FOOD MICROBIOLOGY

FOOD MICROBIOLOGY

K. VIJAYA RAMESH

Head
Department of Microbiology
Dr. M.G.R. Janaki College
Chennai

MJP PUBLISHERS
Chennai 600 005

ISBN 10: 81-8094-019-5
ISBN 13: 978-81-8094-019-4

Cataloguing-in-Publication   Data
Vijaya Ramesh, K (1967 –   ).
    Food Microbiology / by K. Vijaya Ramesh. –
Chennai : MJP Publishers, 2007
    xx, 802p. ; 23 cm.
    Includes Glossary, References and Index.
    ISBN 81-8094-019-5 (pbk.)
    1. Microbiology, Food. I. Title.
    664.001 579 VIJ MJP 030

ISBN 978-81-8094-019-4   **MJP PUBLISHERS**
© Publishers, 2007   47, Nallathambi Street
All rights reserved   Triplicane
Printed and bound in India   Chennai 600 005

Publisher : J.C. Pillai
Managing Editor : C. Sajeesh Kumar
Project Editor : P. Parvath Radha
Assistant Editors : B. Ramalakshmi, S. Revathi
Composition : N. Yamuna Devi, Lissy John, M. Uma
Cover Designer : N. Yamuna Devi
CIP Data : Prof. K. Hariharan

To
all my teachers

PREFACE

The purpose of this book is to provide basic and advanced theoretical information about various facets of food microbiology. Every aspect of food microbiology has been effectively dealt with, be it the tracing of evolutionary aspects or discussing food as an ecosystem with its microbial consortia. These have been dealt with in the introductory unit of this book.

The second unit deals with preservation strategies and I have tried to give a detailed account of various traditional and advanced systems of food preservation. Since food has already been discussed as an ecosystem, the various microbial interactions that take place have been well-documented in the third unit, food spoilage. The significance of microbial spoilage of food has been thoroughly laid down with many examples. ood is a major vehicle of disease in human beings giving way to a number of food-borne infections and intoxications.

I have tried to give a comprehensive account of the various pathogenic microorganisms causing food poisoning, their survival characteristics in food and their detection methods.

A brief account of food-hygiene laws and standards is also available along with details investigating epidemiology of food-borne diseases. Fermented foods are gaining more importance in daily life and I have traced the importance of various classes of fermented foods and their microbiology.

I would appreciate suggestions from students and teachers to improve the work subsequently.

K. Vijaya Ramesh

ACKNOWLEDGEMENTS

During the course of writing this book, I have been very fortunate that great scientists have obliged my requests to part with their findings and reviews.

Any amount of thanking will not suffice my gratefulness towards their support for this work.

I would like to place on record, voluminous help provided by **Dr. Umesh Kumar**, Deputy Director and Head, Food Microbiology Department, Central Food Technological Research Institute, Mysore.

I would also extend my thanks to the following scientists around the globe who have their share in my work.

Professor (Dr.) Gerit Smit, Research Director, Department of Flavour Generation and Delivery, Unilvere Food and Health Research Institute, Netherlands.

Luca Cocolin, Associate Professor, DIVAPRA University of Turin, Italy.

Prof. Dr. Helmut K. Mayer, Head of Food Chemistry Division, Department of Food Science and Technology, BOKU- University of Natural Resources and Applied Life Sciences, Vienna, Austria.

Roger McFeeters, Research Leader and Professor of Food Science, Food Science Research Unit, USDA-ARS. NC State University, Raleigh.

Victoria Enever, Springer, Corporate Reprint Sales Manager, NY, USA.

Dr. M.J.R. Nout, Laboratory of Food Microbiology, Wageningen University, Netherlands.

Dr. Gustavo V. Barbosa Canovas, Department of Biologial Systems Engineering, Washington State University, Pullman, WA.

Professor Lone Gram, Department of Seafood Research, Danish Institute for Fisheries Research, Technical University of Denmark, Denmark.

K. Vijaya Ramesh

CONTENTS

Unit III   Food Spoilage

Unit V   Microbial Food Fermentation

Unit I

FUNDAMENTALS OF FOOD MICROBIOLOGY

The Evolution of Food Microbiology

Food as an Ecosystem

A Consortium of Microorganisms in Food

1

THE EVOLUTION OF FOOD MICROBIOLOGY

ORIGIN OF FOOD MICROBIOLOGY AS A SCIENCE

Food spoilage and food poisoning were the main causes that led man to preserve food and prevent diseases due to food. A study of archaeological sites dated between 18,300 and 17,000 years ago revealed that barley flourished in the Nile valley. The practice of animal husbandry originated about 8,000 to 10,000 years ago. These facts reveal the age of food as a science even without the proper knowledge about it.

The spread of agricultural practices from one society to another made it more possible to assure a stable food supply and encourage community development. By 3,000 BC the people of Iraq had developed a sophisticated agricultural economy.

The production of bread, alcoholic beverages and a variety of acid-fermented foods, the preservation of meat and fish products by drying or adding salts and the production of other indigenous foods were critical to the development of stable societies.

These microbial processes originated in different parts of the world at various times. People began to understand that foods should be kept away from contact with air, light and moisture. Some foods were preserved in early times by coating them with clay and olive oil. Salt became an especially valuable commodity because it was essential to human health and useful for food preservation, the availability of which influenced the course of history.

For thousands of years, people recognized that diseases could be spread by foods. Prohibitions on eating pork had their origins in medical doctrine. Rice, which had turned sour on being left to stand overnight, ready-made food from the market, and dishes which had been sullied by insects or mice or sniffed by an animal were all regarded as unfit for eating. It was not until the

10th century AD that microbiological food poisoning was recognized in civil law. Poisoning by spoiled grains was recognized by the ancient Greeks and Romans, and many epidemics of major proportions occurred through the middle ages in Europe, Russia and elsewhere. Ergotism, caused by growth of a mould, *Claviceps purpurea*, in grain, was recognized in 1582 and reported again around 1600. In the mid-century, epidemics were associated with scabrous rye and other grains infested with *C. purpurea* and precautions were taken to avoid the contaminated grains. The last major outbreak of ergot poisoning in the United States was in 1825.

Because its underlying causes were unknown, microbiological food poisoning was recurrent. Botulism reappeared many times. In 1793, 13 were affected and six died in Germany after consuming blood sausage. It was believed that the illness was caused by a fatty acid.

A. Gartner, in 1888, isolated a bacterium, subsequently named *Salmonella enteritidis* from meat incriminated in a large food-poisoning outbreak. The genus *Salmonella* was named in 1900 after Dr. Salmon, a bacteriologist of the U.S. Department of Agriculture, who first described a member of the group *Salmonella cholerae-suis*, which he thought to be the causative agent of hog cholera. Salmonellosis remains a major problem. K. Shiga, in 1898 discovered an enteric pathogen closely related to the salmonellae. Shiga's bacterium, which causes bacillary dysentery was later named *Shigella dysenteriae*.

In 1906, an aerobic, spore-forming bacillus was recognized as a cause of food poisoning. The true significance of this discovery did not become apparent until the taxonomy of the genus *Bacillus* was clarified by N.R. Smith and R.E. Gordon in 1946 and S. Hauge identified from among the numerous *Bacillus* species that the pathogen was always *B. cereus*.

Staphylococci were first recognized by Pasteur, in pus. They were not associated with food poisoning until studies by T. Denys in 1894 and M.A. Barber in 1914. Barber used himself as a test subject to show that milk obtained from a farm in the Philippines contained staphylococci that induced vomiting. The existence of an exotoxin was confirmed by G.M. Dack and co-workers who studied cream-filled christmas cakes in 1930.

In 1939, J. Schleifstein and M.B. Coleman described gastroenteritis caused by a bacterium that, in 1965, was named *Yersinia enterocolitica* by R. Sakazaki. It was not until 1969 that B. Nilehn documented a food-borne outbreak in 1971 when T. Dadisman and co-workers documented the first United States outbreak, which occurred in New York.

Clostridium perfringens, the causative agent of human gas gangrene infections, was first implicated as a cause of food-borne illness by E. Klein in 1885. This type of food poisoning went almost unrecognized until

R. Knox and E.K. McDonald in England in 1943 and L.S. McClung in the United States in 1945 alerted the scientific community about perfringens food poisoning. In 1953, B. Hobbs and co-workers in England reported that perfringens food poisoning was common but had been overlooked by most laboratories because anaerobic techniques were not used in food microbiology unless botulism was suspected. C.L. Duncan and D.W. Strong demonstrated in 1969 that perfringens food poisoning was caused by an enterotoxin.

Milk had been implicated in the transmission of several diseases. In 1915, G. Jubb reported milk as the vehicle responsible for a small outbreak of poliomyelitis in England. Raw milk was also implicated in the transmission of infectious hepatitis virus by M.D. Campbell in 1943 in England and an outbreak caused by contaminated shellfish in Sweden was described by B. Roos in 1956.

In 1951, T. Fujino showed that *Vibrio parahaemolyticus* caused food poisoning associated with seafood consumption in Japan. Food poisoning caused by moulds attracted attention once again during the 1960s when *Aspergillus flavus* was isolated from groundnut meal and was shown to produce a toxin that caused acute hepatitis at high levels. Later, it was shown that the toxin was carcinogenic at lower levels. Within 10 years many different mycotoxigenic moulds were identified.

Various nematodes and cestodes had been known, for many decades, to be transmitted by foods. Anton van Leeuwenhoek, in 1681, examined his own stools during a bout of diarrhoea and observed a protozoan, *Giardia lamblia*, in large numbers. He subsequently observed similar microorganisms in the guts of rodents and frogs, but did not associate animal reservoirs with disease transmission. This association was not fully appreciated until 1965, when a major waterborne outbreak of giardiasis occurred in Aspen.

Since the difficulties with transportation and storage of foods were aggravated by war, in 1795 the French government offered a substantial award for a new preservation method. Nicholas Appert, a Paris confectioner, accepted the challenge and developed a method whereby wide-mouthed glass bottles were filled with food, corked and heated in boiling water. Appert won the prize in 1805 and the same year, P. Durand of England patented the use of tin cans for thermally processed foods. Neither Appert nor Durand understood why thermally processed food did not spoil. Appert only recognized that the container must be sealed and heated to eliminate what he called "agents of putrefaction" and "fermentable principles". I. Solomon, a Baltimore canner developed in 1860 a simple process that enabled his cannery to increase its output from 2,500 to 20,000 cans per day. In this process, calcium chloride was added to the water in which cans were heated to increase the water temperature from boiling temperature to

116°C. This modest increase in temperature reduced the cooking time from 6 hours to only 25 to 40 minutes. R. Chevallier-Appert was issued a patent in 1853 for sterilization of food at even higher temperatures by using steam under pressure in an autoclave, and by 1874 commercial retorts had been introduced.

Between 1854 and 1864, Louis Pasteur placed heat preservation methods on a scientific basis. The first use of what we now know as "pasteurization", the heating of wine to destroy undesirable organisms was introduced commercially in 1867–1868. In the 1800s, methods to cultivate microorganisms in pure culture and to associate specific bacteria as the causative agents of specific diseases were developed by R. Koch.

Many other pioneering developments on other food preservation processes occurred in the 1800s. In 1842, an English patent was issued to H. Benjamin for a process in which an ice–salt mixture was used to depress the freezing point to freeze foods more rapidly. In 1861, United States patent on freezing fish was issued to E. Piper of Maine. These patents were not used extensively because refrigeration was in its infancy and there were problems in keeping the foods frozen. It was not until the 1950s, that rapid frozen foods gained popular acceptance. Powdered milk was produced in England in 1855. Pasteurized milk was sold in Germany by 1880 and in the United States by 1890. Commercially dried fruits and vegetables appeared in 1886.

Studies on the use of ionizing energy to preserve foods were initiated in 1825 by F. Ludwig and H. Hopf in Germany. However, 70 years after its inception, food irradiation remains an underutilized technology. If the period from the 1890s to the 1940s was described as the era of food preservation, then the era extending from the 1950s through the 1980s can be characterized as the era of "food science" based on chemistry and engineering. As the importance of water activity became clear, intermediate moisture foods were introduced. Spray-dried and freeze-dried foods soon appeared in many areas.

When microbiological standards and regulations were being formulated, it was realized that tests could not be conducted for each and every enteric pathogen that might be present in a sample. Instead, a surrogate must be selected. In 1892, F. Schardinger suggested that *Escherichia coli* would be a useful indicator of faecal pollution. At that time, methods to detect *E. coli* among a large group of related bacteria termed "coliforms" were not readily available. C. Eijkman, in 1904, determined that incubating test samples at 46°C would differentiate "faecal coliforms" that arose from other environments and do not grow at 46°C.

Hazard Analysis Critical Control Point (HACCP) involving complete control over raw materials, process, environment, personnel, distribution and storage was developed during the mid-1900s. The HACCP system was first made public in 1971 but it was not seriously considered by the food industry until it was recommended in 1985 by the National Academy of Sciences Subcommittee on Microbiological Criteria for Foods and Food Ingredients.

The International Commission on Microbiological Specifications for Foods (ICMSF), a standing committee of the International Association of Microbiological Societies, was formed in 1962. The ICMSF establishes internationally acceptable microbiological criteria and attempts to reach agreement on the essential supporting methods from among the plethora of methods in the literature.

Until the 1960s the practice of food microbiology was relatively unchanged since the time of Pasteur. It was a descriptive, qualitative science that focused on *what* happens with relatively less emphasis on or understanding of the *why* or underlying mechanisms. Molecular biology was born in the 1940s and 1950s when scientists from the physical sciences entered the field of biology. They brought with them a more quantitative and mechanistic approach to science which now permeates all areas of biology, including food microbiology.

Only within the last 30 years, fundamental concepts of biology such as the genetic code, the structure–function relationship of proteins, the chemiosmotic coupling of energy-generating and requiring reactions, and the transfer of genetic information have been developed. In most cases, the general principles have been developed using relatively simple, well-studied bacteria such as *E. coli*. Frequently, food-borne microorganisms turn out to be quite different. The *lac* operon is frequently used to teach the concepts of induction, de-repression, carbon catabolite repression, and the role of protein kinases in the synthesis of the beta-galactosidase which is ultimately excreted and cleaves lactose to glucose and galactose. But from a practical standpoint, the most important application of lactose catabolism is in the dairy industry, where the lactose in milk is fermented by lactic acid bacteria such as *Lactococcus lactis*. Lactose catabolism is a completely different mechanism from the *lac* operon model.

The evolution of the dairy industry from a farm-based "art" to a highly technological industry provides other excellent examples of how basic science affects food microbiology. A fundamental understanding of the plasmid biology of fermentative organisms has reduced the incidence of "stuck" fermentations that have lost the ability to metabolize lactose. An understanding of the complex process by which bacteriophage attack and kill starter culture bacteria has generated many strategies for development of phage-resistant fermentations. Through the use of recombinant DNA technology, *E. coli* is rapidly replacing the fourth stomach of a milk-fed calf as the source of rennet (chymotrypsin) used to make many cheeses. Advances in molecular biology and genetics have revolutionized analytical food microbiology. Plate count methods are replaced by ELISA readers, thermocyclers and gel boxes which allow direct quantification of pathogens and their toxins. Rapid salmonella tests have reduced analysis time from 5 days to less than 48 hours.

The history of food microbiology is rich and exciting. It has taken us from the slow realization that certain diseases are caused by microorganisms that grow in food to the empirical control of these microorganisms using physical, chemical and biological manipulation. A mechanistic understanding of microbial physiology and metabolism has provided new approaches to food preservation and laid the foundation for genetic control of food-borne pathogens. Perhaps one day there will be salmonella-resistant chicken or listeria-resistant milk. Food microbiology stands on a scientific footing not more than 100 years old. Its current practice is being transformed by knowledge and tools generated by molecular biology and genetics.

SCOPE OF FOOD MICROBIOLOGY

Food and microorganisms are inseparable, either microorganisms go into the production phase of food where we get fermented oriental foods, or they remain as a contaminant in foods depending on whether they are plant-based or animal-based. Microorganisms can be detrimental to foodstuff when they cause food spoilage leading to heavy economic loss in the production phase or in the consumption phase. Microorganisms can be nutritionally beneficial in relation to food but can also cause dangerous food-borne diseases if left unchecked, considerably affecting the economy class.

Fermented foods were discovered before mankind had any knowledge of microorganisms other than visual proof to their activities. Originally the most important of these changes must have been an improvement in the shelf life and safety of the product along with varieties in consumption. In food fermentation, conditions of treatment and storage produce an environment in which certain types of organism can flourish and have a beneficial effect on the food rather than spoiling it. Fermented foods can be either traditional or conventional. One of the oldest traditional fermented food is bread, nearly 6 million years old. Certain traditional methods have been replaced by biotechnological advances on a commercial scale since oriental foods are much sought after since they represent each country's indigenous palate.

In addition to its undoubted value, food has a long association with the transmission of disease. According to the WHO, food-borne disease is perhaps the most widespread health problem in the contemporary world and an important cause of reduced economic productivity. Outbreaks of food poisoning involve a number of people and a common source, and are consequently more intensively investigated than the numerous sporadic cases that occur. Valuable information is derived from these investigations about contributory factors and the common faults in food hygiene that can occur. Information on outbreaks are collected by the Public Health Organization, from microbiologists and environmental health officials around the country and published as annual reports which gives a clear idea on sensitive epidemic areas and can help trace out legal guidelines through which food safety is given prime importance.

Another aspect of paramount importance regarding food science is the spoilage that can lead to heavy economic loss to the country. Microbial spoilage is very sudden, reflecting the exponential nature of growth as is its metabolism. Once the threshold of spoilage is reached, maintaining the quality of food as acceptable to the consumer is difficult and the food becomes spoilt. Spoilage microflora of any food increasingly depends upon the contaminating flora which is influenced by various factors both external and internal to the food. A thorough knowledge about the possible contaminants of any food can pave the way to develop predictive preservation methods to increase the shelf life of food.

Food microbiology has attained greater heights in applied aspects where quality has to be maintained in terms of microbiological assurance, for example, in every star hotel a food microbiologist is appointed with his team who is very particular about the microbiological criteria of food as well as the hygiene of the personnel working there and the equipment used regarding maintenance of food quality. In terms of microbiology of foods, quality comprises three aspects:

1. *Safety*   A food should be free from pathogen or its toxin that is likely to cause illness when the food is consumed.

2. *Acceptability/shelf life*   A food must not contain levels of microorganisms sufficient to render it organoleptically spoiled in an unacceptably short time.

3. *Consistency*   A food must be of consistent quality both with respect to safety and shelf life. The consumer will not accept products which display large batch-to-batch variations in shelf life and is certainly not prepared to play with his/her health every time he/she eats a particular food.

REVIEW QUESTIONS

1. Trace the origin of food microbiology as a science.
2. List the factors that influence microbial growth within food.

2

FOOD AS AN ECOSYSTEM

Food microbiology is a branch of microbial ecology. The importance of ecological concepts in understanding the occurrence and growth of microorganisms in foods is well recognized by food microbiologists. These ecological principles are the foundations upon which modern quality assurance, predictive modelling and risk analysis strategies have been developed to prevent outbreaks of food spoilage and food-borne disease. They form the bases for the functional use of microorganisms in the production of fermented foods and beverages, and for their use as probiotic and biocontrol agents. Guided by commercial objectives, food microbiology has evolved into a field of study with a strong focus on groups of microorganisms and groups of commodities but, increasingly less focus on the microbiology of the ecosystem as a whole. This trend to compartmentalize our knowledge needs to be balanced against the "big picture" and, ironically, is occurring at a time when other branches of microbial ecology (e.g. water, soil, phyllosphere) are strengthening the totality of their microbiological studies and advancing fundamental understanding of their ecosystems. Food microbiologists must remain aware of the totality of the ecosystem. The growth, survival and activity of any one species or strain, whether it be an unwanted spoilage or pathogenic organism, or a desirable biocontrol or probiotic organism, will, in most cases, be determined by the presence of other species.

WHAT ECOLOGICAL INFORMATION IS NEEDED?

To effectively manage the growth and activities of microorganisms in foods, the following "layers" of information and understanding are needed:

- Reliable data about the diversity and taxonomic identity of the species and strains that contaminate and colonize the food at every stage of production, from the raw material to the time the product is consumed.

- Quantitative data that describes the growth cycle and changes in populations of these species and strains throughout the production and retailing chain.
- Information about the spatial distribution of microbial species throughout the product.
- Biochemical and physiological explanation of the colonization process.
- Impact of the so-called intrinsic, extrinsic, processing and implicit factors on microbial growth, survival and biochemical activity.
- Correlations between the growth and activity of individual organisms, and product quality and safety. Obtaining this information is a challenging task and, most likely, the food microbiologist will need to collaborate with other specialists in chemistry, biochemistry, electron microscopy and sensory evaluation.

LIMITATIONS OF METHODOLOGY

The aim of microbiological analysis, either to profile the diversity of species occurring in a product or to determine the presence or absence of specific pathogens, should be to obtain the ecological truth. Is this achievable with the current portfolio of analytical methods? Despite many innovations in recent years, the examination of foods for total or specific microflora follows the basic operations of

- i. Maceration/blending of the sample
- ii. Dilution of the homogenate
- iii. Plating of dilutions onto appropriate agar media, and
- iv. Isolation and identification of colonies.

Pre-enrichment and selective enrichment culture before plating will be needed to recover species present at low populations (e.g. less than 100–500 cells/g). While this basic approach has had long-term acceptance and general success, there are inherent limitations that are worthy of re-emphasis.

Maceration

Microbial cells occurring on the surface of products suddenly become exposed to a vastly different chemical environment on maceration for microbiological analysis. Tissue extracts generated by maceration could be toxic or inhibitory to some species, thereby giving erroneous data about the ecological composition of the natural product. The assumption that maceration is an ecologically sound prelude to microbiological analysis requires more rigorous scrutiny, since the extracts of vegetables, herbs and spices are toxic to some microorganisms. This question becomes especially relevant when analysing heterogeneous products (e.g. mixed salads and pastas), and when attempting to increase the sensitivity of detection by macerating greater quantities of sample.

Dilution

To many microbiologists, the dilution stage is an innocuous or harmless operation with respect to an ecological outcome. It should facilitate the dispersion of cell clumps and should not affect the cell viability. Many years ago, bacteriologists realized the influence of diluent composition and the time span between dilution and plating on viable plate counts and, without rigorous trialling, more or less standardized this operation using 0.1% peptone as the general diluent. For the isolation of yeasts and moulds, no such analytical standardization has yet occurred and diluents commonly used range from distilled water, saline, phosphate buffer and 0.1% peptone, with various outcomes. In an effort to standardize this operation, the International Commission on Food Mycology has undertaken an international collaborative trial on diluents used for the analysis of yeasts in foods. Although this work is still going on, it has shown that the response of yeasts to any one diluent was repeatedly inconsistent. It was concluded that, apart from diluent composition and timing between dilution and plating, other factors such as stage of cell life cycle, cell stress prior to dilution, degree of cell clumping and aggregation, shear forces during shaking, presence of contaminating metal ions, pH and temperature could all impact on the survival of the yeast cells during the dilution operation. These conclusions could also apply to bacteria and filamentous fungi. In essence, the dilution operation may not be ecologically innocuous, as frequently assumed.

Enrichment Cultures

Enrichment cultures are widely used in food analyses to enhance the cell population and detectability of minority species, especially pathogens such as *Salmonella*, *Listeria monocytogenes*, *Escherichia coli* and *Campylobacter jejuni*. The goal is to amplify populations as low as 1 cell/25 g of product to minimum levels of 10^{10} cells/ml of culture for routine detection by plating, ELISA or nucleic acid probe technologies. Failure of the enrichment culture to give this minimum population within the prescribed incubation time leads to a false negative result. Given the serious consequences of false negative data in managing food safety, it is surprising to find very few studies on the growth kinetics of these pathogens in commonly used enrichment media. These growth kinetics and achievement of a detectable population are determined by many factors such as medium composition, time and temperature of incubation, degree of aeration, interference from food components, competition from non-target flora in the food, the possible influences of bacteriophages, and extent of any sublethal injury. We have frequently experienced occasions where *Salmonella* or *L.monocytogenes* have not reached 10^5–10^6 cells/ml in approved enrichment media. Ecological surveys of foods for the presence of food-borne pathogens are almost invariably based on the use of enrichment cultures but the reliability and limitations of this methodology are rarely questioned.

Anaerobes

It can be expected that obligate anaerobes, in addition to facultative anaerobes, will contribute to the microflora of many foods and beverages. Attention must be given to good anaerobic methodology to successfully isolate these organisms. The literature would suggest that food microbiologists have not been particularly rigorous in meeting these requirements and it is likely that significant anaerobic microflora have been overlooked in the ecology of many products. An illustration of this point is the recent discovery of the strictly anaerobic bacterial species, *Pectinatus cerevisiphilus*, *P. frisingensis*, *Selenomonas lacticifex* and *Megasphaera cerevisiae*, in spoiled, packaged beer.

Unknown and Non-culturable Species

It is now well-established that many natural ecosystems such as soil, water, sediments and sludge harbour microbial populations that greatly exceed those measured by culturing on agar media. Indeed, it is estimated that plate culturing techniques reveal only 10% (or less) of the true microbial population in those environments. This anomaly is explained by two phenomena: (i) the presence of unknown, novel species that are not culturable by existing methods, and (ii) the presence of known species that are metabolically active and viable but have entered a non-culturable state. Understanding these phenomena has evolved from the use of molecular methods that can detect the non-culturable species. The principal strategy is based on analysis of the total DNA extracted from the ecosystem. Using PCR technology, microbial ribosomal DNA (rDNA) in the extract is specifically amplified, cloned, and then sequenced. The sequenced data is compared with sequences in rDNA data bases to give genus or species identification. Another approach uses fluorescently labelled rDNA probes that allow detection and spatial location of targeted species in situation. By combining these methods with denaturing gradient gel electrophoresis to separate PCR amplified rDNA fragments, it is even possible to profile specific changes in the composition of complex microflora as the ecosystem evolves with time. To date, these molecular technologies have received almost no application in studying the microbiology of food ecosystems. Consequently, we may not fully know the microbial composition of some foods, especially fresh products such as vegetables and meats, or complex fermented products (e.g. mould-ripened soft cheeses, meat sausages, cocoa bean fermentations) where a diversity of species may be present. Food microbiologists are, however, familiar with the viable but non-culturable (VBNC) phenomenon, where adverse conditions such as nutrient depletion, low temperature and other stresses can cause healthy, culturable cells to enter a phase which does not produce colonies on media that normally support their growth. Such cells, nevertheless, remain metabolically active and capable of causing infection. Given appropriate conditions, they can recover from their debilitated condition. While their presence is not evident by colony culture, they are detectable by assay with fluorescent stains that can measure membrane, DNA, RNA and other physiological functions. The VBNC state has been experimentally induced in most food-borne pathogens, including *Salmonella, C. jejuni, Vibrio vulnificus, Vibrio parahaemolyticus* and *E. coli*, but there is debate as to whether it occurs in nutrient-rich food environments. However, it is not inconceivable that VBNC forms could occur in bottled waters or on the surfaces of fresh fruits and vegetables where nutrients may be limiting. The VBNC state should not be confused with the concept of sublethal injury. The main difference between the two phenomena is that sublethally injured cells will not grow on selective media but grow on non-selective media, whereas the cells in the VBNC state will not grow on either type of media. Both types of cells are capable of repair and resuscitation to the healthy state.

Quantitative Data

The economic and social consequences of microorganisms in foods depend not only on the species present but, most importantly, on their quantitative populations. It is the number of microbial cells that ultimately determines whether or not the product will cause an outbreak of disease or develop an off-flavour. Also, populations are not static and can change both qualitatively and quantitatively throughout the production and retail chain. In many products, sequential development of species and strains occurs, with each organism impacting upon the chemical composition of the ecosystem according to its biochemical reactivity and, importantly, its maximum population. Reliable, confident assessments of public health and spoilage risks require quantitative ecological data that take into consideration the dynamic nature of microorganisms in food ecosystems. Unfortunately, the vast majority of ecological studies in food microbiology fall significantly short of providing this quantitative knowledge. Many studies simply describe the isolation and identification of the "most predominant" species at one point in the product's history, while others have provided semiquantitative data by reporting the frequency of isolation of specific organisms. Population changes are mostly described in reference to microbial groups (e.g. total plate count, coliforms, psychrotrophs, lactic acid bacteria) rather than data for particular species or strains. Ecological surveys for pathogens such as *Salmonella, L. monocytogenes* and *E. coli* continue to be reported as isolation frequencies (e.g. % sample positive) with population levels (cells/g) rarely being mentioned. Further advances in understanding and managing the microbial ecology of foods will require more quantitative knowledge. An obstacle here is methodology, especially for the pathogens where enrichment culture forms the basis of the analyses. For some years, we have been using a centrifugation-plating technique which avoids cultural enrichment and enables direct enumeration of cells down to a detection limit of 1 cell/25 g of sample. This strategy not only gives fast, quantitative data, but also reveals the failures of enrichment methods.

ECO-BIOCHEMISTRY AND PHYSIOLOGY

Microbial growth in food ecosystems requires biochemical and physiological explanation to understand how specific microorganisms impact on food quality and are able to develop processes for managing this growth. Excellent progress has been made in understanding the biochemical reactions associated with microbial growth in many foods and in explaining how food properties regulate the growth response. However, much of our knowledge is derived from pure culture studies of microbial isolates in laboratory media, and more specific attention is needed to describe the unique reactions of the *in situ* environment which will simultaneously harbour microbial cells in a diversity of physiological states—growing, non-growing, dead and autolysed. In most food ecosystems, the *in situ* environment will mean association of microbial cells with a solid substrate either through attachment or entrapment or both. In essence, the cells are immobilized and localized in high densities. Not much is known about the specific biochemical and physiological properties of high densities of microorganisms associated with solid systems. However, studies with microbial cells that have been specifically immobilized in alginate, polyacrylamide and other substrates clearly demonstrate that their properties of growth, survival, tolerance to extremes, biochemical activity and even cell composition can be significantly different from those of cells growing freely in liquid culture. Thus, the *in situ* solid-phase associations found in food ecosystems could induce unique biochemical and physiological reactions. Because of the physical closeness of cells in these environments, cell–cell signalling and other communication mechanisms could also influence these responses. For many foods, the *in situ* association of microbial cells can extend over a long time frame, thereby providing good opportunity for these cells to adapt to specific stresses of the environment and to adapt to an existence which, for the greater part, is essentially a resting or non-proliferating state. The growing or exponential state probably presents only a short time span throughout the total association with the ecosystem and, indeed, cyclical movement between exponential and non-proliferating stages is likely to occur, depending on nutrient availability and environmental factors. Nutrient limitation and end product accumulation will induce transition to the non-proliferating phase. While the availability of carbon and nitrogen substrates may not be an issue in many foods, there could be limiting concentrations of sulphur, phosphorus, trace metals and vitamins. Also, nutrient availability must be considered in relation to the concepts of nutrient uptake and transport into the cells. It is well known in the fermentation of alcoholic beverages by yeasts that sugar transport into the yeast cells can be a significant rate-limiting reaction. The physiology and molecular biology of substrate transport will emerge as simultaneously important issues in explaining the growth responses of microorganisms in food ecosystems. Environmental factors (pH, temperature, water activity, preservative concentrations) that limit growth have been described for many organisms, especially the food-borne pathogens. Nevertheless, there are species in many important groups of microorganisms such as lactic acid bacteria, pseudomonads and yeasts where this information is not complete and more studies are needed. These limiting values for microbial growth are the bases of predictive and risk assessment strategies now used in quality assurance programs. The uncertainties of the *in situ* response can diminish confidence in these initiatives. A key issue, here, is the adaptive reactions of microorganisms on exposure to *in situ* stresses. Through a range of molecular mechanisms, they can change their limits of tolerance to environmental pressures. The problem is compounded by the fact that *in situ* food environments rarely present a single stress, and microbial cells may be simultaneously exposed to a combination of stresses (e.g. low pH, low temperature; high NaCl concentration, low temperature, low pH; ethanol, acetic acid). The effect of the combined stresses on growth and survival may be additive or synergistic (interactive) where the outcome is significantly greater (or less in some cases) than the additive response. Further complications are introduced by the changing state of the ecosystem which, in many cases, is not static. The severity of the stress can change with time as will the physiological state of the microbial cells. Cells in the stationary phase of growth are, generally, more tolerant of stresses than exponential phase cells. A related but overlooked outcome of the stress response is the changing chemical composition of the microbial cells (e.g. changes in protein and lipid composition, increased concentrations of some amino acids, glycerol, trehalose) which could impact on sensory and other acceptability criteria for the food. As noted already, weakly or non-proliferating cells in a resting or stationary phase of growth, probably dominate in many foods (e.g. consider the 10^6–10^8 cfu/g for many vegetable salads; the 10^6–10^8 cfu/g in many soft cheeses). These cells are still metabolically active and can conduct biochemical reactions that are distinctively different from cells growing exponentially. The flavour-impacting reactions that occur during the production of many fermented foods and beverages are generally the consequences of microbial activity under non-proliferating conditions.

The non-proliferating phase is eventually followed by cell death and cell autolysis. Consequently, most "long-term" food ecosystems will harbour a substantial population of dead, autolysed cells. Autolysis is characterized by extensive loss of cellular structure and function, and enzymatic breakdown of cell proteins, nucleic acids, lipids and polysaccharides. These degradation products become part of the ecosystem, serving as nutrients or antagonists for other microorganisms and, also, they impact on food sensory properties. Microbial autolysis is an important *in situ* phenomenon, the biochemistry, physiology and significance of which have been very much overlooked in studies on the microbial ecology of foods.

SPATIAL HETEROGENEITY

With few exceptions, most foods present an environment that is heterogeneous in physical structure and chemical composition. Consequently, microbial growth throughout the ecosystem is likely to be spatially heterogeneous. That is, different locations within the same food product could have significantly different microflora. Thus, the microbial ecology on the surfaces of cut or damaged vegetables and fruits will be different from that of uncut or undamaged product which have intact cuticle and waxy layers. Similarly, the microbial populations on the outer surfaces of meat and dairy products will be different from those of the inner parts of the foods. Electron microscopy has revealed the potential for microbial cells to attach to food surfaces, become entrapped within the food structure and to grow as microcolonies and biofilms. Cells within microcolonies and biofilms have increased resistance to processing conditions, as well as altered biochemical behaviour. Consequently, it is relevant to know their occurrence and location. More significantly, it is important to know where particular species are located throughout the product. Until recently, it was not possible to obtain such information. The combined uses of fluorescence microscopy and confocal scanning laser microscopy with fluorescently labelled antibodies or nucleic acid probes now enable the *in situ* localization of specific organisms. Using these technologies, it has been possible to demonstrate that *E. coli* 0157:H7 does not necessarily attach to the outer surfaces of radish sprouts or lettuce leaves and can become associated with cut surfaces and the inner parts of stomata and other tissue. Similarly, it has been possible to pin-point the location of the yeast-like fungus, *Aureobasidium pullulans* on leaf surfaces. Further application of these new methods will significantly advance and refine our *in situ* knowledge of food ecosystems.

MICROBIAL INTERACTIONS

With the exception of highly processed products, most foods harbour a mixture of microorganisms which includes different species of bacteria, yeasts and filamentous fungi as well as strains within these species. In addition, bacteriophages and killer yeasts with virus-like particles will also constitute a part of the microflora. In the natural pursuit of survival, growth and dominance, interactions will occur between these different strains and species, the outcome of which will determine the population levels of any particular organism at any given time during the production and retailing time frame. Ecological theory describes the range of interactive associations as competitive, amensalism or antagonism, commensalism, mutualism and parasitism or predation and these could occur both within and between different microbial groups (e.g. bacteria–bacteria; yeast–yeast; bacteria–yeast; bacteria–fungi, etc.). There are many examples of these types of interactive associations scattered throughout the food microbiology literature, but only a few will be discussed here.

Antagonism is probably the best known microbial interaction in food ecosystems because it can be applied as a natural biocontrol strategy to enhance food quality and safety. The production of bacteriocins and their use to control spoilage and pathogenic bacteria have been extensively studied in recent years. While this is a classic example of "bacteria–bacteria" interaction, there are circumstances where bacteriocins will inhibit yeasts. The production of killer toxins by yeasts is somewhat analogous to the bacteriocin production by bacteria. These toxins are extracellular proteins or glycoproteins that disrupt cell membrane function in susceptible yeasts. While these antagonistic interactions were originally thought to be species-specific, there is now clear evidence that they occur across species in different yeast genera and, indeed, they can kill various filamentous fungi. Moreover, there is no doubt that killer interactions between yeasts naturally occur in food ecosystems. A less recognized form of antagonism is the production of cell wall lytic enzymes. Examples include the production of β-1,3-glucanases by bacterial and yeast species that destroy the β-1,3-glucans in the cell walls of fruit spoilage fungi such as *Penicillium expansum* and *Botrytis cinerea*. Another less familiar form of microbial interaction that could be significant in food systems is the ability of yeasts and bacterial cells to agglutinate and aggregate. Most species of Enterobacteriaceae and some lactic acid bacteria will agglutinate *Saccharomyces cerevisiae* by reaction with the surface mannoproteins of the yeast. In addition to antagonism, commensalistic microbial interactions frequently occur in food environments. Examples include the degradation of complex proteins and carbohydrates by some species to produce simple substrates for the growth of other species, utilization of organic acids by yeasts and moulds to favour the growth of bacteria, the autolytic release of nutrients by dead cells, and the production of vitamins, specific amino acids, carbon dioxide and other micronutrients by some species that will assist the growth of other species. Greater consideration needs to be given to the role of bacteriophages in food environments. Most studies concern their ability to destroy starter cultures of lactic acid bacteria used in milk fermentations, but this is a very specialized case. By analogy to other ecosystems, such as the marine environment, foods are likely to harbour an enormous diversity of bacteriophages that could have a significant impact on the *in situ* bacterial ecology. Generally, bacteriophages occur in the same habitats as their bacterial hosts. There are only a few reports on the isolation of bacteriophages from foods, including phages against *Pseudomonas* spp. from refrigerated meat, *Leuconostoc oenos* from wines, *Propionibacterium* spp. from cheese and *V. vulnificus* from oysters. Phages against *Enterococcus durans* from cheese and against *Pseudomonas fluorescens*, *P. viridiflava* and *P. corrugata* from broccoli have been readily isolated. Phage activity would certainly cause bacterial cell lysis and nutrient release in ecosystems and probably accounts for the variable population data often obtained from food samples.

DIVERSITY IN THE MICROBIAL ECOLOGY OF FOODS: CASE STUDIES

When the ecological principles outlined in the previous sections are applied to specific food commodities, two conclusions become apparent: (i) the microbial ecology of most foods is more diverse and complex than generally thought; and (ii) there remain many gaps in knowledge and understanding. These points will be illustrated by reference to two products: cheese and wine.

CHEESE

Cheese manufacture is a vast, economically important industry that produces a diverse range of products. It is an excellent example to consider in the context of microbial ecology, since it embraces all of the practical interests in food microbiology—fermentation, spoilage, safety, biocontrol, probiotics and, also, the issues and controversies surrounding the use of genetically modified microorganisms in food production.

Cheese Processing

The basic steps in cheese manufacture are:

i.   collection and processing of the milk which includes pasteurization in most, but not all, cases.

ii.   conversion of the milk into a cheese curd by the action of proteolytic enzymes and by fermentation with lactic acid bacteria that are usually added as a starter culture but, in some cases, are allowed to develop naturally.

iii. processing of the curd by heating, cutting, addition of sodium chloride and moulding.

iv.   maturation (ripening) of the curd, generally, by storage under controlled humidity and temperature for periods ranging from a few weeks to many months.

    v.   packaging and retailing.

To many microbiologists, the microbial ecology of the process is, principally, fermentation of the milk by lactic acid bacteria. Unfortunately, this is a gross underestimation of the total picture. Rather, it is the microbiological and biochemical changes that occur during maturation where distinctive and unique cheese character is developed and where the important issues of spoilage and safety emerge.

Microbiology of Milk Fermentation

Most research on cheese microbiology has focussed on the role of lactic acid bacteria in fermentation of the milk. Today, this process is largely accomplished by the inoculation of commercially produced starter cultures of *Lactococcus lactis* and, in some cases, *Leuconostoc mesenteroides*, *Streptococcus thermophilus*, *Lactobacillus helveticus* and *Lactobacillus delbrueckii* ssp. *bulgaricus*. Since these 6 species are added to milk at initial populations of 10 cfu/ml or more, the milk is pasteurized in most cases, and since the time of action is relatively short (1–4 hours), the ecology of this fermentation is rather unremarkable. The fermentation starts as an aqueous system, but the curd which develops harbours very high populations (10^{10} cfu/g) of entrapped, essentially immobilized lactic acid bacteria in a non-proliferating phase of growth. The main ecological issue is the intervention of bacteriophages which destroy the lactic acid bacteria and disrupt the fermentation. It is important that the fermentation commences rapidly and produces sufficient lactic acid, bacteriocins and other antagonistics to prevent the growth of spoilage and pathogenic bacteria. Most research is now directed towards understanding the biochemistry, physiology and molecular biology of the fermentation, especially the metabolism of lactose into lactic acid and the production of components that contribute to cheese flavour and texture.

Microbiology of Maturation

It has been known for more than 50 years that many cheeses, especially the soft and semi-soft varieties, contain high populations (10^6–10^9 cfu/g) of microorganisms that are not the lactic acid bacteria added as starters to ferment the milk. These microorganisms have been loosely referred to as the secondary or adventitious microflora, and it is generally considered that they positively contribute to the maturation process. In some cases, microorganisms are deliberately added as part of the maturation process.Well known examples are use of the filamentous moulds *Penicillium camemberti* and *P. roqueforti* and the yeast like mould, *Geotrichum candidum*, in the maturation of Camembert, Brie and blue-veined cheeses, and use of *Propionibacterium shermanii* during maturation of Swiss, Emmenthal and Gruyere cheeses. However, even in these cases, the maturation microflora comprises a complex mixture of wild bacteria, yeasts and bacteriophages as well as any added organisms. Such complexity develops whether the cheese is produced from pasteurized or non-pasteurized milk, or whether or not starter cultures are used to ferment the milk.

These organisms originate as natural contaminants of the process coming from the milk, added proteolytic enzymes, brine (NaCl) solutions, surrounding air and contact with equipment. When conditions within the curd become favourable, they initiate growth. Over the years, there has been significant progress in identifying the main species that comprise the maturation microflora, although the data are far from complete and are extremely variable, even for the one type of cheese. The bacteria associated with maturation are diverse and include: (i) the so-called non-starter lactic acid bacteria that comprise various species of *Lactobacillus, Pediococcus, Leuconostoc* and *Enterococcus*; (ii) micrococci and staphylococci (e.g. *Micrococcus varians, Staphylococcus xylosus*); (iii) corynebacteria and brevibacteria (e.g. *Brevibacterium linens*); (iv) propionibacteria (e.g. *Propionibacterium shermanii, P.freudenreichii*), and (v) various Enterobacteriaceae. Occasionally, spoilage species of clostridia (*Clostridium tyrobutyricum*), coliforms and *Bacillus*, and pathogenic species of *Salmonella, L. monocytogenes* and *E. coli* can develop during maturation. Process failure generally accounts for such problems but natural "biocontrol" of these adverse species by bacteriocins produced by other microflora (e.g. *Enterococcus* spp., *B. linens*) is an important but underestimated function of the maturation process. There are suggestions that the yeast *Debaryomyces hansenii* can inhibit the growth of spoilage clostridia. Yeasts have emerged as significant organisms in the maturation process, although their precise role is not understood. Predominant species include *D. hansenii, Yarrowia lipolytica* and *Kluyveromyces marxianus*, which are often, but inconsistently, present in cheeses at populations of 10^6–10^9 cfu/g. Cheese isolates of *D. hansenii* have been found to possess killer activity, the action of which is enhanced by the presence of NaCl. While the significance of bacteriophages in milk fermentation is well recognized, their potential impact on the bacterial ecology of maturation has not been considered. The main intrinsic factors affecting the growth, survival and biochemical activities of the maturation flora are the moisture content, salt (NaCl) concentration and pH of the curd as well as availability of oxygen. These properties are not uniform throughout the curd and change with time. For example, in many cheeses the salt content is initially higher at the outer surface but progressively equilibrates as it diffuses into the inner parts of the curd. Although microorganisms are distributed throughout the curd, substantially higher populations are located on the outer surface because of the availability of oxygen. Thus, oxidative microorganisms (brevibacteria, micrococci, *D. hansenii, Y. lipolytica*, moulds) are more prevalent on the curd surface, while fermentative species (lactic acid bacteria, *K. marxianus*) are more predominant within the curd. Some key biochemical reactions of the maturation microflora are fermentation of residual lactose, utilization of lactic acid and enzymatic degradation of curd proteins and lipids, but these activities are moderated by the intrinsic properties of the curd, especially pH. The *Penicillium* spp. as well as yeasts such as *D. hansenii* and *Y. lipolytica* are particularly strong utilizers of lactic acid, especially at the surface of the curd, causing its pH to significantly increase over time. This decrease in acidity is also assisted by the strong proteolytic activity of many of the species present. Thus, the pH of the curd can increase from less than pH 5.0 at the beginning of maturation to values near neutrality during maturation and retailing. Accordingly, bacterial species that could not grow in the original curd because of low pH, can now grow. Such species could include spoilage clostridia and pathogenic strains of *E. coli* and *L. monocytogenes*. Also, this bacterial growth is assisted by the greater availability of nutrients such as amino acids originating from proteolysis, and glycerol formed during lipolysis. Autolysis of yeasts and moulds is considered to be another important source of nutrients for bacterial growth. Microbial autolysis, generally, is emerging as a most significant reaction in cheese maturation. Its occurrence is not restricted to yeasts and moulds, but also extends to lactic acid bacteria. It should be recalled that the curd harbours high populations of the starter lactic acid bacteria (e.g. *L. lactis*) that, generally, are not salt tolerant and quickly die off after brining. Autolysis of these cells and release of their intracellular contents, including active enzymes, will have profound impact on the chemical and physical properties of the curd during maturation. A further consideration in relation to autolysis would be the altered composition of microbial cells when exposed to the acidic, high-salt stresses of the curd. In response to the salt stress, yeasts accumulate significant concentrations of glycerol whereas bacteria are likely to produce amino acids. These substances will be released during cell autolysis.

Summary

Most cheeses harbour a diversity of wild microflora which evolves in a successional process throughout maturation and retailing. The subtleties of cheese character, as well as cheese shelf life and safety, are uniquely determined by the composition and evolution of this flora, yet this ecology remains poorly described. Obtaining this information is important so that processes can be managed to encourage growth of the desired species and prevent the growth of undesirable species. This challenging task is complicated by the very complexity of the microflora, itself, and the limitations of cultural methodologies in analysing complex ecosystems. Molecular ecological techniques will be needed to help unravel the microbiological mysteries of cheese and, indeed, are likely to reveal the presence of a greater diversity of organisms. The dynamics of growth, survival and biochemical activity of this microflora will reflect an array of stress reactions in response to the changing conditions of salt and pH, but will be moderated by the phenomenon of cell immobilization within the curd. Cell–cell interactions will play an important role in shaping the ecological profile, and current thinking must be expanded beyond the concept of bacteriocins to include the influences of bacteriophages, killer yeasts and cross-group (e.g. yeast–bacteria, bacteria–fungi, etc.) responses.

WINE

Almost 150 years ago, Louis Pasteur showed that wine was the product of an alcoholic fermentation of grape juice by yeasts. This section will show that the microbial ecology of the process now extends far beyond a simple alcoholic fermentation and involves complex interactive contributions from yeasts, filamentous fungi, lactic acid bacteria, acetic acid bacteria, other bacterial groups, and even bacteriophage. It will demonstrate how recent approaches of quantitative ecological analyses and molecular methods have been significant in advancing this knowledge.

Wine Processing

The basic operations in wine preparation are

    i. crushing of the grapes and extraction of the juice
    ii. alcoholic fermentation of the juice by yeasts
    iii. optional malolactic fermentation (MLF) of the wine by lactic acid bacteria
    iv. bulk storage and ageing of the wine in cellars
    v. packaging and retailing

The production of fortified wines (ports, sherries) and sparkling wines involves additional specialized operations. Microbial growth and activity can be significant at all stages of wine production.

The process is further discussed in later chapters.

Microbiology of Grapes

Grapes represent a principal source of microorganisms in wine production but, surprisingly, their ecology remains poorly researched and diminished by studies that have used inadequate sampling and cultural (enrichment) methods.

Nevertheless, it is generally agreed that the surfaces of mature sound grapes harbour microbial populations at levels of 10^3–10^5 cfu/g consisting mostly of yeasts and various species of lactic acid bacteria and acetic acid bacteria. Contrary to many early reports, the principal wine yeast, *Saccharomyces cerevisiae*, is not prevalent (50 cfu/g) on sound grapes, thereby raising questions about its true origins in wine fermentation. Many intrinsic and extrinsic factors that affect the occurrence and growth of microorganisms on the surfaces of grape berries, include rainfall, temperature, grape variety, berry maturity, location of berry in the bunch, physical damage due to bird, insect and mould attack, and the application of chemicals such as fungicides and insecticides. Moreover, the outer surface of the berry is covered by a waxy, cuticular layer which will affect the adherance of microbial cells and their ability to colonize the surface. Scanning electron micrographs suggest localized colonization of the surface by yeast cells, especially where the surface layers are damaged. Unfortunately, there are no definitive studies on how microorganisms contaminate and colonize the surfaces of grapes. However, damaged grapes quickly develop populations of 10^6–10^8 cfu/g and harbour high populations of filamatous fungi (e.g. *Botrytis cinerea*) and acetic acid bacteria. These grapes have altered chemical composition and fungal enzymes that adversely affect wine flavours and colour, and can negatively impact on yeast growth during alcoholic fermentation and the growth of lactic acid bacteria during MLF. Control of fungal growth on grapes is a key issue in wine preparation. Mycotoxin production and carryover into wine is a possibility that has not been adequately addressed. Fungicide residues on grapes can impact on the yeast ecology of alcoholic fermentation. Genetically modified grapes with anti-fungal levels of glucanase and chitinase are sound in principle, but these enzymes also destroy wine yeasts. Biocontrol of grape fungi with selected, antagonistic species of yeasts is an interesting initiative.

Microbiology of Alcoholic Fermentation

The alcoholic fermentation is dominated by the growth of yeasts because of their ability to rapidly develop at the low pH (3.0–3.5) of the juice, and produce ethanol that inhibits the growth of filamentous fungi and bacteria. Many qualitative ecological studies over the past 100 years have shown that *S. cerevisiae* predominates in almost every wine fermentation, and collectively, they have given a misleading impression that this is the only species of relevance. The important contribution of other species to the overall fermentation has become evident only in recent years when more quantitative studies of yeast growth were undertaken.

The first 2–4 days of the fermentation are characterized by the growth of various species of *Kloeckera, Hanseniaspora, Candida, Metschnikowia, Pichia* and *Kluyveromyces* which achieve populations of about 10^7cfu/ml before progressively dying off according to their tolerance of accumulating concentrations of ethanol. By this time, they have utilized sufficient sugars and amino acids in the juice, and generated sufficient amounts of end products to have an imprint on wine character. *S. cerevisiae* also grows during these early stages but because of its unique and implicit tolerance of ethanol, continues to grow and predominate as the only species during the mid-to-final phases of the fermentation. The application of molecular techniques to the study of this ecology has revealed even further complexity. It is now evident that each of the yeast species may be represented by several strains, and that successive strain evolution and death is characteristic of the ecological profile. Strains with killer activity are commonly isolated from wine fermentations and also contribute to the changing profile. This profile and, hence, wine quality, can be moderated by a range of intrinsic, extrinsic and processing factors including grape juice composition, pesticide and fungicide residues, addition of sulphur dioxide, and degree of juice clarification and temperature. The temperature of fermentation, in particular, can have a profound impact. Low temperatures (10–15°C) increase the ethanol tolerance of *Kloeckera, Hanseniaspora* and *Candida* species to a point that they do not die off and become dominant contributors along with *S. cerevisiae*. Red wines are fermented in contact with grape skins during the early stages and present the interesting possibility of biofilm development at the solid–liquid interface. The origins of the yeasts responsible for the fermentation have attracted significant controversy. Many wines are produced by traditional, natural fermentation, where the yeasts originate from the grapes and winery equipment (e.g. crushers, pumps, hoses, fermentation tanks). For many years, it was believed that grapes were the principal source of *S. cerevisiae* that dominated the fermentation. It is now evident that this species largely originates from winery equipment. This equipment accumulates a residential microflora that is dominated by strains of *S. cerevisiae* because of selection through its ethanol tolerance. Using pulsed field gel electrophoresis and restriction analysis of DNA, it has been possible to type the strain profile of these *S. cerevisiae* and demonstrate the specificity of their winery association and carry over into fermentations from one vintage to the next. Because of their predominance, strains of *S. cerevisiae* have been commercialized as starter cultures for the inoculation and induction of wine fermentation. Inoculated fermentations tend to proceed more rapidly and predictably than their natural counterparts and are now practiced by many winemakers.

There is a general assumption that the inoculated strain will overwhelm and suppress the growth of the natural flora and dominate the fermentation. Various quantitative and molecular ecological studies have now shown that these assumptions are not necessarily correct: the yeasts naturally present continue to contribute to the fermentation, and indeed there are many examples where the inoculated *S. cerevisiae* did not even dominate the fermentation. With respect to biochemistry, it is pertinent to note that the greater part of the fermentation occurs after the yeast cells have entered the stationary phase and when a substantial proportion of the ethanol-sensitive yeast species have died and entered autolysis.

Microbiology of Malolactic Fermentation

Almost 100 years ago, it was observed that many wines underwent a natural secondary fermentation about 2–4 weeks after completion of the alcoholic fermentation. This fermentation has been called the malolactic fermentation (MLF) and its ecology and biochemistry have been extensively studied. It is conducted by acid and ethanol tolerant strains of lactic acid bacteria that survive the alcoholic fermentation or come from winery equipment. They are generally present at low or non-detectable levels ($<$10–100 cfu/ml) in the wine but, over 2–4 weeks, grow to populations of 10^7–10^8 cfu/ml. The most notable feature of this growth is the stoichiometric decarboxylation of L-malic acid to L-lactic acid and carbon dioxide, causing a deacidification of the wine and an increase in pH by about 0.3–0.5 unit. L-malic acid is a major component of most wines and originates from the grape.

For high-acid wines, such as those with an initial pH of 3.0–3.5, this deacidification gives a most desired improvement in wine sensory quality, but for low-acid wines (e.g. pH 3.5–4.0), the deacidification depreciates sensory quality and, also gives the wine a final pH of 4.0 or higher, which makes it more prone to bacterial spoilage. Wines that have not undergone the MLF at the winery have a probability that this reaction will occur in the bottle. In such cases, the wine becomes gassy and cloudy, and is spoiled.

Leuconostoc oenos is the main species that conducts MLF, and it is uniquely found in the winery environment. Molecular taxonomic studies have confirmed its uniqueness and this organism is now considered to be a new genus and species, *Oenococcus oeni*. Other species of lactic acid bacteria, namely *Pediococcus parvulus*, *P. pentosaceus*, *P. damnosus*, and various *Lactobacillus* spp., can also conduct the MLF but, in addition, they give unpleasant off-flavours, and their growth is undesirable. Unlike *L. oenos*, these species do not grow in wines below pH 3.5, so that control of pH is an effective mechanism for preventing their occurrence. There are interesting microbial interactive factors that affect the growth and ecology of lactic acid bacteria during MLF and these include the action of bacteriophages, bacteriocin production by different strains of bacteria, autolysis of yeasts which provide nutrients for bacterial growth, but also the production of inhibitory substances by some strains of yeasts associated with the alcoholic fermentation. The MLF is a key process in modern winemaking. Its successful completion within a practical time frame or its complete prevention have emerged as major challenges. The unpredictability of naturally occurring MLF has led to the commercial availability of cell concentrates of *L. oenos* for inoculation into wines to induce this reaction, but even this strategy can fail. Other innovations include the use of bioreactors charged with high densities (10^9–10^{10} cfu/ml) of immobilized cells of *L. oenos*, and the development of genetically engineered yeasts with malolactic activity.

Wine Spoilage

Even after alcoholic and malolactic fermentations, wines may contain sufficient nutrients to support the growth of a range of spoilage yeasts and bacteria. Generally, these species will be tolerant of the combined effects of low pH (3.0–4.0) and high ethanol concentrations (10–15% w/v). Yeasts include oxidative species of *Pichia* and *Candida*, and fermentative species of *Zygosaccharomyces*, *Saccharomycodes* and *Brettanomyces/Dekkera*. Lactic acid bacteria include species of *Lactobacillus* and *Pediococcus* as mentioned already. The acetic acid bacteria, *Acetobacter pasteurianus* and *A. aceti*, are well known for their ability to oxidize ethanol at the wine–air interface to give vinegary (acetic acid) spoilage, but their involvement is probably more complex, since these species are frequently isolated from the middle of barrelled wines where little oxygen is present. The application of molecular taxonomic methods will probably reveal the occurrence of novel species of acetic acid bacteria in wines. The potential for acid-tolerant, ethanol-tolerant species of *Bacillus* and *Clostridium* to grow in wines should not be undersurface estimated. Various species of *Actinomyces*, *Streptomyces* and filamentous fungi can grow within the cracks and pores of wooden barrels and lenticels of corks, producing metabolites that give overpowering, deleterious taints when leached into the wine.

Summary

Thus, despite major advances in understanding the microbiology of wine fermentations, there are many areas where further information is required. The microbial ecology of the grape, especially the issue of fungal contamination and control, needs systematic study since it impacts on the rest of the process. It is now clear that yeast species other than *S. cerevisiae* are significant in alcoholic fermentation and more data are needed about their ecological and biochemical contributions, especially at the strain level. Bacteria associated with winemaking demonstrate unique tolerances to the harsh stresses of the wine environment and require more thorough study of their taxonomy and physiology using molecular methods. These studies are likely to reveal the presence of novel species.

1. 'Food—an ecosystem of spatial heterogeneity', explain.
2. Write a short note on non-culturable microbes in food.

3

A CONSORTIUM OF MICROORGANISMS IN FOOD

INTRODUCTION

Viable organisms may be found in a very wide range of habitats from the coldest of brine ponds in the frozen waters of polar regions, to the almost boiling water of hot springs. They play an important role in the maintenance of the stability of the biosphere. The surfaces of plant structures such as leaves, flowers, fruits and especially the roots, as well as the surfaces and the guts of animals, all have a rich microflora of bacteria, yeasts and filamentous fungi. This normal flora may affect the original quality of the raw ingredients used in the manufacture of foods, the kinds of contamination which may occur during processing, and the possibility of food spoilage or food- associated illness. Thus, in considering the possible sources of microorganisms as agents of food spoilage or food poisoning, it will be necessary to examine the natural flora of the food materials themselves, the flora introduced by processing and handling, and the possibility of chance contamination from the atmosphere, soil or water.

BACTERIA IN FOODS

Gram-negative Aerobic Rods and Cocci

Organism	Features	Source	Importance in foods
Pseudomonas *P. aeruginosa* *P. cepacia*	Grows in distilled water; produces pigments which protect them from UV rays; produces enterotoxin in food.	Soil, plants, animals, raw vegetables	• Proteolytic spoilage of proteinaceous food. • Important spoilage agent in refrigerated food and fresh animal food (meat).

(Contd.)

Table   (Continued)

Organism	Features	Source	Importance in foods
	• Utilizes non-carbohydrate energy source and produces a variety of products affecting flavour.		
	• Uses simple nitrogenous foods to synthesize its own growth factors/vitamins.		
	• Proteolytic/lipolytic, psychrophilic		
	• Resistant to disinfectants and sanitizers.		
	• Readily killed by heat.		
Xanthomonas	• Plant pathogen • Excretes xanthan (gum)	Plants	Jelly, used in food industry.
Gluconobacter *G. oxydans*	• Producers of acetic acid	Vegetables, fruits, beer, wine, vinegar	Souring of fruits.
Halobacterium *Halococcus*	• Extreme halophiles • Produces the red pigment bacterioruberin.	Solar salt	Spoilage of salted commodities.
Alcaligenes		Soil, water, decaying matter, intestinal tract of animals	Spoilage of eggs and dairy products.
Acetobacter	Oxidizes ethanol to acetic acid	Fruits and vegetables	Souring of fruit juice, beer and wine.

(Contd.)

Table   (Continued)

Organism	Features	Source	Importance in foods

Brucella
  B. melitensis
  B. abortus
  B. suis
	Susceptible to heat	Milk, air, uncooked meat	Brucellosis, a food-borne disease

Campylobacter *C. jejuni*	Grows in reduced oxygen tension	Faecal matter	Gastroenteritis
Desulfotomaculum	Sulphur reducer	Soil, fresh water, rumen	Sulphide stinker in canned food

Gram-negative Facultative Anaerobic Rods

Organism	Features	Source	Importance in foods
Escherichia *E. coli*	• Heat-sensitive cells • Indicator organism of faecal contamination	Soil, water, plants, intestinal tracts of animals	Spoilage in food with acid and gas. "Barny" flavour in foods.
Edwardsiella *E. tarda*	Non-lactose fermenting	Humans and animals	Spoilage
Citrobacter	Confused with *Salmonella*	Animal products like eggs	Enteritis
Salmonella	Heat-sensitive readily destroyed by pasteurization	Ubiquitous—soil, water, sewage, animals, humans, processing equipment, feed, human faeces	Food-borne infections
Shigella	Encapsulated opportunistic pathogen	Water, sewage, soil, flora of mouth, pharynx and intestinal tract, grains, frozen foods.	Gastroenteritis

(Contd.)

Table   (Continued)

Organism	Features	Source	Importance in foods
Enterobacter			
E. cloacae	Indicator organism	Soil, water, sewage, intestinal tract	Spoilage of cream-filled pastries, milk.
E. aerogenes			
Haffnia	Non-faecal origin	Soil, water, faeces	Spoilage of dairy products.
Serratia	Red-pigment-(prodigiosin) producing bacteria	Soil	Common spoilage flora.
S. marcescens			
Proteus	Swarming growth on agar plates	Gastrointestinal tract, sewage, soil, decomposed animal protein	Enteric infection.
Yersinia	Produces heat-stable entertoxin	Ice creams, chocolates	Gastroenteritis in humans.
Y. enterocolitica			
Erwinia	Produces pectinases that degrade pectin causing soft rots	Plants, soil	Spoilage of vegetables.
E. carotovora			
Vibrio	Growth at alkaline pH	Diseased fish and cured meats	Gastroenteritis in humans.
V. anguillarum			
V. costicola		Seafoods	
V. parahaemolyticus			
Aeromonas	Misidentified as coliforms	Water, fish/seafood	Gastroenteritis, spoilage of fish.
Chromobacterium	Produces violet and dark blue pigmentation, possesses antibiotic properties	Water, soil	Spoilage of food.
C. violaceum			
C. lividum			
Flavobacterium	Produces yellow and orange pigment	Water, soil, animals, refrigerated fish, meat	Discolouration in food.

Gram-positive Cocci

Organism	Features	Source	Importance in foods
Micrococcus	Grows in the presence of 5% salt.	Soil, water, skin of man and animals, milk, dairy products, meat products	Potential spoilage organisms
Staphylococcus S. *aureus* S. *epidermidis*	Grows at 7.5–15% salt, sensitive to chlorine, chloramines, iodine, iodophore; moderately resistant to radiation, sensitive to heat	Human nasal cavity, skin, non-hygienically handled food, fingertips, pimples, acne, boils, wounds	Acute gastroenteritis, food poisoning organism
Streptococcus	Susceptible to phage attack, heat-resistant, survives at 60°C for 30 minutes, survives freezing hence an important organism in frozen foods.	Air, water, sewage, plants, animals, milk	Production of milk products, yoghurt, butter milk, curd, sauerkraut, spoilage of fresh milk.
Leuconostoc L. *mesenteriodes* L. *dextranicum*	• Requires complex nutrients • Non-pathogenic • Produces dextrans causing slime in sugar solutions • Lactic starters for buttermilk, butter and cheese • Production of diacetyl, tolerance to salt concentration • Initiates fermentation in vegetable products • Tolerates increased sugar concentration (55–60%)	Plant surfaces, dairy equipment.	Flavour defect in orange concentrate Manufacture of fermented vegetables and dairy products.

(Contd.)

Table   (Continued)

Organism	Features	Source	Importance in foods
	• Produces carbon dioxide to cause openness in cheese and leavening in bread • Heavy slime production		
Pediococcus *P. damnosus* *P. cerevisiae*	• Microaerophile, poor surface-growth, homofermentative, • Salt-tolerant and acid-tolerant	Pickles, wine, beer	Spoilage of beer
Aerococcus *A. viridans*	• Similar to *Pediococcus* • Grows in blood agar with a green zone surrounding the colony	Air, dust, humans, raw vegetables	Spoilage of vegetable products and meat products

Gram-positive Endospore-forming Rods

Organism	Features	Source	Importance in foods
Bacillus	Psychrotrophs, mesophiles and thermophiles (–5 to 45°C)	Soil, water, faecal matter, decaying material, sugar, starch	Proteolytic enzymes useful to clot milk in cheese.
B. acidocaldarius *B. alcalophilus*	• pH 2 • pH 7.5–8 • Salt tolerant (2–25%) spores resistant		
B. subtilis	• Decompose pectin in plant tissue • Low-acid foods, quality control organism for sterility testing		Spoilage of fresh plant products
B. stearothermophilus			Spoilage
B. coagulans *B. cereus* *B. pumilus*	• Sterility testing organism		Spoilage of tomatoes, food- borne pathogen

(Contd.)

Table (Continued)

Organism	Features	Source	Importance in foods
Clostridium *C. botulinum* *C. sporogenes* *C. perfringens* *C. thermosaccharolyticum* *C. putrefaciens*	• Tolerates salt (2.5–6.5%) • Thermophilic, proteolytic and saccharolytic	Soil, intestinal tract of animals and humans, feed and manure	Food-borne disease and spoilage organism.
Lactobacillus *L. bulgaricus* *L. helveticus* *L. lactis* *L. acidophilus* *L. thermophilus* *L. casei* *L. plantarum* *L. brevis* *L. trichodes* *L. buchneri*	Homofermentative and hetero-fermentative, heat-resistant, thermoduric in nature	Plants, animals, manure, dairy products	• Useful in the production of milk-based fermentation products • Green discolouration in meat products, pink sauerkraut (*L. viridescens*) • Bloaters in cucumber • Spoilage of vinegar products
Listeria *L. monocytogenes*	Psychrophilic pathogen	Water, air	Food-borne disease
Microbacterium *M. lacticum*	• Homofermen-tative, heat-resistant • Thermodurics found in high counts in pasteurized milk and dairy products	Raw milk, dairy products	Spoilage of meat products, eggs, dairy products
Photobacterium *P. phosphoreum*	Luminescent, not widespread	Marine environ-ment, water	Phosphorescence in meat and fish

(Contd.)

Table (Continued)

Organism	Features	Source	Importance in foods
Propionibacterium *P. freudenreichii*	Aero-tolerant rods, ferments lactic acid, carbohydrates, polyalcohol to propionic acid and carbondioxide	Humans, intestinal tract of animals	• Production of cheese, causes colour defects in cheese • Large amount of vitamin B_{12} is synthesized by the organism
Brochothrix *B. thermosphacta*	• Long filamentous chain forming knots • Growth temperature is 0–45°C at pH 5–9 and salt concentration 6.5–10%	Environment	Spoilage of meat

Actinomycetes in Food

Organism	Features	Source	Importance in foods
Streptomyces	• Growth temperature ranges from 10–50°C with pH requirements of 5–10 and salt concentrations of 1–2% to 10%. • Produces antibiotics	Soil, air, flour, wheat whole corn	Musty or earthy odour in food, production of SCP and protease enzyme

Rickettsiae in Food

Organism	Features	Source	Importance in foods
Coxiella *C. burnetti*	• Gram-negative short bacilli, non-motile, grows in vacuole of host cell • Resists drying and high temperatures • Does not grow on agar, but only in chick yolk • Temperature of 62.8°C for 30 minutes or 71.7°C for 15 seconds eliminates the organism	Air, cattle shed environment, raw milk	Causes Q-fever in humans

FUNGI IN FOODS

Yeasts in Food

Yeasts are divided into true yeasts (ascomycetes) and false yeasts or yeast-like organisms which produce no ascospores (deuteromycetes). These yeasts can be oxidative (forming a film (pellicle), called film yeasts), fermentative (yeasts found throughout the liquid medium producing carbon dioxide) and yeasts of both the nature.

Another type of yeast classification is based on the products formed, i.e., brewer's yeast and baker's yeast. Top yeasts undergo rapid fermentation at 20°C, producing carbon dioxide allowing the yeasts to rise up causing clumping of yeast cells at the surface.

Bottom yeasts do not show clump formation and they ferment at a lower temperature (10–15°C).

Organism	Features	Source	Importance in foods
Debaryomyces *D. hansenii* Film-forming yeasts	Spherical to globose, multilateral budding, high salt tolerance (18–20%)	Soil brines	Brined food, salted meat, mushrooms, cheese, tomato puree
Hanseniaspora	Diploid cells, lemon shaped, shows vigorous fermentation, low tolerance to alcohol	Soils of orchards, vineyards, grapes	Fermenting/spoilage of fruits.
Hansenula	Multilateral budding, pseudomycelium may be formed, colonies are white, grayish white or cream	Grains, fruits, brine of cucumber and olives	Production of kanji and rice wine, single cell protein
Kluyveromyces *K. lactis*	• Multilateral budding, red pigmented colonies, shows vigorous fermentation (5–46°C) • The organism is a osmophile	Fruits, preservatives, corn meal, grape must, milk, dairy products	Spoilage of figs, dairy products.

(Contd.)

Table   (Continued)

Organism	Features	Source	Importance in foods
Pichia	Cells are multishaped with pseudomycelium	Various sources	Spoilage of fruits, beer, dairy, sorghum, wine.

Saccharomyces
  S. uvarum
  S. cerevisiae
  S. aceti
	• Sugar fungus, vigorous fermentative with production of alcohol and carbon dioxide	Sugar, honey, dairy products	• Brewing and baking, production of yeast extract, source of SCP
			
			• Assay for vitamins, source of enzyme invertase, production of acetic acid

S. baili var osmophilus	• Grows on 40–60% sugar		• Spoilage of honey, dates
S. rouxii, S. bisporus		Beer, sugar	
Torulopsis T. utilis	• Spherical, oval or elongate cells • White cream colonies, grows in 2–21% NaCl	Pickles, salted bread, butter, cheese	Surface slime on cottage cheese, cream, butter and production of SCP
Trichosporon		Butter, cheese, fruit juice, rice, part of flora used in idli	Source of protein, spoilage of crab and shrimp
Schizosaccharomyces	• Spheroidal and cylindrical shaped cells • No budding reproduction seen, only fission • Closely related to Saccharomyces	Low-acid wine	Spoilage of figs, raisins, wine

False yeasts/yeast-like organisms

Organism	Features	Source	Importance in foods
Candida	Pseudomycelium	Soil, water, air, plants, animals, sewage, humans, processing equipments	Fodder yeast, SCP
C. vini	Film-forming yeasts		"Flowers" on wine
C. mycoderma			Food yeast
C. utilis			Fodder yeast
C. albicans			Pathogenic species
Rhodotorula	Red, yellow cells producing carotenoid pigments	Water, air, foods	Spoilage of foods, source of lipids, cysteine and methionine, SCP

Moulds in Foods

Organism	Features	Source	Importance in foods
Mucor *M. rouxii* *M. racemosus*	Zygomycetes	Ubiquitous	Ripen cheese, oriental food preparation, saccharification of starch by "amylo" process
Rhizopus *R. stolonifer*	Mycelia differentiated into stolon, sporangiophore and rhizoids	Ubiquitous	Spoilage flora Manufacture of enzymes (amylase)
Absidia	Pear-shaped sporangia	Ubiquitous	Spoilage flora
Thamnidium *T. elegans*	Grows well in low temperatures (6–7°C) producing sporangia, at higher temperatures (15–20°C), it produces sporangiola	Animal products	Improves flavour of beef, causes defect in meat called "whiskers"

(Contd.)

Table   (Continued)

Organism	Features	Source	Importance in foods
Geotrichum G. *candidum*	White, yellowish, orange, red-coloured firm yeast-like colonies Called the "dairy mould", hyphae are septate, dichotomously branched, reproduction is by oidia	Dairy products, vegetables	Spoilage organism, produces "watery rot" in tomatoes
Aspergillus A. *niger* (brewery waste) A. *flavus* A. *parasiticus* A. *oryzae* A. *fumigatus*	Over 100 species, fungi of storage grains Contains over 29% crude protein Toxinogenic	Ubiquitous	• Spoilage organism of cereal grains, production of citric acid, gluconic acid, amylase, pectinase • Breaks down rice starch to glucon and useful in the manufacture of "sake" • The proteolytic enzyme of *Aspergillus* is used for clotting milk—a substitute of rennet in cheese making • Useful source of fungal protein, • Feed supplement • Produces aflatoxins in grains

(*Contd.*)

Table   (Continued)

Organism	Features	Source	Importance in foods
Botrytis *B. cinerea*	Conidiophore is a sclerotium, irregularly branched Conidia on short sterigmata, position and numbers of sterigmata cause a grape like conidial arrangement- "gray mould"	Common soil contaminant	Spoilage of vegetables, grapes
Penicillium	• Related to aspergilli, branched conidiophore		Meat and refrigerated food spoilage Soft rot in fruits
P. expansum	• Blue-green conidia	Fruits	Soft rot in citrus fruits
P. digitatum			
P. italicum	• Yellowish green	Fruits	Spoilage of fruits
P. camemberti and *P. roqueforti*	• Blue	Fruits Cheese	Ripening of cheese
P. cyclopium and *P. viridicatum*	• Grow at low level of moisture and temperatures	Grains	Spoilage fungi of grains
P. martensii	"Blue eyes" in corn	Corn grains	Spoilage fungus
P. islandicum and *P. citrinum*	Mycotoxin producing fungi	grains	Yellow-rice disease

Sporotricum
  S. thermophile
 	
  S. carnis
	Saprophyte, grows at 40°C producing cellulose
	Grows at –5 to –8°C
		Grains, animal-based products	Spoils meat causing white spot

(Contd.)

Table   (Continued)

Organism	Features	Source	Importance in foods

Alternaria
  A. tenuis
  A. citri
  A. brassicae
	Dark brown mycelia	Vegetables, soil, air	Spoilage of vegetables (tomato), rancid odour in dairy products

Cladosporium *C. herbarum*	Conidia is one- or two-celled	Air, vegetables	Spoils refrigerated meat causing black spots Spoilage of stored grains, dairy products
Helminthosporium *H. maydis*	Dark mycelium Conidia is more than two celled	Rice stub and other cereal crops	Spoilage of cereal crops

1. Differentiate between oxidative and fermentative yeasts.

Unit II

FOOD PRESERVATION STRATEGIES

INTRODUCTION

Microorganisms are present everywhere since they are spread easily through the air, human beings, insects and such like. The microbiologist makes a distinction between bacteria, moulds, yeasts and viruses. Viruses often need a living host and therefore do not enter products directly. If ambient conditions are favourable, microorganisms can multiply very rapidly indeed. A single microbe can multiply into countless microorganisms within a very short period of time. This is a highly desirable phenomenon in the production of wine, yoghurt or beer, but is highly undesirable in other cases, where there is microbial contamination. One speaks of microbial contamination if the number of undesirable microorganisms and/or the waste products generated by them render the product unacceptable. Undesirable microorganisms can be distinguished into contamination-causing and illness-causing (pathogenic) microorganisms. Pathogenic microorganisms can spoil the product without human beings coming to know of their existence through the five senses. Microbial contamination can render the taste of food products unacceptable, or may be injurious to health if ingested. Known examples of taste spoilage of products are mould formation on cheese and the souring of milk due to the waste products of lactic acid bacteria.

The microbiological stabilization of a product can be achieved through the following:

- By restricting the growth of undesirable microbes or
- Inactivating microbes by destroying them

Microbiological stabilization is not concerned with the removal of all existing microbes but with the reduction in the number of undesirable microorganisms to a value below a specific critical value and thus prevent them from rising above this critical value during the shelf storage period of the product. Specific bacterial counts are used as an indication for these purposes. The bacterial count represents the number of microorganisms per gram of product which is capable of multiplying if favourable conditions arise. This provides an indication of the contamination level. Just as for human beings, microorganisms also need living conditions in order grow, for example the presence of water, optimum temperature, oxygen and nutrition. If one or more of these elements is removed, the number of microorganisms can be limited to below spoilage levels. Microbiological stability can be achieved through a play-off between product characteristics, packing, manufacturing processes and distribution chains. However, other factors also have an effect, such as the quality of the raw materials and other additives, product treatment and the method of operation of the operator.

The prevention of microbiological spoilage is possible through finding an optimum balance between preservation methods and packing systems. Preservation consists of using a treatment process for extending the shelf life of the product. The most suitable preservation method depends upon the properties of the product, the desired shelf life and the desired final quality of the product.

The food and beverage industry still suffer from significant loss due to food poisoning and spoilage microorganisms. Even in developed countries like the United States, there exists a threat to the economy due to food-borne pathogens and food spoilage organisms. Figures from the United States Ministry of Agriculture showed that *Escherichia coli* O157:H7 and non-O157:H7 shiga-toxic *E. coli* cost the US yearly > \$900 M with more than 2000 registered cases in 2000. When such is a state of the developed world, a fast-developing nation like India needs to do a lot of research in this area. Hence, there is an urgent need to minimize the risk of food contamination.

In order to ensure that in an optimal way, the risk on food contamination should be considered in a quantitative way. There is a need to predict the behaviour of undesirable microorganisms and thus get an insight into their cellular functioning under conditions generally encountered under relevant food manufacturing conditions. Optimal use of the mechanistic knowledge will be made only if these data can be put in the context of single cell versus cellular population behaviour and in the context of the physico-chemical parameters that determine food taste, flavour and its nutritional value. It needs to cover the translation of mechanistic cellular stress models at the population level to the molecular events and linked cellular physiology occurring in single cells. In doing so, the stochastic events in terms of modelling, e.g. the amount of transcripts and, hence, molecules of certain transcription factors regulating the onset of stress response routes will play a prime role. The distribution of transcription factors over a population of cells determines the likelihood of a given cell displaying resistance under a given set of conditions occurring in raw materials and during food processing. For example, under conditions of nutrient starvation, as spoilage cells present in the food chain will often experience, bacilli are known to opt for either sporulation, general stress response through αB or specific stress responses. Thus the physiology of microbial cell plays a major role in predicting the method of preservation that would be best suited for maintaining the integrity of food.

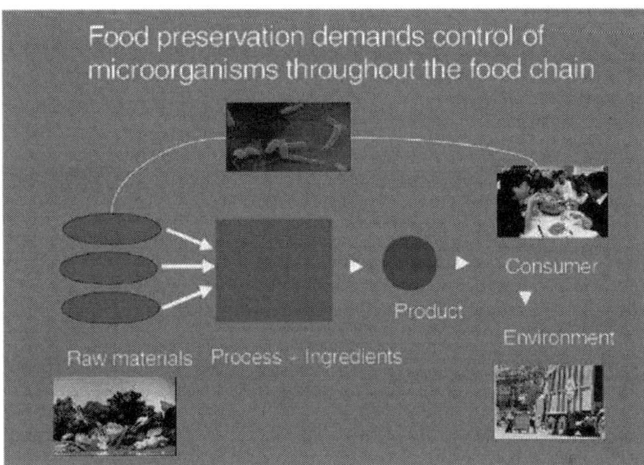

Figure 4.1 A schematic illustration of the point that proper food preservation strategies demand a control of microorganisms throughout the entire food chain, from the raw materials to the consumer up to disposal of the remains of the product

PHYSIOLOGY OF CELLS

The physiology of cells can be studied under two headings:

Genomics

The large-scale sequencing of microbial genomes has opened the way to a full analysis of microbial behaviour. The genomes available at various sites include the model organism for gram-negative bacteria *E. coli*, the model organism for gram-positive bacteria, *Bacillus subtilis* and the model organism, for fungi *Saccharomyces cerevisiae*.

S. cerevisiae has long been and still an organism (eukaryote) of choice for many different types of cellular physiology studies. It was among the first organisms to be fully sequenced and analysed at the genome level for its gene expression under environmental changes, e.g. leading to the diauxic shift and sporulation. Subsequently, in bacteria, gene expression of *E. coli* was analysed upon growth in rich and poor media.

The work on *B. subtilis* has lead to the identification of cellular events that take place during carbon catabolite repression and protein secretion at the proteome level.

Cellular Homeostasis

Cells strive to maintain an optimal balance between those metabolic processes needed for growth and those needed for survival stress response. From an application point of view, studies on growth regulatory systems have been the focus, particularly of those who aimed at increasing the yield of a fermentation process. Fundamentally, interest in how cells adjust their control of cellular metabolic performance can be found at three levels (Figure 4.2). Do cells adjust their metabolism, their protein composition and/or even their gene expression pattern upon applying stress? The issue of

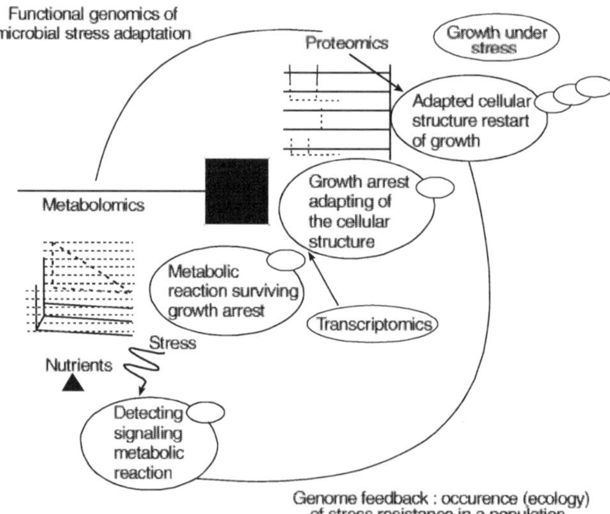

Figure 4.2 A schematic outline on the application of metabolomics, transcriptomics and proteomics in a functional genomics (global physiology) study of stress response. Initial stress survival, stress adaptation and stress resistance are indicated. Note the feedback loop from stress resistance and outgrowth to the ecological distribution of stress sensitivity in a given population of cells.

which step in cellular metabolism exert most control on the overall performance then complements this type of questions. Important in that respect is the notion that it is virtually never only one "rate" controlling step that is key in this. Stress is imposed on cells in various ways. The environment may change in terms of water availability, acidity, temperature profile, presence of antimicrobial compounds (preservatives), absence of nutrients, etc. In our perspective, the change from a normal physiological situation to a situation of stress is a gradual one. If one takes the analogy with humans walking outdoor in winter time, it goes from walking outdoors and feeling the chillness and "freezing to death". There is a gradual scale. Similar to human populations, microbial cells also generally have two types of defence responses to these events. First, there is an intrinsic distribution of resistance against a particular stress among a population. Second, there is a metabolic signalling leading to stress adaptation through production of stress-adaptive proteins.

Indeed in microbial inactivation curves, very often there are shoulders and tails indicative of more resistance of a subpopulation in the isogenic start population. Studies linking this concept to cellular physiology and molecular biology have been, however, scarce. Recently, it has been shown that fluorescent staining methods have allowed a physiological classification of individual cells (*E. coli, Rhodococcus* sp. and *S. cerevisiae*) based on metabolic activity, reproductive activity and membrane integrity. At least four subpopulations were distinguished based on this analysis. This was brought one step further when it was shown that with respect to resistance to heat stress, there was a varying level of induction of the molecular response in the population analysed. The induction of heat stress response at the molecular level is correlated thereby with a low membrane damage. In stress-adapted cells, the inactivation curve did not change in shape but was raised a few decimals indicating that both intrinsic and acquired (induced) stress response is distributed heterogeneously in the population. The cellular defence against hostile environments generally consists of fortification of the cell wall and membrane, the "outer and inner walls" of microbes where it concerns the action of antimicrobial agents. Membrane adaptations in *Clostridium botulinum* cells resistant to heat and the antimicrobial peptide nisin have been reported. Also, acid-adapted *Listeria monocytogenes* displays enhanced tolerance against the antibiotics nisin and lacticin 3147, which were presumably at least partially correlated with observed changes in the fatty acid composition of the bacterial membrane. Activity at and induced changes in the bacterial cell wall have also been reported as a response to the presence of antibacterial enzymes and peptides such as nisin. In yeast cells, studies on resistance development against membrane-active peptides have shown that this generally leads to a significant increase in the chitin and cell wall protein levels, specifically of cell wall mannoproteins such as Cwp1p and Cwp2p.

The resulting cellular changes as a function of stress adaptation to weak organic acids have been described extensively in yeast. The notion that a proton-pumping ATPase plays a crucial role has been established since long. However, these studies received a strong push at the end of 1990s as a molecular characterization of the involvement of a multidrug resistance pump in resistance development against weak organic acids was established. Further, it has been shown that this pump transports preservatives, sorbic acid, benzoic acid and acetic acid from the cytosol to the extracellular environment. Recently, microarray and proteomic analyses of weak organic-acid-resistant cells have been performed. This has indicated important additional resistant mechanisms such as the activation of heat-shock proteins and the activation of the cell integrity pathway.

CELLULAR STRESS SIGNALLING SYSTEMS

Genomic Transcript Profiling

Genomic transcript profiling combined with cellular physiology has also opened the way to assess the cellular signalling systems involved in transmitting stress and regulating the defence systems. This has a crucial spin-off to food preservation research that truly predictive models of microbial growth inhibition through hurdle technology can be generated. Of the classical preservatives (weak organic acids) little is known about the initial signal that sets off a cellular response. Besides lowering the internal pH, the acids also have an effect at the cell membrane. Particularly, sorbic acid was thought to exert a large part of its antimicrobial effect in this way. Furthermore, in order to prevent the development of a futile ATP-consuming metabolic cycle in which the anions and the protons are extruded from cells while re-influx occurs when the proton and anion reassociate at an extracellular pH equal to the pK, adjustments of the cell envelope were thought to take place. Indeed, the recent genome-wide analysis has indicated that cells responding to sorbic acid stress activate their cell integrity pathway signal transduction system. Upon inspection of the set of induced genes and proteins, it is evident that events at the cellular plasma membrane must be the start of the signalling cascade. Indeed, recently, it has been shown that the membrane sensor Wsc2p, belonging to the Wsc sensor family involved in cellular response against membrane perturbation through heat via the protein kinase C1 (PKC1) pathway, regulates the membrane H-ATPase. The latter is known to form a crucial part of the sorbic acid stress response output. Acid stress response in enteric bacteria involves an orchestrated stress response of partially overlapping arrays of acid stress response proteins. Stress response against stressful environmental conditions has also been studied recently genome-wide in *E. coli* as indicated earlier. Such transcription profiling showed that the general stress regulator RpoS, which is normally only induced upon reaching stationary growth phase, is induced in cells experiencing low nutrient levels already in the logarithmic phase of growth (Figure 4.3). In *B. subtilis*, the response of cells towards environmental conditions in terms of forming high or low heat-resistant spores was studied. A regulatory role for the small acid soluble spore proteins (Sasps) was proposed. This was confirmed by experiments using knockouts of Sasps and assessment of heat stress resistance in the resulting spores. Stress response against membrane active antimicrobial agents, often of natural origin, has also recently been the subject of extensive investigation particularly in yeast and vegetative cells of gram-positive bacteria. In yeast, it has been shown that the incorporation of different amounts of cell wall proteins 1 and 2 and the increase in cell wall chitin content in response to nisin and synthetically modelled antimicrobial peptides are presumably also regulated at the cellular membrane via the PKC1 pathway. In gram-positive bacteria, the concentration of cell wall precursor lipid II is crucial in determining nisin resistance. Nisin binds to this molecule and in this way, prevents normal cell wall biosynthesis next to having a direct membrane perturbing effect. In fact, recently, it has been shown that in spontaneous nisin-resistant mutants, the expression of a putative penicillin-binding protein was significantly increased, further indicating the link between cell wall and membrane homeostasis. Changes in the membrane phospholipid fatty acid and head-group composition were proven to be crucial in increasing resistance of *B. cereus* against this compound. In quite all cases in bacterial stress response reactions, one or more alternative factors are involved that direct the RNA-polymerase to the specific stress response genes. The link between cell wall and membrane homeostasis was recently also evident from studies on the cellular response systems against cell wall degrading enzymes in yeast. Treating cells with cell wall lytic enzymes resulted in the activation of membrane localized stretch-sensitive receptors from the Wsc and mid-protein family.

Integrating Physiology with Molecular Biology

Cells like machines can use so many options, as there is energy available to drive the systems. Recent experiments in yeast aim at documenting the exact energy requirements of the cells, when actively growing, when initiating stress response systems and when recovering from the growth lag after stress application. First results with a physical stress, heat, in yeast show that applying a continuous temperature stress on wild type cells leads to an arrest of cell growth, a temporal increase in the glucose flux and ethanol production followed by activation of the cell integrity protein kinase C1 pathway. The latter alludes to the likelihood that the cells stop cell division and alter their carbon metabolism upon activating their stress response systems. Stress response upon heat stress was shown to activate at the molecular level the transcription factor Slt2 (Mpk1) by directly measuring the protein phosphorylation status using specific antibodies.

Figure 4.3 shows a schematic outline of cellular signal transduction and response mechanisms in yeast cells towards environmental factors. The combination of a given nutrient availability and a set of environmental (stress) conditions will determine whether a cell can restart growth. Depicted are the response against hyper- and hypoosmotic stress, heat stress and stress with the food preservative sorbic acid. A few future scenarios are indicated for the system analysis of cellular behaviour against (food preservation) stresses.

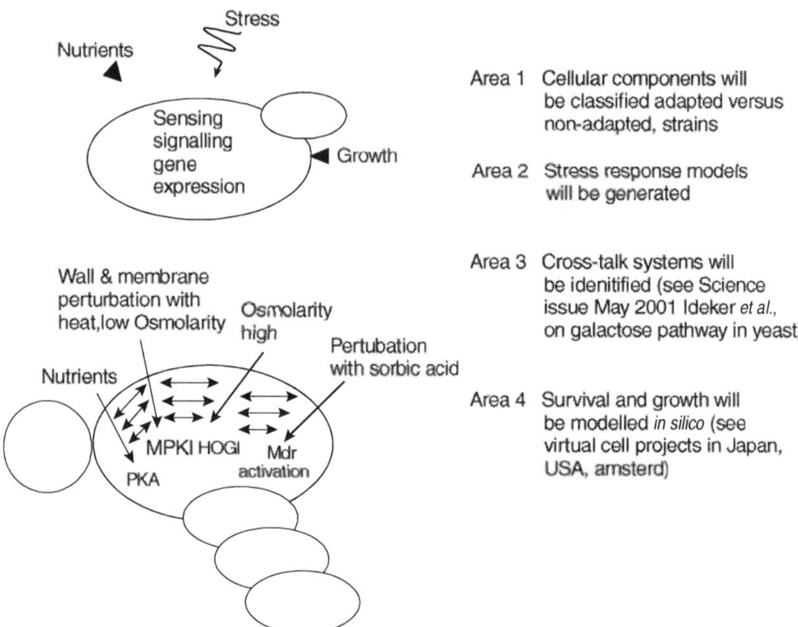

Figure 4.3 Integrated genomics or systems analysis of cellular behaviour against (preservation) stress: The future

OUTLOOK

The understanding of cellular response at the level of the molecular events opens up the way to integrally assess microbial response to environmental conditions beyond the level of growth–no-growth or survival and death. Thus, this allows for the development of mechanistic growth and inactivation models that will also have a predictive power outside the measured data points models that will be able to act as guides in identifying novel combinations of environmental stress conditions through their level of detail at the "wiring" of the cellular machinery, and models that will look like chemical engineering tools describing the flow of metabolites through the cellular regulatory systems. With such predictive models at hand, the development of food preservation systems can be approached more and more quantitatively and can be handled by integrating with other food processing unit operations in one process model. The technology, when applied in an integrated way in food processing, should lead to a significant shortening of the evaluation of novel processing (preservation) systems. This may be achieved in a number of cases through the full identification of relevant cellular targets, but may also be at the level of generating fingerprints of given treatments. In this way, a very fast comparison of cellular response towards a large variety of processing/preservation treatments will be made possible. In the pharmacology area, such a pharmacogenomics approach has been used in characterizing novel antifungal compounds. In addition, strain comparisons in terms of responses to treatments can easily be made (strain fingerprints). In this way, equivalence of treatments and strains can unequivocally be assessed in a relatively short time frame. The following are the advantages of a better understanding of the physiological actions of preservative systems in the food manufacturing industry as seen over the next 5–10 years.

- Improved food quality and wholesomeness through lower thermal treatment
- Possibilities for new products (mildly preserved, organic foods); energy-saving through better controlled processes
- Less use of waste and cooling water
- Reduction in the use of cleaning and disinfecting agents
- Less waste through better process control

Thus food preservation is based on the following:

1. ***Delaying microbial decomposition***   This can be achieved by

- Keeping away from microbes (asepsis)
- Removal of microbes (filtration, centrifugation, washing, trimming, blanching)
- Hindering the growth of microbes (refrigeration, drying, adding chemical preservatives)
- Killing the microbes (irradiating and heat processing)

2. ***Delaying self-decomposition of foods***   This can be achieved by

- Blanching
- Addition of antioxidants

The various traditional preservation techniques (Table 4.1) include :

1. Asepsis/keeping away from microbes
2. Removal of microbes
3. Use of high-temperature
4. Use of low-temperature
5. Drying
6. Chemical preservatives

7. Biological preservatives
8. Irradiation
9. Mechanical destruction of microbes (grinding)
10. A combination of any of these

Some of the technological innovations in food processing and preservation include:

1. Pulsed electric field
2. Ohmic heating
3. Oscillating magnetic field
4. Light pulses
5. High ultrasonic hydrostatic pressure
6. Active packaging
7. Natural antimicrobial compounds from animals or plants
8. Supercritical CO_2
9. Polycationic polymers

If one considers the bacterial growth curve pattern, one can divide the growth phase into various subdivisions as follows:

1. Lag phase
2. Phase of positive acceleration
3. Log phase
4. Negative acceleration
5. Stationary phase
6. Accelerated death phase
7. Death phase
8. Survival phase

Lag phase is the inital growth phase during which there is no increase in cell number, only increase in cell mass occurs.

During the next phase of positive acceleration the rate of growth continuously increases. The next phase is the logarithmic phase of growth during which the rate of multiplication is most rapid and is constant. The phase of negative acceleration shows a decrease in the multiplication rate. Next is the stationary phase where the cell numbers remain constant.

During the accelerated death phase and the death phase, cell numbers decrease at a faster rate than the rate at which new cells are formed. The last phase is pictured as the survival phase during which no cell division occurs but remaining cells survive on endogenous nutrients.

Table 4.1   A chronological classification of food processing and preservation

Old processes (since pre-history)	Current processes
Sun drying	Spray drying
Oven drying	Freeze drying
Smoking	Canning
Salting	Aseptic processing
Pickling	UHT pasteurization
Fermentation	UHT sterilization
Freezing	Extrusion cooking
	Irradiation
	Microwave heating
	Reverse osmosis
	Osmotic dehydration
	Modified atmosphere packaging
	Freezing or chilling

The main point in preservation from the point of view of a food microbiologist is to lengthen the lag phase. This can be accomplished by reducing the amount of contamination, (fewer the organism, longer is the lag phase), avoiding addition of actively growing organisms as can be seen in unclean utensils, by changing one or more environmental factors like pH, temperature, water activity, relative humidity, etc. and by causing actual damage to organism by heat or radiation.

1. Define cellular homeostasis.
2. Give a brief account of the cellular stress signaling systems.
3. List the principles of food preservation.

5

ASEPTIC PACKAGING

INTRODUCTION

Packaging is an essential component in the complex distribution system which moves agricultural products from their point of origin to their point of consumption. Along with our culture, the packaging methodology has become increasingly specialized and complex. Sterile boxes of fruit juice, plastic food pouches that can tolerate pressure canning, and microwave-safe containers are a few examples of modern specialized food packaging methods. Food packaging has several functions in addition to the primary function, which is protection. Packaging protects food from microbial and other environmental contaminants, and from damage during distribution. It also provides the consumer the information on nutrition and ingredients, cooking instructions, product weight, brand identification and pricing. Finding the right method for extending the shelf life of highly perishable products without affecting their quality is indeed a difficult challenge. There are a number of factors which are to be taken into consideration in this connection. Stringent hygienic standards have to be maintained during the manufacturing process, and during storage and distribution, the net effect of which is to improve the shelf life of the product to a small extent. It has also been found that the general pattern for food products is to raise the product to the intermediate shelf life zone (which can extend from a few weeks to a few months). The shelf life is therefore oriented more towards the conventional distribution and consumption pattern. Since milder preservation methods can be used, the product quality is also higher and the energy consumption and the environmental hazards are therefore correspondingly reduced. The contamination of the product is avoided since the packing is done according to more hygienic standards.

Recent years have brought increased environmental pollution, decreased natural resources and landfill space. But improper reuse of food packaging materials can threaten health. So it is important to follow certain guidelines. For example, it is generally safe to reuse glass that has been used for food, but not to reuse plastic or paper. Understanding the uses, specialized functions and limitations of packaging materials used for protecting consumer foods can help consumers take safe decisions regarding the reuse of packaging materials.

CAUSES OF SPOILAGE

Extending the shelf life actually means delaying the spoilage of the products. One can speak of spoilage when a product has an unacceptable quality, or is no longer suitable as a product of that type. This may happen due to the following:

1. Biological damage such as microbial or enzymatic spoilage or through external damage such as through insects and vermin.
2. Chemical damage such as change in taste, colour and odour of modifying agents, through reaction with oxygen.
3. Physical damage such as dehydration, absorption of moisture, rupture or abrasion.

In principle, there are always various kinds of potential spoilage-causing elements present in the product. These various influences gain the upper hand on their way to the ultimate consumer. For extending the shelf life of the product, attention is focused on the most important factors and elements causing spoilage. For example, for countering the effect of moisture absorption or colour change occurring due to the absorption of oxygen, one may opt for packing materials which offer a higher physical barrier. To counter enzymatic spoilage, it is sufficient to subject the product to heat for a brief period of time.

MANAGING MICROBIAL ACTIVITY

In conventional stabilization techniques, the product is packed and processed in the sealed packing container itself. It is however, also possible to first process the product and then pack it. This trend is gaining ground at the present moment. In aseptic packing, the sterilized product is placed inside the sterilized packing material in a sterile environment. The most important advantages of this technique are a greatly lowered thermal load on the product along with reduced energy consumption as well.

Stabilizing without Loss of Quality

Consumers and commercial users are constantly demanding higher quality in products. For maintaining quality, the chemical loading (preservatives) or physical loading (intensive sterilization) to which the product can withstand is quite limited. Using a light product-stabilization process followed by hygienic packing, is perhaps the best solution for a large number of products and application, taking into account the variety of parameters which have to be met. Increasingly, process equipment will be chosen on the basis of microbiological requirements. The equipment has a very important effect on the quality of the product from the microbiological point of view. This is the case where the equipment causes re-contamination of the product. Depending on the hygiene related options selected, one can distinguish three types of packing equipment.

1. *GMP equipment* GMP (Good manufacturing practice) is a preventive system with specifications, standards and procedures according to which the manufacturing process is to be carried out. This ensures the achievement of highest level of quality, consistency and constancy. The process is also quite safe, and manufacturing also takes place very efficiently, and the legal rules and regulations relating to the product in question are also adequately satisfied thereby. The specifications for product quality are translated into specifications for raw materials, the production process, personnel and so on. If all the production factors satisfy the specifications, the chances of achieving a consistent product quality are the highest.

GMP equipment is an adequate option because it does not have any adverse effect on the product. After cleaning the equipment, there are no product remnants left behind, although the surface may still contain the relevant microbes, for example, if used when the product is pasteurized in the packing container, or if the product itself has life-inhibiting conditions for microorganisms. If during packing, re-contamination takes place, the microbes can be killed at a later stage or cannot proliferate further.

2. *Hygienic equipment* Hygienic equipment is used because heavy recontamination may have adverse effect on the ultimate product quality. After cleaning the equipment, there must be no microorganisms remaining behind in the product contact area. Hygienic equipment is used, for example, in connection with packing perishable products which are to be stored under refrigeration. Hygienic equipment is impervious to microorganisms. For example, it is used for packing of milk intended for extended storage periods and UHT milk, which can be stored at room temperature. Even the smallest amount of re-contamination will lead to unacceptable levels of product spoilage.

3. *Aseptic equipment* Aseptic equipment is used if after the stage in which microbes are killed, further microorganisms are to be prevented from re-entering the product. The product characteristics (pH, the absence of hydrogen and the life-limiting conditions), the production process (whether the microorganisms are killed or suppressed), and the conditions of transport and storage (whether the microorganisms can proliferate or not), are important factors for microbiological stability and the selection of process equipment. Here is an example, by way of illustration: if the product has a pH value of more than 4.6, various types of pathogenic microorganisms can multiply easily. It is therefore necessary to kill the microorganisms and to prevent re-infection from taking place. This can be done by packing the product aseptically. If the pH is less than 3.0, there are no known spoilage causing microorganisms which can multiply within the product. The product is then microbiologically stable. Even other living conditions for microorganisms such as a low a_w (water activity) due to a high salt content, can make the product microbiologically stable. It will be adequate to pack done according to GMP specifications. Products with a pH value between 3.0 and 4.6 are microbiologically vulnerable and therefore need a hygienic approach before extending their shelf life.

DESIGN OF ASEPTIC PACKAGING

Thus aseptic packing means free from disease causing germs. Aseptic packing consists of filling a sterile product into a sterilized packing container at room temperature followed by sealing the packing in a sterile environment. Aseptic production is done in two stages:

1.   *Aseptic manufacturing* which gives the product commercial sterility.
2.   *Aseptic packing* which ensures that the product remains sterile.

In general, the quality of an aseptic product is better than a conventionally sterilized product. A major advantage of successful product stabilization is that the quality of the packed product becomes independent of the size of the packing container. This is often very important for sensitive products or very large packing containers. Another advantage of aseptically packed products are the relatively less stringent requirements regard to temperature and shape consistency of the packing material in comparison to the conventional stabilization processes. This often offers additional design possibilities and cost price advantages.

Establishing microbiological stability is of primary importance for most food products. Early food packaging had the ability to tolerate heat so that the product could be canned (terminally sterilized in the container). Glass and tin cans were the original processed food packages because they could tolerate sterilization temperatures. Because food flavours are prone to change during heating, a variety of sterilization techniques have been designed to kill or remove microbes from food products using little or no heat processing. These techniques include the use of hydrogen peroxide treatments, aseptic packaging (packaging in a sterile environment), ultra-high temperature (UHT) sterilization (which uses high temperatures for very short times) and gamma irradiation. A variety of foods are "hot-filled" at 170°F, so the chemically sterilized packaging must be able to tolerate temperatures slightly above 170°F. Packaging materials that tolerate these processes will not always tolerate temperatures over 180°F without substantial breakdown.

When manufacturers design a packaging system, they consider the type of food (high-acid versus low-acid), the susceptibility of the food to light and oxygen (milk, fruit juice), the amount of physical protection needed (eggs) and the amount of product visibility desired (fresh meats). They also consider the amount of heat to be used during sterilization, the container size, and the type of "home processing" to which the product will be subjected (for example, microwaveable pizza). Other considerations include the cost and availability of the packaging material, the cost of transporting the finished product (plastic versus glass soda bottles) and recyclability (plastic versus paper). Plasticizers increase the flexibility of plastics. They intersperse around the polymer molecules and prevent them from bonding to each other so tightly that they form a rigid substance. Food packaging materials are used to provide only barrier and protective functions. However, various kinds of active substances can now be incorporated into the packaging material to improve its functionality and give it new or extra functions. Such active packaging technologies are designed to extend the shelf life of foods, while maintaining their nutritional quality and safety. Active packaging technologies involve interactions between the food, the packaging material and the internal gaseous atmosphere. The extra functions they provide include oxygen scavenging, antimicrobial activity, moisture scavenging, ethylene scavenging, ethanol emitting and so on.

ACTIVE FOOD PACKAGING

In recent years, the main driving force for innovation in food packaging technology has been the increase in consumer demand for minimally processed foods, the change in retail and distribution practices associated with globalization, new consumer product logistics, new distribution trends (such as internet shopping), automatic handling systems at distribution centers, and stricter guidelines regarding consumer health and safety. Active packaging (AP) technologies are being developed as a result of these driving forces. Active packaging is an innovative concept in which the package, the product, and the environment interact to extend shelf life or enhance safety or sensory properties, while maintaining the quality of the product. This is particularly important in the area of fresh and extended shelf life foods. Antimicrobial (AM) packaging is one of the most promising versions of an AP system.

The general principles of AP and AM packaging concepts including oxygen scavenging, moisture absorption and control, carbon dioxide and ethanol generation, AM migrating and non-migrating systems are discussed in this section.

Oxygen Scavenging Systems

The most promising active packaging systems are oxygen scavenging systems which absorb oxygen gas in the package and prevent rancidity of food. They are being developed as forms of sachets or polymer additives and antimicrobial systems.

The presence of oxygen in food is often a key factor that limits the shelf life of a product. Oxidation can cause changes in flavour, colour, and odour, as well as destroy nutrients and facilitate the growth of aerobic bacteria, moulds, and insects. Therefore, the removal of O_2 from the package headspace and from the solution in liquid foods and beverages, has long been a target of the food-packaging scientists. The deterioration in quality of O_2-sensitive products can be minimized by recourse to O_2 scavengers that remove the residual O_2 after packing. Existing O_2 scavenging technologies are based on oxidation of one or more of the following substances: iron powder, ascorbic acid, photo-sensitive dyes, enzymes (such as glucose oxidase and ethanol oxidase), unsaturated fatty acids (such as oleic, linoleic and linolenic acids), rice extract, or immobilized yeast on a solid substrate. These materials are normally contained in a sachet. Oxygen scavenging is an effective way to prevent growth of aerobic bacteria and moulds in dairy and bakery products. Oxygen concentrations of 0.1% v/v or less in the headspace are required for this purpose.

Packaging of crusty rolls in a combination of CO_2 and N_2 (60% CO_2) has shown to be an effective measure against mould growth for 16 to 18 days at ambient temperature. However, such an "anaerobic environment" is not totally effective without the incorporation of an oxygen scavenger into the package to ensure that the headspace O_2 concentration never exceeds 0.05%. Under such conditions the rolls remain mould-free even after 60 days. Many researchers have expressed concern about the safety of modified atmosphere packaged (MAP) foods, especially with respect to the growth of psychrotrophic pathogens such as *Listeria monocytogenes* and anaerobic pathogens such as *Clostridium botulinum*. They monitored the physical, chemical, microbiological, textural, and sensory changes in surimi nuggets inoculated with *L. monocytogenes*, packaged in either air or 100% CO_2 with and without an oxygen scavenger and stored at 4°C and 12°C. They found that MAP was not effective in controlling the growth of the pathogen in either raw or cooked nuggets and also that the pathogen overcame competitive inhibition and pH reduction caused by lactic acid bacteria. They concluded that nuggets packaged under these conditions and contaminated with this pathogen could pose a risk to the consumer. More importantly, it was found that the product retained the normal odour and appearance at the above storage temperatures, even though the level of the pathogen increased considerably. The latter is indeed a cause for concern, since the contaminated product may appear safe from the sensory point of view.

Oxygen scavenging is advantageous for products that are sensitive to O_2 and light. One important advantage of AP over MAP is that the capital investment involved is substantially lower; in some instances, only the sealing of the system that contains the oxygen absorbing sachet is required. This is of extreme importance to small-and medium-sized food companies for which the packaging equipment is often the most expensive item. An alternative to sachets involves the incorporation of the O_2 scavenger into the packaging structure itself. This minimizes negative consumer responses and offers a potential economic advantage through increased outputs. It also eliminates the risk of accidental rupture of the sachets. Since the share of polymers in primary packages for foods and beverages increases constantly, they have become the medium for incorporation of active substances such as antioxidants, O_2 scavengers, flavour compounds, pigments, enzymes, and AM agents.

BP Amoco Chemical (U.S.A) is marketing Ambsorb® 2000 and 3000, which are polymer-concentrates containing iron-based O_2 scavengers. These can be used in polyolefins and in certain polyester packaging applications for wines, beers, sauces, juices, and other beverages. Others in the line are organic-based, UV light-activated O_2 scavengers that can be tailored to allow them to be bound into various layers of a wide range of packaging structures. Oxbar™ is a system developed by Carnaud-Metal Box (now Crown Cork and Seal) that involves cobalt-catalysed oxidation of a MXD6 nylon that is blended into another polymer. This system is used especially in the manufacturing of rigid PET bottles for packaging of wine, beer, flavoured alcoholic beverages, and malt-based drinks. Another O_2 scavenging technology involves using directly the closure lining. Darex® container products use ethylene vinyl alcohol with a proprietary oxygen scavenger. In dry forms, pellets containing unsaturated hydrocarbon polymers with a cobalt catalyst are used as oxygen scavengers in mechanical closures, plastic and metal caps, and steel crowns (both PVC and non-PVC lined). They can prolong the shelf life of beer by 25%. Oxygen scavengers have opened new horizons and opportunities in preserving the quality and extending the shelf life of foodstuff. However, much more information is needed on the action of O_2 scavengers in different environments before optimal, safe, and cost-effective packages can be designed. The need for such information is especially acute on O_2 scavenging films, labels, sheets, and trays that have begun to appear in recent years.

Moisture-absorbing and Controlling Systems

In solid foods, a certain amount of moisture may be trapped during packaging or may develop inside the package due to generation or permeation. Unless it is eliminated, it may form a condensate with attendant spoilage and low consumer appeal. Moisture problems may arise in a variety of circumstances, including respiration in horticultural products, melting of ice, temperature fluctuations in food packs with a high equilibrium relative humidity (ERH), or drip of tissue fluid from cut meats and products. Their minimization via packaging can be achieved either by liquid water absorption or humidity buffering.

Liquid water absorption   The main purpose of liquid water control is to lower the water activity of the product, thereby suppressing the growth of microorganisms on the foodstuff. Temperature cycling of high a_w foods has led to the use of plastics with an antifog additive that lowers the interfacial tension between the condensate and the film. This contributes to the transparency of the films and enables the customer to see clearly the packaged food although it does not affect the amount of liquid water present inside the package. Several companies manufacture drip-absorbent sheets such as Thermarite or Peaksorb, or Toppan™ (Japan) for liquid water control in high a_w foods such as meat, fish, poultry and fresh produce. Basically, these systems consist of a superabsorbent polymer located between 2 layers of a microporous or non-woven polymer. Such sheets are used as drip-absorbing pads placed under whole chickens or chicken cuts. Large sheets are also utilized for absorption of melted ice during air transportation of packaged seafoods. The preferred polymers used for this purpose are polyacrylate salts and graft copolymers of starch.

Humidity buffering   This approach involves interception of moisture in the vapour phase by reducing the in-pack relative humidity and thereby the surface-water content of the food. It can be achieved by means of one or more humectants between two layers of a plastic film that is highly permeable to water vapour or by a moisture-absorbing sachet. An example of this approach is the Pichit™ film manufactured by Showa Denko (Japan). It is marketed in Japan for wrapping fish and chicken and reduces the ERH in the vicinity of the product, but has not been evaluated experimentally. Pouches containing NaCl have also been used in the US tomato market. Desiccants have been successfully used for moisture control in a wide range of foods, such as cheeses, meats, chips, nuts, popcorn, candies, gums and spices. Silica gel, molecular sieves, calcium oxide (CaO) and natural clays (such as montmorillonite) are often provided in sachets.

Carbon Dioxide Generating Systems

Carbon dioxide is known to suppress microbial activity. Relatively high CO_2 levels (60 to 80%) inhibit microbial growth on surfaces and, in turn, prolong shelf life. Therefore, a complementary approach to O_2 scavenging is the impregnation of a packaging structure with a CO_2 generating system or the addition of the latter in the form of a sachet. Since the permeability of CO_2 is 3 to 5 times higher than that of O_2 in most plastic films, it must be continuously produced to maintain the desired concentration within the package. High CO_2 levels may, however, cause changes in taste of products and develop undesirable anaerobic glycosis in fruits. Consequently, a CO_2 generator is only useful in certain applications such as fresh meat, poultry, fish and cheese packaging. In food products for which the volume of the package and its appearance are critical, an O_2 scavenger and CO_2 generator could be used together in order to prevent package collapse due to absorption. An oxygen-free environment alone is insufficient to retard the growth of *Staphylococcus aureus*, *Vibrio* species, *Escherichia coli*, *Bacillus cereus* and *Enterococcus faecalis* at ambient temperatures. For complete inhibition of these microorganisms in foods, a combined treatment involving O_2 cavenging with thermal processing, or storage under refrigeration, or using a CO_2 enriched atmosphere is recommended. An O_2 and CO_2 absorber inhibited the growth of *Clostridium sporogenes* while an O_2 absorber and a CO_2 generator enhanced the growth of this microorganism, which is quite a surprising result. This result indicates the importance of selecting the correct scavenger to control the growth of *Clostridium* species in MAP foods.

Ethanol Generating Systems

Ethanol is used routinely in medical and pharmaceutical packaging applications, indicating its potential as a vapour phase inhibitor. It prevents microbial spoilage of intermediate moisture foods (IMFs), cheeses, and bakery products. It also reduces the rate of staling and oxidative changes. Ethanol has been shown to extend the shelf life of bread, cake and pizza when sprayed onto product surfaces prior to packaging. Sachets containing encapsulated ethanol release its vapour into the packaging headspace thus maintaining the preservative effect. Many applications of ethanol-generating films or sachets have been patented and marketed, including an adhesive-backed film that can be taped on the inside of a package to provide AM activity. Mitsubishi Gas Chemical Co. patented a sachet containing encapsulated ethanol, glucose, ascorbic acid, a phenolic compound and an iron salt , thereby achieving the combined effect of O_2 scavenging and ethanol generation. An ethanol-generating technology was originally developed in Japan whereby food grade ethanol is encapsulated in a fine inert powder inside a sachet. The rate of ethanol vapour release can be tailored by controlling the permeability of the sachet. Several Japanese companies manufacture this type of ethanol generator, the most widely used being Ethicap® or Antimold Mild® produced by the Freund Industrial Co. These systems, approved for use in Japan, extend the mould-free shelf life of various bakery products. The usefulness of ethanol vapour in extending the shelf life of apple turnovers was effectively demonstrated where the shelf life was found to be 14 days for the product packaged in air or in a CO_2/N_2 gas mixture (60% CO_2) and stored at ambient temperature. Afterwards, visible swelling occurred as a result of *Saccharomyces cerevisiae* growth and additional CO_2 production. When encapsulated ethanol was incorporated in the package, yeast growth was totally suppressed and the shelf life was extended to 21 days. On the other hand, this solution caused the packages to contain 1.5% ethanol at the end of the storage period as compared to only 0.2% when packed without ethanol. Consequently, the final products may be unacceptable to the consumer due to elevated ethanol contents. This problem can be partially resolved by heating the contents of the package prior to consumption, thereby evaporating the ethanol.

Gas Emission or Flushing

Gas emission or flushing controls the growth of mould. Typical spoilage moulds include *Botrytis cinerea*, *Penicillium*, *Aspergillus* and *Rhizopus* species commonly found in citrus and berry fruits. To extend the storage period of these fruits, fungicides or antimycotic agents can be applied. Sulphur dioxide (SO_2) is known to be the most effective material in controlling the decay of grapes and is superior to the gamma irradiation and heat-radiation combination methods. However, a SO_2-releasing material entails a number of problems, including bleaching and SO_2 residues. In Australia, two different SO_2 release sheets were tested for packaging of the white "Thompson Seedless" and the purple "Red Globe" grapes. SO_2 in the surrounding air is absorbed into the grapes and initially converted to sulphite and then metabolized into the sulphate form. At the end of the experiment (after 4 days at 21°C), the sulphite levels in the "Red Globe" were found to be lower than those in the "Thompson Seedless", even though the former was subjected to a higher SO_2 level. This reflects the different metabolic rates of the two varieties. The development of a controlled-release polymer that would apply the fungicide at a sufficient level to retain satisfactory fungistatic action was suggested, while minimizing undesirable effects. Another volatile compound exhibiting AM effects is allyl isothiocyanate (AIT), the major pungent component of black mustard (*Brassica nigra*), brown mustard (*Brassica juncea*) and wasabi (*Eutrema wasabi* Maxim.).

Antimicrobial Migrating and Non-migrating Systems

Antimicrobial food packaging materials have to extend the lag phase and reduce the growth rate of microorganisms in order to extend shelf life and to maintain product quality and safety. Alternatives to direct additives for minimizing the microbial load are canning, aseptic processing and MAP. However, canned foods cannot be marketed as "fresh". Aseptic processing may be expensive and hydrogen peroxide, which is restricted in level by regulatory agencies, is often used as a sterilizing agent. In certain cases, MAP can promote the growth of pathogenic anaerobes and the germination of spores, or prevent the growth of spoilage organisms which indicate the presence of pathogens. If packaging materials have self-sterilizing abilities due to their own AM effectiveness, the need for chemical sterilization of the packages may be obviated and the aseptic packaging process simplified. Food packages can be made AM active by incorporation and immobilization of AM agents or by surface modification and surface coating. Present plans envisage the possible use of naturally derived AM agents in packaging systems for a variety of processed meats, cheeses, and other foods, especially those with relatively smooth product surfaces that come in contact with the inner surface of the package. This solution is becoming increasingly important, as it represents a perceived lower risk to the consumer. Table 5.1 lists a number of substances, which can be bound to polymers to impart AM properties. Such substances can be used in AM films, containers and utensils. Antimicrobial materials have been known for many years. However, antimicrobial packages have had relatively few commercial successes, except in Japan. Antimicrobial films can be classified in to 2 types: (1) those that contain an AM agent that migrates to the surface of the food, and (2) those that are effective against surface growth of microorganisms without migration.

Table 5.1 Examples of antimicrobial agents of potential use in food packaging materials

Class	Examples
Acid anhydride	Benzoic anhydride
	Sorbic anhydride
Alcohol	Ethanol
Amine	Hexamethylenetetramine (HMT)
Ammonium compound	Silicon quaternary ammonium salt
Antibiotic	Natamycin
Antimicrobial peptide	Attacin
	Cecropin
	Defensin
	Magainin
Antioxidant phenolic1	Butylated hydroxyanisole (BIIA)
	Butylated hydroxytoulene (BHT)
	Tertiary butylhydroquinone (TBHQ)
Bacteriocin	Bavaricin
	Brevicin
	Carnocin
	Lacticin
	Mesenterocin
	Nisin
	Pediocin

(Contd.)

Table 5.1   (Continued)

Class	Examples
	Sakacin
	Subtilin
Chelator	Citrate
	Conalbumin
	EDTA
	Lactoferrin
	Polyphosphate
Enzyme	Chitinase
	Ethanol oxidase
	β-Glucanase
	Glucose oxidase
	Lactoperoxidase
	Lysozyme
	Myeloperoxidase
Fatty acid	Lauric acid
	Palmitoleic acid
Fatty acid ester	Monolaurin (lauricidin)
Fungicide	Benomyl
	Imazalil
	Sulphur dioxide
Inorganic acid	Phosphoric acid
Metal	Copper
	Silver
Miscellaneous	Catechin
	p-Cresol
	Hydroquinones
Natural phenol	Reuterin

Coating of Films with Antimicrobial Agents

Appropriate coatings can sometimes impart AM effectiveness. A polymer-based solution coating would be the most desirable method in terms of stability and adhesiveness of attaching a bacteriocin to a plastic film. It was found that low-density polyethylene (LDPE) films coated with a mixture of polyamide resin in *iso*-propanol/*n*-propanol and a bacteriocin solution provided AM activity against *Micrococcus flavus*. The migration of bacteriocins reached equilibrium within 3 days, but the level attained was too low to affect several bacterial strains spread on an agar plate media. When the films were in contact with a phosphate buffer solution containing strains of *M. flavus* and *L. monocytogenes*, a marked inhibition of microbial growth of both strains was observed. LDPE film was successfully coated with nisin using methylcellulose (MC)/hydroxypropyl methylcellulose (HPMC) as a carrier. Nisin was found to be effective in suppressing *S. aureus* and *L. monocytogenes* respectively. The efficacy of nisin coated polymeric films such as PVC, linear low-density polyethylene (LLDPE), and nylon was studied in inhibiting *Salmonella typhimurium* on fresh broiler drumstick skin. As anticipated, the more hydrophobic LLDPE film repelled the aqueous nisin formulations to a greater extent than the other films and caused coalescence of the treatment solution droplets. The repulsion between the LLDPE film and the treatment solution may have affected the overall inhibitory activity of the formulations by causing more localized inactivation of the target. An agarbased film containing nisin was also studied. It was found that in this film, the degree of cross-linking depends on the agar concentration, which may affect the migration of nisin to the surface of a broiler drumstick skin. Thus, 0.75% w/w compared with 1.25% w/w gels formed a more open and elastic network, allowing greater migration of the treatment components over time.

Incorporation of Antimicrobial Additives

The direct incorporation of AM additives in packaging films is a convenient mean by which AM activity can be achieved. Several compounds have been proposed and/or tested for AM packaging using this method. The incorporation of 1.0% w/w potassium sorbate in LDPE films was studied. A 0.1-mm thick film was used for physical measurements, while a 0.4-mm thick film was used for AM effectiveness tests. It was found that potassium sorbate lowered the growth rate and maximum growth of yeast, and lengthened the lag-period before mould growth became apparent. The results of this study, however, contradict some others who studied with LDPE films (0.05 mm thick) containing 1.0 % w/w sorbic acid. In the latter case, the films failed to suppress mould growth when brought into contact with inoculated media. Further studies confirmed that ethylene vinyl alcohol/linear low-density polyethylene (EVA/LLDPE) film (70 μm thick) impregnated with 5.0% w/w potassium sorbate was unable to inhibit the growth of microorganisms on cheese and to extend its shelf life. Very limited migration of potassium sorbate into water as well as into cheese cubes occurs, probably because of the incompatibility of the polar salt with the nonpolar LDPE. The choice of an AM agent is often restricted by the incompatibility of that agent with the packaging material or by its heat instability during extrusion.

While polyethylene (PE) has been widely employed as the heat-sealing layer in packages, in some cases the copolymer polyethylene-co-methacrylic acid (PEMA) was found to be preferable for this purpose. A simple method for fabricating PEMA films (0.008–0.010 mm thick) with AM properties by the incorporation of benzoic or sorbic acids was formulated. The experimental results suggest that sodium hydroxide and preservative-treated films exhibit dominantly AM properties for fungal growth, presumably due to the higher-amount of preservatives released from the films (75 mg benzoic acid or 55 mg sorbic acid per g of film) than hydrochloric acid and preservative-treated films. Chitosan films made from dilute acetic acid solutions block the growth of *Rhodotorula rubra* and *Penicillium notatum* if the film is applied directly to the colony-forming organism. Since chitosan is soluble only in slightly acidic solutions, production of such films containing the salt of an organic acid (such as benzoic acid, sorbic acid) that is an AM agent is desirable. However, the interaction between the AM agent and the film-forming material may affect the casting process, the release of the AM agent and the mechanical properties of the film. Anhydrides may serve as appropriate additives to plastic materials for food packaging. LDPE films impregnated with benzoic anhydride completely suppressed the growth of *Rhizopus stolonifer*, *Penicillium* species and *Aspergillus toxicarius* on potato dextrose agar (PDA). Similarly, LDPE films that contained benzoic anhydride delayed mould growth on cheese. Studies on shelf life of packaged cheese and toasted bread demonstrated the efficiency of LDPE film containing benzoic anhydride against mould growth on the food surface during storage at 6°C). The migration of benzoic anhydride, ethyl paraben (ETP) and propyl paraben (PRP) in LDPE films was studied. It was found that the incorporation of these parabens in the polymer was more difficult than that of benzoic anhydride due to their higher volatilities.

No single AM agent can cover all the requirements of food preservation. A range of anhydrides was investigated for use in food packaging. It is known that for mould growth inhibition, the effectiveness of sorbic anhydride (10 mg sorbic anhydride/g of PE initial concentration) incorporated in PE films (0.10 mm to 0.12 mm thick) is much better with slow-growing (*Penicillium* species) than with fast-growing mould (*Aspergillus niger*). This is due to the time required for the PE to release sorbic acid to an inhibitory concentration. The potential of incorporating nisin directly into LDPE film was highlighted for controlling food spoilage and enhancing product safety. Hexamethylene-tetramine (HMT) was the first AM packaging agent. The AM activity of the latter is believed to be due to the formation of formaldehyde when the film comes into contact with an acidic medium. It was found that an LDPE film containing 0.5% w/w HMT exhibited AM activity in packaged cooked ham and therefore this agent is a promising material for food packaging applications. In Japan, the ions of silver and copper, quaternary ammonium salts, and natural compounds such as Hinokitiol are generally considered as safe AM agents. Silver-substituted zeolite (Ag-zeolite) is the most common agent with which plastics are impregnated. It retards a range of metabolic enzymes and has a unique broad antimicrobial spectrum. As an excessive amount of the agent may affect the heat-seal strength and other physical properties such as transparency, the normal incorporation level used is 1 to 3% w/w. Application to the film surface (that is increasing the surface area in contact with the food) is another approach that could be investigated in the future.

Another interesting commercial development is triclosan-based antimicrobial agents such as Microban. LDPE films containing 0.5 and 1.0% w/w triclosan exhibited antimicrobial activity against *S. aureus*, *L. monocytogenes*, *E. coli* O157:H7, *Salmonella enteritidis* and *Brochothrix thermosphacta* in agar diffusion assay. The 1.0% w/w Triclosan film had a strong antimicrobal effect in *in vitro* simulated vacuum-packaged conditions against the psychrotrophic food pathogen *L. monocytogenes*. However, it did not effectively reduce spoilage bacteria and growth of *L. monocytogenes* on refrigerated vacuum packaged chicken breasts stored at 7°C. This is because of ineffectiveness towards microbial growth. Other compounds with AM effects are natural plant extracts. Recently, Korean researchers developed certain AM films impregnated with naturally-derived AM agents. These compounds are perceived to be safer and were claimed to alleviate safety concerns. It was reported that the incorporation of 1% w/w grapefruit seed extract (GFSE) in LDPE film (30 mm thick) used for packaging of curled lettuce reduced the growth rate of aerobic bacteria and yeast. It was found that LDPE films (48 to 55 µm thick) impregnated with either 1.0% w/w *Rheum palmatum* and *Coptis chinensis* extracts or silver-substituted inorganic zirconium retarded the growth of total aerobic bacteria, lactic acid bacteria and yeast on fresh strawberries. However, a study showed that LDPE films (48 to 55mm thick) containing 1.0% w/w *R. palmatum* and *C. chinensis* extracts or Ag-substituted inorganic zirconium did not exhibit any AM activity in a disk test against *E. coli*, *S. aureus*, *Leuconostoc mesenteroides*, *S. cerevisiae*, *A. niger*, *Aspergillus oryzae*, *Penicillium chrysogenum*. A film containing sorbic acid showed activity against *E. coli*, *S. aureus*, and *L. mesenteroides*. During diffusion assays, the AM agent is contained in a well or applied to a paper disc placed in the centre of an agar plate seeded with the test microorganism. This arrangement may not be appropriate for essential oils, as their components are partitioned through the agar due to their affinity for water. Accordingly, broth and agar dilution methods are widely used to determine the AM effectiveness of essential oils. The AM activity of 5.0% w/w Propolis extract, Chitosan polymer and oligomer, or Clove extract in LDPE films (0.030–0.040 mm thick) against *Lactobacillus plantarum*, *E. coli*, *S. cerevisiae*, and *Fusarium oxysporum* is best determined through viable cell counts. Overall, LDPE films with incorporated natural compounds show a positive AM effect against *L. plantarum* and *F. oxysporum*. Preliminarily studies with LLDPE films (45 to 50 mm thick) containing 0.05% w/w linalol or methyl chavicol showed a positive activity against *E. coli*. Edible films and various AM compounds incorporated in edible food packages have also been investigated recently. Edible AM materials produced by incorporating lysozyme, nisin and ethylenediamine tetracetic acid (EDTA) was investigated in whey protein isolate (WPI) films. Such lysozyme or nisin-containing films are effective in inhibiting *Brochothrix thermosphacta* but failed to suppress *L. monocytogenes*. The incorporation of EDTA in WPI films improved the inhibitory effect on *L. monocytogenes* but had a marginal effect only on *E. coli* O157:H7.

Immobilization of AM Substances

Besides diffusion and sorption, some AM packaging systems utilize covalently immobilized AM substances that suppress microbial growth. The efficiency of lysozyme immobilized on different polymers was studied. It is known that cellulose triacetate (CTA) containing lysozyme yields the highest AM activity. The viability of *Micrococcus lysodeikticus* was reduced in the presence of immobilized lysozyme on CTA film. PE/polyamide (70:30) film formed a stable bond with Nisin in contrast to lacticin 3147. Nisin-adsorbed bioactive inserts reduced the level of *L. innocua* and *S. aureus* in sliced cheese and in ham.

Surface modification   Functional groups possessing AM activity was introduced (by chemical means) into polymer films to prevent the transfer of the AM agents from the polymer to the food. A new biopolymer containing a chito-oligosaccharide (COS) side chain was also synthesized. The COS was introduced on polyvinylacetate (PVA) by cross-linking with the bifunctional compound, *N*-methylolacrylamide (NMA). It was found that the growth of *S. aureus* was almost completely suppressed by this means. Surface amine groups formed in polymers by electron irradiation were also shown to impart AM effectiveness. Another AM film has recently been developed using a UV excimer laser. Nylon films irradiated in air by a laser at 193 nm exhibited AM activity, apparently due to a 10% conversion of the amide groups to amines on the nylon surface bound to the polymer chain. In contrast, irradiation at 248 nm did not change the surface chemistry or initiate the conversion of amide groups. A decrease in all bacterial cells, including *S. aureus*, *Pseudomonas fluorescens*, and *E. faecalis* in bulk fluid was observed when using an AM nylon film. The results indicate that this decrease is may be due to the bactericidal action than surface adsorption. Although the mechanism of the reduction in the bacterial population remained uncertain, electrostatic attractive forces between the positively charged film surface and the negatively charged *E. coli* and *S. aureus* were presumed to be the reason for this effect. Further research is needed to characterize the AM active groups on the irradiated film surface and the mechanism of AM action.

Ionomers, with their unique properties such as a high degree of transparency, strength, flexibility, stiffness and toughness, as well as inertness to organic solvents and oils, have also drawn much attention as food packaging materials. Benomyl fungicide was successfully incorporated into ionomer films via its carboxyl groups. Unfortunately, Benomyl is not an approved food preservative. Application of AM ionomers combined with approved food preservatives was investigated. Anhydride linkages in the modified films were formed by reaction of acid/or base-treated films with benzoyl chloride. The AM activity was characterized in terms of the release of benzoic acid, which was higher in the base-treated version indicating the superiority of the latter. The AM effect of modified ionomer films was further demonstrated by their ability to inhibit the growth of *Penicillium* species and *A. niger*.

FACTORS TO CONSIDER WHILE
PREPARING ANTIMICROBIAL FILMS

It is clear that the selection of both the substrate and the AM substance is important in developing an AM packaging system. Furthermore, when an AM agent is added to a packaging material, it may affect the inherent physico-mechanical properties of the latter.

Process conditions and residual antimicrobial activity   The effectiveness of an AM agent applied by impregnation may deteriorate during film fabrication, distribution and storage. The chemical stability of an incorporated AM substance is likely to be affected by the extrusion conditions, namely, the high temperatures, shearing forces and pressures involved. To minimize these problems, master batches of the AM agent in the resin was recommended for the preparation of AM packages. Also, all operations such as lamination, printing and drying as well as the chemicals used (adhesives and solvents) in the process may affect the AM activity of the package. In addition, some of the volatile AM compounds may be lost during storage. All these parameters should be evaluated.

Characteristics of antimicrobial substances and foods   The mechanism and kinetics of growth inhibition are generally studied in order to permit mathematical modelling of microbial growth. Foods with different biological and chemical characteristics are stored under different environmental conditions, which, in turn, may cause different patterns of growth of microflora. Aerobic microorganisms can exploit headspace O_2 for their growth. The pH of a product affects the growth rate of target microorganisms and changes the degree of ionization of the most active chemicals, as well as the activity of the AM agents. LDPE film containing benzoic anhydride was more effective in inhibiting moulds at low pH values. The diffusion of sorbic acid decreased with an increase in pH. The food a_w may alter the microflora, AM activity, and chemical stability of active ingredients applied by impregnation. It was shown that the diffusion of potassium sorbate through polysaccharide films increases with a_w; this has a negative impact on the amount available for protection. It was found that potassium sorbate diffusion rates in MC/HPMC film containing palmitic acid were much higher at higher values of a_w.

Chemical interaction of additives with film matrix   During incorporation of additives into a polymer, the polarity and molecular weight of the additive have to be taken into consideration. Since LDPE itself is non-polar, additives with a high molecular weight and low polarity are more compatible with this polymer. Furthermore, the molecular weight, ionic charge and solubility of different additives affect their rates of diffusion in the polymer. The diffusion of ascorbic acid, potassium sorbate, and sodium ascorbate in calcium-alginate films were compared at 8, 15, and 23°C. It was found that ascorbic acid had the highest and sodium ascorbate the lowest diffusion rate at all studied temperatures. These findings were attributed to the different ionic states of the additives.

Storage temperature   The storage temperature may also affect the activity of AM packages. Several researchers found that the protective action of AM films deteriorated at higher temperatures, due to high diffusion rates in the polymer. The diffusion rate of the AM agent and its concentration in the film must be sufficient to remain effective throughout the shelf life of the product. Low amounts of benzoic anhydrides in LDPE at refrigeration temperatures might be as effective as high levels at room temperature.

Mass transfer coefficients and modelling   Mathematical modelling of the diffusion process could permit prediction of the AM agents release profile and the time during which the agent remains above the critical inhibiting concentration. With a higher diffusivity and much larger volume of the food component compared to the packaging material, a semi-infinite model in which the packaging component has a finite thickness and the food component has infinite volume could be practical. The initial and boundary conditions that could be used in mass transfer modelling have been identified.

Physical properties of packaging materials   AM agents may affect the physical properties, processability or machinability of the packaging material. No significant differences in the tensile properties before and after the incorporation of potassium sorbate in LDPE films, but the transparency of the films deteriorated as the sorbate concentration increased. No noticeable differences in clarity and strength of LDPE film containing 0.5 and 1.0% benzoic anhydride was reported. Similar results were reported for naturally-derived plant extracts such as propolis at 5.0%, clove at 5.0%, *R. palmatum* at 1.0% and *C. chinensis* at 1.0%. On the other hand, LDPE film coated with MC/HPMC containing nisin was difficult to heat-seal. Statistically significant differences were found between the physical properties of films without AM agents and with different agents at concentrations of 5 g/kg and 10 g/kg. It was found that the tensile and sealing strengths were lower in all samples containing AM agents including benzoic anhydride, ethyl paraben (ETP) or propyl paraben (PRP). In all studied cases, the coefficient of friction increased with the addition of AM substances, water vapour permeability declined by the incorporation of PRP, and oxygen permeability decreased by the impregnation of benzoic anhydride or PRP.

Cost The cost of films impregnated with AM agents, are more expensive than their basic counterparts. Commercialization of such films could therefore become viable for high-value food products only.

ANTIMICROBIAL PACKAGING SYSTEMS

Most food packaging systems consist of the packaging material, the food, and the headspace in the package. If the void volume of solid food products is assumed as a kind of headspace, most food packaging systems represent either a package/food system or a package/headspace/ food system (Figure 5.1).

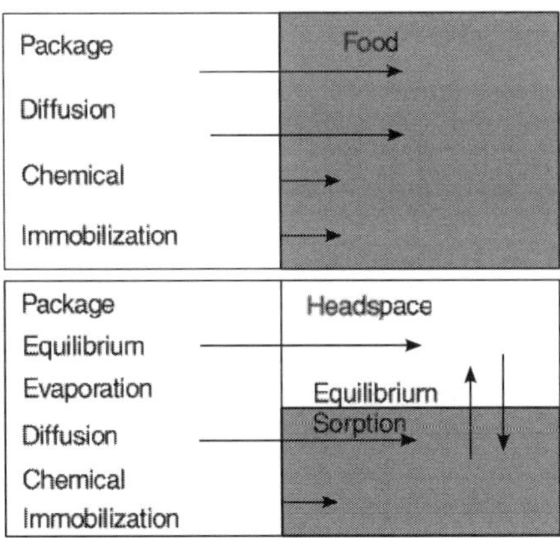

Figure 5.1   Food packaging systems

Figure 5.2 reflects migration mechanisms to estimate the interfacial distribution of the substance. Compared to a nonvolatile substance, which can only migrate through the contact area between the package and the food, a volatile substance can migrate through the headspace and air gaps between the package and the food. Besides diffusion and equilibrated sorption, some antimicrobial packaging uses covalently-immobilized antibiotics or fungicides. This utilizes surface suppression of microbial growth by immobilization of the non food-grade antimicrobial substance without diffusional mass transfer. Figure 5.2 shows the mass transfer phenomena of active substance with different applications. Control of the release rates and migration amounts of antimicrobial substances from food packaging is very important. Biochemical factors affecting the mass transfer characteristics of antimicrobial substances include antimicrobial activity and the mechanism/ kinetics of selected substances to target microorganisms. Studies on these problems provide information on the exact amount and release rate of the antimicrobial substances required to achieve a given effect. The release kinetics have to be designed to control the growth kinetics and maintain the antimicrobial concentration above the critical inhibitory concentration.

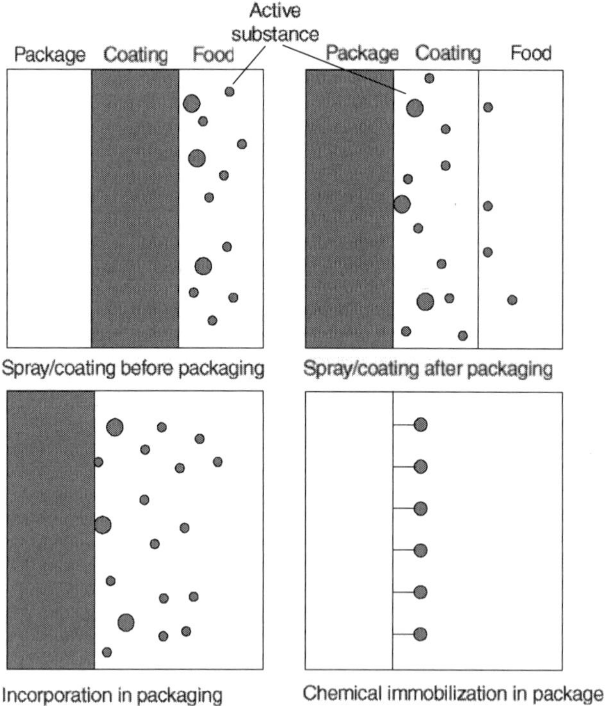

Figure 5.2   Migration of active substance

DESIGNING OF THE AM PACKAGING SYSTEM

A number of factors that affect the design or modelling of an antimicrobial film or package are as follows:

Casting process conditions and residual antimicrobial activity   Antimicrobial activity of an incorporated active substance may be deteriorated during casting (film or container), converting, and/or storage and distribution of the packaging material. The residual antimicrobial activity is the effective activity of the antimicrobial agents utilized for the antimicrobial packaging of final films after the casting and converting processes. Most activity deterioration may occur during the extrusion process using high temperature, shearing force, and pressure. During extrusion of plastic resins, the high pressure and temperature conditions in the extruder affect the chemical stability of incorporated antimicrobial substances and reduce their residual antimicrobial activity. The temperature, pressure, and residence time in the extruder have to be quantified mathematically to predict the residual antimicrobial activity.

The effects of converting operations such as lamination, printing, drying, and storage processes and the effects of adhesives and solvents should also be characterized quantitatively. Besides chemical degradation, loss of volatile antimicrobial compounds is also a reason for antimicrobial activity loss during casting (extrusion or coating) and storage of the packaging materials.

Characteristics of antimicrobial substances and foods   The growth-inhibition mechanism and kinetics are the important factors to be considered in designing the antimicrobial packaging system. The mathematical model of microbial growth can be built up from the inhibition kinetics and mechanism. The release kinetics of antimicrobial agents have to be designed to maintain the concentration above the critical inhibitory concentration with respect to the growth kinetic studies. Because foods have different chemical and biological characteristics such as pH, water activity, carbon source, nitrogen source, partial pressure of oxygen, and temperature, they provide different environmental conditions to microorganisms and included antimicrobial agents. For example, the pH of food affects the microflora and growth rate of target microorganisms and alters the ionization (dissociation/association) of most active chemicals, which could change the antimicrobial activity. Water activity also alters the antimicrobial activity and chemical stability of incorporated active substances as well as microflora. Oxygen in the package headspace can be utilized by aerobic microorganisms, and the oxygen permeability of the packaging materials can alter the headspace oxygen concentration. Therefore, it is essential to study the pH and water activity of the food, oxygen permeability of the packaging material, and microbial profile to design antimicrobial packaging systems.

Storage temperature   Storage temperature can change the antimicrobial agents in the packaging material. A multilayer structure for antimicrobial-release packaging systems was suggested: outer antimicrobial barrier layer (optional); antimicrobial-containing matrix layer; release-control layer; and food. The outer layer prevents the loss of active substance to the environment and gives a uni-direction of diffusion forward to the inner layer. The matrix layer contains active substances and has a very fast diffusion of the antimicrobials and a very large portion of amorphous structure to provide a space for the antimicrobials. The control layer is the key layer to control the initial lag period and the flux of penetration of the antimicrobials. It has a customized thickness and diffusibility with respect to the characteristics of microbial spoilage and food products. This multilayer design has advantages of easy customization by deciding on the thickness and material of the control layer which controls the diffusivity of antimicrobial agents. Selected control-layer film will be laminated to the pre-made antimicrobial base-film consisting of an antimicrobial-matrix layer and an outer layer. Containers as well as films can be formed from the multilayer sheet of antimicrobial material by vacuum or press moulding processes. Papers as well as plastics can be used as antimicrobial packaging materials. Since paper has a porous structure, the antimicrobial agents plugged in the pores may improve the paper material's performance, such as water vapour and gas permeability, physical strength, optical properties, and surface properties as well as the antimicrobial activity.

Physical properties of packaging materials   When an antimicrobial agent is added to packaging materials to reduce microbial spoilage, it may affect general physical properties and process/machinability of the packaging materials. General properties of packaging materials include mechanical properties such as tensile strength, elongation, burst strength, tearing resistance, stiffness, and physical properties such as oxygen (and other gas) permeability, water vapour permeability, wettability, water absorptiveness, grease resistance, brightness, haze, gloss, transparency, etc. The performance of the packaging materials must be maintained with the addition of the active substances, even though the materials contain more heterogeneous formulations. In the case of plastics, the active agents are usually very-low-molecular-weight chemicals compared to the size of the polymeric structure and are added in small amounts. In a properly designed antimicrobial packaging system, the chemicals will position themselves in the amorphous structural ions of the polymeric structure and may not affect the mechanical strength of the polymeric packaging materials. Considering the huge size of the amorphous area of the polymeric materials (or porous area of papers) to the relatively small size of the antimicrobial molecule, a large amount of the antimicrobial substance may be needed to show any effect on the tensile strength the packaging materials. However, after the amorphous area of the polymers and the porous area of papers are saturated by a high concentration or large powder particles of antimicrobial chemicals, the tensile strength of the antimicrobial materials could be adversely affected. Besides mechanical strength changes, incorporation of antimicrobial agents usually reduces optical properties of plastic films such as transparency. For example, the transparency of LDPE films decreased with increasing potassium sorbate concentration. This may result in serious disadvantage while using antimicrobial plastic film for packaging.

Verification of Anti-microbial Activity

The antimicrobial activity of the packaging materials could be measured by microbiology experiments. Before food samples are packaged in the antimicrobial packaging materials, target microorganisms may be inoculated onto the surface of the food or mixed into the food samples. Measuring growth by counting the microorganisms with incubation time can provide the characteristic values of growth rate at the exponential growth phase, maximum growth at the stationary phase, and initial lag period. These values can be compared to conventional packaging or reference samples without packaging. Reduced growth rate and reduced maximum growth population could indicate improved microbial safety, and the extended lag period would show the prolonged shelf life with microbial quality assurance. The growth rate can be easily obtained from the slope of the plot of logarithmic transformed growth against time, using the data from the exponential growth period.

Sorbate releasing LDPE film (plastic film) for cheese packaging is a good example of successful research and development of antimicrobial packaging. To develop the film, the diffusivities of the food preservative potassium sorbate through various plastic films and cheeses were first determined. Plastics have diffusivities ranging from 1×10^8 to 1×10^{14} cm^2/s for potassium sorbate, while cheeses have diffusivities ranging from 1×10^6 to 1×10^7 cm^2/s at 25°C. Cheeses and solid foods usually have diffusivities that are 10^2–10^6 times greater than those of plastics for potassium sorbate. High moisture foods and liquid foods may have much higher diffusivities than dry and solid foods. Since diffusivity in plastics is smaller than that in foods, release of most antimicrobial agents, including potassium sorbate, highly depends on the lowest diffusivities of the active agents in the packaging materials. Because cheese has surface microbial (usually fungal) contamination due to post-process contamination, the potassium sorbate concentration on the cheese surface is critical and must be obtained above 0.1% to inhibit mould growth.

Antimicrobial Edible Coatings and Films

Edible coatings and films have a variety of advantages such as biodegradability, edibility, biocompatibility, aesthetic appearance, and barrier properties against oxygen and physical stress. They can also serve as a carrier for edible antimicorbial agents and preservatives. For example, whey-protein coatings and films can incorporate adequate amounts of edible antimicrobial agents (e.g. lysozyme, nisin, potassium sorbate, EDTA) as well as a plasticizer (e.g. glycerin or sorbitol).

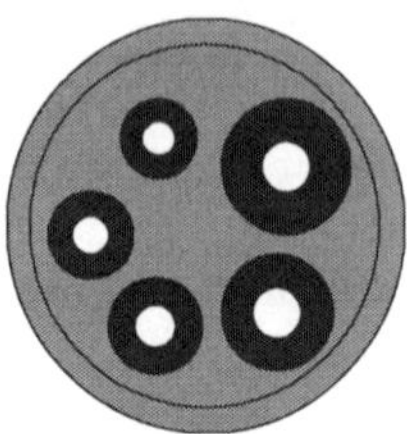

Figure 5.3 Clear zones of inhibition of the spoilage bacteria *Brochothrix thermosphacta* around discs of whey-protein film containing lysozyme

The effectiveness of lysozyme-impregnated whey-protein film against a spoilage microorganism, *Brochothrix thermosphacta* was studied. The lysozyme, slowly released from the film, effectively inhibited the growth of the microorganism as indicated by the clear inhibitory zone around the film discs (Figure 5.3). The lysozyme-impregnated whey-protein film showed great clarity and maintained the tensile strength at concentrations of up to 100 mg of lysozyme/g of dried film. This film may have great potential as a microbial hurdle against gram-positive spoilage and pathogenic bacteria. Antimicrobial packaging is a promising form of active food packaging. Even though most packaged perishable food products are heat-sterilized or have a self-protecting immune system, microbial contamination could occur on the surface or damaged area of the food through package defects or restorage after opening. The antimicrobial substances incorporated into packaging materials can control microbial contamination by reducing the growth rate and maximum growth population and extending the lag period of the target microorganism. They exclude moisture and most gases (including oxygen), and they can be either optically clear or opaque. Because of their unique characteristics, plastics are becoming the most important packaging material for food products. The packaging industry is the largest user of plastics; more than 90 per cent of flexible packaging is made of plastics.

HYGIENE OF OPERATING STAFF

Operators should be given instructions before commencing work. Such instructions should include the basics of personal hygiene, hygienic handling of foods, with emphasis on precautions necessary to prevent contamination of food. Hands must be washed frequently and thoroughly using a suitable hand cleansing preparation and running, warm, potable water. Hands must be washed before entering a food handling area, immediately after using the toilet, and after handling any material which might be capable of transmitting disease. There must be adequate supervision to ensure that this is done. The use of gloves for product safety purposes is of dubious benefit. Thorough and frequent hand washing has been shown to be just effective, provided the nails are well trimmed. In case of a cut or minor infection or any condition which might pose a threat to the product, effective covering of the hands is essential and gloves may be appropriate in these cases. It should be noted that wearing of gloves does not reduce the need for frequent hand washing. Operators should be supplied with a complete set of hygienically laundered protective coating or with disposable clothing for one-time use. This must be worn only for the intended purpose and must not be worn in other departments. Clothing must be changed daily. It is the company's responsibility to provide clean clothing, which may be facilitated by using a laundry service. In the changing room, there must be physical separation of work clothes and outdoor clothes. Every operator has to wear disposable hair and, if needed, beard protection. There should be an area designed with suitable drainage for boot washing operations. Manual cleaning (preferably during the cleaning shift) and industrial washing machines are satisfactory boot washing methods.

HEATING, VENTILATION AND CONDITIONING OF THE PACKAGING ENVIRONMENT

Requirements for room air are

- Complete air-conditioning (with room temperature at 12°C), operating at overpressure, with all air circulating through filters 5–25 Pa, a complete volume change every hour. A pressure difference between high risk to lower care zones should be 5–15 Pa or have an air velocity 1.5 m/s or greater through openings. Supplied air should comply with the desired standard, typically final filtration F9 for room air and H11 for process equipment air.

- During cleaning operations, the air used has to go directly to the exhaust and after cleaning operations ambient air has to be recycled to increase environmental drying as necessary. Air should have a relative humidity of 60–70%.

PACKING MACHINE REQUIREMENTS

Hygienic food processing equipment should be easy to maintain, easy to clean and protect the products from contamination and must prevent the ingress of microorganisms.

Considerations

1. The equipment must be installed such that it will not cause contamination of the ingredients, raw foods and end products.

2. Separation between product contact and non-product contact areas prevents cross-contamination during operations. Indirect product contact zone areas are designed as if they were product contact zone areas.

3. Product contact surfaces are made to prevent build-up of product residues during operations.

4. Separation between product contact areas and non-product contact areas has to be determined by a risk analysis.

5. All parts of the equipment should be installed at a distance of at least 1 m from walls, ceilings and adjacent equipment to be accessible for transport systems for ingredients and packaging material and for easy access of operating staff (for inspection, cleaning and disinfecting, maintenance and to solve breakdowns).

6. Surfaces with direct and indirect product contact are cleanable as measured by <1 cfu per 25 cm^2, <1 cfu per 10 ml when the item is rinsed, and/or negative for residual protein or carbohydrate when using swabs to detect residual protein or carbohydrate (measured post-installation).

Product Contact Surfaces and Coatings

Surfaces of the machine parts that come in direct contact with high-risk ingredients must be made from non-toxic materials and must have finishes which are impervious, non-absorbing, washable, smooth and crack-free to resist microbial settlement and to be easily cleaned and disinfected. Surfaces must have:

- cleanability to a microbial level
- accessibility for inspection, maintenance and cleaning
- no niches and hollow areas
- no product or liquid collection
- hygienic operational performance

Conveyors with product contact   Conveyor belts must be made from non-toxic materials and have finishes which are impervious, non-absorbing, washable, smooth, crack-free in order to resist microbial settlement and facilitate cleaning and disinfection. The quality of the surface finish must be such that, under the conditions of the food packing process, formation of biofilm cannot occur or is reduced to a minimum whereby the film can be removed. All belting should be easily removable or the belt tension reduced easily without tools, so that the surfaces underneath can be cleaned. Belt tension should be adequate throughout operations to prevent water pooling on belts.

Monitoring and control   Protocols for hygiene monitoring and control of packing systems can only be defined once the system design meets the requirements of this guideline.

A GLOSSARY OF FOOD PACKAGING TERMS

Acrylonitrile monomer (AM) The term refers to acrylonitrile that is polymerized with styrene or methyl acrylate. AN films, coatings, and semi-rigid containers are cleared for use only for single use food-contact surfaces. They have not been tested for use above 120°F. AN films are used primarily in the pharmaceutical industry.

Cellulose acetate (CA) This material is manufactured like cellophane, but there is an additional reaction that includes acetic anhydride. It is easily deformed by heat, acids, bases, and some solvents but is resistant to fats and oils. It builds up static. Maximum continuous service temperature for CA is 175°F.

Cellophane Regenerated cellulose-based flexible packaging material. Cellulose is dissolved in sodium hydroxide, then extruded into an acid-salt bath to produce filaments of regenerated cellulose. It can be coated with saran for use with oily or greasy products. Cellophane was introduced in the 1920s as a bread wrap. It is semi-moisture-proof and heat-sealable. The heat-sealing range is 200 to 300°F. Cellophane is flammable. Food-grade plastics often have information embossed into the bottom of the container. The type of plastic material may be referred to by an acronym such as HDPE or ethylene vinyl alcohol.

ABS Poly(acrylonitrile-butadiene-styrene) copolymer. Heat distorts it at 200°F. Maximum continuous service temperature is 165°F. ABS is resistant to acids, bases, fats, and oils. It does not resist solvents and yellows if exposed to sunlight. ABS has been cleared for food use with all foods except those containing alcohol and those subjected to thermal processing. It can be used with foods that are filled and stored at room temperature, and stored refrigerated or frozen.

Acetyl tributyl citrate (ATBC) ATBC is a derivative of citric acid. ATBC is used in flexible packaging films and has been cleared for use as a plasticizer for food contact surfaces of resinous and polymeric coatings and in paper and paperboard for use with fatty foods. It has been sanctioned by the FDA and falls into the "relatively harmless" toxicity category as a consumable chemical. Estimated daily human intake from current uses of plastic film is around 0.0008 mg per kg of body weight. ATBC is considered safe at this level.

Ethylene-vinyl acetate copolymer (EVA) EVA is a flexible packaging material formed from low-density polyethylene (LDPE) and vinyl acetate. It is more flexible than PE but more permeable to water and gases. EVAs are unstable at high temperatures but stable at low temperatures. EVA is cleared for use with fatty foods and for treatment with irradiation as a sterilant (up to 8.0 megarads total), so it need not be heat-sterilized. It is very stretchy, so it can be used as a shrink wrap.

Ethylene vinyl alcohol (EVOH)   A copolymer of vinyl acetate and ethylene. It may be modified by blending it with another organic polymeric phase (polystyrene or nylon). EVOH can be formed into films, sheets, and rigid containers. It provides an excellent barrier to flavours and gases (such as carbon dioxide). EVOH has excellent chemical barrier properties, so it can be used with a wide variety of foods. There are three types of EVOH:

1. An ethylene content of 55 per cent to 30 per cent vinyl alcohol can be used with acid aqueous and non acid aqueous foods with no free surface oil; dairy oil in water emulsions; oil-free beverages; bakery products and dry solids with no free surface oil; and foods that are hot-filled or pasteurized at 150°F or less, and room-temperature, refrigerator, or frozen-stored.

2. An ethylene content of 55 per cent to 15 per cent vinyl alcohol can be used with aqueous acid and nonacid foods with free surface oil; dairy oil in water emulsions; beverages; low-moisture fats and oils; moist bakery products and dry solids with free surface oil; and foods that are refrigerated or frozen-stored with no thermal treatment.

3. An ethylene content of 20 to 40 per cent, to 60 to 80 per cent vinyl alcohol is approved for food use with all foods except those containing alcohol and those which are boiling water-sterilized, hot-filled, or pasteurized above 150°F. This material can be used on refrigerated or frozen readymade foods to be reheated in the container at the time of use, provided the specific film has been tested and has been shown not to release substances exceeding 0.15 mg per square inch of food-contact surface at 212°F.

Food-simulating liquids (FSL)   These are substances which are used to simulate the possible reactions between food and materials with which they come into contact. The chemical characteristics of FSLs are specific, so a particular food may be simulated with several food-simulating liquids including distilled water, 3 per cent acetic acids (acid foods), 8 per cent or 50 per cent ethanol (alcohol containing foods), and n-heptane (fat- or oil-containing foods).

High-density polyethylene (HDPE)   HDPE has a maximum continuous service temperature of 160°F but distorts at 140°F. It is resistant to moisture, gases, acids, bases, solvents, and fats and oils. It will build up static. HDPE is used for milk jugs, cleaning supply bottles, and trash bags.

High-oxygen-barrier polymers   Polymers are used to prevent almost all oxygen transfer. These materials include vinyl alcohol, vinylidene chloride, and acrylonitrile. Pure polymers of these substances are not melt-processable, so they are made into copolymer forms such as PVDC, EVOH, or AN.

HP polyester High-performance PET. This material has a melting point of 275°F and can be melt-blown and hot-filled. The final container is yellow but has excellent barrier properties. For instance, it retains carbon dioxide in soda.

Hydrogen peroxide An oxidizing agent used for cold sterilization of packaging such as films and bottles before filling them with hot or sterilized food. Residual hydrogen peroxide is limited by the FDA to 0.1 parts per million (ppm) after packaging (FDA 21CFR 178.1005). Hydrogen peroxide is permitted as a sterilant for PET, olefins, and EVA.

Polyethylene terephthalate (PET) A polyester resin used for high-impact resistant containers. When melt-blown, it provides a good barrier for both flavours and hydrocarbons (fat). It is not transparent. PET is resistant to acids, bases, some solvents, and oils and fats. It is difficult to mould. PET is used for soda, mouthwash, pourable dressings, edible oils, and peanut butter. Mono-layer films will hold a crease, and are heat-sealable and transparent. The film is used for cereal box liners, soda bottles, boiling-the-bag pouches and microwave food trays. It can be heated in a microwave or in a conventional oven at 350°F for 30 minutes. There has been a moderate amount of concern that additives from these trays may migrate into foods, particularly if the trays are reused in a microwave oven.

Low-density polyethylene (LDPE) Maximum continuous service temperature is 140°F; heat distortion at 104°F. LDPE is resistant to moist-vapour, acids, bases, and fats and oils. It has poor resistance to solvent and builds up static. LDPE is used primarily for packaging films and for bread-wrapping bags.

Oriented polypropylene (OPP) OPP has good light and oxygen barrier characteristics and protects against moisture. It is clear but will accept printing ink. OPP can be manufactured to be semi-rigid to flexible. It can be overlaid with metalized polyester (silver, gold, or coloured). It is used for controlled atmosphere snack foods in which air has been flushed out and replaced with nitrogen to prevent development of rancid flavour. Examples of these snack food are chips in "bubble packs" and airline peanuts.

Polycarbonate (PC) PC is a rigid plastic. It is very clear and break-resistant. It is stable to acids and fats but has poor resistance to bases and solvents. It will not tolerate continuous boiling in water. Maximum continuous service temperature and heat distortion occur at 275°F. This material is used for baby bottles.

Polyethylene (PE) Polyethylene is cleared for food contact surfaces as an "olefin polymer." The structure of this polymer is relatively loose, so it is a less effective barrier than PVDC or PVC. It was introduced in the food industry in the 1950s. PE can be produced by two processes: low pressure produces high-density PE, while high pressure produces low-density PE. Low-density PE is strong and clear. It is a good moisture barrier but a poor oxygen barrier, and it will heat-seal itself. PE builds up static and tends to cling to itself. Its heat-sealing range is from 250 to 350°F. In general PE is cleared for use for packaging but not for cooking. Copolymers have been tested and approved for use with aqueous acid and nonacid foods containing free oil that are heat-sterilized at temperatures over 212°F if the material is less than 0.02-inch thick, or hot-filled or pasteurized at 150°F if the material is between 0.004- and 0.02-inch thick. PE has been cleared for irradiation sterilization up to a total of 6.0 mega-rads. Certain polymers have restricted uses, especially with respect to temperature.

Plasticizers   Substances used to increase the flexibility of polymeric (plastic) films. They prevent formation of some of the bonds between polymer molecules and make the films less rigid. Common plasticizers include dioctyl adipate (DOA), which is used with PVC; butylated hydroxyanisol (BHA), which is used with polystyrene and styrene; low-density polyethylene (LDPE); high-density polyethylene (HDPE); and high-molecular weight hindered phenols, which are used with ethylene-vinyl acetate (EVA).

Polyamide (Nylon-6, Nylon-12)   A flexible packaging material made from amino acids or dimerized vegetable oils. Polyamides are good oxygen and water barriers, but they have low melting points. Polyamides are approved for use as food contact coating materials not to exceed room temperatures, as a component of paperboard, as sealing gaskets for food containers, and as film coatings. Certain polyamides are approved for higher temperature applications. Nylon-66 has a very high melting point but is difficult to heat-seal. Heat distortion occurs at 400°F; maximum continuous service temperature is 175°F.

Polypropylene (PP)   Polypropylene was introduced in the 1950s. It is very transparent but cracks and breaks at low temperatures (15°F). It was used successfully in the mid-1960s as a copolymer with PE but was eventually replaced by plain PE for bags. It is more rigid, stronger, and lighter than PE. It is resistant to water vapour, grease, acids, bases, and solvents, and some polypropylenes are resistant to high temperature. Heat distortion occurs at 113°F; maximum continuous service temperature is 100°F. PP is cleared for irradiation sterilization up to 1 mega-rad total. Polypropylene is used for yoghurt containers, margarine tubs, and bottles for some pourable foods such as syrup.

Polystyrene (PS)   A rigid or flexible packaging material. PS is brittle. It heat distorts at 170°F. It is resistant to water, oxygen (but not carbon dioxide), weak acids, bases, and fats and oils, but it is damaged by solvents and alcohol. PS films are approved for irradiation sterilization up to a total of 1 megarad. PS can be formed into foam that is less brittle than solid PS film. PS can be foamed using *n*-pentane, iso-pentane, or toluenes the "blowing agent"; the use of chlorfluorocarbons (CFCs) has been discontinued in the manufacture of items used by the food service industry.

Polyvinyl chloride (PVC)   It is manufactured by polymerizing vinyl chloride monomer. Plasticizers must be added to give it flexibility. The structure of this polymer is relatively tight, but some air will pass through it. PVC has a heat distortion temperature of 165°F and a maximum continuous service temperature of 150°F. Since its shrink and melt points are so low, PVC can be used to shrink wrap foods which tolerate very little heat. It is resistant to water, acid, bases, some solvents, and fats and oils. PVC is approved for use as the film to wrap fresh red meats because it allows enough air to pass through the package to make the meat pigments "bloom" bright red. PVC is prior sanctioned for use in general food contact applications. The heat-sealing range is from 200 to 350°F. PVC is approved for irradiation sterilization up to 1 megarad.

Polyvinylidene chloride (PVDC)   A copolymer of vinylidene chloride and vinyl chloride or methyl crylate. VDC was invented in 1941 to protect equipment from outdoor elements. After world War II, it was approved for food packaging, and it was prior sanctioned in 1956 (Society of the Plastics Industry). PVDC is cleared for use as a base polymer in surfaces of materials that contact food, including food package gaskets, coverings that come in direct contact with dry foods, and paperboard coating in contact with fatty and aqueous foods. The copolymerization results in a film with molecules bound so tightly together that very little gas or water can get through. The heat distortion temperature of the polymer is 113°F, and the maximum continuous service temperature is 160°F. PVDC films soften at 250°F, so they can be used with food that is boiling, but not necessarily with high-sugar or high-fat foods that may get much hotter than 212°F. PVDC and PVC can be copolymerized to produce a family of flexible "Sarans." Sarans are clear and have excellent barrier properties. Their heat-sealing range is from 280 to 400°F. PVDC is resistant to oxygen, water, acids, bases, and solvents.

Susceptor   A thin metalized film joined to a container (often paperboard) with an adhesive. Its purpose is to "heat up" in the microwave oven to cause browning or crisping of microwaveable foods such as pizza or to help generate enough heat to cause popcorn to pop. Susceptors attain temperatures of 450 to 500°F and have the potential for volatilizing the adhesives and plasticizers in the packaging material, causing them to migrate into the food product. First-generation susceptors used aluminum-coated polyester film or paperboard. Second-generation susceptors are currently being developed; they will use nickel, cobalt, iron, and stainless steel to improve the susceptor's heating capacities.

Vinyl chloride monomer (VCM)   PVC can have residues of vinyl chloride monomer (VCM) in the finished plastic film. These levels have been reduced to insignificant amounts since the late 1970s. The Codex Committee on Food Additives and Contaminants has proposed guideline levels of 1ppm for VCM in PVC packages and 0.01 ppm for the monomer in food.

THE FUTURE

Antimicrobial packaging is a rapidly emerging technology. The need to package foods in a versatile manner for transportation and storage, along with the increasing consumer demand for fresh, convenient, and safe food products presages a bright future for AM packaging. However, more information is required on the chemical, microbiological and physiological effects of these systems on the packaged food especially on the issues of nutritional quality and human safety. So far, research on AM packaging has focused primarily on the development of various methods and model systems, whereas little attention has been paid to its preservation efficacy in actual foods. Research is essentisal to identify the types of food that can benefit most from AM packaging materials. It is likely that future research into a combination of naturally-derived AM agents, biopreservatives and biodegradable packaging materials will highlight a range of the merits of AM packaging in terms of food safety, shelf life and environmental friendliness. The reported effectiveness of natural plant extracts suggests that further research is needed in order to evaluate their antimicrobial activity and potential side effects in packaged foods. An additional challenge is in the area of odour/flavour transfer by natural plant extracts to packaged food products. Thus, research is needed to determine whether natural plant extracts could act as both an antimicrobial agent and as an odour/flavour enhancer. Moreover, in order to secure safe food, amendments to regulations might require toxicological and other testing of compounds prior to their approval for use.

REVIEW QUESTIONS

1.   Give a detailed account of aseptic packaging.

2.   Give examples of antimicrobial agents for potential use in food packaging materials.

3.   Name a few antimicrobial additives.

4.   Explain how surface modification plays a role in maintaining asepsis.

5.   What are the factors to consider in the manufacturing of antimicrobial films?

PRESERVATION BY HIGH TEMPERATURE

INTRODUCTION

New food preservation systems are a fascinating area, particularly for the microbiologist, many of the developments resulting from growing consumer demand for "fresher" foods. There are major differences in the effects on microorganisms of newer processes compared with the traditional canned product. Canned foods would receive a substantial heat sterilization process compared to a short shelf life chilled ready-meal which receives a lower heat process. One of the key parameters to be considered in any thermally processed foods is the target organism, which could be a pathogen or spoilage organism, which should be adequately reduced by the process. While a full canning sterilization process has been designed to achieve at least 12 log reductions of key spore-forming pathogens (mesophilic *Clostridium botulinum*), a short shelf life chilled product would have been designed to achieve 6 log reductions of key vegetative pathogens (*Salmonella, Listeria, E. coli*). The main reasons for this difference relate to temperature and length of storage before consumption. For example, a sterilized canned product would be expected to remain microbiologically stable for up to 2 years at ambient temperatures whereas a short shelf life chilled meal would be stable for 10 days only.

Heat causes membrane damage, loss of nutrients and ions, ribosome aggregation, DNA strand breaks, inactivation of essential enzymes, protein coagulation, etc. In other words, almost every cellular structure is somehow affected by elevated temperatures and it is very difficult to discern which events are leading to cell death. In fact, this is only possible if a direct relationship between the degree of inactivation and the degree of modification of a given target under different environmental conditions is found.

RANGE OF FOOD PRODUCTS

The range of food products that are thermally processed is very diverse; it can include low- medium- and high-viscosity liquids, some with particulates. The effect of these different substrates on the heat resistance of a microorganism can be quite marked, with some proteins, fats and high total solids increasing the heat resistance by a factor of 2 or 3 when compared to the standard heat resistance of a similar microorganism in a broth. The influence of these components on the heat resistance of microorganisms must therefore be carefully considered during product development. Changes in design of food products in recent years have been driven by consumer demand, in particular for convenience and speed of food availability. The time spent preparing a meal has been reduced by over 50% over the past 10 years. This has been attributed to the busy life-styles of modern society. The requirement for fast convenient food is coupled with the desire to improve the "fresh-like" qualities of such foods, often by reducing the heat process applied. The major challenge to the thermally processed food manufacturer has therefore been to prepare food, which is safe in terms of the pathogenic organisms, microbiologically stable over its shelf life, but minimally processed to preserve the quality.

One of the very noticeable changes to product formulation and design over this time period has been the combination of different types of solid particles within foods and the trend towards one-pack meals. The microbiologist and food process engineer have to consider the viscosity of the carrier fluid—low (e.g. thin soup), or medium (e.g. gravy) or high (e.g. concentrated starch). This fluid has a great effect on heat penetration into the product. The presence and the numbers of microorganisms in a product before processing will depend on the quality of the raw materials used and how they have been stored. This will be true for both the ingredients of the carrier fluid and any particulates, therefore the effective heat process must be designed to adequately reduce these organisms.

The type of food particle in a product can pose a serious challenge for the design of a safe process. The most simple particle would be of uniform composition, shape and size; however, in reality food ingredients such as vegetables, rarely grow uniformly and indeed many manufactured food particulates have a wide range of shapes and sizes. Heat penetration into particles will vary with size, and so will microbial death kinetics.

Certain constituents of foods directly affect the ability of microorganisms to survive heat treatments. If the food product contains a high proportion of protein, fats or high total solids it is very likely that the basic heat resistance characteristics of contaminating organisms will be greatly increased and so the published or accepted rates of thermal inactivation, which are often based on studies performed in broth cultures may be underestimates of that required in such foods. In some cases, the increase in process required may be very difficult to achieve as it may begin to denature the remaining constituents of the product, making it unacceptable to the consumer. The current trend to reduce salt or acids in foods would also affect the inhibitory properties of the food and therefore the required heat process to maintain stability. This point must be borne in mind by any manufacturer making seemingly minor changes to product recipes and such apparent minor change to a salt or acid level could have a major effect on heat resistance of contaminating organisms rendering foods unstable or unsafe.

There is no doubt that the heat resistance characteristics of the contaminating microorganisms will have a major effect on the overall design of the manufactured food. Recent advances in the use of curve fitting equations have given technologists an extra tool to assess inactivation curves that are nonlinear. This is particularly useful if you require a process time to reduce a target population by a certain factor but the death kinetics follow a sigmoid pattern or have an extended tail (Figure 6.1). These types of curves have been found in laboratories with both spore formers and vegetative organisms heated in a very wide range of substrates and temperatures. When designing and setting any thermal process it will be designed to inactivate a particular type or group of organisms (e.g. spore formers and vegetative pathogens); this tends to be known as the target organism for the process. For example, if the product was pasteurized and chilled a typical target organism may be *Listeria monocytogenes* where it has been established that a typical heat process at 70°C for 2 minutes would be required to achieve 6 log reductions. However, if the product was acidified and pasteurized the choice of target organism could be *Clostridium butyricum* where it is recognized that if the product is below pH 4.2 the process required may be 95°C for 5 minutes in order to achieve this reduction.

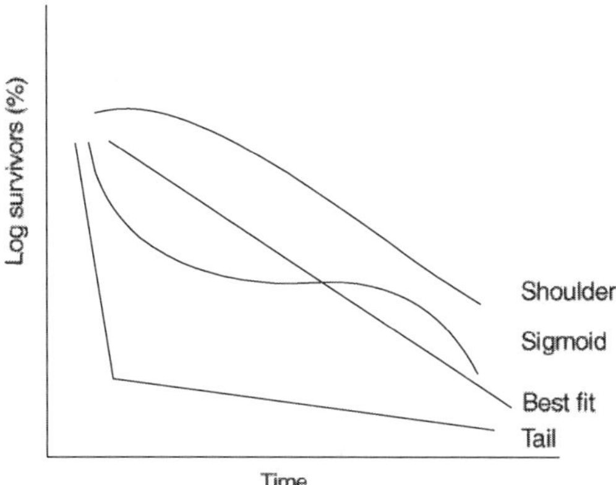

Figure 6.1   Death kinetics of a target population

Considering a commercially sterile, hermetically sealed container or indeed an aseptically processed product at ultra-high temperature, the target organism would be proteolytic *Clostridium botulinum* and the traditional minimum of F_o3 to achieve 12 log reductions still applies.

The target organism selected for a given process will vary according to the raw material used in the product and similarly the heat resistance characteristics will vary according to the behaviour of the organism when heated in the food. The diversity and variety of food product composition and its resultant effects on microbial composition and heat resistance has meant that a far greater understanding of thermal microbiology is needed to ensure multicomponent food products are manufactured safely.

THERMAL PROCESSING EQUIPMENT

In recent years, there have been considerable advances in the range of processing equipment available for the manufacture of food products. Foods may be processed in static vessels, or as a continuous flow. They may be processed in container or may be processed first and then aseptically filled or hot-filled. The effective temperature monitoring of such processes can be very difficult and it is in these situations that biological process verification is essential. The accuracy and reliability of the technique has improved greatly and forms a key part of many innovative process technologies.

Static Processing Method

The static heating methods tend to be used for in-pack processing, and these will follow the principles of canning retorts where the process would be designed to achieve the required temperature in the slowest heating container. The traditional approach to these investigations includes the use of thermocouple probes, which are inserted into packs and placed at specific places in the processing equipment in order to determine the slowest heating points. These investigations will be designed to establish the slowest heating point of the product (which may be within specific particles) within individual containers and the processing equipment (many vessels have cold spots). It is often necessary to include numerous probes in any one trial and these can effectively be simultaneously registered using a data logger, which could record and store at least twenty different sets of data at the same time. It is then possible to manipulate the data after processing in order to determine the overall process lethality and the slowest heating points.

When multicomponent packs are considered, the slowest heating point in the pack will vary considerably depending on the number of components in the pack, the recipe and pack size. The thermal diffusivity or speed at which the heat will penetrate throughout the pack will strongly influence the ability of contaminating organisms to survive. Thermal diffusivity is influenced by particle size of the food. It is evident that considerably more time is required to achieve process temperatures in the larger particulates. This is a very important factor to be considered if microbial contamination is likely to be present in the body of the particle itself.

It is also important to note that historically only the process 'hold time' was used for lethality calculations not the 'come up' and 'cool down' time. More recently, some manufacturers do consider the heating up and cooling down times in process calculations, particularly for long come up times. Any reductions in these margins of safety in order to produce 'fresher' products and higher throughput could cause serious microbiological problems.

Continuous Processing Methods

Continuous flow processing methods are much more difficult to monitor with regard to the minimum process, which will be based on the fastest moving particulate through the heating system. The establishment of heating profiles and the heat processing of certain foods in heat exchangers can be very difficult. Some products can be sticky and may change phases as they are heated to different temperatures. The microbiological complexity is often compounded in a product when continuous flow processing is then aseptically filled. Aseptic filling requires not only adequate thermal processing, but also a chemical decontamination step for both the filler head and each individual container.

Many products are now also manufactured with a cold or hot clean fill where the potential for cross-contamination must be very carefully assessed to avoid serious spoilage and food poisoning issues. There are many problems that the manufacturer of such products faces when the question of "proof of process" is asked. In many cases, it is impossible to obtain accurate thermocouple readings in order to assess the temperatures achieved during the filling process.

When undertaking a validation of a continuous flow thermal process, the most important data with regard to microbial reduction would be the overall effect of the process on microorganisms present in particulates as they flowed through the system. Food research association has developed a method for the validation of such processes using alginate to immobilize marker organisms in particles that simulate those in the food product. In this way survival data can be calculated and the overall process lethality assessed biologically.

Process validation techniques

Immobilization   The basic principles of the alginate technique are to immobilize the marker organism in a homogenate of the food to be tested and an alginate mixture; this is made into the appropriate shape and size to mimic a food particle in that product and allowed to set. A minimum number of these alginate particles (based on a statistical population) are then added to the bulk product and given the scheduled process in production equipment. These particles are often coloured with charcoal to identify them among the bulk food. Once retrieved, the microbiological assessment will determine the number of surviving organisms and consequently the lethality of the process. When using this technique, the choice of marker organism is of critical importance: pathogens are never used, instead a non-pathogenic marker having a similar heat resistance to a pathogen is employed.

T-T indicators   Another process validation technique for use in the evaluation of specific-pasteurization processes that has been recently developed is the use of time–temperature indicators. The potential applications of such a technique utilizing enzyme activity was discussed to indicate process lethality. The enzyme used must have a known denaturation rate when heated and measurement of this can give an indication of applied process.

In principle, the time–temperature indicators comprise silicone tubes of length 10 mm and bore 3 mm which are prepared by injecting 15–20 µl of amylase inside the tube and sealing either end with a silicone plug. These capsules can then be included in the container to be processed, heated and cooled as normal, then removed and the solution extracted in order to determine the remaining enzyme activity. A control experiment is always included in order to confirm the enzyme denaturation rate at the test temperature; this can be expressed as a D value, which allows a comparison to be made with microbial levels. These data can then be used to demonstrate the process lethality achieved in the container.

There are a number of new innovative heating technologies that may apply to the pasteurization and sterilization of food products. Some of these use techniques other than direct heat to achieve inactivation of contaminating organisms. It is very important to understand the microbial inactivation kinetics achieved by these techniques, for example high pressure, pulsed electric field, Ohmic heating and microwaves. The assurance of adequate microbial inactivation in these processes will be a constant challenge. The ever-growing trends for minimally processed (pasteurized) foods will require validation techniques that assure the safety of these treatments. The consequences of combining too many changes to a safe design could be severe. It is essential that the implications of changes to the manufacturer's process and product recipe are fully understood and that safety margins are not reduced to levels that compromise stability. It is essential that all due care and diligence be given during the manufacture of these products to ensure safety.

PROCESS METHODS IN FOOD INDUSTRY

There are two fundamentally different process methods by which heat sterilization is accomplished in the food industry. These two methods are known as retort processing and aseptic processing. In retort processing, foods to be sterilized are first filled and hermetically sealed in cans, jars, or other retortable containers; then heated in their containers using hot steam or water under pressure so that heat penetrates the product from the can-wall inward, and both product and can-wall become sterilized together. In aseptic processing, a liquid food is first sterilized outside the container by pumping it through heat exchangers which deliver very rapid heating and cooling rates. Then, the cool sterile product is filled and sealed in a separately sterilized package equipment under a sterile environment at room temperature. Thus, retort processing can be thought of as in-can sterilization; and aseptic processing can be thought of as "out-of-container" sterilization. Each of these methods will be described in more detail.

Retort Processing

In retort processing (in-can sterilization), the food to be sterilized is first filled and hermetically sealed in rigid, flexible, or semi-rigid containers such as metal cans, glass jars, retort pouches or plastic bowls or trays which are then placed within large steam retorts (pressure vessels that work like giant pressure cookers). Once the retorts are full of containers to be sterilized, the retort doors are closed tightly and the air is replaced by hot steam under pressure to achieve temperatures above the atmospheric boiling point of water. A common retort temperature for sterilizing canned foods is 121°C (250°F), at approximately one atmosphere of added internal pressure. After the containers have been exposed to the sterilizing temperature for sufficient time to achieve the desired level of sterilization, the steam is shut off and cooling water is introduced to cool the containers and reduce the pressure, thus ending the process. Once the retort pressure has returned to atmospheric pressure, the doors can be opened, and the processed containers are removed for labelling, case-packing, and warehousing.

Batch retort systems   As would be expected, considerable time, effort, and labour is required to repeatedly unload and reload the retorts after each retort load or batch of containers has completed the sterilization cycle. This is known as a batch retort operation which can become very labour-intensive for large-scale production. Modern food canning plants that produce large volumes of canned foods operate with great efficiency by using continuous retort systems.

Continuous rotary retort systems   In continuous rotary retort systems, filled and sealed containers travel in single file along automated conveying-tracks into a series of continuous retorts. They enter through a rotating pressure-seal valve that works like a revolving door to maintain the steam pressure inside the retort while introducing container after container from the outside atmosphere at speeds approaching 500 units per minute. Once inside the continuous retort, the containers travel slowly along a rotating helical path (much like being pushed along by riding within the groove of a rotating screw) until they exit the opposite end of the retort through a similar rotating pressure-seal valve directly from the high-pressure steam retort into a cooling retort which is filled with cooling water instead of hot steam to accomplish the cool-down portion of the process. The cool sterilized containers are then conveyed automatically to the labelling, case packing, and warehousing operations.

Continuous hydrostatic retort system   An alternative to the continuous rotary retort system described above, is the continuous hydrostatic retort system, which makes use of two U-shaped columns of water over 20 metres high separated by a pressurized steam chamber in which the containers are sterilized. Both columns of water are open at the top and are open to the steam chamber at the bottom. One column serves as the water entrance leg while the other serves as the exit leg. Meanwhile a chainlink-driven conveyor travels continuously through the system carrying cradles of incoming containers up, along the outside wall to the top of the inlet water leg and then down the inlet water leg into the steam chamber. The conveyor speed and length of its path within the steam chamber are designed so as to deliver a sufficiently long residence time of exposure to the hot steam. So the containers are fully sterilized before they are carried up the exit leg and down the other side where they automatically transfer to the conveying tracks that take them away for labelling, case packing and warehousing.

Aseptic Processing

The retort temperatures and time required to make food microbiologically safe may also cause unavoidable quality degradation. For these heat-sensitive products, sterilization temperatures and times are chosen which will produce a commercially sterile product, but with minimum quality degradation. Because in a retort the container is heated from the outside, it is apparent that the food next to the container wall reaches the process temperature much sooner than the material in the centre of the container particularly with solid foods that heat by conduction. As a result, the major portion of the product must be held at an elevated temperature much longer than is necessary to render it sterile while time is being used to assure that sufficient sterilization is reached at the slowest heating point of the container. Obviously the container size, the product consistency, and its ability to conduct heat exerts a very measurable effect on the required time–temperature relationship. Large containers require much longer processing times. Thus, the quality of solid food is generally poorer in larger containers as compared to the same food processed in small containers.

Following the above logic further, it can be reasoned that if the food can be taken out of the container altogether, the distance for heat penetration can be reduced to a minimum if the food can move as a liquid through very narrow tubes or channels in heat exchangers, or exposed directly to hot steam in a fine spray or thin liquid film. This is the concept behind out-of-container sterilization or aseptic processing.

Limitations Aseptic processing is essentially limited to liquid foods that can be pumped or sprayed through heat exchangers which are capable of heating the product almost instantaneously to the sterilizing temperature, and cool it down just as quickly. The exposure time that is needed at the sterilizing temperature is achieved by letting the heated product flow through an insulated holding tube of sufficient length before entering the cooling section of the heat exchanger. The main drawback to this processing concept has been the difficulty in avoiding recontamination of the cool sterile product from subsequent exposure to the atmosphere or package when attempting to package it for long term storage. These drawbacks have since been overcome by the development of sophisticated aseptic packaging and filling systems which are capable of forming, filling, and sealing sterile packages with sterile liquid products under controlled sterile environments.

MICROBIAL GROWTH AND INACTIVATION

Most of the kinetic models of microbial growth and inactivation have been developed, exclusively, for either growth or inactivation. This is regardless of whether the conditions are stationary (e.g. isothermal, constant chemical agent concentration, etc.) or transient (e.g. non-isothermal temperature profiles, dissipating chemical agent, etc.) However, there are situations where a change in the conditions will halt the growth and start inactivation or will stop an ongoing inactivation and allow the survivors to grow. An example of the former, with practical implications in the operation of fermenters, is the effect of a rising temperature on microorganisms. In general, the higher the temperature, the better is the fermenter's performance (in terms of either production yield or growth rate). But if the temperature is too high, the microorganism will be destroyed. An example of the latter is a resumed bacterial growth after sublethal thermal or chemical treatment (including antibiotics). Published reports on the transition from growth to inactivation and vice versa are rather scarce in the food literature In most of the experimental studies, the objective was to develop a mathematical model that will be able to account and eventually predict growth/inactivation curves of a particular organism. The model is therefore consistent with the fact that in a closed system, death must inevitably follow the microbial population's growth. According to the model, when the temperature rises to a sufficiently high level, the inactivation becomes dominant and the F growth curve is turned into a F survival curve, Although there is no agreed upon combined model of growth and inactivation, one can still ask what would be the mathematical implications of extending growth models to the inactivation regime or inactivation models into the growth regime. If, on a pertinent time scale, both the growth and inactivation follow the first-order kinetics, then modelling in the transition from growth to lethality is rather simple. All one has to do is to find or determine experimentally how the isothermal rate constant varies with temperature. The transition from growth to inactivation or vice versa will be only manifested by a sign change, i.e., if growth is characterized by a positive logarithmic rate constant, then inactivation will be characterized by a negative one Figure 6.2.

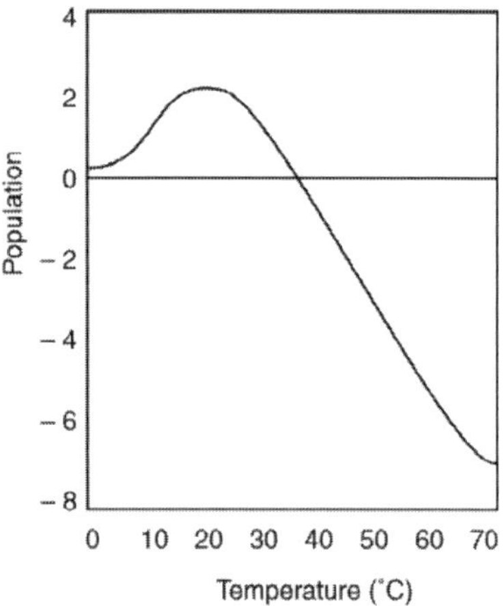

Figure 6.2 Simulated transition between growth and inactivation of a hypothetical microorganism whose growth and inactivation both follow the first order kinetic model.

From a modelling viewpoint, a more complicated situation arises when neither the organism's isothermal growth nor its inactivation follows the first-order kinetics. In such a case, the isothermal growth and inactivation pattern must be characterized by at least two temperature-dependent parameters and if the growth curve has an asymmetric sigmoid shape, by at least three. The transitions themselves have direct relevance to food and water safety. Microbial growth in partially or insufficiently processed foods is a potential cause of food poisoning and in certain types of slowly cooked meats, substantial growth can occur before inactivation ensues, thus affecting the efficacy of the heat treatment. A similar kind of danger may arise in chemically disinfected water if the agent is not replenished in time or at all, thus allowing the microbial population to bounce back. A health hazard may also be created if the disinfectant administration is delayed, thus enabling the microbial population to grow unchecked, which might decrease the efficacy of the belated chemical treatment.

FUNDAMENTALS OF THERMOBACTERIOLOGY

Heat Resistance

It is the capacity to withstand heat and is generally expressed as thermal death time (time taken at a constant temperature to kill all the microbes in food). It varies with different organisms. Generally heat resistance or sensitivity is a term associated with the decimal reduction time (D value is time taken to kill 90% of the microorganisms at a certain temperature).

Calculation of the heat treatment required to kill a given number of bacteria is based on the assumption that destruction of a pure bacterial culture by wet heat at constant temperature exhibits a negative exponential kinetics, i.e., equal time periods at same temperature (lethal temperature) results in equal percentage of death independent of population size. One can calculate the heat resistance of a particular organism by calculating the D values for the same.

Method for Measuring the Heat Resistance

Standard spore/cell suspension is prepared (using standard strains) e.g. *Bacillus stearothermophilus* ATCC 7953 and *Bacillus subtilis* ATCC 6633.

Inoculum should contain 10^6 cells/ml/g of food (6 vials)

↓

Vials heated in a water bath ($<100°$C)/oil bath ($> 100°$C)
at a selected temperature

↓

Vials removed periodically (at constant time intervals),
immediately cooled and subcultured in a fresh medium

↓

Incubated at $37°$C for 24 hours

↓

Observed for growth/no growth

For example, from a population of 300 cells, population is reduced by one log cycle in 10 min. There is a relationship between the temperature employed for sterilization and the time period with the number of survivors.

Table 6.1   Calculation of D values

Temperature	Time	No. of survivors
110°C	1 min	300
	2 min	265
	3 min	200
	4 min	170
	5 min	155
	6 min	100
	7 min	85
	8 min	60
	9 min	52
	10 min	46
	11 min	31

D value at 110°C–10 min.

Hence death rate is exponential. D value can be calculated for various temperatures for the same organism and a graph plotted. As temperature increases, D value decreases. From the plotted values of D, another important factor, the Z value can be obtained. Z value is the temperature required to reduce the D value by tenfold. For moulds, bacteria (vegetative cells) and yeasts, Z value normally ranges between 5–8°C and for spores, 6-16°C. Z values can be used to determine the equivalent D values at different temperatures.

For example, *Listeria monocytogenes* has a D_{60} of 8.3 minutes and a Z of 5°C. What is the D value at 71.7°C (pasteurization temperature)?

Solution

Given

$$D_1 = 8.3 \text{ min.}, \quad Z = 5, \quad T_1 = 60°C, \quad T_2 = 71.7°C$$

$$\frac{D_1}{D_2} = 10^{\frac{T_2 - T_1}{Z}}$$

i.e., $$\frac{8.3}{D_2} = 10^{\frac{71.7-60}{5}}; \quad = \frac{8.3}{D_2} = 10^{2.34}$$

i.e., $$\frac{8.3}{D_2} = \frac{2.34 \log 10}{2} = \text{Antilog of } 2.34 = 218.7$$

i.e., $$D_2 = \frac{8.3}{218.7} = 0.0370 \text{ minutes i.e., } 2.22 \text{ seconds}$$

Thus D value for 71.7°C for 2.2 seconds (HTST).

If the Z value is known, it can be used at any other temperature above the maximum for the heat treatment for eliminating a particular organism.

There is another value of importance (F value) in the canning industry, which will be discussed.

Heat Resistance of Various Microbes

Heat resistances (D values) of various microorganisms and effects of temperature changes on the death rate are different and are influenced by many factors. There are inherent differences among species, among strains within the same species, and between spores and vegetative cells.

Psychrotrophs are less heat-resistant than mesophiles which are in turn less resistant than thermophiles. Gram-positive bacteria are more heat-resistant than gram-negative bacteria. Most of the vegetative cells get killed at 100°C. For example, D_{65} of *Salmonella* is 0.02–0.25 minutes and for *Staphylococcus*, D_{65} is 0.2–2 minutes. Spores are more resistant to heat than vegetative cells, especially thermophilic spores. Yeast ascospores and the asexual spores of moulds, show slight increase in heat resistance than the vegetative cells.

Heat Resistance of Spores

Heat resistance, probably the spore resistance most familiar to food microbiologists, has the most implications for the food industry and is probably the best studied form of resistance in spores. Spores of many species can withstand 100°C for several minutes. An often overlooked feature of spore heat resistance is that the extended survival of spores at elevated temperatures is paralleled by even longer survival times at lower temperatures. Spore D values increase 4–10 fold for each 10°C fall in temperature. Consequently, a spore with a D_{90} of 30 minutes may have a D_{20} value of many years.

The identity of the target(s) whose damage results in heat killing of spores has not been clearly understood even though reports say that DNA damage, mutagenesis, protein damage may be the targets but these have not been proved. There are several factors that modulate spore heat resistance.

1. *Sporulation temperature* Elevated sporulation temperatures increase spore heat resistance. Hence, spores of thermophiles generally have much higher heat resistance than spores of mesophiles. The reason being that the total macromolecular content of spores from thermophiles could be more heat-stable than that from mesophiles, accounting for the higher heat resistance of spores from thermophiles. Another reason could be due to the reduced core water content in spores. In growing bacteria, adaptation to heat stress involves the proteins of the heat shock response. The levels of these proteins would increase with increasing sporulation temperature.

2. α *and* β *type small acid-soluble proteins (SASP)*   Spore DNA is remarkably well protected against heat damage. The major cause of spore DNA protection against heat damage appears to be the saturation of spore DNA by α/β type SASP. The above facts hold good for spores treated with heat in water, but spores treated with dry heat are somewhat different. First, wild type spores are much more resistant to dry heat than to aqueous heat as D values are 2–3 orders of magnitude higher in dry versus hydrated spores. Second, wild type spores exhibit a rather high level of mutagenesis upon killing by dry heat and this mutagenesis is associated with generation of damage in spore DNA. α/β–type SASP are major factors increasing the dry heat resistance of wild type spores over that of vegetative cells and they also provide significant DNA protection against dry heat damage.

3. *Spore mineralization*   Spores accumulate large amounts of divalent cations late in sporulation. Spore mineralization is implicated in heat resistance. Both the amount and type of mineral ions accumulated can affect spore heat resistance. The order of spore heat resistance with different cations are: $H^+ < Na^+ < K^+ < Mg^{2+} < Mn^{2+} < Ca^{2+} <$ untreated. Alteration of spore mineralization can alter spore core water content which presumably causes a significant effect on heat resistance.

4. *Spore core water content*   Low core water content is a major factor causing spore heat resistance. The dehydration of the spore begins in the stage III–IV transition and continues throughout stages IV and V, with final dehydration taking place approximately in parallel with acquisition of spore heat resistance. Synthesis of the spore cortex is essential both for affecting this dehydration and for maintaining the dehydrated state of the spore core. This is undoubtedly due to the ability of peptidoglycan to change its volume markedly upon changes in ionic strength and pH. The important role of the spore cortex in heat resistance is shown by the inverse correlation between spore heat resistance and the volume occupied by the spore cortex relative to that of the core.

Sensitive Screening Method for Heat Resistance of Spore-forming Organisms

Bacterial spores are common contaminants of food products, and their outgrowth may cause food spoilage or food-borne illness. They are extremely resistant to heat and other preservation treatments in comparison to vegetative cells. The inactivation of spores requires high temperatures and long heating times, which are costly and detrimental to the nutritional and organoleptic quality of most food products. To minimize the required heat treatment, there is an urgent need in the food industry for tailored preservation procedures, based on models that accurately predict the presence of viable cells at every step of the food production process. To assess the required heat inactivation procedure for the most resistant cell type, the bacterial spore, a rapid and sensitive screening method was developed to determine the heat resistance of their spores. Under nutrient-limited conditions, vegetative cells of *Bacillus* species undergo the cell differentiation process of sporulation. The resulting spores are metabolically dormant and show, besides resistance to heat, resistance to other potentially lethal treatments that include radiation, high pressure, chemicals, and desiccation. Although spore dormancy and associated resistance are very stable, spores may survive over hundreds and even millions of years. These properties are lost within minutes during the process of germination, which is triggered by the presence of nutrients. There is a considerable amount of information on the factors that modulate the heat resistance of spores, but the exact nature of the damage that actually kills the spore is still obscure. Heat resistance factors include the protection of spore DNA by small acid -soluble proteins, the accumulation of divalent cations, such as Ca^{2+} and Mn^{2+}, and the dehydration of the spore core. In addition, there is a role in heat resistance for dipicolinic acid (DPA or pyridine-2,6-dicarboxylic acid), to which the divalent cations are chelated in the core of the spore. Dipicolinic acid was first identified in bacterial spores by Powell. This compound has been identified exclusively in bacterial spores and is involved in their dormancy, wet-heat resistance, and germination.

Synthesis of DPA   The synthesis of DPA occurs in a sporulating cell in one step from dihydroxydipicolinic acid, an intermediate in lysine biosynthesis. DPA is transported from the mother cell compartment over the outer and inner membranes of the forespore. The proteins involved in DPA transport are most probably encoded by the *spo*VA operon ; the DPA synthase is encoded by the two genes of the *spo*VF operon. Mutations in the *spo*VF locus show significantly increased spore core water content and decreased heat resistance. The addition of exogenous DPA to sporulating cells of these mutants rescues the heat resistance of their spores. However, DPA is not indispensable for full heat resistance, as mutants generating DPA-less spores with restored heat resistance have been isolated. The DPA content of wild-type spores is approximately 10% of the dry weight of the spore, and DPA is usually present in a 1:1 molar ratio with Ca^{2+}. No clear correlation has been found between heat resistance and the total amount of DPA present in the core of wild-type spores. In contrast, many studies have shown that differences in the amount and type of cation strongly affect spore heat resistance. Previously, an assay for the release of DPA during spore germination was developed using absorption of DPA in the UV region; this assay allowed the detection of DPA concentrations down to 0.5 μM. An alternative method for the detection of DPA was obtained by the use of strongly enhanced fluorescence of the lanthanide ion Tb^{3+} upon DPA binding. This fluorescent DPA assay was initially developed as a method for the detection of bacterial spores. The assay has been optimized to a detection limit of 2 nM DPA, which corresponds to 10^4 spores/ml.

Release of DPA   Release of DPA from bacterial spores occurs under a number of different conditions. First, DPA is excreted from spores during the first minute of germination, when nutrients bind to the germinant receptors. The release of DPA is one of the first events in the process of spore germination and occurs simultaneously with the release of cations, the uptake of water, and the loss of the phase-bright appearance of the spore. Second, DPA release occurs during the process of spore activation by a sublethal heat treatment that breaks spore dormancy and leads to an increase in the number of germinating spores. The fraction of DPA released from the spores during heat activation differs strongly among several published studies and depends in part on the nature of the heat treatment and the *Bacillus* species involved. Third, DPA is also released during wet-heat-induced spore inactivation. The relationship between the release of DPA and the heat resistance of spores has been studied for spores from a number of *Bacillus* species, all showing that DPA release is slower than the loss of viability of the spores during heating. Although the correlation between DPA release and the spore death rate is complex, higher rates of death were associated with higher rates of DPA release. A more recent study on heat-induced DPA release from *Bacillus stearothermophilus* spores showed that the rate of spore death has a higher temperature dependence than the rate of DPA release.

Heat inactivation and counting of spores   The wet-heat inactivation of spores carried out using the screw-cap tube method. A spore suspension (0.5×10^8 to 5×10^8 spores per ml) of 200 is injected with a Hamilton syringe into a preheated (15 min. of equilibration) metal screw-cap tube containing 9.8 ml of inactivation medium. The inactivation medium used is either sterile trypticase soy broth or sterile physiological salt solution. Control experiments do not show significant differences in the rate of spore inactivation between these two media. Heating carried out with the metal tubes completely immersed in a glycerol bath. Sampling after a desired incubation time is done through immediate transfer of a tube to ice water. Spore suspensions are diluted 10 times and counted using a haemocytometer. Ten randomly selected squares are used for counting with a surface area of 0.0025 mm^2 and a depth of 0.01 mm each. Heat inactivation of spores is determined by the loss of their ability to germinate and to form colonies (i.e., viability counts). Dilution series of spore suspensions are prepared in 0.1% peptone/0.85% NaCl and added to trypticase soy agar pour plates.

The number of colonies are counted after 4 days of incubation at 37°C. All heat inactivation experiments and viability counts are carried out in duplicate.

Factors Affecting the Heat Resistance

(a) *Temperature–time relationship*   As temperature of exposure increases, time of exposure decreases.

(b) *Initial concentration of spores*   More spores/cells in a food commodity, greater is the heat resistance, hence higher temperature is needed to kill them.

No. of spores/ml	TDT (minutes)
50000	14
500	9
50	4

(c) *Previous history of vegetative cells/spores*   Media and environmental conditions under which the cells have been grown will influence the heat resistance.

   i. *Medium*   Better the medium for growth, more resistant are the cells/spores. Presence of growth factors increases heat resistance. More amount of sugar may lead to increased heat resistance but large concentrations will lead to acid production and thereby decreases heat resistance.

   ii. *Temperature of incubation*   When the incubation temperature is at the optimum, heat resistance for the cell is maximum. Even a slight increase in temperature increases heat resistance. For example *E.coli* is more heat-resistant at 38.5°C than at 28°C.

Bacillus subtilis in peptone water (shown below):

Temperature (°C)	Time to kill at 100°C (minutes)
21–23	11
27 (opt.)	16
41	18

(d) *Desiccation*   Dried spores are more heat-resistant than those kept moist.

(e) *Substrate in which spores are heated*

   i. *Moisture*   Moist heat sterilizes faster than dry heat

   ii. *pH*   spores/cells are more heat-resistant at neutral pH. An increase or decrease in pH decreases the heat resistantce. Acidity of the medium is more effective in killing spores than alkalinity of medium.

iii.   *Other constituents in the substrate*   NaCl at low concentrations has protective effect on spores. Sugars also protect spores but concentrations and effects vary. High concentration of sugar is necessary for osmophilic yeasts. Decrease in a_w is evident as a reason for heat resistance. Glucose protects *E. coli* against heat than NaCl. Glucose is harmful to *Staphylococcus aureus* whereas NaCl is protective to it. Proteins and fats are protective against heat. Germicidal substances in the substrate decreases the heat resistance. For example, hydrogen peroxide, in addition to heat reduces the bacterial content and hence their heat resistance.

Heat Resistance and Enzymes

Enzymes present in food may be responsible for product deterioration during storage. During heat process, the vegetative cells may perish but their extracellular enzymes may remain active at that temperature and spoil the food during storage. Some proteinases and lipases retain activity after an ultra-high temperature. Bovine phosphatases in processed milk indicates that milk is not properly pasteurized. If microbial phosphatases are present this becomes a more complicated problem.

Heat Penetration

This property of heat is important in determining the thermal process necessary for preservation. Canning is a thermal process of heating food along with their hermetically sealed containers. This process is dependent on property of heat penetration or ability to transfer heat to every part of the food such that it becomes sterile inside the container. The transfer of heat into food depends on the thermal properties of food, geometry of the container of food and thermal processing conditions. Normally, heat penetrates from the external source of a can to the centre of the can. At a particular time, the centre of the can may not be heated adequately, i.e., that part of container which gets heated slowly is the critical one.

Heat is transferred by two ways:

1. Conduction
2. Convection

Conduction is slow in foods and heat transfer in convection depends on the opportunity of currents in the liquid and rate of flow of these currents. Sometimes both conduction and convection are involved in foods like solid food suspended in liquid. Solid pieces are heated by conduction whereas liquid parts by convection. Whatever be the heat process, the main aim in canning is to maintain the high temperature in the cold point. Since heating processes are not uniform, for describing a uniform heating process for canned foods, a value F is used. F value depends on the Z value of the organism.

F value is the time necessary to destroy a given number of microorganisms at a standard temperature (usually 121°C for spores, 60°C for vegetative cells) having a specific Z value.

For example, F_{60} is TDT at 60°C and F_{121} is the TDT at 121°C

$Z = 10°C$, if D_{111} is 1 minute and F_{121} is 0.1 min.

Since F values represent the number of minutes to diminish a homogeneous population with a specific Z value at a specific temperature, and because Z values as well as temperatures vary (for many of the thermophilic spores Z value is 18°C), it is necessary to designate a reference F value. This reference is the F_o value, which is the number of minutes at 121°C required to destroy a specific number of cells whose Z value is 10°C. The Fo value of a heat treatment, i.e., its sterilization value, is a measure of the lethality of a given heat treatment.

Since heat sensitivities of microorganisms and the characteristics of the thermal death curve are affected by many factors, thermal death curves should be established in the specific food for which a heat process is designed and it is necessary to have a knowledge about the Z values and F values before deciding the heating process.

Because the spores of *C. botulinum* types A and B are the most heat-resistant spores of a food-borne pathogen, commercial sterilization of low-acid foods must be sufficient to destroy these spores.

12D concept   A 'botulinum cook' is a heat process which reduces the population of the most resistant *C. botulinum* spores by an arbitrarily established factor of 12 decimal values, i.e., killing of cells by 12 log cycles. (A DRT kills by 1 log cycle).

For example, if D_{121} of *C. botulinum* is 0.21 minutes,

Botulinum cook shows $F_o = 12 \times 0.21 = 2.52$ minutes, i.e., a time exposure of 2.52 minutes at 121°C is necessary for assuring the removal of all spores of *C. botulinum* in the can. Assuming every can will contain 1 spore of *C. botulinum*, (initial spore concentration), at the heat treatment of 12D, will ensure that one spore will survive in one can out of 10^{12} cans. This provides a substantial margin of safety in low-acid canned foods.

The botulinum cook will produce a safe product but not necessarily commercially sterile. Other heat-resistant spores, which may cause spoilage but are not pathogenic, may be present and viable. A minimum safe public health sterilization value for a low-acid canned food is:

F_o = 3 minutes at pH 7

2.3 minutes at pH 5.5

1.6 minutes at pH 5

1.2 minutes at pH 4.6

Since the pH decreases, the D value also decreases.

Factors Affecting Heat Penetration

1. *Material of container*   Glass containers show slow heat penetration than metal cans.

2. *Size and shape of container*   In larger cans, cold point increase in temperature takes a longer time. Long, slim cylindrical cans will heat faster than compact cylinders.

3. *Initial temperature of food*   Food at a lower initial temperature, (before actual heat processing begins) heats faster since transfer of heat is greatest with greater differences in temperature, but not so in foods at higher initial temperatures. A higher initial temperature is necessary to kill microbes since cold points get heated up faster in foods with higher initial temperatures than otherwise.

4. *Retort temperature*   Heating is faster in hottest retort.

5. *Consistency of can contents (size and shape of pieces)*

    (a) Pieces that retain their identity (e.g. peas, plums, beets,)—small pieces heat faster than large pieces.

    (b) Pieces that cook apart (e.g. squash, pumpkin)—heat penetrates slowly in the food.

    (c) Pieces that layer (e.g. asparagus that layers vertically and spinach that layers horizontally)—heating is slower in asparagus than in spinach.

6. Rotation and agitation (10–12 rpm)—helpful in foods that layer.

HEAT TREATMENTS IN FOOD PROCESSING

Heating is employed to kill, if not all organisms, at least the pathogenic ones from food stuffs. Various degrees of heating used are:

1. Heating below 100°C/pasteurization

2. Heating at 100°C

3. Heating above 100°C.

HEATING BELOW 100°C (PASTEURIZATION)

First used by Louis Pasteur while handling wine, nowadays it is used for increasing the keeping quality of milk and milk products. Pasteurization is a term given to heat processes in the range of 60–80°C for few minutes followed by immediate cooling. This treatment is used for two processes:

1. Elimination of a specific pathogen/pathogens associated with the product. For example, in milk, liquid egg, ice cream mix.

2. To eliminate a large proportion of potential spoilage organisms thus extending its shelf life. For example, as in beer, fruit juice, pickles, sauces.

Thermoduric spore formers, gram-positive vegetative cells of *Enterococcus, Microbacterium* and *Arthrobacter* can survive pasteurization hence refrigerated storage is an additional requirement. Heating by pasteurization may be by means of steam, hot water, dry heat or electric currents.

When is pasteurization used?

1. When more rigorous heat treatments might harm the quality of product.

2. When the aim is to kill the pathogens (as in milk).

3. When the main spoilage organisms are not so heat-resistant as in yeasts in fruit juices.

4. When competing organisms are to be killed, during a desirable fermentation by a starter culture organism, as in cheese making.

The preservation methods that should supplement pasteurization include:

- Refrigeration (as in milk)
- Asepsis (packaging in containers)
- Maintenance of anaerobic condition (evacuated sealed containers)
- Addition of high amount of sugar (as in condensed milk)
- Presence/addition of chemical preservatives (organic acids in pickles)

Various pasteurization temperatures include:

- Low temperature long time/Holding temperature (LTH)–62.8°C for 30 minutes
- High temperature short time (HTST)–71.7°C for 15 seconds
- Ultra-high temperature (UHT)–137.8°C for 2 sec (to eliminate *Coxiella burnetti*)

The pasteurization temperature of various products are given in Table 6.2.

Table 6.2   Pasteurization temperature of various products

Product	Pasteurization temperature	Time
Ice cream mix	71.1°C	30 minutes
	82.2°C	16–20 seconds
Grape wine	82–85°C	1 minute
Beer	60°C	Varying time
Dried fruit	65.6–85°C	30–90 minutes

HEATING AT 100°C

Done by home canners with the help of pressure cookers. The different processes of heating at 100°C are given below:

- Baking—oven temperature is high enough to kill microbes but the internal temperature of cake/bread never reaches 100°C if there is moisture.
- Simmering—gentle boiling
- Roasting—internal temperature reaches 60–85°C.
- Frying
- Cooking
- Warming

HEATING ABOVE 100°C

This is called canning with a temperature of heating, 100°C for high-acid foods and 121°C for low-acid foods. The heat processing depends on the heat resistance of microbes and variety of foods. Heating is done on retorts and food may be sterilized separately, then packaged into cans, sealed and filled in retorts and given a milder heating through :

- High pressure steam.
- Direct gas flame
- 'Flash 18' method in a high-pressure chamber.

History of Canning

During the early Revolutionary Wars, the notable French newspaper Monde, prompted by the government, offered a hefty cash award of 12,000 Francs to any inventor who could come up with a cheap and effective method of preserving large amounts of food. The massive armies of the period required regular supplies of quality food, and so preservation became a necessity. In 1809 the French confectioner Nicholas Appert developed a method of vacuum-sealing food inside glass jars. However, glass containers were unsuitable for transportation, and soon they had been replaced with cylindrical tin or steel cans (tin-openers were not invented for another thirty years—at first, soldiers either had to cut the cans open with bayonets or smash them open with rocks to get the food out!). The French Army began experimenting with issuing tinned foods to its soldiers, but the slow process of tinning foods and the even slower development stage, along with the difficulties of loading wooden wagons with tons of metal canisters, prevented the army from shipping large amounts around the Empire, and the war ended before the process could be perfected. Unfortunately for Appert, the factory which he had built with his prize money was burnt down in 1814 by Allied soldiers invading France. Following the end of the Napoleonic Wars, the process was gradually put into practice in other European countries and in the United States. Based on Appert's methods of food preservation the packaging of food in sealed airtight tin-plated wrought-iron cans was first patented by an Englishman, Peter Durand, in 1810. Initially, the canning process was slow and labour-intensive, making the tinned food too expensive for ordinary people to buy. Only the rich could afford it, and rich people generally preferred fresh food to tinned alternatives. However, increasing mechanization of the process, coupled with a huge increase in urban populations across Europe, resulted in a rising demand for tinned food.

A number of inventions and improvements followed, and by the 1860s, the time required to process food in a can reduced from six hours to 30 minutes. Thomas Kensett established the first U.S. canning facility for oysters, meats, fruits and vegetables in New York in 1812 and also patented an improved tin canister method. Urban populations in Victorian era demanded ever-increasing quantities of cheap, varied, good-quality food that they could keep on the shelves at home without having to go to the shops every day for fresh products. In response, companies such as Nestle, Heinz and others appeared to provide shops with good-quality tinned food for sale to ordinary working-class city-dwellers. Demand for tinned food skyrocketed during the First World War, as military commanders searched for cheap, high-calorie food which could be transported safely, would survive trench conditions, and which would not spoil in between the factory and the front lines. Complete meals in a tin appeared in 1916, but throughout the war, soldiers generally subsisted on very low-quality tinned foodstuffs, such as the British "Bully Beef" (cheap corned beef) and the notoriously disgusting "Pork and Beans" produced by the MacConnaughy Corporation. The tinned food issued to French soldiers was by far the worst in any army, whilst shortages of tinned food in the British Army, in 1917, led to the government issuing soldiers with cigarettes and even amphetamines to suppress their appetites. After the war, companies that had supplied tinned food to national militaries improved the quality of their goods for sale on the civilian market. Canned foods were soon commonplace, and today tin-coated steel is the material most commonly used. Some food firms are currently dabbling with self-heating cans.

Procedure of Canning

The various steps followed in canning is given in Figure 6.3.

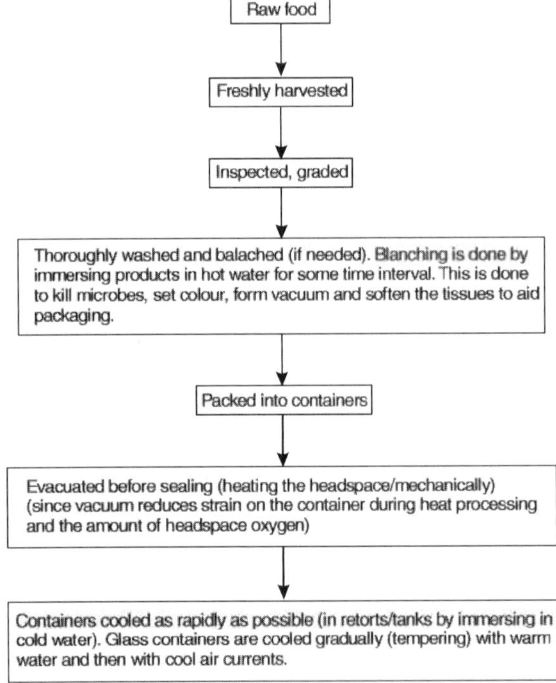

Figure 6.3   Food canning procedure

Heat Processing and Aseptic Packaging

In recent years, considerable advances have been made in the type and design of packaging materials used for heat-processed foods. Packaging materials are not necessarily rigid; flexible-laminates can be formed into pouches and heat-sealed. This in turn brings a number of potential issues in relation to microbial leakage into packs, particularly if seal and pack integrity is not sound. In addition, there is a growing trend to develop multicomponent and multicompartment foods; these also present very demanding microbiological issues.

Multicomponent products may be in one pack and mixed together or in special packaging with several different compartments within one overall pack. It is important to be aware that many of these developments may be market-driven as the packaging materials lend themselves to far more adventurous labelling and direct on-pack advertising. There are many microbiological criteria to consider for these types of products. For example, for the multicomponent product, each compartment may contain different types of microorganisms and while each may be stable individually, when these are all placed together, the presence or the by-products of one microorganism may influence the ability of other organisms to grow or survive. In addition, the heat resistance characteristics of different organisms heated in the different food components could be markedly increased.

The thermal process design must therefore be structured to ensure that the most heat-resistant pathogen is adequately reduced. In addition it should be noted that the presence of contaminating organisms in one component, for example, an acidic sauce, may not pose a threat. If, however, these organisms leak from one compartment of the package to an another neutral pH food component, it is possible that growth could occur. One of the most popular type of development of pasteurized foods has been ready-meals. One particular process known as 'Sous vide' employs the vacuum packing of the product followed by cooking. Provided the minimum safest heat process is achieved, this technique can produce very high quality dish that can be preserved for several weeks at chilled temperatures. Again one of the key parameters for safety would be to establish where the slowest heating point would be in the pack in order to ensure microbial stability. The question of pack integrity must also feature highly at the design stages, particularly as the product is *in-pack* pasteurized and any leakage of microorganisms into the packs will compromise safety. Considerable research has been done on the use of antimicrobial packaging, the main principle of which is to deposit coatings onto packaging materials and assess the inhibitory effects on specific organisms. There are a number of microbiological questions that must be considered when designing and developing a new packaging format. They are

1. How is the packaging material itself decontaminated?

2. Are there opportunities for the packaging to become contaminated with pathogenic microorganisms?

3. Can microbial growth be sustained on the packaging material?

4. Can contamination occur during filling?

5. What is the maximum temperature that can be achieved during filling and processing without causing distortion to the packaging material?

6. Do the suggested packaging design and subsequent required heat process maintain product quality?

7. Are there any risks of breach of container integrity throughout the product life, from manufacture, storage and distribution?

Thermal Processing and Allergenicity

It has been observed that food processing, and particularly thermal processing, could also reduce the allergenicity of some foods as an indirect effect and it has been proposed that this might serve for a better management of the allergenic risk of foods.

The rationale is apparently quite simple. Heat treatments alter the structure of proteins, i.e., the allergenic constituents of a food and, as a consequence, alter their allergenic potential and therefore the allergenicity of the whole food. Although no clear and general relationship between the structure of a protein and its allergenicity has been established, this view assumes, albeit not explicitly, that allergenicity is an intrinsic property of a protein because of or strictly related to particular structural features. This view, however, does not take into account the qualitative and quantitative variability of the allergen repertoire of a whole food, the multiplicity of epitopes on a given allergen or the genetic/geographic variability of the immune response in atopic human beings.

Significant alterations in protein structure do occur during heat treatments, the nature and extent of such changes being dependent on the temperature and duration of the thermal processing as well as on the intrinsic characteristics of the protein and the physico-chemical conditions of its environment (e.g. pH). Typically loss of tertiary structure is followed by (reversible) unfolding, loss of secondary structure (55–70°C), cleavage of disulphide bonds (70–80°C), formation of new intra- and intermolecular interactions, rearrangements of disulphide bonds (80–90°C) and the formation of aggregates (90–100°C). These modifications reflect a progressive passage to a disorganized structure with denaturation of the proteins that adopt an unfolded, random-coil conformation. The denatured molecules associate to form aggregates and then gels resulting in a modification of the surface properties and an increase in size.

Besides those physical transformations, chemical modifications of the protein may also occur at high temperatures (100–125°C and higher). These may involve formation of covalent bonds between the lysine residues of a protein and other constituents of the food matrix leading to various adducts. Advanced glycation end-products, carboxymethyl lysine, malondialdehyde and 4-hydroxynonenal formed through protein–sugar interactions (i.e. by the Maillard reaction) and the cross-linking of oxidized lipid products with proteins are commonly found and may contribute to the formation of new immunologically reactive structures.

It appears that there are no general rules regarding the consequences of thermal treatment on allergenicity. Some allergens or, more properly, some allergenic foods, are described as heat-stable (e.g. milk, egg, fish, peanuts, and products thereof), while others are considered partially stable (e.g. soya bean, cereals, celery, etc.) or labile (fruits of the Rosaceae family and carrots). In addition, thermal processing can create new allergenic epitopes as well as destroying existing epitopes. Whether and how heat treatments may significantly alter the allergenicity of a food is thus a complex question.

Microwave Heat Treatment

Heating due to electromagnetic radiowaves generate internal heat. Water molecules are dipolar—oxygen atoms have a slight negative charge and hydrogen atoms bear a slight positive charge. When rapidly oscillating microwave is applied to foods with rich water content, the molecules reorient with each change in field direction. This increases intermolecular friction and produces heat.

Effect on microbes   Electromagnetic fields cause ion shifts in cell membrane leading to permeability changes, functional disturbances and cell rupture. Microbial cells are differentially heated than the food particles that cause specific inactivation of microbes. Advantage of this method is that it reduces process timings.

Factors affecting microwave heating are

- geometry of food product
- thermal properties (physical, radiation frequencies).

Ohmic Heating

This makes use of direct electric heating. Electric current is passed through the food material as it is passed through a tube. Heating rate depends on electrical conductance of solid and liquid phases of food. If both have similar conductance, both the phases gets heated at the same rate. This is still at the development stage and microbial death appears to be caused solely by thermal effect.

Ohmic heating, sometimes called electrical resistance heating or Joule heating, uses electrical power to be transformed into heat energy. When an alternating electrical current is passed through food, it heats up the food system due to its electrical resistance. For the process to work efficiently, it requires electrodes to be kept in close contact with the food.

The idea of using electrical current to generate heat in food is not new. Ohmic heating is applied to pasteurize milk. A resurgence of interest in ohmic heating has emerged due to two major events: first, the availability of improved non-fouling electrodes; second, consumer demand for "fresh-like" or minimally-processed foods. Although ohmic heating cannot be truly classified as a minimal process, with careful design and operating conditions, the heat generated may be controlled to cause less damage to the food than conventional thermal processing. During conventional heat processing of viscous food, heating occurs from the surface to the interior, with considerable lag time in heat transfer between liquids and solids. When a solid–liquid mixture is conventionally heated, the temperature of the liquid phase will increase more rapidly than that of the particles. In marked contrast, with ohmic heating the solid particles gain heat faster than the liquid. This makes it an attractive process for HTST sterilization or pasteurization of particulate foods. Ohmic heating has been successfully used to process proteinaceous foods such as egg, cheese, etc. It is also a useful method to thaw frozen fish products.

REVIEW QUESTIONS

1. Explain the principle behind high temperature preservation.
2. What is commercial sterilization? Explain.
3. What is aseptic processing?
4. Explain the fundamentals of thermobacteriology.
5. Explain heat resistance in spores.
6. List out the factors that contribute to the heat resistance or heat sensitivity of certain cells.
7. What is heat penetration?
8. Explain the 12D concept.
9. List the factors affecting heat penetration.
10. Explain the term pasteurization.
11. Brief out on microwave heat treatment.
12. What is ohmic heating?

7

LOW TEMPERATURE AS A PRESERVATION AGENT

INTRODUCTION

Refrigeration is the most common means of preserving food, either alone or in combination with other methods such as addition of preservatives. Therefore, an understanding of the response of food spoilage and food-poisoning microorganisms to the stress imposed by low temperature is fundamental to the design of effective preservation strategies. This is particularly relevant in the context of modern demands for foods containing lower levels of preservatives for a more natural flavour and wholesomeness, since many of the major spoilage (e.g. *Brochothrix thermosphacta*, *Pseudomonas* spp., *Micrococcus* spp.) and poisoning (e.g. *Listeria monocytogenes*, *Yersinia enterocolitica*) microorganisms of concern are psychrotrophic (psychrotolerant) (Table 7.1). They are notable for having a particularly broad growth temperature range, often approaching 40°C, in contrast to psychrophiles which have much narrower ranges. Although psychrotrophs cannot match the sub-zero growth of psychrophiles, they are nevertheless able to grow at low temperatures approaching 0°C (e.g. 1–3°C for strains of *L. monocytogenes*) as well as being capable of growing rapidly when temperatures rise to (warm) room temperature and, for pathogens, the human body temperature of 37°C. It is this wide thermal capability that makes them specially significant in terms of food quality and safety. In relation to refrigeration, the temperature ranges that are most relevant are 4–6°C (refrigerators) and 10–12°C (open chiller display units). This thermal range from 4–12°C is one over which psychrotrophs are capable of growing at rates that may be only two-to-four-fold lower than their optimum rates at 20–30°C. Therefore, they pose a particular threat to chilled foods because bacterial populations can reach levels that are capable of serious spoilage or are above the threshold for causing food poisoning.

Table 7.1 Thermal characteristics of cold-adapted food spoilage and food poisoning bacteria

Bacterium	Lower growth limit (°C)	Comments
Pseudomonas fluorescens	0	Food spoilage, non-pathogenic
Micrococcus spp.	1	Food spoilage, non-pathogenic
Listeria monocytogenes	1	Most strains will grow at refrigeration temperatures
Clostridium botulinum	3	Only some non-proteolytic type E strains are psychrotropic
Salmonella spp.	5	Very slow growth of some strains at refrigeration temperatures; most species grow slowly at chilling temperatures
Lactic acid bacteria (LAB)	5	May cause unwanted deterioration of LAB fermented foods
Staphylococcus aureus	6	Some strains will grow at refrigeration temperatures in processed foods
Bacillus cereus	10	Psychrotrophic strains can spoil milk or poison (emetic toxin) in chilled foods
Clostridium perfringens *Clostridium botulinum* type A	13	May grow slowly at chill cabinet temperatures

There are two prominent effects of low temperature as a preservation technique:

1. It retards chemical reaction and action of food enzymes.

2. It stops or retards growth and activity of microorganisms in food.

Each microbe has an optimal temperature for growth and its growth retards with a change in temperature. But low temperatures are preferred by psychrotrophs and psychrophiles.

Table 7.2 Growth temperature ranges of psychrotrophic organisms

Organism	% spoilage flora at each temperature		
	1°C	10°C	15°C
Pseudomonas	90%	37%	15%
Acinetobacter	7%	26%	34%
Enterobacteriaceae	3%	15%	27%
Aeromonas	–	4%	6%

Cladosporium and *Sporotrichum* survive at 6.7°C, whereas *Penicillium* and *Monilia* survive at 4°C and yeasts survive at 3.4°C.

METHODS OF STORAGE USING LOW TEMPERATURES

Common/Cellar Storage

In common or cellar storage the temperature of storage is not lower than 15°C. It is used for storing root crops and fruits (potatoes, cabbage, celery, apples, etc.). However, does not prevent deterioration of the fruits and vegetables by their own enzymes.

Chilling/Cold Storage/Refrigeration

It refers to storage at temperatures above freezing, i.e., from 16°C to –2°C. Most foods do not freeze until a temperature of –2°C or lower is reached. One can store food at cold storage for a few days to several weeks. This storage is based on the fact that reducing the temperature decreases the rate of chemical reactions and growth of microorganisms. The temperature of cold storage is less than the minimal growth temperature of most food-borne pathogens. It is used for storing perishable foods like eggs, dairy products, meat, seafood and fruits. Main organisms of concern in refrigerated food are the psychrophiles (which can grow at –15°C) having an optimum at 10°C, i.e., food will spoil 4 times as fast at 10°C and twice as fast as at 5°C than at 0°C. Food-borne psychrophilic organisms will not grow or produce toxins below 4.4°C. None of the food-borne organisms grow at temperature less than 1.7°C. Because growth rates of microorganisms increase significantly even with a slight increase in temperature, variation of storage temperature should be avoided. The minimum growth temperature of a microorganism is determined by the inhibition of solute transport. Hence the growth temperature range of an organism depends on how well the organism can regulate its lipid fluidity within a given range of temperature. For example, psychrotrophs contain increased amounts of unsaturated fatty acids in their lipids when grown at low temperature. This increase in unsaturated fatty acids leads to a decrease in lipid melting point (lipid melts at lower temperature) thus the increased synthesis of unsaturated fatty acids at low temperature acts to maintain the lipid in a fluid and mobile state thereby allowing membrane proteins to continue to function. The transport permeases of psychrotrophs are more operative at low temperatures than those of mesophiles and transport system of psychrophiles is cold resistant.

Factors affecting cold storage

1. *Temperature*   Lower the temperature of refrigeration, higher the shelf life of storage. But certain foods like banana undergoes spoilage

at temperatures less than 5°C (best temperature for storage is at 1–16°C).

2. *Relative humidity*   It varies with food and environmental factors like temperature, composition of atmosphere, etc., i.e., if relative humidity is low, it leads to loss of moisture from food further leading to loss of weight and wilting; on the other hand, if relative humidity is high, there is an increase in moisture content in food which favours microbial spoilage. A condition called "sweating" is evident when there is precipitation of moisture on the food.

3. *Ventilation*   Control of air movement directly influences the relative humidity thus affecting the environment of chilled storage.

4. *Composition of storage atmosphere*   In the presence of optimal concentration of carbon dioxide, food remains unspoiled for a longer period as in eggs (2–5%), chilled beef (10%), pig meat (100%), etc.

5. *Irradiation*   Combination of UV rays with chilling storage increases the shelf life of food products.

Freezing/Frozen Storage

Freezing lowers the temperature of a food to −18°C, stores it at that temperature or below. Freezing is a method to preserve food without causing major changes in their consumer quality, but they are energy intensive. Freezing of foods occurs over a broad range of temperature. Water in food starts freezing at −1°C to −3°C. This increases the solute concentration in the water that is not yet frozen which results in decrease in the freezing point of the solution (the solution takes longer time to freeze than water). At a particular temperature, called the "eutectic temperature", the solutes precipitate and the residual water freezes. A totally frozen state results at −15°C to −20°C for fruits and vegetables and at less than −40°C for meats (as water content of food is more, total frozen temperature is lowered).

Growth of microorganisms under freezing conditions   During freezing, both temperature and water activity decreases. Thus in frozen foods, only those microorganisms which are cold tolerant and xerotolerant can grow. For example, while the spoilage of refrigerated non-frozen meat is due to bacteria, on frozen meats xerotolerant moulds will be problematic. Yeast *Debaryomyces* grows at −12.5°C. During freezing, microorganisms suffer multiple damage which may lead to their inactivation immediately or later. Gram-negative bacteria, specifically mesophiles are more susceptible to freezing than gram-positive bacteria (since gram-negative bacteria lack unsaturated fatty acids).

Effect of freezing on microbial cells (cryoinjury)   There are several reasons listed for the cryoinjury of cells. They are

1. Thermal shock
2. Toxic action of concentrated intracellular solutes
3. Effect of concentrated extracellular solutes
4. Dehydration
5. Internal ice formation
6. Attainment of a minimum cell volume

Chilling injury   Microbial cells can be damaged when they are cooled from ambient to chill temperature, a phenomenon called "chilling injury". The rapid cooling of mesophilic bacteria from the normal growth temperature brings about immediate death to a proportion of cultures. Freezing lowers the water activity through the removal of water in the form of ice crystals and involves a temperature shock causing metabolic injury (damage to plasma membrane). Rapid chilling results in membrane phase transition (from liquid to gel state) without allowing phase separation of phospholipids and protein domain, resulting in loss of permeability control by the plasma membrane. Cold shock sensitizes cells to various other forms of oxidative stress.

There are two main types of chilling injury namely, cold shock (direct chilling injury) and indirect chilling injury.

Cold shock or direct chilling injury is associated with the process of cooling foods from an ambient temperature to chill temperature and the level of injury depends on the rate at which food is cooled.

Slow freezing   When cooling is slow (as in domestic freezers), ice crystals form outside the cell. This leads to an increase in the concentration of solutes in the environment outside the cell followed by plasmolysis, cell shrinkage and eventually death.

Fast freezing   When cooling is fast (commercial freezing), ice crystals form inside the cells. During freezing of water inside a bacterial cell, an osmotic shock occurs (increase in the solute concentration inside the cell) leading to the development of a hypertonic environment. This changes the concentration of cellular liquids further leading to change in cellular pH and ionic strength thereby inactivating enzymes, denaturing proteins, and hampering function of DNA, RNA and cellular organelles.

Ultra-fast freezing   When cooling is ultra fast (freezing rates produced by plunging cells into liquid nitrogen at $-196°C$), water freezes to form a glass like substance thus preventing the formation of intracellular ice crystals thus reducing its damage. Here most of the injury to cells appear to be associated with thawing rather than the freezing process.

Indirect chilling injury is associated with holding food at chill temperatures for prolonged periods (several days) and is independent of the rate at which the food has been cooled. This type of injury seems to be caused by lack of exchange of materials with the environment leading to accumulation of toxic metabolic products and/or the depletion of important cell metabolites such as ATP, resulting in cell starvation and eventually death.

Effect of thawing on microbial cells   During thawing, ice crystals melt and the liquid is either absorbed back into the tissue or leaks out from the food. If the thawing process is rapid there is less injury but if the thawing process is gradual, due to osmotic differences, permanent injury may result inside the cells leading to cell death. Hence the extent of injury depends on the rate of freezing and thawing. Slow freezing is more injurious than quick freezing and thawing at a slow rate is hazardous than rapid thawing.

Factors Affecting Survival of Microorganisms Under Freezing Conditions

1. *Type of organisms*   Gram-negative bacteria are susceptible whereas gram-positive bacteria are resistant. Spores and viruses remain unaffected.

2. *Age of cells in population*   Actively growing cells are more sensitive than stationary phase cells due to higher lipid content in the cell membrane of stationary phase cells.

3. *Rate at which food is cooled by freezing*   In the initial period, microbial cells show chilling injury—faster the rate of freezing, less damage to the cells.

4. *Composition of food*   Acid foods appear to increase the damaging effects of freezing. Certain compounds enhance and others diminish the effects of freezing—sodium chloride reduces the freezing point of solutions thereby extending the time period during which the cells are exposed to high concentration of solutions before freezing occurs. Other compounds like glycerine, saccharose, gelatin have a "cryoprotective effect".

5. *Time of storage in the frozen state*   There is a decline in the number of living cells with time as seen noticeably in gram-negative rods. Often the death of survivors is fast initially and then slows gradually and finally the survival level stabilizes.

6. *Change during storage*   During frozen storage, death rate of microorganisms is usually lower than during the process of freezing. Death is probably due to the not yet frozen, very concentrated residual solution formed by freezing. The concentration and composition of this residual solution may change during the course of storage. There is less loss of viability during frozen storage when the storage temperature is stable rather than fluctuating.

7. *Treatment before freezing*   Cells already damaged by processes like blanching are more likely to be killed during freezing.

8. *Rate of thawing*   This depends on the original rate of freezing. With slow freezing, the thawing rate has little effect on survival because less ice crystals will be produced. But with fast freezing and slow thawing, ice crystals formed inside the cells, increase in size during the thawing process leading to cell injury. During slow thawing, ice melts slowly outside but the water at lower temperature inside the cells, remain crystalline. Due to the osmotic imbalance water moves into the cells, crystallizes inside due to low temperature and thus ice crystals increase in size leading to bursting of cells. Fast freezing followed by rapid thawing minimizes the formation of ice crystals inside the cell thus preventing damage to the cells.

BACTERIAL MEMBRANES AND THE EFFECT OF CHILL STORAGE

Cold-adapted Enzymes

The fact that cold-adapted bacteria grow at chill temperatures at rates that are either equivalent to or not much slower than mesophiles at room or body temperatures means that they must contain proteins (enzymes) that are adapted to function at low temperatures.

This adaptation has evolved over many generations and is fixed in the genome. The resulting amino acid sequence of each enzyme gives a protein its three- dimensional structure that remains conformationally flexible and thus catalytically active in the cold. Different enzymes have evolved different mechanisms for achieving cold activity, but some common evolutionary adaptations have been identified. These include a reduction in the number of hydrogen bonds, salt bridges, proline and arginine contents, aromatic interactions and hydrophobic clustering, together with increase in solvent interactions and additional surface loops. Not every kind of change is found in each enzyme, but the overall effect decreases the number of enthalpy-driven interactions is more flexible at low temperatures. A corollary of the enhanced activity at low temperatures is the fact that cold-adapted enzymes are more thermolabile than their mesophilic counterparts, so that at quite moderate temperatures (typically 40–50°C), they become too flexible, lose catalytic efficiency and eventually denature. This means that psychrotrophic bacteria are usually killed by mild heat treatment, which could be an advantage in preservation regimes such as those of sous-vide foods in which mild heating is followed by refrigerated storage. Enzymes are found either free within the cytoplasm or in the membrane, but all of the structural data on cold-active enzymes come from studies of soluble cytoplasmic ones. Nothing is known about the structure of membrane-bound cold-adapted enzymes compared to their mesophilic or thermophilic counterparts, but they will presumably also have α–helical sections that span the hydrophobic core of the membrane where they interact with the fatty acyl chains of membrane lipids. Therefore, it is certain that they too will be specifically adapted in order to function at low temperatures and that this adaptation will depend not only on their intrinsic protein structure but also on the physical properties of the surrounding lipids.

Low Temperature and Membrane Lipids

It is well known that a change in temperature alters the lipid composition of membranes. The main changes are in the fatty acyl components of membranes. Changes in the head-group composition of the lipids are much less pronounced and have much less influence on the thermal properties of the membrane.

Increasing the extent of fatty acyl unsaturation, *cis/trans*-unsaturation ratio, methyl branching or the ratio of *anteiso-* to *iso*-branched acyl chains, or shortening the average acyl chain length, all lower the temperature of transition from a liquid-crystalline to a gel phase and so preserve membrane fluidity that is necessary for survival and growth. The term "membrane fluidity" is a convenient one to summarize a multifaceted phenomenon that has contributions from molecular packing (order) and molecular motions (viscosity).

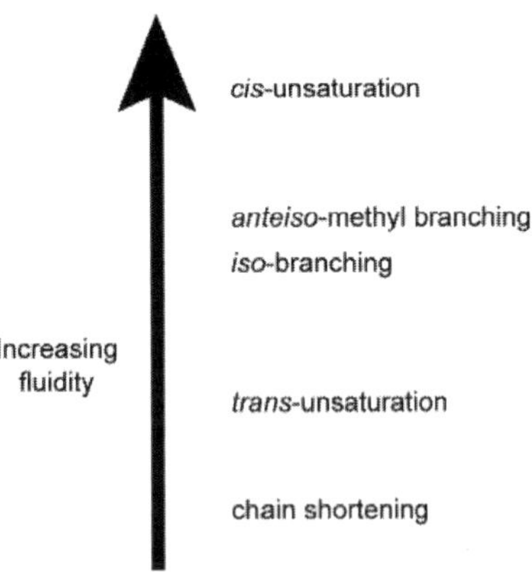

Figure 7.1 A figurative representation of the proportional increases in membrane fluidity given by different fatty acyl changes relative to saturated lipids

The changes in fatty acyl composition may alter either or both of these aspects of fluidity. For example, the introduction of a *cis*-unsaturated bond introduces a "kink" into the acyl chain, which therefore occupies a greater molecular profile. Similarly, a methyl group disrupts the packing of acyl chains by occupying more space, *anteiso*-branches more so than *iso*-branches. *Trans*-unsaturated double bonds alter the orientation of the acyl chain much less than *cis*-unsaturated double bonds and so have a correspondingly smaller effect on fluidity. Disruption of acyl chain packing will not only change the packing order but also the strength of interaction between individual acyl chains. Shortening acyl chains will lessen the van der Waals, intermolecular forces and therefore make the membrane more fluid, particularly if only one of the pair of acyl chains in each lipid molecule is modified, leaving more space for molecular motion of the longer acyl chain. The magnitude of the effect that the different temperature-dependent fatty acyl changes have on membrane fluidity is visualized in Figure 7.1. Each microorganism will have a lipid fatty acyl composition that is adapted for its particular growth temperature range. The same level of fluidity can be achieved by many combinations of different fatty acids and hence, even though different cold-adapted bacteria may have similar low temperature growth abilities, they will almost certainly have quite different fatty acyl compositions. Differences will be influenced by phylogenetic distinctions, different metabolic capabilities and by specific protein–lipid interactions. For example, *Salmonella* adapts to temperature almost entirely by changing lipid unsaturation whereas *Listeria*, which contains predominantly branched fatty acyl chains, modifies the *anteiso*/isobranched ratio and the acyl chain length; bacilli use a combination of unsaturation and changes in branching pattern. As temperature falls, membrane fluidity will decrease, and membrane-associated metabolic processes mediated by enzymes, cytochromes and permeases will slow down. These events will trigger compensatory changes in fatty acyl composition so as to make the membrane more fluid. As long as the temperature change is within the normal growth temperature range, the fatty acyl changes will more or less compensate for the kinetic loss of activity. If the temperature is shifted to or just beyond the lower or upper limits, then cold shock or heat shock, respectively, will occur.

The rate at which the lipid changes occur on transfer to or from the cold will depend on the biosynthetic mechanisms used to modify lipid acyl composition. Changes in unsaturation brought about by desaturase enzymes are usually rapid because they generally occur *in situ* in the membrane through the modification of intact lipids without concomitant growth, i.e., the desaturase enzymes are located within the membrane and interact directly with lipids that are their potential substrates. In contrast, changes in methyl branching and acyl chain length take longer because they require *de novo* synthesis of the whole lipid molecule by cytoplasmic enzymes that are usually linked to growth. In most cases, the effect of temperature is a direct one, acting on the key regulatory enzymes to modulate the overall rate of the reaction. Induction of new enzyme synthesis is much less common, although some desaturases in bacteria and yeasts are cold inducible. Activation (or inhibition) by temperature will be virtually instantaneous, since microorganisms are too small to insulate themselves against thermal effects, and enzyme induction and synthesis in bacteria take only a few minutes to complete. Therefore, the effects of changes in temperature will be reflected in an altered fatty acyl composition within minutes. However, changes that involve *de novo* synthesis of fatty acids will require further steps of fatty acid activation and incorporation into membrane lipids that require cellular growth. Modifications involving desaturases are an exception because existing membrane lipids are the substrate, although such changes only serve for a short term and other mechanisms based on cell growth take over after the initial adaptation phase. Therefore, whatever the strategy, the effects of membrane lipid modification require cellular growth (lipid and membrane synthesis) for their influence to be exerted and so it is significant that if the decrease in temperature is sudden, the bacteria will suffer a cold shock and stop growing for a period of up to several hours due to a block in the initiation of protein synthesis. If the temperature is shifted to or just below the lower growth limit, cold shock will occur.

Cold Shock and Cold Acclimation

Sudden changes in temperature will induce the synthesis of stress proteins—heat-shock proteins for a rise and cold-shock proteins for a fall in temperature. The cold-shock response (CSR) has been identified in food-associated mesophilic and psychrotrophic bacteria that cause spoilage (e.g. *Pseudomonas fluorescens*, *P. fragi* and lactic acid bacteria) or poisoning (e.g. *L. monocytogenes*, *S. typhimurium*, *S. enteritidis*, *Staphylococcus aureus*, *Escherichia coli*, *Y. enterocolitica* and *Bacillus cereus*). Interestingly, cold-shock protein (CSP) genes are not present in *Campylobacter jejuni* which may explain why this food-poisoning bacterium has a very narrow growth temperature range and is unable to grow below 30°C. The CSR involves the differential expression of genes for up to 50 different CSPs, depending on the species. Many of these are concerned with the major functions of CSPs, (Table 7.2), which reflect the importance of ensuring that protein synthesis continues at an appropriate rate at low temperature to give balanced growth. These functions also reflect the importance of the ribosome sensing temperature changes and the fact that the cellular function most sensitive to cold shock is the initiation of translation. In addition to the common features of the CSR given as in Table 7.3, the expression of other genes is involved and these differ among species. At present, there is no complete picture of CSPs in different microorganisms—only less is known about their functions. The regulation of CSP synthesis occurs at several levels, both transcriptional and translational, involving both protein and mRNA stabilities. There are common regulatory sequences, upstream of (e.g."the cold-shock box") and downstream within the coding region, which together coordinate the expression of cold-shock genes. A key question in relation to psychrotrophs and food storage is "How large does the temperature fall have to be, and at what rate it must decrease, for cold shock to occur?'' This is an important consideration because cold shock enables the food-associated microorganisms to survive better and grow at the new lower temperature and, therefore, to spoil and/or poison the product.

Table 7.3 Major characteristics of the cold-shock response

Effect of low temperature	Cellular response
Block in initiation of protein synthesis	Synthesize proteins to stabilize interaction of mRNA with 30s ribosomal subunit
Disruption of ribosome structure	Synthesize proteins to stabilize protein–protein and protein–rRNA interactions within the ribosome
Formation of secondary structures	Synthesize RNA chaperones to maintain mRNA in linear form ("hairpins")
Increased negative supercoiling of DNA	Induction of DNA unwinding enzymes and stabilizing (histone-like) proteins

Significantly, compared to mesophiles, after cold shock in psychrotrophs, there is no concomitant suppression of the expression of so-called housekeeping genes that encode, for example, enzymes of central metabolic pathways, so growth lag times are likely to be shorter or non-existent. Moreover, the number of CSPs and the extent of their synthesis depend on the depth of the cold shock, and one particular class of proteins (the cold-acclimation proteins) is permanently induced during constant growth at low temperature. For example, the food-spoilage bacterium *P. fragi* makes 15 CSPs on shifting from 20 to 5°C, but 24 CSPs when shifted from 30 to 5°C. It is capable of growing after 3–5 hour lag following such temperature shift. The functions of the extra CSPs, whether they increase the ability to grow and survival in foods are not known, but significantly, the temperatures and time scales involved are the ones that might be relevant. For instance, considering the retail purchase and transfer of foods from a supermarket chill cabinet (10–12°C) to a car that might be then left in the sun (30°C) before the food is placed in a domestic refrigerator (5°C). Another practical scenario might be a cooked food left inadvertently overnight in a warm kitchen (20°C) before refrigeration. The time scales involved could easily be a few hours, which would be sufficient to give several generations of growth and large enough numbers of warm-adapted spoilage psychrotrophic bacteria that would then experience cold shock on being placed in the refrigerator and subsequently would cold-adapt and grow. Whether the same considerations apply to a food-poisoning bacterium such as *L. monocytogenes* is arguable, as there is disagreement about the length of the lag times following cold shock. Thus far, changes in lipid composition and cold shock have been discussed separately. However, in the cell, they must be linked because balanced growth requires coordination of intracellular and extracellular events, as well as those occurring within the membrane matrix. The membrane is likely to have a role in the sensing of temperature, and a number of two-component signalling systems involved in global regulatory phenomena are well known. An inducible desaturase has been identified as a cold-shock protein in *B. subtilis*, thus linking the cold-stress response to lipid changes in the membrane. The links between lipid changes mediated by the cytoplasmic fatty acid synthetase and those of the CSR are more difficult to determine. A second aspect of membrane structure that is less commonly taken into account in discussion of thermal adaptation is the need to preserve the bilayer (lamellar) phase, i.e., prevent the formation of non-bilayer phases such as hexagonal, which destroy the selective permeability properties of the membrane. Even more so than changes of temperature, the presence of salt(s) has a large effect on the transition between bilayer and non-bilayer phases, and there is an interplay of effects between temperature and solute concentration, which is relevant to products such as minimally processed foods that rely heavily on chilling for extension of their shelf life. Taking into account what we know about the membrane lipid changes that are triggered separately by low temperature and the presence of (preservative) salts, in combination they may act antagonistically. Since for a given lipid composition, lowering growth temperature will reduce the likelihood of formation of non-bilayer phases whereas lowering the water activity (i.e., raising the salt concentration) will have the opposite effect. Therefore, at lower temperatures, bacteria should grow better in salt (i.e., have higher optimum salt concentrations). Just such an effect has been demonstrated for a moderately halophilic bacterium, but has not been explored for relevant food-spoilage or food-poisoning bacteria. Ignoring any other growth inhibitory effects of the preservative, the reduction of salt in chilled foods would have beneficial effect as far as changes in membrane lipid composition are concerned. In practice, the effect is likely to be small and it would be more effective to find a means of disrupting membrane stability so that lipid changes were less effective in adapting the spoilage/poisoning microorganisms to grow at low temperatures. One approach is to combine physical methods of membrane disruption with cold storage. The available methods include treatment with ultrasound, high pressure or pulsed electric fields.

1. *Ultrasound and membranes*   Ultrasound disrupts biological membranes, probably by a combination of cavitation phenomena and associated shear disruption, localized heating and free radical formation. Typical treatments are for 1–30 seconds using 20–40 kHz ultrasound. Ultrasonication in combination with mild heating (e.g. 50–60°C) is more effective at inactivating a range of vegetative food-spoilage and food-poisoning microorganisms, as well as spores, but no satisfactory explanation exists for the synergy of so-called thermosonication. If the rapid pressure changes that occur during cavitation are responsible for the lethal effect of ultrasound, then raising the temperature and hence membrane fluidity (i.e., weakening the intermolecular forces) would enhance the disruption. However, it is not known if membrane lipid composition of the target organisms is a determining factor in ultrasound sensitivity.

2. *High pressure and membranes*   High hydrostatic pressure, in the order of 100–1000 MPa (i.e., 1–10 kbar), inactivates enzymes and causes a variety of structural changes in the morphology, cell wall and membranes of microorganisms; and membrane fluidity may be a factor in the pressure-sensitivity of an organism. Gram-positive bacteria are less sensitive than gram-negative bacteria, probably due to the thicker cell wall of the former. It has also been noted for *L. monocytogenes* and *E. coli* that there can be considerable differences in sensitivity between strains of the same organism. For *E. coli,* it has been suggested that the differences in resistance between strains are related to their susceptibility for membrane damage, but the relationships between pressure sensitivity, phase of growth of batch cultures and whether the applied pressure is high or low are complex. It had been suggested previously that bacteria with a more fluid membrane are more resistant to high pressure but no direct membrane fluidity measurements were made. Membrane fluidity may exert its influence through control of the ion pumps in the membrane that are essential for maintaining pH homeostasis. This is consistent with the fact that pressurization is more effective at inactivating microorganisms when it is combined with mild heat treatment, since membrane repair of pressure-induced pores would be harder to accomplish if intermolecular forces are weakened by warming. Such co-treatment may further be combined with ultrasound treatment, in a process known as "manothermosonication".

3. *Pulsed electric field and membranes*   A drawback of high pressure or ultrasound processes is that they all give inactivation curves with "tails" of surviving microorganisms, and the use of higher temperatures or other operational parameters to reduce the number of survivors would have adverse effects on the organoleptic qualities of the food. In contrast, it has been demonstrated that treatment of either *L. monocytogenes* or *S. typhimurium* with pulsed electric field (PEF) is an "all or nothing" effect in that the bacterial cells are either killed or survive normally following exposure to effective doses. This is in marked contrast to other novel physical preservation methods, such as high-pressure treatment, which result in a proportion of bacteria that are damaged but are capable of recovery under favourable growth conditions. Thus, predictions of food shelf life should be more reliable after PEF treatment compared with ultrasound or high-pressure treatments. It is generally agreed that PEF treatment leads to the permeabilization of biological membranes. It was found that increasing the applied electric field gave greater microbial inactivation, which was matched by increased leakage of UV-absorbing cellular material and loss of the ability to maintain pH homeostasis, although, in contrast with high-pressure treatment, PEF efficacy did not correlate with the inhibition of membrane H^+–ATPase activity.

Pulsed electric field is a modification of the original "Electropure process" in which an alternating current was used to pasteurize milk. The current was not pulsed and the lethal effect was derived from the heating that occurred. If, instead, the electric field is delivered in pulses with proper control of their strength, number and format (e.g. bipolar, square waves are better than monopolar, exponential waves), then microbial inactivation will not be due to thermal effects. A number of theories have been put forward to explain the membrane-disrupting action of PEF, but they are similar and all are based on the fact that lipid molecules are dipolar and the membrane bilayer has a net electric charge. Application of a PEF causes reorientation of the lipids, stressing the lipid bilayer and eventually causing pores to form. This is consistent with the observation that electric field strength has a more profound influence than treatment time on the lethal effect, because it is the field strength that overcomes the intermolecular forces responsible for maintaining the lipid bilayer core of the membrane. It is also consistent with the fact that raising the treatment temperature increases the efficacy of PEF. The viscoelastic properties of the bilayer oppose these disruptive forces, and the pores will cyclically reseal themselves as soon as they are formed, unless the PEF treatment is sufficient in strength to overcome the repair process. Therefore, membrane lipid composition will influence this balance so as to make the cells more or less sensitive to PEF treatment. Growth at low temperatures will give bacteria with a membrane lipid composition that will be more fluid at room temperature than the membranes of bacteria grown at moderate temperatures. Therefore, one can predict that cultures grown in the cold would have increased PEF sensitivity (at room temperature) compared with those grown at room temperature, because the cold-adapted membranes have less ability to repair electropores. Indeed, found that cultures of *L. monocytogenes* or *S. typhimurium* grown near the upper limit of their temperature ranges (37 and 45°C, respectively) were more resistant to PEF treatment than those grown near the lower limit (4 and 10°C, respectively), as measured by the number of bacterial survivors and loss of UV-absorbing material, but not by the ability to maintain pH homeostasis. It was shown that the growth temperature-triggered alteration in PEF sensitivity was correlated with changes in membrane lipid composition, particularly the lipid fatty acyl composition. In *L. monocytogenes*, the major growth temperature-dependent alteration was in the ratio of the two major *anteiso*-branched fatty acids, *anteiso* 15:0 and *anteiso* 17:0, whereas in *S. typhimurium*, it was a change in the proportion of total unsaturated fatty acids (mainly 16:1D9 and 18:1D11). It is hypothesized that alteration in membrane lipid fatty acyl composition gives a membrane with changed viscoelastic properties, which modify the ability of the bacterial cell to immediately repair damage due to the formation of PEF-induced electropores.

Support for this hypothesis comes from the observation that ethanol, which is a membrane-fluidizing agent, increases microbial inactivation; and phenethyl alcohol, a more potent membrane fluidizer, has a greater effect. Further support comes from a comparison of different strains of *L. monocytogenes*, including two nisin-resistant strains, which have altered lipid compositions and PEF sensitivities. Most recently, using defined growth media as an independent means of giving membranes that have altered membrane fluidity, the ratio of the major *anteiso*-branched fatty acids in *L. monocytogenes* has been manipulated for example, having membranes in which the *anteiso*-branched fatty acyl content is reduced from 78% to 51% and the *anteiso/iso*-branched ratio is lowered from 7.1 to 1.4. One would predict that such a change in lipid composition should give a membrane that was less fluid and therefore less sensitive to PEF, if the hypothesis is correct. However, the opposite result was found. Therefore, either the fluidity prediction is wrong or the role of membrane fluidity in PEF sensitivity is more complex. Other features of lipid organization may be involved that, in order to elucidate them, will require direct biophysical examination of the modified membranes. Refrigeration remains one of the most effective means of extending the shelf life of fresh and processed foods, because lowering temperature reduces the growth rate of even cold-adapted microorganisms, particularly if they have been damaged by use of a physical retreatment. In order to optimize the efficacy of such combined preservation regimes and provide a synergistic extension of the safe shelf life of the food, a better understanding of the molecular basis of the lethal action of the physical treatments is needed. In particular, the role of membrane lipid physico-chemical properties and organization requires investigation.

What Happens to the Injured Cells After Thawing?

Cells that are injured but not killed can recover after thawing as long as there is an ample supply of nutrients and the environment does not contain inhibitors. Injured cells will recover quite readily in thawed foods. For example, cells of food poisoning organisms that are rather than getting killed are still likely to be infective or recover and grow if the food is held at a suitable temperature. This poses a problem in food analysis.

Cryoprotective Agents

There are many substances that protect the microbial cell during freezing and thawing. For example, glycerol and dimethyl sulphoxide penetrate the cells. Other protective agents include egg white, carbohydrates, peptides, serum albumin, malic acid, milk, glutamic acid, yeast extract, diethylene glycerol and between 80. Cryoprotectants act by reducing the amount of ice formed in the cells or by increasing the time required for the water to leave the cells or by increasing the viscosity of the extracellular solution. Glycerol reduces damage to the cell wall and cell membrane.

Preparation of Food for Freezing

The preparation of food for freezing involves the following steps.

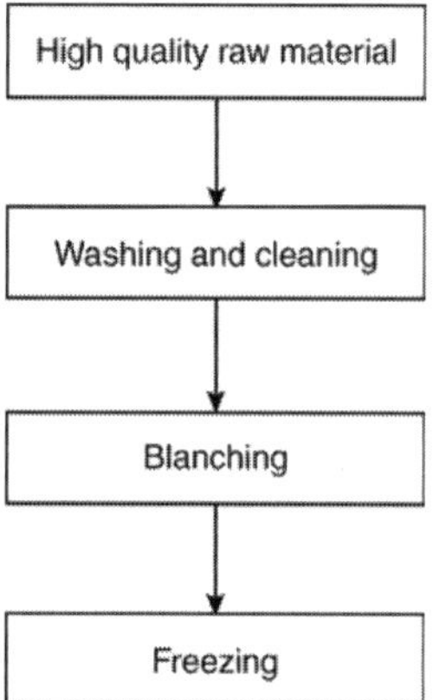

Systems for Freezing

Heat can be removed from food by convection, conduction, evaporation or radiation. Products are frozen after or before packaging. Packaged foods freeze more slowly due to insulation provided by the packaging material.

1. *Still air (sharp/slow freezing)*   This is freezing in air with only natural air circulation or electric fans. Temperature is usually -23°C (ranging from -15 to -29°C) and the process takes 3–72 hours.

2. *Air blast (quick freezing)*   Freezing is done by one or more methods to minimize the time of freezing. Time taken is generally 30 minutes. Temperature achieved is -17.8 to -45.6°C, much lower than slow freezing. In air blast, frigid air (cool air) is blown across the materials being frozen.

3. *Fluidized bed (quick freezing)*   It is used for freezing small particulate materials. The food to be frozen is placed on a mesh belt and very cold air is blown upwards through the mesh. The food particles seem to float over the mesh due to velocity of air. Food is frozen in

a few minutes. Food particles are frozen individually and then packaged.

4. *Liquid immersion freezing*   It is used to freeze poultry foods. Since liquid is a better conductor of heat than air, more rapid freezing is achieved. Brine is one of the liquid freezants.

5. *Direct contact freezing*   It uses a fluorocarbon; dichloro difluoro methane (freon) at $-22°C$ as a spray to freeze food.

6. *Liquified gases*   It is called cryogenic freezing. Surface oxidation is prevented since food is frozen in an inert atmosphere of nitrogen or carbon dioxide. Rapid freezing and low temperatures prevent microbial growth.

Changes in Food During Freezing

There is expansion in volume of frozen food with the formation of ice crystals. This effect is more pronounced in slow freezing than in quick freezing since in slow freezing, large ice crystals are formed (more ice is accumulated between cells), water is drawn from the cells to form ice with a resultant increase in concentration of solutes in the unfrozen liquor leading to dehydration and tissue disruption, whereas in quick freezing, small ice crystals are formed. Quick freezing rapidly slows chemical and enzymatic reactions in food and stops microbial growth.

Changes in Food during Frozen Storage

Chemical and enzymatic reactions proceed slowly. Red myoglobin of meat is oxidized at surfaces to brown *"metmyoglobin"*. Fats of meat may become oxidized and hydrolysed. The unfrozen concentrated solutes of sugars and salts may ooze from packages as a viscous material called *"metacryotic fluid"*. During prolonged storage at a particular temperature, desiccation of the food takes place on its surface leading to evaporation of ice crystals causing *"freezer burn"*. The spot looks dry and brownish.

FUTUROLOGY IN FREEZING TECHNOLOGY— ULTRASOUND FREEZING AND FREEZE DRYING

Ultrasonics is a rapidly growing field of research and development for the food industry. Ultrasound can be classified into two types—high-frequency low-energy diagnostic ultrasound in the MHz range and low-frequency high-energy power ultrasound. The former is usually used as an analytical technique for quality assurance, process control and non-destructive inspection, which has been applied to determine food properties, to measure flow rate, to inspect food packages, etc. However, the application of the latter in the food industry is relatively new and has not yet been profoundly explored until recent years. Various areas have been identified with great potential for future development, e.g. crystallization, drying, degassing, extraction, filtration, homogenization, meat tenderization, oxidation, sterilization, etc. One of the basic components of freezing a food system can always be simply pictured as ice crystals distributed across the unfrozen aqueous phase. The transmitting of sound waves across the aqueous phase can cause the occurrence of cavitation if its amplitude exceeds certain level. The negative pressure during the rarefaction will cause liquid to fracture, leading to the formation of bubbles or cavities. During the negative pressure portion of the sound wave, bubbles (including bubbles that are inherently present in the liquid) will grow rapidly and create a vacuum, causing gases dissolved in the liquid to diffuse into them. As the rarefaction portion of the sound wave passes, the negative pressure is reduced and when atmospheric pressure is reached, the bubbles will start to shrink under surface tension. When the compression cycle starts and while the positive pressure lasts, gas that diffused into the bubbles will be expelled into the fluid. The diffusion of gas out of the bubbles will not take place until after the bubbles are compressed. However, once the bubble is compressed, its boundary surface area available for diffusion is decreased, therefore, the amount of gas that is expelled is less than the amount that is taken up during the rarefaction cycle. Consequently, these bubbles will grow bigger over each ultrasound cycle. These cavitation bubbles can serve as nuclei for ice nucleation once reaching the critical nucleus size. Experiments have shown that power ultrasound can significantly increase the nucleus number in a concentrated sucrose solution. Microstreaming is another significant acoustic phenomenon associated with cavitation, which occurs when the oscillating bubbles produce a vigorous circulatory motion, and thus setting up strong eddy currents in the fluid surrounding them.

The diffusion of gases into and out of the bubbles can also create microcurrents around themselves and further spread into the liquid. Computational fluid dynamics (CFD) simulation revealed that an average acoustic velocity of 3 mm/s could be obtained with 500 kHz ultrasound. The turbulence (violent agitation) that microstreaming provides has been used to enhance heat and mass transfer in many processes. Due to its ability to provide violent agitation in the liquid phase, microstreaming can therefore also benefit the freezing process by reducing both the heat and mass transfer resistance at the ice/liquid interface and thus increasing the freezing rate. Ice crystal is another major component of the food freezing system. Similar to other dense and practically incompressible materials, ice crystals will fracture when they are subjected to sound waves. This effect has been demonstrated by, an observation that when a pulse of ultrasound of approximately 3 s was applied to a freezing sucrose solution every 30 seconds for a duration of 10 minutes, the front of the dendritic ice formed on the cold surface was clearly seen to fracture and the ice fragments were dispersed into the unfrozen bulk liquid. Fragmentation of ice crystals leads to crystal size reduction. It was found that inside a frozen sucrose solution treated with power ultrasound during freezing, 32% of the water exists as crystals with diameter of 50 mm or larger, compared to 77% for the one without acoustic treatment. The effect of power ultrasound on fragmenting crystals has already been successfully used in the production of a crystalline drug.

Applications

Traditionally, power ultrasound has been applied to accelerate the ice nucleation of many chemical processes. Compared to other methods, for instance, the usage of chemicals such as silver iodide, amino acids, ice nucleating bacteria or seed crystals, power ultrasound offers several advantages. It is a very efficient treatment, since one or two pulses of ultrasound can fulfill the requirement. Also, the initial nucleation temperature of the liquid can be dictated. Unlike nucleating agents, it does not require direct contact with the products. Furthermore, it is not chemically invasive and thus unlikely to encounter legislative difficulties. Therefore, power ultrasound has been recently studied in assisting and/or accelerating various freezing processes. These applications could be extended to freezing of high-value food (ingredients) and pharmaceutical products.

1. Explain the growth of microorganisms under freezing conditions.
2. What is cryoinjury?
3. Explain chilling injury.
4. Write down the effects of thawing on microbes.
5. Explain the factors affecting survival of microbes under freezing conditions.
6. Give an account of bacterial membranes and the effect of chill storage.
7. List a few cryoprotective agents.
8. Give an account of changes in food during freezing.
9. Discuss the futurology in freezing technology—ultrasound freezing and freeze drying.

8

PRESERVATION BY DRYING

INTRODUCTION

Water activity, a_W, is a physical property that has a direct implication on microbiological safety of food. Water activity also influences the storage stability of foods as some deteriorative processes in foods are mediated by water. Storage life of dry foods such as biscuits is generally longer than that of moist foods such as meat at the same temperature. In this connection, freezing of foods is equivalent to drying. Water is removed from the food matrix although it is still in the food as ice. Water activity is an important factor affecting the stability of dry and dehydrated products during storage. Dry and dehydrated products have a high level of popularity among today's consumers. Dry mixes are economical and convenient with increased shelf life, reduced packaging, decreased cost (via weight and/or volume reduction) and improved handling properties. Controlling water activity in a dry product maintains proper product structure, texture, stability, density and rehydration properties.

Water activity affects the textural properties of dry cereal-based foods and starch-based snack products. Crackers, potato chips, puffed corn curls and popcorn, each loses its sensory crispness with increasing water activity. The crispness intensity and overall texture of dry snack food products are a function of a_W. Critical water activities are found where the product becomes unacceptable from a sensory standpoint. These fall into the a_W range where amorphous to crystalline transformations occur in simple sugar food systems and mobilization of soluble food constituents begins. Excessive and rapid drying or moisture re-absorption by a glassy material can cause the undesirable consequence of product loss by cracking and excessive breakage. To preserve the initial quality as much as possible during dehydration and storage, the chemical and biochemical reactivity and stability must be considered. Water activity influences non-enzymatic browning, lipid oxidation, degradation of vitamins, enzymatic reactions and protein denaturation (Figure 8.1). The likelihood of non-enzymatic browning increases with increasing a_w, reaching a maximum at an a_w range of 0.6 to 0.7. Generally, further decrease in water activity will hinder browning reactions. Lipid oxidation has a minimum in the intermediate a_w range and increases at both high and low a_w values, although due to different mechanisms. This type of degradation results in the formation of highly objectionable flavours and odours, and loss of fat-soluble vitamins. Water-soluble vitamin degradation in food systems increases with increasing a_w values. Enzyme and protein stability are influenced significantly by water activity due to their relatively fragile nature. Most enzymes and proteins must maintain their conformation to remain active. Therefore, maintaining critical a_w levels to prevent or entice conformational changes is important to food quality. Most enzymatic reactions are slowed down at water activities below 0.8, but some reactions occur even at very low a_w values. Knowledge of the water activity of powders as a function of moisture content and temperature is essential for the control of water content during processing, handling, packaging and storage to prevent the deleterious phenomenon of caking, clumping, collapse and stickiness.

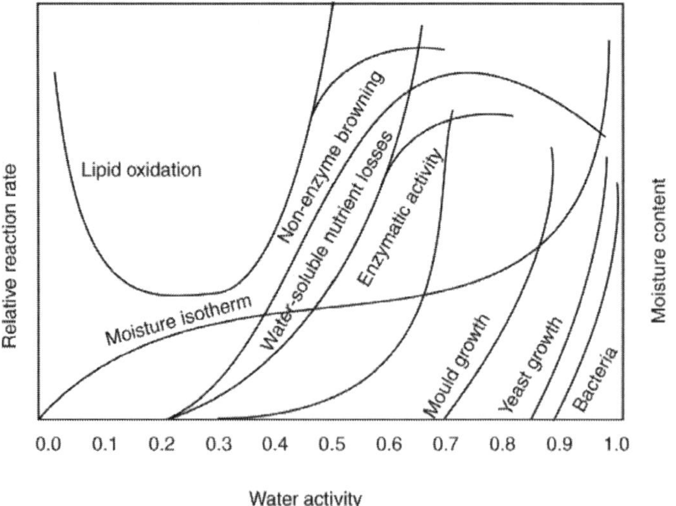

Figure 8.1   Relation between a_w and cellular metabolism

Caking is a deleterious phenomenon by which a low-moisture, free-flowing powder is transformed into lumps and eventually into an agglomerated solid, resulting in loss of functionality and lowered quality.

This problem is ubiquitous in the food and pharmaceutical industries. Caking is dependent on water activity, time and temperature, and is related to the collapse phenomena of the powder under gravitational force. Stages in caking involve bridging, agglomeration, compaction and liquefaction. Factors known to affect caking kinetics may be divided into those related to the powder itself (particle size distribution, hygroscopicity and charge of particles, state of the material, presence of impurities) and external factors such as temperature, relative humidity and mechanical stress applied to the substance. Water activity is an important factor affecting the stability of dry and dehydrated products during storage. Controlling water activity in a dry or dehydrated product maintains proper product structure, texture, stability, density and rehydration properties. From the physicist's point of view, water activity is defined in terms of thermodynamic concepts such as chemical potential and is related to the osmotic pressure of an aqueous solution. When a substance such as salt (sodium cloride) is dissolved in water, the water activity is reduced. This is how salting preserves food. The a_W of a food or solution is the ratio of the water vapour pressure of the food or solution (p) to that of pure water (p_0) at the same temperature and is given by

$$a_W = \frac{p}{p_o}$$

The a_W is related to the boiling and freezing points, equilibrium relative humidity (ERH) and osmotic pressure. The a_W of a solution is a colligative property, i.e., dependent upon the number of particles (molecules or ions) in solution. Increases in solute concentration decreases a_W. Microorganisms require water for solution of cell contents and metabolic processes. The cell membrane is semipermeable (or more correctly selectively permeable), and decreases in the a_W of the suspending medium below a certain maximum value (dependent upon the specific organism) will withdraw water from the cell, concentrating the cellular contents until the internal and external a_W values are in balance. This concentrating effect slows down the metabolic processes until at a limiting value, growth ceases. Many microorganisms under osmotic stress (low a_W) can accumulate or synthesize compatible solutes to relieve the stress. These solutes generally interfere little with the metabolic functions of the cell, and may include accumulation of K^+ ions, accumulation or synthesis of proline, glutamine, betaine, certain sugars or sugar alcohols (e.g. trehalose in yeasts), etc. However, this activity also requires energy, diverting some of the metabolic activities from growth to accumulation of solutes, and resulting in lowering of growth rates. Microorganisms generally grow best between a_W values 0.995–0.980 (Table 8.1), while most cease to grow at $a_W < 0.900$. However, halophiles (salt-loving) are unable to grow in salt-free media and often have an obligatory requirement for substantial concentrations of salt (NaCl). For example, *Halobacterium halobium* will not grow in salt concentrations below 14% w/w, a_W 0.89. Halotolerant organisms, while capable of growth at low a_w/high salt concentrations, grow best at high a_W values, e.g. *Staphylococcus aureus* will grow at a_W 0.90. Xerophilic organisms grow best at low a_W values adjusted with sugars, for example, the mould *Xeromyces bisporus* grows best at a_W 0.92, although it is capable of growth at a_W 0.70 (10% of maximum rate), but ceases growth at a_W 0.96, when a_W is adjusted with sucrose. Microorganisms react not only to a_W, but also to the solute adjusting the a_W. Minimum a_W values for growth are often very different for different solutes.

Food poisoning is the result of ingesting a pre-formed toxin in food. These toxins may result in vomiting (e.g. *Staphylococcus aureus* or *Bacillus cereus* enterotoxins), or other systemic effects, e.g. botulinal neurotoxin paralysis of the nerve–muscle junction. Food-borne infections result from ingesting an organism capable of surviving in the acidic environment of the stomach and growing in the intestinal tract, e.g. *Salmonella* spp. Gastrointestinal symptoms, e.g. diarrhoea, result from toxic metabolites produced in the gut. Microbial spoilage of foods results from changes in the food composition, and/or appearance or structure as a result of growth and metabolism of microorganisms. Commonly, the evolution of obnoxious odours is the cause for rejection of foods, e.g. fresh meats, although the appearance of mould colonies on semi-dry foods, e.g. bread, cheeses, is also common. A wide range of organisms can be responsible for spoilage, and therefore a wide range of changes in foods may be regarded as spoilage. Certain controlled spoilage by microorganisms is used to produce a food different from the starting ingredients, e.g. yoghurt or cheese from milk, fermented sausages from raw meats or sauerkraut from shredded cabbage.

SOLUTE EFFECTS ON MICROBIAL GROWTH AND/OR DEATH

Cell Membrane Phenomena

The cell membrane is semipermeable, or rather selectively permeable. Thus glycerol penetrates the membrane readily, glucose penetrates poorly, sucrose very poorly, and NaCl is almost non-penetrating. When an organism is grown or exposed to low a_W conditions, the cells may accumulate from the environment or synthesize compatible solutes, e.g. glutamine, proline, betaine in bacteria, trehalose in yeasts. These internal solutes interfere little with the metabolism of the cell, although metabolic energy must be diverted for synthesis, but increase resistance to low external a_W conditions, and also increase resistance to other injurious treatments, e.g. heat. This effect differs with different external solutes, e.g. *Staphylococcus aureus* synthesizes compatible solutes at high NaCl levels, but not in the presence of sugar. If the partially dehydrated cell is exposed to a high temperature, then the microorganism displays a greater thermal resistance than when grown at a higher a_w. Proteins and other essential cellular components are more resistant to thermal damage in the partially dehydrated state. Water activity plays an important role in the heat resistance of microorganisms (Table 8.1). Death curves are not always linear and interpolation of D values (and Z values) into application of thermal processes may not always be safe. Similarly, the ratio of effects of the different solutes on D values (Table 8.2) differs for each organism. Thus D values in low a_w solutions or foods must always take into account the actual solute controlling the a_w, and if necessary be determined in that solute, interactions with other physico-chemical parameters. Since extra energy is required to combat the harmful effects of low a_w, other conditions also require expenditure of extra energy, e.g. low pH, presence of preservatives, will result in an additive or synergistic effect in limiting microbial growth. Thus even moderate reductions in a_w in combination with low levels of preservative or pH values, can be sufficient to inhibit growth. One good example is that of inhibition of *Clostridium botulinum*. Under ideal conditions 10% NaCl is required to inhibit the proteolytic species; at the pH values typical of meats (pH 5.4–5.8) and in the presence of 100 ppm nitrite, only 3.5% NaCl is required to produce botulinal-stable cured meats.

Table 8.1 Water activities of common food-borne pathogens

Food poisoning organisms	Minimum a_w for growth*	Food-borne infectious organisms	Minimum a_w for growth
Bacillus cereus	0.95	*Clostridium perfringens*	0.95
Campylobacter coli	0.97	*Escherichia coli*	0.95
Campylobacter jejuni	0.98	*Salmonella* spp.	0.95
Clostridium botulinum		*Vibrio parahaemolyticus*	0.94
type A	0.95	*Yersinia enterocolitica*	0.96
type B	0.94		
type E	0.97		
Listeria monocytogenes	0.92		
Staphylococcus aureus	0.86		

*, the minimum a_w for growth of bacteria is generally by addition of salt. Minimum a_w for growth with other solutes may be different. For toxin production, minimum a_w values may be rather higher.

Table 8.2   Effects of solutes on the D values of *Salmonella* species

Solute	%w/w	D_{65} values (minutes)	
		S. typhimurium	*S. senftenberg*
Sucrose	30	0.7	1.4
	70	53	43
Glucose	30	0.9	2.0
	70	42	17
Fructose	30	0.5	1.1
	70	12	8.5
Glycerol	30	0.2	0.95
	70	0.9	0.7

FUNDAMENTALS OF DRYING

The basic operation of drying converts a solid, semi-solid or liquid feedstock into a solid product by evaporation of the liquid into a vapour phase via application of heat. In the special case of freeze drying, drying occurs by sublimation of the solid phase directly into the vapour phase. This definition thus excludes conversion of a liquid phase into a concentrated liquid phase (evaporation), mechanical dewatering operations such as filtration, centrifugation, sedimentation, supercritical extraction of water from gels to produce extremely high porosity aerogels (extraction) or so-called drying of liquids and gases by use of molecular sieves (adsorption). Phase change and production of a solid phase as end product are essential features of the drying process. Drying is an essential operation in the chemical, agricultural, biotechnology, food, polymer, ceramics, pharmaceutical, pulp and paper, mineral processing and wood processing industries. Drying of various feedstock is needed for one or several of the following reasons: need for easy-to-handle free-flowing solids, preservation and storage, reduction in cost of transportation, achieving desired quality of product, etc. In many processes, improper drying may lead to irreversible damage to product quality and hence a non-salable product. Before proceeding to the basic principles, it is useful to note the following unique features of drying which make it a fascinating and challenging area for Research and Development:

1. Product size may range from microns to tens of centimetres (in thickness or depth)
2. Product porosity may range from 0 to 99.9 per cent
3. Drying times range from 0.25 seconds (drying of tissue paper) to five months (for certain hardwood species)

4. Production capacities may range from 0.10 kg/h to 100 t/h
5. Product speeds range from zero (stationary) to 2000 m/s (tissue paper)
6. Drying temperatures range from below the triple point to above the critical point of the liquid
7. Operating pressure may range from fraction of a millibar to 25 atmospheres
8. Heat may be transferred continuously or intermittently by convection, conduction, radiation or electromagnetic fields

Clearly, no single design procedure that can apply to all or even several of the dryer variants is possible. It is therefore essential to revert to the fundamentals of heat, mass and momentum transfer coupled with a knowledge of the material properties (quality) when attempting design of a dryer or analysis of an existing dryer. Mathematically speaking, all processes involved, even in the simplest dryer, are highly non-linear and hence scale-up of dryers is generally very difficult. Experimentation at laboratory and pilot scales coupled with field experience and know-how is essential to the development of a new dryer application.

BASIC PRINCIPLES AND TERMINOLOGY

Drying is a complex operation involving transient transfer of heat and mass along with several rate processes, such as physical or chemical transformations, which, in turn, may cause changes in product quality as well as the mechanisms of heat and mass transfer. Physical changes that may occur include shrinkage, puffing, crystallization and glass transitions. In some cases, desirable or undesirable chemical or biochemical reactions may occur leading to changes in colour, texture, odour or other properties of the solid product. In the manufacture of catalysts, for example, drying conditions can yield significant differences in the activity of the catalyst by changing the internal surface area. Drying occurs by effecting vapourization of the liquid by supplying heat to the wet feedstock. As noted earlier, heat may be supplied by convection (direct dryers), by conduction (contact or indirect dryers), radiation or volumetrically by placing the wet material in a microwave or radiofrequency electromagnetic field. Over 85 per cent of industrial dryers are of the convective type with hot air or direct combustion gases as the drying medium. Over 99 per cent of the applications involve removal of water. All modes except the dielectric (microwave and radiofrequency), supply heat at the boundaries of the drying object so that the heat must diffuse into the solid primarily by conduction. The liquid must travel to the boundary of the material before it is transported away by the carrier gas (or by application of vacuum for non-convective dryers). Transport of moisture within the solid may occur by any one or more of the following mechanisms of mass transfer.

- Liquid diffusion, if the wet solid is at a temperature below the boiling point of the liquid
- Vapour diffusion, if the liquid vaporizes within the material
- Knudsen diffusion, if drying takes place at very low temperatures and pressures, e.g. in freeze drying
- Surface diffusion (possible although not proven)
- Hydrostatic pressure differences, when internal vaporization rates exceed the rate of vapour transport through the solid to the surroundings
- Combinations of the above mechanisms

It is to be noted that since the physical structure of the drying solid is subject to change during drying the mechanisms of moisture transfer may also change with elapsed time of drying.

Thermodynamic Properties of Air–Water Mixtures and Moist Solids

As noted earlier, a majority of dryers are of direct (or convective) type. In other words, hot air is used both to supply the heat for evaporation and to carry away the evaporated moisture from the product. Notable exceptions are freeze and vacuum dryers, which are used almost exclusively for drying heat-sensitive products because they tend to be significantly more expensive than dryers that operate near to atmospheric pressure.

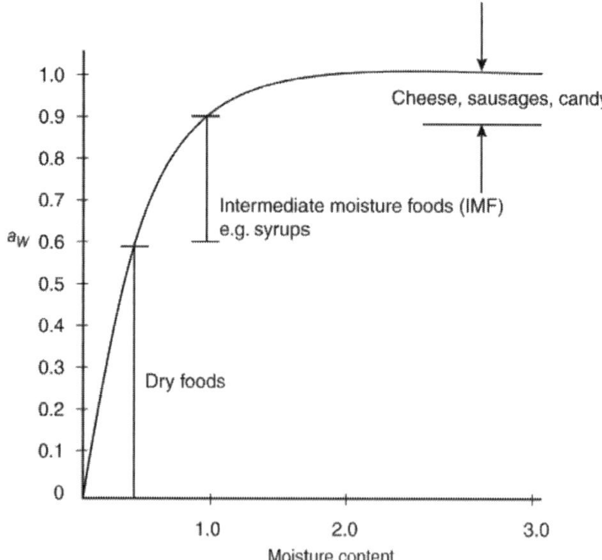

Figure 8.2　Water activity versus moisture content plot for different types of food

Another exception is the emerging technology of superheated steam drying. In certain cases, such as the drum drying of pasty foods, some or all of the heat is supplied indirectly by conduction. Drying with heated air implies humidification and cooling of the air in a well-insulated (adiabatic) dryer. Figure 8.2 shows the water activity versus moisture content plot for different types of food.

QUALITY OF DRIED FOODS AND DETERIORATIVE REACTIONS DURING DRYING

The quality of dried foods is dependent in part on changes occurring during processing and storage. Some of these changes involve modification of the physical structure. These modifications affect texture, rehydrability and appearance. Other changes are due to chemical reactions, but these are also affected by physical structure, primarily due to effects on diffusivities of reactants and of reaction products. The most commonly examined properties of dried products can be classified into two major categories: engineering and quality properties. The engineering properties of the dried products involve effective moisture diffusivity, effective thermal conductivity, drying kinetics, specific heat and equilibrium moisture content. Figure 8.3 shows the deterioration rates as a function of water activity. In addition there are

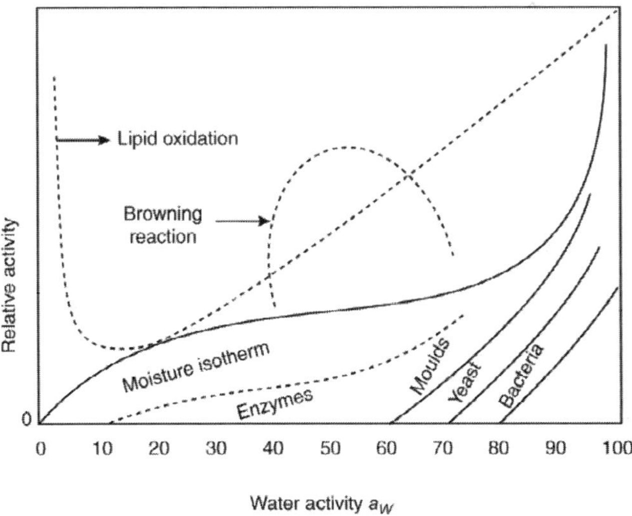

Figure 8.3   Deterioration rates as a function of water activity for food systems

properties related to product quality. These properties are necessary for the determination and the characterization of the quality of dried products and can be grouped into:

- Thermal properties—state of product: glassy, crystalline, rubbery
- Structural properties—density, porosity, pore size, specific volume
- Textural properties—compression test, stress relaxation test, tensile test
- Optical properties—colour, appearance
- Sensory properties—aroma, taste, flavour
- Nutritional characteristics—vitamins, proteins and
- Rehydration properties—rehydration rate, rehydration capacity

During the last few decades, much attention is paid on the quality of dehydrated foods. The specific drying method as well as the physico-chemical changes that occur during drying seems to affect the quality of dehydrated products. More specifically, drying method and process conditions affect significantly the drying constant, colour, texture, density and porosity and sorption characteristics of materials. The increasing need for producing efficiently high quality and convenient products at a competitive cost has led to the employment of several drying methods in practice.

METHODS FOR DRYING

Conventional Air-drying

Conventional air-drying is the most frequently used dehydration operation in food and conventional chemical industry. Dried products are characterized by low porosity and high apparent density. Significant colour changes occur during air-drying, and most frequently the dried product has low sorption capacity.

Microwave Drying

Microwave drying is an alternative drying method that has recently been used in food industry. Applying microwave energy under vacuum combines advantages of both vacuum drying and microwave drying as far as improved energy efficiency and product quality are concerned. Vacuum-dried materials are characterized by higher porosity, depending on the level of vacuum, and less deterioration of colour and volatile aroma.

Osmotic Dehydration

Osmotic dehydration minimizes the heat effects on colour and flavour, prevents enzymatic browning and thus limits the use of sulphur dioxide, increasing in this way the retention of nutrients during subsequent convective drying. Osmotic dehydration greatly affects apparent density and porosity.

Freeze Drying

Freeze drying is one of the most sophisticated dehydration methods. It provides dried products their porous structure, little or no shrinkage, superior taste and aroma retention and better rehydration properties, compared to products of alternative drying processes. However, its advantages are directly weighed against its corresponding high treatment cost.

Reactions occurring during drying can result in quality losses, particularly nutrient losses and other deteriorations caused by browning reactions. Reactions during drying may be classified as browning reactions and nutrient losses. Moreover, structural changes also occur, which affect quality of dried fruits and vegetables.

CHEMICAL FACTORS INFLUENCED BY DRYING

Browning Reactions

Browning reactions, which are some of the most important phenomena in food processing and storage, represent an interesting research area for the implications in food stability and technology, as well as in nutrition and health. They can involve different compounds and proceed through different chemical pathways. Browning reactions in foods are of widespread occurrence, and become evident when food materials are subjected to processing or mechanical injury. They are important in terms of alteration of appearance, flavour and nutritive value. Browning is considered to be desirable if it enhances the appearance and flavour of a food product in terms of tradition and consumer acceptance like in the cases of coffee, maple syrup, beer and in toasting of bread. However, in many other instances, such as fruits, vegetables, frozen and dehydrated foods, browning is undesirable as it results in off-flavours and colours. Therefore, it is important to know the mechanisms and inhibition methods of browning reactions. Another significant adverse effect of browning is the lowering of the nutritive value of the food article. Rate of browning reactions depends on temperature of drying, pH and moisture content of the product, time of heat treatment, and the concentration and nature of the reactants. Rate increases with increasing temperature, and the increase is faster in systems high in sugar content. For moisture contents above 30%, a decrease in reaction rate is caused by dilution, whereas below 30%, decrease is caused by the intrinsic ability of sugars to lower water activity.

Browning reactions change colour, decrease nutritional value and solubility, create off-flavours, and induce textural changes. There are two important forms of browning—enzymatic and non-enzymatic (Maillard reactions, caramelization, ascorbic acid oxidation). This colour development is usually undesirable, but with knowledge of the type of reaction involved, it is easier to work out methods for controlling this change.

Enzymatic browning   A group of enzymes, collectively called "phenolase" is responsible for browning of some fruits and vegetables, such as potatoes, apples and banana. When the tissue is bruised, cut, peeled, diseased, or exposed to any number of abnormal conditions, the colour of the fruits or vegetables is changed. The injured tissue rapidly darkens on exposure to air, due to the conversion of phenolic compounds to brown melanins. This enzyme group includes diverse such enzymes as phenoloxidase, cresolase, dopa oxidase, catecholase, tyrosinase, polyphenoloxidase, potato oxidase, sweet potato oxidase and phenolase complex. This type of browning is a serious problem during the dehydration process where any injury to the plant tissue, sustained through the use of heat or through poor handling procedures, can result in phenolase activation. The enzymatic browning of foods is usually undesirable because it cuts down the acceptability of the food in question for two reasons: (1) the undesirable development of off-colour and (2) the formation of off-flavours.

Non-enzymatic browning   During manufacturing process changes in the structure of derivative fruit products are produced, therefore these modify the colour and final aspect of the product. Although most non-enzymatic browning in food materials is undesirable because it indicates deterioration in flavour and appearance of the product involved, the development of brown colour in some products is entirely acceptable. Examples are the development of brown colours in baked goods during the baking process, in beer, molasses, coffee and substitute cereal beverages, many breakfast foods, and the roasting and other forms of heat preparation of meat. However, the brown colours developing in most other products are not desirable, and methods to prevent or retard such changes are in use. There are three main non-enzymatic reaction pathways: (i) Maillard reaction, (ii) Caramelization, (iii) Ascorbic acid oxidation.

The Maillard reaction   The Maillard reaction has been named after the French chemist Louis Maillard (1912) who observed the formation of brown pigments or melanoidins when heating a solution of glucose and glycine. The Maillard reaction is the action of amino acids and proteins on sugars. The carbohydrate must be a reducing sugar because a free carbonyl group is necessary for such a combination. The end product is the melanoidins, which are brown pigments. The mechanism of reaction has three stages:

1. Initial stage (colourless)
   (a) sugar–amine condensation
   (b) Amadori rearrangement
2. Intermediate stage (colourless to yellow)
   (a) sugar dehydration
   (b) sugar fragmentation
   (c) amino acid degradation

3. Final stage (highly coloured)

    (a) aldol condensation

    (b) aldehyde–amine polymerization, formation of heterocyclic nitrogen compounds.

Caramelization   This process is another example of non-enzymatic browning involving the degradation of sugars in the absence of amino acids or proteins. When sugars are treated under anhydrous conditions with heat, or at high concentration with dilute acid, caramelization occurs, with the formation of anhydrous sugars. Caramels for commercial use are made from glucose syrups, but usually caramelization is the result of reactions that take place when sucrose is heated. There are three stages in this process (at 200°C), during which water is lost; and isosacchrosan and other anhydrides are formed. The first stage starts with the melting of sucrose, followed by foaming, which continues for 35 minutes during which one molecule of water is lost from a molecule of sucrose. The foaming then stops. Shortly after this, a second stage of foaming starts which lasts 55 minutes. During this stage, about 9% of the water is lost, and the compound formed is caramelan, a pigment with the formula of $C_{24}H_{36}O_{18}$. Caramelan melts at 138°C, is soluble in water and ethanol, and is bitter in taste. The pigment caramelen is formed during the third stage of foaming which starts after about 55 minutes. The formula of this pigment is $C_{36}H_{50}O_{25}$. Caramelen melts at 154°C and is soluble in water. The main disadvantage of this reaction is the production of unpleasant, burned and bitter products, which can arise if this process is allowed to proceed uncontrolled. This reaction may be slowed down by bisulphites, which react with sugar to decrease the concentration of aldehydic form.

Ascorbic acid oxidation   A further mechanism appears to operate during the discolouration of dehydrated vegetables in which ascorbic acid is involved. The formation of dehydroascorbic acid and diketogluconic acids from ascorbic acid is thought to occur during final stages of the drying process and is capable of interacting with the free amino acids, non-enzymatically, producing the red-to-brown discolouration. This reaction may involve Strecker degradation.

Lipid Oxidation

Lipid oxidation is responsible for rancidity, development of off-flavours, and the loss of fat-soluble vitamins and pigments in many foods, especially in dehydrated foods. Factors that affect oxidation rate include moisture content, type of substrate (fatty acid), extent of reaction, oxygen content, temperature, presence of metals, presence of natural antioxidants, enzyme activity, ultraviolet light, protein content, free amino acid content and other chemical reactions. Moisture plays an important role in the rate of oxidation.

The elimination of oxygen from foods can reduce oxidation, but the oxygen concentration must be very low to have an effect. The effect of oxygen on lipid oxidation is also closely related to the product porosity. Freeze-dried foods are more susceptible to oxygen because of their high porosity. Air-dried foods tend to have less surface area due to shrinkage and thus are not much affected by oxygen. Minimizing the oxygen level during processing and storage, and addition of antioxidants as well as sequesterants, have been recommended in the literature to prevent lipid oxidation.

Colour Loss

The colour of foods is dependent upon the circumstances under which food is viewed, and the ability of the food to reflect, scatter, absorb or transmit visible light. Drying changes the surface characteristics of food and hence alters the reflectivity and colour. Carotenoids are fat-soluble pigments present in green leaves and red and yellow vegetables. Chemical changes in the carotenoid and chlorophyll pigments are caused by heat and oxidation during drying. In general, longer drying times and higher drying temperatures produce greater pigment losses. Oxidation and residual enzyme activity cause browning during storage. This is prevented by improved blanching methods and treatment of fruits with ascorbic acid or sulphur dioxide. Many studies indicate that the bulk of carotene destruction occurs during storage rather than as a result of the dehydration process. Pigment retention in dried foods decreases as temperature and moisture increases. Thus it was found that the beet pigments were most stable in the powders, than slices, and least stable in solution. The natural green pigment of all higher plants is a mixture of chlorophyll a and chlorophyll b. The retention of the natural green colour of chlorophyll is directly related to the retention of magnesium in the pigment molecules. In moist heating conditions, the chlorophyll is converted to pheophytin by losing some of its magnesium. The colour then becomes olive green rather than grass green. The interaction of amino acids and reducing sugars (Maillard reaction) occurs during conventional dehydration of fruits. If the fruits are sulphured, enzymatic browning can be inhibited, and the Maillard reaction retarded.

PHYSICAL FACTORS INFLUENCED BY DRYING

Rehydration, Shrinkage and Food Porosity

Rehydration is a complex process aimed at the restoration of raw material properties when dried material is contacted with water. Pre-drying treatments, subsequent drying and rehydration induce many changes in structure and composition of plant tissue, which result in impaired reconstitution properties. Hence, rehydration can be considered as a measure of the injury to the material caused by drying and treatments preceeding dehydration.

Rehydration of dried plant tissues is composed of three simultaneous processes: the imbibition of water into dried material, swelling and leaching of soluble. It has been shown that the volume changes (swelling) of biological materials are often proportional to the amount of absorbed water. It is generally accepted that the degree of rehydration is dependent on the degree of cellular and structural disruption. There are a large number of research reports in which authors measure the ability of dry material to rehydrate.

The ratio between the dry material mass and water mass varies from 1:5 to 1:50, temperature of rehydrating water is from room temperature to boiling. Time of rehydration varies from 2 minutes to 24 hours. The degree to which a dehydrated sample will rehydrate is influenced by structural and chemical changes caused by dehydration, processing conditions, sample preparation and sample composition. Rehydration is maximized when cellular and structural disruptions such as shrinkage are minimized. Several researchers have found that freeze-drying causes fewer structural changes and fewer changes to product's hydrophilic properties than other drying processes. Most of the shrinkage occurs in the early drying stages, where 40 to 50% shrinkage may occur. To minimize shrinkage, therefore, low-temperature drying should be employed so that moisture gradients throughout the product are minimized. Many drying techniques or pretreatments given to food before drying are aimed at making the structure more porous so as to facilitate mass transfer and thereby speed drying rate. Porous sponge-like structures are excellent insulating bodies and generally will slow down the rate of heat transfer into the food. Porosity may be developed by creating steam pressure within the product and a case hardened surface through rapid drying. Porosity also can be developed by whipping or foaming a food liquid or puree prior to drying. The porous product has the advantages of quick solubility or reconstitution and greater volume appearance, but the disadvantages of increased bulk and generally shorter storage stability because of increased surface exposure to air, light, etc.

Solubility

Many factors affect the solubility, including processing conditions, storage conditions, composition, pH, density and particle size. It has been found that increasing product temperatures is accompanied by increasing protein denaturation, which decreases solubility. A low bulk density is required for good dispersibility of non-fat dry milk. It was found that particle agglomeration, which increases particle size, increases sinkability. However, some scientists found that larger particles were less soluble. This was attributed to the longer drying time required to dry large particles. Thus more protein was denatured and solubility decreased. This shows that the heat treatments as well as the particle size must be considered when determining solubility.

Texture

Texture is one of the most important properties connected to product quality. Texture change of solid foods is an important cause of quality deterioration. Factors that affect texture include moisture content, composition, variety, pH, product history (maturity) and sample dimensions. The chemical changes associated with textural changes in fruits and vegetables include crystallization of cellulose, degradation of pectin and starch gelatinization.

Texture is also dependent on the method of dehydration. High air temperatures (particularly with fruits, fish and meats) cause complex chemical and physical changes to the surface, and the formation of hard impermeable skin. This is termed "case hardening". It reduces the rate of drying and produces a food with a dry surface and a moist interior. It is minimized by controlling the drying conditions to prevent excessively high moisture gradients between the interior and the surface of the food. On rehydration the product absorbs water more slowly and does not regain the firm texture associated with the fresh material. There are substantial variations in the degree of shrinkage with different foods.

Drying is not commonly applied to meats in many countries owing to the severe changes in texture compared with other methods of preservation. These are caused by aggregation and denaturation of proteins and a loss of water-holding capacity, which leads to toughening of muscle tissue.

The rate and temperature of drying have a substantial effect on the texture of foods. In general, rapid drying and high temperatures cause greater changes than do moderate rates of drying and lower temperatures. As water is removed during dehydration, solutes move from the interior of the food to the surface. Evaporation of water causes concentration of solutes at the surface.

In powders, the textural characteristics are related to bulk density and the ease with which they are rehydrated. These properties are determined by the composition of the food, the method of drying and the particle size of the product. Low-fat foods, (for example fruit juices, potato and coffee) are more easily formed into free-flowing powders than whole milk or meat extracts. Powders are "instantized" by treating individual particles so that they form free-flowing agglomerates or aggregates, in which there are relatively few points of contact. The surface of each particle is easily wetted when the powder is rehydrated, and particles sink below the surface to disperse rapidly throughout the liquid. These characteristics are respectively termed wettability, sinkability, dispersibility and solubility. For a powder to be "instant", it should undergo these four stages within a few seconds.

Aroma Loss

There is often decrease in the quality of the dried products because most conventional techniques use high temperatures during the drying process. Processing may also introduce undesirable changes in appearance and will cause modification of the natural "balanced" flavour and colour. The dehydration technologies should be focusing on the production of dried products with little or no loss in their sensory characteristics together with the advantages of added convenience.

The properties of dried vegetables are influenced by chemical and physical changes. Chemical changes mainly affect sensory properties such as colour, taste and aroma, whereas physical changes mainly influence the handling properties such as swelling capacity and cooking time. Heat treatment of fruits and vegetables often reduces the number of original volatile flavour compounds, while introducing additional volatile flavour compounds through the autoxidation of unsaturated fatty acids and thermal decomposition, and/or initiation of Maillard reactions. Volatile organic compounds responsible for aroma and flavour have boiling points at temperatures lower than water. Volatiles, which have a high relative volatility and diffusivity, are lost at an early stage in drying. Fewer volatile components are lost at later stages. Control of drying conditions during each stage of drying minimizes losses. Foods that have economic value due to their characteristic flavours, herbs and spices, are dried at low temperatures. A second important cause of aroma loss is oxidation of pigments, vitamins and lipids during storage. The open porous nature of dried food allows access of oxygen. The storage temperature and the water activity of the food determine the rate of deterioration.

In dried milk the oxidation of lipids produces rancid flavours owing to the formation of secondary products including b-lactones. Most fruits and vegetables contain only small quantities of lipid, but oxidation of unsaturated fatty acids to produce hydroperoxides, ketones and acids, causes rancid and objectionable odours. Vacuum or gas packaging, low storage temperatures, exclusion of ultraviolet or visible light, maintenance of low moisture contents, addition of synthetic antioxidant or preservation of natural antioxidants reduce these changes.

The technical enzyme, glucose oxidase, also protects dried foods from oxidation. A package, which is permeable to oxygen but not to moisture and which contains glucose and the enzyme, is placed on the dried food inside a container. Oxygen is removed from the headspace during storage. Flavour changes, due to oxidative or hydrolytic enzymes are prevented in fruits by the use of sulphur dioxide, ascorbic acid or citric acid, by pasteurization of milk or fruit juices and by blanching of vegetables.

Other methods that are used to retain flavours in dried foods include:

1. Recovery of volatiles and their return to the product during drying,
2. Mixing recovered volatiles with flavour fixing compounds, which are then granulated and added back to the dried product (for example, dried meat powders), and
3. Addition of enzymes, or activation of naturally occurring enzymes, to produce flavours from flavour precursors in the food (for example, onion and garlic are dried under conditions that protect the enzymes that release characteristics flavours). Maltose is used as a carrier material when drying flavour compounds.

NUTRITIONAL FACTORS INVOLVED IN DRYING

Nutrient Losses

In drying, a food loses its moisture content, which results in increase in the concentration of nutrients in the remaining mass. Proteins, fats and carbohydrates are present in larger amounts per unit weight in dried foods than in their fresh counterpart. Large differences in reported data on the nutritive value of dried foods are due to wide variations in the preparation procedures, the drying temperature and time, and the storage conditions. In fruits and vegetables, losses during preparation usually exceed those caused by the drying operation. The water-soluble vitamins can be expected to be partially oxidized. The water-soluble vitamins are diminished during blanching and enzyme inactivation. Some vitamins during the drying process, (for example, riboflavin) become supersaturated and precipitate from solution. Losses are therefore small. Others (for example, ascorbic acid) are soluble until the moisture content of the food falls to very low levels and react with solutes at higher rates as drying proceeds. Ascorbic acid is sensitive to high temperatures at high moisture contents.

Several studies have shown that the maximum rate of ascorbic acid degradation occurs at specific (critical) moisture levels. The critical moisture level appears to vary with the product being dried and/or the dehydration process. Short drying times, low temperatures, and low moisture and oxygen levels during storage are necessary to avoid large losses. To optimize ascorbic acid retention, the product should be dried at a low initial temperature when the moisture content is high since ascorbic acid is most heat-sensitive at high moisture contents.

The temperature can then be increased as drying progresses and ascorbic acid is more stable, due to decrease in moisture. Thiamine is also heat-sensitive, but other water-soluble vitamins are more stable to heat and oxidation, and losses during drying rarely exceed 5–10%. Fruits can be sun dried, dehydrated or processed by a combination of the two. Sun drying causes losses in carotene content. Dehydration especially spray drying, can be accomplished with loss of this nutrient. Vitamin C is lost in great proportions in sun-dried fruits. Freeze drying of fruits retains greater portions of vitamin C, and other nutrients. The retention of vitamins in dehydrated foods is generally superior in all counts than in sun-dried foods.

Vegetable tissues dried artificially or in the sun tend to have losses in nutrients in the same order of magnitude as fruits. The carotene content of vegetables is decreased as much as 80% if processing is accomplished without enzyme inactivation. The best commercial methods will permit drying with losses in the order of five per cent for carotene. Thiamine content reduction can be anticipated to be in the order of 15% in blanched tissues, while unblanched may lose three-fourths of this nutrient. With ascorbic acid, rapid drying retains greater amounts than slow drying. Generally the vitamin C content of vegetable tissues will be lost in slow, sun-drying processes. In all events the vitamin potency will decrease on storage of the dry food.

With milk products, the nutrient level of the raw milk and the method of processing will dictate the level of vitamins retained. Vitamin A is retained in good proportions in drum-dried and spray-dried milk. Vacuum-packed dry milk can be stored with good retention of vitamin A. Thiamine losses occur during both spray and drum drying, but losses are of a lower order of magnitude than with fruit and vegetable drying. Similar results are obtained with riboflavin. Ascorbic acid losses occur during the drying of milk. Being sensitive to heat and oxidation, vitamin C may be totally lost in a drying process. With careful processing, vacuum drying and freeze drying, ascorbic acid values can be retained in the same order of magnitude as fresh raw milk. The vitamin D content of milk is generally greatly decreased by drying. Fluid milk should be enriched with vitamin D prior to drying. Other vitamins such as pyridoxine and niacin are not materially lost. Usually dried meat contains slightly less vitamins than fresh meat. Thiamine losses occur during processing, greater losses occurring at high drying temperature. Vitamin C is in most part lost in dried meat. Small losses of riboflavin and niacin occur. Oil-soluble nutrients (for example, essential fatty acids and vitamins A, D, E and K) are mostly contained within the dry matter of the food and they are not therefore concentrated during drying. However, water is a solvent for heavy metal catalysts that promote oxidation of unsaturated nutrients. As water is removed, the catalysts become more reactive, and the rate of oxidation accelerates. Fat-soluble vitamins are lost by interaction with the peroxides produced by fat oxidation. Losses during storage are reduced by low oxygen concentration and storage temperatures and by exclusion of light.

Influence of drying on protein The biological value of dried protein is dependent on the method of drying. Prolonged exposures to high temperatures can render the protein less useful in the diet. Low temperature treatments of protein may increase the digestibility of protein over native material. Milk proteins are partially denatured during drum drying, and these result in the reduction in solubility of the milk powder, aggregation and loss of clotting ability. At high storage temperatures and at moisture contents above approximately 5%, the biological value of milk protein is decreased by Maillard reactions between lysine and lactose. Lysine is heat-sensitive and losses in whole milk range from 3–10% in spray drying and 5–40% in drum drying.

Influence of drying on fats   Rancidity is an important problem in dried foods. The oxidation of fats is greater at higher temperatures than at lower temperatures of dehydration. Protection of fats with antioxidants is an effective control.

Influence of drying on carbohydrates   Fruits are generally rich sources of carbohydrates, poor sources of proteins and fats. The principal deterioration in fruits is in carbohydrates. Discolouration may be due to enzymatic browning, or to caramelization types of reactions. In the latter instances, the reaction of organic acids and reducing sugars causes discolourations noticed as browning. The addition of sulphur dioxide to tissues is a means of controlling browning. The action is one of enzyme poisoning and antioxidant power. The effectiveness of this treatment is dependent upon low moisture contents. Carbohydrate deterioration is most important in fruit and vegetable tissues being dried. Slow sun drying permits extensive deterioration unless the tissues are protected with sulphates, or suitable agents. Burning sulphur is the least expensive method of obtaining such protection, and is done prior to drying.

MICROBIOLOGICAL QUALITY

Since microorganisms are widely distributed throughout nature, and foodstuff at one time or another are in contact with soil and dust, it is anticipated that microorganisms will be active whenever conditions permit. One obvious method of control is in the restriction of moisture for growth. Living tissues require moisture. The amount of moisture in food establishes which microorganisms will have an opportunity to grow. Reducing the water activity of a product below 0.85 inhibits growth but does not result in a sterile product. The heat of the drying process does reduce their numbers, but the survival of food-spoilage organisms may give rise to problems in the reconstituted food. Recommendations for the control of microorganisms during processing are often very basic. The highest possible drying temperatures should be used to maximize thermal death even though low drying temperatures are best for maintaining organoleptic characteristics. If a process is optimized for other quality factors, there are constraints on the maximum allowable water content.

Sodium chloride is commonly employed in conjunction with drying. Salt is useful in controlling microbial growth during sun drying and dehydration processes, i.e., meat and fish drying. The most positive control would be to start with high quality foods having low contamination, pasteurize the material prior to drying, process in clean factories, and store under conditions where the dried foods are protected from infection by dust, insects, rodents and other animals.

Storage Stability

When discussing storage stability, one is concerned with the organoleptic, physical and chemical changes that take place in the dried fruits and vegetables during storage and the rates at which these changes occur. Darkening and loss of flavour are the major types of deteriorations of dried fruits and vegetables in storage. Sulphur dioxide content, storage temperature, light, packaging material, moisture content, antimicrobial treatment and trace elements are major factors affecting storage stability. Only free sulphite is effective in retarding the formation of pigment materials. During storage, the loss of sulphur dioxide determines the practical shelf life of the dried product with respect to spoilage through non-enzymatic browning. Storage of products at semitropical or summer temperatures requires residual sulphites to prevent darkening and flavour bittering, and to make the dried fruit less favourable medium for growth of microorganisms. Sulphur dioxide helps to maintain a light, natural colour during storage. Darkening rates during storage is inversely proportional to sulphur dioxide content. Therefore, any condition accelerating sulphur dioxide loss, in turn, accelerates the darkening of the product. One way to retard sulphite loss, thereby darkening, is the addition of oxygen scavenger pouch to the sealed, packed sulphured dried fruit.

Storage temperature is of vital importance in relation to maintenance of quality. Storage of dried fruits and vegetables should be at relatively low temperatures to maximize storage life. There is an important effect of temperature on loss of sulphur dioxide from the dried product during storage. A 20°F increase in temperature increases the rate of sulphur dioxide loss approximately 3 times. Moreover at higher temperatures, the rate of change in flavour also increases. Light, during storage, is detrimental for quality. It causes a reduction in carotene content, increases the rate and amount of sulphur dioxide loss, and thereby increases the rate of darkening. In addition, it also affects riboflavin content. Packaging material used and the package environment are other major factors in terms of storage stability. The type of package used varies with expected storage conditions. Packaging may be done under vacuum, nitrogen or atmosphere. Dried foods have moisture content below 20% and a water activity of 0.7 or below. They are hard and firm, resistant to microbial deterioration. There are critical water activities for some products below which browning is minimized. Storage stability increases with decreasing moisture content. But, it was also reported that the maximum rate of deterioration of dried fruits occurs at a moisture content of 5–8% moisture.

Dried fruits and vegetables must be protected from rodents and insects during storage. Fumigation is often used to prevent insect infestation during storage and before packaging. In addition to fumigation, antimycotic agents (fungistats) are used to stabilize most prunes and figs against mould growth at 30–35 % moisture. Sorbic acid and sorbate salts are used as dips or sprays to prevent moulding; sulphur dioxide or sulphite salts are used to preserve fruits during drying from colour changes and browning, and to ward off insects. Potassium sorbate dip is the most effective one. The effectiveness depends on pH of the product. Some salts and metals are detrimental to nutritive value, flavour and storage quality. Raw materials may be exposed to these trace elements during washing or pretreatment. Calcium has a firming effect on texture; iron and copper combine with tannins to cause blackening and may accelerate degradation of ascorbic acid. Sodium, magnesium and calcium sulphates impart bitter flavour. Certain salts of zinc, cadmium and chromium have toxic effects.

CALCULATION OF DRYING EFFICIENCY

The majority of artificial drying operations are based on hot air drying, where air is heated by the combustion of fossil fuels prior to being forced through the product. This type of drying requires high energy inputs, due to the inefficiencies of such dryers. Often, the exhaust air is simply released to the surrounding ambient air. Some systems allow for the recycling of exhaust heat, which can greatly increase the overall energy efficiency of the dryer. With increasing pressures to reduce environmental degradation, both from the public and from governments, it is necessary to improve drying processes to reduce energy consumption and greenhouse gas (GHG) emissions, while still providing a high-quality product with minimal increase in economic input. In order to achieve these goals, much work needs to come from the advances in novel technologies for drying. There is a great debate about the proper method to calculate energy efficiencies for the purpose of providing an objective comparison between different dryers and drying processes. The basic approach to calculating any energy efficiency, η, is to take the ratio of energy required, E_r, to energy supplied, E_s

$$\eta = \frac{E_r}{E_s} \tag{1}$$

However, the energy efficiency for the drying process can be calculated as a whole (total energy required and total energy supplied), instantaneous efficiency (energy required and energy supplied at given time), or it may be only for the drying chamber, not including other peripheral energy requirements. Typical convective dryers account for about 85% of all industrial dryers. The drying medium is generally hot air or direct combustion gases. The energy efficiency for convective dryers, η_{cov}, is usually calculated based on the temperature of the drying medium at the inlet, T_{in}, outlet, T_{out} and the ambient air temperature, T_{amb}:

$$\eta_{cov} = \frac{T_{in} - T_{out}}{T_{in} - T_{amb}} \tag{2}$$

The temperature of the heating medium cannot drop below the wet bulb temperature, T_{wb}. This will result in the maximum efficiency of a convective dryer as

$$\eta_{cov,\,max} = \frac{T_{in} - T_{wb}}{T_{in} - T_{amb}} \tag{3}$$

This is, however, only the efficiency of the dryer itself, not including any energy inputs or losses that are not directly associated with the drying chamber (inputs to blowers, heat loss prior to entry into the drying chamber), though these other inputs and losses are generally quite small.

Another drawback of equations (2) and (3) is that the cumulative energy efficiency only holds true if all the temperatures remain constant, and if the outlet temperature is representative of the drying process. Thus, the instantaneous energy efficiency is calculated as

$$\eta_{ins} = \frac{\text{energy used for evaporation at time } t}{\text{input energy at time } t} \tag{4}$$

The cumulative energy efficiency, η_c can be calculated by integrating equation (4) with respect to time

$$\eta_c = \frac{1}{t} \int_0^t \eta_{ins}(t)\, dt \tag{5}$$

It is an overall efficiency, but does not describe the ability of the heat to remove moisture from the product. The instantaneous drying efficiency, ε_{ins} is designated as

$$\varepsilon_{ins} = \frac{\text{energy used for evaporation at time } t}{(\text{input energy } - \text{ output energy with outlet gas}) \text{ at time } t} \tag{6}$$

This equation can also be integrated to give the cumulative drying efficiency, ε_c. An alternative indicator of the energy efficiency often used for heat pump dryers, is the specific moisture extraction ratio, SMER (kg/kW/h):

$$SMER = \frac{\text{amount of water evaporated}}{\text{energy used}} \tag{7}$$

The *SMER* can be calculated either as an instantaneous value or as an average value during drying. During the drying process, the *SMER* value invariably decreases as the removal of moisture becomes more difficult due to smaller water vapour deficits at the surface of the product. In theory, the maximum value for SMER in a conventional dryer is 1.55 kg/kW/h, which is based on the latent heat of water evaporation at 100°C. For dryers with heat recovery systems, such as heat pump dryers, the *SMER* value can be above the theoretical maximum value.

ENERGY SOURCES

Increasing concern of global warming and related environmental problems are causing changes in all sectors of industry. Increasing legislation and tighter environmental policies are requiring that companies and corporations evaluate their energy use and GHG emissions. GHG emissions are generally calculated based on their CO_2 equivalent emissions. Thus, a change in the fuel source can increase or decrease the GHG emissions. Calculations for dryers that use electrical energy input (microwave dryers, heat pump dryers) can be a little more difficult, as the CO_2 emissions produced in the production of electricity vary to a large degree, due to the method of production and the different fuels used in the production. An attractive alternative energy source is biomass, which is considered to be CO_2-neutral when grown in sustainable conditions.

COMBINED TECHNOLOGIES

Hybrid or combined drying technologies include implementation of different modes of heat transfer, two or more stages of the same or different type of dryer. The efficiency of drying concerning both energy and time of process dictates intensive research in this area, and some of the most promising drying methods include the electromagnetic waves and sonic-assisted drying.

Electro Technologies

Electromagnetic waves can penetrate deep into material causing volumetric heating "targeting" mostly water, offering higher energy conversion rates and therefore shorter process time. Unfortunately, this technology is still only being accepted in industry, mainly because of high capital costs and a lack of documented energy savings. Nevertheless, the implementation of electromagnetic waves could be one of the most promising drying techniques in the future, especially because of its potential for savings in energy and time, and providing high quality products. Dielectric heating is based on volumetric heat generation throughout the material being dried. Radio frequencies cover the electromagnetic spectra with frequencies of 1 to 300 MHz, but special care should be taken here: only specific frequencies have been designated within both radio frequency (RF) and microwave (MW) spectra for industrial applications 13.56, 27.12, and 40 MHz for RF, and 915 MHz and 2450 MHz in the MW region. RF drying can be appropriate for large loads such as paper and timber drying, where high power and short duration are required. Drying employing RF has a few applications in food industry such as biscuit post-baking, drying and meat/fish tempering. Microwaves cover electromagnetic wave spectrum with wavelengths from 1 mm to 1 m. They are the most extensive researched electromagnetic waves in drying so far, but as with all other methods, there is a lack of documented energy analysis and the majority of work dealing with MW are related to the influence of MW-assisted drying on the quality of specific product. Biological materials are very attractive for MW drying, because they are heat-sensitive and can benefit from the "targeted" heating obtained with MW. MW was used to dry grapes, and specific energy consumption calculated (defined as the total energy in MJ used to evaporate a unit mass of water) for convective drying, and for MW-convective drying. For convective drying, specific energy consumption ranged between 81.2 and 90.4 MJ/kg, whereas under similar convective conditions and with implementation of MW, this consumption decreased radically—7.1 to 24.3 MJ/kg (depending on process conditions). Enhancement of MW-assisted drying can be achieved by introducing intermittent MW power exposure, as opposed to continuous exposure. Discontinuous MW power can substantially reduce the energy loss. The total drying time was increased, but the total MW exposure was shorter, giving the product a higher quality. It was also demonstrated that energy consumption (defined as MJ per kg of evaporated water) was influenced by both MW cycling period and MW power density in intermittent MW drying. Further improvement of MW drying can be achieved in drying under vacuum.

Sonic Drying

Sonic drying is usually used for viscous products that are difficult to dry with other methods, and is especially suitable for temperature-sensitive products because it is a non-thermal drying (separation or dewatering) process. Sound used in industrial applications has low frequencies (20 to 40 kHz). Unfortunately, there are only a few works which consider the energy requirements for sonic drying, stating that consumption of sonic energy per kg of evaporated water is approximately the same or even higher as in convective drying. However, when sound effect is used in pulse-combustion drying, which combines heat and sound, this vibrating airstream removes water much faster than conventional air-drying, and uses 3.53 MJ/kg, when conventional air-drying uses between 5.8 and 7.0 MJ for one kilogram of evaporated water.

NEW METHODS

Superheated Steam

Superheated steam (SHS) serves as a drying medium, supplying heat to a drying product and carrying off evaporated moisture. This method has been industrially implemented, but so far on a very small scale. The advantages are that no oxidative or combustion reactions take place in or near the dryer, higher drying rates (in some cases), and it can permit pasteurization of food products. The disadvantages are that it is a more complex system, heat-sensitive materials are prone to damage, and there is limited documented experience about this method. It was shown that substantial energy savings (more than 80% in some instances) could be made by substituting air with SHS. These savings were made by heat recovery from exhausting SHS and eliminating the need to heat from ambient temperature.

Heat Pump-Assisted Drying

A heat pump works on the principle of refrigeration to cool an airstream and condense the water contained in it. This renders the air dry and also recovers the latent heat of evaporation through water vapour removal which permits air recirculation. Heat pump drying (HPD) had the lowest operating cost when compared to electrically heated convective dryers and direct-fired dryers. One important disadvantage of this method is that it uses an expensive energy source—electricity, needed to run the compressor, but it could be economically feasible during the initial stages of drying for high moisture products. HPD has higher drying efficiency, offers better product quality, and it is environmental-friendly. HPD is useful for materials with high initial moisture content and in regions with high humidity of ambient air. HPD can be combined with MW.

FREEZE DRYING AND ITS APPLICATIONS IN FOOD INDUSTRY

Vacuum freeze drying of biological materials is one of the best methods of water removal, which results in final products of highest quality. Freeze drying is based on dehydration by sublimation of the ice fraction of a frozen product where water passes from a solid to gaseous state. Due to the absence of liquid water and the low temperatures required for the process, most of the deterioration reaction and microbiological activities are stopped, giving a final product of excellent quality. The solid state of water during freeze drying protects the primary structure and shape of the products with minimal volume reduction. Freeze-dried products have a long shelf life without refrigeration—usually two years for a 2% residual moisture content product. This technique has been used successfully for several biological materials such as meats, coffee, juices, dairy products, etc. Despite its many advantages, freeze drying has always been recognized as the most expensive process for manufacturing a dehydrated product.

Traditionally, the application of the freeze drying process to dehydrate fruits and vegetables has been reduced to the production of space shuttle goods and military or extreme-sport/outdoor living foodstuff. Recently, however, the market for 'natural'and 'organic' products and the demand for foods with minimal processing and high quality has been strongly increasing. In this regard, the market for freeze-dried fruits and vegetables is not only increasing in volume but also diversifying.

Freeze Drying of Plant Products

Numerous plant products have interesting bioactive compounds (i.e., isoflavones, carotenoids, anthocyanins, etc.) that are the centre of attention of consumers today, due to their action in preventing diseases such as cancer and neurological and coronary diseases. Some of these valuable compounds can be deteriorated during processing by high temperatures, oxygen and light. But freeze drying is always not the best method to preserve certain target bioactive compounds such as folic acid or ascorbate oxidase, vitamin C and volatile compounds.

QUALITY OF FREEZE-DRIED PRODUCTS

Since the last decade, the quality of foods became one among the main pre-occupations in food research. The advantages and disadvantages of hot drying and freeze drying helps us to understand the ways to improve the quality of the foodstuff. This is because of taking into account some important features such as shrinkage, glass transition temperature (T_g), process quality interaction, drying kinetics, costs and new improvements.

T_g can be defined as the temperature at which an amorphous system changes from the glassy to the rubbery state. When the temperature of some processes exceeds the T_g, the quality of foodstuff is seriously altered. Thus process quality relationships based on the T_g concept could be a solid basis to optimize the dehydration methods.

On this basis, two temperature limits are shown to be essential in order to avoid quality problems during freeze drying of a particular product. The frozen core has to be below the ice melting onset temperature, and the temperature of the dry matrix has to be lower than the T_g of the dry solids. The first thermal limit is difficult to corroborate in practice due to the limited data available about the ice melting on set temperature for foods and the difficulty of precisely measuring the frozen core temperature during freeze drying of solid foods. Theoretically, if the vacuum level in the freeze dryer is low enough, then the first thermal limit can be achieved. The second thermal limit is accomplished when the final temperature of the product is lower than its T_g. Shrinkage and T_g are interrelated, in that, significant changes in volume and collapse of structure can be noticed only if the temperature of the process is higher than the T_g of the material at that particular moisture content.

The porous structure created during freeze drying is interesting for manufacturing food powders, preparing instant foods or for adding freeze-dried fruits to cereal mixtures due to improved and quick rehydration. However, this feature is no longer beneficial during long-term storage of freeze-dried products that contain bioactive compounds since the increased porosity increases certain deterioration reactions that depend on the exposed area (i.e., oxidation). Chlorophyll in freeze-dried spinach deteriorated during storage compared to controlled low-temperature vacuum dehydrated samples. The loss of chlorophyll can be attributed to the high porosity of freeze-dried products. Thus porosity is a key parameter for understanding rehydration and moisture sorption of freeze-dried products. In general, moisture sorption is higher for freeze-dried vegetables compared with other dehydration methods. Rehydration of freeze-dried samples is almost instantaneous and depends on bath temperature, the product's open or close porosity and rehydration media viscosity. The effect of bath temperature on rehydration of freeze-dried materials has a strong correlation with the effect of the different drying methods on food structure. Hot air drying usually destroys food cells with considerable shrinkage from the high drying temperatures. The final air-dried products have a compact structure and a reduced volume and thus increasing the rehydration temperature would help moisture diffusion. On the other hand, the solid state of water during freeze drying protects the primary structure and the shape of the products with minimal volume reduction, keeping the food structure and cells almost intact, with a high porosity end product. Pores and pore-size distribution in solid foods have an enormous impact on mass transfer during rehydration, which logically makes freeze-dried materials of rapid rehydration. Therefore, when dealing with high porosity materials, such as freeze-dried products, capillarity rather than diffusion could be the primary mode of water transport. However, this high porous structure could easily collapse if the temperature of the rehydration media is not well set.

HYBRID REHYDRATION METHODS

With regards to freeze drying, some developments have been made recently through combination of hot air drying followed by freeze drying. This has been studied with regards to vegetables. This also showed that this combination produces dehydrated carrot and pumpkin products having similar quality as freeze-dried products and more superior than hot air-dried products. The drying time and total energy consumption was favourably 50% lower than freeze drying alone.

EFFECT OF DRYING ON MICROBES

This process is not lethal to microorganisms. Bacteria usually do not grow since they need an a_w of >0.9. At a_w levels of 0.9, organisms most likely to grow are the yeasts and moulds. Bacterial endospores survive drying as do yeasts, moulds and many gram-negative and gram-positive bacteria.

Food-borne parasites like *Trichinella spirallis* survive the drying process. Death or injury from drying results from denaturation in the undried portion and removal of bound water. When death occurs during drying, it is highest during the early stages of drying. Young cultures are more sensitive to drying. During initial warming up period of drying, temperature is still low and relative humidity of the food is high. Length of this phase depends on the size of the food particles. If this phase is long, microbes may grow during a slow increase in temperature (20–40°C) under the existing high a_w. During the phase of drying, temperature exceeds 50–70°C. There is no opportunity for growth but destruction due to heat is neither significant since the heat is dry (a_w decreases parallel to an increase in temperature) and wet heat is more destructive than dry heat.

But microbes can be destroyed completely using time–temperature relationship (at least for 30 minutes). With certain drying technologies, surface temperature reaches 100°C when internal temperature remains lower. Microbes on the surface are inactivated. The a_w on the surface decreases, thereby decreasing the heat sensitivity of microbial cells. A combination of high temperature and high relative humidity is necessary for killing all microbes (since moist heat is more effective than dry heat) but this is rare in drying technologies. Hence the major microbicidal effect can be due to:

- High temperature and low humidity conditions which can last for a long time.
- Loss of viability continues during storage since reversibly damaged cells are unable to regenerate at low a_w and they gradually die.
- Further loss of viability may occur during rehydration of dried foods especially when relative humidity is rapid due to large difference in osmotic pressure.

Effect of Freeze Drying on Microbes

Freeze drying is the combination of decreased a_w by freezing and further reduction of a_w by sublimation of ice. Extent of cell damage may depend on the temperature and rates of freezing and sublimation. Microbial survival also depends on the composition of freeze-dried foods. For example, carbohydrates, proteins and colloidal substances have a protective effect on microbes. The gram-positive bacteria survive better than gram-negative bacteria.

FACTORS CONTROLLING DRYING

1. *Temperature*   It varies with the food and method of drying.
2. *Relative humidity*   It varies with food and method of drying and also the stage of drying.
3. *Velocity*   Velocity of air indirectly affects the relative humidity and thus the drying process.
4. Time process of drying.

If any of these factors are not controlled, it leads to a condition called "case hardening". (More rapid evaporation of moisture from the surface than diffusions from the interior with a resulting hard impermeable surface film that hinders further drying).

TREATMENT OF FOODS BEFORE DRYING

Blanching of vegetables and fruits (apricots) and sulphuring of light-coloured fruits by exposing to SO_2 gas produced by burning sulphur this helps maintain an attractive light colour, conserves vitamin C and vitamin A and helps to repel insects and also kills many microbes.

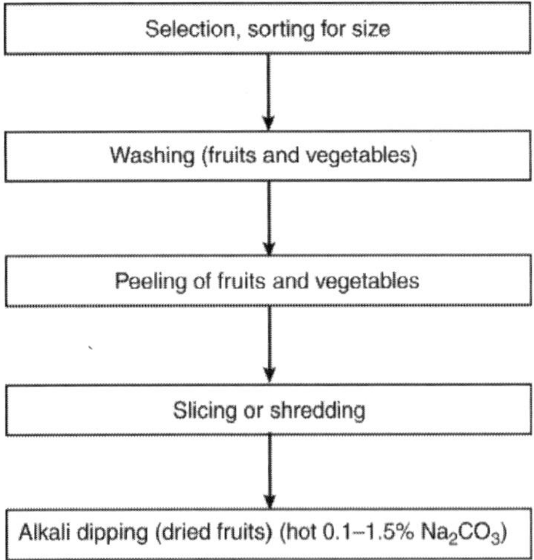

TREATMENT OF FOODS AFTER DRYING

Sweating   It is a process related to storing dried food in boxes for re-addition of moisture to a desired level—used with almonds, walnuts.

Packaging   Most foods are packaged after drying for protection against moisture, protection from contamination with microbes and protection from insects.

Pasteurization Fruits are pasteurized in the package and heated at 65.6–85°C for 30–70 minutes at 70–100% relative humidity.

MICROBIOLOGY OF DRIED FOODS

Before Receiving at the Processing Plant

Microflora is present on the fruits and vegetables (natural flora of soil and water). Spoiled parts of fruits and vegetables add microflora to the same. When vegetables are piled, due to heat and moisture, it may support the surface growth of slime forming or rot producing organisms.

In the Plant Before Drying

Equipments and workers (hands) contaminate the food. Some pretreatments like blanching, washing, peeling may reduce the microflora. Selection, sorting will influence the kinds and numbers of microbes present. For example, washing eggs prove to be harmful, adding moisture to favour growth of bacteria. Dipping dried foods in alkali and sulphuring reduces the number of organisms.

Microbiology During Drying Process

Heat causes a reduction in the total number of microbes but the efficiency varies with kind, number of microbes and the drying process employed. Spores of bacteria and moulds survive the drying process. Vegetative cells of certain heat-resistant bacteria also survive drying.

Microflora After Drying

In adequate conditions, no growth of microbes is seen. During storage, there is slow decrease in number of microbes. There is a high percentage of survivability seen among heat-resistant microbes. Pasteurization reduces number of microbes. During repackaging contamination may be seen.

MICROBIOLOGY OF SPECIFIC DRIED FOODS

Dried fruits	Dried vegetables	Dried eggs	Dried milk
Spores of bacteria and moulds	*E.coli, Enterobacter, Bacillus, Clostridium, Micrococcus, Pseudomonas and Streptococcus, Lactobacillus and Leuconostoc*	Mostly bacteria. Very few in fresh eggs. micrococci, streptococci, coliforms, moulds	Thermoduric streptococci, micrococci and spore formers

To conclude, the need for drying biological materials is very important in the agro-food industry, producing high-quality and shelf stable products. However, there is a downside to the process, as it is a high energy consuming process. This has two drawbacks, the first is the cost of energy, and the second is the environmental degradation that is associated with some types of energy production. In response to these concerns, there has been much work on novel drying techniques to improve energy and drying efficiencies. Some of these novel drying techniques, with the hope that these techniques, along with future research, will produce dryers and drying process that are more economical and less harmful to the environment.

INTERMEDIATE MOISTURE FOODS (IMF)

The systematic investigation and the acquired new knowledge and understanding of water–food interactions and the importance of water activity in food has led to a rediscovery of old techniques of preservation and a resurgence of interest in foods that are made shelf stable by a reduction in their water activity. New methods of production designed to utilize this new understanding have resulted in several generations of new foods requiring less energy for production and distribution. Such foods are generally classified as intermediate moisture foods (IMF). IMF are foods that have a moisture content higher than that of dry foods and are edible without rehydration. Despite their higher moisture, they are designed to be shelf stable without needing refrigeration in distribution. Thermal processing to the extent needed for canning is not required, although some may be pasteurized. IMF have no precise definition based on water content or water activity. Generally, their moisture content is in the range of 10–40%, and their a_W is 0.60–0.90. They have non-refrigerated shelf stability.

Some traditional IMF foods are

- Cakes and pastries
- Jams
- Honey
- Soft candies
- Jellies

These products are low in moisture and can be eaten without preparation or rehydration. They are also called "reduced water activity products". The characteristic properties of IMF offer a number of advantages over conventional dry or high-moisture foods. IMF processing, as well as distribution, generally is significantly less energy-intensive than drying, refrigeration, freezing or canning. Also, IMF technology can potentially lead to higher retention of nutrients and quality than that result from more vigorous processes such as some methods of dehydration and thermal processing. Being easily masticated without an oral sensation of dryness, IMF are suitable for direct consumption with no preparation, offering convenience and further savings of energy. Because of their relatively low moisture, IMF are concentrated in weight and bulk and have high nutrient and caloric density. Because of their plasticity, they can be moulded into blocks of uniform geometry for easy packaging and storage, and they can be formed conveniently into individual servings. IMF can be stored without special precautions for several months. Although appropriate packaging is a factor in prolonged shelf life, packaging requirements for IMF are not as strict as for many other categories of foods. Completely impervious packages are not necessary, and loss of package integrity may not pose a health hazard, especially in environments with average humidity.

These advantageous characteristics of IMF are particularly compatible with the needs of modern consumers for convenient foods of high nutrient density. IMF are especially attractive where the food supply load, ability to resupply and preparation time are limiting factors, as in military settings, space travel, exploration and mountaineering. IMF technology can be an alternative to energy-intensive methods of drying for preservation and storage. Thus, in tropical climates, especially in countries of the Third World where refrigeration is scarce and food spoilage a vital problem, IMF technology will become increasingly important. The Current Good Manufacturing Practice regulations include a provision for IMF. Foods (such as IMF) that rely on the control of a_W for preventing the growth of undesirable microorganisms shall be processed to and maintained at a safe moisture level. Compliance with this requirement may be accomplished by an effective means, including: (1) monitoring the a_W of the food, (2) controlling the ratio between water and soluble solids in the finished food and (3) protecting the finished food from moisture pickup, so that the a_W does not increase to an unsafe level. IMF production methods, technological aspects of IMF processing, and applications and limitations of the IMF technology are presented in the following sections.

Technology of Intermediate-moisture Foods

Humectants   The basic step in the production of an intermediate moisture food is the reduction of water activity of the product to a_W values in the IMF zone. This reduction generally cannot be achieved by simply drying the food, as the resulting texture is too dry for direct consumption. For example, most vegetables and meats would have to be dried until their moisture content is below 15% to achieve an a_W less than 0.85. At this moisture level, rehydration would be required before consumption. This problem is partially solved by the addition of humectants, materials that lower water activity but also allow products to retain their moist properties and give a plastic texture. The use of humectants is a fundamental and characteristic step in the production of IMF, regardless of the specific manufacturing process applied. A number of hygroscopic chemical compounds are presently employed or are being considered for use by the food industry as humectants. Most of them can be classified in to one of the four general categories: salts (mineral and organic), sugars, polyols and protein derivatives. The effectiveness of a humectant in lowering water activity depends on its ability to lower the mole fraction of water as well as interact with and alter the structure of the water in the food system. The water activity of an ideal solution is a direct function of the mole fraction of the solved component. According to Raoult's law

$$a_W = X_w$$

$$= \frac{n_w}{(n_w + n_s)}$$

where X_w is the mole fraction of water, n_w is the total moles of water and n_s is the total moles of the solute or solutes. Thus, the water activity of the aqueous ideal solution depends only on the total number of solute molecules (kinetic units) and not on the nature of the solutes. The smaller the molecular weight of the humectant (solute), the greater its water activity lowering effect (humectancy) per unit of weight dissolved. Dilute aqueous solutions of non-electrolytes exhibit ideal solution behaviour. For example, the water activity of solutions of glucose and glycerol in concentrations up to 4 M and of sucrose in concentrations up to 2 M is predicted accurately (with a deviation of less than 1%) by Raoult's law.

In more concentrated systems, substantial deviations from ideality can be observed. These deviations can be attributed to several factors. The most prevalent are interactions between solute molecules; the unavailability of some of the water in the food (e.g. monolayer water) to act as a solvent; and the binding of solutes to insoluble food components (e.g. proteins), which prevents them from getting into actual solution. These theoretical approaches, although useful for predicting the a_W of solutions, become cumbersome for multisolute mixtures and can give only approximate values for a real food system. To minimize development time for IMF products, a prediction equation is needed for the water activity-lowering effect of humectants in a complex food system containing solids that do not go completely into solution. This need has led to the development of a number of semitheoretical and empirical equations for predicting a_W. The Margules–Van Laar equations, expresses the water activity of a binary system as:

$$\ln a_w = \ln X_w = K_1 X S_1$$

where K_1 is the interaction constant between the solute (S_1) and water; this constant is determined by the slope of $\ln(a_w/X_w)$ versus $X S_1$. Unfortunately, the use of these humectants at levels required to achieve a water activity in the IMF zone results in undesirable flavour. Combinations of these humectants and the addition of less effective or less common humectants are the usual approaches in product development in alleviating the taste problem.

Polymeric humectants, such as high-molecular-weight polyols and water-soluble gums, result in water activity that deviates strongly from that predicted by Raoult's law. This is because they are more effective per unit of hydrophilic groups than compounds with lower molecular weights. The problem with high-molecular-weight humectants is the high viscosity of their solutions. An alternative way of producing soft, pliable textured foods, especially fruits and vegetables, at a shelf stable a_W of 0.6–0.7 is to build up or alter the polymeric matrix of plant material by adding materials that could interact with the natural matrix and lead to increased water retention at a given a_W. Vegetable pieces (green snap beans), which would normally have a poor texture when dried to a_W levels of IMF, were cooked in a number of solutions of naturally occurring substances, equilibrated overnight, drained, and dried to an a_W of 0.6–0.7. The water retention of these IMF pieces was compared to that of control samples of water-cooked pieces. Pieces treated with algin, gum arabic, tapioca starch, glycerol and glycerol oligomers, lactic acid, aloe vera gel and yucca extract had increased water retention and plasticity. Other starches and gums had a negative effect. A combination of such naturally occurring humectants, at concentrations below their off-flavour threshold, can be considered for use in the production of IMF. Similarly, mild lactic acid fermentation was shown to have a very beneficial effect on the texture of the further-processed intermediate-moisture vegetable.

Several oligoglycerols and polyglycerols and their esters were evaluated for their potential as humectants. Although effective in lowering a_W, most of them resulted in unacceptable odour and taste. This could be a result of impurities that were due to the synthesis method used. The economic feasibility of pure polyglycerols having better flavour is a subject of further research.

Novel ingredients could also be successfully used as humectants. Neosugar is an example of such a material. It is a non-nutritive sweetener composed of glucose attached in a α (2-1) linkage to two, three or four fructose units and is accordingly designated GF2, GF3 or GF4. It is produced by the action of a fungal enzyme, fructosyltransferase, on sucrose. The sweetness of neosugar is 0.4–0.6 times that of sucrose. Thus, GF2 with about half the sweetness and 1.5 times the molecular weight of sucrose, would probably result in greater lowering of a_W than an amount of sucrose of equal sweetening power. Its non-caloric property is also desirable in a market of increased demand for lower-calorie foods. Another method, alternative or complementary to the use of humectants, is the addition of surface-active agents or semi-solid fats to plasticize the texture of an IMF. Gelling or emulsion formation in the mixture of an IMF does not lower the water activity but substantially improves the texture of the product. The plasticizing effect of fats is fundamental in the confectionery industry, and the softening of texture by surfactants is widely practiced in the baking industry and has been applied in the production of intermediate-moisture pet foods.

Methods of production   Various production techniques can be used in the manufacture of IMF. They can be classified into four main categories:

Partial drying   It can be used in the production of IMF only if the starting materials are naturally rich in humectants. This is the case with dried fruits (raisins, apricots, prunes, dates, apples and figs) and maple syrup. The a_w of these products is in the range of 0.6–0.8.

Moist infusion or osmotic drying   It involves soaking solid food pieces in a water–humectant solution of lower water activity. The difference in osmolality forces water to diffuse out of the food into the solution. Simultaneously, the humectant diffuses into the food, usually more slowly than water. Salt or sugar solutions are usually employed. This is the method for the production of candied fruits. Also, novel meat and vegetable IMF have been produced by infusion in solutions of salt, sugar, glycerol or other humectants.

Dry infusion   It involves dehydrating solid food pieces and then soaking them in a water–humectant solution of the desired water activity. This process is more energy-intensive than the others, but it results in high-quality products. It has been used extensively in the preparation of IMF for the U.S. Army and the National Aeronautics and Space Administration.

Blending   It consists of weighing and direct mixing of food ingredients, humectants and additives, followed by cooking, extrusion or other treatment that results in a finished product of the desired water activity. This method is fast and energy-efficient and offers great flexibility in formulation. It is used for both traditional IMF (confections, preserves) and novel IMF (pet foods, snacks).

Microbial stability   Microbial stability is of primary concern when the potential success of an IMF formulation is considered. As mentioned above, the ability of microorganisms to grow on a substrate is a function of a_w, pH, temperature, oxidation–reduction potential, preservatives and existing microflora. Most, but not all pathogenic microorganisms are inhibited in the water activity zone of IMF. Numerous microorganisms, of significance to both spoilage and health, have been shown to be able to grow with a_w in the range of 0.6–0.9 when other conditions are favourable (Table 8.5). Thus, additional precautions, besides the adjustment of a_w, must be taken to inhibit or limit the proliferation of these microorganisms in IMF. Numerous pathogenic microorganisms of major concern in foods are effectively inhibited by the reduction of water activity to the IMF zone. Thus, the growth of clostridia is prevented by such reduced water activity, regardless of storage temperature and pH. Nevertheless, growth could conceivably occur during formulation and storage before the reduction of a_w, and therefore good hygienic and manufacturing practices are essential. *Bacillus* species require a minimum a_w of 0.89–0.90 for growth. At IMF water activities, salmonellae cannot multiply, but their resistance to heat is greatly increased, and they may persist in IMF for long periods. Pasteurization of the ingredients before formulation is generally necessary.

One of the major concerns for intermediate moisture foods is *Staphylococcus aureus*. It has been shown that the organism is able to grow at a_w above 0.84–0.85 if the pH is favourable. Formulation of IMF at the highest possible moisture content, for improved texture and palatability, requires additional measures for the inhibition of *S. aureus*. The same is true for moulds. The most often encountered ones, for example, the common *Aspergillus* and *Penicillium* species, can grow at a_w above 0.77–0.85. The minimum a_w for mycotoxin production by these moulds is usually higher (Table 8.3). Several xerophilic and xerotolerant moulds can grow at a_w in the range 0.62–0.64.

Table 8.3   Microorganisms growing within the range of water activity (a_w) of intermediate-moisture foods

a_w	Microorganisms generally inhibited by lowest a_w in the range	Examples of traditional foods with a_w in this range
0.80–0.91	Many yeasts (*Candida, Torulopsis, Hansenula*)	Fermented sausage (salami), sponge cake, dry cheeses, margarine, foods containing 65% (w/w) sucrose (saturated) or 15% NaCl
0.80–0.87	Most moulds (mycotoxigenic penicillia), *Staphylococcus aureus*, most *Saccharomyces* spp., (e.g. *S. bailii*), *Debaryomyces*	Most fruit juice concentrates, sweetened condensed milk, chocolate syrup, maple and fruit syrups, flour, rice, pulses containing 15–17% moisture, fruitcake, country-style ham, fondants, high-sugar cakes
0.75–0.80	Most halophilic bacteria, mycotoxigenic aspergilli	Jam, marmalade, marzipan, glace fruits and some marshmallows

(Contd.)

Table 8.3   (Continued)

a_W	Microorganisms generally inhibited by lowest a_W in the range	Examples of traditional foods with a_W in this range
0.60–0.75	Xerophilic moulds *(Aspergillus chevalieri, A, candidus, Wallemia sebi), Saccharomyces bisporus*	Rolled oats containing about 10% moisture, grained nougats, fudge, marshmallows, jelly, molasses, raw cane sugar, some dried fruits and nuts
0.60–0.65	Osmophilic yeasts *(Saccharomyces rouxii)*, few moulds *(Aspergillus echinulatus, Monascus bisporus)*	Dried fruits containing 15–20% moisture, some toffees and caramels and honey

With *S. aureus* and moulds being of primary concern in IMF, it is not surprising that the majority of microbial studies and tests in these foods have used these types of microorganisms as indicators and challengers.

Combinations of pH, water activities and preservative concentrations that offer adequate protection can be established on the basis of tests with these microorganisms. Effective mould inhibitors often used are sorbates and propionates. Very few inhibitor systems prevented the growth of all three organisms at a pH above 5.4 and an a_W of 0.86–0.90. Only propylene glycol, a humectant with specific antimicrobial activity, achieved complete inhibition. At a high a_W, acidification to pH 5.2, in conjunction with the mould inhibitors, was effective for inhibition of *S. aureus*. At pH 5.2–6, a combination of propylene glycol (4–6%) with either potassium sorbate or calcium propionate (0.1–0.3%) was required. As was pointed out by the researchers, the effectiveness of a microbial protection combination is very system-dependent, and extrapolations to other systems are not always valid.

Organic acids are generally more effective in inhibiting *S. aureus* at high a_W, but an inorganic acid, phosphoric acid, is the most effective at lower a_W, at the upper limit of IMF range. Besides staphylococci, bacteria such as *Streptococcus faecalis* and a *Lactobacillus* species can grow at the upper limit of a_W in IMF. Both can grow at a_W above 0.87–0.88. Sorbate was less effective on these bacteria than on staphylococci. Propylene glycol had an equally inhibitory effect on staphylococci and *Streptococcus faecalis* but had no specific antimicrobial effect (besides the lowering of a_W) on lactobacillus. Yeast growth is another potential problem in IMF. Osmophilic yeasts can grow at a_W down to 0.60. Good manufacturing practices, pasteurization of the mixtures, and the use of chemical preservatives such as sulphites, benzoates, *para*-hydroxybenzoates, sorbates and diethyl pyrocarbonate are the usual control measures. The means used to achieve the final water activity is another factor affecting microbial stability. In one study, IMF was prepared by blending method (desorption) and the dry infusion method (adsorption). In desorption systems, the minimum a_w for the growth of microorganisms, as reported in the literature, could be used to evaluate microbial stability. However, in adsorption samples, the minimum a_w for microbial growth was much higher under the same conditions. *S. aureus* grew in desorption samples but died in adsorption samples of the same a_w. Desorption–adsorption samples having similar moisture contents (i.e., different a_w) showed similar growth and inhibition behaviour. Also, the specific antimicrobial effect of the humectant should be taken into account. The inhibitory effect of propylene glycol and of the aliphatic diols had already been mentioned. A specific bacteriostatic effect on *S. aureus* was shown for low concentrations (2–4%) of ethanol.

A novel approach to microbial stabilization of IMF, with a minimum amount of chemical preservatives, is the use of an optimum distribution of a preservative throughout the food. The more susceptible part of the food (namely, the surface) should have a higher concentration of the preservative than the interior. Temperature changes during distribution and storage can result in local condensation of water on the surface, leading to microbial outgrowth on the surface. Two methods for improving surface stability by maintaining a high concentration of preservatives were demonstrated. The first method involved zein, an impermeable, edible food coating. The second was based on the maintenance of a pH differential between the surface and the bulk of the food. The reduction of surface pH increases the surface availability of the most active form of sorbic acid and other lipophilic acids used as preservatives. A negatively charged macromolecule was immobilized in the form of a component of a surface coating, whereas other molecules, particularly electrolytes, moved freely. A deionized mixture of carrageenan and agarose resulted in a pH differential of up to 0.5 pH units. Both methods were tested and very substantially increased the microbial stability of an IMF with $a_w = 0.88$.

Chemical stability At water activities in the IMF range, chemical reactions increase rapidly and reach a maximum. Because enzymatic activity is usually prevented by enzyme inactivation with an initial thermal treatment, lipid peroxidation and non-enzymatic browning are the major deterioration reactions in IMF. Water has a dual effect on the rate of lipid peroxidation. It can retard oxidation by hydrating or diluting heavy metal catalysts or even precipitating them as hydroxides. Water forms hydrogen bonds with hydroperoxides and slows down the steps of peroxide decomposition. By promoting radical recombination, it can terminate the chain reaction.

On the other hand, water can speed up the reaction by lowering the viscosity, thereby increasing the mobility of reactants and bringing catalysts into solution. It also swells the solid matrix of the system, with the result that new surfaces are exposed for catalysis. These contrary effects, occurring simultaneously, result in a minimum oxidation rate at a water activity close to that corresponding to the monolayer moisture content. In the IMF zone, the promoting action predominates; therefore, the oxidation rate increases with water activity. In systems high in trace metal catalysts, a maximum rate is reached at a_w of 0.75–0.80, followed by a decline at higher a_w, the dilution effect again predominating.

The method used to achieve a given level of water activity must also be considered. Foods prepared by dry infusion (adsorption) oxidize much more rapidly than those prepared by blending (desorption), at the same a_w. Thus, the actual water content is important, as the systems prepared by adsorption have higher moisture contents than those prepared by desorption, because of sorption hysteresis. Lipid peroxidation is a serious problem in IMF, leading to unacceptably rancid products if control measures are not taken. Oxidation can be prevented by the elimination of oxygen through vacuum packing and oxygen-impermeable packaging materials, by antioxidants, or by oxygen scavenger sachets as used in Japan. Fat-soluble free radical scavengers, such as butylated hydroxytoluene (BHT) or butylated hydroxyanisole (BHA), or water-soluble metal chelators, such as ethylene diaminetetra acetic acid (EDTA) or citric acid, may be used as antioxidants. Chelators, although more effective in model systems of high water activity, proved less effective than BHA in actual IMF systems, probably because of binding to proteins. The maximum reaction rate for non-enzymatic browning occurs in the IMF water activity range, usually at a_w of 0.65–0.70. The observed maximum rate of browning can be attributed to a balance of viscosity-controlled diffusion, dilution and concentration effects. At low water activities, the slow diffusion of reactants limits the rate. At higher water activities, faster diffusion enables reactions to occur faster, until dilution of the reactants again slows them down. Also, the higher concentrations of water retard the reversible reaction steps that produce water, e.g. the initial condensation stage. Up to 3.5 mol of water are formed per mole of sugar consumed in the reaction. On the other hand, water may increase deamination reactions, such as the production of furfural or hydroxymethyl furfural, in the browning reaction sequence. The maximum browning rate occurs at different a_w, depending on the humectant used to reduce the water activity. The overall effect of liquid humectants is to shift the maximum to a lower a_w. Liquid humectants influence the rate of browning by acting as solvents and thus increasing reactant mobility at lower moisture contents. However, increasing the viscosity by the addition of viscosity agents, such as sorbitol, can dramatically decrease the reaction rate at all a_w. Non-enzymatic browning, although sometimes desirable, as in the production of confectionery and bakery products, has deleterious effects on the shelf life of IMF. It results in loss of protein quality and the undesirable production of off-flavours and dark pigments. Loss of protein quality refers mainly to the loss of the essential amino acid lysine via reactions involving its free α-amino group. When an IMF formulation is considered, the reactivity of the ingredients with respect to non-enzymatic browning must also be considered. The use of reducing sugars (especially pentoses) and amino acids as humectants should be avoided when the Maillard reaction is a major concern. Methods of controlling non-enzymatic browning, besides the use of low-reactive ingredients, include lowering the pH, maintaining low storage temperatures and adding sulphites. The use of sulphiting agents is a subject of debate, because of their potentially harmful effects on sulphite-sensitive individuals, and alternative anti-browning agents are being investigated.

Future of Intermediate Moisture Food

The chief areas for future research in the IMF field can be summarized as follows:

1. The development of new formulations containing humectants of high organoleptic acceptability

2. The development of new antimicrobial agents suitable for IMF and new methods to limit chemical deterioration and loss of quality during storage

3. The development of improved processes for large-scale production

No sudden growth or major breakthroughs in the commercialization of IMF products are expected in the near future. With the possible exception of foods produced by freeze-flow technology, new IMF products will be introduced slowly but steadily. With the life style in the United States demanding more foods that are easy to eat, easy to store and prepare, individually sized and nutritionally dense, IMF technology can successfully fill a real need. It is up to the product development scientists to innovatively use known principles and come up with improved or new IMF products. IMF technology will most likely be used to produce foods of higher a_w (0.9–0.96) where only a short shelf life (four to six weeks) is needed. This concept is already being used by the Japanese.

1. Explain the term 'water activity'.
2. Discuss the solute effects on microbial growth and/or death.
3. Explain the fundamentals of drying.
4. Give an account of the prevalent and alternative methods of drying.
5. What are the major food deterioration reactions pertaining to drying?
6. What is freeze drying?
7. Explain the effect of drying on microbes.
8. Explain the Intermediate Moisture Foods in detail.

9

IRRADIATION

INTRODUCTION

Many processing methods have been developed to prevent food spoilage and improve safety. The traditional methods of preservation, such as drying, smoking and salting have been supplemented with pasteurization (by heat), canning (commercial sterilization by heat), refrigeration, freezing and chemical preservatives. Food irradiation is another technology that can be added to this list. It is not new; interest was shown in Germany in 1896 and it began in the early 1920s, while in the 1950/60s the US Army Natick Soldier Centre (NATICK) experimented with both low-dose and high-dose irradiation for military rations. In the UK, at the same time, the Low Temperature Research Station programme concentrated on low-dose pasteurization. Irradiation is extensively used in the medical field for sterilizing instruments, dressings, etc. Food irradiation is the process of exposing food to a carefully controlled amount of energy in the form of high-speed particles or rays. These occur widely in nature and are included among the energy reaching earth all the time from the sun. While the knowledge of how to produce them originated from research into nuclear energy many years ago, modern methods are available which are straightforward and safe.

The discovery of X-rays by W.K. Roentgen in 1895 and the discovery of radioactive substances by H. Becquerel in 1896 led to intense research of the biological effects of these "radiations." Initially, most of the irradiations made use of X-rays, which are produced when electrons from an electron accelerator are stopped in materials. These early investigations laid the foundation for food irradiation. Ionizing radiation was found to be lethal to living organisms soon after its discovery. The use of this lethality to control spoilage and other organisms that contaminate foods was demonstrated in the early decades of the 20th century.

However, no commercial development of this use occurred then, due to the difficulty to obtain ionizing radiation in quantities needed and high costs.

In the mid 1940s, the interest in food irradiation was renewed when it was suggested that electron accelerators could be used to preserve food. However, the accelerators in those days were rather costly and too unreliable for industrial application. From 1940 through 1953, exploratory research in food irradiation in the United States was sponsored by the Department of the Army, the Atomic Energy Commission, and private industry. Early research in the late 1940s and early 1950s investigated the potential of 5 different types of radiation (ultra-violet light, X-rays, electrons, neutrons, and alpha particles) for food preservation. Researchers concluded at that time that only cathode ray radiation (electrons) had the necessary characteristics of efficiency, safety, and practicality. They considered X-rays to be impractical because of the very low conversion efficiency from electron to X-ray that was possible at that time. Ultraviolet light and alpha particles were considered to be impractical because of their limited ability to penetrate matter. Neutrons exhibited great penetration and were very effective in the destruction or inactivation of bacteria, but were considered inappropriate for use because of the potential for inducing radioactivity in food.

In the 1940s, as described, sources of proper kinds of ionizing radiation became available. The first sources were machines that produced high-energy electron beams of up to 24 million electron volts. This energy was sufficient to penetrate and sterilize a 6-inch No. 10 can of food when electron beams were "fired" from both sides of the can. Also in this same decade, man-made radionuclides such as Cobalt-60 and Cesium-137 (which in their radioactive decay emit gamma rays) became available through the development of atomic energy. The availability of these sources stimulated research in food irradiation for commercial process. The early development of food irradiation is given in Table 9.1.

Table 9.1 Food irradiation—some major milestones

1895	Von Roentgen discovered X-rays
1896	Becquerel discovered radioactivity. Minsch published proposal to use ionizing radiation to preserve food by destroying microorganisms.
1904	Prescott published studies at MIT on bactericidal effects of ionizing radiation
1905	US and British patents issued for use of ionizing radiation to kill bacteria in foods.
1905–1920	Much research conducted on the physical, chemical and biological effects of ionizing radiation

(Contd.)

Table 9.1 (Continued)

1921	USDA researcher Schwartz published studies on the lethal effect of X-rays on *Trichinella spiralis* in raw pork
1923	First published results of animal feeding studies to evaluate the wholesomeness of irradiated foods
1930	French patent issued for the use of ionizing radiation to preserve foods
1943	MIT group, under US army contract, demonstrates the feasibility of preserving ground beef by X-rays
Late 1940 and early 1950s	Beginning of era of food irradiation development by US Government (among Atomic Energy Commission, industry, universities and private institutions) including long-term animal feeding studies by US Army and Swift and Company
1950	Beginning of food irradiation program by England and numerous other countries

Source *Irradiated Food* (3rd edition), American Council on Science and Health. 1988.

WHAT IS FOOD IRRADIATION?

This refers to microbial destruction without the generation of high temperature and is also referred to as *'cold sterilization'*. Much of the food preservations use ultraviolet, ionizing radiation and microwave heating. In particular, radiations can be divided into electromagnetic and particulate. Food irradiation preserves meat, spices, seasonings, potatoes, fresh fruits and vegetables and poultry with high-energy gamma rays to improve product safety and shelf life. This method of preservation prevents growth of food poisoning bacteria, destroys parasites, and delays ripening of fruits and vegetables. Food irradiation could be used to reduce or replace chemical preservatives used in foods. More than 40 years of research on food irradiation has established that foods exposed to low-levels of irradiation are safe and wholesome, and they retain high quality. Table 9.2 provides the different electromagnetic radiation techniques available.

Table 9.2 Electromagnetic radiation and its particulates

Electromagnetic radiation	Particulates
γ-rays	α-rays
X-rays	β-rays
UV rays	Neutrons
Visible	Protons
Infra red	High speed electrons
Microwave	

Radiation is measured in Rad and Megarad. 1 Rad is the unit of radiation dosage equivalent to absorption of 100 ergs/g of irradiated material. A dose of one Rad is obtained when 0.01 Joules of radiation energy is absorbed/kg of material. Recent formulation is Gray. 1 Gray = 100 Rads. Radappertization is beneficial when refrigeration and frozen storage are not available.

Types of Irradiation

There exist different types of irradiation techniques. They are:

1. *Radappertization/Radiation sterilization*   Implying high dose treatments for increasing shelf life. It is the destruction of all of the organisms. Spores of *C. botulinum* are very resistant to irradiation hence the 12 D concept is used to produce commercially sterile food.

2. *Radurization*   The use of low doses of a radiation to destroy a sufficient number of microbes and enhance the storage life of the product. It is also referred to as Radiation pasteurization. The doses of radiation used ranges from 100–1000 K Rad and destroys 90–99 % of microbes. Some heat-resistant vegetative cells and spores survive. Hence refrigeration is essential to retard microbial growth. Dose used is the maximum amount of irradiation which produces no detectable change in the product. Psychrophilic pseudomonads are very sensitive to this kind of treatment. *C. botulinum* type E (fishes) survive this irradiation treatment.

3. *Radicidation*   Low level irradiation treatment to destroy viable non-spore forming pathogenic organisms to reduce the problem of food-borne illness. Spores of *C. botulinum* or *C. perfringens* are not destroyed. This term was suggested for elimination of salmonellae from food and feed.

4. *Thermoradiation*   Combination of heat and radiation. Enzymes and spores of *C.botulinum* are resistant to radiation but are more sensitive to heat. In contrast, thermophilic spores are very heat-resistant but sensitive to radiation. Hence a combination of heat and radiation is of great value in food industry. Preheating sensitizes vegetative cells to irradiation. And preirradiation sensitizes spores to heat treatment.

IONIZING RADIATION USED FOR FOOD IRRADIATION

Energy exists in the form of waves and is defined by its wavelength. As the wavelength gets shorter, the energy of the wave increases. Electric power, radio and television, microwaves (radar) and light have longer wavelengths and lower energies. They cause molecules to move but cannot structurally change the atoms in those molecules. Ionizing radiation (gamma rays, X-rays) has a very short wavelength and higher energy, enough to change atoms by knocking-off an electron from them to form an ion, but not enough to split atoms and cause exposed sources to become radioactive. Therefore, the sources of radiation allowed for food processing (Cobalt-60, Cesium-137, accelerated electrons and X-rays) cannot make food radioactive.

Electron beam irradiation   High-voltage electron beams (accelerated electrons) generated from linear accelerators are an alternative to radioisotope generators. They lack the penetration depth of gamma irradiation (about 0.5 cm.) per 1,000,000 electron volts (MeV) of energy, however, they require much shorter exposure times (seconds vs. hours for gamma irradiation) to be effective. Electron beam irradiation is currently being used to disinfest grain at 1.4 MeV and to pasteurize frozen mechanically separated meat products with 10 MeV. Irradiation research facilities at Iowa State University's Meat Irradiation Technology Centre features a 10 MeV linear accelerator that can switch from electrons to X-ray beams.

X-rays   X-rays are generated when electrons from an electron beam bombard a heavy metal target such as tungsten, have a greater penetration depth but are less desirable because of the low energy conversion efficiency of electrons to X-rays.

Gamma rays   Gamma rays used for irradiation processing of food are radioactive fission products of Cobalt-60 and Cesium-137. Gamma rays have good penetration, as do X-rays. With "cross-firing", they can easily deliver a uniform (less than 25% overdose) energy. Cobalt-60 is not a waste product from the nuclear industry. It is specifically manufactured for use in radiotherapy, sterilization of medical products, and the irradiation of food. Cesium-137 is one of the fission products contained in used fuel rods. It must be extracted in reprocessing plants before it can be used as a radiation source. Currently, almost all radiation facilities in the world use Cobalt-60 rather than Cesium-137.

Gamma rays are more powerful than the rays emitted by a microwave oven. Rays from a microwave oven cause food to heat rapidly, whereas gamma rays, with much shorter wavelengths and higher frequencies, penetrate through the food so rapidly that no heat is produced. After food is irradiated, it is stored and may be transported back to the processing plant for further handling and packaging. Once the food has been irradiated, it must be handled appropriately to prevent recontamination.

Comparison of Gamma Rays, X-rays and Cathode Rays

Gamma rays, X-rays, and electron beams are equally effective in sterilization for equal quantities of energy absorbed. The greatest drawback at present to the use of X-rays in food preservation is the low efficiency and high cost of production. For this reason, most research has concentrated on the use of gamma photons and electron beams. Gamma rays have a maximum of

10 to 25% utilization efficiency, while the maximal efficiency of electrons from electron beam generators ranges between 40 and 80% (depending on the shape of the irradiated material). Radioactive sources of gamma rays (Cobalt-60 or Cesium-137) decay steadily and hence weaken with time. This requires constant replenishment which is expensive.

Like X-rays, e-beams are machine-generated using ordinary electricity and can be powered on and off at the touch of a switch. E-beams offer extremely rapid and cost-effective processing, but in some cases sacrifice penetration depth depending on product density. Treatment of food using either X-rays or electron beams are occasionally referred to as "electronic pasteurization" or "electronic irradiation" methods because they are derived from electricity. The use of electrons from electron beam generators presents fewer health problems than the use of gamma rays, since electron beams are directional and less penetrating, can be turned off for repair or maintenance work, and present no hazard of radioactive materials after a fire, explosion, or other catastrophe. Gamma rays are emitted in all directions, are penetrating, are continuously emitted, and come from radioactive sources. Gamma rays require more shielding to protect workers.

The one overriding requirement for an energy source to be employed in food irradiation is that the energy levels must be below those that could possibly cause the food to become radioactive. After that requirement is met, sources are considered on the basis of their practical and economic feasibility. Machine sources must produce radiation with relatively simple technology. Isotopes must have sufficient long life and emit penetrating radiation.

Regardless of the source of ionizing energy, the food is treated by exposing it to the energy source for a short time period. In the case of e-beam, food is irradiated in just a few seconds, while it takes more time with gamma and X-rays. The food is never in contact with the energy source; the ionizing energy merely penetrates into the food but does not stay in the food. It takes very little energy to destroy harmful bacteria. At these levels there is no significant increase in temperature or change in composition. Irradiation does not make food radioactive nor does it leave any residues. The levels of ionizing energy used to treat foods for pathogen reduction or disinfestation are measured in kiloGrays (kGy). A low-to-medium dose of 1–10 kGy is usually sufficient to render a product safe from harmful bacteria or insects such as fruit flies, while causing little or no effects on product quality or nutrition.

The most significant public health benefit of food irradiation is that it stops the spread of food-borne disease. It greatly reduces or eliminates the number of disease-causing bacteria and other harmful organisms that threaten us and our food supply. Many of these organisms, including *Salmonella, Escherichia coli* O157:H7, *Staphyloccoccus aureus, Listeria monocytogenes, Campylobacter jejuni* and *Toxoplasma gondii* have caused many outbreaks of food-borne illness. When food is irradiated, the penetrating energy breaks down the DNA molecules of the harmful organisms. The food is left virtually unchanged, except that it is much safer because the number of harmful organisms is greatly reduced or eliminated. An added advantage is that food can be irradiated in its final packaging—fresh or frozen, which prevents the possibility of contamination in the distribution system, at the store, or even in the home, prior to the package being opened. Although reduction of disease-causing bacteria is of greatest importance to public health and safety, there are other significant benefits of food irradiation. Irradiation can also help keep meat, poultry and seafood fresh longer by reducing the level of spoilage-causing microbes. It also allows consumers to keep certain fruits and vegetables fresh longer. For example, irradiated strawberries stay unspoiled for up to three weeks, while untreated berries stay unspoiled for 3–5 days. For many developing countries, food spoilage is an ever-present and costly reality, often causing spoilage rates in excess of 40 per cent. In these countries, irradiation stands to benefit millions by providing more nutritious fruits and vegetables to consumers. When grains and spices, fresh and dried fruits, legumes and condiments are irradiated, the process eliminates any insects that might be present and can replace the use of chemical fumigants, which could leave residues or harm the environment. For example, irradiation is used as an alternative to chemical fumigation or vapour heat processes for treating fruits, to meet quarantine requirements for international trade in fresh fruits and vegetables.

It is important to note that toxins, viruses or bacterial spores are resistant to irradiation. Therefore, it is essential that irradiation be used in conjunction with all other established safe food handling and good manufacturing practices.

Table 9.3 Irradiation conversion table

1000000 rads	1 megarad (Mrad)
1 gray (Gy)	100 rads
1 kilogray (kGy)	100000 rads
1 kGy	100 kilorads (Krads)
1 kGy	0.1 Mrad
10 kGy	1 Mrad

The irradiation dose applied to a food product is measured in terms of kilograys (kGy) (Table 9.3). One kilogray is equivalent to 1,000 grays (Gy), 0.1 megarad (Mrad), or 100,000 rads. The basic unit is the gray, which is the amount of irradiation energy that 1 kilogram of food receives.

he amount of irradiation applied to a food product needs to be carefully controlled and monitored. The irradiation dose applied to a food product will depend upon the composition of the food, the degree of perishability, and the potential to harbour harmful microorganisms. The amount of radiation that a food product absorbs is measured by a dosimeter.

FOODS CURRENTLY BEING IRRADIATED

Internationally, foods such as apples, strawberries, bananas, mangoes, onions, potatoes, spices and seasonings, meat, poultry, fish, frog legs, and grains have been irradiated for many years. In Japan, more than 20,000 pounds of potatoes are irradiated each year to prevent sprouting. In the Netherlands, more than 18,000 pounds of foods such as strawberries, spices, poultry, dehydrated vegetables, and frozen products are irradiated daily. Belgium irradiates more than 8,000 tons of food per year. Canada has approved the irradiation of potatoes, onions, wheat flour, fish fillets, and spices and seasonings. In the United States, spices and seasonings have been approved by the Food and Drug Administration (FDA) to be irradiated up to 30 kGy to reduce the number of microorganisms and insects. Today more than 35 countries have approved irradiation of nearly 40 different food products.

SENSITIVITY AND RESISTANCE OF MICROBES
TOWARDS IONIZING RADIATIONS

This is given by the D value which denotes the radiation dose (kGy) necessary to reduce the viable cell number by 90%. Bacterial spores are generally most resistant. Most resistant vegetative cell is that of *Deinococcus radiodurans* other than prions. Among the clostridial spores, *C. botulinum* type A and B are the most resistant. Type E is highly sensitive. Among the *Bacillus* spp., *B. pumilus* is the most radiation resistant. In general, multicellular organisms are more sensitive to radiation than the unicellular organisms.

Gram-negative bacteria are more sensitive than the gram-positive bacteria. Viruses exhibit high resistance to radiation and prions with no nucleic acid are extremely resistant (D value of scrapie agent is approximately 50 kGy; lethal dose to humans <10Gy)

Enzymes, pyrogens, toxins, antigens of microbial origin are very radiation resistant compared to living cells. Hence the number of microbes present prior to a radiation sterilization is of importance when dealing with medical products.

Mechanism of lethal action

- Direct ionization of a vital cellular molecule (DNA)

- Indirectly through the reaction of the free radicals produced in the cellular fluids.

Main target site

- Nucleic acids
- Enzymes involved in nucleic acid repair and replication

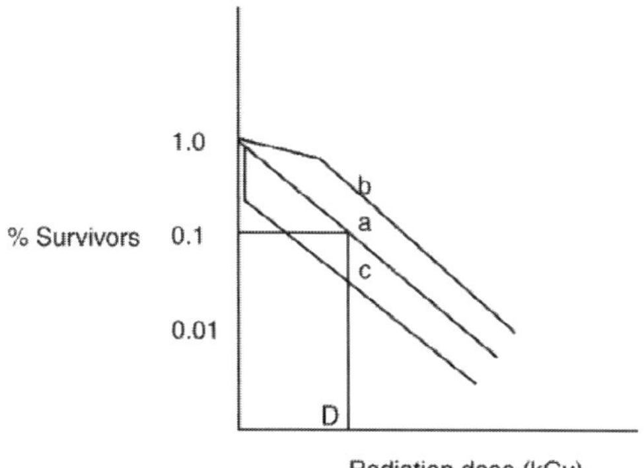

Figure 9.1   Death kinetics due to irradiation

In the graph (Figure 9.1) curve 'a' indicates that a single 'hit' on the sensitive target site, DNA is responsible for cell death. An initial lag period before an exponential rate of killing 'b' suggests that multiple hits on DNA are needed to inactivate DNA or that DNA repair mechanisms protect the organism. Curve 'c' represents exponential death followed by tailing off, i.e., decrease in sensitivity due to development of mutant strains—seen rarely in spore forming bacteria.

DNA damage   Ionizing radiation induce structural damage in microbial DNA which unless repaired will inhibit DNA synthesis or cause some error in protein synthesis leading to cell death. These radiations are highly reactive, producing short-lived OH^+ radicals in water within microbes which cleave phospodiester bonds in DNA leading to single-stranded breaks (SSB) or double-stranded breaks (DSB). Damage to the sugars and bases may also occur as for example, the production of 5,6-dihydroxy 5,6-dihydrothymine from thymine.

DNA repair   Mechanisms involved in DNA repair involve excision of modified bases and scaling SSBs and DSBs through recombination with undamaged daughter strands. DSBs are more difficult to repair than SSBs. Very few DSB are tolerated in organisms like *E. coli* but they occur readily in radiation-resistant bacteria such as *D. radiodurans*. Lethality due to ionizing radiation, as proposed by the target theory, occurs when the irradiated microorganisms are destroyed by the passage of an ionizing particle or quantum of energy through, or in close proximity to, a sensitive portion of the cell. This direct "hit" on the target causes ionization in this sensitive region of the organism or cell and subsequent death. It is also assumed that much of the germicidal effect results from the ionization of the surroundings, especially water, to yield free radicals, some of which may be oxidizing or reducing and therefore helpful in the destruction of the organisms. This effect is reduced if food is irradiated in the frozen state. Irradiation may also cause mutations in the organisms present.

Bacterial spores are more resistant to ionizing radiation than vegetative cells. Gram-positive bacteria are more resistant than gram-negative bacteria. The resistance of yeasts and moulds varies considerably, but some are more resistant than most bacteria.

The bactericidal efficacy of a given dose of irradiation depends on the following:

1. The kind and species of the organism (Table 9.4).
2. The numbers of organisms (or spores) originally present. The more organisms there are, the less effective a given dose will be.
3. *The composition of the food*   Some constituents [e.g. proteins, catalase, and reducing substances (nitrites, sulphites, and sulphydryl compounds)] may be protective. Compounds that combine with the SH groups would be sensitizing.
4. *The presence or absence of oxygen*   The effect of free oxygen varies with the organism, ranging from no effect to sensitization of the organism. Undesirable "side reactions" are likely to be intensified in the presence of oxygen and to be less frequent in a vacuum.
5. *The physical state of the food during irradiation*   Both moisture content and temperature affect different organisms in different ways.
6. *The condition of the organisms*   Age, temperature of growth and sporulation, and state (vegetative or spore) may affect the sensitivity of the organisms.

Table 9.4   Approximate killing doses of ionizing radiations in kilorays (kGy)

Organisms	Lethal dose (kGy)
Insects	0.22 to 0.93
Viruses	10 to 40
Yeasts (fermentative)	4 to 9
Yeasts (film)	3.7 to 18
Moulds (with spores)	1.3 to 11

(Contd.)

Table 9.4 (Continued)

Organisms	Lethal dose (kGy)
Bacteria (cells of pathogens)	
Mycobacterium tuberculosis	1.4
Staphylococcus aureus	1.4 to 7.0
Corynebacterium diphtheriae	4.2
Salmonella spp.	3.7 to 7.8
Bacteria (Cells of saprophytes)	
Gram-negative	
Escherichia coli	1.0 to 2.3
Pseudomonas aeruginosa	1.6 to 2.3
Pseudomonas fluorescens	1.2 to 2.3
Enterobacter aerogenes	1.4 to 1.8
Gram-positive	
Lactobacillus spp.	0.23 to 0.38
Streptococcus faecalis	1.7 to 8.8
Leuconostoc dextranicum	0.9
Sarcina lutea	3.7
Bacteria (spores)	
Bacillus subtilis	12 to 18
Bacillus coagulans	10
Clostridium botulinum type A	19 to 37
Clostridium botulinum type B	15 to 18
Clostridium perfringens	3.1
Putrefactive anaerobe 3679	23 to 50
Bacillus stearothermophilus	10 to 17

Source *Food Microbiology* (4th edition), W.C., and Westhoff D.C. 1988. McGraw-Hill. New York.

IMPORTANCE OF SURVIVING BACTERIA IN LOW-DOSE IRRADIATED FOOD

It is known that some microorganisms grow in foods and spoil or deteriorate them. The growth of these microorganisms also inhibits the growth of some pathogenic microorganisms. Therefore, there must be some "spoilage survivors" to inhibit the outgrowth of pathogens from surviving spores. For example, radiation survival curves were determined for 7 strains of

Enterococcus faecium, 10 strains of *Enterococcus faecalis,* and 8 strains of the proteolytic variety of *E. faecalis.* The D values (i.e., the doses giving 90% reduction of viable counts) ranged from 5.0 to 47 kGy for *E. faecium* strains, 3.5 to 21 kGy for the *E. faecalis* strains, and 3.0 to 4.5 kGy for the proteolytic variants of *E. faecalis* strains. The survival curves were linear for most strains but some exhibited non-linear trends. Useful radiation-resistant strains of group D streptococci, may find application in low-dose irradiated foods for preventing toxin formation by pathogenic microorganisms such as *Clostridium botulinum* in bacon and other foods.

Although low-dose irradiation (3 kGy or less) is effective in destroying most harmful bacteria, it does not prevent the growth or toxin production of *Clostridium botulinum,* the organism that produces the deadly toxin that causes botulism. Much higher irradiation doses, up to 30 to 60 kGy, are needed to destroy this organism in foods. Sodium nitrite, a food additive used in cured meats to prevent botulism, could be reduced as much as 66 per cent (120 mg/kg to 40 mg/kg) with an irradiation dose of 7.5 kGy. These products were found to have an excellent shelf life of more than 90 days at 39°F and exhibit good odour, flavour, texture, and colour. The dose of irradiation needed to eliminate microorganisms in food will depend upon the type, amount, and growth stage of the microorganisms present and the properties of the food including moisture, pH, temperature, oxygen present, and nutrient composition. Irradiation suppresses the microbiological contamination of foods and cannot be used to cover up spoiled foods. Thus irradiation combined with good food-handling practices would reduce the incidence of food-borne disease.

USES OF FOOD IRRADIATION

Ionizing radiation can be used to process food. Its effect on the food is dependent on the dose level (amount) of irradiation to which the food has been subjected. High-dose levels of irradiation (20 to more than 70 kGy) can be used to sterilize foods by eliminating all vegetative microorganisms and spores in the food. Very low doses of irradiation (less than 0.1 kGy) can be used to inhibit sprouting in potatoes, onions and garlic. Irradiated foods are used by astronauts during space travel.

Low doses have also been shown to be as effective as pesticide fumigants for deinfesting grain products prior to shipment and storage, and for reducing microbial and insect contamination on fresh fruits and vegetables.

Not all fresh products is suitable for irradiation. The shelf life of mushrooms, potatoes, tomatoes, onions, mangoes, papayas, bananas, apricots, strawberries, and figs can be extended with low-dose irradiation with no loss in quality. However, the quality of some foods (some citrus fruits, avocados, pears, cantaloupes, and plums) is actually degraded by irradiation.

Pasteurizing doses of irradiation can kill or reduce the populations of both food spoilage and pathogenic microorganisms in food. For example, *Salmonella* spp. and *Campylobacter jejuni* can be eliminated from poultry, and trichinae from pork. *Escherichia coli* O157:H7, *Salmonella, C. jejuni, Listeria monocytogenes,* and *Staphylococcus aureus* can be eliminated in uncooked ground beef.

Different doses (levels) of radiation are used for different purposes as shown in Table 9.5.

Table 9.5 Applications of food irradiation

Type of food	Radiation doses in kGy	Effect of treatment
Meat, poultry, fish, shellfish, some vegetables, baked goods, prepared foods	20 to 71	Sterilization. Treated products can be stored at room temperature without spoilage. Treated products are safe for hospital patients who require microbiologically sterile diets.
Spices and other seasonings	Up to 30	Reduces number of microorganisms and insects. Replaces chemicals used for this purpose.
Meat, poultry, fish	0.1 to 10	Delays spoilage by reducing the number of microorganisms in the fresh, refrigerated product. Kills some types of food poisoning bacteria and renders harmless disease-causing parasites (e.g. trichinae).
Strawberries and some other fruits	1 to 5	Extends shelf life by delaying mould growth.
Grain, fruit, vegetables and other foods subject to insect infestation	0.1 to 2	Kills insects or prevents them from reproducing. Could partially replace post-harvest fumigants used for this purpose.
Bananas, avocados, mangoes, papayas, guavas and certain other non-citrus fruits	Up to 0.05	Delays ripening
Potatoes, onions, garlic, ginger	0.05 to 0.15	Inhibits sprouting
Grain, dehydrated vegetables, other foods	Various doses	Desirable changes (e.g. reduced rehydration times)

Source Irradiated Food (3rd edition), American Council on Science and Health. 1988.

EFFECT OF IONIZING RADIATION ON NUTRIENTS IN FOOD

When foods are exposed to ionizing radiation under conditions envisioned for commercial application, no significant impairment in the nutritional quality of protein, lipid and carbohydrate constituents was observed. Irradiation is no more destructive to vitamins than other food preservation methods.

It was noted that there were small losses of vitamin E and thiamine. According to a study, thiamine in pork is not significantly affected by the FDA-approved maximum radiation dose to control *Trichinella*, but at larger doses, it is significantly affected. Protection of nutrients is improved by holding the food at low temperature during irradiation and by reducing or excluding free oxygen from the radiation environment. This is accomplished by irradiating vacuum-packaged foods at temperatures below 0°C (32°F).

The effect of irradiation on retention of vitamin E (alpha tocopherol) in chicken breasts was determined when the chicken breasts were irradiated in air with a Cesium-137 source at 0, 1, 3, 5.6, and 10 kGy at 0° to 2°C (32.0° to 35.6°F). The fresh muscle tissue was saponified and the total tocopherols were isolated and quantitated using normal phase high performance liquid chromatography with a fluorescence detector. Gamma irradiation of the chicken resulted in a decrease in alpha tocopherol with increasing dose. At 3 kGy and 2°C, the radiation level approved by the FDA to process poultry, there was a 6% reduction in alpha tocopherol level. No significant changes were observed for gamma tocopherol. Free radical scavengers were tested for their ability to reduce the loss of thiamine and riboflavin in buffered solutions and in pork during gamma irradiation. In aqueous solution, the tested compounds were twice as effective for the protection of riboflavin as for the protection of thiamine. The presence of nitrous oxide doubled the rate of loss of thiamine and riboflavin in solution, indicating a predominance of reactions with hydroxyl radicals. In buffered solutions, niacin was not affected by gamma radiation unless either thiamine or riboflavin was present, in which case, the niacin was destroyed rather than the other vitamin. Ascorbate, cysteine, and quinoid reductants were demonstrated to be naturally present in sufficient quantities to account for the lower rates of loss of thiamine and riboflavin observed during irradiation of pork meat, as compared to irradiation in buffered solution.

A study was made of thiamine content of the skeletal muscles and livers of pork, chicken, and beef after gamma irradiation. Gamma irradiation from a Cesium-137 source was used to irradiate the samples with doses of 0, 1.5, 3, 6, and 10 kGy at 2°C (35.6°F). Samples were also titrated with dichlorophenoindophenol to determine the reducing capacity of the tissue. The rate of loss of thiamine upon irradiation was found to be about 3 times as fast in skeletal muscle as in liver and to be a function of the reducing capacity of the tissues, the loss decreasing with increasing reductant titer. For the same amount of thiamine loss, liver could be irradiated to 3 times the dose as could muscle.

PACKAGING FOR IRRADIATED FOODS

Since food is often packaged before irradiation, it is possible that irradiation might either affect barrier properties or that radiolytic products formed in the package might be absorbed into the product. Ionizing radiation rarely creates new compounds and any novel compounds are generally of low molecular weight. These possibly already existed in the unirradiated polymer but are desorbed from the polymer over time. Any product unique to irradiation will only occur at very low levels. A benchmark level of 0.5 ppb for non-carcinogenic compounds has been set; below which level the compounds are considered too insignificant to warrant regulatory concern. The original packaging materials (polyolefins, polystyrene, cellophane, vinylidene chloride copolymers, etc.) were approved back in 1964. In 2001 the FDA determined gamma rays, X-rays and e-beam to be equivalent in terms of types and levels of radiolytic products generated in the packaging materials used in packaged foods before irradiation.

EFFECT OF IONIZING RADIATION ON MEATS

Various parasites cause human illnesses and food is an important vehicle for some of these parasitic diseases.

Treating fresh or frozen meats with ionizing radiation is an effective method to reduce or eliminate several of the food-borne human pathogens such as *Salmonella, Campylobacter, Listeria, Trichinella*, and *Yersinia*. It is possible to produce high-quality, shelf stable commercially sterile meats. Irradiation dose, processing temperature, and packaging conditions strongly influence the results of irradiation treatments on both microbiological and nutritional quality of meat. These factors are especially important when irradiating fresh non-frozen meats. Radiation doses up to 3.0 kGy have little effect on the vitamin content, enzyme activity, and structure of refrigerated non-frozen chicken meat, but have very substantial effects on food-borne pathogens.

Ionizing (gamma) radiation can be used as an alternative to thermal processes for the preservation of food. Research studies of the uses of ionizing radiation to extend the safety of processed meats were undertaken. Radiation research studies of meat products included: bacon, ham, frankfurters, corned beef and pork sausage, and beef, chicken and pork. Beef, chicken and pork were organoleptically acceptable after irradiation *in vacuo* at −30°C ± 10°C. 12 D doses for the process were 4.12, 4.27, and 4.37 Mrad for beef, chicken, and pork loin, respectively. It was also reported that while sublethal radiation doses enhance the safety and storage life of raw beef, chicken, and pork, it may actually increase the spoilage rate in cured meat products.

Gamma radiation doses of 0.26 kGy and 0.36 kGy, administered *in vacuo* at 0°C (32°F), destroyed 90% of log-phase and stationary-phase colony forming units (cfu) of *S. aureus*, respectively, in mechanically deboned chicken meat (MDCM). Samples inoculated with 103.9 cfu/g of *S. aureus* were treated with gamma radiation *in vacuo* at 0°C (32°F) and then held for 20 hours at 35°C (95.0°F) (abusive storage). Enterotoxin was not detected in irradiated MDCM. A predictive equation was developed for the response of *S. aureus* in MDCM to radiation dose and irradiation temperature.

E. coli O157:H7 had a significantly higher D value when irradiated at –17° to –15°C (1.4° to 5.0°F), compared to treatment at 3° to 5°C (37.4° to 41.0°F). At a given temperature, the level of fat in beef did not have an effect on D values. *Salmonella* behaved similarly to *E. coli* O157:H7 in low-fat beef, but temperature had no effect on D values when the pathogen was irradiated in high-fat beef. Significantly higher D values were calculated for *C. jejuni* in frozen compared to refrigerated low-fat beef. The pathogen was more resistant to irradiation in low-fat beef when treatment was done at –17° to –15°C (–1.4° to 5.0°F). Neither the level of fat nor the treatment temperature significantly affected the D values for *L. monocytogenes* and *S. aureus*. Considering all combinations of fat level and treatment temperature, the ranges of D values (kGy) (Table 9.6) were, in ascending order of irradiation resistance.

Table 9.6   D values of certain pathogens

Campylobacter jejuni	0.175 to 0.235 kGy
Escherichia coli O157:H7	0.241 to 0.307 kGy
Staphylococcus aureus	0.435 to 0.453 kGy
Listeria monocytogenes	0.507 to 0.610 kGy
Salmonella	0.618 to 0.800 kGy

A study was reported on the sensitivity of *E. coli* O157:H7 suspended in beef or mechanically deboned chicken meat (MDCM) to gamma irradiation and also to determine the influence of processing parameter such as atmosphere or temperature on that sensitivity. Undercooked and raw meat has been the reasons for the outbreaks of haemorrhagic diarrhoea due to the presence of *E. coli* O157:H7; therefore, treatment with ionizing radiation was investigated as a potential method for the elimination of this organism. Response-surface methods were used to study the effects of irradiation dose (0 to 2.0 kGy), temperature [–20° to +20°C (– 4.0° to 68.0°F)], and atmosphere (air and vacuum) on *E. coli* O157:H7 in mechanically deboned chicken meat. Differences in irradiation dose and temperature significantly affected the results. 90% of the viable *E. coli* in chicken meat was eliminated by doses of 0.27 kGy at +5°C (41.0°F) and 0.42 kGy at –5°C (23.0°F). Small but significant differences in radiation resistance by *E. coli* were found when finely ground lean beef rather than chicken was the substrate. Unlike non-irradiated samples, no measurable verotoxin was found in finely ground lean beef that had been inoculated with 104.8 cfu *E. coli* O157:H7 per gram, irradiated at a minimum dose of 1.5 kGy, and temperature abused at 35°C (95.0°F) for 20 hours. The study concluded that radiation is an effective method to control this food-borne pathogen.

Vacuum-packaged ground fresh pork samples absorbed gamma radiation doses of 0, 0.57, 3.76, 5.52, or 7.25 kGy at 2°C (35.6°F). Samples were analysed after 1, 7, 14, 21, 28, or 35 days storage at 2°C (35.6°F) for the presence and number of aerobic and anaerobic mesophiles and endospore formers and aerobic psychrotrophs. Conventional plate counts did not detect surviving microflora in any sample that received an absorbed dose of 1.91 kGy or higher, even after refrigerated storage for up to 35 days. The microflora in the control were predominantly gram-positive for the first 21 days; however, *Serratia* predominated at 28th and 35th days. *Staphylococcus, Micrococcus*, and other yeast species predominated in samples that received 0.57 kGy.

The gamma radiation resistance of 5 enterotoxic and 1 emetic isolate of *Bacillus cereus* vegetative cells and endospores was tested in mechanically deboned chicken meat (MDCM), ground turkey breast, ground beef round, ground pork loin, and beef gravy. The D values for *B. cereus* were 0.184, 0.431, and 2.56 kGy for logarithmic-phase cells, stationary-phase cells, and endospores at 5°C (41.0°F) on MDCM, respectively. Neither the presence nor absence of air during irradiation significantly affected radiation resistance of vegetative cells or endospores of *B. cereus* when present on MDCM. Irradiation temperature [–20° to +20°C (-4.0° to 68.0°F)] did affect the radiation resistance of stationary-phase vegetative cells, and to a limited extent that of spores on MDCM. Impedance studies indicated that surviving vegetative cells were severely injured by radiation. A dose of 7.5 kGy at 5°C (41.0°F) was required to eliminate a challenge of 4.6×10^3 *B. cereus* from temperature-abused MDCM [24 hours at 30°C (86.0°F)]. The radiation resistance of a mixture of endospores of 6 strains to gamma radiation was 2.78 kGy in ground beef round, ground pork loin, and beef gravy, but 1.91 kGy in turkey and MDCM. The results indicate that irradiation of meat or poultry can provide significant protection from vegetative cells but not from endospores of *B. cereus*.

The effects of water content, activity, NaCl, and sucrose content on the survival of *Salmonella typhimurium* on irradiated MDCM and ground pork loin were investigated. The effects of NaCl and sucrose concentration were investigated by adding various amounts to MDCM or ground pork loin with NaCl solutions with various degrees of saturation. The effects of water content were investigated by rehydrating freeze-dried ground pork loin with different quantities of water. Inoculated samples were irradiated at 5°C (41.0°F) *in vacuo* to doses up to 6.0 kGy. The survival of *S. typhimurium* was effected by water content, water activity, and NaCl content, but not by sucrose content. The failure of sucrose to provide the same protection for *S. typhimurium* in meat against radiation argues against reduced water activity being a primary mechanism of protection. The results indicate that the survival of food-borne pathogens on irradiated meat with reduced water content or increased NaCl levels may be greater than expected.

Research has demonstrated that ionizing radiation can inactivate parasites, eliminate or greatly reduce the populations of microbial pathogens, and extend the shelf life while preserving the desired nutritional and sensory properties of refrigerated poultry and red meat. Food-borne pathogens can be greatly reduced in population and sometimes completely eliminated from foods by low doses of ionizing radiation. The shelf life of poultry, pork, and beef can be significantly extended by treatment with ionizing radiation. Combination treatments with vacuum packaging or modified atmosphere packaging and ionizing radiation have produced better than predicted results. Additional research is needed on the combined processes.

The American Meat Institute (AMI) is actively involved in the investigation of viable pathogen-preventing technologies which can be applied to the meat and poultry industry. Irradiation is one of those technologies and it is one that has gathered support from governments and scientists worldwide. In AMI's view, reducing pathogens in the food supply and preventing food-borne illness will demand a multi-faceted, farm-to-table approach. While irradiation may be helpful, it alone will not solve public health problems related to food-borne pathogens.

ULTRAVIOLET RADIATION

The UV radiations are light with wavelength range from 200 nm to 400 nm.

- UV–A range 315–400 nm causes changes in skin (tanning)
- UV–B range 280–315 nm causes sun burns
- UV–C range 200–280 nm leads to cell mutation and cell death. This is the germicidal range and very effective at inactivating bacteria and viruses.

UV light cannot penetrate solid, opaque, glass and its main use is to disinfect surfaces, air, equipments, water etc. Actual use of UV light include control of microbial growth in bulk, stored meat, sterilization of packaging materials used in aseptic packaging, control of surface growth on bakery products, treatment of sea water used for washing shell fish, sanitation of equipments and air purification, prevention of growth of film yeasts on pickles and killing of spores on sugar crystals.

Factors Influencing Effectiveness

1. *Time* Longer the time of exposure, more effective is the treatment.
2. *Intensity* Intensity of the rays reacting on object depends on power of the lamp, distance from lamp to the object, kind and amount of interfering material in the path of rays. It increases with the power of the lamp and is measured in mW/sq.cm. A lamp is about 100 times as effective in killing microbes at 5 inches than at 8 ft from the irradiated object. Dust reduces the effectiveness as does atmospheric humidity.

Sensitivity of Microbes to UV Radiation

Bacterial spores are more resistant to UV radiation than vegetative cells. Viruses are inactivated by UV rays and they are less resistant than spores but more resistant than non-sporulating bacteria. Non-enveloped viruses are more resistant to UV radiation than enveloped ones. Among enteroviruses, human adenovirus type 2 is the most UV resistant. Cysts of water borne protozoa like *Giardia lamblia* and *Cryptosporidium parvum* are more UV resistant than non-sporulating bacteria.

UV radiation is equally effective against yeasts, gram-positive and gram-negative bacteria. *Micrococcus radiodurans* is the highly resistant organism to UV due to very efficient DNA repair mechanism. Some moulds are protected by fatty or waxy secretions that inhibit penetration of UV radiations and pigments in some microbes has a protective effect.

Target Site and Inactivation

DNA is the target site and inactivation results are due to

1. Accumulation of photoproducts, i.e., dimers between adjacent thymine residues in the same DNA strand. These dimers are most stable for they inhibit DNA synthesis.
2. Induces DNA intrastrand links and NA–protein cross links.
3. In *Deinococcus radiodurans*, another type of photoproduct, i.e., 5,6-dihydroxy-5,6-dihydrothymine (TDHT) is also induced.

Effect of UV Radiation on Bacterial Spores

Spores are more resistant than vegetative cells. During germination, they become much more resistant. When spores are irradiated, TDHT accumulate which are removed by photoproduct lyase, a Fe-S protein. 10-20% of protein in the core of dormant spore is in the form of small acid-soluble protein (SASP). These proteins coat the DNA thus protecting it from the ill effects of UV radiation. During germination, SASPs are rapidly degraded and are not found in vegetative cells of spore forming species.

To conclude, food irradiation can be used to combat food-borne diseases, including the emergence of disease-causing organisms such as *C. jejuni, E. coli,* and *L. monocytogenes.* Food irradiation is not a substitute for proper handling, cooking, and storage of food. Care must be taken to ensure that irradiated foods do not become recontaminated. Also, food irradiation could be used instead of fumigants to kill moulds and insects on products and grains. Food irradiation has been studied more extensively than any other food additive, yet there is only limited application in this country.

Food irradiation has been endorsed by FAO, WHO, USDA, the American Medical Association (AMA), and the Institute of Food Technologists (IFT) as a safe and practical method for preserving a variety of foods and reducing the risk of food-borne diseases. International imports and exports of fresh foods could be expanded, increasing the abundance of food worldwide. Food irradiation makes food safer to eat, improves quality, and extends shelf life.

REVIEW QUESTIONS

1. Explain the term 'Cold Sterilization'.
2. Define radicidation.
3. What are the types of irradiation used in food preservation?
4. Give a brief account of foods currently being irradiated.
5. Throw some light on sensitivity and resistance of microbes towards ionizing radiations.
6. Discuss the uses of food irradiation.

10

PRESERVATION OF FOOD USING CHEMICALS

INTRODUCTION

The quality of a food product decreases from the time of harvest or slaughter until it is consumed. Quality loss may be due to microbiological, enzymatic, chemical or physical changes. The consequences of quality loss caused by microorganisms are consumer hazards due to the presence of microbial toxins or pathogenic microorganisms and economic loss due to spoilage. Many food preservation technologies, some in use since ancient times, protect foods from the effects of microbial and inherent deterioration. Microorganisms may be inhibited by chilling, freezing, water activity reduction, nutrient restriction, acidification, modification of packaging atmosphere, or fermentation or through addition of antimicrobial compounds. Food antimicrobials are traditionally used for their ability to inhibit spoilage microorganisms and, thus, prolong shelf life and preserve food quality. Recently, however, antimicrobials have been used increasingly as primary interventions to inactivate or inhibit the outgrowth of pathogenic microorganisms in foods.

Food antimicrobials are chemical compounds added to or present in foods that retard microbial growth or kill microorganisms thereby resisting deterioration in safety or quality. They are different than therapeutic antibiotics (e.g. penicillin, tetracyclines) used to treat human or animal disease. Food antimicrobials are sometimes called "preservatives." The term "preservative" often includes antioxidants in addition to antimicrobials. "Antimicrobial" is the more specific term in the context of this discussion and, therefore, is used throughout this chapter. Antimicrobials may be classified as "traditional" or "naturally occurring". A number of traditional antimicrobials, e.g. acetic acid and benzoic acid, are approved for use in foods by most international regulatory agencies. The major targets for antimicrobials are food poisoning microorganisms (infective agents and toxin producers) and food spoilage microorganisms whose metabolic end products or enzymes cause off-odours, off-flavours, texture problems, discolouration, slime, or haze. Food antimicrobials are sometimes referred to as food preservatives. However, food preservatives include not only antimicrobials but also antibrowning agents (e.g. citric acid) and antioxidants (e.g. butylated hydroxyanisole). The use of food antimicrobials is decreasing because of consumer trends favouring consumption of natural foods.

Most food antimicrobials are only bacteriostatic or fungistatic and not bactericidal or fungicidal. Because they are generally bacteriostatic or fungistatic, they will not preserve a food indefinitely. Depending on storage conditions, the food product eventually spoils or becomes hazardous. Food antimicrobials are often used in combination with other food preservation procedures.

Cellular targets of food antimicrobials include the cell wall, cell membrane, metabolic enzymes, protein synthesis systems, and genetic systems (Figure 10.1). Since all are essential, inactivation of any one can inactivate cells. The exact mechanisms or targets for food antimicrobials are often not known or well defined. There are several possible reasons for this. It is difficult to pinpoint a target when many interacting reactions take place simultaneously. For example, membrane disrupting compounds could cause leakage, interfere with active transport or metabolic enzymes or dissipate cellular energy in the form of ATP. Food antimicrobials generally have multiple targets with concentration dependent thresholds for inactivation or inhibition. A given target is important only if its sensitivity is within the inhibitory concentration range of the antimicrobials.

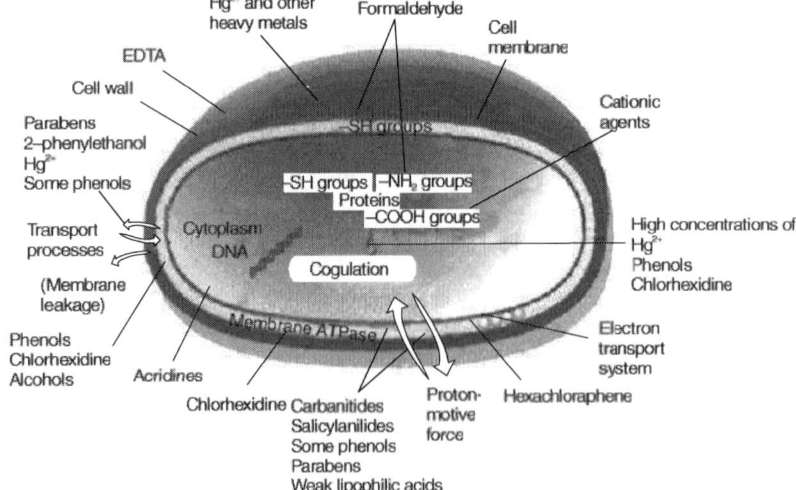

Figure 10.1   Cellular targets of food antimicrobials

FACTORS AFFECTING ACTIVITY

The effectiveness of food antimicorbials depends on many factors associated with the food product, its storage environment, its handling, and the target microbes. Food preservation is best achieved when the antimicrobial type and concentration, storage time and temperature, food pH and buffering capacity and presence of other agents which may influence shelf life are known and taken into account.

Microbial factors that affect antimicrobial activity include inherent resistance (e.g. vegetative cells versus spores, strain differences), initial number and growth rate, interaction with other microbes (e.g. antagonism), cellular composition (gram stain reaction) and cellular status (injury). Intrinsic factors affecting antimicrobial activity are those associated with a food product and include composition, pH buffering capacity, oxidation–reduction potential, and water activity. Extrinsic (environmental) factors affecting antimicrobial activity include temperature and time of storage, atmosphere, and relative humidity. Processing factors include changes in food composition, shifts in microflora, changes in microbial numbers and changes in microbial structure. Most factors influence microorganisms in an interactive manner.

pH

pH is the most important factor influencing the effectiveness of most food antimicrobials. Most food antimicrobials are weak acids and are most effective in their undissociated form. This is because weak acids are able to penetrate the cytoplasmic membrane of a microorganism more effectively in the protonated form. Therefore, the pH value of these compounds is important in selecting a particular compound for an application. The lower the pH of a food product, the greater the proportion of acid in the undissociated form and the greater the antimicrobial activity. Some researchers have suggested that only the undissociated form of a weak acid has antimicrobial activity.

Polarity

Another important factor affecting antimicrobial activity is polarity. This relates both to the ionization of the molecule and contribution of any alkyl side groups of hydrophobic parent molecules. An antimicrobial must be lipophilic to attach and pass through the cell membrane and also be soluble in the aqueous phase, because most food antimicrobials are at least partially hydrophobic. The presence of lipid in food may decrease the activity as a result of solubilization or binding of the compound. Proteins in foods may also decrease activity of antimicrobials through hydrophobic interactions.

TRADITIONAL ANTIMICROBIALS

Organic Acids and Esters

Many organic acids are used as food additives, but not all have antimicorbial activity. The most active antimicrobials are acetic, lactic, propionic, sorbic and benzoic aids. Citric, caprylic, malic, fumaric and other organic acids have limited activity but are used for flavourings.

The activity of organic acids is highly pH dependent. Therefore, in selecting an organic acid for use as an antimicrobial food additive, both the product pH and the pK, of the acid must be taken into account. The use of organic acids is generally limited to foods with pH less than 5.5, since most organic acids have dissociation constants (pK) of pH 3.0–5.0.

The mechanisms of action of organic acids and their esters have some common elements. In the undissociated form, organic acids can penetrate the cell membrane lipid-bilayer more easily. Once inside the cell, the acid dissociates because the cell interior has a higher pH than the exterior. Bacteria maintain internal pH near neutrality to prevent conformational changes of the cell structural proteins, enzymes, nucleic acids, and phospholipids. Protons generated from intracellular dissociation of the organic acid acidify the cytoplasm and must be extruded to the exterior. The cytoplasmic membrane is impermeable to protons, and the protons must be transported to the exterior. This proton extrusion creates an electrochemical potential across the membrane called the proton motive force (PMF). Since protons generated by the organic acid inside the cell must be extruded using energy in the form of ATP, the constant influx of these protons will eventually deplete cellular energy. This same phenomenon could be caused by interference with membrane permeability as well. The short chain organic acids interfere with energy metabolism by altering the structure of the cytoplasmic membrane through interaction with membrane proteins. This reduces ATP regeneration by uncoupling the electron transport system or by inhibiting active transport of nutrients into the cell. These short-chain fatty acids act as uncouplers of amino acid carrier proteins from the electron transport system. This inhibition of active transport was due to destruction of the PMF, which in turn cause active transport to cease. The organic acids and their esters have significant effects on bacterial cytoplasmic membranes, interfering with metabolite transport and maintenance of membrane potential. Many organic acids and esters also affect the activities of microbial enzymes.

Acetic Acid and Acetates

Acetic acid, the primary component of vinegar, and its sodium, postassium and calcium salts are some of the oldest food antimicrobials. Only *Acetobacter* species, lactic acid bacteria and butyric acid bacteria are tolerant to acetic acid. Bacteria inhibited by acetic acid include *Bacillus* species, *Staphylococcus aureus*, *Clostridium* species, *Listeria monocytogenes*, salmonellae, *E. coli*, *Campylobacter jejuni* and pseudomonads. Moulds and yeasts are generally more resistant to acetic acid than bacteria. Yeasts and moulds sensitive to acetic acid include *Aspergillus, Penicillium* and *Rhizopus* species and some strains of *Saccharomyces*.

Acetic acid and its salts have shown variable success rate as antimicrobials in food applications. Acetic acid can increase poultry shelf life when added to cut chicken parts in cold water at pH 2.5. Addition of acetic acid at 0.1% to scald tank waster used in poultry processing decreases the heat resistance of *Salmonella newport*, *Salmonela typhimurium* and *Campylobacter jejuni*. Acetic acid has shown variable effectiveness as an antimicrobial for use as a spray sanitizer on meat carcasses.

Sodium acetate at 1.0% increases the shelf life of catfish fillets by 6 days when stored at 4°C. Sodium acetate is also an effective inhibitor of rope forming bacteria *B. subtilis* in bakery goods and of the moulds *Aspergillus flavus, A. fumigatus , A. niger, A. glaucus, Penicillium expansum* and *Mucor pusilus* at pH 3.5–4.5. It is useful in the baking industry because it has little effect on the yeast used in baking. Sodium diacetate is inhibitory to *L. monocytogenes, E. coli, Pseudomonas fluorescens, Salmonella enteritidis* and *Shewanella putrefaciens*.

Acetic acid is used commercially in baked goods, cheeses, condiments and relishes, dairy products, fats and oils, gravies and sauces, and meats. The general mechanism by which acetic acid inhibits microbes is related to that or other organic acids. It inhibits oxygen uptake and resultant ATP production by 76–77%. It inhibits the growth by uncoupling substrate transport and oxidative phosphorylation from the electron transport system. This inhibits uptake of metabolites into the cell.

Benzoic Acid and Benzoates

Benzoic acid and sodium benzoate were the first antimicrobial compound permitted in food by the U.S. Food and Drug Administration. Benzoic acid occurs naturally in cranberries, plums, prunes, cinnamon, cloves and most berries. Sodium benzoate is highly soluble in water while benzoic acid is much less soluble.

Benzoic acid and sodium benzoate are used primarily as antifungal agents. *Zygosaccharomyces bailii* is particularly resistant to benzoic acid. Sodium benzoate is used as an antimicrobial at up to 0.1% in carbonated and still beverages, syrups, jams, jellies, fruit salads and in the storage of vegetables. There are multiple cellular targets for benzoic acid. Only the undissociated form is taken up by the cells because of its lipophilic character and the ability to cross the cytoplasmic membrane. When the cells are exposed to >60°C the uptake rate decreases. Heat inactivation of this uptake process suggests enzymatic inactivation. Benzoic acid uncouples both substrate transport and oxidative phosphorylation from the electron transport system. It destroys the PMF by continuous transport of protons into the cell, causing disruption of the transport system. Benzoates inhibit various microbial enzymes and enzyme systems. Acetic acid metabolism and oxidative phosphorylation, α-ketoglutarate and succinate dehydrogenases and trimethylamine-N-oxide reductase activity of *E. coli* are inhibited by benzoic acid. Even aflatoxin production by *A. flavus* is inhibited by the benzoates.

Lactic Acid and Lactates

Lactic acid is produced naturally during fermentation of foods by lactic acid bacteria, while the acid and its salts act as preservatives in food products, their primary uses are as pH control agents and flavourings. Lactic acid inhibits spore-forming bacteria, *S. aureus* and *Y. enterocolitica*. The antimicrobial activity of lactic acid depends on the food application and the target microorganism. Lactic acid is more effective than malic, citric, propionic, or acetic acid in inhibiting growth of *Bacillus coagulans* in tomato juice. Lactic acid at 1–2% reduces Enterobacteriaceae and aerobic mesophilic microorganisms on beef, veal, pork, and poultry and delays growth of spoilage microflora during long term storage of products. Sodium lactate inhibits *Clostridium botulinum, C. sporogenes, L. monocytogenes* and spoilage bacteria in various meat products.

Presumably, it functions similar to other organic acids and has a primary mechanism involving disruption of the cytoplasmic membrane PMF. The lactate salts are less efficient than the organic acids.

Propionic Acid and Propionates

Propionic acid is produced naturally in Swiss cheese by *Propionibacterium freudenreichii* subsp. *shermanii*. The activities of propionates depend on the pH of the substance to be preserved, with the undissociated acid is the most active form. Propionic acid and its salts are used primarily against moulds.

Sodium propionate is used as a fungicide and for mould prevention. Most uses of sodium propionate are as a food additive, in particular for use in baked goods, confections, and gelatin. It is also used in cosmetics. It is used as a topical antifungal agent in livestock, and also as a preservative for hay and silage. Propionic acid is also used as an inert ingredient in various biorational pesticides and is used as a preservative in various fertilizer products.

Propionic acid and its salts, including sodium propionate, are toxic to moulds and certain bacteria based on the inability of the affected organisms to metabolize the three-carbon chain. It is most effective at an acid pH. Propionic acid also interferes with cytoplasmic membrane or cellular enzymes. Propionic acid inhibits amino acid uptake and inhibits growth of various bacteria. Propionic acid neutralized the PMF of the microbial membrane by passing through the membrane as the undissociated molecule and dissociating intracellularly. This leads cells to utilize energy to pump out the excess protons and eventually depletes cellular ATP.

Sorbic Acid and Sorbates

Sorbic acid is a *trans–trans* unsaturated monocarboxylic fatty acid which is slightly soluble in water at 20°C. The potassium salt of sorbic acid is readily soluble in water. The antimicrobial activity of sorbic acid is greatest when the compound is in the undissociated state.

They inhibit fungi and certain bacteria. Food related yeast inhibited by sorbates include species of *Brettanomyces, Byssochlamys, Candida, Cryptococcus, Debaryomyces, Hansenula, Pichia, Rhodotorula, Saccharomyces, Torulaspora* and *Zygosaccharomyces*. Food related mould species inhibited by sorbates belong to genera *Fusarium, Geotrichum, Alternaria, Aspergillus, Botrytis, Cephalosporium, Helminthosporium, Mucor, Penicillium, Pullularia, Sporotrichum* and *Trichoderma*.

Sorbates inhibit the growth of yeasts and moulds in microbiological media, cheeses, fruits, vegetables and vegetable fermentations. Sorbates inhibit growth and mycotoxin production by the mycotoxigenic moulds and many food-borne pathogenic and spoilage bacteria including salmonellae and *S. aureus* in cooked, uncured sausage. Sorbates are effective anticlostridial agents in cured meats and other meat and seafood products. Sorbate prevents spores of *C. botulinum* from germinating and forming toxin in beef, pork, poultry, etc. Potassium sorbate is a strong inhibitor of both *Bacillus* and *Clostridium* spore germination at pH 5.7.

Sorbate is applied to foods by direct addition, dipping, spraying, dusting or incorporation into packaging materials.

One of sorbic acid's primary targets in vegetative cells appears to be the cytoplasmic membrane. It also inhibits amino acid uptake, which in turn eliminates the membrane PMF through nutrient depletion. Sorbic acid inhibits dehydrogenases involved in fatty acid oxidation. Addition of sorbic acid results in the accumulation of unsaturated fatty acids that are intermediate products in the oxidation of fatty acids by fungi. This prevents the function of dehydrogenases and inhibits metabolism and growth. Sorbic acid also inhibits sulphydryl enzymes like fumarase, aspartase, succinic dehydrogenase and alcohol dehydrogenase.

Fatty Acid Esters

Glyceryl monolaurate (monolaurin) is more effective against gram-positive bacteria than the gram-negative bacteria.

Nitrites

Nitrites find a special place in cured meat products. Meat curing utilizes salt, sugar, spices and ascorbate in addition to nitrites. As nitric oxide, it reacts with the meat pigment, myoglobin, to form the nitrosomyoglobin, It also contributes to the flavour and texture of cured meats and serves as an antioxidant.

The primary use for sodium nitrite as an antimicrobial is to inhibit *C. botulinum* growth and toxin production in cured meats. Nitrite inhibits bacterial spore-formers by inhibiting outgrowth of the germinated spores.

Nitrites effectiveness depends on several environmental factors. Nitrites are more effective at lower pH and under anaerobic conditions. Storage and processing temperatures, salt concentration and initial inoculum size also influences the antimicrobial effectiveness of nitrite.

Nitrite has variable effects on microorganisms other than *C. botulinum*, *Clostridium perfringens*, *E.coli*, *Achromobacter*, *Enterobacter*, *Flavobacterium*, *Micrococcus* and *Pseudomonas*. Meat products that may contain nitrites include bacon, ham, fermented sausages, perishable canned cured meat. Nitrite is also used in a variety of fish and poultry products.

Nitrites inactivated a variety of enzymes associated with respiration. In spore- forming bacteria, nitrite causes a reduction in intracellular ATP and excretion of pyruvate in bacterial cells. Two enzymes in the system, pyruvate ferredoxin oxidoreductase (PFR) and ferredoxin are susceptible to nitrite. Nitric oxide is the active antimicrobial principle of nitrite. Mechanism of inhibition against non-spore forming microbes is different from that for spore formers. Nitrite inhibits active transport, oxygen uptake, and oxidative phosphorylation by oxidizing ferrous ion of an electron carrier, such as cytochrome oxidase to the ferric form. Thus nitrites inhibit aerobic bacteria by binding the haem ion of cytochrome oxidase.

Parabens

Parabens are produced by the esterification of the carboxyl group of benzoic acid. They are thus phenolic derivatives. Methyl, propyl and heptyl parabens are directly added to foods as antimicrobials. The antimicrobial activity of parabens is direcly proportional to the chain length of the alkyl component. As the alkyl chain length of the parabens increases, inhibitory activity also increases. Parabens are generally more active against moulds and yeasts than bacteria. Among bacteria, the parabens are more active against gram-positive genera. Parabens are incorporated into foods by being dissolved in water, ethanol, propylene glycol or the food product itself. Parabens are used in a variety of foods including baked goods, beverages, fruit products like jams and jellies, fermented foods, syrups, salad dressings, wine and fillings.

These compounds cause physical damage to the membrane or permeability barrier causing leakage of cellular materials. The amount of leakage is proportional to the alkyl chain length of the paraben. Parabens also inhibited serine uptake and the oxidation of a-glycerol phosphate and NADH in membrane vesicles of bacteria thus inhibiting both membrane transport and electron transport system.

One paraben compound, butylated hydroxytoluene (BHT), causes leakage by disrupting the symmetry of the membrane and increasing fluidity of lipid alkyl chain in phospholipids. Thus to conclude, phenolic compounds do not have the same mechanism of action and there may be several targets which lead to inhibition of microorganisms by these compounds.

Sodium Chloride

Sodium chloride, common salt, is the oldest known food preservative. Salt is primarily used as an adjunct to their processing methods such as canning and pasteurization. In general food-borne pathogenic bacteria (except *S. aureus*) are inhibited by a water activity of 0.92 or less. Fungi are more tolerant to low water activity than bacteria. The minimum for growth of xerotolerant fungi is 0.61 to 0.62 but most are inhibited by 0.85 or lower. Sodium chloride inhibits microorganisms primarily by its plasmolytic effect. The antimicrobial activity of sodium chloride is related to its ability to reduce water activity and create unfavorable conditions for microbial growth. As the water activity of the external medium is reduced, cells are subjected to osmotic shock and rapidly lose water through plasmolysis. During plasmolysis, a cell ceases to grow and either dies or remains dormant. In order to resume growth, the cell must reduce its intracellular water activity. Aside from the osmotic influence on growth, other possible mechanisms include limiting oxygen solubility, alteration of pH, toxicity of sodium and chloride ions, and loss of magnesium ions. Certain isolated enzymes are more susceptible to inhibition by sodium chloride than is the growth of the host species.

Sulphites

Use of sulphur dioxide as a food preservative is an age old practice. Salts of sulphur dioxide include potassium sulphite, sodium sulphite, potassium bisulphite, sodium bisulphite, potassium metabisulphite, sodium metabisulphite. Sulphites are used primarily in fruits and vegetable products to control three groups of microorganisms: spoilage and fermentative yeasts and moulds on fruits and fruit products, acetic acid bacteria and malolactic bacteria. In addition to use as antimicrobial, sulphites act as antioxidants and inhibit enzymatic and non-enzymatic browning in a variety of foods.

Important factor that affects the antimicrobial activity of sulphites is pH. The inhibitory effect of sulphites is most pronounced when the acid is in the undissociated form. Increased effectiveness at low pH is likely due to the ability of unionized sulphur dioxide to pass across the cell membrane. The bisulphites are very reactive, forming addition compounds with aldehydes and ketones. These bound forms have much less antimicrobial activity compared with the free forms. Sulphur dioxide is fungicidal even in low concentrations against yeasts and moulds. The inhibitory concentration range of sulphur dioxide is 0.1–20.2 µg/ml for *Saccharomyces, Zygosaccharomyces, Pichia, Hansenula* and *Candida* species. Around 25–100 µg of sulphur dioxide per ml is required to inhibit *Byssochlamys nivea* in apple juices.

Sulphite may be used to inhibit acetic acid and lactic acid producing bacteria in wines and fruit products and spoilage bacteria in meat products. Sulphur dioxide is more inhibitory to gram-negative rods than to gram-positive rods.

Sulphur dioxide is used to control the growth of undesirable microbes in fruits, fruit juices, wines, sausages, fresh shrimp and acid pickles. During fermentation, sulphur dioxide also serves as an antioxidant, clarifier and dissolving agent. The optimum level of sulphur dioxide is maintained to prevent post-fermentation changes by microorganisms.

Because of their extreme reactivity, it is difficult to pinpoint the exact antimicrobial mechanism of sulphites. This reactivity is due to the ability of sulphites to act as reducing agents or take part in nucleophilic attack. Sulphites react with disulphide bonds of proteins and with glutathione forming thiosulphonates thus inactivating enzymes which have disulphide links and effect conformation of proteins. Sulphites react with coenzymes and enzyme prosthetic groups. The coenzymes NAD and the related NADP are inactivated by sulphite addition. Prosthetic groups including thiamine, folic acid, pyridoxal, flavins and haemes are all susceptible to sulphite inactivation.

The targets for inhibition by sulphites include the cytoplasmic membrane, DNA replication, protein synthesis, membrane bound or cytoplasmic enzymes or individual components in metabolic pathways. Sulphite dissipates the PMF and inhibit solute active transport. Once inside the cell, sulphur dioxide acts on enzymes and individual components of metabolism.

Dimethyl Dicarbonate (DMDC)

This is a colourless liquid which is slightly soluble in water. The compound is very reactive with many substances, including water, ethanol, alkyl and aromatic amines and sulphydryl groups. The primary target microorganisms for DMDC are yeast, including *Saccharomyces, Zygosaccharomyces, Rhodotorula, Candida, Pichia, Torulopsis, Torula, Endomyces, Kloeckera* and *Hansenula* species. The compound is also bactericidal to a number of species, including *Acetobacter pasteurianus, E.coli, Pseudomonas aeruginosa, S. aureus* and *Pediococcus cerevisiae*. Moulds are generally more resistant to DMDC than yeasts or bacteria. It acts by inactivating the enzymes.

Phenolic Antioxidants

Phenolic antioxidants are used in foods primarily to delay autoxidants of unsaturated lipids by interrupting the free-radical chain mechanism of hydroperoxide formation during the autoxidation process. Gram-positive bacteria are more susceptible to phenolic compounds. Tertiary butyl hydroquinone (TBHQ) is an extremely effective inhibitor of gram-positive bacteria including *S. aureus* and *L. monocytogenes*.

Many environmental factors influence the antimicrobial activity of the phenolic antioxidants. The presence of lipid or protein dramatically decreases the activity of phenolic antioxidants. Calcium increases the antimicrobial activity of the phenolic antioxidants.

Phosphates

Phosphates have important uses in food processing including buffering or pH stabilization, acidification, alkalization sequestration or precipitation of metals, formation of complexes with organic polyelectrolytes, deflocculation, dispersion, peptization, emulsification, nutrient supplementation, antimicrobial preservation and leavening.

Gram-positive bacteria are generally more susceptible to phosphates than gram- negative bacteria. The ability of polyphosphates to chelate metal ions appears to play an important role in their antimicrobial activity. The polyphosphates inhibited gram-positive bacteria and fungi by removal of essential cations from binding sites on the cell walls of these microbes. Polyphosphates may also interfere with RNA function or metabolic activities of cell.

RESISTANCE AND ADAPTATION TO FOOD ANTIMICROBIALS AND OTHER PROCESS CONTROLS

Most antimicrobials used in food manufacture have been in use for about 50 to 100 years. A few antimicrobials, e.g. sulphites and nitrites, have been in use for an even longer period of time. Similarly, sanitizing agents, used to reduce microorganisms on processing equipment, have been in use for nearly 100 years. Concern has been recently raised, about the potential for target pathogenic microorganisms to develop resistance to these compounds. Despite the considerable period of time that food antimicrobials and equipment sanitizers have been used in the food industry, there is little data about the development of microbial resistance to these compounds. This lack of data might be viewed as a good indication that resistance development is probably not a major problem.

Concern remains, however, for three reasons. One concern is the increasing incidence of microorganisms exhibiting resistance to antibiotics used for therapeutic purposes in human and animal medicine. A second concern is the increasing reliance on antimicrobials and sanitizers as primary tools for controlling the outgrowth of pathogens in foods. A third concern is the evidence indicating that tolerance to antimicrobials, sanitizers, and other preservation processes may be generated within microorganisms exposed to certain stresses.

If antimicrobials and sanitizers are to play a major role in effective control of food-borne pathogens, food manufacturers and others within the food industry must know more about the potential for development of resistance among target microorganisms.

Antimicrobials approved for use in food are shown in Table 10.1.

Table 10.1 Various chemical preservatives and their mode of action

Name	Effect on microbes	Mode of action	Foods used
CHEMICAL PRESERVATIVES			
Organic acids			
Acetic acid (Ca, K, Na salts)	*Tolerant* *Acetobacter* Lactic acid bacteria and Butyric acid bacteria *Inhibitory* *Bacillus, Clostridium, L. monocytogenes, Salmonella, S. aureus, E. coli, C. jejuni, Pseudomonas Aspergillus, Penicillium Rhizopus*	Inhibits oxygen uptake, ATP production and transport mechanisms, acts on cellular enzymes by reducing intracellular pH.	Chicken, poultry processing (0.1–1%), commercially baked foods, cheese, dairy products, sauce, breakfast cereals, jams, jellies, soft candy.

(Contd.)

Table 10.1 (Continued)

Name	Effect on microbes	Mode of action	Foods used
	Resistant Moulds and yeasts		
Benzoic acid Sodium benzoate—first antimicrobial compound permitted by FDA. Naturally occurring in cranberries, plums, prunes, cinnamon, clove.	Antifungal Yeasts, moulds Higher concentrations affect bacteria (0.1% reduces viable *E. coli* O157:H7).	Affects the structure of cell membrane and enzymes Uncouples substrate transport and oxidative phosphorylation. Destroys proton motive force.	Carbonated beverages (0.1%), syrups, olives, pickles, soysauce, jams, jellies, pastry, fruit salads.
Lactic acid Sodium lactate—natural product of fermented foods.	Inhibits Spore-forming bacteria, *S. aureus,* *Y. enterocolitica,* *B. coagulans*, 1-2% reduces enterobacteriaceae and aerobic mesophiles. Sodium lactate 2.5-5% inhibits *C. botulinum,* *C. sporogenes* and *L. monocytogenes.*	Disruption of CM and PMF. Reduction of a_w.	Tomato juice, beef, poultry, meat.
Propionic acid Naturally present in Swiss-cheese (1%) available as Na, K, Ca salts	Inhibitory to moulds, yeasts and some bacteria. 0.1–5% retard growth of *E.coli, S. auerus, Sarcina lutea, Salmonella* sp. *Proteus vulgaris, Candida* and *Saccharomyces*	Affects CM and inhibits amino acid uptake, depletes cellular ATP.	Bakery foods, cheese, bread dough (0.4%)
Citric acid	Retards growth and toxin production by *A. parasiticus, A. versicolor, Salmonella* and *C. botulinum*	Chelation	Animal food, tomato products

(Contd.)

Table 10.1   (Continued)

Name	Effect on microbes	Mode of action	Foods used
Sorbic acid Na, Ca, K naturally occurring in rowanberry. Best antimicrobial	Inhibitory to fungi, certain bacteria, yeasts like *Candida, Cryptococcus, Rhodotorula* and moulds like *Alternaria, Botrytis, Aspergillus, Fusarium, Geotrichum, Mucor, Byssochlamys fulva.* Inhibits mycotoxin production. Inhibits food-borne pathogens, anti-clostridial agent	Affects CM, inhibits amino acid uptake, eliminates PMF, inhibits dehydrogenases involved in fatty acid oxidation, inhibits sulphydryl enzymes forming stable thiohexenoic acid complexes. Decreases the rate of cell division of germinated spores of *Bacillus* and *C. botulinum*	Cakes, pastries, doughnut icing, fruit fillings. Jams, jellies, margarine, chocolate syrup, salads, dried fruits.
Fumaric acid Flavouring agent	Prevents malolactic fermentation in wines		
Fatty acid esters (Glyceryl monolaurate)	Active against gram-positive bacteria and inactive against gram- negative bacteria		
Nitrites $NaNO_3$, KNO_3	Inhibits *C. botulinum, E. coli, Achromobacter, Micrococcus, Pseudomonas.*	Reacts with meat pigment myoglobin to yield nitrosomyoglobin. Serves as an antioxidant. As a nitrite, it is an effective anti-microbial agent. Inhibits outgrowth of germinated spores and inactivates enzymes associated with respiration. Reduction of cellular ATP in non-spore-forming organisms and inhibits active transport of oxygen, oxidative phosphorylation since it oxidizes the ferrous ion of the electron carrier, cytochrome oxidase.	Cured meat products, salmon fish.

(Contd.)

Table 10.1   (Continued)

Name	Effect on microbes	Mode of action	Foods used
Parabens (esterification of carboxyl groups of benzoic acids yield paraben) Methyl, propyl and heptyl parabens have antimicrobial activity.	Active against yeasts and moulds. Active against gram-positive bacteria.	Attacks the cell membrane and affect the release of cell constitutents. Leakage of cellular RNA Inhibits serine uptake. Inhibits membrane transport and electron transport system.	Baked foods, beer, non-carbonated soft drinks, jam, jellies and syrups.
Sulphites K_2SO_2, Na_2SO_2, $KHSO_3$, $K_2S_2O_3$	Inhibits spoilage and fermentative yeasts and moulds on fruits. Inhibits acetic acid bacteria and malo-lactic bacteria. It is fungicidal at low concentrations (0.1–20.2?g/ml)	Extreme reactivity. SO_2 reacts with disulphide bonds of proteins giving rise to glutathion further to thiosulphonates which inactivates enzymes having disulphide links thus changing the conformation of proteins. Affects cell membrane structure, DNA replication, protein synthesis.	Fruits and vegetable products.
Phosphates	Active against gram-positive spore formers	Chelates metal ions	Food processing, meat curing.
Sodium chloride Oldest preservative used as an adjunct to canning and pasteurization.	Food-borne pathogens are inhibited (3%,0.9 a_W) *S. aureus* and *L. monocytogenes* are salt tolerant.	Exerts a plasmolytic effect. Reduces a_W leading to osmotic shock. Decreases oxygen solubility and alters the pH. The Na^+ and Cl^- ions are toxic leading to the loss of Mg^{++} ions.	Raw meat and raw fish

(Contd.)

Table 10.1 (Continued)

Name	Effect on microbes	Mode of action	Foods used
Dimethyl dicarbonates Colourless liquid	Very reactive with water and ethanol. Inhibit yeasts, *E. coli*, *P. aeruginosa*. Moulds are resistant.	Inactivation of enzymes by reacting with histidyl groups of proteins	
Phenolic antioxidants Butylated hydroxyl anisole, propyl gallate.	Effective against *E. coli*, *S. typhimurium*, *S. aureus* and other gram-positive bacteria.	Interrupts the free radical formation.	
Tertiary butyl hydroquinone (TBHQ)	Extremely effective against gram-positive bacteria *S. aureus* and *L. monocytogenes*.		

BIOLOGICAL PRESERVATIVES (ANIMAL SYSTEMS)

Name	Effect on microbes	Mode of action	Foods used
Lactoperoxidase systems	Gram-negative bacteria are sensitive when compared to the gram-positive. Inhibits food-borne pathogens	LP+thiocyanate+ $H_2O_2 \rightarrow$ Hypo-thiocyanate (anti-microbial compound) Thiocyanate ion oxidizes the sulphydryl groups in proteins, inactivates enzymes involved in Kreb's cycle.	Enzyme found naturally in raw milk, colostrum and saliva. Increases shelf life of raw milk in countries with poor refrigeration facilities.
Lactoferrin	Inhibits *B. subtilis*, *B. stearothermophilus*, *Micrococcus*	Produces an iron-deficient environment. Responsible for loss of LPS in gram-negative bacteria. Chelates cations which stabilizes the LPS in CM.	Milk and eggs. In milk it is referred to as transferrin (iron-binding protein)
Ovotransferrin	Gram-positive are more sensitive. Inhibits *E. coli*, *Shigella*, *Bacillus*, *Micrococcu.s*	Iron chelator.	Raw eggs
Avidin	Inhibits bacteria and yeasts.	Binds to cofactor biotin which is involved in TCA cycle and fatty acid biosynthesis.	Egg albumen

(Contd.)

Tabel 10.1   Continued

Name	Effect on microbes	Mode of action	Foods used
Lysozyme	Gram-positive bacteria are more sensitive. It inhibits *C. botulinum,* *C. thermosaccharolyticum,* *B. stearothermophilus,* *B. cereus* and *L. monocytogenes.*	Catalyses the hydrolysis of glycosidic bonds between NAM and NAG of the peptidoglycan of bacterial cell walls causing cell wall degradation and lysis in hypotonic solutions.	Avian eggs, mammalian milk, insects, fishes, dried egg white
BIOLOGICAL PRESERVATIVES (PLANT SYSTEMS)			
Spices, essential oils	Effective against spores of *B. anthracis,* *B. subtilis.* *S. aureus* is also sensitive	Interference with the function of CM, PMF and active transport.	Cinnamon, cloves, thyme, rosemary, sage.
Allicin	Inhibits *B. subtilis, Serratia marcescens, C. botulinum,* salmonellae, shigellae, *S. aureus* and fungi like *A. flavus, A. parasiticus, Candida albicans, Cryptococcus, Penicillium, Trichosporon*	Inhibition of sulphydryl containing enzymes involved in Kreb's cycle and other metabolic reactions.	Onion and garlic
Flavanoids	Bacteriostatic effect on *S. aureus.* Inactivates poliovirus, reovirus and herpes simplex virus	Antimicrobial activity	Various plants

Naturally occurring antimicrobials include compounds that originate from microbial, plant and animal sources (Table 10.1). A subgroup of naturally occurring antimicrobials is the bacteriocins, proteins produced by lactic acid bacteria, e.g. *Lactococcus, Lactobacillus,* and *Pediococcus* species, and a few other bacteria. Only a few naturally occurring antimicrobials, such as nisin, natamycin, lactoferrin and lysozyme, have regulatory approval for application to foods. Many additional antimicrobials, especially those derived from microorganisms, hold the potential for regulatory approval in the future. Although food antimicrobials have been used for many years, few of these substances are used exclusively to control the growth of specific food-borne pathogens. Examples of those used exclusively to control specific pathogens are nitrite to inhibit the growth of *Clostridium botulinum* in cured meats, selected organic acid sprays to reduce pathogens on beef carcass surfaces, nisin and lysozyme to inhibit growth of *C. botulinum* in pasteurized process cheese, and lactate and diacetate to inactivate *Listeria monocytogenes* in processed meats. Generally, these compounds serve as the primary microbial controls among a combination of inhibitors and inhibitory conditions (e.g. low pH and low temperature). Such use of combinations of several microbial controls (multiple interventions) is sometimes called "hurdle technology". If a population of microorganisms is exposed to a sufficiently high concentration of an antimicrobial compound, susceptible cells will be killed. However, some cells may possess a degree of natural resistance or they may acquire it later through mutation or genetic exchange and will, therefore, survive and grow. To fully understand antimicrobial resistance, one must understand the mechanisms of action and/or the specific targets of an antimicrobial within a microorganism. For example, antibiotics used for therapeutic purposes often have specific target sites in a microbial cell and the development of resistance to these compounds is the result of changes in these target sites. These changes may include inactivation or modification of the antibiotic by enzymes within the cell, absence of or bypassing of an enzymatic or metabolic step targeted by the antibiotic, impaired uptake or efflux of the antibiotic, modification of the antibiotic target site, or overproduction of a target molecule. Unfortunately, while we know a great deal about the mechanisms of action and resistance to antibiotics used therapeutically, the precise mechanisms and targets of most food antimicrobials and sanitizers remain a mystery. Therefore, we are less able to predict and/or understand potential resistance to these groups of compounds. The resistance responses of microorganisms to antimicrobials or sanitizers may be innate, apparent, or acquired. Innate resistance is a chromosomally controlled property that is naturally associated with a microorganism. Differences in resistance to antimicrobials occurring among different types, genera, species, and strains of microorganisms under identical environmental conditions and antimicrobial concentrations are most likely controlled innately. Mechanisms of innate resistance may include cellular barriers preventing entry of the antimicrobial (e.g. the outer membrane of gram-negative bacteria and teichoic acids contained within gram-positive bacteria), cellular efflux (i.e., mechanisms that pump compounds out of the cell), lack of a biochemical target for antimicrobial attachment or microbial inactivation, and inactivation of antimicrobials by microbial enzymes.

Apparent resistance is related to assay or application conditions. As with any preservation technique, susceptibility to antimicrobials is dependent upon the conditions of the application.

Bacterial Stress Responses

Food preservation processes are designed to either inhibit the growth of or inactivate bacteria, depending upon the type and severity of the process used. Thus, food preservation exposes bacteria to both lethal and sublethal stresses. Bacteria may have different mechanisms for surviving these external environmental stresses. For example, the formation of endospores in response to stress is a survival strategy for *Bacillus* and *Clostridium* species. Bacteria that cannot form endospores undergo other significant physiological changes that enhance their ability to survive environmental stressors. Regardless of the specific microbial strategy, genetic regulatory modification is involved. Common genetic regulatory factors, called sigma (s) factors, are frequently involved in enhanced stress resistance. Sigma factors produced in response to a stress bind to core microbial RNA polymerase, conferring different promoter specificities and leading to the production of stress proteins which protect the cell from the stress. *RpoS*, for example, is a regulatory factor required for transcriptional activation of a large number of genes required for tolerance to environmental stresses, including growth phase-dependent acid tolerance. The *rpoS*-deficient mutants were highly sensitive to food processing conditions compared with non-mutants.

Resistance of Microorganisms to Traditional Antimicrobials

Benzoic acid and its salts were one of the first groups of antimicrobials approved for application to foods in the United States. The primary application of benzoic acid and benzoates is to inhibit yeasts and moulds in acidic foods. Differences in microbial resistance to benzoates occur as a result of differences in innate tolerance. Because benzoates are used primarily as antifungal agents, one might conclude that bacteria are generally more resistant to these compounds than fungi. In fact, however, bacteria are quite variable in their resistance to benzoates. Benzoates are used primarily as antifungals because: (1) they function best in the undissociated state, which is the predominant form of the compound at low pH in high-acid foods; and (2) fungi are the primary spoilage microorganisms in acidic foods. Therefore, the innate resistance of yeasts and moulds to benzoates is of greater concern than that of bacteria. A number of yeasts, including *Schizosaccharomyces pombe* and *Zygosaccharomyces bailii*, have been observed to grow in the presence of about 500 µg/ml benzoic acid. Other yeasts, including *Pichia membranefaciens* and *Byssochlamys nivea*, are also known to be resistant to benzoates.

The mechanism by which yeasts develop resistance to weak acidic antimicrobials, including propionic as well as benzoic acids, is related to membrane permeability and the ability of the cells to continuously pump antimicrobials out of the cell. Some microorganisms on the other hand have innate resistance to benzoates because they metabolize the compounds. These bacteria (*Bacillus*, *Pseudomonas*, *Corynebacterium*, *Micrococcus*, and the mould *Aspergillus*) degrade benzoic acid through their β-ketoadipate pathway, in which benzoic acid is converted to succinic acid and acetyl coenzyme A. Few studies examine the potential for acquired resistance to benzoic acid. In a study, a variety of yeasts were incubated, including *Candida krusei, Hansenula anomala, Kluyveromyces fragilis, Kloeckera apiculata, Saccharomyces cerevisiae, Saccharomycodes ludwigii, S. pombe* and *Z. bailii,* in the presence of either 0.25 mM (31 µg/ml) or 2 mM (244 mg/ml) benzoic acid. The minimum inhibitory concentration (MIC) or lowest concentration preventing growth for unexposed cells was significantly lower for cells exposed to these concentrations of benzoic acid than for cells previously   exposed to subinhibitory concentrations of benzoic acid. Pre-exposure to benzoic acid **caused a 1.4 to 2.2 fold increase in MIC, with** *Z. bailii* and *S. pombe* exhibiting the greatest MIC increases. The proposed resistance mechanism was an increased cellular efflux. There was no evidence to indicate any increased resistance due to mutation nor any evidence that the resistance was stable.

Sorbic acid has been used as an antimicrobial in foods in the United States since the 1940s when it was patented for use in foods and on packaging to retard spoilage by moulds. Innate resistance to sorbate is demonstrated by bacteria, including catalase-negative lactic acid bacteria, *Sporolactobacillus,* some *Pseudomonas,* yeasts (including *Brettanomyces, Candida, Saccharomyces, Torulopsis,* and *Z. bailii),* and moulds (including *Aspergillus, Fusarium, Geotrichum, Mucor,* and *Penicillium).* As with benzoic acid, some microorganisms can metabolize sorbic acid. Moulds isolated from cheese, including seven *Penicillium* species, exhibited growth in the presence of and degradation of 0.3 to 1.2% sorbate. *Penicillium puberulum* and *Penicillium cyclopium* were the most resistant species evaluated. A study demonstrated that *Penicillium* species isolated from cheese produced 1,3-pentadiene, which has a kerosene off-odour, from sorbic acid. Sorbic acid is also degraded by *Mucor* species to 4-hexenol and by *Geotrichum* species to 4-hexenoic acid and ethyl sorbate. High numbers of lactic acid bacteria can produce ethyl sorbate, 2,4-hexadien-1-ol, 1-ethoxyhexa-2,4-diene, 5-hexadien-1-ol, and 2-ethoxyhexa-3,5-diene in sorbic acid-treated red wine. The 2,4 hexadien-1-ol metabolic product can cause "geranium" type off-odours in wines and fermented vegetables. For benzoic acid, there is a little evidence of acquired resistance to sorbic acid. *Z. bailii* grown in the presence of sorbic acid acquired resistance to subsequent exposure to the compound. There was little or no increase in the resistance of *Penicillium digitatum* or *Penicillium italicum* when exposed to increasing concentrations of sorbic acid according to one study. To combat the effects of sorbic and other organic acids, yeasts have several mechanisms by which they can develop resistance. One mechanism for acquired resistance that has been demonstrated among yeasts is the triggering of an inducible, energy-requiring system that increases the sorbic acid efflux. However, resistance of yeasts to sorbic acid and other weak acids probably involves more than one system.

The mechanism by which organic acids inhibit microorganisms involves passage of the undissociated form of the acid across the cell membrane lipid bilayer. Once inside the cell, the acid dissociates because the cell interior has a higher pH than the exterior. Protons generated from intracellular dissociation of the organic acid then acidify the cytoplasm and must be extruded to the exterior. Yeasts use the enzyme, H^+-ATPase, along with energy in the form of ATP to remove excess protons from the cell. Inhibition and/or inactivation may be due to eventual loss of cellular energy or inactivation of critical cellular functions due to low intracellular pH.

Another mechanism used to prevent depletion of energy pools involves the induction of a membrane protein that can decrease the activity of the ATPase to conserve energy. In addition, exposure of *S. cerevisiae* to sorbic acid can strongly induce a membrane protein ATP-binding cassette transporter (Pdr12), which is a "multidrug resistance pump" that confers resistance by mediating energy-dependent extrusion of anions. Mutants without the transporter are hypersensitive to sorbic, benzoic, and propionic acids. One problem with extruding anions and protons is the potential for recombination in the extracellular medium, thus allowing them to re-enter the cell. To prevent the futile cycle allowing the acid back into the cell, adapted yeasts apparently reduce diffusion and passage of the weak acids into the cell, most likely by altering cell membrane structures. Similar mechanisms likely also exist for bacteria that are capable of developing resistance to sorbic or other organic acids. Considering the length of time that sorbic and benzoic acids have been applied to food products it would seem, however, that the development of acquired resistance by spoilage and pathogenic microorganisms is very rare or non-existent. Pre-exposure to sub-inhibitory concentrations of other food antimicrobials has demonstrated varying resistance responses by microorganisms. For example, the effectiveness of methyl paraben and potassium sorbate was compared on the growth of four psychrotrophic food-borne bacteria—*Aeromonas hydrophila*, *L. monocytogenes*, *Pseudomonas putida* and *Yersinia enterocolitica*. They observed little or no adaptation when cells were exposed to subinhibitory concentrations of antimicrobials. Similarly, the relationship between lipid composition of *S. aureus* and resistance to parabens was compared. Differences in total lipid, phospholipids, and fatty acids were found for *S. aureus* strains that were relatively resistant and a strain that was sensitive to parabens. The paraben-resistant strain had a higher percentage of total lipid, higher relative percentage of phosphatidyl glycerol, and decreased cyclopropane fatty acids compared with the sensitive strains. It was suggested that these changes could influence membrane fluidity and, therefore, adsorption of the parabens to the membrane. Thus, a correlation exists between lipid composition of the microbial cell membrane and susceptibility to antimicrobial compounds.

Resistance of Microorganisms to Naturally Occurring Antimicrobials

Most of the attention on acquired resistance to naturally occurring antimicrobials has been focused on microbiologically derived antimicrobials. The probable reason for this is the similarity in form and/or ability to kill target cells that some of these compounds have to be medically important antibiotics. Because of the similarities, it has been suggested that use of microbiologically derived antimicrobials in foods may result in the development of acquired resistance to the compounds themselves or possibly cross resistance to antibiotics used in human medicine. In contrast to antibiotics used for therapeutic purposes, microbiologically derived antimicrobials generally have a much narrower spectrum of activity, i.e., affecting limited types of target microorganisms, and often having different mechanisms which may reduce chances for acquired resistance.

Two microbiologically derived antimicrobials that have been studied for their impact on the development of acquired resistance are natamycin and nisin. Natamycin, formerly called pimaricin, is an antifungal produced by *Streptomyces natalensis* that is effective against nearly all moulds and yeasts but has little or no effect on bacteria. Natamycin has no medical uses; however, it is used primarily as an antifungal agent on cheese. Nisin is a polypeptide composed of 34 amino acids that is produced by certain strains of *Lactococcus lactis* ssp. *lactis*. Nisin has a narrow spectrum of activity affecting primarily vegetative cells and spores of gram-positive bacteria. Susceptible strains are found among lactic acid bacteria, *Bacillus*, *Clostridium*, *Listeria*, and *Streptococcus*. The peptide alone generally does not inhibit gram-negative bacteria, yeasts, or moulds. The mechanism of antimicrobial action of nisin against vegetative cells includes binding to the anionic phospholipids of the cell membrane and insertion into the membrane, resulting in pore formation. Disruption of the cytoplasmic membrane causes efflux of intracellular components and eventual depletion of the proton motive force (PMF). Microorganisms exhibiting resistance to nisin may inactivate the peptide via enzymatic action or they may alter their membrane susceptibility. *Streptococcus thermophilus, Lactobacillus plantarum,* and certain *Bacillus* species that produce the enzyme nisinase neutralize the antimicrobial activity of the polypeptide. In addition, spontaneous nisin resistant mutants, including *L. monocytogenes, C. botulinum, Bacillus* species, and *S. aureus,* could occur via exposure of wildtype strains to nisin or transfer of strains in media containing increasing concentrations of nisin. *L. monocytogenes* resistant mutants, which are stable, may occur at a rate of 10^6 to 10^8 or even lower. It was observed that nisin-resistant strains of *L. monocytogenes* (NisR) had altered phospholipid composition, including decreased anionic phospholipid (cardiolipin and phosphatidylglycerol) and increased phosphatidylethanolamine in the cell membrane resulting in a decreased net negative charge that could hinder binding of cationic compounds such as nisin. In addition, the cell membranes of NisR strains exhibited increased long-chain fatty acids and reduced ratios of C15/C17 fatty acids, suggesting reduced fluidity and stabilization caused by reduced effect on PMF. These and other changes suggest an alteration of the cytoplasmic membrane to prevent access by nisin.

The obvious implication of the emergence of pathogenic microorganisms resistant to bacteriocins is the potential hazard in foods that are preserved exclusively by a single compound. To overcome this potential hazard, some researchers suggest using combinations of bacteriocins or combinations of bacteriocins with other antimicrobials or preservation methods. Combinations of bacteriocins could be successfully applied if the mechanisms of action of the bacteriocins are different. However, even this strategy must be validated for each combination since one study demonstrated that *L. monocytogenes* ATCC 700302 was both nisin- and pediocin-resistant. Cross resistance among bacteriocins, however, is variable. There was no reported cross resistance between nisin- and pediocin-resistant strains of *L. monocytogenes* but pediocin and bavaricin cross resistance was observed. Because microorganisms in foods are often exposed to some variation of acidic environmental conditions it is interesting to speculate whether acid adaptation of a microorganism could alter its sensitivity to bacteriocins. On investigation of acid adaptation at pH 5.5 and bacteriocin sensitivity during a study, it was found that acid-adapted *L. monocytogenes* was more resistant to nisin and lacticin 3147. The difference in resistance between the acid-adapted and non-adapted cells was more noticeable with nisin than with lacticin 3147. The potential for this change in resistance in a food system remains uncertain, however, because the experiment was done in a microbiological medium (tryptic soy broth supplemented with 0.6% yeast extract).

The most important question concerning the potential for microorganisms to acquire resistance to bacteriocins is whether such resistance conveys a natural advantage over non-resistant strains in food systems. It was demonstrated that nisin-resistant strains of *L. monocytogenes* and *C. botulinum* were not as resistant as wild-type strains to other traditional food antimicrobials including sodium chloride, sodium nitrite and potassium sorbate. Leucocin- and sakacin-resistant *L. monocytogenes* B73 had a reduced growth rate in a microbiological growth medium (brain heart infusion broth) without bacteriocin than bacteriocin sensitive strains. In addition, NisR strains failed to compete with bacteriocin-sensitive strains when grown in mixed populations, even at a 1:1 ratio. Thus it was concluded that the bacteriocin-resistant phenotype of *L. monocytogenes* B73 was not likely to become stable in natural populations. It was also observed that pediocin-resistant *L. monocytogenes* frequently exhibited a reduced growth rate and extended lag phase in a microbiological broth medium compared to wild-type cells. However, nisin-resistant *L. monocytogenes* strains had fewer and less pronounced growth rate reductions. Interestingly, pediocin- and nisin-resistant strains were no more stress susceptible (pH, salt, low temperature) than sensitive strains, and they grew equally well in a model sausage system as the parent strains. It was demonstrated that nisin-resistant *C. botulinum* 169B spores had similar heat resistance patterns as wild type spores. Therefore, while acquired resistance to a single bacteriocin neither automatically confer resistance to other antimicrobials or preservative treatments nor any natural advantage for a population in the absence of the inhibitor, more research in food systems is definitely warranted.

Resistance of Microorganisms to Other Processing Conditions

It has been shown repeatedly in laboratory situations that bacteria can become resistant to certain environmental factors under conditions that would normally be considered lethal to the organism. For example, *E. coli* O157:H7, *Salmonella typhimurium*, and *L. monocytogenes* can become more acid-resistant and possibly more resistant to other stresses (e.g. heat, osmotic pressure), if subjected to relatively mild acidity before exposure to more acidic conditions. Developed resistance is referred to as tolerance, adaptation, or habituation depending upon how the microorganism is exposed to the stress and the physiological conditions that lead to enhanced survival. In addition, production of acidic conditions by the microorganism itself can produce acid tolerance. For example, growth of *E. coli* in an acidogenic broth (acid generating) (e.g. tryptic soy broth [TSB] + glucose) produced cells that expressed an acid resistance response while cells grown in a non acidogenic broth (TSB without glucose) did not. Whether this occurs in actual food environments during processing situations or in the product itself is intriguing, but largely unanswered. If pH resistance is the issue, then pH levels typically found in foods of concern should be evaluated. In a conducted study, *L. monocytogenes* exhibited an acid tolerance response when it was acid-adapted to pH 5.5 with lactic acid and then challenged in acidified skim milk at pH 3.5 and 4.0. When the challenge pH of 4.5 was used, however, there was no adaptive acid tolerance response. Because the pH of 4.5 is more closely related to pH levels that might occur in fermented products made from skim milk, results based on this pH may be the most meaningful. In apple, orange, and white grape juices, it was found that the thermal resistance of *E. coli* O157:H7, *Salmonella,* and *L. monocytogenes* increased after acid-adaptation. The typical pasteurization process applied to these types of fruit juices provides sufficient thermal inactivation of these pathogens, regardless of whether or not they have enhanced thermal resistance due to acid adaptation. Many studies of acid shock and acid adaption of bacteria have been conducted. Most studies have evaluated these responses over a relatively short period of time (typically a few hours). Within foods and on food contact surfaces, adaptive alteration of cells might be of more concern if the resistance is sustained by the adapted cells. The greater the degree of severity of the antimicrobial challenge, the less likely it will be for the microorganism to survive for extended periods, even if it is adapted. Numerous product and antimicrobial combinations are possible in foods. Under certain scenarios, there may be reason for concern when considering whether or not bacteria can acquire some degree of resistance to a particular antimicrobial. Direct acidification of a food or food ingredient may shock microflora. So they become more acid-resistant. Fermentation of foods, however, may lead to somewhat different situations. Lactic acid bacteria can lower the pH of a substrate gradually over time, likely resulting in a pH gradient rather than a sharp change in pH as would be expected with direct acidification. An acid adaptive response in *E. coli* O157:H7 was reported which enhanced its survival in fermented sausage (pH 5.6). The use of acidic antimicrobial sprays on the surfaces of meat carcasses has become very common. One could ask if bacteria on the meat surface become more acid-resistant when they are exposed to a low concentration of a weak acid (e.g. <3%) solution. Whether or not the exposure of microorganisms to acids on substrates such as meat is sufficient to result in acquired resistance is largely unknown at this time. One study demonstrated a lack of increased resistance to lactic acid for *E. coli* O157:H7, *S. typhimurium, S. aureus,* and *C. jejuni* when acid-adapted cells were inoculated on pork bellies and treated with 2% lactic acid as a sanitizer. However, most other studies related to acquired stress response have been done either in laboratory culture media with lower than typically encountered concentrations of antimicrobials, or in foods under conditions which might be only marginally similar to actual situations. In another study, *E. coli* O157:H7, which was adapted to acid under mild acidic conditions, acquired cross protection to heat in an acidic environment. Thus it was concluded that this might have implications for fermented meat production and similar processes, which are also given a thermal process. However, since this study was done in culture broth, it does not fully mimic what could occur in the more complex food matrix. Several studies involving acid-adapted microbes in heated and or dried meat products have been conducted. In many of these, acid adaptation did not offer any enhanced protection to other environmental stresses to which the microorganism was subsequently exposed in an actual food product. In addition to looking at the possibility of *L. monocytogenes* becoming more acid-resistant through an acid tolerance response, in a study conducted to determine if acid adaptation of the microorganism by cross protection enhanced survival in the presence of an activated lactoperoxidase system, a naturally occurring antimicrobial system in milk, it was observed that the survival rates were similar for the acid-adapted and non-adapted cells at pH 4.5 both in the presence and absence of an activated lactoperoxidase system, indicating that no cross protection was afforded the acid-adapted cells. In contrast, a cross protection to an activated lactoperoxidase system was reported in another study with acid-adapted *S. typhimurium* in a laboratory culture medium. The activity of the lactoperoxidase system can vary depending on the medium, a possible explanation for the contrasting results. Another possible explanation for the difference in the studies may be the greater degree of acid tolerance exhibited by *Salmonella* than *Listeria*. The lactoperoxidase system may also be a less effective antimicrobial towards *Salmonella* compared with *Listeria*. It was also reported that acid-adapted *S. typhimurium* offered cross protection against heat, salt, and selected surface active agents. It was found in a study that acid-adapted *L. monocytogenes* also exhibited increased resistance to heat, cold, salt, selected surface active agents, and ethanol. However, another study also reported that when populations of *S. typhimurium* were acid-adapted at a pH of 5.0 to 5.8, the sensitivity of the cells to hypochlorous acid and iodine increased. The mechanism of the inactivation was also investigated by hypochlorous acid and was concluded that, whether the cells were acid-adapted or not, the mechanism involved changes in membrane permeability, inability to maintain or restore energy charge, and probably oxidation of essential cellular components. It was proposed that acid pretreatment in a food plant sanitation program may thus enhance the efficiency of halogen sanitizers. Using a *L. monocytogenes* isolate from a food processing plant drain, the survival and heat resistance of the microorganism after exposure to alkaline pH was investigated (up to pH 12.0) and after exposure to chlorine (up to 6.0 mg of free chlorine per litre). The alkaline stress enhanced the resistance of *L. monocytogenes* to thermal processing conditions of 56 and 59°C. In contrast, exposure of the cells to chlorine resulted in populations that were more sensitive to heating at 56°C. Based on these results, there are differences in the cross protection offered after exposure to alkaline pH depending on whether an alkaline- or chlorine-based sanitizer is used in sanitation routines.

Because evidence exists that microorganisms can acquire varying levels of resistance or tolerance to environmental stresses, there is some concern that this might provide protection for food-borne pathogens against antimicrobials and preservation processes. Development of resistance to manufacturing and processing treatments could occur at numerous points in a food production system and could influence preservation treatment efficacy. However, before decisions can be made with confidence concerning the development of tolerance or resistance among food-borne pathogens, it is critical to acquire data that is relevant to real food processing situations. In addition, conditions most representative of those in actual products or food processing situations must be included in experimental protocol. For example, as was noted in a study, the physiological state of food-borne pathogens used in challenge studies in food and in evaluating Hazard Analysis Critical Control Point programs is an important consideration because the effectiveness of control measures may vary with varied microbial physiological states. A major problem with the use of antimicrobial combinations, however, is the general lack of knowledge about food antimicrobial mechanisms. Without this information, antimicrobials cannot be applied effectively to achieve synergistic interactions. More research is needed to determine: (1) the frequency and mechanisms of resistance to food antimicrobials and process stresses in food systems, (2) mechanisms of action of traditional and naturally occurring food antimicrobials, and (3) application strategies to minimize tolerance or resistance development. Simple methods for overcoming the potential for development of acquired resistance include using appropriate antimicrobials, avoiding the use of sublethal concentrations of antimicrobials, using combinations of antimicrobials for environmental or process controls (such as hurdle technology) and using combinations of antimicrobials that have different mechanisms. Although data concerning development of resistance to food antimicrobials is scarce, antimicrobial resistance does not appear to be a phenomenon that would have a major negative impact on public health.

REVIEW QUESTIONS

1. Define food antimicrobials.
2. Give a detailed account of mycotoxic antimicrobial additives.
3. Give a detailed account of bactericidal antimicrobial additives.
4. Discuss in detail about bacteriocins and their use as food preservatives.

11

MICROBIOLOGY OF PRESSURE-TREATED FOODS

INTRODUCTION

High hydrostatic pressure has the potential to produce high-quality foods that are microbiologically safe with an extended shelf life. Microorganisms vary in their response to high pressure. Bacterial spores are the most resistant group and they cannot be significantly inactivated by pressure alone. Combination treatments using high pressure and heat have been proposed as a method of producing shelf-stable low-acid foods. Viruses are less resistant than bacterial spores and their infectivity can be abolished without destroying their ability to elicit antibodies, leading to the possibility of vaccine production. Yeasts, moulds and vegetative bacteria vary in their response to pressure, depending on factors such as species, strain, processing temperature and substrate. A knowledge of how these factors interact is necessary in order to select the optimum processing conditions for food.

Consumers, today demand high-quality foods that are free from additives, have fresh taste, microbiologically safe with an extended shelf life. A food technology that has the potential to meet these demands is high pressure processing. High pressure processing (HPP), also known as high hydrostatic pressure (HHP) or ultra-high pressure processing, uses pressures up to 900 MPa (9000 atmospheres, 135, 000 pounds per square inch) to kill many of the microorganisms found in foods, even at room temperature. These pressures are immense. [A mid-range food processing pressure of 500 MPa is equivalent to the weight of three elephants on a strawberry].

The idea of using high pressure in food processing is not new. The first report of high pressure being used as a food preservation method was in the year 1899. It was reported that milk "kept sweet for longer" after a pressure treatment of

600 MPa for 1 hour at room temperature. It was also reported in 1914 that while pressure could be used to extend the shelf life of fruits, it was less successful with vegetables. It was concluded that fruits and fruit juices responded well to high pressure because the "yeasts and other organisms responsible for decomposition are very susceptible to pressure". In vegetables, this treatment is not successful due to the presence of spore-forming bacteria that survived the pressure treatment and could grow in the low-acid environment. The problem of pressure-resistant spores still remains one of the challenges for the technology today. Much experimental data has been produced over the last 100 years but it was not until the early 1990s that the first commercial food applications of the technology were seen. There are considerable engineering problems involved in repeatedly generating and containing the immense pressures in a vessel suitable for food products. However, within the last 25 years a range of specialized high-pressure vessels, based on those used routinely in the production of polymers, ceramics and artificial diamonds, became available and reopened the possibility of commercial production of pressure-treated foods.

HIGH PRESSURE PROCESSING EQUIPMENT

A typical pressure treatment system consists of a pressure vessel, the pressure transmission fluid (usually water) and one or more pumps to generate the pressure. It is traditionally a batch process and pressure vessels used for commercial food production having capacities of 35–350 L.

Food packages are loaded into the vessel, the top is closed and the pressure transmission fluid is pumped into the vessel from the bottom. Once the desired pressure is reached, pumping is stopped, valves are closed and the pressure can be maintained without further need for energy input. The pressure is transmitted rapidly and uniformly throughout the pressure-fluid and the food. The high pressure is applied in an isostatic manner so that all parts of the food are subjected to the same pressure at exactly the same time, unlike heat processing where temperature gradients are established. As it is equal from all sides, the pressure does not significantly affect the product shape. The pressure is released after the desired treatment time and the food packages can be unloaded. In the case of liquids, such as fruit juices, the whole vessel can be filled with the juice, which itself becomes the pressure transmission fluid. After treatment, the juice can be transferred to an aseptic filling line, similar to that used for UHT (ultra-high pressure treatments) liquids. A series of these vessels can work in sequence, with a vessel filling with juice, a vessel pressurizing and another emptying, all operating simultaneously, so the overall system can become semi-continuous.

High-pressure equipment suitable for food use is specialized and the capital equipment cost is relatively high, although running costs are relatively low. The cost of a typical commercial vessel varies depending on its size.

FUNDAMENTAL EFFECTS OF PRESSURE ON MICROBIAL CELLS

Effect on Cell Membranes

The cell membrane is generally acknowledged to be a primary site of pressure damage in microorganisms. Evidence of physical damage to the cell membrane has been demonstrated as leakage of ATP or UV-absorbing material from bacterial cells subjected to pressure or increased uptake of fluorescent dyes such as propidium iodide that do not normally penetrate membranes of healthy cells. Stationary-phase cells are normally more pressure-resistant than exponential-phase cells. A study proposed that exponential-phase cells are inactivated under high pressure by irreversible damage to the cell membrane. In contrast, stationary-phase cells have a more robust cytoplasmic membrane that can better withstand pressure treatment. This proposal was based on the fact that exponential-phase cells showed changes in their cell envelopes that were not seen in stationary-phase cells. These changes included physical perturbations of the cell envelope structure, a loss of osmotic responsiveness and a loss of protein and RNA to the extracellular medium. Loss of membrane functionality resulting from pressure treatment has also been described in *Lactobacillus plantarum*. The ability to maintain an internal pH was also reduced and the acid-reflux was impaired.

Effect on Cell Morphology

The cell wall is less affected by high pressure than the membrane and generally no morphological changes can be observed in prokaryotes and lower eukaryotes while observed under a light microscope. However, intracellular damage can be observed using electron microscopy. Using scanning electron microscopy (SEM), it was reported that bud scars appeared on the cell surface of *Listeria monocytogenes* after a 10 minutes pressure treatment at 400 MPa in citrate buffer. Similarly, the effect of pressure was studied on the ultrastucture of *L. viridescens*. Nodes on cell walls of organisms treated at 400 MPa and above for 5 minutes at 25°C were observed using SEM. Transmission electron micrographs indicated empty cavities between the cytoplasmic membrane and the cell wall after this treatment.

Effect on Biochemical Reactions

High pressure treatment favours biochemical reactions that lead to a volume decrease while it can inhibit or retard reactions that lead to a volume increase. Most biochemical reactions result in a volume change and are therefore affected by pressure. Studies carried out on volume changes in proteins have shown that the main targets of pressure are hydrophobic and electrostatic interactions while hydrogen bonding, which stabilizes the α-helical and β-pleated sheet forms of proteins, is not significantly influenced by pressure. Enzymes vary greatly in their ability to withstand pressure. Certain microbial enzymes, such as *Bacillus subtilis* *a*-amylase, can withstand pressures of 500 MPa, while others, such as *L. monocytogenes* phosphoglucomutase and aconitase, are inactivated by 200 MPa. However, in the latter case, *L. monocytogenes* was little affected at this pressure treatment, suggesting that the inactivation of these enzymes was not critical to survival.

Covalent bonds are generally unaffected at the pressures used in food processing. This means that many of the components responsible for the sensory and nutritional quality of foods, such as flavour components and vitamins, are not destroyed by high pressure. This is an advantage of pressure treatment.

Effect on Genetic Mechanisms

Nucleic acids are relatively resistant to high pressures and as the structure of the DNA helix is largely the result of hydrogen bond formation, it is also stable under pressure. However, the enzyme-mediated steps involved in DNA replication and transcription are disrupted. It has been reported that pressure causes a condensation of nuclear material in *L. monocytogenes*, *Salmonella typhimurium* and *L. plantarum*. It was postulated that at elevated pressures, DNA comes into contact with endonucleases that cleaves the DNA.

HIGH PRESSURE INACTIVATION OF MICROORGANISMS

Inactivation Kinetics

High pressure inactivation of microorganisms is complex, and plotting the log of surviving numbers against time does not always form a linear relationship (first-order kinetics). Often there is an initial linear decrease in numbers followed by a decrease in the killing rate leading to a pressure-resistant 'tail'. Studies have shown that when this 'tail' population is isolated, grown and again exposed to pressure, there is no significant difference in pressure resistance between it and the original culture. Such tails are also found with heat processing but the phenomenon seems to be more pronounced with high pressure processing. The tailing effect is not fully understood. It may be because of inherent phenotypic variation in pressure resistance in some cells. Experimental conditions, such as the substrate and growth conditions, may also be a factor. Tailing phenomena can make calculation of pressure D values difficult and have to be taken into account when doing future studies designed to optimize processing conditions for various foods.

Pressure Injury

A high pressure treatment may not always completely inactivate microorganisms but rather may injure a proportion of the population. Recovery of the injured cells will depend on the conditions after treatment and this has implications for microbiological enumeration. Compounds such as sodium chloride added to plating media can act as selective agents and inhibit growth of injured cells. Using selective media can, therefore give inaccurate estimates of numbers of survivors.

MICROBIAL RESPONSES TO PRESSURE

Microorganisms vary in their response to pressure.

Vegetative Bacteria

In general terms gram-positive bacteria tend to be more resistant to pressure than gram-negatives and cocci are more resistant than rod-shaped bacteria. It has been suggested that the cell membrane structure is more complex in gram-negative bacteria, making it more susceptible to environmental changes caused by pressure. However, there are many exceptions to these general rules. Certain strains of *Escherichia coli* O157:H7, for example, can be exceptionally pressure-resistant. However, *Salmonella* have shown only a weak or no correlation between pressure resistance and other stresses.

Bacterial Endospores

Bacterial endospores can be extremely resistant to high pressure, just as they are resistant to other physical treatments such as irradiation and heat, and can survive treatments of more than 1000 MPa. There can be significant variation between spores of different species and also between strains of the same species. *Clostridium botulinum* spores are among the most pressure resistant, especially non-proteolytic type B. However, relatively low pressures (below 200 MPa) can trigger spore germination. This has led to the suggestion that spores could be killed by applying pressure in two stages. The first pressure treatment would germinate the spores while the second treatment, at a higher pressure, would kill the germinated spores. This process could be repeated several times leading to the idea that pressure cycling between relatively low and then high pressures could be one way of overcoming the problem of spore resistance. However, the extent of the inactivation can be highly variable and a small proportion of each spore population seems to remain resistant to pressure-induced germination.

Another approach to the problem of the pressure resistance of bacterial spores is to combine high temperature along with pressure treatment. There have been many reports indicating that this can be very successful. This approach is now being actively considered for the commercial production of shelf-stable foods and is the subject of a number of patents designed to achieve the commercial sterilization of foods that have a pH greater than 4.5. One such patent describes a process involving two or more cycles of high heat (>70°C) and high pressure (>530 MPa) with a pause between the cycles. The temperature, pressure level, treatment time and time interval between the cycles can be varied depending on the product but are designed to give greater than the equivalent of a 12D process for *C. botulinum*. These treatments use an initial temperature of below 100°C and rely on the fact that adiabatic heating occurs when the product is pressure-treated. Adiabatic heating results from the work of compression during pressure treatment leading to an increase in the temperature of food. The extent of the temperature increase varies with the composition of the food but is normally 3–9°C/100 MPa. The overall treatment conditions are less severe than conventional retorting. This results in shelf-stable products which are of higher quality in terms of texture, flavour and retention of nutrients than those obtained by conventional processing.

Viruses and Prions

There is relatively little information on pressure inactivation of viruses compared with other microorganisms but it does appear that viruses can vary significantly in their response to treatment. Polio virus in tissue culture medium appears to be relatively resistant with 450 MPa for 5 minutes at 21°C giving no reduction in plaque-forming units (pfu). The same treatment conditions with hepatitis A resulted in a $6\log_{10}$ pfu/ml stock culture being reduced to undetectable levels. However, treatment in seawater increased the pressure resistance of hepatitis A virus, suggesting a protective effect of the salts. Feline calicivirus, a Norwalk virus surrogate, and human rotavirus were more pressure-sensitive than hepatitis A when treated in tissue culture medium. There are also reports that HIV-1 is relatively sensitive to pressure with 400–600 MPa for 10 minutes at 25°C resulting in $4–5\log_{10}$ reduction in viable particles when treated in tissue culture medium, although different strains varied in their pressure resistance. Foot-and-mouth disease virus (FMDV) is also relatively sensitive to high pressure. A treatment of 250 MPa at –15°C for 1 hour destroyed FMDV infectivity but maintained the integrity of capsid structure. The treated virus could also elicit neutralizing antibody production in rabbits. These results suggest that high pressure could be a safe, simple, cheap and reproducible method of producing viral vaccines. Evidence is emerging that high pressure may have some effect on prions. Further work still needs to be done but the early work suggests the possibility of producing safe products for specialized, high added-value markets, such as baby foods.

Yeasts and Moulds

Yeasts are generally not associated with food-borne disease but are important in spoilage, especially in acidic foods. They are relatively sensitive to pressure and this is one reason why pressure treatment of fruit products to extend shelf life is particularly successful.

There is relatively little information on the pressure sensitivity of moulds but it has been shown that vegetative forms are relatively sensitive, while ascospores are more resistant. The effect of pressure on pre-formed mycotoxins is thought to be limited as the treatment has little effect on covalent bonds. However, one study reported that patulin, a mycotoxin produced by several species of *Aspergillus, Penicillium* and *Byssochlamys*, was found to be degraded by pressure. The patulin content in apple juice decreased by 42, 53 and 62% after 1 hour treatment at 300, 500 and 800 MPa, respectively, at 20°C.

EXTRINSIC FACTORS AFFECTING THE SENSITIVITY OF MICROORGANISMS TO HIGH PRESSURE

High pressure is not different from other physical preservation methods in that its effectiveness against microorganisms is influenced by a number of factors. These all interact and contribute to the lethal effect and therefore have to be considered when designing process conditions to ensure the microbiological safety and quality of pressure-treated foods.

Effect of Substrate

The chemical composition of the substrate during treatment can have a significant effect on the response of microorganisms to pressure. Certain food constituents such as proteins, carbohydrates and lipids can have a protective effect. Inactivation data obtained using buffers or laboratory media, therefore, should not be extrapolated to real food situations where a more severe pressure treatment may be needed to achieve the same level of inactivation. For example, a treatment of 375 MPa for 30 minutes at 20°C in phosphate buffer (pH 7) gave a 6 $\log_{10}$ inactivation of a pressure-resistant strain of *E. coli* O157:H7. However, the same treatment gave a 2.5 $\log_{10}$ reduction in poultry meat and only 1.75 $\log_{10}$ reduction in milk. Cations, such as Ca^{2+}, can be baroprotective and this may explain why many microorganisms appear more pressure-resistant when treated in certain foods, such as milk.

A low water activity protects microorganisms against the effects of pressure. It was reported that reducing a_W of the medium from 0.98–1.0 down to 0.94–0.96 resulted in better survival of *Rhodotorula rubra* when it was subjected up to 200–400 MPa for 15 minutes at 25°C. However, the nature of the solute is important. At the same a_W, cells were more pressure-sensitive in glycerol than in monosaccharides and disaccharides. Trehalose is reported to confer most protection.

The pH of acidic solutions decreases as pressure increases and it has been estimated that in apple juice, there is a pH drop of 0.2 per 100 MPa. To date, the pH change which occurs during pressure treatment cannot be measured directly in solid foods, but methods have been developed for *in situ* pH measurement during pressure treatment of liquids. When the pressure is released, the pH reverts to its original value but it is not known whether these sudden changes in pH affect microbial survival in addition to the effect of pressure. It is known that pH and pressure can act synergistically leading to increased microbial inactivation. It was shown in a study that initial pH had a significant effect on inactivation rates of *E. coli* O157:H7 in orange juice. As pH was lowered, the cells were more susceptible to pressure inactivation and sublethally injured cells failed to repair and died more rapidly during subsequent storage of the juice.

Food additives can have varying effects on microbial resistance to pressure. Pressure has been used to sensitize gram-negative bacteria such as *Salmonella*, as well as gram-positive bacteria such as *L. monocytogenes* to nisin and lysozyme. The combination of high pressure and nisin was also used to increase the inactivation of *B. cereus* spores in cheese. However, in this case the pressure-treatment conditions were relatively severe. The treatment (60 MPa at 30°C for 210 minutes to germinate the spores followed by 400 MPa at 30°C for 15 minutes to kill the vegetative cells) was carried out in the presence of 1.56 mg/ml nisin in the cheese. This resulted in approximately 2.4 $\log_{10}$ inactivation of the spores.

Pediocin AcH also works synergistically with pressure. A combination of 345 MPa for 5 minutes at 50°C in the presence of 3000 AU/ml pediocin AcH gave at least a 7 $\log_{10}$ inactivation of a range of bacteria including *L. monocytogenes*, *S. typhimurium*, *Staphylococcus aureus*, *E. coli* O157:H7, *L. sake* and *Pseudomonas fluorescens*. This level of inactivation could not be achieved by pressure. Similarly, the combination of high pressure with other antimicrobial agents such as lacticin 3147, lactoperoxidase and carvacrol can work synergistically to enhance the killing of microorganisms, including pathogens.

Effect of Temperature

Temperature during pressure treatment can have a significant effect on microbial survival. Increased inactivation is usually observed at temperatures above or below 20°C.

High temperatures (>70°C) can be particularly effective in helping to achieve high pressure sterilization, as discussed above. The combination of elevated temperatures (<50°C) with pressure has also been suggested as a practical way to overcome the problem of pressure-resistant strains of vegetative cells. Approximately a 6 $\log_{10}$ inactivation of a pressure-resistant *E. coli* O157:H7 was reported in poultry mince and a 5 $\log_{10}$ inactivation in milk using a treatment of 400 MPa at 50°C for 15 minutes. Neither heat nor pressure alone could achieve this level of inactivation.

Refrigeration temperatures can also enhance pressure inactivation. It was reported in a study that ewes' milk pressurized at 450 MPa at 2°C for 15 minutes was more effective at inactivating *L. innocua* than the same treatment at 25°C but less effective than the pressure treatment at 50°C.

PRESSURE TREATMENT TO IMPROVE THE MICROBIOLOGICAL QUALITY OF FOODS

Pressure-treated fruit jams and sauces first became commercially available in Japan in the early 1990s. Treatment of fruit jams with around 400 MPa for up to 5 minutes at room temperature can significantly reduce the number of microorganisms, especially yeasts and moulds. Refrigeration of the jam after processing is necessary due to browning and flavour changes caused by enzymatic activities. These products have a shelf life of around 30 days and have superior sensory quality compared with those prepared in a conventional manner.

Fruit juices are normally processed at 400 MPa or greater for a few minutes at 20°C or less. This can significantly reduce number of yeasts and moulds and so extend shelf life for up to 30 days. Pathogens, such as *E. coli* O157:H7 can also be destroyed by this treatment. Pressure-treated orange and grapefruit juices have been available in France since 1994, while pressure-treated apple juice is available in Italy.

Pressure treatment of vegetable products is problematic because of their relatively high pH along with the possibility of survival and growth of pathogenic spore-forming organisms.

Sliced-cooked ham and other delicate meat products, in flexible pouches, can be successfully treated using 500 MPa for a few minutes. The sensory properties of ham are preserved, and shelf life can be extended to 60 days under chilled storage. Cooked delicate products have a risk of post-processing contamination from pathogens such as *L. monocytogenes*. High pressure treatment as a final preservation step, after packaging, can give additional microbiological safety assurance.

Another example of commercially successful pressure-treated foods available in the USA are oysters. The initial aim of the pressure treatment was to eliminate *Vibrio* spp. from oysters, which are often eaten raw or only lightly-cooked. *Vibrio* spp. are relatively sensitive to high pressure, although there can be species variation. Typical treatments of 250–350 MPa for 1–3 minutes at ambient temperature are used commercially without significantly affecting sensory quality. An additional benefit of pressure-treated oysters is the mechanical shucking effect it causes, releasing the adductor muscle from the shell. For this reason, a heat shrink plastic band is placed around each oyster prior to processing so that the shell is kept shut until the meat is required. Pressure will successfully shuck other shellfish, such as mussels, *Nephrops* and crabs as well as improving their microbiological quality and it is likely that the technology will be more widely used for this purpose. There is also increasing interest in pressure-treating finfish to improve microbiological safety and quality as well as using the technology to produce a range of novel food products.

Pressure treatment of milk and dairy products to improve microbial safety and quality has been of interest since the early work in 1899. However, it is likely that the technology will only be used commercially for niche applications, where it can provide a commercial advantage over existing, usually heat-treated products. For example, pressure treatment may be of value in treating milk that is to be used in the manufacture of raw milk cheese, where it could reduce the numbers of pathogens such as *L. monocytogenes*. However, some pathogens, such as certain strains of *E. coli* O157:H7, are known to be extremely pressure-resistant in milk. Therefore, this approach will not solve all the microbiological safety problems associated with raw milk cheese. High pressure treatment of yoghurt has also been investigated. The range of products now being considered for high pressure treatment is increasing steadily.

REVIEW QUESTIONS

1. Give an account of the effects of pressure on microbial cells.
2. Explain how different microorganisms react to pressure.

12

NEW PRESERVATION TECHNOLOGIES

INTRODUCTION

The increasing consumer demand for 'fresh-like' foods has led to much research effort in the last 20 years to develop new mild methods for food preservation. Non-thermal methods allow microorganisms to be inactivated at sublethal temperatures thus better preserving the sensory, nutritional and functional properties of foods. The mechanisms of inactivation, sensitivity of different microbial groups and factors affecting it and kinetics of inactivation are discussed. Microorganisms are the main agents responsible for food spoilage and food poisoning and therefore food preservation procedures are targeted towards them. Food preservation methods currently used by the industry rely either on the inhibition of microbial growth or on microbial inactivation. Methods which prevent or slow down microbial growth cannot completely assure food safety, as their efficacy depends on the environmental conditions such as, for instance, the maintenance of the chill chain. Thermal treatment is the most widely used procedure for microbial inactivation in foods. However, heat causes unwanted side effects in the sensory, nutritional and functional properties of food. These limitations together with increasing consumer demand for fresh-like foods has promoted the development of alternative methods for microbial inactivation, among which ionizing irradiation, ultrasound under pressure, high hydrostatic pressure (HHP) and pulsed electric field (PEF) are very important interest. The irradiation process involves the application of electromagnetic waves or electrons to foods. Radiation sources are either gamma rays from cobalt-60, electron beams or X-rays, and the amount of irradiation absorbed by a food is measured in kGy (1 Gy 1/4 1 J kg^{-1}). Commercial application of ionizing radiation (IR) treatment on foods was started at the beginning of the 1980s, but its success has been prevented by consumer concerns.

Nowadays, social perception of IR is changing and this technology is being re-examined. Ultrasound is defined as sound waves with frequencies above the threshold of human hearing (>16 kHz). Although ultrasound was initially discarded for food preservation because of its weak lethal action, the application of an external hydrostatic pressure of up to 600 kPa [manosonication (MS) increases substantially the lethality of the treatment. In addition, a combination of MS with temperature manothermosonication (MTS)] has been proposed. The HHP involved the application of pressures from 100 to 1000 MPa. The initial studies on the lethal effect of HHP were conducted at the end of the 19th century, but the commercial applications of this procedure have started. Finally, PEF technology uses the application of short duration (1–100 ls) high electric field pulses (10–50 kV cm^{-1}) to a food placed between two electrodes. Like ultrasound under pressure, PEF technology is not yet being used to preserve food commercially.

Food preservation is a continuous fight against microorganisms spoiling the food or making it unsafe. Within the disposable arsenal of preservation techniques, the food industry investigates more and more for the replacement of traditional food preservation techniques (intense heat treatments, salting, acidification, drying and chemical preservation) by new preservation techniques due to the increased consumer demand for tasty, nutritious, natural and easy-to-handle food products. The most investigated new preservation technologies are non-thermal inactivation technologies such as high hydrostatic pressure (HHP) and pulsed electric fields (PEF), new packaging systems such as modified atmosphere packaging (MAP) and active packaging, natural antimicrobial compounds and biopreservation. In spite of the intensive research efforts and investments, very few of these new preservation methods are until now implemented by the food industry.

NEW PRESERVATION TECHNOLOGIES

NON-THERMAL INACTIVATION TECHNOLOGIES

Non-thermal inactivation techniques include ionizing radiation, HHP, pulsed electrical fields, high pressure homogenization, UV decontamination, pulsed high intensity light, high intensity laser, pulsed white light, high power ultrasound, oscillating magnetic fields, high voltage arc discharge and streamer plasma.

High Hydrostatic Pressure

HHP is the technology by which a product is treated at or above 100 MPa. For the last 15 years, the use of HPP has been explored extensively in food industry. In recent years, HPP has been extensively used in Japan and a variety of food products like jams and fruit juices have been processed.

There have been 10 to 15 types of pressurized foods on the Japanese market, but several have disappeared, and those that remain are so specific that they would have little interest to European or American markets. Nevertheless, interest in HPP derives from its ability to deliver foods with fresh-like tastes without added preservatives.

The industrial equipment used at present are discontinuous (from 10 to 500 L of capacity) for solid, viscous and particulated foods: batch processing, and semicontinuous (from 1 to 4 ton per hour of production) for liquid foods: bulk processing. The processing cost of some recent equipment has been estimated on 10–15 Eurocent per kg of product, inclusive of investment and operation costs. Commercial production of pressurized foods has been reported for fruit jams, jellies, sauces, juices, rice wine, cake, avocado pulp, guacamole and cooked ham. For dairy applications, the technique has been studied, among others, to improve the shelf life of goat's cheese to reduce the ripening time of cheese to 3 days at 250 MPa and to prevent over-acidification of yoghurt, increasing the shelf life to more than 2 weeks at 4°C when treated at 200 MPa for 15 min. at 20°C. The possibility of using HHP to reduce milk allergenicity of milk has been demonstrated by specific hydrolysis of β-lactoglobulin at 250 MPa. Many reports have demonstrated the inactivation effect of HHP on microorganisms, extending in this way the microbial shelf life and improving the microbial safety of food products. An overview of the sensitivity of microorganisms at several processing conditions and in several substrates was given by several scientists. Substantial count reductions ($>4 \log_{10}$ units) of most vegetative microorganisms are realized when a pressure treatment of 400–600 MPa at room temperature is applied. However, pressure treatment alone is often not sufficient for substantial reduction of viable spore counts. Spores of some species survive pressures above 1000 MPa, when the temperature is not higher than 45–75°C. Pressure-induced germination of spores has been examined and subsequent pressure treatment of the germinated/germinating spores was shown to be an effective means of reducing spore counts. However, 'superdormant' spores which may be present in the tail of a germination curve, may also be more resistant to germination induced by pressure, making HHP not suitable as such for sterilization of food products. To cope with this important drawback of HHP, combination treatments have been suggested from which the combination with an increased temperature was mostly investigated. An overview of the investigated combinations to inactivate bacterial spores by HHP is given in the Table 12.1. Several studies indicate that it is possible to reduce bacterial spores through combinations of mild heat and HHP. *Bacillus stearothermophilus* was inactivated with 6 decimals by a treatment at 500 MPa at 70°C during 6 cycles of 5 minutes demonstrated a 8-decimal reduction of *Bacillus subtilis* spores when treated at 500 MPa at 70°C during 10 cycles of 1 minute. A mild heat treatment can encourage spores to germinate, resulting in them being more susceptible to pressure treatments. Therefore, a preheat treatment followed by pressurization is generally more effective at inactivating spores than heating during pressurization. Other proposed combination treatments are the use of HHP together with the addition of nisin, lactoperoxidase or lysozyme.

Inactivation of *Listeria innocua* exceeding 7 decades was achieved for all strains with a mild treatment (400 MPa, 15 minutes, 20°C) in the presence of lactoperoxidase, which in the absence of the lactoperoxidase system caused only 2–5 decades of inactivation depending on the strain. In contrast, for none of the tested *Escherichia coli* strains, the lactoperoxidase system increased the inactivation as compared to a treatment with high pressure alone. Compared with that of phosphate buffer, the complex physico-chemical environment of milk exerts strong protective effect on *E. coli* MG1655 against HHP inactivation, reducing inactivation from 7 logs at 400 MPa to only 3 logs at 700 MPa in 15 min. at 20°C. An increase in lethality was achieved by addition of high concentrations of lysozyme (400 mg/mL) and nisin (400 IU/mL) to the milk before pressure treatment. The additional reduction amounted maximally to 3 logs in skim milk at 550 MPa but was strain dependent and significantly reduced in 1.55% fat and whole milk. One of the major concerns about HHP for the food industry is the occurrence of pressure-resistant vegetative bacteria after successive pressure treatments. Pressure-resistant strains of *Listeria monocytogenes* and *E. coli* have also been described. When pressure resistance is so easily occurring in food-contaminating bacteria, a selection of pressure-resistant strains into a food processing environment is also not imaginary. More knowledge is necessary on the mechanisms behind the introduction of resistance towards HHP to enable the evaluation of possible risks associated with the introduction of this technique.

Mechanism of HPP

Any phenomenon in equilibrium (chemical reaction, phase transition, change in molecular configuration), accompanied by a decrease in volume, can be enhanced by pressure (Le Chatelier's Principle). Thus, HPP affects any phenomenon in food systems where a volume-change is involved and favours phenomena which result in a volume decrease. The HPP affects non-covalent bonds (hydrogen, ionic, and hydrophobic bonds) substantially as some non-covalent bonds are very sensitive to pressure, which means that low-molecular-weight food components (responsible for nutritional and sensory characteristics) are not affected, whereas high-molecular weight components (whose tertiary structure is important for functionality-determination) are sensitive. Some specific covalent bonds are also modified by pressure. The other principles which govern HPP are as follows:

i. The *Isostatic Principle* which implies that the transmittance of pressure is uniform and instantaneous (independent of size and geometry of food), however, transmittance is not instantaneous when gas are present, and

ii. *Microscopic Ordering Principle* which implies that at constant temperature, an increase in pressure increases the degree of ordering of the molecules of a substance.

iii. Another interesting rule concerns the small energy needed to compress a solid or liquid to 500 MPa as compared to heating to 100°C, because compressibility is small.

HPP offers several advantages: reduced process times; minimal heat penetration/heat damage problems; freshness, flavour, texture, and colour are well retained; there is no vitamin C loss; multiple changes in ice-phase forms resulting in pressure-shift freezing; and functionality-alterations are minimized compared with traditional thermal processing.

Table 12.1 Combination treatments to inactivate bacterial spores by means of high hydrostatic pressure

Treatment combination	Target microorganism	Process conditions	Reduction
HHP+heat	*B. stearothermophilus*	500 MPa, 70°C, 6 cycles of 5 min.	6D
		700–900 MPa, 70°C, 5 min.	5D
		700 MPa, 70°C, 5 min.	5D
		400 MPa, 70°C, 25 min.	3D
		200 MPa, 85°C, 11 min.	1D
		350–400 MPa, 70°C, 10 cycles of 1 min.	6D
		300 MPa, 60°C, 360 min.	2.5D
	B. subtilis	500 MPa, 70°C, 10 cycles of 1 min.	8D
		404 MPa, 70°C, 15 min., pH 6.0.	5D
		600 MPa, 40°C, 20 min., pH 7.0, followed by heat treatment of 10 min, 60°C.	1.8D
	B. coagulans	400 MPa, 70°C, 30 min., pH 7.0.	4D
	C. sporogenes	404 MPa, 70°C, 15 min., pH 6.0.	<0.5 D

(Contd.)

Table 12.1   (Continued)

Treatment combination	Target microorganism	Process conditions	Reduction
HHP+nisin	*B. subtilis*	404 MPa, 70°C, 30 min., pH 7.0 without nisin	<0.5D
		With 0.6 IU/mL nisin	2.5D
	C. sporogenes	404 MPa, 25°C, 30 min., pH 7.0 without nisin.	<0.5D
		With 0.6 IU/mL nisin.	2.6D
HHP + acid environment	*B. subtilis*	404 MPa, 25°C, 30 min.	
		pH 7.0	<0.5D
		pH 4.0	<0.5D
		600 MPa, 40°C, 20 min.	
		pH 3–8	<0.5D
	C. sporogenes	404 MPa, 25°C, 30 min.	
		pH 7.0	<0.5D
		pH 4.0	2.5D

APPLICATIONS OF HPP

Retention of sensory and nutritional characteristics and extension of shelf life of foods and yielding value-added products are the main advantages of HPP.

As mentioned above, HPP does not depreciate the nutritional and sensory characteristics of food, and yet it maintains shelf life. Total inactivation of microbes and polyphenol oxidase activity occurred at 20°C (using dilute citric acid solution at 0.5 or 1.0% as immersion medium). Water-blanched and high pressure-treated potato cubes had similar softness but potassium leaching was reduced by 20%, in addition, ascorbic acid was better retained (90% at 5°C to 35% at 50°C) in high-pressure-treated vacuum-packaged samples. However, they did not investigate use of different pressure-ranges, which may have resulted in different functional characteristics. In studies with tomatoes it was reported that colour, sugar content, and pH were affected by pressure, and that viscosity decreased with increasing blanching temperature and increased with increasing pressures. It was also found that pressure has a significant inactivating effect on poly-galacturonase type pectic enzymes, but had only little effect on pectin esterases (PE). It was concluded that HPP may control various quality parameters of chopped tomatoes.

The quality (volatile flavour components, anthocyanins, browning index, furfural, sucrose, and vitamin C content) of pressure-treated and heat-treated jams was compared in a study during storage at 5 and 25°C for 1–3 months. Immediately after processing, the pressure-treated jams had better fresh quality than heat-treated jams, and the quality was maintained in both at low temperature storage, but not at room temperature for pressure-treated jams. The presence of dissolved oxygen and enzymes was believed to have resulted in deterioration of pressure-treated jam held at ambient temperature. However, pressure-treated jam could be stored at refrigeration temperatures with minimal loss in sensory and nutritional characteristics for up to 3 months. The rate of microbial inactivation was directly proportional to pressure used. Stabilization of white wine must occurred at 507 MPa for 3 minutes, whereas complete microbial inactivation was not achieved in red must (even at 811 MPa for 5 minutes), which could have been due to its high concentration of suspended solids. A substantial inactivation of microbes (< 10 cfu/mL) was observed in pressure-treated thermally pasteurized "guava puree" at 600 MPa, and the pressure-treated samples showed no colour-change, no degradation of pectin, no cloud formation, and had the same ascorbic acid content as fresh samples. However, enzyme inactivation was more pronounced in thermally-treated samples. The pressure-treated "guava puree" (600 MPa) maintained good quality (similar to freshly extracted "guava puree") for 40 days when stored at 4°C. It is to be noted that "guava puree" is particularly sensitive to enzymatic browning reactions which are inhibited during HPP to some extent.

The effect of HPP on the shelf life and sensory characteristics of *Angelica keiskei* juice was studied by subjecting it to a pressure of 558 MPa for 7 minutes and storing it at 4°C. *Pseudomonas* spp., *E. coli*, and coliforms were totally inhibited by HPP. During storage, microbiological quality and sensory characteristics were monitored and it was found that HPP did not significantly influence freshness, sweetness, and bitterness; but after 8 days (at 4°C), pressure-treated juices had better freshness ratings than controls (non-pressurized juice). Preservation of fresh fruits, vegetables, fruit- and vegetable-juices was studied by subjecting them to pressures and subsequently storing them at 6°C. HPP had no beneficial effects on keeping quality of fruits and vegetables, whereas immediately after pressurization, and after 55 days of refrigerated storage, pressure-treated juices had better aroma, flavour, and microbiological quality than untreated controls, and vitamin C content remained the same or declined slightly.

The use of high-pressure throttling (HPT) as an alternative to thermal processing of milk before acidification to yoghurt was investigated. It was reported that when HPT was used to increase milk temperature to 80°C (above minimum pasteurization temperature) followed by rapid cooling to 40°C, this resulted in a 4 log reduction of microbial numbers and increased viscosity, yielding a thick and creamy yoghurt (requiring no further addition of polysaccharides).

The effects of high pressure on the lipid oxidation of extra virgin olive and seed oils were studied. Peroxidase-value, *p*-anisidine value, rancimat test, and volatile hydrocarbons were the analytical parameters measured to study the oxidative stability of the pressure-treated oils. Pressure treatment changed *p*-anisidine values, but not others (peroxidase values and volatile hydrocarbons). Other parameters affecting high pressure treatment were origin, composition, initial quality, and age of the oils, and it was found that olive oils were more resistant to oxidation which suggests need for replacement of seed oil with extra virgin olive oil during HPP to extend the shelf life of foods. The study was done using a pressure, temperature and time period. Further studies should be conducted using different combinations of pressure, temperature, and time intervals, which may result in other value-added products and opportunities.

Microbial Inactivation

Application of HPP as a method for microbial inactivation has stimulated considerable interest in the food industry. The effectiveness of HPP on microbial inactivation has to be studied in great detail to ensure the safety of food treated in this manner. Currently, research in this area has concentrated mainly on the effect of HPP on spores and vegetative cells of different pathogenic bacterial species. Detectable effects of HPP on microbial cells include an increase in the permeability of cell membranes and possible inhibition of enzymes vital for survival and reproduction of the bacterial cells. To design appropriate processing conditions for HPP of food materials, it is essential to know the precise tolerance levels of different microbial species to HPP and the mechanisms by which that tolerance level can be minimized. A knowledge of critical factors that affect the baro-resistance of different bacterial species will help in the development of more effective and accurate high pressure processors. Inappropriate use of a variety of parameters like pressure range, processing temperature, initial temperature of sample, holding time, and packaging type may adversely affect the outcome of HPP. Thus a thorough understanding of the effect of a variation in critical factors on the intracellular changes undergone by pressure-treated microbial species is essential for documentation of a safe HPP. The physico-chemical environment can adversely change the resistance of a bacterial species to pressure. In most cases, the effect of HPP on gram-positive bacteria is less pronounced than on gram-negative species. Factors such as the water activity and pH also influence the extent to which foods need to be treated to eliminate pathogenic microorganisms.

Most bacteria are baroduric, i.e., they are capable of enduring high pressures but grow well at atmospheric pressures. HPP resistance development was studied in *Escherichia coli* MG1655 mutants. Three barotolerant mutants (LMM1010, LMM1020, LMM1030) were isolated and pressure-treated. Mutants showed 40–85% survival at 200 MPa for 15 minutes (ambient temperature) and 0.5–1.5% survival at 800 MPa for

15 minutes (ambient temperature). In contrast, survival of the parent strain (MG1655) decreased from 15% at 220 MPa to 2×10^{-8} % at 700 MPa. It should be noted that pressure sensitivity of the mutants increased from 10 to 50°C, as opposed to the parent strain which showed minimum sensitivity at 40°C. This research indicated that the development of high levels of barotolerance should be properly understood in order to predict the safety of HPP. Similar studies are needed to document the barotolerance of other potentially pathogenic bacterial species.

The effects of HPP on the heat resistance of fungi was studied in apricot nectar and distilled water. Complete inactivation of *T. flavus* ascospores was reported at 900 MPa for 20 minutes at 20°C; a two log cycle reduction in *N. fischeri*, but no effect was seen on *B. fulva* and *B. nivea* populations. In contrast, pre-heating apricot nectar (50°C) followed by pressure treatment of 800 MPa for 1–4 minutes resulted in complete inactivation of all four species. Also, lower pressure resistance was observed in distilled water samples.

Factors Affecting Microbial Inactivation

Processing conditions Several theories on the effect of HPP on bacterial species have been proposed over the years to explain the mechanism behind microbial inactivation and to better optimize HPP of foods. The processing conditions (initial sample temperature, circulating water temperature, pressurizing medium, holding time) under which high pressures are applied significantly influence the level of inactivation as well as the overall effect on the nutritional and sensory characteristics of food. The pressure tolerance of *Saccharomyces cerevisiae*, *Escherichia coli*, and *Staphylococcus epidermidis* was stuided in various media (agar, broth, apple jam, and juice), by subjecting the inocula to a pressure of 300 MPa at 5–25°C for 1–20 minutes. It was reported that media pH played a very important role in the destruction of microbes; *S. epidermidis* was inhibited > 90% at 300 MPa in 11.2 minutes at pH 7.2 and in 4.8 minutes at pH 4.0. Variations in the pH and water activity of foods can result in different levels of lethality to a particular bacterium for the same high pressure processing parameters. Studies conducted on *B. subtilis* showed that the pressure resistance of the bacterium (when subjected to 483 MPa for 30 minutes) was decreased as the pH in milk medium was lowered or raised from a pH value of 8. This value is not a constant for all microorganisms and the survival of *B. subtilis* at a specific pH can vary with the pressure and temperature of treatment. The type of culture medium used for growing the microbial species can also have a significant impact on the pressure and heat resistance of any microorganism. In general, the richer the growth medium, the better the baroresistance of the microorganisms. This is thought to be because of the increased availability of essential nutrients and amino acids to the stressed cell. It must be kept in mind that the parameters governing pressure tolerance are not constant for every bacterial species, they vary from one bacterium to another, and may also be different for a single species grown under different conditions or in different growth media. It is very likely that the application of pressure affects a multitude of functions in a cell, thus interacting to retard or even kill the cell. Therefore, studies related to the inactivation of bacterial species during HPP should specifically describe and document the processing conditions under which the inactivation took place. Also, specific information should be given about the variation in sample temperature during HPP.

The pressure resistance of spores of six *Bacillus* strains was compared. The spores were cultivated on nutrient agar and suspended in cold sterile distilled water with the filtrate being heated for 30 minutes at 80°C to destroy any vegetative cells. Spores of these six strains were then treated under pressures ranging from 196 to 981 MPa at 5–10°C for holding times of 20–120 minutes. It was found that *B. stearothermophilus* IAM 12043, *B. subtilis* IAM 12118 and *B. licheniformis* IAM13417 had the most resistance to pressure, but *B. coagulans* was actually activated when treated with high pressures. There was no actual correlation between pressure and heat resistance, although they chose rightly a number of spore-forming bacteria varying widely in heat resistance. This could be due to applications of very low pressure in most cases. It should also be noted that, in most cases, the reference heat treatment is much more inactivating that the pressure treatment to be compared, therefore, direct comparisons between heat resistance and pressure resistance is often not fair.

The sensitivity of vegetative pathogens was studied (*Yersinia enterocolitica* 11174, *Salmonella typhimurium* NCTC 74, *Salmonella enteritidis*, *Staphylococcus aureus* NCTC 10652, *Listeria monocytogenes*, *Escherichia coli* O157:H7) in buffer (pH 7.0), UHT milk, and poultry meat to high pressures up to 700 MPa at 20°C. A 10^5 reduction in numbers was obtained in all cases when pressures in the range of 275–700 MPa for 15 minutes were applied at 20°C. Different strains of *L. monocytogenes* and *Escherichia coli* O157:H7 showed significant variation in pressure resistance, which were further used to examine the effect of substrates on pressure sensitivity, and indicated that substrate affected the baro-resistance of the microorganisms significantly.

pH The pH of a food material plays a very important role in determining the extent to which HPP affects the microorganisms under study. Several studies have documented and analysed changes in heat resistance of organisms grown under different pH conditions, but there has not been very many studies on the pressure resistance of spores at different pH values. The effects of changes in pH values combined with a variety of other factors on the pressure-resistance of *B. coagulans* ATCC 7050 was investigated.

An increase in the effectiveness of pressurization was reported as the pH of the buffer was lowered. A decrease of an additional 1.5 log was observed as the pH was decreased from 7.0 to 4.0. On the basis of these results, it is highly probable that, other factors remaining constant, a neutral pH value is most conducive to high pressure resistance in the cells. Earlier, it was observed by some workers that at high pressure the spores were most resistant at neutral pH and at low pressure, most sensitive to neutral pH. That is perfectly in line with other findings that a pressure between 50 and 200 MPa enhances germination followed by direct killing at high pressures above 900 MPa. In contrast, another study reported that the inactivation of bacterial spores by pressurization was maximum when the buffer was at a pH near neutral and was lowest at extreme values of pH. The change in pH was thought to affect membrane ATPase and intracellular functions of the spore thereby destabilizing the microorganisms. This effect of pH on the pressure resistance of any microorganism is accentuated by other factors like the addition of salts, temperature conditions, and general process parameters. A better understanding of the exact process by which the variations in the pH affects the stability of the spore is still awaited to give a better overall picture of the effect of high pressures. Therefore, studies to examine the effect of pH variation on the inactivation of different types of spores by HPP need to be done.

Water activity The water activity (a_w) of cells also affects the pressure resistance. It is reported that the lower the a_w, the higher the pressure resistance of cells. The combined effect of HPP and a_w was studied on *Zygosaccharomyces bailii* inhibition. Complete inhibition of yeast was reported at $a_w > 0.98$, and an increase in the surviving fraction was reported with a decrease in a_w thus concluding that addition of sucrose (to decrease a_w) acts as a baro-protective layer, preventing the inhibition of yeast even at high pressures. Such a mechanism can be utilized to prevent inhibition of favourable microbes during HPP. More work is also needed in this area.

It is clear that HPP represents another interesting and promising dimension for food processing because it not only inactivates microorganisms but also provides opportunities for the development of new "value-added" food products. The need for an alternative to thermal processing as the primary means of eliminating pathogenic and spoilage microorganisms is substantial. High pressure processing holds promise since food materials treated by this method retain their natural flavour, colour, and texture without loss in vitamin or nutrient content. Furthermore, predictable changes in functional characteristics of proteins and complex carbohydrates (where little work has been done), mean that there are some exciting avenues of work in HPP treatment of foods that remain to be explored. Although a lot of research has been conducted in the area of high pressure processing, a lot remains to be done in terms of understanding the critical limits of the process and the extent to which this might ensure appropriate treatment of food materials. Research has been done in various research institutions, universities, and food industry R&D laboratories, yet a direct comparison of the data obtained is not possible. There can be no possibility of extensive implementation of this technology unless specific processing parameters are established for each food material treated by HPP. The possibility of commercialization of HPP also depends on its economic viability. Therefore, a detailed economic analysis of HPP needs to be done by comparing the process costs with the present costs of processing by conventional means and at the same time factoring in the added value of the improved sensory characteristics of HPP-treated foods. Some indications on investment and operational costs of HPP are available from the machine manufacturers, although perhaps not fully reliable.

Pulsed Electric Fields

The use of PEF is gaining considerable popularity since it represents an alternative to thermal processing of food. PEF involves the passage of high voltage, in short bursts, into foods placed between two electrodes. The electric field is normally applied at ambient or refrigerated temperature for less than 1 second. In this process, there is little or no heating of food, leaving the product 'fresh-like' and retaining its physico-chemical and nutritional properties. It is suggested that PEF may be the most important 'new' development in non-thermal processing of foods. Bacterial inactivation is believed to be due to structural changes to the cell membrane and pore formation.

The application of PEF requires two components: first a pulsed power supply, and second a treatment chamber. Normally, low voltage is converted to high voltage and stored in a capacitor. The energy stored can then be discharged instantaneously. In the chamber, liquid food is passed between two electrodes spaced 5 mm apart. The liquid to be processed (e.g. milk) flows at the rate of 200–500 ml/minute with a PEF intensity of 80 kV/cm, pulse width 0.5–5 s, and the pulse repetition rate of 0.1–10 Hz. Using this technique, the following foods have been successfully processed: orange juice, milk, apple juice, apple-juice concentrate, green pea soup and liquid eggs. A non-thermal inactivation technology gaining commercial interest and evolving more and more from laboratory and pilot scale to industrial scale is pulsed electric field technology (PEF). PEF technology is based on a pulsing power delivered to the product placed between a set of electrodes that connect the treatment gap of the PEF chamber. The destruction of microorganisms is realized by the application of short high-voltage pulses between the set of electrodes, causing disruption of microbial cell membranes. Only pumpable food products can be treated. Industrial PEF equipment is at the moment rare and expensive, having a limited capacity of around 1800 l/h. However, many efforts are made from different angles to increase the availability of high capacity PEF equipment. Studies about the effect of PEF on dairy products have been conducted on skim milk, whole milk and yoghurt. Generally, gram-positive, vegetative cells are more resistant for PEF than gram-negative bacteria while yeasts show a higher sensitivity than bacteria. Most reports describe the inactivation of vegetative cells by PEF, but only some reports can be found on the inactivation of spores, describing a limited effect of PEF. Mould conidiospores showed to be very sensitive for PEF in fruit juices but *Neosartorya fischeri* ascospores were resistant to PEF treatments. Probably, factors other than process parameters such as the preparation of the spores and the type of medium in which the treatment was performed can influence the inactivation of bacterial spores and explain therefore the high variation in reported results. It is indeed known that the electrical conductivity of the medium influences greatly the inactivation effect of PEF. Little information is available about the ability of microorganisms to become resistant to PEF treatments after consecutive treatments. Similar to HHP treatment, the inactivation effect of PEF can be increased by applying it in combination with other stressing factors such as the presence of antimicrobial compounds such as nisin and organic acids, increased water activity, pH and mild heat treatments. All of them have a synergistic effect on the inactivation by PEF treatments but more research is necessary to understand the mechanisms behind these synergisms, more specifically when the effects on spores are concerned.

New Packaging Systems

New packaging systems have greatly contributed the last decade in extending the shelf life of chilled minimally processed food products. MAP has been widely applied by the food industry, often without the appropriate knowledge about the mechanisms behind its preservation effect. MAP may be defined as "the enclosure of food products in gas-barrier materials, in which the gaseous environment has been changed." One of the bottlenecks in MAP is defining the optimal gas atmosphere for a food product in a specific packaging design. This optimal atmosphere depends on the intrinsic parameters of the food product (pH, water activity, fat content, type of fat) and the gas/product volume ratio in the chosen package type. The intrinsic parameters will determine the sensitivity of the product for specific microbial, chemical and enzymatic degradation reactions. Products that are susceptible to microbial spoilage due to the development of gram-negative bacteria and yeasts (e.g. yoghurt, cheese) will be packaged in a CO_2-enriched atmosphere as the growth of those microorganisms is significantly retarded by CO_2. In general, oxygen is excluded from the gas mixture. For prolonging the shelf life of products which are spoiled due to mould growth (e.g. hard cheese) or due to oxidation, it is essential to package in oxygen-free atmospheres. The use of CO_2 is however limited due to its solubility in water and fat. This high solubility can cause collapsing of the package when too high CO_2 concentrations are applied. This will specially be the case for food products containing high amounts of unsaturated fat. The influence of lard content, water activity, pH, temperature and gas/product ratio on the CO_2 solubility was quantified, concluding that especially the gas/product ratio and temperature are influencing the amount of dissolved CO_2 in the water phase of a packaged product. Moreover, too high CO_2 concentrations in the atmosphere can lead to an increased drip loss during storage. This can be explained by the pH drop induced by CO_2 dissolving in the water phase of the product, causing a decrease in the water binding capacity of the proteins. Active food packaging concepts provide some additional functions in comparison to traditional passive packaging materials which are limited to protection of the packaged food product against external influences. Active packaging materials change the condition of the packaged food product to extend shelf life and/or improve microbial food safety and/or improve sensorial properties. The most applied active packaging system is the oxygen scavenger, removing the residual oxygen from the headspace and/or absorbing oxygen diffusing through the packaging material during storage. One promising type of active packaging is the incorporation of antimicrobial substances in food packaging materials to control undesirable growth of microorganisms on the surface of foods. Nowadays, research focuses on the incorporation of natural antimicrobial or antioxidant compounds such as plant extracts and bacteriocins. An overview of natural compounds that are investigated for use in active antimicrobial packaging materials is given in the Table 12.2. The main cause of spoilage of many refrigerated foods is microbial growth on the product surface. The application of antimicrobial agents to packaging materials could be useful to prevent the growth of microorganisms on the product surface and hence may lead to an extension of the shelf life and/or improved microbial safety of the product. The antimicrobial packaging materials can be prepared by incorporating the antimicrobial agent either into the packaging material, by coating the active compound on the surface of the packaging or by adding sachets into the package from which the antimicrobial compound is released during further storage. For most antimicrobial active packaging concepts, intensive contact between the active material and the food product is required. Therefore, potential food applications include especially vacuum packaged products. Another possibility is to incorporate the antimicrobial compound into an edible film or coating, applied by dipping or spraying onto the food.

However, these coatings, which are supposed to be eaten, are not considered or legally defined as packaging materials. Major drawbacks for the application of active packaging systems are the incompatibility with the European legislation, instability of active compounds during extrusion of packaging materials and the limited amount of mass available in the packaging material to be transferred to the food product. It is also important to mention to potential users of advanced packaging systems, that packaging cannot improve the quality of the packaged product but will only slow down further degradation. To realize a substantial shelf life extension with an advanced packaging type, it is essential that the initial quality of the product is optimal.

Table 12.2   Natural components in active antimicrobial packaging

Active component	Polymer/carrier	Substrate
Lysozyme	PVOH, nylon, cellulose acetate	Culture media
	Edible film	Culture media
Nisin	Cellulose film, PE, SPI, corn zein film	Cheese, ham Culture media Beef carcass tissue
	HPMC	Culture media
	Silicon coating	Culture media
	Corn zein film	Culture media
	PVC, LDPE, nylon	Culture media
	Corn zein film	Culture media
	SPI, WPI, WG, EA	Phosphate buffer
Lacticin	Cellulose film, PE	Phosphate buffer
Pediocin	Cellulose	Cooked meats
Glucose oxidase	Alginate	Fish
Acetic acid	Chitosan	Water
	Chitosan	Bologna, cooked ham, pastrami
Lactic acid	Alginate	Lean beef muscle
Sorbic acid	WPI	Culture media
Propionic acid	Chitosan	Water
	Chitosan	Bologna, cooked ham, pastrami
Grapefruit seed extract	LDPE	Lettuce, soybean sprouts
Hinokitiol	LDPE	Vegetables, fruits
Bamboo powder	Not stipulated	Fishery products, vegetables, etc.
Tocopherol	LDPE	Beef

Abbreviations: PVOH—polyvinyl alcohol, PE—polyethylene, SPI—soy protein isolate, HPMC—hydroxypropyl methyl cellulose, PVC—polyvinyl chloride, LDPE—low density PE, WPI—whey protein isolate, WG—wheat gluten, EA—egg albumen.

BIOPRESERVATION

An increasing number of consumers prefer minimally processed foods, prepared without chemical preservatives. Many of these ready-to-eat and novel food types represent new food systems with respect to health risks and spoilage association. Against this background, and relying on improved understanding and knowledge of the complexity of microbial interactions, recent approaches are increasingly directed towards possibilities offered by bio(logical) preservation. In biopreservation, storage life is extended and/or safety of food products is enhanced by using natural or controlled microflora, mainly LAB and/or their antibacterial products such as lactic acid, bacteriocins and others.

NATURAL ANTIMICROBIAL COMPOUNDS

Numerous antimicrobial agents exist in animals, plants as well as microorganisms where they are often evolved as host defence mechanisms. Typical examples of antimicrobial compounds are lactoperoxidase (milk), lysozyme (egg white), saponins and flavonoids (herbs and spices), bacteriocins (lactic acid bacteria (LAB) and chitosan (shrimp shells).

All these compounds show a substantial antimicrobial character that has been demonstrated mainly *in vitro* in liquid or solid media. However, two aspects, that are crucial for the practical application of natural compounds, are often overlooked: (1) the changes in the organoleptical and textural properties of the food when added and (2) the interaction of these compounds with food ingredients and the influence of this interaction on its efficacy. In many cases, concentrations of the antimicrobial compounds in herbs and spices are too low to be used effectively without adverse effects on the sensory characteristics of a food product. Moreover, the lipophilic character of many of these phyto-antimicrobials hampers the practical use of them. Moderate concentrations of NaCl present in many food products, are often already responsible for the neutralization of the antimicrobial character of natural compounds. This is the case for lysozyme, bacteriocins and chitosan. Also other food compounds such as proteins, fat and starch can alter the antimicrobial activity of natural compounds.

Antagonistic cultures that are only added to inhibit pathogens and/or prolong the shelf life, while changing the sensory properties of the food product as little as possible, are termed protective cultures. LAB, considered as "food-grade" organisms, show special promise for implementation as protective cultures. Their antagonism refers to the inhibition of other microorganisms through competition for nutrients and/or by the production of one or more antimicrobially active metabolites such as organic (lactic and acetic) acids, hydrogen peroxide, antimicrobial enzymes, bacteriocins and reuterin. With regard to safety of LAB, it was found that, with the exception of enterococci, the overall risk of LAB infection is very low, particularly in view of their occurence in substantial numbers on fresh foods such as meats and vegetables, and their use since antiquity in the production of a wide range of fermented products where they are present at high numbers.

MICROBIAL SOLUTION TO MICROBIAL PROBLEMS

Lactococcal Bacteriocins

Bacteriocins are antibacterial peptides produced by a wide range of bacteria. The majority of gram-positive bacteriocins identified to date are produced by lactic acid bacteria (LAB), and a number of these are produced by members of the genus *Lactococcus*. Because lactococci and other LAB are generally regarded as safe (GRAS) organisms, their bacteriocins have attracted particular attention in recent years in an attempt to develop their potential commercial applications. Although the bacteriocins identified to date are diverse, both structurally and genetically, they are usually assigned to one of the three broad subgroups. All lactococcal bacteriocins characterized thus far belong to either group I or group II.

Group-I bacteriocins These are termed lantibiotics. These are ribosomally synthesized peptides characterized by post-translational modification of primary residues into modified amino acids such as lanthionine (Ala-S-Ala) and β-methylanthionine (Abu-S-Ala). Members of this group include three lactococcal bacteriocins; nisin, lacticin 481 and lacticin 3147, in addition to those produced by other gram-positive genera, which include subtilin, epidermin and mersacidin.

Nisin It is a prototype lantibiotic which was first discovered in the 1920s. It is a 34- amino acid peptide produced by some strains of *Lactococcus lactis* with a wide spectrum of inhibition against gram-positive microorganisms in favourable conditions. Nisin has a wide range of applications because of its broad bactericidal mode of action and because it can be easily broken down by digestive proteases so as to not disturb gut biota. It is the only lantibiotic to date that has been approved for commercial use. Nisin is now approved for use as a food preservative in approximately 50 countries. Although nisin is the only commercially exploited lantibiotic to date, there has been an extensive effort devoted to developing applications for other lantibiotics.

Lacticin 3147 It is also a broad-spectrum lantibiotic with potential uses in the food industry and in medicine. It is a two-component lantibiotic (two separate peptides act in concert for full activity) produced by *L.lactis* subsp. *lactis* (DPC3147), which was first isolated from an Irish kefir grain. *Lactococcus lactis* (IFPL105) is another lacticin 3147 producer and was isolated from goats milk in Spain. The genes required for DPC3147 to produce lacticin 3147 are carried on a conjugative plasmid, pMRC01 whereas IFPL105 relies on a non-conjugative plasmid, pBAC105. Lacticin 3147 has a broad spectrum of inhibition exhibiting a bactericidal mode of action against all gram-positive bacteria tested to date; including food spoilage bacteria such as *Clostridium* sp., food pathogenic organisms such as *Listeria monocytogenes*, the mastitis-causing pathogen *Staphylococcus aureus* and *Streptococcus dysgalactiae* and human pathogens such as methicillin-resistant *S. aureus*, vancomycin-resistant *Enterococcus faecalis*, penicillin-resistant *Pneumococcus*, *Propionibacterium acne* and *Streptococcus mutans*. It is also extremely heat-stable, active over a broad pH range and, as the genetic determinants are encoded on a self-transmissible plasmid, pMRC01, it can be conjugally transferred to different strains. These features make lacticin 3147 attractive for developing potential applications in the food industry, veterinary medicine and possibly in the treatment of human diseases.

Lacticin 481   It is another lantibiotic produced by certain *L.lactis* strains and exhibits a medium spectrum of activity, inhibiting a broad range of other LAB and *Clostridium tyrobutyricum*. This peptide can have a bacteriolytic effect on sensitive organisms. It has been studied for its potential use in cheese ripening to induce lysis of starter strains and therefore deliver lactococcal enzymes during cheese manufacture to improve both flavour and quality.

Group II bacteriocins   The Group II bacteriocins are small, unmodified, cationic, hydrophobic peptides. Their inhibitory spectrum is usually narrow, limited to those with a low G + C content such as *Listeria* sp., *Clostridium* sp. or other LAB. In some cases, a single strain can produce multiple bacteriocins. For example, some *L. lactis* strains can produce lactococcins A, B and M (three class II bacteriocins).

Although they have a narrow spectrum of activity, inhibiting only other lactococci, they have potential applications in the dairy industry as, like lacticin 481, they have a bacteriolytic mode of action which can be exploited for the accelerated lysis of starter cultures during cheese manufacture and ripening.

Applications of Lactococcal Bacteriocins in the Food Industry

Lacticin 3147 has been investigated for use in the dairy industry to control NSLAB during the fermentation of dairy products. NSLAB are primarily lactobacilli which, although not deliberately added to the fermentation process, can enter through cheese vats and become the dominant flora in a six-month ripened cheese. The role of NSLAB in the development of cheese flavour is still not clear. Although they can contribute positively to the overall quality of the cheese, they are also associated with off-flavours and with the formation of calcium-lactate crystals and slit-defects often found in cheese.

The control of NSLAB in cheese ripening would allow for a more predictable end product. The most efficient method of introducing lacticin 3147 to a fermented dairy product is to use a lacticin-producing culture as the starter or as a starter adjunct in the fermentation process. DPC3147, the natural lacticin 3147 producer, is unsuitable as a starter-culture as the strain is associated with an off-flavour. However, this problem can be eliminated by taking advantage of the conjugative nature of pMRC01. This approach, which is a frequently used method for genetic improvement, allows the directed transfer of the plasmid responsible for producing the bactcriocin, pMRC01, to a commercially used lactococcal starter (the newly producing strain is termed a transconjugant). Over 30 lacticin 3147 transconjugants have been created to date. An important consideration when creating and using transconjugants as starters is to ensure that their important industrial traits have been retained and that they remain suitable for cheese manufacture.

Nisin-producing starters, for example, have often been associated with slow acid production, reduced proteolysis and poor heat resistance when compared with a commercial starter. It was shown that a lacticin 3147-producing transconjugant starter (DPC4275) produced acid at comparable rates with the parental commercial starter. It was also shown that the level of bacteriocin produced was constant throughout the ripening process and at the end of cheese production the NSLAB population had been reduced by at least 100-fold. A reduction in the NSLAB population was seen when DPC4275 was used to make low-fat cheese with increased ripening temperatures. This offers the cheese-maker far greater control over the developing adventitious flora without the need to resort expensive alternatives such as cold-ripening. However, as mentioned earlier, NSLAB may contribute positively to flavour and quality of the cheese and therefore it may be desirable that certain NSLAB increase in number during the ripening process. This objective was achieved when a lacticin 3147-resistant variant of *Lactobacillus paracasei* subsp. *paracasei* DPC5336 was isolated after repeated exposure to low levels lacticin 3147. This strain, which still remains sensitive to high levels of lacticin, was then used in conjunction with lacticin 3147-producing starter in Cheddar cheese manufacture. While other NSLAB were inhibited during ripening, the resistant mutant could tolerate the levels of lacticin 3147 and became the dominant microflora in the cheese.

Developments like this give the cheese-maker greater control of the microbial biota. Another advantage of using lacticin 3147 transconjugants as dairy starters is that the plasmid pMRC01 encodes bacteriophage resistance, specifically an abortive infection mechanism. The susceptibility of lactococcal starter cultures to phage attack is a serious problem in the dairy industry and can result in production and economic losses. Thus, the introduction of the lacticin 3147 genetic determinants to a starter culture has the added benefit of conferring protection from lactococcal phage attack as a result of the linked abortive infection mechanism. This is contrast to nisin-producing starters which are associated with phage susceptibility. Bacteriocins such as lacticin 3147 and nisin, which inhibit a broad range of food spoilage and food pathogenic bacteria, have obvious applications in the food industry.

Class II bacteriocins, such as lactococcin ABM, only inhibit other lactococci and therefore their associated applications are limited. However, it has been shown that lactococcin ABM can have both a bactericidal and bacteriolytic mode of action on target cells. In fact, sensitive strains undergo complete lysis following exposure to an ABM producer, resulting in the release of intracellular enzymes. The lytic abilities of these bacteriocins may have an application in the acceleration of cheese ripening as explained below. Cheddar cheese usually has a maturation time of at least six months, during which gradual autolysis of the starter cultures occurs. Lysis results in the release of intracellular enzymes such as lactate dehydrogenase (LDH) and post-proline dipeptidyl aminopeptidase (PepX) which break down the casein in the cheese to small peptides and amino acids. The amino acids released are the precursor compounds responsible for flavour development in cheese. As cell lysis is a slow process and often a limiting step in cheese maturation, controlled early lysis is known to be advantageous for improved flavour development. The potential of using lactococcin ABM to induce starter cell lysis has been investigated. Addition of these bacteriocins to growing sensitive cells is associated with a gradual decrease in optical density (O.D.) in laboratory media and a concomitant increase in the release of intracellular enzymes. Cheese making trials using *L. lactis* subsp. *lactis* (DPC3286) (an overproducer of lactococcin ABM) as a starter adjunct with the cheese-making strain *L. lactis* subsp. *cremoris* demonstrated increased starter cell lysis, elevated enzyme release and an overall reduction in bitterness when compared with the control cheeses. However, a problem that may be encountered when using bacteriocins to lyse starters is that the rate of acidification may be compromised if the target lactococcal strain is also the primary acidifier. To overcome this problem with lactococcin ABM, a three-strain system was developed.

Three-strain System

This system included a lactococcin-producing adjunct *L. lactis* subsp. *lactis* (DPC3286), a bacteriocin-sensitive starter, *L. lactis* subsp. *cremoris* as a target and *Streptococcus thermophilus*, a bacteriocin-resistant species. *Streptococcus thermophilus* is included as it is not inhibited by lactococcin ABM, is insensitive to lactococcal phage and is an efficient acid producer. As this strain continued to grow throughout ripening, overall acidification was not affected and elevated enzyme release was achieved by lysis of the starter. This indicates a potential use for lactococcin ABM as a tool in cheese manufacture to increase enzyme release, thereby accelerating the rate of ripening while not affecting the quality of the cheese. Lacticin 3147, produced by *L. lactis* (IFPL105) can also have a bacteriolytic effect on sensitive cells. *Lactococcus lactis* (IFPL105) and the lacticin 3147-producing transconjugant *L. lactis* (IFPL3593) have been successfully used in the acceleration of cheese ripening. Lacticin 481, produced by a number of lactococcal strains, also has applications in the ripening of cheese. This is a medium-spectrum lantibiotic, inhibiting other LAB and *C. tyrobutyricum* and also has a bacteriolytic mode of action. It was demonstrated that LDH release caused by the addition of lacticin 481 (*L. lactis* subsp. *lactis* DPC5552) to target cells was higher than levels found with either lacticin (3147 DPC3147) or lactococcin ABM (DPC3286). It was also shown, in laboratory-scale cheese trials, that use of a lacticin 481-producing culture as a starter adjunct results in elevated levels of LDH from the starter. Another interesting and novel phenomenon with lacticin 481 was that the O.D. of exposed cells did not decrease; in fact it increased even while intracellular enzymes were released. This suggests that lacticin 481 could be used to induce early cell lysis but also the target cell would continue to grow leaving its acid producing abilities unaffected. Furthermore, as lacticin 481 is effective against other LAB, it may have an application in inhibiting NSLAB during cheese manufacture. This indicates a potential dual role for lacticin 481 in cheese manufacture. To conclude, given that Cheddar cheese ripening can be a long and costly process, it would be economically advantageous to accelerate the process. Results suggest that the use of bacteriocins as tools to induce early cell lysis would appear to be a possible option. Also, as the use of bacteriocin-producing cultures involves the natural alteration of starter cultures, there would not be any additional technology costs or genetic modification events associated with the process. As lacticin 3147 inhibits a large number of food pathogenic organisms, it would appear to be particularly suited to use as a biopreservative in foods.

Delivering lacticin 3147 into foods may be accomplished in a number of ways. As discussed, a lacticin 3147-producing strain may be used as a starter or a starter adjunct in fermented foods. This approach has the advantage that lacticin 3147 can be produced by food grade strains and is not considered an additive, and has no cost implications for the manufacturer. Lacticin 3147 may also be added in the form of powdered whey or a milk fermentate (whey or milk is fermented with a lacticin producer and subsequently spray-dried) or as a concentrated lacticin 3147 preparation. Powdered lacticin fermentate would be added as a food ingredient (most probably replacing skim milk or whey powder in an existing formulation). The use of concentrated or purified lacticin preparations may also be problematic as such preparations would be regarded as food additives and, furthermore, may not be economically feasible. One of the food pathogens that lacticin 3147 controls is *L. monocytogenes*. This organism can be a difficult pathogen to control because of its ubiquitous distribution, tolerance to high levels of salt and its ability to grow at a relatively low pH and at refrigeration temperatures. While fermented foods are generally considered to be relatively free of pathogens, *Listeria* has been found to survive and possibly grow in fermented foods made with raw materials containing the organism. The susceptibilities of a large range of *L. monocytogenes* strains were tested to varying levels of lacticin 3147. Importantly, while the level of sensitivity varied among different isolates, a completely resistant strain was encountered.

NON-BACTERIOCINOGENIC CULTURES

Despite the potential of bacteriocinogenic strains as biopreservatives in foods, only a few bacteriocinogenic strains are marketed and mainly in the meat sector. An alternative to overcome the disadvantages of bacteriocinogenic cultures is the use of non-bacteriocinogenic but nevertheless very competitive cultures, e.g. *Lactobacillus alimentarius* BJ-33 or *Lactobacillus sake* TH1 and others.Their antagonistic character may be due to the production of lactic acid and the accompanying acidification controlling growth of spoilage and/or pathogenic bacteria. The full explanation of the inhibition is probably more complex and may be due to the combined effects such as production of antimicrobials and competition for or depletion of specific nutrients. In vacuum-packaged cooked- meat products, the growth of *L. monocytogenes* stops when the number of LAB reaches 10^7 cfu g/1. This inhibition cannot be explained by pH-reduction and lactic acid production as the pH-reduction at the moment of inhibition is too limited due to the low glucose content and the large buffering capacity of cooked meat products.

A mechanism involving nutrient depletion was suggested to explain the ability of *Carnobacterium piscicola* to suppress *L. monocytogenes*. Antagonism of pseudomonads towards *Shewanella putrefaciens* on fish is caused by the competition for iron, mediated by bacterial siderophore production. It was found that the effect of Enterobacteriaceae on *L. monocytogenes* in a medium simulating the composition of endives (a food product common in the west) was presumably due to a competition for glucose and/or amino acids. Competitive effects of LAB towards mainly *L. monocytogenes* have been studied on several food products including meat products and MAP minimally processed vegetables. Adding a combined culture of a *Pediococcus acidilactici*, a *Lactobacillus casei* and a *Lactobacillus paracasei* strain to frankfurters, co-inoculated with *L. monocytogenes*, resulted in a bactericidal effect towards *L. monocytogenes*; numbers of *L. monocytogenes* were 4.2–4.7 $\log_{10}$ lower than controls after the 28-day-refrigerated storage observed an inhibitory effect of different *Lactobacillus* and *Lactococcus* strains on growth and sporulation of *B. cereus* in milk. *Lactococcus* strains were more inhibitory to *B. cereus* than the *Lactobacillus* strains.

Other non-LAB protective cultures are, *Pantoea agglomerans* CPA-2, an effective antagonist for the control of the major post-harvest pathogens on fruits. Well characterized, psychrotrophic, homofermentative, mildly acidifying LAB are ideal candidates for biopreservation of (refrigerated) food products. Relatively high levels of these cultures may be required for protection against some pathogens. Most protective cultures are LAB, not surviving commercial cooking processes, and should therefore be added by dipping or spraying after heat treatment. Therefore, heat resistance is a desirable characteristic for a protective culture to minimize cost attributed to product supplementation with viable cells experimented with the addition of bacteriocinogenic cultures through the curing brine, whereas cultures may also be added prior heating in a spray-dried or freeze-dried form. The production of competitive and bacteriocinogenic LAB may provide an additional hurdle to improve preservation by natural means. However, application of a protective culture should be considered as an additional measure to good manufacturing, processing, storage and distribution conditions.

Mechanisms of Inactivation

The successful implementation of a novel technology for food preservation relies on the progress in the field of mechanisms of inactivation. An adequate knowledge of the physiological behaviour of microorganisms towards inactivation agents is essential for the development of safe foods.

It is necessary for understanding the effect of environmental factors on resistance and also for identifying critical factors. It would help to interpret kinetics of inactivation and to develop mathematical models based on parameters with a biological meaning and therefore more useful and able to predict microbial inactivation in a wider range of conditions. A better knowledge of the effect of the preservation agents on the microorganisms would lead to a more rational design of processes.

Inactivation Targets and Mode of Action

Cell death has been associated with either structural damage or physiological dysfunctions. Among structural damage, disruption of the envelopes, DNA conformational changes, ribosome alterations or protein aggregation are the most frequently described. Also physiological disorders, such as membrane selective permeability alterations or loss of function of key-enzymes have been proposed as events leading to cell death. Perhaps the most important difficulties that researchers encounter in this area is that several of these lesions may occur simultaneously when the cells are subjected to an agent and it is therefore difficult to attribute the loss of viability of the cell to a single event. Nevertheless, it must be kept in mind that also multitarget inactivation is feasible, this being the addition of several lesions that together cause death. It is also possible that the key target is only affected when a secondary structure is previously damaged. For instance, heat causes membrane damage, loss of nutrients and ions, ribosome aggregation, DNA strand breaks, inactivation of essential enzymes, protein coagulation, etc. In other words, almost every cellular structure is somehow affected by high temperatures, and it is very difficult to discern which events are leading to cell death. In fact, this is only possible if a direct relationship between the degree of inactivation and the degree of modification of a given target under different environmental conditions is found. Regarding novel inactivation technologies, progress on inactivation mechanisms research is heterogeneous. In this way, sound hypotheses have been put forward to describe the effect of IR and ultrasound on microorganisms, but much research is still needed to completely explain the way PEF and HHP can inactivate bacterial cells. Table 12.3 summarizes the relevant events related to bacterial inactivation by the four technologies.

Table12.3 Lethality of irradiation (IR), manosonication (MS), high hydrostatic pressure HHP) and pulsed electric field (PEF) treatments

Microbial group	IR*	MS#	HHP@	PEF$
Bacterial spores	0–1	0–1	0	0
Gram-positive (vegetative cells)	1–4	1–3	<1–2	<1–3
Gram-negative	3 to > 9	4–6	1–7	3–4
Yeasts and Moulds	0–1	ND	2 to > 8	3–5

*—Log cycles after 1.5 kGy   #—Log cycles after 5 minutes treatment at 117mm and 200 KPa

@—Log cycles after 15 minutes treatment at 300 MPa $—Log cycles after 200 minutes treatment at 22 kVcm^{-1}.

It is well established that the critical target for irradiation is the chromosome. The effects of ionizing irradiation on bacterial cells are classified as direct and indirect. Direct actions comprise the events caused by the absorption of radiation energy by the target molecules, whereas indirect actions are those derived from the interaction between the reactive species formed by the radiolysis of water, such as the hydroxyl radical, and the target molecules. The hydroxyl radical OH is able to react with the sugar-phosphate backbone of the DNA chain giving rise to the elimination of hydrogen atoms from the sugar. This causes the excision of the phosphate ester bonds and subsequent appearance of single strand breaks. Double strand breaks occur when two single strand breaks take place in each chain of the double helix at a close distance. Bases are also attacked by the free radicals generated by radiolysis, but it is not clear whether this is relevant to cell death. The mechanism of inactivation of bacterial cells by ultrasound under pressure has also been described. Most authors agree that the cavitation phenomenon is responsible for the lethal effects of ultrasound. When bubbles implode under an intense ultrasonic field, very high pressures and temperatures are generated, and consequently strong mechanical forces and free radicals are formed. Free radicals could therefore inactivate bacterial cells in a similar mode as that described for IR. However, experimental data using free radical scavengers have lead to the conclusion that the possible effect of free radicals is negligible in comparison with that of the strong mechanical effects generated by cavitation. The effect of equivalent heat, MS and MTS treatments (99% of inactivation) on the degree of cell disruption was evaluated through phase contrast microscopy and it was observed that heat-treated cells maintained full cellular integrity, whereas MS-treated cells were completely broken. MTS-treated cells showed a medium degree of disruption. Therefore it is confirmed that ultrasound inactivate microbial cells through envelope breakdown in an all-or-nothing type phenomenon. There are however some unclear aspects still not solved regarding the mechanism of action of ultrasound. In some experimental conditions, and with some bacterial species, a synergistic effect of MS and heat (MTS) has been observed. This is the case of *Enterococcus faecium*, heat-shocked cells of *Listeria monocytogenes* and cells suspended in a low water activity media.

The reasons for the increased sensitivity of these cells to a combined MS-heat treatment are still not known. It has been suggested that moderately high temperatures (55–60°C) would cause a weakening effect on cell envelopes, facilitating the mechanical disruption of the cell by ultrasonic waves. This weakening effect would have no relevance for thermo-sensitive cells, as they are killed by heat at relatively low temperatures. In addition, a synergistic effect of MS and heat has been described for bacterial spores. Ultrasonic treatments cause the release of some low-molecular weight polypeptides and dipicolinic acid from the spore. It has also been found that MS treatments sensitize spores of *Bacillus subtilis* to lysozyme. Therefore, it has been suggested that ultrasonic waves could damage the external layers of the spore, facilitating its rehydration and consequently reducing its extreme heat-resistance.

In contrast to the clear mechanisms of inactivation proposed for irradiation and ultrasound, it is much more complicated, in high hydrostatic pressure inactivation. Much research has been carried out in this matter, but there is still controversy about the key-target leading to cell death by high pressure. Initial investigations on mechanism of inactivation of HHP suggested the cytoplasmic membrane as the key-target. Evidences of bacterial membranes being a target for HHP inactivation are clear. High pressure causes tighter packing of the acyl chains within the phospholipid bilayer of membranes and promotes membrane transition from liquid crystalline to gel phase, in a similar way as a temperature downshift. Although phase-transition of membrane lipids is not necessarily lethal to bacteria, it has been demonstrated that the composition and state of the bacterial cell membrane prior to pressure-treatment affect bacterial resistance to HHP. Cells with a more fluid membrane, i.e., with a higher degree of unsaturation are more barotolerant. It is not clear how a more fluid membrane renders a more resistant cell to HHP. The pressure at which phase transition occurs would be higher in cells with a more fluid membrane, but it is not known in which circumstances, if any, cell damage is linked to phase transition. Damage to the cytoplasmic membrane after pressurization has also been repeatedly reported, through loss of osmotic responsiveness, uptake of vital dyes, loss of intracellular material, and formations of buds and vesicles of lipidic origin. The loss of function of some proteins including the membrane ATPase or multidrug efflux pumps has also been described. Even a direct relationship between loss of membrane integrity and loss of viability has been found for pressure-treated exponentially-growing cells. However, it has also been demonstrated that both outer and cytoplasmic membrane permeabilization is transient to some extent and that pressure-treated stationary-phase cells of *Escherichia coli* may maintain a physically intact cytoplasmic membrane upon decompression even in dead cells. Therefore, other structures inside the cell have also been proposed as potential key targets for inactivation by HHP. Some study have reported similarities between cell inactivation and protein denaturation kinetics by HHP, and changes in the conformation of the nucleoid, ribosomes and cytoplasmic protein have been described. A direct relationship between loss of viability in *E. coli* and ribosome damage, evaluated by differential scanning calorimetry. Further incubation of treated cells in a magnesium-rich medium, which is known to have a stabilizing effect on ribosome structure, aided the ribosomes to recover the initial conformation. The study concluded that other factors together with ribosome initial destabilization accounted for cell death and suggested that the loss of essential ions like magnesium through a damaged membrane could trigger the ribosome destabilization. It seems clear that some of these cellular lesions like DNA and protein condensation are not necessarily lethal and are repairable if the cell keeps functional membrane and the environmental conditions suitable.

HHP inactivation seems to be multitarget in nature. Membrane is a key target, but in some cases additional damaging events such as extensive solute loss during pressurization, protein coagulation, key enzyme inactivation and ribosome conformational changes, together with impaired recovery mechanisms, seem also needed to kill bacteria. Bacterial spores are extremely resistant to HHP, being able to withstand up to 1000 MPa for long treatment times, unless they are in the germinated state. Pressure itself at a moderate level induces spore germination. This has been the basis for the design of a cyclic combined treatment in which spores are induced to germinate in a first step and inactivated in a second step, generally by a combination of mild heat and pressure. Membrane structural or functional alteration is generally accepted as the cause of cell death by PEF. It was demonstrated that the inactivation was the result of the direct effect of PEF on the membrane (electroporation) rather than because of the temperature increase or the electrolysis products. Electroporation can be defined as the formation of pores in cells and organelles, and the most accepted theory to explain it is that proposed by Zimmermann *et al*. They compared the cell membrane to a capacitor. Free charges tend to accumulate in the inner and outer surface of the membrane generating a transmembrane potential of about 10 mV. When an external electric field is applied, as in PEF-treatment, a higher amount of free charges of opposite charge accumulate at both membrane surfaces, resulting in compression of the membrane. When the external electric field exceeds a critical value or threshold, the membrane is unable to withstand the electrocompression and pores are formed. The size and amount of pores depend on the electric field strength and the duration of the treatment. Alternative theories propose that the permeabilization is the consequence of dipolar reorientation of the membrane phospholipids under an electric field. It was also suggested that the formation of hydrophilic pores would lead to a localized Joule heating phenomenon that could be responsible for the denaturation of proteins and phase changes in the membrane. Studies on electron microscopy of several bacteria and yeasts have shown morphological alterations like surface roughness, disruption of organelles, ruptures in the membrane, etc. However, no correspondence between the frequencies of appearance of morphological alterations with loss of viability has been proved. Many attempts have been made to establish whether there is a relationship between membrane damage and microbial inactivation by PEF. Using vital staining and detection of UV-absorbing material leakage various researchers have shown the occurrence of membrane permeabilization in relation to cell death. They have described a linear relationship between the percentage of permeabilized cells and the intensity of the PEF treatment, supporting the hypothesis of membrane permeabilization being the cause of cell inactivation. An interesting and almost nonexplored aspect of PEF treatments is the occurrence of reversible pores. The same way as it happens with high pressure, a proportion of cell membranes could become leaky during PEF treatment but reseal to a certain extent after it. Experiments using the addition of propidium iodide to the treatment medium as a marker of nonpermanent permeabilization of the cytoplasmic membrane, have shown that the degree of staining of *Salmonella* serotype *senftenberg* 775 W cells was approximately twice as that observed when the propidium iodide was added after PEF treatment. These results would indicate that a percentage of cells were able to reseal their pores just after PEF treatment.

Sublethal Injury

Microorganisms surviving the lethal action of preservation agents may be sublethally injured: able to repair the damage and outgrow only if the environmental conditions are suitable. The occurrence of sublethal injury has two main consequences. First, injured cells might not be detected when selective conditions are used for enumeration of survivors. This can lead to an overestimation of the lethality of the treatment. Secondly, the occurrence of sublethal injury means that a delay time between treatment and outgrowth of survivors takes place. Alternatively, if repair is adequately prevented by the combination of additional preservation agents (hurdles) that interfere with cellular homeostasis maintenance, the cell might not be able to outgrow, and the inactivation level attained might be higher. Thus, detection and characterization of sublethal injury by novel preservation technologies is essential for the optimization of mild combined methods with a higher lethal effect on microbes. Survival curves of MS treatments of gram-positive and gram-negative cells recovered in media with sodium chloride added are virtually identical to those recovered in a non-selective medium. This indicates the total absence of repairable membrane damage and is in concordance with the hypothesis of the all-or-nothing lysis of the cell as the mechanism of inactivation of ultrasonic waves on bacterial cells. Sublethal or repairable injury has been detected in irradiated, pressurized and PEF-treated cells. For each agent the mechanism of inactivation is different, and so is the nature and the magnitude of the sublethal injury.

Studies on sublethal cell injury and repair of irradiated cells have focused on DNA. It seems clear that the relative sensitivity of the different microbial groups depends not only on their susceptibility to the direct and indirect action of irradiation itself, but also on their capability of repairing the single and double strand breaks through several enzymatic actions. Irradiation, however, does not discriminate among molecules in a sample, and virtually all the molecules in an irradiated cell may be affected. Irradiation damage to other structures has been scarcely studied. Some studies have reported the sensitization of irradiated cells to selective media but it was shown that irradiation did not induce membrane damage as detected by vital dye staining. It is not clear whether the secondary hurdle added (salt, acid, CO_2) would exert its inhibitory action either towards the damaged DNA as suggested by some researchers with heat, interfering with DNA repair systems or at any other level.

Bacterial membranes are the main target for sublethal injury in HHP treated cells. Permeabilization of the outer and cytoplasmic membrane has been described and combined treatments of HHP and antimicrobial peptides in the pressurizing medium, such as lysozyme, nisin, pediocin AcH, lacticin, lactoferrin and lactoferricin with increased efficacy for both gram-positive and gram-negative microorganisms have been proposed. However, the permeabilization of the outer membrane of gram-negative cells is transient and a fully functional membrane is recovered automatically shortly after decompression. On contrary, cytoplasmic membrane damage repair is highly demanding, and requires energy, RNA and protein synthesis. An intact cytoplasmic membrane is essential for the maintenance of the homeostasis under unfavourable environmental conditions, and thus, combinations of HHP with alkali, salt or acidic environment also attain a higher lethal effect on microorganisms.

The occurrence of sublethal cell damage in PEF-treated cells is a matter of controversy. Most authors have not observed the occurrence of sublethal injury using the selective medium plating technique so it was generally accepted that bacterial inactivation by PEF was an all-or-nothing effect. However, recent results obtained in some laboratories have shown the occurrence of sublethal cytoplasmic membrane damage in a large proportion of the population of gram-negative cells estimated by differential plating. Unpublished results of a research group indicate that the discrepancies in published data may arise from the fact that the occurrence of sublethal membrane damage by PEF treatments depends on the pH of the suspending medium and on the bacterial species. In this way, several gram-negative bacteria showed a higher resistance to PEF at acidic pH (when compared with neutral pH) that was correlated to the capability to repair the cytoplasmic membrane, which was extensively damaged. For gram-positive cells the reverse was true. They showed a greater resistance to PEF, the occurrence of membrane damage and the subsequent repair ability at neutral pH. At acidic pH the cells became more sensitive and irreversibly damaged. It is worthy to note that PEF-sublethally injured cells stored under acidic conditions had lost its viability during storage time. This means, from a practical point of view, that if adequate post-treatment holding conditions are selected, the intensity of treatments, both for HHP and PEF, could be diminished without affecting the microbial quality of the product. Alternatively, a higher degree of safety could be attained. The role of the outer membrane in PEF inactivation needs further attention. These observations provide new useful data that contribute to the understanding of mechanisms of membrane electroporation in bacterial cells. They also contribute to clarify the environmental circumstances under which PEF might act synergistically with other hurdles for food preservation.

Stress Adaptation and Resistance

It is well known that microorganisms can develop adaptive responses and resistances when exposed to sublethal stresses, which may have serious implications for food safety. In the last 15 years, much research interest has been directed towards the elucidation of bacterial stress adaptation mechanisms and gene regulation behind it. The modification of sigma factors (*r*) bound to core RNA polymerase, conferring promoter specificity, is possibly the most important regulatory mechanism in bacterial cells. Sigma factor (*r*) regulates, in gram-negative bacteria, the transcription of more than 50 genes involved in resistance to osmotic, heat, oxidative and acid stress, among others. The induction of this sigma factor occurs in response to starvation, generally when cells enter the stationary phase of growth, and also when exponentially growing cells are subjected to stresses other than starvation. In gram-positive bacteria (*B. subtilis*, *L. monocytogenes* and *Staphylococcus aureus*), an alternative sigma factor with equivalent physiological functions has been described (*sigB*). Therefore, it seems that a parallel mechanism for the acquisition of multiple stress resistance exists in gram-positive and gram-negative cells. The influence of the activation of these regulatory networks on bacterial resistance to novel preservation processes has not yet been sufficiently studied. Nevertheless, it is known that the higher resistance to HHP of stationary-phase cells is partly the result of the presence of the *RpoS* protein in *E. coli* and *sigB* in *L. monocytogenes*.

A range of morphological and physiological changes dependent on *RpoS* expression has been described for *Salmonella* and *Escherichia* cells that might possibly account for the increase in pressure resistance. It can be foreseen that these changes might also have an influence on bacterial resistance to PEF, ultrasound and irradiation, but till now no data are available on the influence of the *RpoS/sigB* regulon on the resistance to these technologies. Sigma factor *r32*, encoded by the *rpoH* gene, controls the heat shock response, which consists of a rapid and transient overexpression of chaperons and proteases. The application of sublethal HHP treatments promotes the expression of several heat shock proteins, suggesting a direct role of the heat-shock response in HHP resistance. In fact, it has been reported that a previous heat-shock may protect cells against HHP. In addition, the basal expression of several heat-shock proteins like *dnaK*, *lon* or *clp* is increased in pressure-resistant mutants. The influence of a sublethal heat-shock on PEF resistance has been scarcely investigated but it was shown that survival seems to be increased. Preliminary results suggest that the ability to recover from PEF damage is also increased. More research is needed in this field. The application of a heat-shock does not protect bacteria to a subsequent MS treatment. This indicates that the possible changes that heat-shock may induce in cell envelopes, if any, are not relevant to ultrasound resistance. Although scattered information is available about the influence of other types of stresses on bacterial resistance to novel preservation methods, from published data it can be deduced that in most cases cross resistance is induced. Certain studies have described how the exposure to hydrogen peroxide, ethanol and cold-shock induces baroresistance in *Saccharomyces cerevisiae*. Increase of irradiation resistance of *E. coli* O157:H7 was observed when cells were previously acid-adapted. Likewise, acid, cold and heat adaptation also seems to protect this microorganism to PEF treatments. Adaptation of microorganisms to adverse environmental conditions during processing poses a risk that should not be underestimated. Extensive research is still needed to characterize the physiology and genetics of microbial stress responses involved in survival to food preservation processes.

FACTORS AFFECTING MICROBIAL RESISTANCE

Microbial inactivation by irradiation, ultrasound under pressure, HHP and PEF has been found to depend on many factors. The factors affecting resistance are classified into three groups: process parameters, microbial characteristics and product parameters.

Process Parameters

Some process parameters are intrinsic to each technology and no general conclusions can be drawn. For instance, the intensity of an irradiation treatment is given by the irradiation dose absorbed, as the radiation energy is normally fixed. Critical inherent parameters for ultrasound under pressure are treatment time, amplitude of the ultrasonic waves and external pressure applied. Whereas HHP efficacy depends on treatment time and pressure, PEF lethality varies with parameters such as electric field strength, pulse characteristics and frequency, apart from treatment time. There are, however, some process parameters that are common for the four technologies. This is the case for treatment temperature. As a general conclusion, as the temperature is raised, the lethality of the four technologies increases. The lethal effects of irradiation on microorganisms are more pronounced when the treatment is carried out at elevated temperatures. When irradiation takes place in the frozen state the sensitivity of the microorganisms is reduced by a factor of 2–5 , and this has been attributed to the reduced mobility of free radicals. The lethality of ultrasound under pressure treatments is almost not modified by an increase in temperature unless lethal temperatures are reached (MTS treatments), in which case an additive lethal effect is generally attained although in some cases the total lethal effect has been found to be synergistic. The effect of temperature on high-pressure inactivation is complex. Combinations of HHP with mild treatment temperature lead to a higher lethal effect, and this has been attributed to a higher degree of damage on proteins. Also treatment temperatures below 20–30°C activate cells faster and this effect has been proposed to be the result of the reduced fluidity of the membrane. Regarding the effect of treatment temperature on PEF lethality, increases in temperature, both at nonlethal and lethal values, improve the efficacy of the treatment. This effect has been related to a higher fluidity of the membrane that would make cells more susceptible to pore formation.

Microbial Characteristics

Data on the resistance of representative microorganisms to medium-intensity irradiation, ultrasound under pressure, HHP and PEF treatments are shown (Table 12.3) Maximum inactivation levels attained with each technology will depend on factors such as equipment technical developments and food characteristics. Comparison of data is hampered by different equipment, treatment media, strains, etc. As a general rule, bacterial spores are the most resistant to physical stresses and gram-positive bacteria are more resistant than gram negatives, and this has been attributed to the greater rigidity of their envelopes. Resistance of yeasts and moulds is quite variable. In general, they are more resistant to irradiation than non-sporulated prokaryotic cells. On the contrary, yeasts are more sensitive to PEF than prokaryotic cells, and this difference is specially evidenced at low electric field strengths, as the minimal electric field strength threshold needed to inactivate yeast is lower. Yeasts and moulds are considered to be sensitive to HHP, but wide differences among genus have been detected.

Regarding virus inactivation, they are among the most radiation-resistant microorganisms, and within the practical limits irradiation is considered not to be effective in virus elimination. The HHP resistance of viruses depends on their structure, which is heterogeneous. Human rotavirus and hepatitis A virus are inactivated to safe levels with treatments of 300 MPa/2 minutes and 450 MPa/5 minutes, respectively. The inactivation of viruses by PEF has been scarcely studied, and no decrease in infectious dose of human rotavirus after PEF treatments of (20–29 kV cm^{-1}) for 146 ls was reported. No data are available on the resistance of viruses to ultrasound. In addition, cell size and shape seem to have a role in resistance to physical stresses. The smaller the cell size, the higher the resistance to ultrasound, HHP and PEF. Coccoid cells tend to be more resistant than rod-shaped cells. Nevertheless there are many exceptions to these general rules. It should be noted that in some cases intraspecific variation is of great importance. Reports have shown that there are differences in survival to HHP treatments of more than 6 log cycles among natural isolates of *E. coli* O157. Difference of 1 and 2 log cycles in the fraction of surviving cells among 40 *Salmonella* strains subjected to irradiation and pressure treatments was reported. Intraspecies variability in resistance to PEF has also been demonstrated with *L. monocytogenes* and *Salmonella enterica*. Moreover, the isolation of pressure-resistant strains has been reported. These data raise concern, as an adequate target strain should always be selected for designing safe processes. In addition, the lack of relationship between the resistances of a given strain to various stresses emphasizes the necessity of carefully choosing the most appropriate strain for each agent.

This wide variation of resistance among strains of the same species has not been found for ultrasound under pressure treatments. *Salmonella* strains that differed greatly in their heat resistances were almost identical in their resistance to MS treatments. Microbial resistance to different physical agents depends not only on the intrinsic resistance of the microorganisms but also on their physiological state. It is well known that bacterial heat resistance varies widely depending on the growth phase, growth temperature and exposure to previous stressing environments. Cells in the exponential phase are generally more sensitive to all types of agents. Data about the influence of growth temperature are scattered and in many cases inconclusive. It seems to be an important factor determining HHP resistance, but its effect is different in exponential- and stationary-phase cells.

Product Parameters

Environmental factors, such as composition of the treatment medium, pH, water activity or addition of preservative substances, strongly affect the resistance of microorganisms to heat. The relative influence of such factors on microbial resistance to novel technologies depends on their mechanisms of action. Additional factors of special relevance in a particular technology have been identified. For instance, one of the most important factors influencing irradiation sensitivity is the composition of the treatment atmosphere. The presence of oxygen during irradiation has been found to enhance lethal effect because of oxygen radical formation. As a general rule, rich media exert a protective effect on bacteria against physical agents. This has repeatedly reported for heat treatments and similar results have been found for new technologies, except ultrasound under pressure, whereas for the rest of factors studied, a very small influence of medium composition has been described. Water activity of the suspending medium is possibly the environmental factor that modifies heat resistance in largest magnitude. It has been reported that the survival of *Salmonella* serovar *typhimurium* to irradiation is increased in low water activity meat. Water activity also affects resistance to ultrasound under pressure. *Listeria monocytogenes* suspended in a medium with a_W of 0.93 showed a decimal reduction time to a MS treatment two-fold greater as that obtained in a medium with a_W 0.99. The magnitude of this protective effect is very small when compared with other technologies, but it is worthy of note that a decrease in the a_W of the treatment medium is practically the unique environmental factor capable of protecting bacteria against MS. Regarding HHP inactivation, it was suggested that the protective effect of media of low a_W in *Lactococcus lactis* depends on the solute added. Sucrose protects cells against HHP inactivation through stabilization of membrane protein functionality. However, the addition of sodium chloride exerts a physical effect, increasing the phase transition of membrane phospholipids. Studies on the influence of a_W on PEF microbial inactivation are very limited but all agree that reduced a_W media increase PEF tolerance.

It was observed that a_w decrease from 0.99 to 0.93 increased the PEF survival of *Yersinia enterocolitica* by 3.5 log cycles after an 800 ls at 22 kV/cm. The reasons for this protection are unknown. The pH of the treatment medium is one of the factors modifying bacterial resistance with a more obvious practical side. The combination of acidification with heat has been used for decades to reduce the intensity of sterilization treatments for canned products, as acid pH decreases the thermal resistance of microorganisms. On the contrary, it has been demonstrated that acid pH has little effect on bacterial resistance to irradiation. It does not affect ultrasound under pressure resistance either. Regarding HHP inactivation, from published results it seems that bacterial cells are more sensitive to pressure in acidic media, but the magnitude of the effect is not remarkable. On the contrary, the pH of the treatment medium has a great influence on PEF resistance, but it depends on the species studied.

KINETICS OF INACTIVATION

The application of any new technology in food preservation requires a reliable model that accurately describes the inactivation rate. The model should be able to establish appropriate treatment conditions to achieve known levels of microbial inactivation, allowing the production of stable and safe foods. Models should be simple, and ideally, they should be built on parameters based on the physiological mechanism of inactivation. Thermal processing parameters have generally been calculated through the first order kinetics model. This model assumes that microbial inactivation results from the chance of the lethal agent striking the key target. As a result, a straight survival curve (plot of the logarithm of the fraction of survival cells vs treatment time) is obtained, and decimal reduction times (D values) and Z values are used to calculate treatment parameters.

However in the last 10 years, the suitability of first order kinetics for the modelling of heat inactivation, as well as for novel technologies, is being reconsidered. This is because of the frequent occurrence of deviations, such as shoulders and tails. Several theories have been proposed to explain these deviations, which may be different for spores and for vegetative cells. Shoulders have been mainly attributed to the occurrence of sublethal injury, multitarget inactivation, cell clumping or activation phenomena for spores. Tails are considered to be the reflection of resistance heterogeneity within the population, either inherent to the bacterial cells or acquired during the treatment. They are commonly detected when survival curves go beyond 4–5 log cycles. Typical survival curves of irradiation, ultrasound under pressure, HHP and PEF are shown in the Figure 12.1. Note that irradiation survival curves represent the log fraction of survivors vs irradiation dose, instead of treatment time. Shoulders have been reported for bacterial inactivation by irradiation, and it has been attributed to repairable cellular damage in the low dose range. Tails have also been detected, especially in complex media such as fish meal and crabmeat. First order kinetics inactivation is normally described for ultrasound under pressure. This is in agreement with the existence of a single key target and an all-or-nothing type inactivation mechanism. It also fits with the absence of adaptation phenomena and with the small intra- and interspecies variation in resistance to MS. Survival curves for HHP and PEF treatments show pronounced tails, concave upwards in shape. It is noteworthy that HHP and PEF treatments are homogeneous throughout the whole sample, and therefore every cell is subjected to the same stress. Several approaches have been proposed to explain the reasons for the upward concavity of survival curves. Some authors have considered that this kind of curves are biphasic and reflect the first order inactivation of two distinct microbial subpopulations, each one homogeneous in resistance. A concave upwards curve could also be justified by a continuous distribution of resistance within the microbial population.

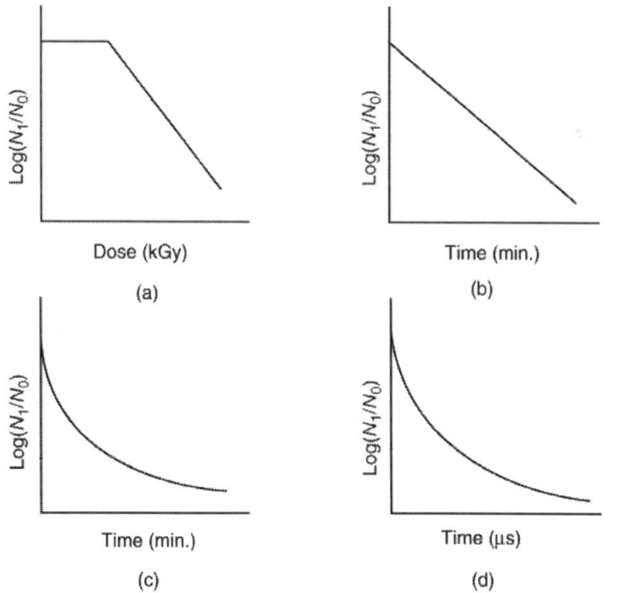

Figure 12.1 Typical survival curves for irradiation (a) manosonication (b) high hydrostatic pressure (c) and pulsed electric field (d) treatments

CONCLUDING REMARKS

The opportunities and limitations of new preservation techniques are listed in Table 12.4. Food scientists are fascinated by the never-ending fight against food spoilage organisms and food pathogens and are often encouraged by promising results of *in vitro* experiments. These *in vitro* experiments are often intensively elaborated but the feasibility of new preservatives or technologies is often too late validated by in *vivo/in situ* experiments resulting in a waste of research funds. One should not be blind for the complexity of food matrixes and the microbial ecology of foods and therefore evolve food products as soon as possible in experimental designs when new preservation methods are investigated. Food components often interact with the mechanism of action of new preservation methods resulting in a moderate to non-significant effect in food products. Moreover, the application of new preservation methods needs the approval of governmental authorities and this approval is often not yet the case. Nevertheless, further research is essential to demonstrate and explain the effect of new preservation technologies on the shelf life and safety of food products.

Table 12.4   Important opportunities and drawbacks of new preservation techniques

New preservative technique	Opportunities	Drawbacks
High hydrostatic pressure	High retention of nutrients and vitamins	Discontinuous for solid, viscous and particulate foods
	High fresh-like organoleptical quality	Semicontinuous for liquid foods
	Applicable for acid foods (as spores will not germinate in acid foods)	Large investment costs
	Spores can be inactivated when combined with heat or lactoperoxidase system or lysozyme	Spores are not sensitive Bacteria have shown to become resistant
Pulsed electric fields	Continuous process is possible	Upscaling of equipment is still under development
	High retention of nutrients and vitamins	Limited to liquid products
	Applicable for acid foods (as spores will not germinate in acid foods)	Spores are not sensitive
	Effect of combinations with other preservation methods to inactivate spores is under investigation	Efficacy depends on electrical conductivity of the food

(Contd.)

Table 12.4 (Continued)

New preservative technique	Opportunities	Drawbacks
Active packaging	Oxygen scavengers in sachets are effective	Oxygen scavengers incorporated in a packaging film show often limited effectivity
	Enables a surface treatment of food products	Compatibility with legislation
	Facilitates processing	Amounts of migrating active compound are often not substantial
Natural antimicrobial compounds	Green labelling	Often expensive
	Natural image	Interaction with food ingredients
		Low water solubility
		Change the organoleptical properties
Bacteriocins	Natural image	Narrow activity spectrum
		Spontaneous loss of bacteriocinogenicity
		Limited diffusion in solid matrixes
		Inactivation through proteolytic enzymes
		Interaction with food ingredients
		Bacteriocin-resistant bacteria
Protective cultures	Green labelling	Sometimes difficult to apply
	Natural image	Heat-labile
		Efficacy is not always proven in food products

Irradiation, ultrasound under pressure, HHP and PEF are effective procedures to inactivate vegetative microorganisms in foods, but the high resistance of spores limits their use as a sole method for food preservation. Therefore, these novel technologies are finding applications as hurdles that assure food safety through microbial inactivation in minimally processed high-quality products. To fully exploit their potential, more research effort is needed to clarify mechanisms of inactivation, especially for HHP and PEF, to better understand the effect of environmental factors, and the occurrence of stress adaptation and sublethal injury, which are aspects of great relevance regarding food safety. Characterization of resistant strains may be useful in mechanistic studies. Identification and selection of target strains for each technology is also necessary. Finally, development of mathematical models based on physiological facts to establish treatment conditions is urgently needed. Ideally, models should consider the lethal event or events leading to death, the heterogeneity within the bacterial population and possible phenomena of adaptation and damage.

REVIEW QUESTIONS

1. Discuss in detail about high hydrostatic pressure as a preservation technology.

2. Discuss the applications of pressure inactivation.

3. Give a brief account of pulsed electric field as a novel preservation technology.

13

NON-THERMAL PRESERVATION OF FOODS USING COMBINED PROCESSING TECHNIQUES

INTRODUCTION

Foods begin to lose their quality the moment they are harvested, through changes resulting from physical, chemical, enzymatic or microbiological reaction. Food preservation prevents these deteriorative reactions, extending a food's shelf life and assuring its safety. Microorganisms and enzymes are the main agents responsible for food spoilage and therefore the targets of preservation techniques. An "ideal" method of food preservation has the following characteristics:

1. It improves shelf life and safety by inactivating spoilage and pathogenic microbes.
2. It does not change organoleptic and nutritional attributes.
3. It does not leave residues.
4. It is cheap and convenient to apply.
5. It is consumer-friendly and adheres to the standards.

Most food preservation procedures inhibit deteriorative agents. Some of these technologies (chilling or freezing) maintain a food's fresh quality. However, applications of others, such as lower a_w or pH, also affect freshness. Overall, these technologies do not ensure complete food safety, since modifications of preservation conditions, such as breakdown in the chill chain or food rehydration, could lead to the growth of pathogenic and spoilage microbes.

Currently, thermal processing is the only food preservation technique to act via inactivation, which is used substantially in the food industry. Inactivation of microbes and enzymes results in a stable and safe product. Thermal processing has most of the characteristics of an "ideal" food preservation method. However, in some foods the high thermotolerance of certain enzymes and microorganisms mainly bacterial spores, needs extreme heat treatments, which affects the nutritional and organoleptic food properties. Therefore, some alternative means of inactivating pathogenic and spoilage microbes are being sought by the food industry.

In the last 20 years, consumers demand for high-quality foods that are safe and stable has led to non-thermal preservation techniques for inactivating microorganisms and enzymes. In non-thermal processing, the temperature level is held below the thermal processing level so that there will be minimal degradation of food. However, non-thermal technologies not only improve food quality, but also enhance the safety levels, when compared to other procedures. Overall, most non-thermal preservation techniques are highly effective in inactivating vegetative cells of bacteria, yeast and moulds. However, it is difficult to inactivate bacterial spores and most enzymes with these procedures. Thus their use is limited to foods where enzymatic reactions do not affect food quality or where spore germination is inhibited by other prevailing conditions such as low pH.

To extend the use of non-thermal processing in the food industry, combinations of these technologies with traditional or emerging food preservation techniques are being studied. This approach, known as "hurdle technology" has already been applied successfully using traditional techniques of food preservation. The antimicrobial effect of preservation treatments can be optimized by combining them as hurdles in an overall preservation strategy.

Hurdle Technology

By placing a number of sublethal stresses (i.e., hurdles) on a microbial cell, the organism expends energy to overcome the hostile environment, potentially leading to metabolic exhaustion and death. Hurdles targeting the same elements within the cell have an additive inhibitory effect only, whereas synergistic effects may result from disturbing several functions of the cell. Attacking various cellular targets will have a synergistic effect by making the organism strain every possible repair mechanism simultaneously, while the activation of stress-shock proteins also becomes more difficult. The multitarget approach may also help to counter stress adaptation associated with sublethal treatments. Two or more novel non-thermal processes and conventional preservative techniques can also be combined to enhance their lethal or inhibitory effects on microorganisms. Such combinations enhance food preservation at lower individual treatment intensities, and microbes that survive processes, such as irradiation and ultrasonication, generally become less-resistant to other factors such as heat, pH changes and antibiotics.

It is debatable whether the theoretical basis of the original hurdle concept can be applied to non-thermal processing combinations, though the modes of action of non-thermal processes are intrinsically different to those of conventional and sublethal preservation hurdles (e.g. low water activity, low-temperature storage, pH manipulation or addition of inhibitory substances).

They are not a series of low-intensity stressors designed to fatigue a cell, but are relatively short-time high-intensity treatments causing irreversible damage to vital cell components. This raises questions about the mechanisms of synergy behind non-thermal processing combinations such as whether effective combinations target the same or different components within the cell. It could be argued that applying a number of treatments with a common target would improve the chances of irreversible damage being inflicted, whereas treatments targeting different sites may cause sublethal damage in several areas of the cell, but fail to inactivate the organism. A good understanding of the modes of action of each individual treatment is crucial to select effective antimicrobial combinations. Unfortunately, the mechanisms of microbial inactivation by non-thermal technologies are currently not well understood although theories have been developed. Food preservation using combined methods involves successive or simultaneous applications of various individual treatments. Combined treatments are advantageous, principally because many individual treatments alone are not adequate to ensure food safety or stability. In some circumstances, combined treatments allow a milder use of single treatments, with consequent improvement in food quality. The combined effect of several preservation techniques can be merely additive, but in terms of food quality and safety a synergistic effect is preferable. However, combining technologies is not always advantageous, because in some circumstances an antagonistic effect may result.

Combining non-thermal methods with the food preservation techniques can

1. Enhance the lethal effects of non-thermal processing
2. Reduce the severity of non-thermal treatment needed to obtain a given level of microbial inactivation
3. Prevent the proliferation of survivors following treatment.

COMBINATIONS WITH HIGH HYDROSTATIC PRESSURE

During high hydrostatic pressure(HHP) processing, foods are subjected to pressure in the range 100 to 1000 MPa. These treatments are capable of inactivating microorganisms and endogenous enzymes, while retaining nutrients and flavours. Initial studies on HHP preservation of foods were conducted at the end of the 19th century. However, this technology has not had commercial application until recent years. The renewed interest in HHP processing in the last decade is not only due to its preservation effects but also its potential to modify the functional properties of foods.

High pressure only affects non-covalent bonds, leaving covalent bonds intact and, consequently, induces alterations in the structure of secondary- and tertiary-bonded molecules. Overall, HHP is very effective in inactivating vegetative cells of microbes, but these alone does not achieve a substantial inactivation of spores and reduction in activity of certain enzymes. The majority of HHP processed products available in the market are high-acid products like fruit juices or sauces. HHP preservation techniques are suitable for these products having low pH that are mainly spoiled by microorganisms relatively sensitive to HHP (yeast, moulds and lactic bacteria) and do not support the germination of pressure-resistant bacterial spores.

Interest in using high-pressure technology to extend the shelf life of low-acid foods (e.g. poultry and red meat, goat-milk cheese, liquid whole egg or cooked ham) is increasing. Preservation of such foods with HHP requires an appropriate combined-treatment that can achieve enough levels of microbial inactivation and shelf life extension.

HHP and Heat

The combined effect of HHP and heat on microbial inactivation has been widely investigated. The resistance of vegetative cells of bacteria to HHP appears to be high at temperatures approximately 20–30°C. At higher and lower temperatures, microorganisms are more sensitive to HHP. The resistance of vegetative cells of bacteria to HHP decreases even when pressure is combined with heat at non-lethal temperatures. This combination allows substantial inactivation ($>6 \log_{10}$ cycles) of spoilage and pathogenic microbes at lower pressures and/or shorter times than that required when pressurization is carried out at room temperature. For example, the inactivation of *Listeria monocytogenes* in UHT milk did not occur at 200 MPa and temperatures up to 45°C. In contrast, at 200 MPa and 55°C, a near $6 \log_{10}$ cycle reduction in cell count was obtained after 15 minutes treatment. Another benefit of this combination is that variation in pressure resistance among strains is much lower when HHP is combined with moderate temperature. Resistance by different strains of *L. monocytogenes*, *Staphylococcus aureus*, *Escherichia coli* 0157: H7 and *Salmonella* species to 345 MPa varied widely at 25°C but these differences were greatly reduced at 50°C.

In general, the kinetics of inactivation of most vegetative cells by HHP at low temperatures shows an initial exponential rate, followed by pronounced tailing. This tail tends to disappear when HHP is combined with heat. Inactivation of vegetative forms of yeast and moulds in combination with HHP and heat has been scarcely investigated, probably because microorganisms involved are very sensitive to mild HHP treatment at room temperatures. Inactivation of ascospores in moulds requires combination of HHP and moderate temperatures (60–70°C).

Unlike inactivation of vegetative bacterial cells, inactivation of bacterial spores with HHP, occurs in two steps. Pressure initially causes the germination of spores and then inactivates the germinated forms. In general, bacterial spores appear to be resistant to HHP treatments at room temperature. It has been reported that they can resist pressures as high as 800 MPa for many hours. However, pressures as low as 10 MPa can initiate germination of bacterial spores, thus sensitizing them to heat, radiation, chemical agents and even HHP treatments.

Initiation of germination and inactivation of bacterial spores by HHP are highly temperature dependent, thus increase at higher temperatures. At low temperatures, the treatments cause spore germination, but in general the pressure is insufficient to cause appreciable inactivation of germinated forms. Combined HHP and heat is especially effective at temperatures allowing inactivation of germinated spores (>60°C) suggesting that spores germinated by HHP are directly inactivated by heat. However, recently it has been observed that inactivation of *B. subtilis* and *B. stearothermophilus* spores by HHP at 70°C and 90°C respectively did not involve spore germination.

Inactivation of *B. stearothermophilus* by combining HHP and heat was more effective when an oscillatory treatment (6 log cycles for 5 minutes) was applied. At 70°C and 600 MPa, 4 log cycles of inactivation were obtained with continuous treatment and more than 6 log cycles of inactivation were achieved with oscillatory treatment.

The effect of HHP on enzymes varies widely for different enzymes probably due to the structural differences of individual enzymes. In general, combined effect of pressure with moderate temperatures increases the level of enzyme inactivation with exceptions.

The high resistance of endogenous enzymes to HHP or to combined HHP heat treatments emphasizes the necessity of combining pressurization with other techniques, such as storage at low temperature, chemical modification of enzymes, and use of enzymes ("killer enzymes") or naturally occurring protein inhibitors to maintain food quality and extend shelf life.

Combining HHP and heat could be of great practical interest, especially in preserving low-acid foods. Combining mild temperatures and pressure allows the pasteurization/sterilization of foods at lower temperatures. Another practical benefit of this combination would be the inactivation of microbes at moderate pressures, which is economically feasible and does not induce quality loss in the treated products.

HHP and Low pH

In general, microbial inactivation of both vegetative cells and bacterial spores with HHP is scarcely affected by the pH of the treatment medium. It has been reported that inactivation of bacterial spores by combining pressure and temperature was high when the pH was near neutrality. However, a higher inactivation of spores in *B. coagulans* and *Clostridium sporogenes* PA 3679 was obtained when the pH of the treatment medium was reduced from 7 to 4. Pressurization narrows the pH range for growth of microorganisms, having a particularly inhibitory effect on membrane ATPase, a very important enzyme in the acid–base physiology of cells.

Different yeasts and moulds showed the same pressure resistance in fruit juice acidified between 2.5 and 4.5 with citric, tartaric, and lactic acid. However, lowering the pH of citrate buffer from 5 to 3 enhanced the lethality in *Sachharomyces cerevisiae* and *Zygosaccharomyces bailii*, 2 and 1.5 $\log_{10}$ cycles respectively.

Although microbial sensitivity to HHP is not increased by lowering pH, the key factor in this combination is the prevention of microbial growth and the germination of spores that can survive HHP treatment at acidic pH.

HHP and Antimicrobials

Interest in natural antimicrobials has expanded in recent years in response to consumer demand for less 'artificial' additives. The potential of these natural antimicrobials is substantial, especially in combination with other food preservation techniques. Recent studies have indicated that HHP inflicts sublethal injury on microorganisms, even at lower pressures required for their death. The primary cause of vegetative cell damage by HHP is permeabilization of the cell membrane due to irreversible changes in the structure of membrane macromolecules such as proteins. Permeabilization of the membrane is also affected by compression of the membrane's bilayer and reduction in the cross-sectional area per phospholipid molecule.

Sublethally injured cells are more susceptible to antimicrobial components. In gram-negative bacteria this fact is predominant because their outer membrane acts as an efficient barrier against hydrophobic solutes and macromolecules, such as bacteriocins. During HHP exposure (>180 MPa) *E. coli* cells became sensitized to lysozyme and nisin, two antimicrobial proteins excluded by the outer membrane at ambient pressure. The high levels of inactivation seen with antimicrobial agents and HHP hurdles are believed to be due to the combined effects of destabilization of membrane structure or function although their specific modes of action are different. High-pressure membrane damage can increase the cell penetrability and activity of antimicrobial agents, while, conversely, treatment with cell wall weakening agents sensitizes pressure-resistant bacteria to HHP. Reversible permeabilization of the outer cell membrane was induced under pressure, enabling temporary access of nisin to the cytoplasmic membrane, or lysozyme to the peptidoglycan layer surrounding the cell. In addition, residual bacteriocins in the food matrix would continue to exert their inhibitory effect after treatment with the non-thermal processing hurdle, thereby inactivating and preventing the recovery of any sublethally injured cells. HHP treatment (320 MPa, 15 minutes, 23°C) reduced viable counts of *E. coli* to 4.06 $\log_{10}$ inactivating 5.7, 5.4 and 6.9 $\log_{10}$ in the population respectively, when nisin (1100 IU/ml), lysozyme (10 µg/ml) or both antimicrobials were present during pressurization. However, adding these antimicrobials immediately after pressure treatment did not cause any viable count reduction. Combining antimicrobials with HHP can enhance the effectiveness of pressurization with advantages in product quality and safety.

HHP and Modified Atmospheres

HHP at low temperatures combined with modified atmosphere packaging (MAP) has been used to extend the shelf life of salmon and prawns. This treatment is more effective in delaying microbial growth during storage at 5°C. However, the best treatment for salmon in terms of shelf-life extension and quality retention was a pressurization treatment of 150 MPa for 10 minutes followed by MAP. This treatment extended the shelf life at least 5 days. However, specific studies should be conducted to discover the best combination for any given food.

HHP and Carbon Dioxide

The antimicrobial effect of carbon dioxide is well known and it has been observed that this effect increases when CO_2 is applied under pressure. In general, pressures used in this combination are below 50 MPa. Microbial inactivation with this combination has been observed in natural flora and with specific spoilage and pathogenic microbes in model systems and foods. Even supercritical CO_2 (at 20 MPa and 35°C) has been shown to cause up to 7 log cycle reductions of bacteria and yeast presumably due to the low-pH inactivation of metabolic enzymes and/or the extraction of cellular substances such as phospholipids. Exposure to supercritical CO_2 has also been used to reduce both viability and pressure resistance of *Bacillus* spores.

However, pressurized CO_2 does not appear to have a significant effect on bacterial spores and fungal ascospores at temperatures below 80°C.

Time, pressure, temperature, water activity and pH are the critical parameters in this combination. An increase in temperature and/or pressure and a decrease in pH enhances the antimicrobial effects of CO_2. On the other hand, inactivation decreases at low water activity. It has been suggested that pressurized CO_2 penetrates cells relatively easily, causing a greater intracellular pH change than other acids. Others have suggested that CO_2 dissolves in the aqueous phase as carbonic acid, then expands upon sudden pressure release, damaging the cell.

The exact mechanism behind microbial inactivation using CO_2 under pressure is still unclear. Some authors suggest that the quick release of carbon dioxide following compression play a significant role in microbial reduction. Some others suggest that the inactivation of microbes occurs during the pressurization step, not during decompression. Under pressure, more CO_2 molecules pass through the membrane and lower the internal pH, thereby affecting the key enzymes in cell's metabolism. The inactivation of pressurized CO_2 has also been related to the extraction of cell wall constituents, such as phospholipids and hydrophobic compounds. Pressurized CO_2 is a promising alternative to improve food quality by the reduction of microbial loads especially present on food surfaces. However, a long treatment period is generally needed to achieve substantial microbial inactivation.

HHP and Irradiation

The use of gamma irradiation was investigated before the HHP treatment of *B. coagulans* spores and found that mildly lethal doses of radiation increased the pressure sensitivity of survivors. It was also found that irradiation sensitized *Clostridium sporogenes* spores to pressure and that mild doses of irradiation and high pressure combined was more effective for inactivating *Clostridium* spores than application of either process alone. Effects of simultaneously applying pressure and ionizing radiation to spores were reported. The two treatments were additive in their sporicidal effect; thus, the intensity of one or both treatments could be lowered while retaining the degree of inactivation. The mechanisms of spore destruction were postulated as either the germination of spores by HHP and consequent sensitization to radiation, or the disruptive effect of irradiation on peptidoglycan in the spore cortex, allowing partial rehydration of the core and increased sensitivity to both radiation and pressure.

HHP and Pulsed Electric Fields

Pressurization may reduce the effectiveness of a PEF treatment, as was found that electric field pulse treatment under high pressure (200 MPa) exerted a protective effect against permeabilization of bacterial cell membranes although each treatment alone cause major damage at this site. The possibility of germinating Bacillus spores using HHP was studied, then the germinated cells were inactivated with a PEF treatment. It was found that germination of more than 5 log cycles of spores was initiated by pressurization, and while the germinated cells did become sensitive to a subsequent heat treatment, they were not sensitized to PEF application below 40°C. It was suggested that spore inactivation by these combined processes could be improved by adding an intermediate holding step to allow germinated spores to outgrow into vegetative cells.

COMBINATIONS WITH ULTRASOUND

Ultrasound is defined as sound waves with frequencies above that of human hearing. These waves can be propagated in a liquid media as alternating compression. If ultrasound has sufficient energy, a phenomenon known as cavitation occurs. Cavitation involves the formation, growth, and rapid collapse of microscopic bubbles. Based on theoretical considerations, extremely high temperatures and pressures are momentarily delivered to the liquid media during the collapse of the bubbles. With extreme conditions, electrical discharge and production of free radicals also occur inside the collapsing bubble. However, the ultimate reason for microbial inactivation via ultrasound is believed to be the mechanical damage caused by cavitation.

Ever since the lethal effects of ultrasound on microbes were first observed in 1929, its use has been continually suggested for disinfection and food preservation. However, this technology has not been adopted, probably due to the long treatment needed for substantial microbial inactivation.

Recently, it was demonstrated that microbial inactivation with ultrasound increases when treatment is applied under pressure (Manosonication, MS). For example, an increment of hydrostatic pressure from 0–500 kPa at constant amplitude (150 µm) decreased the decimal reduction time (D value) of *Yersinia enterocolitica* eight times. The increased lethality of ultrasound under higher static pressure was more remarkable within a given pressure range (0–300 kPa). At constant hydrostatic pressure, microbial inactivation depended on the amplitude of the ultrasonic waves. At 200 kPa, the D value of *Y. enterocolitica* decreased 11 times when the amplitude increased from 21 to 150 µm. An exponential relationship was observed between the lethality of ultrasound and the amplitude. The influence of hydrostatic pressure and the amplitude on the lethality of MS treatments in different gram-negative and gram-positive bacteria was the same.

MS and Low pH

The influence of treatment medium pH on MS resistance to *L. monocytogenes* was studied and the DMS value of this microorganism did not change when pH decreased from 7 to 4. The acidic conditions had a much greater effect on the organism's resistance to heat than its sensitivity to ultrasonication. Thus this combination does not prove to be very effective commercially.

MS and Low a_w

In general, the presence of solutes in the treatment medium prevents microbial inactivation by different lethal agents. However, this influence was greater for heat than for MS treatment. Adding 57% sucrose to the decreased aw in the treatment medium, until 0.94, increased the heat resistance at 62°C of *L. monocytogenes* 25 times, while its MS resistance increased only two times. Therefore, MS could be a very useful alternative to heat treatment for the inactivation of microorganisms in foods with low a_w.

Combining MS and Heat (Manothermosonication-MTS)

Researchers have investigated combinations of heat and ultrasound to decrease the intensity of heat treatments. The heat resistance of *B. cereus*, *B. licheniformis*, *B. stearothermophilus* and thermoduric streptococci decreased following ultrasonication treatment at 20 kHz. This effect was higher when heat and ultrasound were applied simultaneously. The application of MS treatment simultaneously with heat treatment also led to higher microbial inactivation. While in most vegetative cells the lethal effect of MTS was additive, on *Enterococcus faecium* and *B. subtilis* spores, a synergistic effect was observed.

MTS treatments have also been very effective in the inactivation of different enzymes related to food quality such as lipoxygenase, lipase protease or pectin methyl esterase, peroxidase, polyphenoloxidase, etc.

In conclusion, the simultaneous application of ultrasound under pressure with heat treatment results in higher microbial and enzyme inactivation. Therefore, the same inactivation level is achieved over a shorter treatment period or at lower temperature. Consequently, this combination could be advantageous, due to the minimization of heat induced damage in product quality.

MS and Antimicrobials

A study was done with reduced intensity and duration of ultrasound treatment required to inhibit *Zygosaccharomyces rouxii* respectively, by incorporating potassium sorbate and sodium benzoate or eugenol by 67% and 33%,into the recovery media. The study suggested that the different modes of action of the ultrasonication, mild heating (45°C) and antimicrobial hurdles were responsible for the observed inhibition. The combination of ultrasonication and hydrogen peroxide was more lethal to both *Bacillus* and *Clostridium* spores than either treatments done alone. It was believed that the ultrasonic waves improved lethality of the hydrogen peroxide by increasing the permeability of the cells, increasing the rate of reaction between the hydrogen peroxide and cell components, and dispersing cell aggregates to increase the surface area for contact.

MS and High Pressure

At ambient temperature and pressure, ultrasonication has little lethal effect on microorganisms. High treatment intensities may cause some microbial inactivation, but simultaneously produce adverse sensory changes in the food. A much more gentle yet effective treatment, termed manosonication (MS), utilizes moderate doses of ultrasonication under mild pressure. Manothermosonication (MTS) describes an MS process that is carried out at high temperatures. At elevated temperatures, a study proved the inactivation of *Y. enterocolitica* by combining ultrasonication, pressure and heat. The lethal effect of ultrasonication (20 kHz, 150 μm) increased with rising pressure until an optimum pressure of 400 kPa at which maximum inactivation occurred. It was found that lethality of MTS treatments for bacterial cells, spores and yeast to be 6–30 times greater than thermal treatments of equal temperature and concluded that the combined effects of ultrasonication, pressure and heat were synergistic. Studies also noted enhanced microbial inactivation when MS was combined with temperatures above 50ºC. Unlike the synergism between ultrasound and pressure, however, the lethality of MS plus heat was additive only and might be due to two different mechanisms acting independently.

Levels of destruction of *B. subtilis* spores using MTS (20 kHz, 117 μm) followed a similar trend under increasing pressure, with maximum inactivation at a pressure of 500 kPa. In both studies, the lethal effect of pressurization was significantly more pronounced when the amplitude of ultrasonic waves was increased. It was also found that the ultrasonic (20 kHz, 117 μm) inactivation of *L. monocytogenes* increased dramatically when the pressure was raised from ambient to 200 kPa. The increase in the inactivation rate became progressively smaller, however, the pressure was raised from 200 to 400 kPa. The study theorized that the higher lethality of ultrasound under moderate pressure was due to the higher intensity of cavitation. It should be clarified that pressures applied during manosonication (e.g. 200–600 kPa) are not in the lethal range of pressures applied during HHP treatment (e.g. 50–1000 MPa).

MS and PEF

It was observed that ultrasound enhanced the inactivation of *B. subtilis* spores under PEF; however, no details were given in the publications as to the intensity of the ultrasonication treatment, whether it was performed before, during or after PEF, and what the actual levels of inactivation were. 4 log cycles of *B. subtilis* spores were inactivated using 2000 Hz sonication combined with PEF at 30–40 kV/cm.

COMBINATIONS WITH PULSED ELECTRIC FIELDS

Microbial inactivation by pulsed electric fields (PEF) was first observed in the early 1960s. This technology involves applications of high intensity electric field pulses of short duration (microseconds). PEF has received increased attention in the last decade as a food preservation technique because of its potential to inactivate microorganisms at temperatures below those adversely affecting food quality. PEF is highly effective in killing vegetative cells of bacteria, yeast and moulds. However, PEF inactivation of bacterial spores and enzymes as related to food quality is unclear.

PEF technology is not being used to preserve foods commercially at present. However, the feasibility of PEF technology extending the shelf life of different foods without apparent change in product properties, as in apple juice, skim milk, beaten eggs, green pea soup, or orange juice, has been demonstrated on a pilot plant scale. Following PEF treatment, these products must be stored under refrigeration to prevent enzymatic reactions and spore germination. The combinations of PEF with other preservation technologies are being investigated to increase the lethal effect of this non-thermal process and to extend its application to different liquid foods.

PEF and Heat

In general, during PEF treatment the heat generated due to the sample's resistance to current flow is removed. However, it has been observed that lethality of PEF treatments increases with an increase in processing temperature. Several study have demonstrated that PEF treatments applied at higher start temperatures increase the inactivation effect in a batch treatment system or in a continuous treatment system. This increment in the lethality effect of combined treatments has been observed at both lethal and non-lethal temperatures for microorganisms investigated. For example, inactivation of *E. coli* by PEF with a field strength of 36 kV/cm and pulse duration of 2ms, increased from 2 to 3 log cycles when temperature was increased from 7–40°C. The higher susceptibility of microorganisms to PEF has been related to the temperature effect on membrane fluidity properties. While phospholipids are closely packed in a rigid gel structure at low temperatures, at high temperatures they are less ordered, the membrane has a liquid crystalline structure, and its thickness is reduced. Further studies are necessary to determine an optimal heat and PEF combination that can inactivate the maximum level of PEF-resistant microbial species possible with a minimal effect on food quality; studies on the effect of this combination on bacterial spores and enzymes are needed as well.

PEF and Low pH

PEF processing involves passing a high-voltage electric field (10–80 kV/cm) through a liquid food held between two electrodes, in very fast pulses typically of 1–100 ms duration. Destruction of microbial cells by PEF is due to irreversible electroporation of the cell membrane, which leads to leakage of intracellular contents and eventually lysis of the cell. The PEF inactivation of *Escherichia coli* O157:H7 in a 10% glycerol solution was enhanced synergistically by lowering the pH from 6.4 to 3.4 using benzoic or sorbic acid. Similarly, adjusting the pH of skim milk and liquid eggs with organic acids has shown additive and synergistic inactivation of bacteria when combined with PEF treatment.

PEF plus acidification with hydrochloric acid, on the other hand, resulted in no extra inactivation of raw milk's microflora when compared with PEF alone. The synergism between acetic acid and PEF might be due to the fact that both treatments target the cell membrane. In this way, PEF could increase the permeability of the cell wall and membrane, enhancing the entry of undissociated acids into the bacterial cell.

The influence of pH on bacterial inactivation by PEF is unclear. Independent of this pH influence on microbial inactivation by PEF, acid foods such as fruit juices are good candidates for processing with this technology because bacterial spores cannot germinate at low pH.

PEF and Antimicrobials

It is well known that PEF induces reversible or irreversible structural changes in cell membrane, resulting in pore formation and loss of selective permeability properties. On the other hand, some antimicrobials such as lysozyme or nisin act on the cell membranes and others such as organic acids cross the microbial membranes and access the cytoplasm where they act. According to the mechanisms of action, for both preservation techniques, a powerful synergistic effect should be expected when they are combined: antimicrobials could increase the susceptibility of membranes to dielectric breakdown and/or PEF could facilitate the access of antimicrobials to the membranes or cytoplasm.

Nisin is the antimicrobial most studied in combination with PEF. It has been demonstrated that presence of nisin in the treatment medium can increase the inactivation effect of PEF for both gram-positive and gram-negative bacteria. A synergistic effect has been observed when bacteria are incubated with nisin after PEF treatment, when nisin is present during PEF treatment, or when nisin is added to the bacterial suspension after PEF treatment.

Combining PEF with nisin also affects microorganisms suspended in different foods. The combined effect of PEF treatment on *L. innocua* suspended in liquid whole eggs or skim milk, followed by exposure to nisin, was additive or synergistic, depending on the intensity of PEF treatment. Combining organic acids with PEF resulted in higher inactivation of *E. coli* 0157:H7. However, a synergistic effect was observed at pH 3.4 but not at pH 6.4. At the lower pH, the presence of benzoic or sorbic acid (1000 ppm) during a single high voltage electric pulse resulted in an inactivation of around 2 $\log_{10}$ units greater than that with individual treatments. As the undissociated fraction of acids in solution is much higher at pH 3.4 than at pH 6.4, the synergistic effect at low pH may indicate that PEF enhanced the entry of the undissociated fraction but not the dissociated one.

Results indicated that PEF in combination with antimicrobials increases bacterial inactivation and extends the spectrum of action of some bacteriocins. More investigation is needed to understand the mechanisms of action for treatment optimization to enhance the safety and shelf life of foods processed by PEF.

PEF and HHP

The lethality of PEF treatment applied simultaneously with sublethal HHP treatment on *B. subtilis* vegetative cells was lower than the lethality of PEF treatment applied under atmospheric pressure. Thus, the sublethal pressure treatment produced a stabilizing effect on cells of *B. subtilis* against the PEF treatment. However, a synergistic effect was observed when the cells were exposed to a treatment of 200 MPa for 10 minutes followed by PEF treatment immediately before the pressure release. The additional inactivation effect was $2 \log_{10}$ units.

COMBINATIONS WITH IRRADIATION

Irradiation is a non-thermal food preservation process that has been re-examined in recent years. The process involves applications of electrons or electromagnetic waves to foods. Sources of ionizing radiation used to irradiate food include gamma rays, electron beams, and X rays. Although the mechanisms of bacterial inactivation by irradiation are not completely understood, it is thought that irradiation inactivates microorganisms mainly by causing lesions in DNA. Irradiation is considered the only known technique capable of ensuring the hygienic quality of raw food in a fresh or frozen state.

From the beginning, safety and wholesomeness of irradiated foods have been the issues of concern. In 1999, WHO stated that according to established good manufacturing practices, irradiated food products can be considered safe and nutritionally adequate. Currently, many countries have approved irradiation of a variety of foods, such as fruits, vegetables, spice, pork, beef, poultry, fish and sea food. Irradiation enhances the shelf life of foods and ensures their innocuousness because most microorganisms in vegetative form are extremely sensitive to irradiation at low doses. However, in some circumstances combined treatments are more satisfactory since the dose required for complete sterilization can induce undesirable changes in food flavour or is higher than permitted levels.

Irradiation and Low Temperature and Modified Atmospheres

Irradiation of various fresh foods at doses significantly affecting organoleptic properties, in combination with storage at chill temperatures, improves food safety and extends shelf life. To avoid the growth of surviving microorganisms during storage at chill temperatures, the combination of irradiation with modified atmosphere packaging has been investigated in different foods such as pork, beef, poultry, turkey, carrots and lettuce. In general, combining low dose irradiation (< 3 kGY) with modified atmosphere packaging (vacuum or gas packaging) widely extended the shelf life of fresh foods in refrigeration.

As the effects of combining irradiation with MPA on pathogenic and spoilage microorganisms, and sensory quality, depend on food and atmosphere composition, specific studies should be conducted to find an optimal gas composition and irradiation level.

Irradiation and Heat

A combination of irradiation and heat has been proposed for required lethality while preserving food quality, thereby reducing the detrimental effects of heat or irradiation alone on food. While the inactivation effect of heat treatment followed by irradiation is additive, a synergistic effect has been observed when heat treatment is applied after irradiation or when both treatments are simultaneously applied (thermoradiation). Several studies have established that irradiation sensitizes vegetative cells and bacterial spores to a subsequent heat treatment. This heat sensitizing effect has been observed in both vegetative cells and bacterial spores suspended in aqueous or lipid systems or foods.

A synergistic effect on the inactivation of vegetative bacteria and bacterial spores can be achieved with thermoradiation. For examples, it was observed that the D value of *S. aureus* decreased form 0.098 to 0.053 kGy, while the irradiation temperature increased from 35 to 45°C. This synergistic effect of thermoradiation has also been observed in the inactivation of microorganisms in several foods. It was reported that thermoradiation caused greater inactivation of *Salmonella enteritidis* in liquid whole egg than either heat or radiation alone. A synergistic effect was also observed during thermoradiation of *Vibrio vulnificus* in buffer, fresh oyster, and fresh fish at 40°C.

Irradiation and HHP

HHP in combination with irradiation has also been studied. A hydrostatic pressure greater than 500 atm decreased the resistance of *Bacillus pumilus* spores to gamma irradiation. This decrease in resistance was associated with the initial germination of bacterial spores by HHP. Also, an irradiation treatment sufficient to inactivate a proportion of *B. coagulans* spores increased the pressure sensitivity of the survivors. When pressure and radiation were applied simultaneously, the effect obtained in *B. pumilus* was additive, rather than synergistic.

The combined effect of low dose irradiation and HHP on the microbiological quality of different species of meat has also been investigated. Staphylococcal counts in lamb meat were reduced by 1 log cycle in samples subjected to irradiation (1 kGy) or HHP (200 MPa, 30 min.). However, when both treatments combined were applied, an inactivation of more than 4 log cycles was achieved. Therefore, combinations of HHP and irradiation can lower the intensity of treatment required for any process when used alone, and as a consequence can improve the sensorial quality and microbial safety of meat.

Irradiation and Antimicrobials

The shelf life of iced cod fillets were extended from 19 to 25 days when irradiation was supplemented with dipping of the fillets in a 5% potassium sorbate solution. Another study reported a doubling of the shelf life of Dover sole when 0.19% sodium benzoate was added to samples before irradiation. Vacuum and modified atmosphere packaging have also been shown to work well with irradiation treatments to extend the chilled shelf life of fresh beef, pork and fish.

Irradiation and Lowered pH

Irradiation of bulk or pre-packaged foods is achieved by exposing the product to a source of ionizing energy, typically Cobalt-60. The product to be treated is conveyed through a shielded chamber containing the radiation source, and irradiation dosage is controlled by the speed of the conveyor. Inactivation of organisms by ionizing radiation is primarily due to DNA damage, which destroys the reproductive capabilities and other functions of the cell. The lethal effect of ionizing radiation is not greatly enhanced by mild acidification. For example, the destruction of bacteria and the shelf life of irradiated chicken and ground beef were not improved by low concentrations of infused or added acetic acid. In another study, the differences in pH of five commercial orange juice formulations (from pH 3.87 to 4.13) had no influence on the radiation dose required for 90% inactivation of a *Salmonella enteritidis* strain isolated from a citrus juice associated with an outbreak of salmonellosis.

SUMMARY

Most novel non-thermal technologies are still in their early stages of development although some emerging non-thermal processes have now been implemented in industrial-scale systems for commercial and research applications. Other than irradiation, however, it is safe to say that not one non-thermal process has been developed to a point where its use alone can guarantee the safety of low-acid foods. Effective combinations of two or more preservation hurdles may be chosen once the modes of action and cellular targets of each treatment are known. For the intelligent selection of non-thermal processing combinations, therefore, target elements within cell and the effects of treatments on those elements need to be determined. The treatment intensities required for cell inactivation need quantification and standardization also. The small number of scientific studies summarized in this review show that combining non-thermal treatments has great potential for improving the safety and quality of foods although many technological and regulatory barriers still need to be overcome before the food supply can receive these benefits.

1. Explain in detail the possible combinations of preservation technologies with HHP.
1. Define manosonication.
2. Define manothermosonication.

14

MICROBIAL AND BIOCHEMICAL ASPECTS OF FOOD SPOILAGE

OVERVIEW

Food spoilage can be considered as any change which renders a product unacceptable for human consumption. Food spoilage can be obvious, e.g. physical damage, visible growth of microorganisms, slime formation or insect damage. However, when spoilage is due to changes in texture or the development of off-flavours caused by (bio)chemical or microbial reactions, the underlying mechanisms may be difficult to identify. Therefore, evaluation of spoilage will always, directly or indirectly, be related to a sensory assessment. In contrast, biochemical or to a lesser extent, microbial analyses are less expensive, more objective and thus convenient. Consequently, for a limited number of foods, various chemical and biochemical indices for spoilage have been proposed and used as measures of the quality or degree of spoilage. Microbial or biochemical spoilage indices will be discussed in this chapter.

The exact figures of the total economical losses due to food spoilage are unknown but the scarce figures available indicate that it constitutes an enormous financial loss. It is estimated that one-fourth of the world's food supply is lost through microbial activity alone. In the developed countries, spoilage is mainly caused by psychrotrophic microorganisms, yeasts and moulds. In less developed countries, food spoilage due to rodents and other animals is of major concern. Thus, food spoilage is an economical problem that is not yet under adequate control despite modern food technology and the range of preservation techniques available.

Food spoilage is a complex event, in which a combination of microbial and (bio)chemical activities may interact. The microbiology of food spoilage has over the years received considerable attention and the characterization of the typical microflora which develop on different types of foods during storage has been well documented. Therefore, the major problems are to find the relation between microbial composition and presence of microbial metabolites, related to the evaluation and possible prediction of microbial spoilage. Classical microbial evaluation of especially perishable foods is of limited value for predictive prognosis since these foods are sold or eaten before the results of microbiological tests are available. New highly sensitive and possible specific microbial methods based upon immunological and molecular techniques have already been developed for the detection of pathogenic microorganisms. These techniques could also be applied for the early detection of specific spoilage organisms (SSO).

However, before such techniques can be used for the detection of SSO, those microorganisms must be identified for each type of product and their effect on spoilage characteristics must be determined. As yet SSO are only known for a few products and selective or indicative methods for enumeration are not generally available. Table 14.1 summarizes the methods for the evaluation of SSO and some techniques which can be used to determine the spoilage domain of SSO are proposed.

Table 14.1   Methods for the characterization of specific spoilage organisms

Method	Comparison of results from product and model substrate experiments
Spoilage potential (qualitative)	Microorganisms are isolated from products at sensory rejection and the ability of isolates to produce off-odours is determined by inoculation of substrates
Spoilage activity (quantitative)	The concentration of groups of bacteria is determined at the time of sensory rejection and the concentration of these bacteria at the time of off-odour detection is determined in model substrates
Yield factor determination (quantitative)	Number of groups of bacteria and the concentration of selected metabolites are determined in products and the increase in concentration of these bacteria and of special metabolites are then determined in model substrates
Chemical spoilage profiles (qualitative or quantitative)	Chemical spoilage profiles of naturally spoiled products compared to chemical spoilage profiles of isolated microorganisms grown in model substrates

Although the detection levels for the metabolites formed during (bio)chemical spoilage are generally low and thus less accessible, a major disadvantage is that the (bio)chemical processes related to food spoilage appear to be poorly understood. Even less is known about interactions between microbial and chemical spoilage reactions.

Therefore we are still far away from assuring the quality of a food by predicting shelf life on the basis of specific spoilage indicators. A unifying description of the interaction between the microflora developing in the product and the chemical changes in the same product presents a special challenge. Such an integrated understanding of each of the different types of products would indeed be beneficial in relation to the increasing interest in natural preservation systems such as microbially derived antimicrobial agents and antioxidants derived from plants.

A schematical representation of the complex mechanisms of food quality deterioration during storage is presented in the Figure 14.1.

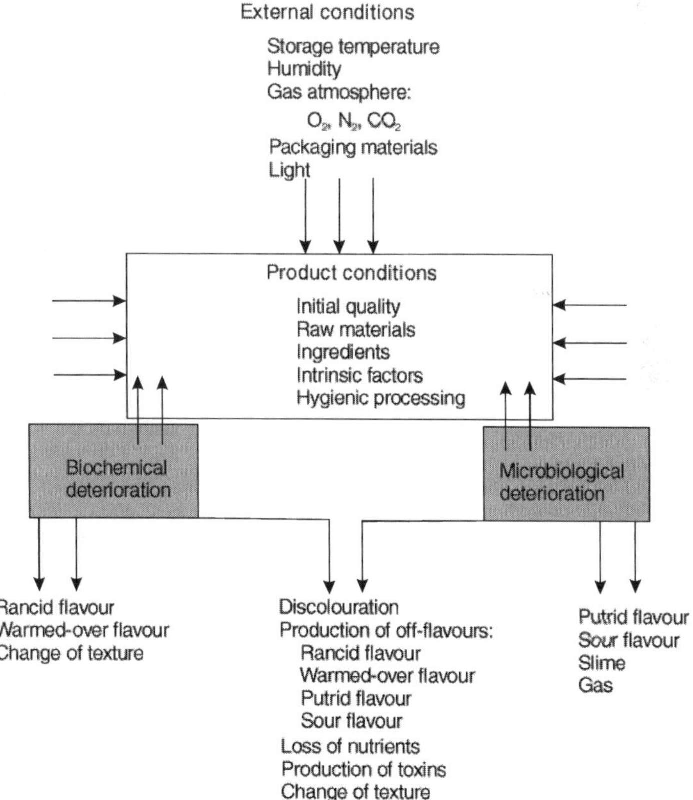

Figure 14.1 Quality deterioration during storage of foods

Factors Determining Microbial Spoilage of Foods

Spoilage of foods and beverages is the result of microbial activity of a variety of microorganisms. The microbial flora that colonizes a particular food or beverage depends highly on the characteristics of the product and the way it is processed and stored. The parameters affecting proliferation of microorganisms in foods can be categorized into four groups:

    i.   intrinsic parameters

    ii. extrinsic parameters

    iii. modes of processing and preservation

    iv. implicit parameters

It must be realized that any of the above-mentioned parameters will influence the effect of others. Therefore the overall effect of a combination of parameters is generally much higher than the perceived effect of each individual parameter.

Intrinsic parameters   Intrinsic parameters are the physical, chemical and structural properties inherent in the food itself. The most important intrinsic factors are water activity, acidity, redox potential, available nutrients and natural antimicrobial substances.

Extrinsic parameters   These are factors in the environment in which a food is stored, notably temperature, humidity and atmosphere composition.

Modes of processing and preservation   Physical or chemical treatments often result in changes in the characteristics of a food product, determining the microflora associated with the product.

Implicit parameters   Implicit parameters are mutual influences, synergistic or antagonistic, among the primary selection of organisms resulting from the influence of the above-mentioned parameters. Thus, implicit parameters are the result of the development of a microorganism which may have a synergistic or antagonistic effect on the microbial activity of other microorganisms present in the food product. Synergistic effects include production or availability of essential nutrients due to the growth of a certain group of microorganisms, allowing development of other organisms which otherwise were unable to grow. Likewise, changes in pH value, redox potential and water activity may enable the development of microorganisms less tolerant to these inhibitory factors, yielding secondary spoilage. Antagonistic processes include competition for essential nutrients, changes in pH value or redox potential or the formation of antimicrobial substances, e.g. bacteriocins which may negatively affect the survival or growth of other microorganisms.

Another important phenomenon, which deserves attention in food preservation is the homeostasis of microorganisms. If the homeostasis of a microorganism, i.e., their internal equilibrium, is disturbed by preservative factors in foods, they will not multiply, i.e., they remain in the lag phase or even die, before their homeostasis is re-established. For instance, in an acid food they will actively expel protons against the pressure of a passive proton influx. Another important homeostatic mechanism regulates the internal osmotic pressure (osmohomeostasis). Cells have to maintain a positive turgor by keeping the osmolarity of the cytoplasm higher than the environment and they generally achieve this using so-called osmoprotective compounds such as proline and betaine. We still do not have enough knowledge about how microorganisms behave under stress situations which regularly occur in foods. Most of our knowledge of how microorganisms respond to changed environments comes from extrapolation from pure cultures in laboratory experiments. However, it is becoming clear that microorganisms possess a series of mechanisms whereby they can adapt rapidly to particular environments, enabling them to colonize and grow on diverse substrates and to adapt to a wide range of host conditions.

One of the most fruitful research themes of recent years has been the discovery that microorganisms are not limited to specific ranges of temperature, pH and water activity, but can adapt to survive at values outside those found in these laboratory experiments.

The above-mentioned accumulation of osmoprotectants is a major adaptive response to an osmotic stress in microorganisms. Microorganisms also possess the ability to adapt and to tolerate low pH environments or higher temperatures by the induction of heat-shock proteins or chaperones.

Another mechanism is that external stimuli stimulate a transmembrane sensor protein. The extracytoplasmic domain senses the environment and transfers the signal to a regulatory protein in the cytoplasm which through phosphorylation leads to specific binding to the genome, DNA supercoiling, alterations in RNA polymerase specificity, etc. which result in major changes in the gene transcription or translation and hence gene expression. These mechanisms have to be understood before we can control and predict the growth of spoilage microorganisms in foods.

We also need to understand how microorganisms react to external stresses, such that we can find ways of interfering with diverse mechanisms whereby they develop resistance to physical and chemical treatments.

In this chapter, first the microorganisms associated with food spoilage are discussed. Subsequently the production of off-flavours by both endogenous enzymes as well as enzymes of microbial origin will be presented. Finally mechanisms involved in chemical spoilage are discussed.

MICROORGANISMS IN FOOD SPOILAGE

Spoilage is most rapid and evident in proteinaceous foods such as meat, poultry, fish, shellfish, milk and some dairy products. These foods are highly nutritious, possess a neutral or slightly acidic pH and a high moisture content and therefore permit growth of a wide range of microorganisms. The pattern of microbial spoilage has been found to be similar for these type of proteinaceous foods (Figure 14.2).

Initially, SSO are present in low quantities and constitute only a minor part of the natural microflora. During storage, SSO generally grow faster than the remaining microflora and produce the metabolites responsible for off-odours, off-flavours or slime and finally cause sensory rejection. The cell concentration of SSO at rejection may be called the "minimal spoilage level" (MSL) and the concentration of the metabolite that corresponds to spoilage can be used as an objective chemical spoilage index (CSI).

Changes in the extrinsic conditions (e.g. refrigeration, MAP) is the only way to delay spoilage. However, storage at adequate low temperatures will not prevent spoilage but will limit spoilage to psychrotrophic microorganisms. In general, they comprise, in addition to some gram-positive rods (lactic acid bacteria) and spore-forming bacteria (clostridia), largely the gram-negative, rod-shaped, non-spore-forming bacteria (Pseudomonadaceae). Although several yeasts and moulds are psychrotrophic, they do not generally compete well with bacteria at low temperatures but they may become important in situations where bacteria have difficulty in growing, i.e., in acid foods, or foods with high sugar or salt concentration.

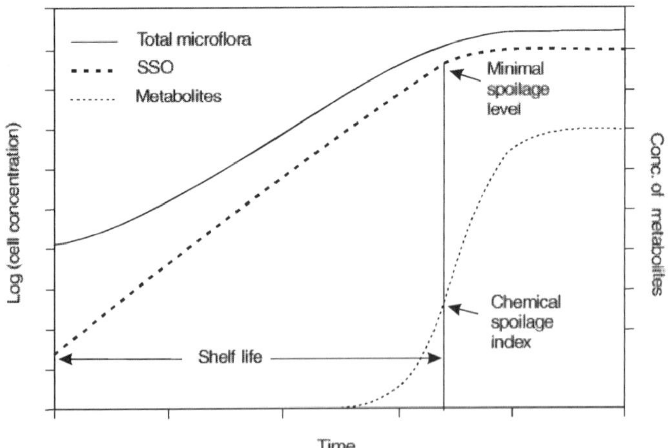

Figure 14.2   General pattern of microbial spoilage; SSO—specific spoilage organisms; MSL—minimal spoilage level; CSI—chemical spoilage index

For convenience, the spoilage microorganisms will be divided into broad categories: gram-negative rod-shaped bacteria, gram-positive spore-forming bacteria, lactic acid bacteria, other gram-positive bacteria (e.g. *Brochotrix thermosphacta),* yeasts and moulds.

Bacteria in Food Spoilage

Gram-negative rod-shaped bacteria   Pseudomonas spp. are the most common spoilage organisms, particularly in aerobically stored foods with a high water content and neutral pH, e.g. red meat, fish, poultry, milk and dairy products. *Pseudomonas* spp. like most of the other gram-negative rod-shaped bacteria, usually comprise only a small proportion of the initial microflora of fresh foods. They are however, widely distributed in the environment and may contaminate foods from many sources and are able to utilize a wide range of materials as substrates for growth.

Food spoilage due to pseudomonads may occur in a number of ways. In foods of animal origin the non-protein nitrogen fraction (NPN) will be first metabolized. Subsequently the production of lipases or proteases will liberate fatty- and amino acids which after metabolism can also result in off-odours, off-flavours and rancidity. At later stages the production of extracellular slime and the development of pigmented growth often becomes visible. A number of other gram-negative rod-shaped bacteria may also grow rapidly at chill temperatures and spoil foods, such as *Aeromonas, Photobacterium, Shewanella* and *Vibrio*. Some or all of the above bacteria have been shown to contribute to the spoilage of chilled red meat, cured meats, poultry, fish, shellfish and milk and dairy products. *Vibrio* spp., are unusual as most are halophilic (salt-loving) and so may cause spoilage of seafish and cured meats. Like the pseudomonads, other gram-negative bacteria also cause food spoilage due to NPN metabolism. Subsequently, foods will spoil by the production of enzymes (resulting in odour and flavour defects), slime production and the formation of visible, often pigmented colonies. At a temperature above 5–10°C Enterobacteriaceae generally dominate over *Pseudomonas* spp. and become responsible for spoilage. Vibrionaceae spoil fish at higher temperatures. Spoilage is characterized by the production of gas, acid, slime, rope, bitter flavours and off-odours. The presence of Enterobacteriaceae is often used as an indication for possible faecal contamination, inadequate processing or post-process contamination.

Gram-positive spore-forming bacteria   Many foods undergo a heating or pasteurization process. Thus microorganisms capable of surviving this process are significant (e.g. *Bacillus* and *Clostridium* spp. particularly if they are able to grow at chill temperatures. Growth of the spore-forming bacteria tends to be much slower than that of the Gram-negative bacteria, but most of these latter bacteria are eliminated by heat processing.

The *Bacillus* spp. are largely aerobic in nature. Perhaps, best recognized is *B. cereus* which may grow at low temperatures (5°C or less) and produce enzymes which results in "sweet curdling" and "bitter cream" in milk. Other *Bacillus* spp., may grow at temperatures of 0–2°C. Most spoilage *Clostridium* spp. are unable to grow at refrigerator temperatures (i.e., 5°C or less), but at slightly higher temperatures may produce gas resulting in "late blowing" of hard cheeses during maturation.

Recently some psychrotrophic *Clostridium* spp. have been observed to spoil vacuum-packed meat and fish with extended shelf life such as vacuum-packed beef and ham and sous-vide cooked beef.

Lactic acid bacteria   Lactic acid bacteria spoil foods by the fermentation of sugars to form lactic acid, slime and CO_2 leading to a drop in pH and off-flavours. These bacteria tend to grow slowly at refrigeration temperatures and are under aerobic conditions generally out-competed by pseudomonads. Generally, they are present in the initial microflora in low numbers and are therefore rarely responsible for the spoilage of fresh proteinaceous foods.

Lactic acid bacteria, however, have been identified as the major spoiling microorganisms of vacuum-packed meat and poultry and are also suggested as possible spoilers of lightly preserved fish products. Cured and fermented meat products may also be spoiled by lactic acid bacteria, as the pH or other preservation methods in the food again prevent the growth of the normal spoilage microflora. Typical lactic acid bacteria are identified as *Lactobacillus, Streptococcus, Leuconostoc* and *Pediococcus* spp.

Other gram-positive bacteria   *Brochothrix thermosphacta* is a gram-positive rod which may be occasionally present on fresh meats. The increased use of modified atmosphere packaging and vacuum packaging will often allow *Brochothrix thermosphacta* to dominate the microflora.

Micrococcus spp. are able to grow in the presence of salt and may be responsible for the spoilage of cured meat products such as bacon producing slime, souring or pigmented growth. These microorganisms also often predominate in freshly collected milk. Many strains are thermoduric and may survive milk pasteurization causing subsequent spoilage, particularly if the other more rapidly growing spoilage organisms have been eliminated by the heat treatment.

BIOCHEMICAL SPOILAGE

Production of Off-flavours

Off-flavours are produced in different foods due to endogenous enzymatic activities.

Off-flavours which arise in foods fall into the three following categories:

(i) off-flavours preformed in the food; (ii) off-flavours formed as a result of cellular disruption and (iii) off-flavours as a consequence of endogenous or microbial enzymes.

Off-flavours may be preformed in the food due to normal biochemical metabolism or as stress metabolites. These type of off-flavours are dependent to a large extent on agronomic factors such as varietal differences, feeding or fertilizer regimes, level of water used, spacing, etc.

Off-flavours in Different Food Products

Milk and dairy products During extended refrigerated storage of milk, heat-stable enzymes of microbial origin may be formed. These can biochemically alter the products, eventually causing spoilage. Two types of enzymes are particularly important in the formation of off-flavours: (i) lipases and (ii) proteinases.

Lipases Lipase activity has been reported for most psychrotrophs isolated from milk and milk products. *Pseudomonas*, *Flavobacterium* and *Alcaligenes* species are the most lipolytic bacteria. Microbial lipases are heat-stable. Lipase activity in milk leads to the preferential release of medium- and short-chain fatty acids from triglycerides. Hydrolysis of as little as 1–2% triglycerides leads to rancid off-flavour. Milk naturally also contains high levels of indigenous lipase. It is therefore extremely likely that indigenous as well as microbial lipases are important in the development of lipolytic rancidity in milk.

Proteinases The major cause of bitterness in milk and milk products is the formation of bitter peptides due to the action of proteinases. Proteinase activity has been detected in many bacterial species, in particular *Pseudomonas*, *Aeromonas*, *Serratia* and *Bacillus* species. Heat stability of proteinases from several bacterial species was investigated. Strict quality control is therefore critical in UHT milk products to ensure that heat-stable proteinases do not cause bitter off-flavours. The most investigated source of bitter peptides is the casein.

Meat and fish Off-flavours which develop due to surface microbial contamination are major causes of spoilage in meat. The first signs of spoilage to be caused are by the formation of fruity, sweet-smelling esters, followed by the formation of putrid sulphur compounds. *Pseudomonas* was identified as the main bacterial contamination. Many putrid odours arise due to decomposition of proteins and amino acids by anaerobic bacteria. Volatiles produced include indole, methanethiol, dimethyl disulphide and ammonia. Rancidity problems which occur are usually due to oxidation of unsaturated lipids and are not associated with microbial growth.

In the case of fish, a four-phase pattern for the changes in flavour quality after harvest is described by the scientists. Initial microbial contamination and growth is by aerobes, which act on carbohydrates giving carbon dioxide and water. As the surface becomes covered and slime builds up, conditions become more favourable to the growth of anaerobes. Reduction of trimethylamine oxide to the unpleasant fish-smelling trimethylamine, catalysed by trimethylamine-N-oxide reductase, is carried out by many bacteria. Many off-flavours are also associated with the breakdown of sulphur-containing amino acids. Typical volatile products are hydrogen sulphide, methyl mercaptan and dimethyl sulphide.

After harvest or slaughter the structural integrity of foods begins to break down due to damage from handling and normal decay processes. This will result in the mixing of compartmentalized enzymes and substrates and the generation of possible flavour compounds. Although in some cases the production of flavour compounds along this way is a typical characteristic of the product (e.g. onions, garlic) and generally this leads to the formation of off-flavours and spoilage of the product.

As by-products of growth and metabolism, microorganisms produce a range of chemicals which alter the quality attributes of foods such as off-flavours, mycotoxins, and biogenic amines ultimately rendering the product inedible or unsafe.

Fruits and vegetables

Citrus fruits   A flavour defect which constitutes a major problem worldwide to the citrus industry is bitterness due to formation of limonin. It was shown that intact fruits were not bitter and did not contain limonin itself but a non-bitter precursor, limonoate A-ring lactone. When juice is extracted, this non-bitter precursor is slowly converted to limonin under acidic conditions and is accelerated by the presence of limonin D-ring lactonase.

Legumes   The enzyme lipoxygenase is believed to be ubiquitous amongst eukaryotic organisms. This enzyme poses a particular problem in legumes such as soy beans, winged beans, lentils, green beans, etc., giving rise to a range of off-flavours described variously as beany, grassy and rancid. There has been considerable interest in lipoxygenase in soy beans, related to the formation of volatile aroma compounds. Soy beans contain high levels of lipoxygenase, constituting 1–2% of the protein. The oil fraction contains 55% linoleic acid and 8% linolenic acid, which are substrates for lipoxygenase.

Brassicas   Perhaps the most important group of compounds to the flavour of brassicas are the volatile degradation products of endogenous glucosinolates.

Brassicas also contain thioglucosidase (myrosinase) enzymes, which are released when the cells of the plant are disrupted and which then come into contact with the glucosinolates. This leads to a range of potent flavour compounds dependent upon (i) the specific glucosinolate involved and (ii) the conditions of reaction.

In many instances this gives rise to desirable flavour notes, e.g. 2-propanylisothiocyanate, derived from the thioglucosidase initiated breakdown of 2-propanyl glucosinolate (sinigrin), is an important flavour component of black pepper. However, the undesirable bitter note in *Brassica oleraceae* cultivars in particular Brussels sprouts has been attributed to the formation of goitrin (5-vinyloxazolidine-2-thione) from the glucosinolate progoitrin (2-hydroxy-3-butanyl glucosinolate).

Wine and beer   Enzymes are essential for the conversion of grape juice into wine, and malt extract into beer. This includes the formation of many flavour compounds, together with the typical ethanol component of these drinks. Off-flavours which arise are invariably associated with fermentation problems or microbial contamination.

Wine   Grass-like tastes which may arise in wines are caused by the formation of hexanal, *cis*-hexen-3-al and *trans*-hexen-2-al. These compounds are formed during the juice extraction phase due to the breakdown of membrane equivalent alcohols. Although sensorically the alcohols are not as potent as the aldehydes, it is believed that they can also contribute to the grass-like aroma. A range of off-flavours, collectively known as "cork taint", may arise in wines and spirits. The major cause of cork taint is believed to be 2,4,6-trichloroanisole (TCA).

Beer   Perhaps one of the most studied off-flavour problems in beer is the formation of the butter-like compound diacetyl α-acetolactate, an intermediate in the biosynthetic pathway from pyruvate to leucine and valine, within a normal metabolizing yeast. However, it is possible for a-acetolactate to pass into the bulk of the wort where chemical oxidative decarboxylation converts it into diacetyl. Given sufficient time, the yeast will absorb the diacetyl and convert it via acetoin to 2,3-butanediol. However, if the yeast is removed, diacetyl will accumulate leading to the off-flavour.

The interrelationship between microbial enzymes, endogenous enzymes and off-flavours is very complex. Although some aspects such as bitterness in milk and diacetyl in beer, have been well-studied, formation of off-flavours due to enzymes, both microbial and endogenous, is often poorly understood.

YEASTS AND MOULDS IN FOOD SPOILAGE

Yeasts and moulds can be found in a wide variety of environments, such as in plants, animal products, soil, water and insects. This broad occurrence can be explained by the fact that yeasts and moulds can utilize a variety of substrates such as pectin and other carbohydrates, organic acids, proteins and lipids. Moreover, yeasts and moulds are relatively tolerant to low pH, low water activity, low temperature and the presence of preservatives. It is also notable that yeasts can utilize food ingredients, such as organic acids like lactic, citric and acetic acids, that are generally considered to have an inhibitory effect on the growth of many microorganisms. Even common preservatives such as benzoate, propionate and sorbate can be utilized by some yeast species.

Contamination of foods and beverages by yeasts and moulds has been extensively reported. Contemporary work has also reported the occurrence of yeasts in fresh seafood, packaged meats, delicatessen salads, and in fresh vegetables.

Changes induced by spoilage of yeasts and moulds can be of a sensory nature, recognizable in the product's appearance by the production of slime, quite often pigmented growth on the surface, fermentation of sugars to produce acid, gas or alcohol or the development of off-odours and off-flavours.

In addition to visible spoilage, moulds can also spoil foods through the formation of mycotoxins. It has now been established that more than 200 different types of moulds do form substances that are orally toxic to man, when growing in certain foods. Although most research has been carried out on the metabolites of *Aspergillus flavus,* it is quite obvious that in addition to the so-called aflatoxins, many other mycotoxins may be of great significance.

CHEMICAL SPOILAGE

Although chemical and physical spoilage processes cannot be totally separated, their main contributions to food spoilage are characterized by flavour and colour changes due to oxidation, irradiation, lipolysis (rancid) and heat. These changes may be induced by light, metal ions or excessive heat during processing or storage.

Chemical processes also may bring about physical changes such as increased viscosity, gelation, sedimentation or colour change.

Lipid Oxidation

Lipid oxidation is one of the most common causes of deterioration of food quality. Unsaturated fats are oxidized by free radical autoxidation, a chain reaction process catalysed by the products of the reaction. The susceptibility to the rate of oxidation increases as the number of double bonds in the fatty acid increases. Oxidation of oils may also be initiated by lipoxygenase or photosensitizers.

Many foods or food ingredients (plants, fruits, roots and meats) contain components which possess the so-called antioxidant properties. Examples of well known natural antioxidants are ascorbic acid, vitamin E (tocopherols), carotenoids and flavonoids. Antioxidants inhibit lipid oxidation by acting as hydrogen or electron donors, and interfere with the radical chain reaction by forming non-radical compounds that will not propagate further radical reaction.

Enzymatic Oxidation

Lipoxygenase occurs in many plants and catalyses the oxidation of unsaturated fatty acids containing a *cis, cis* 1,4-pentadiene system to their corresponding monohydroperoxides. These peroxides have the same structure as those obtained by autoxidation. Lipoxygenase is a metal bound protein with a "Fe"-atom in its active centre. Plant lipoxygenases produce (*cis, trans*)-conjugated monohydroperoxides as primary products. Naturally occurring substrates include linoleic, linolenic and arachidonic acids.

Lipolysis

Fat containing foods, e.g. milk can undergo a number of subtle chemical and physical changes caused by lipolysis. Lipolysis can be defined as the enzymatic hydrolysis of fats by lipases. The accumulation of the reaction products, especially free fatty acids is responsible for the common off-flavour, frequently referred to as rancidity. The lipolytic enzymes could either be endogenous of the food product, e.g. milk, but could also be derived from psychrotrophic microorganisms. In contrast to bacterial lipases, the endogenous milk enzymes have shown to be sensitive to heat.

Discolouration

The red or brown colour of meat depends on the different forms of myoglobin due to different oxygen partial pressure at the meat surface. The depth of the oxymyoglobin layer responsible for the red colour depends on the oxygen penetration into the meat. Also during spoilage, fish flesh may turn yellow due to oxidation of caretenoid pigments and lipids in tissues. The most important colour changes in stored fruits and fruit products are caused by chemical reactions known as browning reactions. Browning of fruits may be enzymatic or non-enzymatic. Polyphenoloxidases (PPO) and peroxidases found in fruit tissues can catalyse oxidation of certain endogenous phenolic compounds to quinones that polymerize to form intense brown pigments.

SUMMARY

Although food spoilage is a major economical loss, the underlying integrated mechanisms are still poorly understood. It is obvious that the presence of high numbers of spoilage microorganisms will eventually lead to deterioration of foods but the time between the moment of reaching these high levels of microorganisms and the actual spoilage may vary considerably depending on the type of food, the actual intrinsic, extrinsic and implicit factors, and the activity of specific spoilage microorganisms (SSO). In order to minimize food spoilage and be able to predict the quality or shelf life of a particular food, a better understanding of the mechanisms underlying food spoilage is essential. As a consequence, there is a need for the identification and control of growth of SSO present on different food commodities. As yet not many SSO have been identified. Therefore, the estimation of the quality of a food product still relies on the quantification of total numbers of microorganisms, which in some cases is a very poor reflection of the actual quality.

In addition to the identification of SSO, a better understanding of the complex interactions between SSO and other microorganisms or their metabolites (synergism/antagonism) is needed. Finally the interaction between microbial spoilage and biochemical spoilage has to be elucidated.

REVIEW QUESTIONS

1. Define microbial spoilage of foods.
2. What are the factors determining microbial spoilage of foods?
3. Discuss the role of bacteria in food spoilage.
4. Discuss the role of microbial enzymes in food spoilage
5. Discuss the role of yeasts and moulds in food spoilage.

15

PHYSIOLOGY OF FOOD-SPOILAGE ORGANISMS

FOOD SPOILAGE

Microbial food spoilage costs the food industry (and indirectly, the consumer) annually and ultimately a great sum of money represents an enormous waste of a valuable resource. Food losses begin on the farm and continue throughout post-harvest storage, distribution, processing, wholesaling, retailing and use in the home and in catering. Although it is technically possible to produce food entirely free of microbial contamination by well-known technologies such as gamma irradiation, such a severe approach is contrary to the current consumer demand for foods that are minimally processed and perceived as "fresh". Therefore, in addition to ensuring the safety of foods, the challenge remains to produce high-quality foods that are easy to prepare, do not require daily trips to the supermarket and rely less heavily on chemically synthesized food preservatives.

IMPORTANCE OF MICROBIAL PHYSIOLOGY

Most traditional food preservation processes have been developed empirically without a full understanding of the mechanisms of action of the antimicrobial agents used. However, with the move away from using high concentrations of individual food preservatives towards increasing reliance on combinations of sublethal levels of antimicrobial compounds or processes, there is a need to re-examine the basics. It could be argued that truly novel combination systems can only be developed logically if they are based on a thorough understanding of the physiology of microorganisms in foods.

Unfortunately, in recent years the accumulation of knowledge on the physiology of spoilage microorganisms has lagged behind that on microbial genetics and molecular biology. For example, the entire genome of *Saccharomyces cerevisiae*, a beneficial and economically-important yeast as well as a spoilage agent of foods, has been sequenced and reported, but only about 50% of the 6000 genes identified are of known function. Important aspects of the regulation of growth and metabolism in many spoilage microorganisms are still poorly understood from a fundamental viewpoint.

RESPONSE OF MICROBES TO PHYSIOLOGICAL FACTORS

Survival and Resistance to Food-associated Stresses

Some of the major stresses encountered by microorganisms in foods include low or high temperatures, acidity, low water activity and modified atmospheres. Whilst the types of stresses encountered by spoilage and pathogenic organisms may be the same, spoilage organisms tend to be more adaptable to harsh environmental conditions, often developing resistance to chemical food preservatives, cleaning and sanitizing agents. Furthermore, food spoilage organisms outnumber food pathogens in terms of both numbers and types.

Maintaining Homeostasis

In general, microbial exposure to low water activity by dehydration or addition of salts or sugars leads to the activation of constitutive transport systems culminating in the accumulation of compatible solutes inside the cell. These are often small, highly soluble molecules and include potassium, proline, glutamate, betaine and trehalose. Notably, accumulation of compatible solutes has been shown, in some microorganisms, to confer increased resistance to heat and improved ability to grow at reduced temperatures. In the last 10–15 years, cross-resistance to different types of stresses has become a recurring theme in studies of microbial physiology.

Spoilage Consortia

The study of individual food spoilage organisms is an attempt to provide guidelines for systematic and predictive determination of shelf life. Unfortunately, most foods contain a mixed microbial flora in which strains compete for survival and/or growth. Therefore, poorly understood microbial interactions in foods often lead to invalidate conclusions drawn on the basis of work done with pure cultures.

FUTURE PROSPECTS

The introduction of entirely new food preservatives will be slow in the future, despite the technological capabilities of genetic engineering, due to regulatory constraints and the high cost of safety testing now required for all new food additives. It is more likely that, by systematic study of the stress responses of microorganisms together with detailed assessment of technological performance of available preservatives and preservation technologies in real food formulations, much more sophisticated usage will emerge, resulting in new processes and products.

1. Give an account of microbial survival and resistance to food-associated stresses.

16

INTERACTIONS BETWEEN FOOD-SPOILAGE BACTERIA

INTRODUCTION

Preservation of foods has, since the beginning of mankind, been necessary for our survival. The preservation techniques used in early days relied without any understanding of the microbiology on inactivation of the spoiling microorganisms through drying, salting, heating or fermentation. These methods are still used today, albeit using less and less preservation and combining various lightly preservation procedures to inhibit growth of microorganisms. Spoilage is characterized by any change in a food product that renders it unacceptable to the consumer from a sensory point of view. This may be physical damage, chemical changes (oxidation, colour changes) or appearance of off-flavours and off-odours resulting from microbial growth and metabolism in the product. Microbial spoilage is by far the most common cause of spoilage and may manifest itself as visible growth (slime, colonies), as textural changes (degradation of polymers) or as off-odours and off-flavours. Despite chill chains, chemical preservatives and a much better understanding of microbial food spoilage, it has been estimated that 25% of all foods produced globally is lost post harvest or post slaughter due to microbial spoilage.

The Concept of Specific Spoilage Organisms (SSO)

Each and every food product harbours its own specific and characteristic microflora at any given point of time during production and storage. This microflora is a function of raw material flora, processing, preservation and storage conditions. Despite the variability in all of the three, some very clear patterns emerge, and based on knowledge of a few chemical and physical parameters it is possible with great accuracy to predict which microorganisms will grow and dominate in a particular product.

At the point of sensory rejection (spoilage), the so-called spoilage microflora (or spoilage association) is composed of microorganisms that have contributed to the spoilage and microorganisms that have grown but not caused unpleasant changes. The former is the so-called specific spoilage organisms (SSO) of the product.

The spoilage potential of a microorganism is the ability of a pure culture to produce the metabolites that are associated with the spoilage of a particular product. In general, several of the organisms isolated from a food product will be able to produce spoilage metabolites when allowed unlimited growth. It is crucial that quantitative considerations are introduced since the spoilage activity of an organism is its quantitative ability to produce spoilage metabolites. Thus it is to be evaluated if the levels of the particular organism reached in naturally spoiling foods are capable of producing the amount of metabolites associated with spoilage. In general, it requires a careful combination of microbiology, sensory analysis and chemistry to determine which microorganisms are the SSOs of a particular food product.

Despite the importance of microbes in food spoilage, the definition and assessment of spoilage relies on sensory evaluation. Specific microbes are the cause of spoilage, however neither the level of "total count", e.g. 10^7 cfu/cm^2 nor the level of SSO can directly predict the sensory quality of a product. In contrast, the level of SSO can be used to predict remaining shelf life of a product under conditions where the SSO is important. Examples of this is the prediction of remaining iced shelf life of cod based on numbers of *Shewanella putrefaciens*, and of modified atmosphere packed cod based on numbers of *Photobacterium phosphoreum*.

Food-spoiling Microorganisms

Almost all groups of microorganisms harbour members that under some conditions can contribute to spoilage of foods. Theoretically, one can assume that all microbes are initially present on a food product where after a selection occurs—based primarily on nutrient composition and on the chemical and physical parameters. Similar microfloras emerge in different food products under the same conditions despite the heterogeneity in the outset. Thus *Pseudomonas* spp. and a few other gram-negative psychrotrophic organisms will dominate proteinaceous foods stored aerobically at chill temperatures. This is true for meat, poultry, milk and fish. If pH is high like in fish and "dark firm dry" (DFD) meat, *S. putrefaciens*-like organisms develop in parallel, and may become the dominant spoilage organisms, as is the case in marine iced fish. The pseudomonads in pasteurized milk originate from post-process contamination but the product may also spoil due to growth of psychrotrophic *Bacillus* of which the spores have survived the heat treatment.

In meat and fish, a change in atmosphere, e.g. by vacuum packing, will inhibit the respiratory pseudomonads and in meats cause a shift in the microflora to lactic acid bacteria (LAB), Enterobacteriaceae and sometimes *Brochothrix thermosphacta*. Also, clostridia may cause a deep-muscle anaerobic spoilage of vacuum-packed meats.

S. putrefaciens, which is capable of anaerobic respiration, also grows and contributes to spoilage in meat with high pH. In fish, vacuum packaging selects for *S. putrefaciens* and for the CO_2 resistant, psychrotolerant marine bacterium *P. phosphoreum*. CO_2 packaging of fish from temperate waters does not select for LAB but allows *P. phosphoreum* to grow and this organism spoils the product. A mild heat treatment of fish results in elimination of vegetative bacteria, and clostridia and *Bacillus* (that both survive as spores) may grow and spoil the product, especially if vacuum-packed.

Further, "selective pressure", e.g. addition of low levels of salt to and drying of fish eventually switch the microflora in the same direction as vacuum-packed meat and meat products. Thus a microflora of LAB, Enterobacteriaceae and to some extent *Brochothrix* develops. Due to the aquatic origin, *P. phosphoreum* is often also present. Identifying the SSO of lightly preserved fish products and of processed meat products has proven difficult and probably different groups of bacteria are important under different conditions. Thus in some trials, the indicators of spoilage parallels the metabolites of *P. phosphoreum*, in other trials that of a LAB flora (*Lactobacillus curvatus*) and in others the combined metabolites of LAB and Enterobacteriaceae.

Increasing the preservation by a decrease in pH (below 5), an increase in the NaCl concentration (above 6%) and by adding sorbate and/or benzoate eliminates the gram-negative microflora. LAB and yeasts are the remaining organisms in semi-preserved fish products. Different *Lactobacillus* spp. including *Lactobacillus alimentarius* have been identified as spoilage organisms of these products. Meats that are cured by salt and low pH are usually shelf stable and are not spoiled by microbial growth. The preservation profile of, e.g. mayonnaise-based salads is similar to the semi-preserved fish products in terms of pH, temperature, atmosphere and chemical preservatives and a microflora of LAB may develop. Also, fruit juices, which are high in sugar and have a low pH, (2.0–4.5) often spoil due to growth of LAB and/or yeasts. In pasteurized juice, the acid-tolerant spore-forming bacteria *Alicyclobacillus acideoterrestris* is a major spoilage organism. Decreasing a_W further eliminates bacterial growth, and only extremophiles (such as halophiles or osmophiles) and filamentous fungi are capable of developing on dried, salted fish, meat and fruit products. Thus, as the preservation profile increases in inhibitory strength, the microflora changes along the following path:

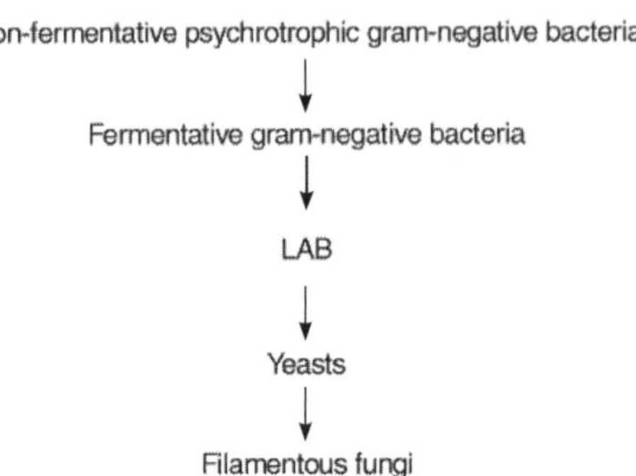

Food products of vegetable origin presents a special case due to the nutrient composition. The high pH will allow a range of gram-negative bacteria to grow, but spoilage is specifically caused by organisms capable of degrading the vegetable polymer, pectin. These organisms, typically *Erwinia* and *Pseudomonas* species are the SSO of several ready-to-eat vegetable products but also pectin degrading fungi can play a role in vegetable spoilage.

The presence of ethanol, as in beer and wine, is similar in "preservation strength" to, e.g. low pH and salted foods and only allows LAB and yeasts to grow. Resistance to the antibacterial compounds found in hop is required for an organism to grow in beer. *Lactobacillus* species are tolerant to these iso-acids and several species have been identified as SSOs in beer. Also, several *Pediococcus* species are important beer spoiling organisms.

Food-spoiling Reactions

As mentioned, microbial spoilage usually manifests itself as slime and/or off-odours and off-flavours. Many studies have described in more chemical terms what compounds characterize the spoilage of particular products and which substrates have been used by the microorganisms to form these compounds. Such knowledge is important, if a chemical spoilage index is to be developed. Examples of this include trimethylamine of iced marine fish. This knowledge may also be used to eliminate a particular spoilage promoting compound, e.g. sucrose as sweetener in brined shrimp. In this case, slime/ropiness is formed by *Leuconostoc* from sucrose and substituting with, e.g. artificial sweeteners or sugar alcohols, may prevent this spoilage.

INTERACTIONS BETWEEN FOOD-SPOILING BACTERIA

The selective conditions imposed on the food microbial community by physical and chemical parameters undoubtedly are the most important in terms of growth and selection of microorganisms. However, microbial food spoilage is a process involving growth of microbes to numbers (10^7–10^9 cfu/g) at which the microbes also must be assumed to interact and influence the growth of one another. The interactions between microbes may be classified on the basis of their effects as detrimental or beneficial. Several types of interactions have been studied in food ecosystems including both antagonistic and coordinated behaviour, and interactions where growth or a particular metabolism of one organism is favoured by the growth of another organism. This is illustrated by three examples:

1. Antagonism caused by the competition for iron as mediated by bacterial siderophore production and subsequent suppression of maximum cell density of less competitive bacteria
2. Change in spoilage profile of an organism by the supply of nutrients from another microorganism (metabiosis)
3. The ability of gram-negative bacteria to coordinate the expression of certain phenotypic traits (e.g. hydrolytic enzymes) through bacterial communication via *N*-acyl homoserine lactones (AHLs).

Antagonism

Changes in environmental conditions, e.g. by lowering of pH, can be a powerful way for a microbe to antagonize other bacteria and create a selective advantage. Also, competition for nutrients may select for the organisms best capable of scavenging the limiting compound(s). Several microbes important in food spoilage have such antagonistic abilities. Thus, the lactic acid bacteria cause a lowering of pH and may produce antibacterial peptides (bacteriocins). The spoilage reactions of certain gram-negative bacteria may produce NH_3 and trimethyl amine, which are toxic to a number of other bacteria and sometimes to the producing organism itself. *Pseudomonas* spp. in particular the fluorescent group, produce a range of antibacterial and antifungal compounds such as antibiotics and cyanide, and at the same time they compete very efficiently for iron.

Competition for iron   Despite the richness of nutrients in most foods, bacteria may be limited in selected compounds such as minerals, amino acids or sugars. Thus, it has been shown that the concentration of iron is low in several fish products. Iron is essential for most microbes and is used in bacterial respiration (as electron shuttler) and redox enzymes. Due to the high oxidative power of Fe^{3+}, iron is mostly bound in insoluble complexes in the environment and in mammals and plants. Most microorganisms have therefore developed highly specific iron chelating systems, and often they produce siderophores, which are iron chelators secreted by the cell. Upon binding of iron, the siderophore–iron complex is taken up by the microbial cell and iron is liberated internally. In particular, pseudomonads are prominent producers of siderophores with high iron binding constants. The iron chelating ability has been of particular interest in the rhizosphere, where the use of pseudomonads as biocontrol agents against fungal diseases has been attributed in part to their competitive advantage vis-a-vis iron.

Bacteria growing on fish produce siderophores that are only induced when the iron concentration is limiting. Siderophores have been extracted from cheeses indicating that this food is iron limited. Also, bacteria spoiling eggs must be able to scavenge iron from the tight iron binding proteins of the egg. *Pseudomonas* spp. isolated from foods produce siderophores, in particular strains isolated from fish. Also *S. putrefaciens* chelates iron by siderophore production but despite this ability, it is strongly inhibited by siderophore producing pseudomonads under iron limited conditions. When grown in co-culture on fish samples, siderophore producing pseudomonads inhibits, e.g. *Shewanella* when the former reach approximately 10^8 cfu/g and cause a suppression of the maximum cell density of the inhibited organism. The depression of maximum cell density of microbes by an "over growing" microflora is also seen for *Listeria monocytogenes* growing in lightly preserved foods with a dominant lactic acid bacterial flora. The latter phenomenon may be caused by bacteriocin production by the LAB, but since several non-bacteriocin producing LAB are as inhibitory, the inhibition may as well be explained by the LAB out competing the *Listeria* on a few essential nutrients. This suppression of maximum population density of a particular organism by an overgrowing microflora is called the Jameson effect.

It is not clear to what extent the competitive activity of, e.g. pseudomonads contribute to their dominance during spoilage of proteinaceous foods. Since *S. putrefaciens* can be isolated from fresh Nile perch and the fish contains trimethylamine oxide, it would be expected that this bacterium contributed to spoilage of the iced fish. However, spoilage is exclusively caused by *Pseudomonas* spp., and it has been suggested that their competitive ability contributes to this dominance.

Removal of iron and other trace metals has been suggested as a way of controlling growth of spoilage microorganisms. Spoilage yeasts did not grow in grape juice in which iron was chelated by resins. It is, however, unlikely that addition of high concentrations of iron chelates will be widely adopted due to the obvious negative effects on the nutritional value of food products limited in iron.

Metabiosis

Innumerable ways of interdependency exists between different organisms. The term metabiosis describes the reliance by an organism on another to produce a favourable environment. For example, the removal of oxygen by a gram-negative microflora allowing anaerobic organisms such as *Clostridium botulinum* to grow or it can be situations where one organism provides nutrients enhancing growth of another. Thus several studies have shown that despite the inhibitory activity of pseudomonads as described above, their presence may also enhance the growth of some microbes. Preinoculation of milk with different gram-negative psychrotrophic bacteria subsequently yielded higher growth and more acid from lactic acid bacteria and growth of *Staphylococcus aureus* may also be stimulated by *Pseudomonas* species. Such nutrient interdependency may also play a role in food spoilage.

Biogenic amines which are formed by bacterial decomposition of amino acids, have been suggested as a quality indicator of vacuum-packed beef. The spoilage microflora of this product typically consists of a mixture of LAB and Enterobacteriaceae. *Hafnia alvei* and *Serratia liquefaciens* as single cultures produce levels of cadavarine similar to the naturally contaminated product, however, the level of putrescine from single cultures does not correspond to the spoiling products. Some LAB degrade arginine to ornithine (a precursor of putrescine) and co-inoculation of the putrescine forming Enterobacteriaceae with arginine degrading LAB resulted in a 6–15 times greater production of putrescine than in single cultures. The changes in biogenic amines and pH during chill storage or vacuum-packed cold-smoked salmon can be combined to a quality index that correlates with sensory evaluation of the product. High levels of biogenic amines may be produced by single cultures of *P. phosphoreum* or by *L. curvatus* but may also arise from co-cultures of lactobacilli and Enterobacteriaceae. Similar to the studies from vacuum-packed meat, production of putrescine was enhanced 10–15 times when ornithine decarboxylase positive Enterobacteriaceae were co-cultured with arginine deaminase positive LAB as compared to single cultures. Other studies have similarly found that interaction between LAB and Enterobacteriaceae can be important in food spoilage. It was found that a mixture of LAB and *H. alvei* produced the spoilage off-odours typical of spoiling vacuum-packed meats whereas growing the organisms as single cultures did not result in the spoilage off-odours, in cold-smoked salmon. Inoculation of these gram-negative bacteria (*Shewanella*, *Photobacterium* and *Aeromonas*) did not cause spoilage of vacuum-packed cold-smoked salmon whereas the coinoculation of the gram-negative bacteria with *B. thermosphacta* and *Carnobacterium piscicola* produced spoilage off-odours.

In pasteurized milk, both Enterobacteriaceae (*Yersinia intermedia*) and *Pseudomonas putida*, may produce off-odours. However, the amount of esters causing fruity off-odours was increased when the two organisms were co-cultured.

Psychrotolerant *Clostridium* species have been identified as the causative agent of "brown pack" spoilage of vacuum-packed meats. To date, there has been no reports on influence of interacting microorganisms on this spoilage. However, aerobic bacteria (such as *S. putrefaciens*) significantly affect (enhance) toxin production by *C. botulinum* type E and one may therefore speculate that aerobic bacteria on the meat through their removal of oxygen may create a more advantageous environment for the psychrotrophic spoilage clostridia.

The term "specific spoilage organism" was originally coined to describe the (assumed) single species being responsible for spoilage. Although Jorgensen introduced the term "metabiotic spoilage association" to describe situations where two or more microbial species contribute to spoilage through exchange of metabolites or nutrients, this scenario could be covered by the "specific spoilage organisms" concept where it should be specified that a consortium of organisms interact to spoil the product.

ACYLATED HOMOSERINE LACTONE-BASED COMMUNICATION AND QUORUM SENSING

The expression of many phenotypic traits in microorganisms is governed by tight gene regulation and influenced by growth phase, nutrients, external stresses and multitude of other factors. Several behavioural patterns are correlated with the density of the population, and the ability to regulate gene expression as function of cell density has been termed "Quorum sensing". Quorum sensing requires that the microorganisms are capable of communicating by means of chemical signals. Peptides serve as the signal molecules to monitor population size in many gram-positive bacteria. In gram-negative bacteria, the most intensely studied signals are the *N*-acyl homoserine lactones. AHL-based quorum sensing systems of the *lux*-homologous family require a minimum of four components for proper function:

i. a diffusible AHL type signal molecule, synthesized by an enzyme (referred to as the Lux I homolog)

ii. an AHL-binding receptor (referred to as the Lux R homolog) which alters the DNA-binding activity in response to binding of its cognate signal molecule

iii. a DNA sequence (typically referred to as the *lux*- box) located in the promoter region of a target gene, acting as target sequence for the regulatory Lux R homolog allowing it to act as transcriptional regulator, and finally

iv. a set of target genes resulting in a particular phenotype, which is either up-regulated or repressed as a function of the Lux R–AHL binding.

The quorum sensing results from the fact, that AHLs are diffusible over the cell membrane and that the regulatory LuxR-protein requires a threshold concentration of the AHLs in order to be activated. Thus even though AHLs are produced at low levels in dilute cultures, binding to the receptor protein and the transcriptional activation of target genes only takes place at high cell densities. The binding of the LuxR–AHL complex to the *lux*-box results in transcription of genes downstream of the *lux* box and this transcription results in an AHL regulated phenotypic response, in some bacteria, the *lux I*-homologue gene is located downstream of the *lux* box. Production of the AHL synthase is therefore also regulated by the LuxR–AHL complex, i.e., it is autoinducible. Thus, the activation of the regulatory protein results in a dramatic up-regulation of AHL production and as a consequence AHL regulated phenotypes. As much as a 1000-fold increase in activity per cell has been reported, e.g. for AHL production and the AHL-regulated bioluminescence in *Vibrio fischeri*. In other bacteria, AHL production is constitutive and it appears that the up-regulation of phenotypes is less dramatic.

The production of AHLs and regulation of different phenotypic traits has been reported in many gram-negative bacteria. The quorum sensing system may be advantageous to the bacteria, as they do not waste energy expressing phenotypes (at low cell densities) that are only required at high densities. Classical examples include not only regulation of symbiotic behaviours (e.g. bioluminescence in *V. fischeri*) or virulence factors (e.g. elastase in *Pseudomonas aeruginosa*, antibiotic production in *Erwinia carotovora* and "Ti" plasmid transfer in *Agrobacterium tumefaciens*) but also more complex behaviour such as surface motility and colonization of *S. liquefaciens* and biofilm formation of *P. aeruginosa* and *Burkholderia cepacia*. Cell density dependent regulation of gene expression probably reflects the need for the invading pathogen to reach a critical population density sufficient to overwhelm host defences and thus establish infection.

AHLs in Foods and Food-spoilage Bacteria

As outlined above, the specific spoilage organism concept requires that a link be made by the qualitative and quantitative production of (spoilage) metabolites in an organism, the impact of these metabolites on sensory impression, and the growth of the organism(s) in naturally spoiling products. In particular, the amount of metabolite per cell, (e.g. the yield) is an important parameter when evaluating spoilage activity. If quorum sensing is used by gram-negative bacteria involved in spoilage and if AHLs (up) regulate traits relevant to the spoilage behaviour, such information is crucial for evaluation of spoilage activity of an organism. Studies have been done to determine if AHL signalling occurs in foods during storage and spoilage, how widespread AHL signalling is amongst gram-negative spoilage bacteria and to what extent AHL signalling plays a role in bacterial food spoilage. AHLs can be extracted from complex samples like foods by homogenizing ethyl acetate. Subsequently, the presence of AHLs can be evaluated using one or several bacterial AHL-monitoring strains. By using the *Chromobacterium violaceum* and the *A. tumefaciens* (with their plasmids) it has been found that many commercial foods contain large amounts of AHLs. Several types of AHLs are produced in foods and this can be visualized by separation of extracted AHLs on thin layer chromatographic plates and subsequent development by AHL monitor strains. Also, AHL production in foods can be visualized directly using reported systems. If an AHL producing strain and an AHL-negative strain carrying an AHL monitor system (e.g. the *luxR*-gene and *luxI*-promoter fused to the *gfp*-reporter gene) are coinoculated on a piece of meat or salmon, the appearance of green fluorescence in single bacterial cells on the meat/salmon signifies *in situ* expression of *luxI* and thereby the production of AHL.

Many gram-negative bacteria isolated from foods produce AHLs, thus, all, except one of 148 Enterobacteriaceae isolated from cold-smoked salmon and vacuum-packed meats produced detectable amounts of AHLs. AHLs are produced by pseudomonads involved in sprout spoilage and strains of *P. phosphoreum* also produce AHL. Also, several pectinolytic *Erwinia* and *Pseudomonas*, which spoil ready-to-eat vegetables like bean sprouts, produce AHLs. Thus, AHL production is a widespread phenomenon in food spoiling bacteria. AHL production can be determined either by random/representative isolation of pure cultures and subsequent determination of AHLs in culture supernatants using the monitors mentioned above. AHL-producing bacteria can also be isolated directly by replica plating from a non-selective aerobic plates to agar media containing AHL monitor strains.

A potential role for AHLs in food spoilage   The mere detection of AHLs in foods before and during spoilage and in gram-negative bacteria that may be involved in spoilage does not allow the conclusion that the AHL production plays a role in spoilage. Storage trials with bean sprouts have demonstrated that inoculation of sprouts with AHL-producing, pectinolytic *E.carotovora* results in a faster spoilage. It is anticipated that the major spoilage reaction is pectin degradation by *Erwinia* spp. and since pectinolytic activity is regulated by AHLs in *E. carotovora*, the hypothesis seems justified.

As mentioned, Enterobacteriaceae isolated from several fish and meat products produce AHLs. These organisms may contribute to the spoilage. Also, in vacuum-packed meats, the detection of AHL correlated with the numbers of Enterobacteriaceae being above 10^6 cfu/g. The spoilage of pasteurized milk is caused by growth of *Bacillus* where spores have survived the heat treatment or by the exoenzymatic activity of *Pseudomonas* species re-contaminating the milk following pasteurization. It has been suggested that quorum sensing is involved in regulation of hydrolytic activities by milk spoiling pseudomonads, however, to date no studies have been conducted on this topic. Enterobacteriaceae which produce AHLs may also occur in pasteurized milk. Due to the common involvement of AHLs either directly in regulation of extracellular hydrolytic enzymes or indirectly through regulation of transport systems, the hypothesis deserves investigation.

Other aspects of AHLs in food   Many of our foods contain AHLs. The major interest in these signalling compounds stems from their role in bacteria–bacteria interactions. However, it has recently been demonstrated that AHLs may act directly on eukaryotic organisms and may suppress reactions important for the immune response of epithelial cells. If AHLs can act directly on gastroepithelial cells, such action could be a component or mechanism facilitating the attack of food-borne pathogens in the gastro-intestinal tract.

To date, little work has been done on stability and degradation of AHLs and it is therefore not known how stable these compounds are in foods and how food preparation procedures affect them. AHLs detected in vacuum-packed meats may disappear during storage and it was found that degradation of AHLs from food derived Enterobacteriaceae may be facilitated by the producing organisms. Two recent studies demonstrate that some bacteria are capable of degrading AHLs and it was demonstrated that an enzyme from *Bacillus* spp. degraded AHLs and it was also found in one study that a strain of *Pseudomonas aureofaciens* degraded its own signal molecules.

Perspectives in Food Preservation

Due to involvement of AHLs in regulation of virulence factors in several opportunistic human and plant pathogenic bacteria, an intense search has been under way during the last 6 years for compounds that specifically could block AHL communication. In the clinical scenario, it is envisioned that such compounds can block expression of virulence without affecting growth, thus eliminating the risk of resistance development. Such quorum sensing inhibitors (QSI) are typically analogues of the AHLs, or compounds that degrade AHLs. One promising group of QSI is the halogenated furanones produced by the Australian red algae, *Delisea pulchra*. These furanones interfere with the receptor proteins and release the AHL signal. Treatment with these compounds reduces expression of AHL-regulated virulence factors and biofilm formation in *P. aeruginosa* and *S. liquefaciens*. QSIs like the *D. pulchra* furanones could also be used as food preservative in selected foods where AHL-regulated traits contribute to product spoilage. The finding of QSI compounds from algal (vegetable) sources raises the possibility of identifying active QSI compounds from a multitude of marine organisms as well as from natural foods or food by-products. Recent studies have shown that a number of foods contain furanone compounds with similarity to the *D.pulchra* compounds. QSI compounds may influence colonization of meat surfaces, toxin formation and perhaps bacterial proliferation. The natural occurrence of QSI is an important consideration for assessment of their toxicological status and may facilitate their use in food.

SUMMARY

Thus many food spoilage occur due to microbial degradation with their metabolites being the cause of the off-flavours or the textural changes resulting in sensory rejection. Despite the wide range of raw materials, the different processing parameters and the diverse array of storage conditions, very similar microfloras will develop in products with similar physical and chemical characteristics. Only a fraction of the microbes present in spoiling food products cause the spoilage characteristics. These organisms are known in some products, but much work is needed on a range of food products to identify the spoilage organisms. Under some conditions, also the interactive behaviour of spoilage bacteria may influence their growth and metabolism. Some examples have been described but it is likely that interactive behaviour (metabiosis and antagonism) is important in any foods in which a mixed flora develops during storage. The detection of chemical signals, acylated homoserine lactones, involved in bacterial quorum sensing regulation, in food and food-spoiling bacteria is of significant interest.

REVIEW QUESTIONS

1. Explain the concept of SSO.
2. Explain in detail the various interactions between food-spoiling bacteria.
3. Discuss the role of AHLs in food spoilage.

17

MICROBIAL SPOILAGE OF FISH

INTRODUCTION

Microbiological spoilage of foods may take diverse forms, but all of them are a consequence of microbial growth and/or activity, which manifests itself as changes in the sensory characteristics as shown in Table 17.1.

Table 17.1 Microbiological spoilage of foods

Microbiological activity	Sensory characteristics
Breakdown of food components	Production of off-odour and off-flavour
Production of extracellular polysaccharide material	Slime formation
Growth per se of moulds, bacteria, yeasts	Large visible pigmented or non-pigmented colonies
CO_2 from carbohydrate or amino acids	Production of gas
Production of diffusible pigments	Discolouration

Raw foods are initially contaminated with a wide variety of microorganisms, but only selective contaminants are able to colonize the food and grow to high numbers. The term "spoilage association" has been coined for such a specific microbial community. The precise mechanism by which one group of bacteria predominates over another, closely related group is not always fully understood. It is well known that only minor changes in processing and packaging of fish products are causing a dramatic change in the development and composition of the spoilage association and a complete different type of spoilage. However, even for the same type of product, spoilage may develop differently, depending on geographical origin and other unknown factors interacting with the microbial development.

FISH AS SUBSTRATE FOR BACTERIAL GROWTH

All food commodities have their own distinctive microbiology. Important factors contributing to the microbiological complexity of seafood are:

- specific as well as non-specific contamination of the live animal from the environment and of products during processing;
- growth conditions for microorganisms due to specific intrinsic and extrinsic factors (temperature, a_W, pH, microbial interactions, etc.).

The wide range of environmental habitats (freshwater to saltwater, tropical water to arctic water, pelagic swimmers to bottom dwellers and degree of pollution) and the variety of processing practices (iced fish products to (sterile) canned products) are all important factors in determining the initial contamination of fish and fish products. The part of the microflora which will ultimately grow on the products will be determined by the intrinsic and extrinsic parameters. There are several important specific intrinsic factors in fish which greatly influence the microbiology and spoilage:

- the poikilotherm nature of the fish and its aquatic environment
- a high post-mortem pH in the flesh (usually > 6.0)
- the presence of large amounts of non-protein-nitrogen (NPN)
- the presence of trimethylamine oxide (TMAO) as part of the NPN fraction

Bacteria establish themselves on the outer and inner surfaces of the live fish (e.g. gills, skin, gastrointestinal tract). The poikilotherm nature of fish allows bacteria with a broad temperature range to grow. Thus, the microflora of temperate water fish is dominated by psychrotrophic gram-negative, rod-shaped bacteria belonging to the genera *Pseudomonas, Moraxella, Acinetobacter, Shewanella, Flavobacterium,* members of Vibrionaceae and Aeromonadaceae, but gram-positive organisms such as *Bacillus, Micrococcus, Clostridium, Lactobacillus* and *Corynebacterium* can also be found in varying proportions. The flora on tropical fish often carries a slightly higher load of gram-positive and enteric bacteria, but is otherwise similar to the flora on temperate-water fish.

An important intrinsic factor related to fish flesh is the very high post-mortem pH (>6.0). Most fish contain only very little carbohydrate (< 0.5%) in the muscle tissue and only small amounts of lactic acid are produced. This has important consequences for the microbiology of fish as amongst other factors it allows the pH-sensitive spoilage bacteria *Shewanella putrefaciens* to grow.

The non-protein-nitrogen (NPN) fraction of the fish flesh consists of low-molecular-weight water-soluble nitrogen-containing compounds such as free amino acids and nucleotides and is a readily available bacterial growth substrate. The decomposition of the sulphur containing amino acids cysteine and methionine is particularly important in spoilage, as it causes off-odours and -flavours due to formation of hydrogen sulphides and methylmercaptan respectively.

Trimethylamineoxide (TMAO) is part of the NPN fraction and its presence in all marine and some fresh water fish species is well established. TMAO is known to cause a high (positive) redox potential (Eh) in the fish flesh however, the significance of this is not clear. The spoilage of fresh fish is certainly influenced by the presence of TMAO, particularly under conditions where oxygen is excluded. A number of well defined spoilage bacteria *(Shewanella putrefaciens, Photobacterium phosphoreum,* Vibrionaceae members) are able to utilize TMAO as the terminal electron acceptor in an anaerobic respiration resulting in off-odours and -flavours due to formation of trimethylamine (TMA). In sugar-salted herring, the presence of TMAO and the high Eh was established as the protective mechanism against the most common type of spoilage (sweet-sour, rotten putrid) as the organism causing this type of spoilage is a strict anaerobe requiring a low Eh for growth.

PRINCIPLES OF BACTERIAL SPOILAGE

In its simplest form, food spoilage is a result of microbiological growth per se and becomes evident as visible growth (moulds, pigmented or non-pigmented, slimy bacterial colonies). In such cases, of course, there is a direct relationship between the total number of microorganisms and the degree of spoilage.

More often spoilage is a result of the production of off-odours and-flavours caused by bacterial metabolism. In this case there is no correlation between total numbers of bacteria and spoilage, since only a fraction of the total flora participates in the spoilage. A clear distinction should be made between the terms "spoilage association" and "spoilage organisms" (bacteria) since the first describes merely the bacteria present on the fish when it spoils whereas the latter is the specific group that produces the off-odours and off-flavours associated with spoilage (Figure 17.1).

It is not an easy task to determine which of the bacteria isolated from the spoiled fish are those causing spoilage, and it requires extensive sensory, microbiological and chemical studies. First, the sensory, microbiological and chemical changes during storage must be studied and quantified, including a determination of the level of a given chemical compound that correlates with spoilage (the chemical spoilage indicator). Second, bacteria are isolated at the point of sensory rejection.

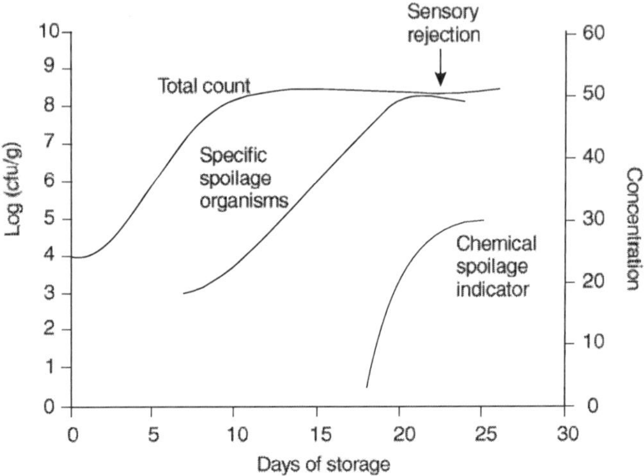

Figure 17.1   Model of changes in total viable count (TVC), specific spoilage organisms (SSO) and chemical spoilage indices during chill storage of a fish product

Pure and mixed cultures of bacteria are screened in sterile fish substrates for their spoilage potential, i.e., their ability to produce sensory (off-odours) and chemical changes typical of the spoiling product. Spoilage potential can be assessed in substrates such as sterile, raw fish juice, heat-sterilized fish juice or on sterile muscle blocks. The latter is the most complicated, but is also the one which is most comparable to the product. Finally, the selected strains are tested to evaluate their spoilage activity, i.e., their growth kinetics and their qualitative and quantitative production of off-odours in the product of concern. The latter step is important, as some bacteria may produce the chemical compounds associated with spoilage, but are unable to do so in significant amounts at the normal conditions prevailing in a product and they are thus not the specific spoilage bacteria. When stored aerobically, levels of 10^8–10^9cfu/g of specific spoilage bacteria are required to cause spoilage of iced fish. The spoilage of iced CO_2-packed fish is seen at a much lower level of 10^7 *P. phosphoreum* per gram. This relatively low cell level is due to the large size (5 mm) of the bacterium resulting in a high amount of TMA per cell and thus a higher spoilage activity. Calculating the amount of TMA per cell surface or volume, the production of TMA by *P. phosphoreum* is, however, not different from that of *S. putrefaciens,* for example. Some organisms may not be able to produce off-odours in sterile fish substrates but only when the substrate is digested by other members of the spoilage association. Therefore, in some products, spoilage potential and activity of a particular strain must be evaluated in competition or symbiosis with the normal spoilage association.

SPOILAGE OF FRESH FISH

Generally the quality deterioration of fresh fish is characterized sensorically by an initial loss of "fresh fish flavour" (sweet, seaweedy). After a period, when the odours and flavours are described as neutral or non-specific, the first indications of off-odours and -flavours are detectable. These will progressively become more pronounced and lead to rejection of the fish. The time to spoilage depends mainly on storage temperature and fish species. The initial quality loss in fish is primarily caused by autolytic changes and is unrelated to microbiological activity. Of particular importance in this respect is the degradation of nucleotides (ATP-related compounds) which is caused by autolytic enzymes. It is now widely accepted that the loss of the intermediate nucelotide, inosine monophosphate (IMP), is responsible for the loss of fresh fish flavour, but apart from this, the autolytic changes are contributing to spoilage mainly by making catabolites available for bacterial growth. The breakdown of IMP and inosine proceeds faster in naturally contaminated fish than in sterile samples and it has been shown repeatedly that several bacteria participate in the degradation.

The off-odours and -flavours developing in aerobically stored fish depend on the fish species and origin of the fish. The spoilage of marine temperate-water fish is characterized sensorically by development of offensive fishy, rotten H_2S-off-odours and -flavours. This sensory impression is distinctly different for some tropical fish and freshwater fish, where fruity, sulphydryl off-odours and -flavours are more typical.

The spoilage association developing in aerobically stored fish consists typically of gram-negative psychrotrophic non-fermenting rods. Thus, under aerobic iced storage, the flora is composed almost exclusively of *Pseudomonas* sp. and *S. putrefaciens*. This is true for all fish and shellfish whether caught or harvested in temperate or subtropical and tropical waters. At ambient temperature (25°C), the microflora is dominated by mesophilic Vibrionaceae and, particularly if the fish are caught in polluted waters, mesophilic Enterobacteriaceae members.

Shewanella putrefaciens is the specific spoilage bacteria of marine temperate-water fish stored aerobically in ice and the number of *S. putrefaciens* is inversely linearly related to remaining shelf life of iced cod. *S. putrefaciens* strains isolated from fish products have similar spoilage potential, however, the group is phenotypically heterogeneous and there may be a clonal selection during a storage trial.

Pseudomonas sp., are the specific spoilers of iced stored tropical freshwater fish and are also, together with *S. putrefaciens* spoilers of marine tropical-water fish stored in ice. *S. putrefaciens* has been isolated from tropical freshwaters, but does not appear to be important in the spoilage of iced freshwater fish from tropical waters. This may be due to occurrence of very low numbers and the inabililty of the organism to compete with high numbers of antagonistic pseudomonads.

At ambient temperature, motile aeromonads are the specific spoilers of aerobically stored freshwater fish.

It was shown that a large proportion of the flora on ambient-stored mackerel consisted of *S. putrefaciens*, indicating that this bacterium may also take part in the spoilage.

Numbers below 10^8 cfu/g are unlikely to be important in spoilage and consequently other organisms must be involved. It was observed that vacuum-packed cod contained some very large, almost yeast-like cells and suggested that these were involved in the spoilage. It was recently shown that these cells are heat-sensitive *Photobacterium phosphoreum*. It is a marine vibrio which has escaped microbiologist as it does not grow when pour plating and incubation at high temperatures are used. It is easily isolated from intestines of various fish. The organism produces 10–100 fold more TMA per cell than *S. putrefaciens* but does not cause off-odours as foul as *S. putrefaciens* probably because it does not produce volatile sulphides. The spoilage of vacuum-packed fish from temperate marine waters is caused by these two bacteria and differences in initial numbers of *S. putrefaciens* and *P. phosphoreum* probably decides which of the two becomes most important. Whilst abundant data exists on marine fish, little is known about the spoilage bacteriology of vacuum-packed freshwater fish. It is unlikely that *P. phosphoreum* plays a major role in the spoilage of freshwater fish as it requires sodium (or halophilic). CO_2 packing of marine fish from temperate water inhibits the development of the respiratory organisms like *Pseudomonas* and *S. putrefaciens* and their numbers rarely exceed 10^5–10^6 cfu/g. Similar changes resulting in dramatic extensions of shelf life are seen for meat products. However, a number of studies have shown that compared to meat products only limited or no extension of shelf life is obtained by packing fish in CO_2 atmosphere. The development of TMA is similar to or delayed only a few days compared to vacuum-packed storage, but increased compared to aerobic storage. The presence and subsequent growth of CO_2-resistant TMA producing *P. phosphoreum* to levels of 10^7–10^8 cfu/g explains these observations. It can further be hypothesized that it is not development and level of TMA per se which causes sensorical rejection of fish, but rather the presence of TMA in combination with other, yet unidentified compounds. Carbon dioxide and vacuum-packing of fish caught in freshwater or warmer water where these particular heat-sensitive, sodium-requiring *P. phosphoreum* are probably not as common, logically result in decrease in TMA production. The microflora becomes dominated by various gram-positive organisms, mainly lactic acid bacteria. However, as TMA can be detected later in the storage, TMAO-reducing organisms must be present at some level. The specific spoilage bacteria of fresh and packed fish are summarized in Table 17.2.

Table 17.2 Specific spoilage bacteria of fresh and packed fish stored chilled (<4°C) or in ice (Gram and Huss, 1996)

Atmosphere	Specific spoilage organisms of fresh, chilled fish			
	Temperate waters		Tropical waters	
	Marine	Fresh	Marine	Fresh
Aerobic	*S. putrefaciens* Pseudomonas spp.	*Pseudomonas* spp.	*S. putrefacnies/* Pseudomonas spp.	*Pseudomonas* spp.
Vacuum	*S. putrefaciens*	Gram-positive bacteria	Lactic acid bacteria/others	Lactic acid bacteria
	P. phosphoreum	Lactic acid bacteria		
CO₂	*P. phosphoreum*	Lactic acid bacteria	Lactic acid bacteria/TMAO reducing bacteria	Lactic acid bacteria/TMAO reducing bacteria

Comparison of the chemical compounds developing in naturally spoiling fish and sterile fish has shown that most of the volatile compounds are produced by bacteria. These include TMA, volatile sulphur compounds. aldehydes, ketones, esters, hypoxanthine as well as other low molecular weight compounds. The substrates for the production of volatiles are TMAO, sulphur-containing amino acids, carbohydrates (e.g. lactate and ribose), nucleotides (e.g. inosine monophosphate and inosine) and other NPN molecules. The amino acids are particularly important substrates for formation of sulphides and ammonia. Most bacteria identified as specific spoilage bacteria produce one or several volatile sulphides. *S. putrefaciens* and some members of Vibrionaceae produce H_2S from the sulphur containing amino acid L-cysteine. In contrast, neither *Pseudomonas* nor *P. phosphoreum* produce significant amounts of H_2S. Thus, hydrogen sulphide, which is typical of spoiling iced cod stored aerobically, is not detected in spoiling CO_2-packed cod. Methylmercaptan (CH_3SH) and dimethylsulphide (($CH_3)_2S$) are both formed from methionine. Taurine, which is also sulphur-containing, occurs as free amino acid in very high concentrations in fish muscle and disappears from the fish flesh during storage but this is because of leakage rather than because of bacterial attack. The development of TMA is in many fish species paralleled by a production of hypoxanthine. Hypoxanthine, which may cause a bitter off-flavour in the fish, can be formed by the autolytic decomposition of nucleotides, but it can also be formed by bacteria; and the rate of bacterial formation is higher than the autolytic. Several of the spoilage bacteria produce hypoxanthine from inosine or inosine monophosphate, including *Pseudomonas* sp., contrary to the iced spoilage by *S. putrefaciens* and the ambient spoilage by Vibrionaceae which is dominated by H_2S and TMA, the spoilage caused by *Pseudomonas* sp. is characterized by absence of these compounds. Fruity, rotten, sulphydryl odours and flavours are typical of the *Pseudomonas* spoilage of iced fish. *Pseudomonas* sp. produce a number of volatile aldehydes, ketones, esters and sulphides. However, it is not known which specific compounds are responsible for the typical off-odours. The fruity off-odours produced by *Pseudomonas fragi* originate from monoamino and monocarboxylic amino acids.

SPOILAGE OF FISH PRODUCTS

Lightly Preserved Fish Products

This group includes fish products preserved by low levels of salt (< 6% NaCl (w/w) in the water phase) and, for some products, addition of preservatives (sorbate, benzoate, NO_2 or smoke). Product pH is high (> 5.0) and they are often packaged under vacuum and must be stored and distributed at chill temperatures (< 5°C). This is a group of high-value delicatessen products (cold-smoked, pickled ('gravad') or marinated fish, brined shellfish) that are typically consumed as ready-to-eat products with no heat treatment. The microbial spoilage developing in these products is not well understood. By comparing the sensory changes in naturally contaminated cold-smoked salmon with changes in samples with much reduced bacterial load, it was concluded in a study that while autolytic enzymes caused texture changes (softening), bacterial activity was indeed the cause of spoilage. The off-odours and flavour developing are variously described as putrid, cabbage-like, sour, bitter, fruity, sweet and the normal shelf life of this product also varied considerably from about three weeks to eight weeks for vacuum-packed, cold-smoked salmon (4–5% NaCl in water phase, pH 6.3–6.4) stored at 5°C.

Several studies have shown that the microflora developing in this type of products is dominated by lactic acid bacteria (LAB). Often high levels (10^7–10^8 cfu/g) of LAB are present for several weeks before the product becomes sensorically rejectable, which demonstrates that the often used "total count of bacteria" is completely useless as a spoilage indicator for this type of products. It was reported that *Lactobacillus curvatus* was the most common species occurring, but also *L. sake/bavaricus*, *L. plantarum*, *Carnobacterium* and *Leuconostoc* sp. were present in smaller numbers. The specific microflora, developing seemed to be related to production environment in the smokehouse (an in-house flora).

Until recently, it was believed that the lactic acid bacteria were of little importance for the sensory changes during storage of fish products and it was shown that no or very faint off-odours were produced by lactic acid bacteria compared to the very obnoxious off-odours produced by gram-negative spoilers.

However, it was also found that several lactic acid bacteria were able to produce some of the off-odours (sour, cabbage, sulphurous) associated with spoilage of cold-smoked salmon. Furthermore, a strain of *L. sake* produced H_2S during growth on cold-smoked salmon.

Although most studies report dominance of lactic acid bacteria in cold-smoked fish, the chemical changes during storage varies as does the composition of the remaining microbial flora. LAB are known not to reduce TMAO to TMA, other bacteria must at times be participating in the spoilage process.

Spoiled lightly preserved fish products may also contain low (10^3 cfu/g) or high (10^6–10^7 cfu/g) levels of Enterobacteriaceae, *Brochotrix thermosphacta*, yeasts and *Photobacterium phosphoreum*. Three different microfloras were present when cold-smoked salmon was sensorically rejected:

- dominated by lactic acid bacteria (10^7–10^9 cfu/g)

- dominated by lactic acid bacteria and Enterobacteriaceae (10^7–10^8 cfu/g), and

- dominated by *Photobacterium phosphoreum* (10^6–10^7 cfu/g) with occasional high levels of lactic acid bacteria.

The Enterobacteriaceae growing in lightly preserved fish products have been identified as psychrotrophic *Hafnia alvei, Serratia liquefaciens* or *Enterobacter* sp. All strains produce spoilage off-odours and reduce TMAO but the ability to produce H_2S from protein substrates varies and little is known about their spoilage activity in lightly preserved fish products.

Salt Curing and Fermentation

There are essentially two types of products where the preservative action of salt is the predominant process, dry salted and wet salted or pickled fish products. Dry-salting is used only for non-fatty fish. There are two types of spoilage of this product. One is growth of the extremely halophilic bacteria which causes a condition known as "pink". These pink halophilic bacteria (*Halococcus, Halobacterium*) are strongly proteolytic and produce off-odours and flavours in the product. The other type of spoilage is moulding by a highly osmophilic type of fungus known as "dun" (*Sporendonema* and *Oospora*).

Wet salting or barrel-salting is used for fatty fish species such as herring and anchovy. The fishes are mixed with salt and kept in a closed container. The waterphase salt in barrel-salted herring is typically 15–20%. Three types of spoilage are known for this product. The most common type is characterized by the presence of sour, sour/sweet and putrid off-odours and -flavours. This type of spoilage is caused by growth of a gram-negative, halophilic, obligate anaerobic rod [(up to 10^6–10^7 cfu/g)]. The growth of this organism is not possible until the general microflora has reduced all the TMAO causing the Eh to drop to negative values. This may take more than one year at chill storage (2–4°C) as the general flora consisted of low levels (10^3–10^5 cfu/g) of mainly gram-negative, halophilic rods able to reduce TMAO. The second type of spoilage is characterized by the development of fruity off-odours and is caused by growth to levels of 10^5 cfu/g of osmotolerant yeast species. Finally the term "ropiness" or "ropy brine" is used to describe the phenomenon that the brine becomes highly viscous or slimy. This third type of spoilage is caused by a gram-negative, halophilic, aerobic, non-motile, rod-shaped (*Moraxella*-like) bacteria.

There are no reports of microbiological spoilage of high-salt fermented fish products such as fish sauce and paste. These products, are despite their name, not really fermented products as very high salt levels (20–30%) are used and the liquefying action believed to be caused by autolytic enzymes. Many types of seafood products receive a heat treatment as part of their processing. Particularly the products receiving only a mild heat treatment and distributed at chill temperatures Refrigerated processed foods with extended durability (REPFEDS) are likely to spoil due to microbial action. Unusual clostridia have been found as spoilers of pasteurized crab meat. It was found that *sous vide* packed cod stored at 5°C spoiled due to growth of a gram-positive spore-forming bacteria producing extremely obnoxious and putrid off-odours.

CONCLUSIONS

A thorough understanding of the spoilage process and knowledge of the specific spoilage organisms are necessary for design of an optimal, product specific quality assurance programme or if microbiological data are to be utilized to predict the shelf life of a product at specific conditions (temperature, packaging). For a number of fish products (heavily preserved as iced or salt cured products) the situation is reasonably well described and understood. One or only a few organisms invariably appear as the spoilage organisms in the same type of product under the same geographical conditions.

However, the situation in, e.g. lightly preserved fish products is much more complicated and the assumption mentioned appears not to hold. Much more work is needed to understand the significance of such factors as the role of bacterial interactions, possible clonal selection of specially active groups within one species and possible changes of bacterial metabolism due to intrinsic and extrinsic conditions.

1. Discuss the role of fish as a substrate for bacterial growth.
2. Brief out the spoilage of fresh fish.
3. Discuss the spoilage of fish products.

18

BACTERIAL SPOILAGE OF MEAT AND MEAT PRODUCTS

VARIOUS PARAMETERS INVOLVED IN SPOILAGE OF MEAT BY MICROBES

The shelf life of meat and meat products is the storage time until spoilage. The point of spoilage may be defined by a certain maximum acceptable bacterial level, or an unacceptable off-odour/off-flavour or appearance. The shelf life depends on the numbers and types of microorganisms, mainly bacteria, initially present and their subsequent growth.

The initial mesophilic bacterial count on meat and cooked meat products is about 10^2–10^3 cfu/cm^2 or gram, consisting of a large variety of species. Only 10% of the bacteria initially present are able to grow at refrigeration temperatures, and the fraction causing spoilage is even lower. Since meat products are heated to a temperature of 65–75°C, most vegetative cells are killed and post heat treatment re-contamination determines the shelf life. The surface contamination of the cut meat and the meat products will determine the potential shelf life.

During storage, environmental factors such as temperature, gaseous atmosphere, pH and NaCl will favour certain bacteria, and affect their growth rate and activity. The shelf life of refrigerated meat and meat products may vary from days up to several months.

Microbial spoilage of meat and seafood generally occurs as a result of the growth of bacteria that have colonized muscle surfaces. The first stage in colonization and growth involves the attachment of microbial cells to the surface. Bacterial attachment to muscle surfaces involves two stages:

(i) the first is a loose, reversible sorption which may be related to van der Waals forces or other physico-chemical factors.

One of the factors that influences attachment at this point is the population of bacteria in the water film.

(ii) The second stage consists of an irreversible attachment to surfaces involving the production of an extracellular polysaccharide layer known as glycocalyx. In addition to cell density, factors such as type of surface, growth phase, temperature, and motility may also influence bacterial attachment to muscle surfaces. Bacteria already present on surfaces may influence the ability of other bacteria to attach to surfaces. *Pseudomonas* species have been observed to attach more rapidly to meat surfaces.

The initial microflora of muscle foods is highly variable, arising from the resident microbes in and on the live animal and from environmental sources such as vegetation, water and soil. Despite this variability and differences between muscle tissues of different species, characteristics of microbial spoilage are remarkably similar.

Refrigeration restricts the growth of mesophiles, generally a major component of the initial microflora, and allows psychrotrophic microorganisms to grow and eventually dominate the microflora. As microbial growth occurs during storage, the composition of the microflora is altered so that it is dominated by a few or often a single, microbial species, usually of the genus *Pseudomonas*, *Lactobacillus*, *Moraxella* or *Acinetobacter* or *Brochothrix thermosphacta*.

The final composition of the spoilage microflora may be influenced by the proportion of specific spoilage microbes in the initial population. High initial numbers of a slowly growing species may compete successfully with lower numbers of a species with a faster growth rate. If the number of spoilage microorganisms in the initial population is high, a slower growth rate may not be an important factor, since less growth may be necessary before spoilage occurs.

Pseudomonas species are typically able to compete successfully on aerobically stored, refrigerated muscle foods for several reasons. The genus *Pseudomonas* is characterized by a competitive growth rate, even at refrigeration temperatures. In addition, pseudomonads are able to grow within the usual pH range of muscle foods, while many bacteria, e.g. *Moraxella* and *Acinetobacter* species, are less capable of competing under the refrigeration temperatures at the lower pH in this range. Since pseudomonads are highly oxidative, they are able to utilize low molecular weight nitrogen compounds as sources of energy. This is a clear competitive advantage for pseudomonads, since meats contain relatively low levels of simple sugars, and more complex energy sources such as protein and fat do not serve as significant substrates for growth until later in spoilage when high bacterial populations are attained.

BACTERIA ASSOCIATED WITH SPOILAGE

The predominant bacteria associated with spoilage of refrigerated beef and pork, are *B. thermosphacta*, *Carnobacterium* spp., Enterobacteriaceae, *Lactobacillus* spp., *Leuconostoc* spp. *Pseudomonas* spp. and *S. putrefaciens*. The main defects of meat are off-odour and off-flavour but discolouration and gas production also occur.

Environmental Influences on Bacterial Growth and Shelf Life

Growth to high numbers is a prerequisite for spoilage. The expected shelf life and growth ability of different bacteria under various environmental conditions, are presented in the Table 18.1.

Table 18.1   Expected shelf life under refrigerated storage, and growth ability of bacterial groups and bacteria on meat and meat products

Product	Storage	Expected shelf life	Growth			
			Pseudomonas spp.	Entero-bacteriaceae	Lactic acid bacteria	*B.thermosphacta*
Meat, normal pH	Air	Days	+++	++	++	++/+++
	High O_2^- MA	Days	+++	++/+++	++/ +++	+++
	Vacuum	Weeks– months	+	+/++	+++	++/+++
	100% CO_2	Months	+	+/++	+++	+
Meat, high pH	Vacuum	Days	+	++/+++	+++	++/+++
	100% CO_2	Weeks– months	+	+/++	+++	+
Meat products	Air	Days	+/++	+	++	+++
	Vacuum	Weeks	+	+	+++	++/+++
	CO_2 + N_2	Weeks	+	+	+++	+

+++, dominant part of microflora; ++, intermediate part of the microflora; +, minor part of the microflora

Packaging

Three different packaging types are in use: air, vacuum and modified atmospheres (MA). MA contain different levels of oxygen and carbon dioxide, balanced with inert nitrogen. Packages containing up to 80% oxygen and

20% carbon dioxide (high-oxygen MA) will reduce the colour deterioration of retail cuts of meat, but will only slightly increase the shelf life, compared to aerobic storage. Pork is generally stored aerobically or in MA, and beef in a vacuum or MA due to the need for tenderization during an extended storage. Transitions between different packaging types may be performed for retail cuts. The shelf life of meat increases in the order: air, high-oxygen MA, vacuum, no-oxygenMA and 100% CO_2.

Pseudomonas spp. dominate on aerobically stored meat, and due to a high growth rate the shelf life is a matter of days. The frequent domination of *Pseudomonas fragi* is suggested to be due to the ability of the bacterium to use creatine and creatinine. In a mixed broth culture of *P. fragi*, *B. thermosphacta* and *Corynebacterium maltaromicus*, all grow at the same exponential rates at 4°C and pH 6.1. *B. thermosphacta* dominated the flora at the early stationary growth phase but it was eventually outgrown by *P. fragi*.

In high-oxygen MA, a variety of bacteria are able to grow to high final numbers such as *B. thermosphacta*, *Pseudomonas* spp., *Leuconostoc* spp., and *Lactobacillus* spp. Most bacteria are more or less inhibited by CO_2, and thus the growth rate is reduced, compared to air, and the shelf life increased.

In a vacuum pack the composition of the gaseous phase changes occur during storage. The concentration of oxygen decreases while that of carbon dioxide increases. The bacterial flora is gradually selected towards a CO_2- tolerant but slowly growing one. Vacuum-packaged beef may have a storage life of 10–12 weeks at 0°C, until the off-flavour becomes unacceptable. The bacterial flora is dominated mainly by lactic acid bacteria, *Corynebacterium* spp., *Lactobacillus* spp. and *Leuconostoc* spp. The film permeability has been shown to affect shelf life. With increasing permeability, the growth rate and the maximum number of *Pseudomonas* spp. increases, the growth rate of *B. thermosphacta* is unaffected, but the maximum count increases; and *Lactobacillus* spp. growth is unaffected.

A long shelf life may be attained in pure CO_2, the time needed to reach 10^7 bacteria/cm^2 and off-odour, was 10 days in air, and 40 days in 100% CO_2 for pork stored at 4°C. The effect of CO_2 is enhanced by a low storage temperature, due to increased solubility of the gas. Shelf life extension by CO_2 results from an immediate selection, as opposed to a gradual one in a vacuum pack, of lactic acid bacteria growing at a reduced rate.

VARIOUS SPOILAGE FEATURES OF MEAT

Off-odours and Off-flavours

Off-odours such as sweet and fruity, putrid, sulphury and cheesy, are characterized in aerobically stored meat. *Pseudomonas* sp. specifically

P. fragi produce ethyl esters coinciding with the early stages of spoilage. Sulphur-containing compounds contribute to the putrid and sulphury odours. The responsible compounds are, for example, hydrogen sulphide formed by Enterobacteriaceae and dimethyl sulphide formed by *Pseudomonas* spp. Cheesy odours are associated with acetoin/diacetyl and 3-methylbutanol formation, presumably by Enterobacteriaceae, *B. thermosphacta* and homofermentative *Lactobacillus* spp.

The off-odour of meat packaged in high-oxygen MA is characterized as cheesy and rancid. In this atmosphere, a variety of bacteria such as *Pseudomonas* spp., *B. thermosphacta* and lactic acid bacteria are able to grow, and contribute to spoilage. The presence of oxygen will increase the spoilage potential of both *B. thermosphacta* and lactic acid bacteria, due to the formation of end products such as acetoin and acetic acid.

The off-odours of vacuum and anaerobic MA packaged meat are less offensive than of aerobically stored meat. The spoilage characteristics are sour and acid, and are typically associated with lactic acid bacteria and the production of lactic acid and acetic acid. Sulphur compounds may also contribute to off-odour. During extended storage of meat, a depletion of glucose is likely to notice. A metabolic switch due to glucose depletion, leads to the formation of end products such as acetic acid and hydrogen sulphide in homofermentative *Lactobacillus* spp. in anaerobic atmospheres.

Discolouration

The bacterial production of hydrogen sulphide converts the muscle pigment to green sulphmyoglobin. Hydrogen sulphide is produced from cysteine and is triggered by glucose limitation. *L. sake* forms hydrogen sulphide, but only when the glucose and oxygen availability is limited. Sulphmyoglobin is, however, not formed in anaerobic atmospheres. Other bacteria able to produce hydrogen sulphide are *H. alvei* and *S. putrefaciens*. Greening is typically associated with high-pH meat, but may also occur in normal-pH meat.

Gas Production

Clostridium spp. have been associated with the production of large amounts of gas (H_2 and CO_2) in vacuum-packaged beef, accompanied by foul off-odours. Gas production (CO_2) by lactic acid bacteria without extensive off-odours may be associated with vacuum-packaged beef and pork.

HEAT-PROCESSED MEAT PRODUCTS

Environmental Influences on Bacterial Growth and Shelf Life

The microbiological stability of cooked, cured meat products depends on extrinsic factors, mainly the packaging methods and storage temperature, and on intrinsic factors, such as the product composition.

Packaging

Cooked meat products are chill stored, usually in vacuum pack or in MA packs, but are also distributed unpacked, i.e., stored in an aerobic atmosphere. Furthermore, in retail shops slicing is performed after the opening of packages, with subsequent storage in an aerobic atmosphere.

During the aerobic storage of cooked, sliced meat products, a mixed flora composed of *Bacillus* sp., *Micrococcus* spp. and *Lactobacillus* spp. is reported to dominate. In addition, *Pseudomonas* spp. may increase up to 10^5 cfu/g. In cured, raw meat products, *B. thermosphacta, Moraxella* spp./ *Psychrobacter* spp., and *Pseudomonas* spp. were retrieved. In addition, good growth of yeast occurred.

Vacuum packaging is frequently used for cooked meat products. The combination of the microaerophilic conditions, the presence of curing salt and nitrite favours the growth of psychrotrophic lactic acid bacteria. *B. thermosphacta* may also be a dominant part of the bacterial flora; this will particularly be the case when the film permeability is high.

Storage under modified atmospheres ($CO_2 + N_2$) is also used for cooked, cured meat products. Comparisons of the shelf life of cooked, cured meat products in vacuum packs and in MA packs have resulted in different findings. While some investigations indicated no extension of the shelf life of MA-packaged meat products, other studies reported an increase in shelf life by MA. For meat products stored in MA, the shelf life with respect to bacterial numbers was prolonged by 75% compared to vacuum-packaged, independent of the proportions of CO_2 and N_2 used. The effects on flavours and slimy spoilage varied, however, with the different gaseous atmospheres used, where 100% N_2 gave the best overall result followed by MA with an initial CO_2 concentration <50%. High concentrations of CO_2 may cause discolouration, off-odours and off-flavours, and a release of liquid from the meat products. The amount of drip from emulsion sausage stored in vacuum or MA decreased in the order: 100% CO_2 > vacuum > 70% N_2 >30% CO_2 > 100% N_2.

The growth of lactic acid bacteria is favoured in atmospheres of CO_2 plus N_2 while the growth of Enterobateriaceae, *B. thermosphacta* and yeast is restricted. However, the growth rate of some lactic acid bacteria is reduced in CO_2 atmospheres compared to aerobic which may explain the prolonged shelf life.

The lactic flora of vacuum or MA packaged cooked, cured meat products consists of *Lactobacillus* spp. predominantly *L. sake* and *L. curvatus*. *Leuconostoc* spp. such as *L. gelidium, L. carnosum* and *L. mesenteroides sensu stricto* group are also frequently found to dominate. Furthermore, *Weissella viridescens, Corynebacterium divergens, C. maltaromicus* and *B. thermosphacta* have been isolated.

Temperature

On vacuum or MA packaged meat products, the dominance of lactic acid bacteria is unaltered by the refrigeration temperature used, but the growth rate is affected. With a decrease in temperature from 7°C to 2°C, the growth of lactic acid bacteria was retarded almost twofold; from 7°C to 0°C about four fold. Thus for meat products the storage temperature is an important factor influencing the shelf life.

Product Composition

Intrinsic factors that contribute to the microbial stability of cooked, cured meat products are mainly salt content, water activity, pH and nitrite concentration. The concentration of NaCl is 3–5% calculated on the water content, pH values are 6.0–6.5, water activity (a_w) values 0.96–0.99 and the residual nitrite level is below 100.

Cooked meat products may, in addition to sodium chloride, also contain other humectants such as phosphates and sodium lactate. These humectants will result in a decrease of the water activity. The addition of 4% sodium chloride in the water phase, will decrease the a_w value from 0.99 to about 0.97. Salt-sensitive microbes such as *Pseudomonas* spp. and Enterobacteriaceae, will not grow at these reduced water activity values, and the microflora developing will shift to more salt-tolerant microbes such as lactic acid bacteria and yeasts. The growth of yeasts is furthermore greatly affected by the gaseous atmosphere, being restricted in an anaerobic atmosphere. The growth rate and the lag phase of lactic acid bacteria are influenced by reduced a_w values, for example, a decrease in a_w value from 0.98 to 0.96 in a sausage resulted in a three-fold increase in the lag time and a two-fold decrease in the growth rate of the lactic acid bacteria that dominated the bacterial flora.

The initial pH value of cooked meat products will not restrict microbial growth. The pH may, however, decrease from pH 6.0–6.5 to pH 5.0–5.3 during storage, due to the activity of lactic acid bacteria. Such a decrease will restrict the growth of *B. thermosphacta*, but not of *Lactobacillus* spp. the pink colour of cooked, cured meat products is the result of the addition of nitrite and or nitrate prior to heating, and the subsequent formation of nitrosohaemochrome. Nitrite has an inhibitory action on the growth of several microbes such as Enterobacteriaceae and *B. thermosphacta* but not on lactic acid bacteria.

SPOILAGE UNDER ANAEROBIC CONDITIONS

The spoilage microflora of muscle foods is dominated by lactic acid bacteria when oxygen is excluded from the storage environment. If the pH of the muscle tissue is high or residual amounts of oxygen are present, other microbes such as *B. thermosphacta* and *S. putrefaciens* may make substantial contributions to product spoilage. The growth rate of bacteria under anaerobic conditions is considerably reduced compared with growth under aerobic conditions. In addition, the maximum cell density achieved under anaerobic conditions is considerably less than that achieved under aerobic conditions. It is reported that maximum populations of spoilage microbes under anaerobic conditions are determined by the rate of diffusion of fermentable substrates, such as glucose and arginine, from underlying tissues. It is not until the maximum density of microflora is reached that obvious spoilage slowly develops.

The sour, acid, cheesy odours or the cheesy and dairy flavours that develop in muscle tissue under anaerobic conditions can be attributed to the accumulation of short-chain fatty acids and amines.

The energy required for muscle activity in the live animal is obtained from sugars (glycogen) in the muscle. In the healthy and well-rested animal, the glycogen content of the muscle is high. After the animal has been slaughtered, the glycogen in the muscle is converted into lactic acid, and the muscle and carcass becomes firm (rigor mortis). This lactic acid is necessary to produce meat, which is tasteful and tender, of good keeping quality and good colour. If the animal is stressed before and during slaughter, the glycogen is used up, and the lactic acid level that develops in the meat after slaughter is reduced. This will have serious adverse effects on meat quality.

Pale Soft Exudative (PSE) Meat

PSE in pigs is caused by severe, short-term stress just prior to slaughter, for example during off-loading, handling, holding in pens and stunning. Here the animal is subjected to severe anxiety and fright caused by manhandling, fighting in the pens and bad stunning techniques. All this may result in biochemical processes in the muscle, particularly in rapid breakdown of muscle glycogen and the meat becoming very pale with pronounced acidity (pH values of 5.4–5.6 immediately after slaughter) and poor flavour. This type of meat is difficult to use or cannot be used at all by butchers or meat processors and is wasted in extreme cases. Allowing pigs to rest for one hour prior to slaughter and quiet handling will considerably reduce the risk of PSE.

Dark, Firm, Dry (DFD) Meat

It is found in carcasses from animals which, before slaughter, have undergone prolonged muscular activity or stress with consequent depletion of glycogen reserves in the muscles. After death, muscle glycogen is converted to lactic acid until a final pH of about 5.5 is attained in unstressed animals, but in DFD meat, glycogen is exhausted while the pH is still above 6.0. The DFD condition is considered to be a problem in both beef and pork. Such meat spoils more rapidly than normal ultimate pH meat. This has been attributed to a faster rate of growth of spoilage bacteria at the higher pH, but the growth rates of most spoilage bacteria on meat are unaffected by pH between 5.5 and 7.0, although the lag phase is extended at the lower end of the range. An alternative possibility is that DFD meat is devoid of glucose, which is normally attacked first by spoilage bacteria. The concentration of glucose in meat varies between species and between different muscles in the same carcass. The average glucose contents of normal ultimate pH beef, mutton, and pork are about 100, 400, and 800 pg/g (wet weight), respectively. In the absence of glucose, pseudomonads, which are the dominant bacteria in aerobic spoilage floras, degrade amino acids with the production of ammonia and off-odours. In DFD meat, this could occur at lower cell densities than normal.

The pH of meat is usually considered to be the most important indicator of the DFD condition. However, the critical factor for keeping quality appears to be the depletion of glucose, and glucose content is highly variable. Analysis of glucose in those muscles most susceptible to glucose exhaustion would seem to be the most reliable guide to the keeping quality of suspected DFD meat.

BACTERIA ASSOCIATED WITH SPOILAGE OF MEAT PRODUCTS

Lactic acid bacteria are the major bacterial group associated with the spoilage of refrigerated vacuum or MA packaged cooked, cured meat products. At the time of spoilage some products contain a "pure" culture of only one species, while in others a mixture of *Lactobacillus* spp. and *Leuconostoc* spp. was found. The great diversity of bacteria isolated from spoiled meat products is confirmed by several studies. The genus or species of lactic acid bacteria responsible for spoilage, depend on the product composition (product related flora) as well as the manufacturing site (house related flora). Lactic acid bacteria spoil refrigerated meat products by causing defects such as sour off-flavours, discolouration, gas production, slime production and decrease in pH.

VARIOUS SPOILAGE FEATURES OF MEAT PRODUCTS

Off-odours and Off-flavours

Off-flavours in vacuum- or MA- packaged cooked meat products are typically described as sour and acid. The dominating bacteria, lactic acid bacteria, produce acids such as lactic acid, acetic acid and formic acid; the levels depending on genus, species and growth conditions. Meat products stored aerobically or vacuum-packaged using a film with a relatively high permeability to oxygen may, in addition to sour and acid flavours, develop a slightly sweet, cheesy obnoxious odour. This is also found in meat products that have initially been stored anaerobically and subsequent to opening the package in an aerobic atmosphere. Anaerobic atmosphere induces the formation of acetoin in *B. thermosphacta*, *Lactobacillus* spp. and *Carnobacterium* spp.

Discolouration

Bacteria producing H_2O_2 may cause a green discolouration through the oxidation of nitrosohaemochrome to choleomyoglobin, frequently seen as green spots. Exposure to air is necessary for the formation of H_2O_2. The bacterial greening in the centre of meat products is caused by bacteria surviving the cooking process which after exposure to air start to produce H_2O_2 due to a high heat resistance. *W. viridescens* has been demonstrated to survive regular heat processing in sausage processing, being able to survive for more than 40 minutes at 68°C. The surface greening is caused by bacteria which contaminate the product after cooking. Homofermentative *Lactobacillus* spp., heterofermentative *Lactobacillus* spp. *Leuconostoc* spp. and *C. divergens* are able to form H_2O_2. Other bacteria that have been associated with greening are *Enterococcus* spp., and *Pediococcus* spp.,

Gas Formation

Accumulation of gas (CO_2) is often associated with the growth of *Leuconostoc* spp. such as *L. mesenteroides*, *L. carnosum* and *L. amelibiosum*.

Slime

Ropy slime formation associated with vacuum-packed, cooked meat products is caused by homofermentative *Lactobacillus* spp. such as *L. sake* and *Leuconostoc* spp. such as *L. amelibiosum*, *L. carnosum*, *L. gelidum*, *L. mesenteroides* subsp. *dextranicum* and *L. mesenteroides* subsp. *mesenteroides* have been associated with slime formation on meat products. Ropy-slime-producing *L. sake* and *L. amelibiosum* were recovered from the processing rooms at meat plants where the meat products were handled after heat processing. Furthermore, slime formation is an early indication of spoilage, often observed before the expiry date. The formation of ropy slime does not require that the meat product contains sucrose.

CHANGES IN THE SPOILAGE-RELATED MICROBIOTA OF BEEF

Owing to its high water content and abundance of important nutrients available on the surface, meat is recognized as one of the most perishable foods. "Spoilage can be defined as any change in a food product that makes it unacceptable to the consumer from a sensory point of view." Apart from physical damage, oxidation, and colour change, the other spoilage symptoms are ascribable to the undesired growth of microorganisms to unacceptable levels. In the case of meat, microbial spoilage leads to the development of off-odours and often slime formation, which make the product undesirable for human consumption. The organoleptic changes may vary according to the microbial association contaminating the meat and to the conditions under which the meat is stored. The development of organoleptic spoilage is related to microbial consumption of meat nutrients such as sugars and free amino acids and the release of undesired volatile metabolites. Microbial loads from 10^7 cfu/cm^2 are usually associated to occurrence of off-odours such as "cheesy" or "buttery" odours; these may evolve into "fruity" odors when the loads increase and putrid smells as the result of free amino acid consumption at loads as high as 10^9 cfu/cm^2. In fact, once the glucose present in the aqueous phase has been utilized, other substrates are sequentially consumed until odorous nitrogenous compounds such as ammonia and dimethylsulphide are released.

Different spoilage-related species and strains can colonize the meat surface through different stages involving adsorption to the meat surface and attachment by glycocalyx formation. The development of these phases depends on the intrinsic and extrinsic ecological factors of a particular meat ecosystem such as pH, meat surface morphology, O_2 availability, temperature, and the presence and development of other bacteria.

Many groups of organisms contain members that potentially contribute to meat spoilage under appropriate conditions. This makes the microbial ecology of spoiling raw meat very complex and the spoilage thus very difficult to prevent. Under aerobic conditions, a few species of the genus *Pseudomonas* are generally recognized to dominate the meat system and to actively contribute to spoilage owing to their capability of glucose and amino acid degradation, even at refrigeration temperatures. *Brochothrix thermosphacta* is a microorganism for which meat is considered an ecological niche, even though it can also occur in spoiled fish. The capability of *B. thermosphacta* to grow on meat during both aerobiosis and anaerobiosis makes it an important and a significant meat colonizer.

Members of the Spoilage-related Flora Due to Off-odour Production

Many members of the Enterobacteriaceae, belonging to the genera *Serratia*, *Enterobacter*, *Pantoea*, *Proteus*, and *Hafnia*, often contribute to meat spoilage. Moreover, lactic acid bacteria (LAB) such as *Lactobacillus*, *Corynebacterium*, and *Leuconostoc* can play an important role in the spoilage of refrigerated raw meat and are also recognized as important competitors of the other spoilage-related microbial groups under appropriate conditions. Packaging can be an effective method for meat shelf life extension that avoids the use of chemical preservatives. These methods of preservation include the use of modified atmosphere packaging (MAP) using gas mixtures containing variable O_2 and CO_2 concentrations in order to inhibit the different spoilage-related bacteria and are often associated with the use of low temperatures during storage. Substantial fractions of CO_2 are used to retard the growth of organisms produced by aerobic spoilage, and a certain concentration of O_2 is employed for red meat MAP to preserve meat colour. This kind of packaging is normally associated with the use of packs made of materials that provide a barrier to the exchange of gases between the pack and the external atmosphere. Improved storage of meat can be achieved by using CO_2, allowing the growth of LAB such as *Lactobacillus* spp. and *Leuconostoc* spp. and thus outcompeting Enterobacteriaceae, *Pseudomonas* spp. and *Brochothrix thermosphacta*. Moreover, the use of high CO_2 concentrations, together with low pH and chill storage, can more readily inhibit the growth of food pathogens than vacuum packaging. In addition, novel technologies of active packaging are currently under exploitation; these techniques can employ natural antimicrobials such as essential oils or bacteriocins to inhibit microbial growth in meat products. Most of the work performed on the changes of microbial associations in meat during storage have focused on comparisons between viable counts of spoilage-related microbial groups. However, microbial analysis alone as a spoilage index may not be exhaustive enough to understand the actual shifts of the microbial ecology of raw meat in response to different storage conditions. Certain flora may be differently influenced by the specific storage conditions, and the different microbial species may unpredictably develop during storage, thus influencing the time and type of spoilage development. There is still a need to assess, within each spoilage-related microbial group, which species are actually involved in the spoilage of meat. Moreover, molecular methods such as PCR and denaturing gradient gel electrophoresis (DGGE) are seldom optimized to monitor changes in spoilage-related microbial flora in food, while they are widely exploited for the characterization of fermented foods. The microbial diversity of the spoilage-related microbial groups that develop during refrigerated storage of beef at the species level under different MAP conditions is assessed by considering quality indices of significant importance for the acceptability of fresh meat by the consumer.

1. Discuss the role of bacteria associated with meat spoilage.

2. Explain discolouration.

3. Discuss meat spoilage under anaerobic conditions.

4. What is Pale Soft Exudative (PSE) meat?

19

WINE SPOILAGE ORGANISMS—YEASTS

YEASTS AND WINE—A PERSPECTIVE

When we look at any subject on food microbiology published during the last 50 years, it appears that food spoilage caused by yeasts receives little attention, even in foods commonly spoiled by yeasts. The following questions should still be asked.

i. What is spoilage yeast?
ii. Does the food industry have adequate information to be sufficiently aware of the microbiological problems of a food commodity?
iii. What are the sources of spoilage yeasts in the food industry?
iv. Does the food industry have the appropriate zymological indicators to assess the quality of foods and to establish fair commercial contracts with retailers and wholesalers?

Concept of "Spoilage Yeast"

In many cases, microbial spoilage is not easily defined, particularly in fermented foods and beverages, where the metabolites produced contribute to the flavour, aroma, and taste of the final products. In fact, for cultural or ethnic reasons, there is little difference between what is perceived as spoilage or beneficial activity. An example of this can be found in the wine industry, where the production of 4-ethylphenol by *Brettanomyces/Dekkera* spp. in red wines is only regarded as spoilage when this secondary metabolite is present at levels higher than about 620 mg/l. At less than 400 mg/l, it contributes favourably to the complexity of wine aroma by imparting aromatic notes of spices, leather, smoke, or game, appreciated by most consumers. Above 620 mg/l, the wines are clearly substandard for some consumers, but remain pleasant for others. Those that can survive in foods but are not able to grow and, for that reason, do not affect the sensory appeal of the food may be termed adventitious or innocent; those responsible for undesirable changes are called spoilage yeasts. However, for food technologists, the concept of spoilage yeast has, in general, a stricter sense. It applies only when a particular species is able to spoil foods which have been processed and packaged according to the standards of good manufacturing practices (GMPs), in spite of the subjective character of these practices. If this is not achieved, many other adventitious yeast contaminants can develop in a product. This distinction is shown by the contamination species where widespread adventitious contaminants are not regarded as spoilers (e.g. *Rhodotorula* spp.) and dangerous spoilage species are not necessarily frequent contaminants (e.g. *Brettanomyces bruxellensis*). The observation that only a small group of about 10 species of yeast is responsible for food spoilage was made by a group of scientists and a second group added species and one genus. These "second-division" yeasts are associated with spoilage of foods, which also allow the growth of bacteria, mainly gram-negative with simple nutritional requirements. In these foods, the factor(s) favouring the growth of yeasts over bacteria cannot be identified. These yeasts should be added to those listed by which usually appear in foods preserved by extreme abiotic stress factors, mainly low water activity (a_w) and high acidity, that inhibit bacterial growth. Although the exact number of species important in food spoilage is debatable, there is no doubt they represent a small proportion of yeasts that can be found in foods.

In fermented alcoholic beverages, the concept of spoilage yeast has a more complex meaning than in non-fermented foods, where any yeast able to change food sensorial characteristics can be regarded as a "spoilage yeast." In fermented drinks or foods, yeast activity is essential during the fermenting process. Detrimental and beneficial activity must therefore be distinguished. In the wine industry where alcoholic fermentation occurs in the presence of many yeast species and bacteria (mainly lactic and acetic), it is very difficult to draw a line between beneficial fermenting activity and spoilage activity. For this reason, spoilage yeasts are rarely sought during wine fermentation, but are required during storage or ageing and during the bottling process. However, many detrimental effects of yeasts occur before fermentation (e.g. ethyl acetate produced by *Pichia anomala* or the early stage of fermentation (e.g. acetate production by *Kloeckera apiculata/Hanseniaspora uvarum*.) Thus, monitoring of spoilage yeasts should include all phases of winemaking. The concept of wine spoilage yeasts *sensu stricto* includes only those species able to affect wines that have been processed and packaged according to GMP. 39 species were listed as the most frequent wine-related yeast contaminant, but spoilage are much fewer. For instance, the yeasts were put in the following groups: (i) fermenting strains (*Saccharomyces cerevisiae*), able to re-ferment sweet-bottled wines; (ii) *Zygosaccharomyces bailii*;

(iii) film-forming yeasts (*Hansenula, Kloeckera, Pichia, Metschnikowia, Debaryomyces*); (iv) *Brettanomyces* spp.

Schizosaccharomyces pombe and *Saccharomycodes ludwigii*, although dangerous spoilers, are not regarded as common contaminants. *D. bruxellensis, Z. bailii*, and *S.cerevisiae* are considered as spoilage yeasts *sensu stricto*. However, this last species appears to be more dangerous than indicated, as some strains isolated from dry white wines seem to be a more potential spoilage yeast than *Z. bailii* due to its sorbic acid and sulphite tolerance at high ethanol levels. Furthermore, strains of *S. cerevisiae* have frequently been associated with re-fermentation of bottled "dry" red wines due to the presence of residual sugars in high ethanol (>13% v/v) wines.

The Increasing Importance of Yeasts in Spoilage

The spoilage of foods and beverages by yeasts has gained an increasing importance in food industry. The reasons for this include the use of modern technologies in food processing, the great variety of new formulations of foods and beverages, the tendency to reduce the use of preservatives, particularly those effective against yeasts (e.g. sulphur dioxide and benzoic acid), and less-severe processing. The increasing importance of yeasts in food spoilage is well illustrated by the case of wine industry. Microbial spoilage of wines may also be due to the activity of lactic and acetic acid bacteria. In fact, most traditional wine "diseases" are bacterial in origin. However, advances in wine technology and improvement in GMPs, e.g. equipment design, sanitation procedures, and use of preservatives, have led to the virtual extinction of these diseases, most of which have never been encountered by today's oenologists.

On the contrary, yeasts are now the most feared cause contaminants leading to wine spoilage. The common spoilage effects are film formation in stored wines, cloudiness or haziness, sediments, and gas production in bottled wines, and off-odours and off-tastes at all stages of wine production. Increasing quality demand by consumers also extended the range of spoilage problems or decreased the tolerance to aspects which were not formerly taken to be defects, most of which are due to yeast activity (e.g. slight haziness in bottled wines, phenolic-tainted wines). For example, *Dekkera/Brettanomyces* spp. have been well known since the beginning of the 20th century, but has only attracted the attention of wine technologists in the last decade. The production of acetic acid and "mousy" off-odours in grape juice, but their main effect–off-flavours were due to volatile phenols (4-ethylguaiacol and 4-ethylphenol). It was reported that these yeasts were the only species isolated from wines with mousy and other ill-defined off-odours. Further the production of volatile phenols by *Brettanomyces* in grape juice was also demonstrated, but lactic acid bacteria were thought to be also responsible for their production in wines. It was demonstrated by various studies that the genera *Dekkera/Brettanomyces* are the sole agents of phenolic off-flavours in wines. In addition, other species capable of producing 4-ethylphenol, with variable efficiency in grapes, grape juice, insects, and cellar equipment were also found. Among these, *Pichia guilliermondii* showed conversion rates of *p*-coumaric acid into 4-ethylphenol similar to *D. bruxellensis*, but apparently, it cannot grow in wines. Therefore, *D. bruxellensis* is today considered to be the main cause of wine spoilage, especially of fashionable premium red wines matured in oat casks, where it can be responsible for serious economic losses. Methods of assessing the presence of spoilage yeasts in food ecosystems are very limited and few improvements in last decades. There are several techniques used to show the presence of spoilage yeasts in foods. However, undoubtedly, the spread plate technique is still the most popular.

The microbiological analysis of a food sample may be compared to taking a photo of the sample, aiming to show the species and size of the yeast population. As in photography, the sharpness depends on the tools and techniques used, which are, for the plate technique (i) sampling, (ii) pretreatment techniques (maceration/ blending of the sample, dilution, and enrichment), (iii) counting techniques (culture media, incubation conditions), and (iv) identification procedures. Traditionally, in most studies of microbial ecology of foods, more attention is paid to the identification of isolated strains than to the previous steps.

ORIGIN OF SPOILAGE YEASTS IN THE WINE INDUSTRY

The wine production environment may be divided in two fundamental parts: the vineyard, which is a natural ecosystem, influenced by cultural practices, and the winery, which is the environment associated with grape fermentation, wine storage and ageing, and bottling. A deep knowledge of these two ecosystems—vineyard and the winery—is essential to establish the origin of wine spoilage yeasts, their routes of contamination, critical points of yeast infection, and their control.

Vineyard

Ecological surveys performed in vineyards and on grape surfaces during ripening are relatively few when compared to those performed on grape musts and on their spontaneous fermentation. Moreover, the majority of them used less optimal sampling, preisolation techniques, enrichment methods, isolation culture media, and incubation times, leading to an insufficient knowledge of grape microbial ecology. In general terms, the available information about the presence of microbial communities in vineyards and on grape surfaces may be summarized as follows:

1. Mature sound grapes harbour microbial populations at levels of 10^3–10^5 cfu/g, consisting mostly of yeasts and various species of lactic and acetic acid bacteria and filamentous moulds.

2. The sources of yeasts and yeast-like microorganisms include all the vine parts, as well as the soil, air, other plants, and animal vectors in the vineyard.

3. Insects are the principal vectors for the transportation of yeasts

4. Yeast colonization on grapes is influenced by the degree of ripeness of the bunch.

5. The occurrence and growth of microorganisms on the skin of the grapes is affected by the rainfall, temperature, grape variety, and application of agrochemicals.

6. Yeasts are mainly localized in areas of grape surface where some might escape in juice and are embedded in a fruit secrete; outer surface of the berries is covered by a waxy layer, which affects the adherence of microbial cells and their ability to colonize the surface.

7. Oxidative Basidiomycetous yeasts, without any ecological interest— *Sporobolomyces, Cryptococcus, Rhodotorula,* and *Filobasidium*—are mostly prevalent in the vineyard environment (soil, bark, leaves, grapes), as well as *Aureobasidium pullulans*, which seems to be a normal inhabitant of grape skin.

8. Apiculate yeasts (*Hanseniaspora* and *Kloeckera* spp.) and oxidative yeasts (mostly *Candida, Pichia,* and *Kluyveromyces* spp.) are predominant on ripe sound grapes.

9. The main wine yeast—*S. cerevisiae*—contrary to many early reports, is virtually absent from sound grapes, being present in one berry among 2016 tested, or about 1 in 1000 berries. Despite the above-mentioned statements, there are still no definitive studies on how microorganisms contaminate and colonize the grape bunch. The controversy on the origin of *S. cerevisiae* is beyond the scope of this book, but illustrates the need for more work on the subject and highlights as well the need to improve appropriate sampling and recovery techniques.

In fact, the dissemination of yeasts on the grape surface is quite variable, the microbial ecology of damaged grapes is poorly studied, or even unknown, and grape rupture is associated with the increasing occurrence of fermentative species. The above-mentioned studies have been addressed to *S. cerevisiae*, but the fact that spoilage species are also fermentative suggests that the knowledge of their dissemination may greatly improve if more attention is given to the microbiology of damaged grapes. Furthermore, as mentioned earlier, selective media and long incubation periods are essential to recover spoilage yeasts from grapes, such as *S. pombe* or *Brettanomyces* spp.

The Microbiology of Damaged and Dried Grapes

Damaged grapes may result from different causes:

1. Increase of berry volume due to rapid rainwater absorption by the wines, especially when the bunches are rather tight and the berry skin is thin;

2. Other meteorological accidents like hail and heavy rain;

3. Attack by *Drosophila* spp., honeybees, wasps, moths and birds;

4. Attacks of phytopathogenic moulds (e.g. downy and powdery mildews, noble or grey rot).

Grapes infected with powdery mildew harbour much higher microbial loads of microorganisms (yeasts, lactic and acetic acid bacteria) and volatile compounds (ethanol, ethyl acetate and acetic acid) than sound grapes. Among yeasts, significant numbers of *Dekkera* and *Kloeckera* were detected, which are probably disseminated by insects attracted to the infected grapes by the volatiles given off by ripening berries. In another work, no particular yeast contaminant species were found with powdery mildew infected grapes, but the resulting wine was scored higher in "yeasty" and 'estery" aromas, probably resulting from undesired fermentation microorganisms.

There are several types of rot, although the most frequent is that caused by the mould *Botrytis cinerea*. In particular climatic conditions, with alternating wet and dry periods, *B. cinerea* induces controlled dehydration of the grapes, leading to the well-known noble rot, which is the base for the production of some of the most famous dessert wines of the world, like Tokay Aszu' and Sauternes. In *B. cinerea*-infected grapes, the presence of *K. apiculata* and *Candida stellata* seems to be favoured when compared with yeast populations of healthy grapes. In one of the few studies on the subject, the presence of *Hanseniaspora osmophila* and a non-culturable fructophilic *Candida* population was shown, besides the expected populations of the genera *Saccharomyces, Hanseniaspora, Pichia, Metschnikowia, Kluyveromyces,* and *Candida.* In addition, *B. cinerea* frequently produces grey rot, which severely damages wine quality and causes serious economic losses. Both rots—noble and grey— change the chemical composition of grape juice dramatically. The first is essentially characterized by a significant increase in sugar concentration as a rule higher than 300 g/l, a slight increase in pH usually about 0.3 pH units, production of gluconic acid, and a significant increase in acetic acid. In grey rot, the sugar concentration is not usually increased, but there is significantly more glycerol, gluconic acid, and due to acetic acid bacteria, acetic acid. *B. cinerea* also produces various antibiotic substances in grapes.

Damage Due to Rots

Grey rot In contrast to noble rot, grey rot yields rather unbalanced wines, with weak maturation ability and with a typical mould odour, which reduces their quality. When grey rot is accompanied by growth of other moulds like *Aspergillus niger, Penicillium* sp. and *Cladosporium* sp., contaminated grapes are frequently extremely bitter and with aromatic flavours, yielding wines with phenolic and iodine odours. It is easy to conclude that besides a sudden increase in microbial load to about 10^6–10^8 cfu/g, deep alterations in yeast microbiota might occur compared with sound grapes. Surprisingly, the studies on microbial ecology of grapes spoiled by *B. cinerea* are so scarce that it is not clear if poor wine quality is a result of changes in the chemical composition of the grape caused by the mould or change in the fermentative microflora or both. Although, there is no information on the alteration of grape microbial community by grey rot; it is plausible that a significant increase of fermentative yeasts, some of which are spoilage yeasts, moulds, lactic and acetic acid bacteria, occurs.

Sour rot Grapes can also be affected by another type of rot, generally known as "sour rot," where yeasts and acetic acid bacteria appear to have a dominating role and where moulds are hardly detected. Sour rot was reported for the first time in Italy and is frequently initiated in the area near berry pedicel or at the level of skin damage. Grape sour rot is easily recognized by browning and desegregation of the internal tissues, detachment of the rotten berries from the pedicel, and a strong ethyl acetate smell.

Another component of the system is *Drosophila*, although their role in the process has not been studied in depth. The yeast species most frequently reported as actively proliferating in rotten berries are *H. uvarum* and its anamorph *K. apiculata, C. stellata, Metshnikowia pulcherrima, Candida krusei, Pichia membranifaciens, Saccharomycopsis vini, Saccharomycopsis crataegensis,* and *Candida steatolytica*. Occasionally, *Zygosaccharomyces* spp. can also be present in high densities, together with other spoilage yeast species, like *Brettanomyces* spices. The contribution of acetic acid bacteria to this disease seems to be well established, and several studies on rotten berries confirm the presence of levels of acetic acid as high as 40 g/l, of ethyl acetate, and of *Gluconobacter* spp. and, less commonly, of *Acetobacter* species. However, in other studies, acetic acid bacteria were rarely or never recovered, suggesting that acetic acid and ethyl acetate may result from the yeast activity. Once more, surprisingly, there are no studies that cast light on the microbial ecology of this type of sour rot. Besides, most studies were performed without using preisolation techniques, selective/differential media or long incubation periods that favour the recovery of slow-growing and minority yeasts. Thus, it is legitimate to think that for many of the most important wine spoilage species, e.g. *Dekkera/Brettanomyces* spp., the main entry to the winery is grapes affected by sour rot, which cannot be eliminated during harvesting. Another example of unbalanced grapes that are used in winemaking with some frequency are those affected by mealy bugs (*Pseudococcus* spp.) excreting honeydew that may not damage grape skin, but a high concentration of sugar is accumulated in the surface. In some regions of Mediterranean countries, including Portugal, this disease can affect more than 10–20% of the crop, so that their yeast population might have an impact on wine quality. As far as we are aware, the microbial ecology of grapes with honeydew has never been investigated, although the typical black colour due to the growth of filamentous fungi on the grape surface and the abundant presence of ants are well known. Bearing in mind that honeydew is essentially plant sap with high sugar content, it is conceivable that damaged grapes are a habitat favourable for yeast growth, especially osmophilic and osmotolerant species, such as *Zygosaccharomyces* species. Similar conditions could occur in dried undamaged grapes used in the production of certain table (e.g. Amarones) or dessert (e.g. Muscat) wines, where the initial sugar concentration may suffer relative increases of 30–40%. However, very limited valid information is available. As a consequence of the above description, many gaps exist in the knowledge of grape microbial ecology, particularly concerning wine spoilage yeasts.

Winery

Essentially, the microorganisms in the winery come from the grapes and vectors, among which, *Drosophila* are likely to be the most important. Conditions enabling colonization of wines and contact surfaces depend on the stringency of GMP. The intrinsic properties of wine are of major importance in influencing the evolution of microbial communities. Considering the winery environment, two sections are relevant: (i) winemaking and bulk wine storage and (ii) the bottling line.

Winemaking and wine storage Studies of microbial ecology in cellars are relatively scarce compared with those of grapes and grape juice. However, all the results obtained seem to confirm that the yeast population of wineries is quite different from that of grapes, particularly due to the high proportion of *S. cerevisiae*. The association between the winery and this species is so close that it is called "the first domesticated microorganism" and claim that it is a result of yeast species evolution in this environment. Besides *S. cerevisiae*, other species frequently recovered from wine or grape juice contact surfaces—tank walls, crushers, presses, floor, winery walls, pipes, etc.—are *P. anomala*, *P. membranifaciens*, *Candida* spp., *Cryptococcus* spp., and more rarely *Rhodotorula* spp., *A. pullulans*, *Trichosporon cutaneum*, *Debaryomyces hansenii*, *K. apiculata*, *M. pulcherrima*, and *T. rosei*. Some of these species, in spite of being common contaminants, are obligate aerobes (e.g. *Rhodotorula* spp., *Cryptococcus* spp., *D. hansenii* and *A. pullulans*) and therefore have little or no ability to grow in or spoil wines. The species able to grow abundantly in wine, with fully aerobic or weakly fermentative metabolism (e.g. *P. membranifaciens, P. anomala,* and *Candida* spp.) are known for film formation on the surface of bulk wines in unfilled containers and with sulphite levels insufficient to prevent their growth. Given their oxidative metabolism and high growth rate, at winery temperature, they rapidly colonize surfaces contaminated with wine residues, being regarded as indicators of hygiene and of the stringency in avoiding wine contact with air. When measures are not taken to prevent their growth, they may affect wine and favour growth of acetic acid bacteria with much more serious consequences.

However, adequate GMP, adequate levels of molecular sulphite, efficient wine protection from air contact with nitrogen, and low storage temperatures (8–12°C) allow high control of these yeast.

Occasionally, they can be tolerant to molecular sulphite levels. *P. anomala, M. pulcherrima* and *H. uvarum (K. apiculata)* are known for producing high levels of ethyl acetate and acetic acid, before and during initial fermentation steps, leading to serious wine deterioration. It seems that ethyl acetate is not produced by *K. apiculata* and *Candida pulcherrima* during fermentation by *S. cerevisiae* . Although these species are common winery contaminants, their activity is especially dangerous when associated with damaged berries, which encourage their growth, leading to high initial populations at the beginning of fermentation. High juice settling temperatures with low protective levels of sulphite can also lead to massive growth, originating from contaminated grapes or poorly sanitized equipments. Surprisingly, the yeast species regarded as the most dangerous to wines, i.e., *Dekkera/Brettanomyces* spp., *Z. bailii* and *S. ludwigii* are seldom detected in yeast studies performed in wineries. Although classical studies on *Brettanomyces* spp. *Z. bailii* and *S. ludwigii* have demonstrated that they may be winery contaminants, and most results from literature suggest that their prevalence is low. In support of this, it was shown that the presence of *Dekkera/Brettanomyces* in grapes and at various sites of grape crushing processing lines by using direct PCR techniques. *D. bruxellensis* was also recovered from air samples of crush, tank, barrel, and bottling line areas using BSM medium (Millipore) followed by a filter-based chemiluminescent *in situ* hybridization technique.

From the technological point of view, the main question is to know the factors and under which conditions they enable slow-growing yeasts like *Dekkera/Brettanomyces* spp. and *Zygosaccharomyces* spp. to become competitive, attain high contamination levels, and cause serious wine defects. Only when this information is available, it can be possible to establish adequate control measures using appropriate culture media and a method developed for the recovery of *Brettanomyces* species from materials heavily contaminated with other species, reported that these yeasts are common contaminants in the winery and its equipment. However, these yeasts are not recovered from husks, pomaces, or fresh grapes, suggesting that the infection of wines and musts by *Brettanomyces* species is due to contamination spreading from foci of infection within the winery. Much later, oak barrels were identified as an ecological niche for *Dekkera/Brettanomyces* spp. which become more dangerous with repeated use. This suggests that barrel sanitation and sulphite utilization (sulphur burning in empty barrels) is not enough to eliminate *Dekkera/Brettanomyces* spp. which develop during the lifetime of the barrel. It was reported that treatment with hot water and steam is not enough to eliminate yeasts and moulds entrapped in barrel staves. It is now generally accepted that control of *Dekkera/Brettanomyces* spp. cannot be achieved by efficient sanitation of all cellar equipment, but demands much more stringent microbiological control and judicious utilisation of sulphite or dimethyldicarbonate (DMDC). However, the primary source of these yeasts remains obscure. One research in a laboratory suggests that rotten grapes, *Drosophila* spp. particularly those frequenting piles of husks, lees and grape leftovers, and wine residues on equipment are foci where *Dekkera/Brettanomyces* spp. can be found by using an appropriate culture medium and MPN enumeration technique. It is then admissible that rotten grapes are the main entry source of these yeasts in wineries and that lees and husks' leftovers are important infection sources, frequently visited by Drosophila flies which carry them into the winery.

Yeasts of the genus *Zygosaccharomyces*, and particularly *Z. bailii*, are very rare in sound grapes and are not regarded as common winery contaminants. However, its presence is well known in wineries processing sweet or sparkling wines using concentrated or sulphited grape juice. A similar situation is observed in other food and beverage industries using processed raw materials such as fruit juices, concentrated juice, glucose syrups, flavouring compounds, and colouring agents. The fact that these yeasts are extremely resistant to preservatives, particularly *Z. bailii*, *Zygosaccharomyces bisporus*, and *Zygosaccharomyces rouxii*, means that addition of high, but sublethal, doses increases their competitiveness and makes them seriously dangerous. Hence, it is regarded as a good manufacturing practice to add the preservative to sweet wines just prior to bottling and to limit the circulation of concentrated grape juice to specific pipes and pumps. This is frequently forgotten in most wineries. It would also be highly desirable to have a strict microbiological control of concentrated grape juice, which is not common in wineries. *S. ludwigii* is another yeast species that may cause serious problems in wines, although it is not regarded as a typical contaminant of winery microflora. Its remarkable tolerance to sulphite makes it a frequent isolate in wineries where high sulphite doses or sulphited grape juices are used. This species was defined as the "winemaker's nightmare" because of the infection of bulk wine which is notoriously difficult to eradicate from a winery. Although it is known that mummified fruits are natural habitats of *Z. bailii* , that *Z. bisporus* can be isolated from fruit tree exudates, and that *S. ludwigii* is present in slime fluxes of *Quercus* spp. The winery contamination routes and vectors for these yeasts are barely known, justifying further studies on this subject.

Bottling line Wine bottling is a critical operation since, with the exception of hot bottling, it is the final contamination source before wine is released to the market. In most dry red wines, yeast contamination during bottling is not serious . However, for wines with residual sugar and for some dry white wines, it can be very serious, being responsible for a major part of the microbiological problems in bottled wines. When sweet wines are processed, either with natural sweetener stabilized with sulphite and sorbate, or alternatively, with concentrated grape juice, the contaminating flora of the bottling line is usually dominated by species that are resistant to chemical preservatives and low a_w, namely, *Z. bailii*, *S. cerevisiae* and *S. ludwigii*. This situation leads to the conclusion that high sulphite levels and sorbate, when used in sublethal doses, play an essential role in favouring the highly resistant yeast. This is mainly observed in bottling lines. The same applies to the use of concentrated grape juice, a well-known source of *Zygosaccharomyces* spp. and other dangerous species.

Critical points Some authors have studied the critical points of bottling lines. The outlet side of the sterilizing filter, the filler, in particular, the bell rubbers and rubber spacers, the corker, in particular, the bells/cork jaws and cork hopper, the bottle sterilizer, the bottle mouth, and the air inside the bottling room are important critical points. Furthermore, the importance of each point is strongly dependent on suitably designed equipment. Packaging materials such as bottles, corks, and rip-caps are generally not significant contamination sources, because they are frequently infected by fungi, spore-forming bacteria, and adventitious yeasts, which do not survive in wine. However, they can be important sources of spoilage yeasts when wine is improperly stored for long periods in a humid and contaminated winery environment. However, observed cork contamination has been observed with dangerous species prior to winery entrance in two separate cases: one with *S. cerevisiae*, due to contamination of the silicone used in cork surface treatment, and the other with *S. ludwigii*, resulting from an inadequate cork routine treatment before cork packaging with sublethal sulphite doses.

Good quality of bottling equipment (Figure 19.1) is also essential to prevent yeast growth.When oxygen is introduced in wines during bottling, it stimulates the growth of *Z. bailii*. In recent years, new bottling equipment, revision of bottling line sanitation programs and overall plant hygiene standards, and the better implementation of HACCP systems in wineries have contributed to a significant improvement of the microbiological quality of wine bottling. However, these improvements have not sufficed to reduce the levels of preservatives used in sweet and dry white wines sterilized by filtration prior to bottling. This could be due to technological and microbiological limitations. The former are related to the design of plant layout and human failures, mainly, the incorrect execution of sanitation programs and the result of cross-contamination. A clear example of this is steam disinfection of the filler, which, is a classical case of a frequently incorrect procedure. In fact, if after steam application sterile air is not injected into the filler during the cooling period, a negative pressure will be formed inside the filler, leading to ingress of air contaminated with potential spoilage yeasts. The limitations of a microbial nature are concerned, once more, with the lack of efficient tools to examine and interpret the contamination of bottling lines on line.

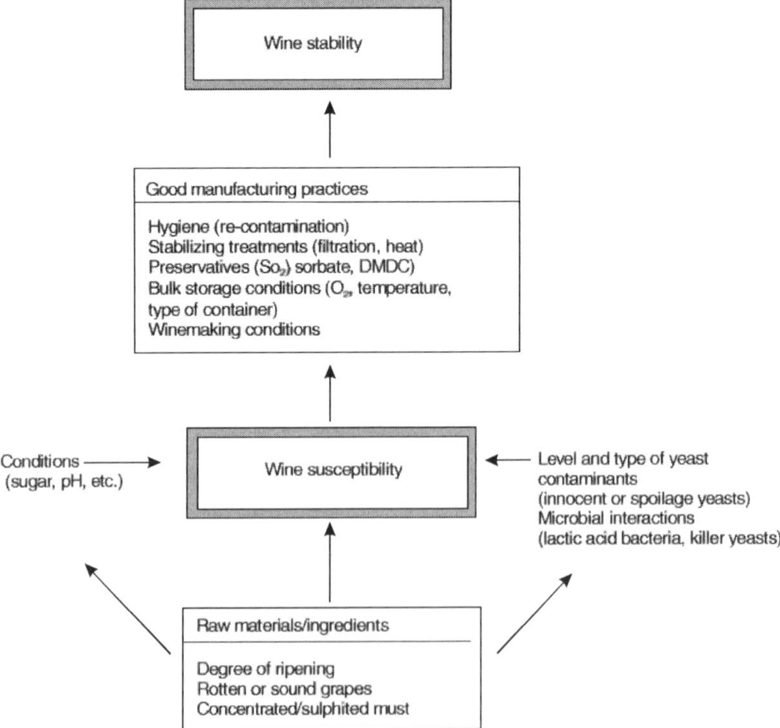

Figure 19.1 Factors affecting wine stability

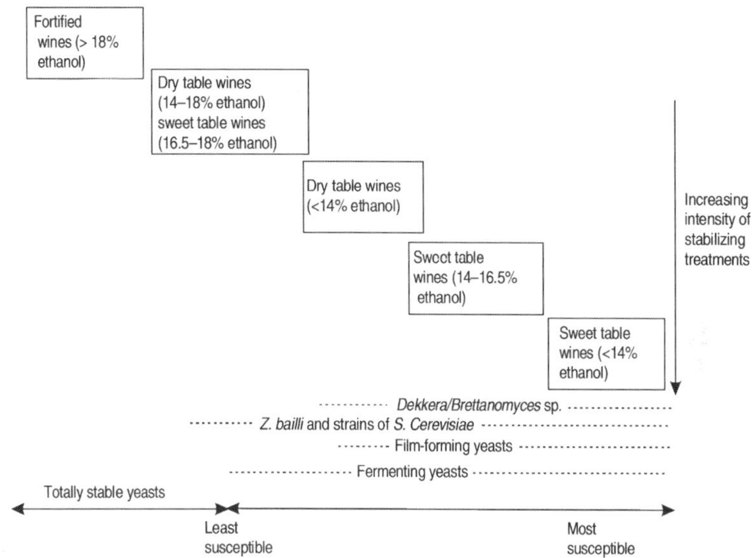

Figure 19.2 The susceptibility of wines to the yeast colonization. (Note : dry table wines have less than 2 g/l reducing sugars.)

SUMMARY

From the several aspects covered in this chapter, the main conclusions may be summarized as follows:

1. Knowledge of the microbial ecology of grapes, particularly damaged grapes, wineries and vectors has many gaps that do not enable a full understanding of the origin and dissemination of spoilage yeasts in wines (Figure 19.2);

   - some yeast species are of great concern in wine industry, namely, *Dekkera/Brettanomyces* spp. in red wines matured in oak barrels and in wines originating from poor sanitary quality grapes
   - microbiological criteria used in industry are, as a rule, old-fashioned, arbitrary, and established without scientific background; scientific research has not yet been able to produce the knowledge needed to solve the problems caused by spoilage yeasts

2. Considering the currently available methods, the wine industry cannot implement efficient and appropriate HACCP systems for spoilage yeasts. Therefore, developments of future research should be aimed to:

   - provide a better knowledge of the yeast microflora of damaged and dried grapes

- elucidate the role of insects as spoilage yeast colonizers of grapes and wineries
- transfer the rapid molecular typing technique from research laboratories to the industry laboratories
- quantify wine susceptibility to yeast colonization
- improve predictive models of wine spoilage

In future, measures should be implemented aiming to improve the wine spoilage risk management:

i. to avoid the dissemination of spoilage yeasts in the winery

ii. to apply adequate zymological control to each type of wine

iii. to standardize microbiological criteria, namely, sampling by attributes, using standard analytical methods and appropriate specifications.

All proposed measures could be much more easily implemented if researchers and industry worked more closely in the future.

1. Discuss the concept of spoilage yeasts.
2. Discuss the microbiology of damaged and dried grapes.

20

SPECIFIC SPOILAGE
ORGANISMS IN BREWERIES

INTRODUCTION

The gram-positive bacteria are generally regarded as the most hazardous beer spoilage organisms in modern breweries, especially (i) the lactobacilli: *L. brevis, L. lindneri, L. curvatus, L. casei, L. buchneri, L. coryneformis, L. plantarum, L. brevisimilis, L. malefermentans* and *L. parabuchneri* and (ii) the pediococci: *P. damnosus, P. inopinatus* and *P. dextrinicus. Micrococcus kristinae* is the only species within the micrococci relevant to brewing. The gram-negative strictly anaerobic bacteria are apparently increasing in importance and include *Pectinatus cerevisiphilis, Pectinatus frisingensis* and *Selenomonas lacticifex,* reported as obligate beer spoilage organisms; *Zymophilus raffinosivorans* as a potential beer spoilage organism; *Megasphaera cerevisiae* is an obligate spoilage organism of low-alcohol beer and *Zymomonas mobilis* as capable of spoiling primed beer. With improved process technology the importance of aerobic bacteria has decreased and the same applies for the gram-negative aerobic bacteria *Hafnia protea* and *Enterobacter cloacae* which are capable of surviving beer fermentation.

Beer spoilage organisms include several so-called wild yeasts, of which *Saccharomyces* species are generally considered the most important. Even though the detection of beer spoilage organisms by cultivation in laboratory media does not always provide the specificity and the sensitivity required, the use of selective media and incubation conditions still appears to be the method preferred by breweries. The media used depends on the type of sample, the specificity required and, for detection of wild yeasts, to some extent, the characteristics of the culture yeast. Among the media reported so far, no single medium can be used to detect all members within a group of specific beer spoilage organisms and further works on the development of improved substrates are required both for bacteria and wild yeasts.

In a brewery, specific spoilage organisms may be defined as any organism which is not deliberately introduced and which is able to survive and proliferate in the environment, i.e., in wort, fermenting wort, beer after filtration or in packaged beer. Many bacterial strains have been shown to retain their viability for extended periods in beer but as long as they are not able to grow in the product they are not harmful and should not be considered as beer spoilage organisms. With the improved technology introduced in modern breweries the importance of various specific spoilage organisms has changed. Well-known spoilage organisms like the aerobic acetic acid bacteria, i.e., *Gluconobacter oxydans* and *Acetobacter* spp. appear no longer to be a problem. Improvements in beer handling and bottling technology have resulted in significant reduction of the oxygen content during the process and in packaged beer which has permitted the growth of strictly anaerobic microorganisms. Typical examples are *Pectinatus* spp. and *Megasphaera cerevisiae* which within the past few decades have caused increasing problems with spoilage of packaged beer.

As mentioned, oxygen content is a major factor in controlling the microflora capable of growth during beer production and storage. However, several other factors are also important for the spoilage potential of beer and include the pH (3.8–4.7), the concentration of hop bitters (approximately 17–55 mg iso-a-acids/l), ethanol (0–8% w/v), CO_2, (approximately 0.5% w/w), SO_2, (approximately 5–30 mg/l), organic acids, acetaldehyde and other metabolites as well as the nutrient contents and storage temperature. The sensitivity of different brands of beer to spoilage by hop-resistant bacteria belonging to the genera *Lactobacillus* and *Pediococcus* has been investigated by predicting the spoilage potential of these bacteria for beer from information on the level of undissociated SO_2, undissociated hop bitter acids, polyphenols, maltotriose and free amino nitrogen as well as colour intensity. The detection of beer spoilage organisms is difficult as they are often present in low numbers, e.g. for wild yeasts the methods used should be capable of detection of one spoilage organism per 10^6 culture yeast cells. Also spoilage organisms are often sub-lethally damaged due to the environmental conditions. In addition some organisms may be very fastidious concerning their conditions of growth (*Lactobacillus* spp. and *Pediococcus* spp.) or may be adapted to the particular product or environment and very reluctant to multiply in other environments including highly nutritious laboratory media. For the detection of wild yeasts the fact that contaminants belonging to the genus *Saccharomyces* are often biochemically and physiologically very similar to the culture yeast is a major problem.

Within the past twenty years, considerable interest has been shown in the development of rapid methods for detection of beer spoilage organisms. These methods include bioluminescence techniques, direct epi-fluorescence filter techniques (DEFT), immunoassays, use of automated turbidimetry, measurement of impedance or conductance, flow cytometry as well as a number of methods involving DNA technologies such as PCR and DNA hybridization techniques. With the exception of the bioluminescence technique which has been used successfully for control of cleaning and disinfection as well as for last rinse water from cleaning in place (CIP), these alternative methods seem not to have found their way into the breweries as they often lack the speed, sensitivity and specificity required or include the use of advanced, expensive equipment and reagents. For these reasons the use of selective media and incubation conditions is still the method preferred in most breweries.

However, due to the diversity of the microflora, several media must be used inorder to ensure the detection of both gram-positive and gram-negative spoilage bacteria as well as *Saccharomyces* and non-*Saccharomyces* wild yeasts. This chapter presents a summary of the specific spoilage organisms important to modern breweries and the selective media and incubation conditions relevant to the detection of these organisms.

SPECIFIC SPOILAGE ORGANISMS

Bacteria

Gram-positive bacteria   The bacteria generally regarded as most hazardous for modern breweries are the gram-positive bacteria belonging to the genera *Lactobacillus* and *Pediococcus*. Among the lactobacilli the most important spoilage organisms according to the brewing literature are *Lactobacillus brevis, L. lindneri, L. curvatus, L. casei, L. buchneri, L. coryneformis* and *L. plantarum*. In addition, the following potential beer spoilage species namely, *L. brevisimilis, L. malefermentans* and *L. parabuchneri. L. delbrueckii, L. fermentum* and *L. fructivorans* have been reported to occur in beer but their spoilage potential is low. Not all lactobacilli reported seem to be recognized as valid species according to recent reviews of the genus and the brewing literature in general appears taxonomically out of date.

Among the pediococci only *Pediococcus damnosus, P. inopinatus* and to some extent *P. dextrinicus* are of importance for spoilage of beer. However, growth of *P. inopinatus* and *P. dextrinicus* is only possible above pH 4.2 and at low ethanol and hop bitter concentrations. It also appears that only some strains of the above species are capable of growth in beer. *P. acidilactici* and *P. pentosaceus* are found on malt and can grow during the early stages of wort production as long as the temperature is below 50°C and hops have not been added, but they have never been reported to cause any defect in the beer produced.

The genus *Micrococcus* includes one species, *M. kristinae* relevant to breweries. It is very sensitive to the concentration of ethanol and hop bitters in beer and only capable of growth above pH 4.5. Unusually for the micrococci, *M. kristinae* is capable of anaerobic growth. The spoilage caused by lactic acid bacteria appears to be dependent on the composition of the beer produced and to be most dangerous during conditioning of beer and in packaged products. Gram-positive bacteria are generally linked to haze or rope formation and acidification. In particular *L. casei* and the *Pediococcus* spp. produce extensive amounts of diacetyl. Honey-like flavours and extended fermentation time have been linked with infections caused by *Pediococcus* spp. and *L. brevis* has been shown to cause super-attenuation due to its ability to ferment dextrins and starch. A fruity atypical aroma has been reported for *M. kristinae* even when present at low numbers. As indicated above, the spoilage potential of lactic acid bacteria is to a large extent dependent on their resistance to hop bitters such as *trans*-isohumulone. The effect of hop bitters can be both bacteriostatic and bactericidal depending on their concentrations and the exposure time, however, the adaptation of cells to hop bitters has been questioned. The antimicrobial activity is related to the undissociated form of the hop bitter acids and thereby pH dependent, e.g. the antibacterial activity of *trans*-isohumulone towards *L. brevis* decreases 800-fold if the pH is raised from 4.0 to 7.0. It appears that hop bitter acids act as ionophores, carrying ions including protons across the plasma membrane and reducing the intracellular pH of the cell. The exact mechanism responsible for the increased resistance of some strains appears not to be known but might be related to the plasma membrane.

Gram-negative bacteria The gram-negative beer spoilage organisms include a number of species belonging to various genera. Among these the strictly anaerobic bacteria *Pectinatus cerevisiphilus, P. frisingensis* and *Selenomonas lacticifex* have been reported as obligate beer spoilage organisms. Within the genus *Zymophilus* which is phylogenetically close to the genus *Pectinatus*, two species have been isolated from brewery samples: *Z. raffinosivorans* and *Z. paucivorans* although only the first mentioned has been reported as a beer spoilage organism. The beer spoilage organisms belonging to the genera *Pectinatus* and *Zymophilus* have been reported to grow in beer at pH above 4.3–4.6 and at ethanol concentrations below 5% (w/v). These strictly anaerobic bacteria have become more important with the increasing production of non-pasteurized and flash-pasteurized beer together with improved bottling technology which results in significantly reduced oxygen content in packaged beers.

The spoilage caused by these organisms includes the production of propionic, acetic, and succinic acids, methyl mercaptan, dimethyl sulphide and hydrogen sulphide as well as turbidity. Another strictly anaerobic gram-negative beer spoilage organism is the coccus *Megasphaera cerevisiae*. It is sensitive to low pH (< 4.1) and growth is inhibited in beer with an ethanol content above 2.8% (w/v), however, growth has been observed in beer with up to 3.8% (w/v) ethanol. As for *Pectinatus cerevisiphilus* it has proved to be quite tolerant towards hop bitters and it may cause spoilage of low-alcohol beer by production of butyric and other fatty acids as well as hydrogen sulphide and development of turbidity.

Of importance for primed (added sugars) beer is the anaerobic but oxygen-tolerant gram-negative bacterium *Zymomonas mobilis* which is resistant to hop bitters and able to grow at pH above 3.4 and ethanol concentrations lower than 10% (w/v). It is not able to ferment maltose and maltotriose but it ferments glucose and fructose, and some strains ferment sucrose. Spoilage of primed beer is primarily caused by the production of high levels of acetaldehyde and hydrogen sulphide.

Zymomonas spp. have not been reported in lager breweries probably due to their stringent carbohydrate requirements. The gram-negative, aerobic acetic acid bacteria, especially *Gluconobacter oxydans. Acetobacter aceti* and *A. pasteurianus* have during the history of brewing been paid a great deal of attention. They are able to convert ethanol into acetic acid and thereby change the flavour of the beer significantly, resulting in vinegary off-flavours. However, being aerobes they are not considered a major problem in modern breweries. The same applies to the Enterobacteriaceae, among which a variety of species have been found in breweries including *Hafnia protea, Hafnia alvei, Klebsiella pneumoniae, Enterobacter cloacae, Enterobacter aerogenes* and *Enterobacter agglomerans,* now recognized as *Rahnella aquatilis*. These members of the family Enterobacteriaceae are considered as wort spoilers as they do not multiply in beer. The two species *H. protea* and *E. cloacae* can survive the fermentation process and since they cosediment together with the culture yeast they may be passed on to subsequent fermentations. Comparing the survival of the two species during yeast storage (10°C) *H. protea* has been shown to have a significantly higher resistance than *E. cloacae*. They can cause beer spoilage by production of fusel alcohols, 2,3–butanediol, dimethylsulphide (DMS) and dimethyl disulphide which are transferred to the final beer. Table 20.1 lists the bacteria mentioned in the brewing literature as capable of growth in beer.

Table 20.1   Bacteria capable of growth in beer

Gram-positive	
Lactobacillus spp.	*Pediococcus* spp.
L. brevis	P. damnosus
L. lindneri	P. inopinatus[a]
L. curvatus	P. dextrinicus[a]
L. casei	
L. buchneri	Micrococcus spp.

(*Contd.*)

Table 20.1   (Continued)

Gram-positive	
L. coryneformis	*M. kristinae*[a]
L. plantarum	
L. brevisimilis	
L. malefermentans	
L. parabuchneri	
Gram-negative	
Pectinatus spp.	*Megasphaera* spp.
P. cerevisiphilus	*M. cerevisiae*[b]
P. frisingensis	
Selenomonas spp.	*Zymomonas* spp.
S. lacticifex	*Z. mobilis*[c]
Zymophilus spp.	
Z. raffinosivorans	

[a] Restricted by high ethanol and hop bitter concentrations
[b] Grows only in low-alcohol beer.
[c] Grows only in primed beers

Yeasts

The diversity of wild yeasts in terms of beer spoilage means that no general description can be given, however, wild yeasts are generally divided into *Saccharomyces* and non-*Saccharomyces* wild yeasts. Often the most severe infections will be caused by *Saccharomyces* spp. which, once isolated, can often be distinguished from lager yeasts by cell morphology and spore formation. Among the *Saccharomyces* wild yeasts, most isolates belong to *S. cerevisiae* with a predominance of strains previously described as *S. diastaticus*, *S. pastorianus*, *S. ellipsoideum* and *S. willianus*. Infections with these yeasts typically cause phenolic off-flavours and super-attenuation of the final beer. The production of phenolic off-flavours is due to the ability of these wild yeasts to decarboxylate different phenolic acids such as ferulic and *trans*-cinnamic acids, while the super-attenuation is due to the production and secretion of glycoamylases with starch debranching activity which enables the wild yeasts to use dextrins not normally fermented by the culture yeast.

The most important non-*Saccharomyces* wild yeasts are *Pichia membranefaciens* and *Hansenula anomala* as well as a number of species belonging to the genera *Torulopsis*, *Schizosaccharomyces*, *Brettanomyces*, *Kloeckera* and *Candida*. Among the *Candida* sp., *C. mycoderma* and *C. krusei* have been reported as capable of beer spoilage. The non-*Saccharomyces* wild yeasts will cause various types of spoilage, e.g. *Pichia membranefaciens* is known to produce film, haze and off-flavours and *Torulopsis* spp. are known to cause super-attenuation. Spoilage caused by wild yeasts belonging to the genera *Pichia*, *Hansenula* and *Debaryomyces* is commonly associated with aerobic conditions even though the yeast species are capable of anaerobic growth.

SUMMARY

Thus in modern breweries, the risk of microbial beer spoilage is associated with growth of specific strains of *Lactobacillus* spp. *Pediococcus* spp. as well as a group of strictly anaerobic gram-negative bacteria which is apparently gaining increasing significance due to improved beer handling and bottling techniques which significantly reduce the oxygen content of beer. Another important factor is the increased production of beers with low alcohol content. In addition, wild yeasts, in particular *Saccharomyces* spp. still constitute a significant risk.

The ability to grow in beer varies between strains of the same species of specific spoilage organisms. Very low levels of infections may eventually result in growth and spoilage. Further, microbes in beer seem to be specifically adapted to this particular environment and are sometimes very difficult to grow in other environments including laboratory media. Microorganisms in beer and brewery samples are often damaged by the hostile environment and difficult to detect even in enriched media.

1. Discuss the specific spoilage microbes capable of growth in beer.

21

MICROBIAL SPOILAGE OF CEREAL AND CEREAL PRODUCTS

INTRODUCTION

The cereal products are often contaminated with soil, air (dust), insects as wells as the natural microflora of the harvested grains. Freshly harvested grains are shown to possess a bacterial count of 10^3–10^6 bacterial cells/g and over 10,000 mould spores/g of grain.

The common bacterial species are from Pseudomonadaceae, Micrococcaceae, Lactobacillaceae and Bacillaceae. Processing methods like washing and milling (microbes in the outer portion of the grain are removed) remove some microbes. Apart from these processing methods, grain bleaching also reduces the number of microbes.

SPOILAGE OF CEREALS

Grains

The spoilage of grains is accompanied with the presence of moulds. The moulds attacking the grains are of two types:

Field fungi and storage fungi Field fungi are well adapted to rapid changing conditions on the surfaces of grains. It has been found that the field fungi commonly found are of the genus, *Cladosporium*, *Alternaria*, etc. They survive hot conditions and desiccation and apparently need a high a_w to survive. On the other hand the storage fungi are well adapted to the more constant conditions of cereals in storage. They generally grow at low a_w. Examples of some storage fungi are *Penicillium*, *Aspergillus*, *Fusarium* and certain xerophilic moulds like *Eurotium* spp. and *Aspergillus restrictus* grow very slowly at a very low a_w (0.71).

Once they start growing and metabolizing, they produce water of respiration and result in rise in a_w allowing more rapid growth of certain mesophilic moulds to germinate and grow. This kind of sequential growth of moulds shows a decrease in germinability of the grain. This is followed by discolouration, production of mould metabolites including mycotoxins, increase in temperature (self heating), production of musty odours, caking and rapid increase in water activity leading finally to the complete decay with the growth of a wide range of microorganisms.

If there is accidental addition of water to the stored cereal grains, the wet mash of grains first undergo acid fermentation with lactic acid bacteria and coliform bacteria normally present on the plant surfaces. Then follows the alcoholic fermentation by yeasts as soon as the acidity has increased enough to favour them. Finally, moulds and film yeasts grow on the top surface. If acetic acid bacteria are present, they will oxidize the alcohol to acetic acid, thus inhibiting the growth of moulds.

Pulse seeds rich in oil, such as groundnuts, have a much lower water content at a particular water activity than cereals, thus groundnuts with a 7.2% water content have a water activity of about 0.65–0.7 at 25°C. Apart from the problem of mycotoxin formation in moulded oilseeds, several mould species have strong lipolytic activity leading to the contamination of the extracted oils with free fatty acids which may in turn undergo oxidation to form products contributing to rancidity. The most important lipolytic moulds are species of *Aspergillus*, such as *A. niger* and *A. tamarii*, *Penicillium* and *Paecilomyces*, while at higher water activities species of *Rhizopus* may also be important.

Flours

In whole wheat flours, spoilage will be similar to grains whereas washing grains, milling, sifting of flours usually reduce the content of microbes. Normally in flours the a_w is very less thus unfavourable to the microbial growth but slight moisturing may permit the growth of moulds and bacteria.

Types of spoilage of flour differs with the different contaminating flora.

If acid forming bacteria are present, an acid fermentation begins, followed by alcoholic fermentation by yeasts and acetic acid production by *Acetobacter* species. Other bacteria include coliforms, species of *Bacillus* and micrococci.

SPOILAGE OF CEREAL PRODUCTS

Bread

Bread is an unstable, elastic, solid foam, the solid part of which contains a continuous phase composed in part of an elastic network of cross-linked gluten molecules and in part of leached starch polymer molecules, primarily amylose, both uncomplexed and complexed with polar lipid molecules, and a discontinuous phase of entrapped, gelatinized, swollen, deformed (wheat) starch granules. *Staling* is a term which indicates "decreasing consumer acceptance of bakery products caused by changes in crumb other than those resulting from the action of spoilage organisms". From the composition of bread, one can predict the typical spoilage pattern that could be found, i.e., mouldiness and ropiness. Of the two, mouldiness is the most common. Moulds usually come from air during cooling, handling or wrapping. The moulds usually initiate growth in the crease of the loaf and between the slices. The most common bread mould is *Rhizopus stolonifer*. Other forms are *R. expansum, Penicillium stoloniferum, A. niger, Monilia sitophila, Mucor* and *Geotrichum* among the rest.

Heavy contamination is evident after baking, slicing, wrapping and storage (warm humid place). Ropiness is common in home-baked bread, during hot weather. It is chiefly due to *Bacillus subtilis* or *B. licheniformis* (formerly *B. mesentericus*). The spores of these bacteria can withstand the temperature of baking (100°C), germinate and grow within the loaves. Ropy condition is a result of encapsulation of the bacillus along with hydrolysis of flour protein (gluten) by proteinases. Hydrolysis of starch by the amylases to sugars further encourage rope formation. The area of ropiness is coloured yellow to brown, and soft and sticky to touch. The slimy material can be drawn into long threads when the bread slice is pulled apart. Odour is unfavourable.

Apart from these, one can see a condition called "red bread" due to the growth of pigmented bacteria—*Serratia marcescens* or the mould *Monilia sitophila*. Sometimes the crumb portion may become red due to the growth of a mould *Geotricum aurantiacum*.

"Chalky bread" is a condition where there are white chalk-like spots in bread principally caused due to the yeast *Endomycopsis fibuligera*.

REVIEW QUESTIONS

1. Discuss about field fungi and storage fungi.
2. Brief out on the spoilage of cereal flours.

22

SPOILAGE OF MILK AND DAIRY PRODUCTS

INTRODUCTION

Being both highly perishable and nutritious, milk has since pre-historic times been subject to a variety of preservation treatments. The various types of spoilage of milk are listed in Table 22.1.

Table 22.1 Types of milk spoilage

Defects caused in milk	Associated microbes	Metabolic product
Bitter	Psychrotrophic bacteria (*Bacillus cereus*)	Bitter peptides
Rancid	Psychrotrophic bacteria	Free fatty acids
Fruity	Psychrotrophic bacteria	Ethyl esters
Coagulation	*Bacillus* sp.	Casein destabilization
Sour	Lactic acid bacteria	Lactic, acetic acids
Malty	Lactic acid bacteria	3-Methylbutanol
Ropy	Lactic acid bacteria	Exopolysaccharides

MICROBES INVOLVED IN SPOILAGE

Psychrotrophic Bacteria in Milk

Psychrotrophic bacteria which spoil raw and pasteurized milk are primarily aerobic gram-negative rods which belong to the family Pseudomonadaceae, with occasional representatives from the family Neisseriaceae and the genera *Flavobacterium* and *Alcaligenes*. It is typical that 65–70% of raw milk psychrotrophic isolates are in the genus *Pseudomonas*. The psychrotrophic spoilage microflora of milk is generally proteolytic, with many isolates able to produce extracellular lipases, phospholipase, and other hydrolytic enzymes but unable to utilize lactose. The bacterium most often associated with flavour defects in refrigerated milk is *P. fluorescens*, with *P. fragi*, *P. putida* and *P. aeruginosa* also commonly encountered.

Soil, water, animals and plant material constitute the natural habitat of psychrotrophic bacteria found in milk. Psychrotrophic bacteria isolated from water are often very active producers of extracellular enzymes and grow rapidly at refrigeration temperatures. As a result, water is an important source of milk spoilage bacteria. The teat and udder area of the cow can harbour high levels of psychrotrophic bacteria, even after washing and sanitizing. These psychrotrophs probably originate from soil.

Milking equipment, utensils, and storage tanks are the major sources of psychrotrophic contamination of raw milk. Milk residues on unclean equipment provide a growth niche for psychrotrophic bacteria which enter milking machines, pipelines, and holding tanks with water rinses or milk.

Pasteurized milk products become contaminated with psychrotrophic bacteria by exposure to contaminated equipment or air. The filling equipment is most often the source of psychrotrophs in packaged milk. Although the levels of psychrotrophic bacteria in air are generally quite low, only one viable cell per container is required to spoil the product.

Defects of fluid milk associated with the growth of psychrotrophic bacteria are related to the production of extracellular enzymes. Sufficient enzyme to cause defects is usually present when the population of psychrotrophs reaches 10^6–10^7 cfu/ml but this depends on the specific product, for example, the shelf life of UHT milk is limited by the presence of less enzyme than is the shelf life of pasteurized milk. Bitter and putrid flavours and coagulation result from proteolysis, rancid and fruity flavours result from lipolysis. The production of extracellular enzymes by psychrotrophic bacteria in raw milk also has implications for the quality of products produced from that milk.

Protease-induced product defects   Proteases of psychrotrophic bacteria cause product defects either at the time they are produced in the product or as a result of enzyme surviving a heat process. These proteases preferentially hydrolyse α-casein, although some show preference for β-casein. Degradation of casein in milk by enzymes produced by psychrotrophs results in the liberation of bitter peptides. Bitterness is a common off-flavour in pasteurized milk that has been subjected to post-pasteurization contamination with psychrotrophic bacteria. Continued proteolysis results in putrid off-flavours associated with lower molecular weight degradation products such as ammonia, amines and sulphides. Bitterness in UHT milk develops when sufficient psychrotrophic bacteria growth occurs in raw milk to leave residual enzyme after heat treatment. The effect of proteases of psychrotrophic origin on quality of cheese and other cultured dairy products is minimal because the combination of low pH and low storage temperature inhibits their activity. In addition, proteases are removed with the whey fraction during cheese manufacture. However, growth of proteolytic bacteria in raw milk lowers cheese yield because proteolytic products of casein degradation are lost to the whey rather than becoming part of the cheese.

Psychrotrophic *P. fluorescens* isolated from milk often produces extracellular lipase in addition to protease. The triglycerides in raw milk are present in globules that are protected from enzymatic degradation by a membrane. Milk becomes susceptible to lipolysis and this membrane is disrupted by excessive shear force. Raw milk contains a mammalian lipase which will rapidly act on the fat if the globule membrane is disrupted. Most cases of rancidity in raw and pasteurized milk result from this process rather than from the growth of lipase producing microbes. Phospholipase C and protease produced by psychrotrophic bacteria can degrade the fat globule membrane, resulting in the enhancement of milk lipase activity. Since milk lipase is heat labile, most milk products will not have residual activity.

Lipase-induced product defects   Sufficient bacterial lipase can be produced in raw milk to cause defects in products manufactured from that milk. Since residual activities are usually low and the reaction environment is less than optimum, usually only products with long storage times or high storage temperatures are affected. Such products include UHT milk, some cheeses, butter and whole milk powder. The rancid flavour and odour resulting form lipase action are usually from the liberation of C4 to C8 fatty acids. Fatty acids of higher molecular weight produce a flavour described as soapy. Low levels of unsaturated fatty acids liberated by enzymatic activity may be oxidized to ketones and aldehydes to produce oxidized or "cardboardy" off-flavour. *P. fragi* produces a fruity off-flavour in milk by esterifying free fatty acids with ethanol.

Residual activity from heat-stable microbial lipases can cause off-flavours in UHT milk, but lipase induced defects are not as common as those resulting from microbial protease.

Rancid defect in butter may result from growth of lipolytic microbes during storage, residual heat-stable microbial lipase originating from the growth of psychrotrophic bacteria in the milk or cream, or milk lipase activity in the raw milk or cream. However, the typical odour of rancid butter is associated with lower molecular weight of fatty acids. Microbial lipases present in butter will exhibit activity even if the product is stored at refrigerated temperature.

Cheese is more susceptible to defects caused by bacterial lipases than those caused by proteases because lipases, unlike most proteases, are concentrated along with the fat in the curd. Camembert cheese is in fact susceptible to defects associated with microbial lipase.

Spoilage by Fermentative Non-spore-formers

Spoilage of milk and dairy products resulting from growth of acid-producing fermentative bacteria occurs when storage temperatures are sufficiently high for these microorganisms to outgrow psychrotrophic bacteria or when product composition is inhibitory to gram-negative aerobic organisms. Fermented dairy foods, though manufactured by using lactic acid bacteria, can be spoiled by the growth of "wild" lactic acid bacteria which produced unwanted gas, off-flavours or appearance defects. Fluid milk, cheese, and cultured milks are the major dairy products susceptible to spoilage by non-spore-forming fermentative bacteria.

Non-spore-forming bacteria responsible for fermentative spoilage of dairy products are mostly in either the lactic acid producing or coliform group. Lactic acid bacteria involved in dairy fermentations can spoil fluid milk, but the strains involved are often environmental types which produce defect-inducing metabolites in addition to lactic acid. Genera of lactic acid bacteria involved in spoilage of milk and fermented products include *Lactococcus, Lactobacillus, Leuconostoc, Enterococcus, Pediococcus* and *Streptococcus*. Coliforms can spoil milk, but this is seldom a problem since they are usually outgrown by either the lactic acid or the pyschrotrophic bacteria. Coliform spoilage is more common with fermented products especially certain cheese varieties. Members of the genera *Enterobacter* and *Klebsiella* are most often associated with coliform spoilage, while *Escherichia* spp. only occasionally exhibit sufficient growth to produce a defect.

The most common fermentative defect in fluid milk products is souring caused by the growth of lactic acid bacteria. Lactic acid by itself has a clean, pleasant acid flavour and no odour. The unpleasant "sour" odour and taste of spoiled milk is a result of small amounts of acetic and propionic acids. Other defects may occur in combination with acid production. Another defect associated with growth of LAB in milk is "ropy" texture. Most dairy associated species of lactic acid bacteria have strains that produce exocellular polymers which increase the viscosity of milk, causing the ropy defect. The defect in non-cultured fluid milk products is usually caused by growth of specific strains of lactococci. The polymer produced by these organisms is a polysaccharide containing glucose and galactose with small amounts of mannose, rhamnose, and pentose. The polysaccharide possibly associates with protein.

Some strains of LAB produce flavour and appearance defects in cheese. Lactobacilli are a normal part of the dominant microflora of aged Cheddar cheese. If heterofermentative lactobacilli predominate, the cheese is prone to develop an "open" texture or fissures, a result of gas production during ageing. Off-flavours are also associated with the growth of these organisms. Gassy defects in aged Cheddar cheese are more often associated with growth of lactobacilli than with growth of coliforms, yeasts, or sporeformers. *Lactobacillus brevis* and *L. casei* subsp. *pseudoplantarum* have been associated with gas production in Mozzarella cheese. *L. casei* subsp. *casei* produces a soft body defect in Mozzarella cheese as a result of its proteolytic ability. The softened cheese cannot be readily sliced or grated and does not melt properly.

Some cheese varieties occasionally exhibit a pink discolouration. Pink spots in Swiss type varieties result from the growth of pigmented strains of propionibacteria. This defect is associated with certain strains of *Lactobacillus delbrueckii* subsp. *bulgaricus* that fail to lower the redox potential of the cheese. Another common defect of aged Cheddar cheese is the appearance of white crystalline deposits on the surface. This may be due to an atypical strain of a facultatively heterofermentative *Lactobacillus* species associated with the deposits. This strain produces an unusually high amount of D-lactic acid during cheese ageing, resulting in the formation of insoluble calcium lactate crystals, the primary component of the white deposits. *L. casei* subsp. *alactosus* and *L. casei* subsp. *rhamnosus* have been associated with the development of a phenolic flavour in Cheddar cheese (similar to horse urine) which develops after 2–6 months of ageing.

Fruity off-flavour in Cheddar cheese is a result of growth of lactic acid bacteria (usually *Lactococcus* spp.) which produce esterase. Fruity flavoured cheeses contain high levels of ethanol, a substrate for esterification. The major esters contributing to fruity flavour in cheese are ethyl hexanoate and ethyl butyrate.

Fermented milk products such as cultured buttermilk, sour cream and cottage cheese rely on diacetyl produced during fermentation for their typical "buttery" flavour and aroma. Lactococci, capable of growing at 7°C produces sufficient diacetyl reductase to destroy diacetyl in cultured milks.

REVIEW QUESTIONS

1. Discuss the role of psychrotrophic bacteria in milk.

2. Brief out on the spoilage of milk by fermentative non-spore-forming bacteria.

23

FOOD SPOILAGE BY MOULDS

MOULDS AND THEIR METABOLITES

Moulds are able to grow on all kinds of food: cereals, meat, milk, fruits, vegetables, nuts, fats and products of these. The mould growth may result in several kinds of food-spoilage: off-flavours, toxins, discolouration, rotting and formation of pathogenic or allergenic propagules. The deterioration of sensorial properties is often due to the production of exoenzymes during growth. Moulds can produce a vast number of enzymes: lipases, proteases, carbohydrases. Once inside the food these enzymes may continue their activities independent of destruction or removal of the mycelium. The enzymatic activities may give rise to flavours like rio coffee beans, musty odours in cork and wine or dried fruits. This is caused by the fungal transformation of 2,4,6-trichlorophenol to trichloroanisol (TCA) by *Penicillium brevicompactum*, *P. crustosum*, *Aspergillus flavus* and other species. Some of these flavours can be detected in very small amounts like TCA or *trans*-1,3 pentadiene produced from sorbic acid by *Penicillium* species. TCA has an odour threshold level of 8 ng/l in coffee. The enzymatic reactions may also lead to complete disintegration of the food structure, like the change of whole pasteurized strawberries into strawberry pulp due to growth of the heat-resistant fungi *Byssochlamys fulva* and *Byssochlamys nivea*.

Moulds can produce volatiles such as dimethyldisulphide, geosmin and 2-methylisoborneol which can affect the quality of foods and beverages even when present in very small amounts. These compounds are produced in large quantities in species specific combinations of different genera such as *Penicillium*, *Aspergillus* and *Fusarium*. The most important aspect of mould spoilage of foods is, however, the formation of mycotoxins. More than 400 mycotoxins are known today, aflatoxins being the best known, and the number is increasing rapidly. Mycotoxins are secondary metabolites which are toxic to vertebrate animals in small amounts when introduced *via* a natural route. The toxicity of these metablities is very different, with chronic term toxicosis being the most important to humans. However, only a few mycotoxins are well described in toxicological terms. The most important toxic effects are different kinds of cancers and immune suppression. Several mycotoxins have very significant antibiotic activity as well, which in time may give rise to bacteria with a cross resistance to the most important antibiotics used today, like penicillins. Further more, it is important to note that some of these mycotoxins act synergistically. The mycotoxins are formed during growth of moulds on foods. Some mycotoxins are only present in the mould, while most of them are excreted in the foods. In liquid foods and in fruits like peaches, pears and tomatoes the diffusion of mycotoxins can be very fast, leaving no part of the product uncontaminated. In solid foods like cheese, bread, apples and oranges the diffusion is slow leaving the major part of the product uncontaminated. Since most of the mycotoxins are very resistant to physical and chemical treatments, a rule of thumb exists—once the mycotoxins are in the food, they stay there during processing and storage. This also means that the use of any mouldy material in the processing of food may contribute mycotoxins to the final product.

Being secondary metabolites, the individual mycotoxins are produced by a limited number of species. Aflatoxin is produced only by the closely related *A. flavus*, *A. parasiticus* and *A. nomius*, while other mycotoxins such as ochratoxin A are produced by few species in different genera: *Petromyces alliaceus*, *Aspergillus ochraceus* and *Penicillium verrucosum*. On the other hand, the individual toxic species are able to produce a considerable number of mycotoxins. *P. griseofulvum* for example produces patulin, griseofulvin, cyclopiazonic acid and roquefortine C consistently. The number of toxic species is large, in fact it is a question whether any naturally existing species can be claimed to be able to produce mycotoxins at all. However, the profile and amount of mycotoxins in the food depends completely on the ecological and processing parameters of the particular foodstuff.

Besides being restrictive to the mycotoxin formation, the food parameters have also proven to be surprisingly restrictive in the spectrum of species, which are able to grow and thus spoil the individual food types. Normally less than ten and often one to three species are responsible for spoilage. On the other hand these critical species are often completely different of each food type. As far as fungi in foods are concerned this discovery is fairly new, and is due to the development of new mycological methods and taxonomy of food-borne mould, especially the genera *Penicillium*, *Aspergillus* and *Fusarium*. The former dominating role of morphology in mould identification has been replaced by the combined use of secondary metabolite profiles, isozyme profiles, physiological and ecological characteristics, DNA patterns and morphology.

EFFECT OF SPOILAGE MOULDS ON FOOD PRODUCTS

Fruits and Vegetables

Citrus fruit   Citrus fruits are non-climacteric fruit. Three spoilage fungi are of paramount importance, *Penicillium digitatum*, *P. italicum* and *P. ulaiense*. *P. ulaiense* only appears when the other two pathogens are inhibited by fungicides and *P. ulaiense* is much more related to *P. italicum* than *P. digitatum*. *Alternaria* species are also spoilers of citrus fruits, but the taxonomy of small-spored *Alternaria* species is complex. *Fusarium* spp. has also been reported on decayed citrus furits.

Germination of *P. digitatum* conidia is stimulated by certain combinations of the volatiles surrounding wounded oranges, notably limonene, α-pinene, sabinene, β-myrcene, acetaldehyde, ethanol, ethylene and CO_2. Other constituents of oranges, such as simple sugars and organic acids also stimulate conidium germination in *P. digitatum*. Thus it can be concluded that the associated fungi can tolerate and is sometimes even stimulated by the acids and other protecting volatile and non-volatile phytoalexins of citrus fruits in combination with the ability to produce pectinases and other citrus skin degrading enzymes. The fungal activities result in serious weight loss, shrinkage and softening of the citrus fruits. Furthermore, a few fruits spoiled by fungi can cause reduced shelf life of the sound fruits due to accelerated ripening or senescence triggered by the releasing ethylene.

Mycotoxins of *Alternaria* have been found in mandarins and they can also be produced in lemons and oranges. The mycotoxins found include tenuazonic acid, alternariol monomethyl ether (AME) and alternariol. *Fusarium* spp. may produce trichothecenes and fusarin C, however *Fusarium* toxins have never been detected in citrus fruits.

P. italicum and *P. digitatum* mycotoxins have not been found in citrus fruits yet these fungi produce compounds that are toxic to bacteria, plants, brine shrimps and chick embryos. *P. digitatum* has been found to produce tryptoquivalins and tryptoquivalons, which are regarded as mycotoxins.

Pomaceous and stone fruit   *Penicillium expansum*, *P. crustosum* and *P. solitum and Alternaria alternata* were reported as organisms able to reproduce rot in apples. Pomaceous and stone fruit and several other berries can be degraded by a large number of pathogenic species including *Monilia laxa*, *M. fructigena* and *Rhizopus stolonifer*. For example the "box rot" of dried prunes, which is soft, sticky, macerated areas on the fruit and slippage of the skin under slight pressure due to the activity of pectinolytic enzymes produced by these fungi. However, these fungi are probably not mycotoxin producers.

Canned fruits like apricots and peaches sometimes suffer from textural changes due to heat-resistant fungal enzymes produced in the raw fruit or to the enzymatic activity of surviving heat-resistant fungi like *Byssochlamys fulva*.

P. expansum is known for its production of patulin and citrinin and these mycotoxins have been found in mouldy fruit. Other mycotoxins produced by *P. expansum*, such as roquefortine C and chaetoglobosin C or by *P. crustosum* such as terrestric acid, roquefortine C and penitrem A have not yet been reported from rotting pomaceous fruits.

Garlic and onions　Apart form *Botrytis aclada*, few species are able to spoil garlic and onions. *Penicillium allii* is a widespread spoiler of garlic while the closely related *P. albocoremium* is more common on the onions. *Petromyces alliaceus*, *Aspergillus niger* and *Penicillium glabrum* are cited as producers of rots in onions. *Penicillium glabrum* appears to grow only in the outer layers.

Petromyces alliaceus is a very efficient producer of ochratoxin A. *P. glabrum* produces the nephrotoxin citromycetin. The presence of allicin and other antimicrobial compounds in onions strongly selects for the associated fungi.

Potato tubers　Dry rot of potatoes is mainly caused by *Fusarium sambucinum* and *F. coeruleum*. The other species frequently reported in relation to dry rot of potatoes, *F. coeruleum*, is a synonym of *F. solani* var. *coeruleum*. *F. cerealis* is also frequently isolated from damaged potatoes, however, its role as a primary pathogen is unclear. The dry rot is normally so pronounced that the tubers are not suitable for consumption. However, as the full extent of the damage is not always visible from the outside, it may be possible that partly rottened potatoes could pass on to further processing in the food industry. In addition to the physical damage of potatoes, mycotoxins may also be produced in the tubers. Diacetoxyscirpenol and related trichothecenes have been detected in tubers inoculated with *F. sambucinum*.

The pathogens are present in soil and tubers and the infection takes place by damage of the periderm. High soil humidity raises the infection rate whereas crop rotation will lower it.

Wheat and Rye Grain

Field condition　*Fusarium*, *Alternaria*, *Cladosporium* and *Claviceps* are very common on grains in the field and can reduce the quality of grains by their growth and mycotoxin production. Cereal plants may be damaged by numerous fungal pathogens. Fusarium ear diseases of cereals is caused primarily by *Fusarium culmorum* and *F. graminearum*. Both species can produce deoxynivalenol, zearalenone and several other biologically active metabolites in the grains. Whereas the fusaria will be eliminated during food processing, a significant carryover of toxins will be possible as they are resistant to cleaning of grains, milling, baking and other cooking processes.

Another important species is *F. avenaceum*, as this species can produce moniliformin, antibiotic Y, and fusarin C. *Fusarium* infections take place by airborne conidia on the heads or by systemic infection. Alternariols and other toxins of *Alternaria* have been detected infrequently in grains. However, together with *Cladosporium* spp., *Alternaria* can cause discolouration of the grains by their abundant presence on the grain, called black (sooty) heads. In some cases these moulds can cause a mild infection, which may result in weakened and undersized grains.

Ergot, *Claviceps purpurea*, occurs mainly on rye but certain wheat lines have also been damaged. The sclerotia, replacing grains, are the visible damage but in addition *C. purpurea* produces a series of alkaloids toxic towards humans. These alkaloids have been detected in rye and wheat (grain and flour).

Stored conditions In temperate-climated storage, the dominating moulds are species within *Penicillium* and *Aspergillus*. *Penicillium* species are of paramount importance in stored cereals. Other species are *P. verrucosum*, the only known ochratoxin A producer in *Penicillium*, *P. hordei* and members of the *P. aurantiogriseum* complex (also named *P. verrucosum* var. *cyclopium* and var. *verrucosum*). Several toxin-producing aspergilli have been reported to dominate on cereals, especially *A. candidus*, *A. flavus*, *A. niger*, *A. versicolor* and *A. penicillioides* and *Eurotium* spp. at lower water activities.

Ochratoxin A, citrinin, xanthomegnin, viomellein and vioxanthin have been found in barley, rye and wheat. Several other possible toxic secondary metabolites are produced by species in the *P. aurantiogriseum* complex such as verrucosidin, penicillic acid, cyclopenin, viridicatol, pseurotins, viridic acid, brevianamide A, nephrotoxic glycopeptides, anacine, auranthine, aurantiamine, terrestric acid, puberulonic acid, verrucofortine, puberuline, roquefortine C, meleagrin, oxaline, viridamine and aspterric acid.

Mycotoxins from penicillia growing in cereals stored in countries with subtropical or tropical climate could include viridicatum toxin (*P. aethiopicum*), citrinin (*P. citrinum*), cyclopiazonic acid, patulin and roquefortine C (*P. griseofulvum*) and secalonic acid D (*P. oxalicum*).

Rye bread The most important species with no preservatives added are *Penicillium roqueforti*, *P. paneum*, *P. carneum*, *P. corylophilum*, *Eurotium repens* and *E. rubrum*. *P. paneum* and *P. carneum* are newly described species based on significant differences in mycotoxin, DNA and morphological characteristics. Isolates belonging to these species have earlier been identified as *P. roqueforti* or *P. roqueforti* var. *carneum*. The growth of yeasts sometimes is a serious problem especially in sliced rye bread, the dominating species being *Endomyces fibuligera*, *Pichia anomala* and *Hyphopichia burtonii*. Species of less importance to the quality of rye bread are: *Paecilomyces variotii, Aspergillus flavus, Penicillium commune, P. solitum, A. niger, P. decumbens* and *Mucor* spp.

Only a few mycotoxins have been detected in rye bread. Aflatoxins, citrinin, ergot alkaloids and patulin.

There are no data on mould spoilage other than mycotoxin contamination in bread. The spoilage is due to species tolerating a lowered water activity (0.95) and the presence of organic acids like acetic acid and propionic acid which are formed during the fermentation or which have been added as preservatives. The infection takes place after the baking process, which obviously kills all fungal propagules and is due to airborne conidia originating from growth of the spoiling species on product wastes in a few specific places in the plant.

Cheese

The most important spoilage species of hard, semi-hard and semi-soft cheese from several countries, without preservatives added, are: *Penicillium commune* and *P. nalgiovense*. Species of less importance are: *P. verrucosum, P. solitum, P. roqueforti, Scopulariopsis brevicaulis* and *Aspergillus versicolor*. It has been shown that important isolates from cheese which have been identified as *P. verrucosum* var. *cyclopium, P. aurantiogriseum, P. cyclopium* and *P. puberulum* can be re-identified as *P. commune*. Recently a new species, *P. discolor* has been isolated from hard cheese. Growth of *P. discolor* on cheese only takes place under very restricted conditions, which have yet to be determined.

The most important mycotoxin found in cheese is sterigmatocystin. Further, mycotoxins which must be considered important in cheese due to the mycotoxin potential of the species are cyclopiazonic acid, rugulovasine A and B and ochratoxin A.

Spoilage of cheese due to fungal growth is also caused by formation of off-flavours. If sorbates are used as preservatives, resistant species are able to metabolize these compounds under formation of a plastic like or "kerosene" off-flavour, which is due to the metabolites *trans*-1,3-pentadiene or *trans*-piperylene.

Fermented Sausages

The associated fungi in naturally fermented sausages are *Penicillium* species: *P. nalgiovense, P. olsonii, P. chrysogenum, P. verrucosum, P. spathulatum, P. solitum, P. oxalicum, P. commune, P.camemberti, P. expansum, P. miczynskii* and *P. simplicissimum*. Dominating species of *Aspergillus* and *Scopulariopsis* have also been reported.

In the beginning of the fermentation process yeasts are dominating the surface fungi, but after a few weeks the above-mentioned naturally occurring moulds take over, *P. nalgiovense* being dominating. This species in some cases is added as a starter culture. The Penicillium species in the associated fungi are known to produce several mycotoxins and antibiotics. Some of these mycotoxins have been detected in fermented sausages after mould inoculation in pure cultures: Citreoviridin, citrinin, cyclopiazonic acid, isofumigaclavine A, ochratoxin A, patulin, roquefortine C, rugulovasine A. Further mycotoxins like viomellein and xanthomegnin are produced by the associated fungi.

SUMMARY

Thus the existence of an associated fungi in foods has great impact on the mycological quality assessment in the food industry. The limited number of species which are shown to be important to the quality of the foods mentioned to a great extent simplifies the preventive and the control actions which must be taken. Knowing the properties of the spoiling species makes it possible to optimize the preservative profile of the food and the hygienic measures during production of the food. Thus prevention of mould spoilage of foods can only be carried out successfully, if the species, which are actually spoiling the food and the associated fungi are known.

1. Give a detailed account of mycotoxin-producing fungal strains and their role in food spoilage.

24

SPOILAGE OF CANNED FOODS

INTRODUCTION

The incidence of spoilage in canned foods is low, but when it occurs it must be investigated properly. Swollen cans often indicate a spoiled product. However, spoilage is not the only cause of abnormal cans. Overfilling, buckling, denting, or closing while cool may also be responsible. Microbial spoilage and hydrogen produced by the interaction of acids in the food product with the metals of the can, are the principal causes of swelling. High summer temperatures and high altitudes may also increase the degree of swelling. Some microorganisms that grow in canned foods, however, do not produce gas and therefore cause no abnormal appearance of the can; nevertheless, they cause spoilage of the product.

CAUSES OF SPOILAGE

Spoilage is usually caused by growth of microorganisms following leakage or underprocessing.

Spoilage by Leakage

Leakage occurs from can defects, punctures, or rough handling. Contaminated cooling water sometimes leaks to the interior through pinholes or poor seams and introduces bacteria that cause spoilage. A viable mixed microflora of bacterial rods and cocci is indicative of leakage, which may usually be confirmed by can examination.

Spoilage by Underprocessing

Underprocessing may be caused by undercooking; retort operations that are faulty because of inaccurate or improperly functioning thermometers, gauges, or controls; excessive contamination of the product for which normally adequate processes are insufficient; changes in formulation or handling of the product that result in a more viscous product or tighter packing in the container, with consequent lengthening of the heat penetration time; or sometimes, accidental bypassing off the retort operation altogether. When the can contains a spoiled product and no viable microorganisms, spoilage may have occurred before processing or the microorganisms causing the spoilage may have died during storage.

Underprocessed and leaking cans are of major concern and both pose potential health hazards. However, before a decision can be made regarding the potential health hazard of a low-acid canned food, certain basic information is necessary.

Table 24.1   Classification of food products according to acidity

Low acid—pH greater than 4.6	Acid—pH 4.6 and below
Meats	Tomatoes
Seafoods	Pears
Milk	Pineapple
Meat and vegetable mixtures and "specialties"	Other fruits
Spaghetti	Sauerkraut
Soups	
Vegetables	
Asparagus	Pickles
Beets	Berries
Pumpkin	Citrus
Green beans	
Corn	Rhubarb
Lima beans	

Spoilage in acid products is usually caused by non-spore-forming lactobacilli and yeasts. Cans of spoiled tomatoes and tomato juice remain flat but the products have an off-odour, with or without lowered pH, due to aerobic, mesophilic, and thermophilic sporeformers. Spoilage of this type is an exception to the general rule that products below pH 4.6 are immune to spoilage by sporeformers. Many canned foods contain thermophiles which do not grow under normal storage conditions, but which grow and cause spoilage when the product is subjected to elevated temperatures (50–55°C).

B. thermoacidurans and *B. stearothermophilus* are thermophiles responsible for flat-sour decomposition in acid and low-acid foods, respectively.

Table 24.2 Spoilage microorganisms that cause high and low acidity in various vegetables and fruits

Spoilage type	pH groups	Examples
Thermophilic		
Flat-sour	≥ 5.3	Corn, peas
Thermophilic* anaerobes	≥ 4.8	Spinach, corn
Sulphide spoilage*	≥ 5.3	Corn, peas
Mesophilic		
Putrefactive anaerobes*	≥ 4.8	Corn, asparagus
Butyric anaerobes	≥ 4.0	Tomatoes, peas
Aciduric flat-sour*	≥ 4.2	Tomato juice
Lactobacilli	4.5–3.7	Fruits
Yeasts	≤ 3.7	Fruits
Moulds	≤ 3.7	Fruits

* spore formers

Naturally, if *Clostridium botulinum* (spores, toxin, or both is found, the hazard is obvious. Intact cans that contain only mesophilic, gram-positive, spore-forming rods should be considered underprocessed, unless proved otherwise. It must be determined that the can is intact (commercially acceptable seams and no microleaks) and that other factors that may lead to underprocessing, such as drained weight and product formulation, have been evaluated.

STAGES OF SPOILED CANS

During spoilage, cans may progress from normal to flipper, to springer, to soft-swell and to hard-swell.

Flat can A can with both ends concave; it remains in this condition even when the can is brought down sharply on its end on a solid, flat surface.

Flipper can A can that normally appears flat; when brought down sharply on its end on a flat surface, one end flips out. When pressure is applied to this end, it flips in again and the can appears flat.

Springer can A can with one end permanently bulged. When sufficient pressure is applied to this end, it will flip in, but the other end will flip out.

Soft-swell can   A can bulged at both ends, but not so tightly that the ends cannot be pushed in somewhat with thumb pressure.

Hard-swell can   A can bulged at both ends, and so tightly that no indentation can be made with thumb pressure. A hard swell will generally "buckle" before the can bursts. Bursting usually occurs at the double seam over the side seam lap, or in the middle of the side seam.

Table 24.3   Spoilage manifestation in acid products

Group of organisms	Manifestation
Low-acid products	
Flat-sour causing organisms	Can becomes flat possibly due to loss of vacuum on storage
	Product appearance not usually altered; pH markedly lowered, sour; many have slightly abnormal odour; sometimes cloudy liquor.
Thermophilic anaerobe	Can swells and may burst
Sulphide spoilage causing organisms	Can becomes flat due to H_2S gas absorbed by product
Putrefactive anaerobe	Can swells and may burst
	Product may be partially digested; pH slightly above normal; typical putrid odour
Aerobic sporeformers	Can becomes flat or swollen; usually no swelling, except in cured meats when nitrate and sugar present; coagulated evaporated milk, black beets
Acid products	
Bacillus thermoacidurans (flat-sour tomato juice)	Can becomes flat due to slight change in vacuum
	Product shows slight pH change; off-odour
Butyric anaerobes (tomatoes and tomato juice)	Can swells and may burst
	Product is fermented; butyric odour
Non-sporeformers (mostly lactic types)	Can swells and usually bursts, but swelling may be arrested
	Product produces acid odour

DETECTION OF SPOILAGE

The preferred tool for can content examination is a bacteriological can opener consisting of a puncturing device at the end of a metal rod mounted with a sliding triangular blade that is held in place by a set screw. The advantage over other types of openers is that it does no damage to the double seam and therefore will not interfere with subsequent seam examination of the can.

The useful descriptive terms for canned food analysis is given in Table 24.4.

Table 24.4   Useful descriptive terms for canned food analysis

Exterior can condition		Internal can condition	
Leaker		Normal	
Dented		Peeling	
Rusted		Slight, moderate or severe etching	
Buckled		Slight, moderate or severe blackening	
Panelled		Slight, moderate or severe rusting	
Bulge		Mechanical damage	
Microleak test		**Product odour**	**Product liquor**
Packer seam		Putrid	Cloudy
Side panel		Acidic	Clear
Side seam		Butyric	Foreign
Cut code		Metallic	Frothy
Pinhole		Sour	
		Cheesy	
		Fermented	
		Musty	
		Sweet	
		Facal	
		Sulphur	
		Off-odour	
Solid product	Liquid product	Pigment	Consistency
Digested	Cloudy	Darkened	Slimy
Softened	Clear	Light	Fluid
Curdled	Foreign	Changed	Viscous
Uncooked	Frothy		Ropy
overcooked			

LABORATORY DIAGNOSIS

The presence of only spore-forming bacteria, which grow at 35°C, in cans with satisfactory seams and no microleaks indicates underprocessing if their heat resistance is equal to or less than that of *C. botulinum*. Spoilage by thermophilic anaerobes such as *C. thermobutylicum* may be indicated by gas in cooked meat at 55°C and a cheesy odour. Spoilage by *C. botulinum*, *C. sporogenes*, or *C. perfringens* may be indicated in cooked meat at 35°C by gas and a putrid odour; rods, spores, and clostridial forms may be seen on microscopic examination. Always test supernatants of such cultures for botulinal toxin even if no toxin was found in the product itself, since viable botulinal spores in canned foods indicate a potential public health hazard, requiring recall of all cans bearing the same code. Spoilage by mesophilic organisms such as *Bacillus thermoacidurans* or *B. coagulans* and/or thermophilic organisms such as *B. stearothermophilus*, which are flat-sour types, may be indicated by acid production in BCP tubes at 35 and/or 55°C in high-acid or low-acid canned foods. No definitive conclusions may be drawn from inspection of cultures in broth if the food produced an initial turbidity on inoculation. Presence or absence of growth in this case must be determined by subculturing.

Incubation at 55°C will not cause a change in the appearance of the can, but the product has an off-odour with or without a lowered pH. Spoilage encountered in products such as tomatoes, pears figs, and pineapples is occasionally caused by *C. pasteurianum*, a spore-forming anaerobe which produces gas and a butyric acid odour. *C thermosaccolyticum* is a thermophilic anaerobe which causes swelling of the can and a cheesy odour of the product. Cans which bypass the retort without heat processing usually are contaminated with non-sporeformers as well as sporeformers, a spoilage characteristic similar to that resulting from leakage.

A mixed microflora of viable bacterial rods and cocci usually indicates leakage. Can examination may not substantiate the bacteriological findings, but leakage at some time in the past must be presumed. Alternatively, the cans may have missed the retort altogether, in which case a high rate of swells would also be expected.

A mixed microflora in the product, as shown by direct smear, in which there are large numbers of bacteria visible but no growth in the cultures, may indicate precanning spoilage. This results from bacterial growth in the product before canning. The product may be abnormal in pH, odour, and appearance.

If no evidence of microbial growth can be found in swelled cans, the swelling may be due to development of hydrogen by chemical action of contents on container interiors. The proportion of hydrogen varies with the length and condition of storage. Thermophilic anaerobes produce gas, and since cells disintegrate rapidly after growth, it is possible to confuse thermophilic spoilage with hydrogen swells. Chemical breakdown of the product may result in evolution of carbon dioxide. This is particularly true of concentrated products containing sugar and some acid foods, such as tomato paste, molasses, minced meats, and highly sugared fruits. The reaction is accelerated at elevated temperatures.

Table 24.5   Laboratory diagnosis of bacterial spoilage

	Underprocessed	Leakage
Can	Flat or swelled; seams generally normal	Swelled; may show normal defects
Product appearance	Sloppy or fermented	Frothy fermentation; viscous
Odour	Normal, sour or putrid, but generally consistent from can to can	Sour, faecal; generally varying from can to can
pH	Usually fairly constant	Wide variation
Microscopic and cultural	Cultures show spore-forming rods only	Mixed cultures, generally rods and cocci; only at usual temperatures
	Growth at 35 and/or 55°C. May be characteristic on special growth media, e.g. acid agar for tomato juice.	
	If the can misses retort completely, rods, cocci, yeast or moulds, or any combination of these may be present.	
History	Spoilage usually confined to certain portions of pack	Spoilage scattered
	In acid products, diagnosis may be less clearly defined; similar organisms may be involved in understerilization and leakage.	

Any organisms isolate from normal cans that have obvious vacuum and normal product but no organisms in the direct smear should be suspected as being a laboratory contaminant. To confirm, aseptically inoculate growing organism into another normal can, solder the hole closed, and incubate 14 days at 35°C. if any swelling of container or product changes occur, the organism was probably not in the original sample. If can remains flat, open it aseptically and subculture as previously described. If a culture of the same organism is recovered and the product is normal, consider the product commercially sterile since the organism does not grow under normal conditions of storage and distribution.

REVIEW QUESTIONS

1. What are the various stages of canned food spoilage?
2. Discuss the spoilage manifestation in low-acid food products.
3. Discuss the spoilage manifestation in acid food products.

25

MICROBIOLOGY OF FOOD TAINTS

INTRODUCTION

Most consumers would readily recognize the odour associated with rotten meat or spoiled milk, and would instantly realize that these materials, if consumed, could be harmful. Such odours are indicative of microbial spoilage, and the compounds responsible are metabolites of some of the microbes present in the foodstuff. These compounds are produced by the metabolic activity of microorganisms exerted on natural components of the food. Because of the association of unpleasant odours and flavours with the possibility of food poisoning, the term "taint" has frequently been used to describe the compound responsible. Today, this term is widely applied to a range of off-odours and off-flavours irrespective of their microbiological or chemical origins.

Microorganisms that can cause food taints are bacteria (including actinomycetes and cyanobacteria), fungi, and yeasts, and their growth in a food is usually limited only by the physical environment and chemical composition of the foodstuff they infect. The most important physical properties that will affect their growth are water activity (a_w), pH and the temperature and the atmospheric composition under which the food is stored. Even so, most food spoilage microorganisms are extremely versatile. For example, bacteria grow in foods at a_w from 0.7–1, pH from 2.5–10 and temperatures from–2 to 75°C, and anaerobic bacteria can grow under a variety of gaseous compositions. Fungi and yeasts are equally versatile; they grow at a_w from 0.62 to 1, pH from 1.5 to 10 and temperatures from –3 to 50°C, and certain species of both of these types of microorganisms can grow in the absence of oxygen. The chemical composition of all foods is usually conductive to the growth of a wide range of microbes: bacteria, fungi and yeasts can all effectively utilize a wide variety of food components, including proteins, carbohydrates (including pectins), organic acids and lipids. The addition of preservatives can control the growth of some organisms, but preservatives can ultimately become the precursors of taints when the food is infected by resistant strains.

MEAT AND MEAT PRODUCTS

Raw Meat

Surface microbial contamination is the major cause of spoilage taints in raw meat, and the main source of such contamination is the soil and faeces that accompany the livestock into the slaughterhouse. These microbes are then transferred to the meat during the subsequent slaughter and processing of the animal's carcass. That such taints occur infrequently is some measure of the quality control standards of most modern abattoirs.

Odours   Off-odours, such as "dairy", "buttery", "cheesy", "sweet", "fruity" and finally "putrid", characterize refrigerated aerobically stored beef. These off-odours usually develop in this sequence, and are related to the changes in microflora and chemical precursors that occur as storage time is increased. The onset of the "buttery" and "cheesy" odours is usually sufficient to cause most cooks to reject the meat as spoilt. The compounds responsible for these odours are diacetyl, acetoin, 3-methylbutanol and 2-methylpropanol. All of these compounds can be formed from glucose by the bacterium, *Brochothrix thermosphacta*, a common microbial contaminant of raw meat. However, other bacteria, including species in the Enterobacteriaceae and homofermentative *Lactobacillus* spp. can also contribute to the production of these "buttery" and "cheesy" odours.

The next stage of the spoilage process is the onset of the "sweet" and "fruity" odours and the compounds principally responsible for these are the ethyl esters of acetic, butanoic and hexanoic acids. These compounds can be formed from glucose by *Pseudomonas* spp. and in particular *P. fragi*. Such organisms are also common microbial contaminants of raw beef. It is of interest that some oxidative strains of *Moraxella* have been shown to produce "sweet", "fruity" odours in minced beef, but the esters involved were those of hexanoic and octanoic acids only. In more recent studies, species of *Moraxella* have been shown to produce ethyl acetate and the ketones acetone and 2-butanone when grown on ground beef. Branched chain esters have also been identified in normal and high pH beef stored at chill temperatures and include the ethyl esters of 2-methylbutanoic, 3-methylbutanoic and 3-methyl-2-butanoic acids, and the acetates of isopropanol, isobutanol and isopentanol. The sources of such branched chain esters have not been established, but are probably the amino acids valine, leucine and isoleucine. Both Enterobacteriaceae and *Brochothrix thermosphacta* have been shown to produce one or more of the acids and alcohols involved, but neither type of organism produces these esters in pure cultures. A possible explanation is that *Pseudomonas* spp., a common bacterial contaminant of beef, catalyse the interaction of the excreted products.

The final stage of beef spoilage is characterized by the appearance of the "putrid" odours. These odours, sometimes described as garlic and onion like, are caused by the formation of the sulphur compounds methanethiol, dimethyl sulphide and dimethyl disulphide. Such sulphur compounds usually appear only after the depletion of glucose from the meat surface and the start of amino acid metabolism. *Pseudomonas* spp. would appear to be the major organism responsible for the production of these compounds on beef under aerobic conditions. However, they are also produced in ground beef by some species of *Moraxella* and *Acinetobacter*. Hydrogen sulphide can also be detected at this stage of spoilage and its presence usually indicates that the beef is contaminated by atypical bacteria such as Enterobacteriaceae. Refrigerated raw beef can also develop sour flavours, caused by the growth of *Lactobacillus* and *Leuconostoc* spp. These bacteria produce lactic and acetic acids from glucose and it is these compounds that are responsible for the sour flavour. Off-odours and off-flavours can also be produced when refrigerated beef is packaged in high-oxygen modified atmospheres (MA). *Moraxella*, *Pseudomonas* spp., and lactic acid bacteria have been identified in meat after 14 days' storage. Although strong off-odours were observed, many of the compounds responsible were not identified, as the headspace extracts were heavily contaminated with components from the packaging material. Compounds that appeared to be produced by the bacteria are 1-hexene, 1-heptene, benzene, ethyl acetate and methyl thiirane.

The spoilage characteristic of normal pH meat stored in vacuum packs are quite different from those of aerobically stored meat. Inoffensive sour acid odours assumed to arise from the acid end products of sugar fermentation are the principal products of spoilage. The compounds responsible are acetic, 2-methylpropanoic, 3-methylbutanoic and lactic acids, and these accumulate in naturally contaminated samples. A "sulphide" type spoilage has been observed in normal pH beef packed in material with high oxygen permeability. The causative organisms were strains of *Lactobacillus sake* that produce hydrogen sulphide from cysteine. With species of *Clostridium*, the off-odour is described as, "sulphurous", "fruity", "solvent-like" and "strong cheese". The compounds responsible for these odours were hydrogen sulphide, methyl mercaptan, dimethyl sulphide, dimethyl disulphide and trisulphide, methyl thioacetate, butanol, acetic acid, butanoic acid, ethyl butanoate and the butyl esters of C1 to C4 fatty acids. Metabolism of cysteine and methionine by the *Clostridium* sp. is the major source of the sulphur compounds, while butanol, acetic and butanoic acids are typical end-products of saccharolytic clostridia. The butyl esters, so characteristic of this spoilage organism, are presumably formed by the interaction of butanol with the acidic components. The esterification reaction is possibly mediated by the *Clostridium* spp. Mustiness or potato-like odours have been reported in a variety of meats, including beef, veal, pork and poultry. Bacteria responsible for such off-odours have been identified as predominantly *Pseudomonas perolens* and *P. taetrolens*. The compound responsible is 2-methoxy-3-isopropylpyrazine. This compound is also produced by *P. perolens*.

Taints of microbial origin in ruminants can also be produced in the digestive system of the living animals and migrate by way of the bloodstream to the muscle or depot fat. A classic example is the presence of "faecal" taints in beef cattle that have fed on pasture containing weeds of the *Lepidium* genus. The compounds responsible for such taints are indole and skatole and these compounds are formed by normal microbial decompostion of tryptophan in the animal's stomach (rumen). However, the presence of the *Lepidium* spp. in the gut interferes with the excretion of indole and skatole and as a consequence, these accumulate in the blood, and eventually in the muscle and depot fat.

Pork

Vacuum-packed pork has a short shelf life than beef, even though lactic acid bacteria dominate on both types of meat. Glycogen and glucose decrease at a faster rate in pork than in beef, leading to an earlier initiation of amino acid degradation in pork. Spoilage is evident as "sour", "cheesy" or "acidic" off-flavours. These taints have been attributed to the presence of short chain fatty acids and other end products of the dominant lactic acid bacteria. However, at high pH (>6.0) the spoilage is characterized by "putrid" and sulphury odours. The compounds responsible for these odours include methanethiol, methyl thioacetate, methyl thiopropionate, dimethyl disulphide, dimethyl trisulphide and bis-(methylthio)-methane. The major organisms present in the high-pH pork are Enterobacteriaceae and *Shewanella putrefaciens*; the latter is unable to grow on meat at normal pH. In a study, *Brochothrix thermospacta* was shown to be the dominant bacterium on refrigerated pork stored in atmospheres enriched with carbon dioxide and oxygen. The spoilage of this meat was characterized by "cheesy" and "acidic" off-odours and the compounds responsible were acetoin, diacetyl and the short-chain fatty acids—acetic, propanoic, 2-methylpropanoic and 3-methylbutanoic acids.

The so-called "boar taint" is attributed to the presence of skatole (3-methylindole) in the back fat of susceptible animals. Skatole is produced in the hindgut of animals by microbial degradation of the amino acid tryptophan, originating from dietary and endogenous protein. Some of the skatole is absorbed from the intestine and transported in the blood to the liver, where the majority is degraded. The undegraded skatole is deposited in the fat and muscle, where it can give rise to an unpleasant flavour. Bacteria found in the gut that can produce skatole include a strain of *Lactobacillus helveticus* and *Clostridium scatologenes*. Studies have shown that *C. scatologenes* can degrade tryptophan directly to skatole, but the strain of *Lactobacillus* requires the tryptophan to be first degraded to indole-3-acetic acid before the odorous compound can be formed. Control of the amount of tryptophan in the animal's diet is seen as one possible way of reducing the incidence of boar taint. An interesting feature here is that this off-flavour develops only in male pigs and can be minimized if males are castrated at appropriate age.

Poultry

Sulphide-like odours are important components of the spoilage aroma of poultry meat, and the bacteria responsible for such spoilage have been extensively studied. Bacteria identified include *Pseudomonas* spp., *Proteus* spp., *Citrobacter* spp., *Shewanella putrefaciens* and coryneforms. *Pseudomonas* spp. produce a wide range of sulphides, aliphatic aldehydes, alcohols and ketones. As previously seen for beef and pork, the sulphur compounds were principally responsible for the "putrid" odour associated with the spoiled chicken. Organoleptical detectable spoilage is noticeable about 6 days after storing the chicken under refrigerated conditions. In fresh and 4-day-old carcass, trace amount of 27 aliphatic alcohols, aldehydes and ketones can be identified. From the sixth day, the appearance of hydrogen sulphide and methanethiol as well as of fatty acid esters can be observed indicating incipient spoilage. By day 8, when spoilage is quite apparent, a total of 11 sulphur containing compounds and 21 fatty acid esters can be identified. It was considered in a study that hydrogen sulphide, methanethiol and the ethyl esters of acetic, hexanoic, 2-methylbutanoic and 2-methyl-3- butanoic acids were prime indicators of chicken carcass spoilage. The bacteria responsible for the production of these odours were principally *Pseudomonas* spp. Generally the autolysis of the bacteria did not play any significant role in the spoilage of refrigerated chicken. Instead, the off-odours associated with the spoilt flesh were the metabolites of the spoilage flora growing on the chicken carcass.

In all of the previous examples of microbial taints in meat and poultry, the compounds responsible were formed either on the surface of the food or in the stomach of the animal. However, in the case of a musty taint in eggs and poultry, the compounds were present in wood shavings used in the bottom of cages to prevent egg damage. The compound responsible was identified as 2,3,4,6-tetrachloroanisole. It was subsequently shown that this compound was present only in the superficial layers of some samples of undressed timber. This led to the speculation that the chloroanisole was derived either from fungal growth on the timber or as a result of the use of a wood preservative. The most important organism isolated was *Penicillium corylophilum and P. brevicompactum* apart from *Paecilomyces rariotii, Aspergillus sydowi* and *A. versicolor*. These results unequivocally showed that 2,3,4,6-tetrachloroanisole was formed in the poultry litter by microbial methylation of the corresponding chlorophenol.

In another outbreak of mustiness in eggs and broilers, an unidentified trichloroanisole was implicated as the cause of the taint. The origin of this compound was again wood shavings together with contaminated chicken feed. This compound was eventually identified as 2,4,6-trichloroanisole and it too was formed by endogenous fungi in wood shavings by microbial methylation. Fungi implicated in the methylation process were *Aspergillus sydowi, Penicillium crustosum* and *Scopulariopsis brevicaulis*. The identification of 2,4,6-trichloroanisole in tainted poultry was to become the first of many examples where this compound was implicated as the cause of mustiness in a variety of foodstuffs.

SEAFOODS

Fish

Most odorous compounds associated with the natural spoilage of fish are produced by bacteria that have been introduced after the animal was caught. Thus fish that are stored aerobically at ice temperatures have a microflora that is dominated by *Pseudomonas* spp. and *Shewanella putrefaciens*, while at ambient temperatures most of the bacteria identified belong to the Vibrionaceae and to a lesser extent, Enterobacteriaceae. However, the off-odours and off-flavours that are produced by such bacteria in aerobically stored fish also depend on the species of fish and its origin. Thus the spoilage of temperate climate marine fish is characterized by the development of ammoniacal, fishy, rotten and hydrogen sulphide type odours and flavours. A four phase pattern of change in flavour quality of fish after harvest have been described. These changes are the initial contamination of the fish by microbes, the growth of aerobic bacteria, development of surface slime and the growth of anaerobic bacteria. Unlike the aerobic bacteria that grow on beef, the aerobes that are found on fish do not produce diacetyl and related compounds from glucose, but break down this sugar to carbon dioxide and water. For most consumers, the first recognizable sign of spoilage is the ammoniacal odour of trimethylamine, a compound formed by the bacterial reduction of trimethylamineoxide. A number of well-defined spoilage bacteria, including *Shewanella putrefaciens, Photobacterium phosphoreum* and species of Vibrionaceae, is able to utilize trimethylamine oxide as part of their anaerobic respiration.

The next stage in the spoilage process is the development of sulphurous and putrid odours and the compounds responsible are principally formed by the microbial decomposition of amino acids. Most bacteria identified specifically as spoilage bacteria produce one or several volatile sulphides from either cysteine or methionine. Thus *Shewanella putrefaciens* and some species of Vibrionaceae produce hydrogen sulphide from cysteine, while *S. putrefaciens, Achromobacter* sp. and *Pseudomonas perolens* produce methanethiol, methyl disulphide and dimethyl trisulphide from methionine. In addition, *Pseudomonas fragi* is known to produce methanethiol, dimethyl sulphide and dimethyl disulphide in sterile fish muscle and *Pseudomonas fluorescens* methanethiol and dimethyl disulphide, presumably also from methionine. It is of interest that neither *Pseudomonas* spp., nor *Photobacterium phosphoreum* produces significant amounts of hydrogen sulphide; accordingly, it has been claimed that this compound is characteristic of iced fish spoiling under aerobic conditions.

Chilled fish muscle can also develop a fruity or ester-like odour during the early stages of spoilage, and the bacterium responsible for the production of this odour is *Pseudomonas fragi*. The compounds responsible for the taint are the ethyl esters of acetic, butanoic and hexanoic acids. In fish, these compounds are thought to be formed from monoamino, monocarboxylic acids, but in raw beef these same three esters are formed by *P. fragi* from glucose. The concentration of glucose in fish muscle is far less than in beef and , as a consequence, this bacterium can presumably adapt to use quite different substrates to produce the same compounds. Another *Pseudomonas* sp., *Pseudomonas perolens*, is responsible for a musty, potatolike odour in chilled fillets of cod during the early stages of spoilage. The compound responsible for this odour is 2-methoxy-3-isopropylpyrazine, an extremely odorous compound responsible for characteristic green odours in many vegetables. A biosynthetic pathway has been proposed for the formation of 2-methoxy-3-isopropylpyrazine by bacteria.

Earthy or muddy off-flavours in aquacultured and freshwater fish are major causes of concern in many countries. Aquatic microbes were first considered to be the source of these off-flavours as early as 1936. The same microbes were also considered to be responsible for similar off-flavours encountered in water supplies. The organisms involved included certain species of actinomycetes, and cyanobacteria, also called blue-green algae. However, it was not until the late 1960's that geosmin and 2-methylisoborneol were identified as the compounds responsible for these off-flavours. Both compounds were initially identified as metabolites of actinomycetes but were subsequently shown to be produced by a variety of cyanobacteria.

Geosmin was first identified as the cause of muddy flavours in rainbow trout that had been grown in saline lakes of Western Canada. It was shown that 10 species of cyanobacteria present in the lakes were capable of producing geosmin; they belonged to *Oscillatoria* spp., *Lyngbya* spp. and *Symploca* sp. Organisms considered to be the most likely cause of the problem were *O. splendida, O. prolifica, O. cortiana and L. aestuary*. In 1980, 2-methylisoborneol was implicated for the first time as a cause of a muddy flavour in commercial fishes. It has led to the identification of a wide range of cyanobacteria, including species of *Anabaena, Aphanizomenon, Microcystis, Raphidiopsis* and *Oscillatoria*. The highest levels of geosmin are associated with blooms of *Anabaena spiroides*.

Studies have shown that both geosmin and 2-methylisoborneol are transferred to the muscle of fish and shrimp by two principal routes: (i) the first involves the release of the compounds by the organism into the water supply, followed by absorption through the gills and skin of the animal, (ii) the second depends on the ingestion of the algae and absorption of the compounds from the small intestine and stomach. By far the major route was absorption through the gills.

Crustaceans

The microbiological origin of off-flavours in crustaceans have received far less attention than those in freshwater and ocean fish. A deep sea prawn, *Hymenopenaeus sibogae*, is known to possess a distinctive garlic-type flavour that, on occasions, renders catches unacceptable to consumers. The prawns are not microbially spoiled and the taint is exclusively concentrated in the animal's gut. The compound responsible was identified as bio-(methylthio)-methane. This compound is a bacterial metabolite that is produced in Gouda cheese and is a metabolite of *Shewanella putrefaciens*. The organism or organisms responsible for its formation in the prawn gut have not been identified; however, a pathway based on the interaction of the bacterial metabolites methanethiol and formaldehyde has been described. Methanethiol is a metabolite of both *Pseudomonas* and *Achromobacter* spp., while formaldehyde is formed by microbial oxidation of trimethylamine by marine *Pseudomonas* spp. bis-(methylthio)-methane has also been identified as the cause of a similar but more intense garlic-like odour in lobsters.

In another study, dimethyl trisulphide was identified as the cause of an unpleasant cooked-onion odour in rejected catches of deep sea prawns. Indole was also identified in this material. The flesh of the prawns was discoloured and in most cases slimy to touch. A microbiological examination confirmed that the prawns were contaminated with a variety of bacteria, but it was not possible to identify the causative organisms. However, it is known that both *Shewanella putrefaciens* and *Pseudomonas perolens* produce dimethyl trisulphide in fish muscle at 0°C and such organisms could be expected to be present in the contaminated product.

In another example, a garlic-like taint in prawns was shown to be due to the presence of trimethylarsine. This compound was found in highest concentration in deep sea prawns but was also present in some shallow water species. Again, the prawns were not microbially spoiled and the trimethylarsine was concentrated in the gut. Environmental arsenic is readily transformed into alkylarsines by a range of microbes under reducing conditions, including the soil mould *Scopulariopsis brevicaulis*, a yeast, *Candida humicola*, from sewage sludge and *Methanobacterium* isolated from ocean mud. In addition, freshwater green algae, such as *Ankistrodesmus* spp., *Chlorella* spp., *Selenastrum* spp. and *Scenedesmus* spp. have all been shown to convert arsenite to trimethylarsine oxide. Prawns are known to accumulate inorganic arsenic and in the presence of adventitious bacteria, such compounds could be converted to trimethylarsine.

MILK AND DAIRY PRODUCTS

Milk

Flavour defects in raw and pasteurized milk as a result of microbial metabolism can occur at any stage of production and processing. Furthermore, the protein, carbohydrate, fat, minerals, vitamins and water content of milk make it an ideal medium for microbial growth. The types of microbial flavour defects that can be produced include, "acid", "malty", "fruity", "unclean", "butter" and "putrid". Each of these is the result usually of complex enzymic activities of microbes in the raw or pasteurized milk.

The principal acid-producing bacteria in milk are *Streptococcus*, *Pediococcus*, *Leuconostoc*, *Lactobacillus* spp. and members of the Enterobacteriaceae. Production of lactic acid from lactose, in combination with short- and medium-chain fatty acids produced by one or more of these bacteria, is the cause of acid flavours in raw and pasteurized milk. Because of the ubiquitous nature of *Streptococcus lactis* and *Streptococcus cremoris* in the environment of milk production, most milk is unintentionally inoculated with these organisms immediately after milking or during processing. If the milk is not cooled to 4.4°C or below, it will eventually develop an acid flavour as a result of proliferation of these bacteria and their conversion of lactose to lactic acid.

Heat-resistant lipases from psychrotrophic bacteria, predominantly *Pseudomonas* spp., also contribute to the flavour of rancid milk by the release of short- and medium-chain fatty acids from milk triglycerides. For example, the lipase isolated from the bacterium *Pseudomonas fluorescens* yields butanoic, hexanoic, octanoic and decanoic acids when added to milk fat slurries. However, normal milk also contains lipases, which can contribute to acid off-flavours, but such contributions usually only occur when milk experiences intense agitation, as on homogenization.

An off-flavour and off-odour described as "cooked", "burnt", "caramel" and "malty" develops in raw milk when it is contaminated by the bacterium *Streptococcus lactis* biovar. *maltigenes*. The compound principally responsible for this off-flavour is 3-methylbutanal and is formed from the amino acid leucine and its precursor a-ketoisocaproic acid. In subsequent studies with skimmed milk cultures, the researchers found that this bacterium also produced 2-methylpropanal, 3-methylpropanol, 2-methylbutanol and 3-methylbutanol. These compounds are formed from the amino acids valine, isoleucine and leucine respectively. All of these compounds contributed to the "malty" off-flavours and off-odour. It is of interest that other strains of *S. lactis* and all strains of *S. cremoris* were incapable of producing these off-flavour compounds in milk cultures.

Fruity odours that occur in pasteurized milk and other dairy products are principally produced by *Pseudomonas fragi* as a result of post-pasteurization contamination. The off-flavour has been described as strawberry-like, ester-like or fruity, and usually develops when pasteurized milk is stored for extended period of time under refrigeration. The compounds associated with this fruity aroma are ethyl esters and those responsible of the odour and flavour are ethyl acetate, ethyl butanoate and ethyl hexanoate. *P. fragi* is strongly lipolytic, removing fatty acids preferentially from the 1- and 3-positions of the triglycerides. Butanoic and hexanoic acids are esterified in the 3-position of constituent triglycerides of milk-fat, and accordingly are the major products of lipolysis by this organism.

Unclean, bitter and putrid flavours are produced by psychrotropic organisms in pasteurized milk. These organisms can multiply at or below 7°C. Although most raw milk psychrotrophs are heat sensitive, some heat-resistant species have been isolated. These latter bacteria can cause spoilage of stored heat-treated milk or milk products either during their growth or by the involvement of specific enzymes that they produce. The two groups of extracellular enzymes that are of most importance are the proteinases and the lipases, which can act directly on micellar casein or on the fat globules in the milk, thus producing unclean, bitter and putrid flavours. Proteolytic strains of *P. fluorescens*, *P. fragi* and *P. putrefaciens* have been shown to degrade caseins and whey products when grown in pure cultures in pasteurized milk at low temperatures. The resultant milk had bitter flavours and peptides were suspected as the cause; however, the actual compounds were not identified. The causes of so-called unclean and putrid flavours in milk also remain unidentified and may be caused by a combination of several compounds, including free fatty acids, formed by different bacteria.

Cheese

Off-flavours of microbiological origin also occur in fermented milk products. For example, "fruitiness" frequently occurs in Cheddar cheese and is attributed to inadequate sanitation during the production of the milk or in the cheese factory. Manufacturing faults can also contribute, such as low acidity or salt content or high moisture content. Comparison of normal and fruity cheddar cheeses showed that the latter were high in ethanol, ethyl butanoate, ethyl hexanoate and ethyl octanoate, but lower in total free fatty acids. Some workers have shown that some strains of *Streptococcus lactis* could impart a characteristic fruity flavour to Cheddar cheese, while others showed that *S. lactis* and *Streptococcus diacetylactis* strains could produce a fruity-fermented flavour with open-textured cheese. Subsequent studies confirmed that strains of *S. lactis*, *S. diacetylactis*, *Lactobacillus casei*, a *Lactobacillus* sp. and two strains of *Pseudomonas* contained esterases capable of esterification of butanoic and hexanoic acids with ethanol. Accordingly, any of these organisms can be involved in the production of fruity flavours in cheese.

Esters have also been implicated as a possible cause of a fermented, yeasty flavour defect in Cheddar cheese. The cheese contained elevated levels of ethanol, ethyl acetate and ethyl butanoate, and was contaminated with large numbers of yeasts, especially *Candida* species.

A potato-like off-flavours in smear-coated cheese was shown to be caused by the presence of 2-methoxy-3-isopropylpyrazine. *Pseudomonas* spp. were shown to be responsible for the production of this compound.

Comte cheese was shown to be caused by the presence of 3-methoxy-2-propylpyridine; however, the organism responsible for this novel odorous compound was not identified. The organisms responsible for the production of a kerosene-like taint in commercial Feta cheese preserved with sorbic acid were also not identified; however, the compund responsible was shown to be *trans*-1,3-pentadiene. Earlier studies had shown that sorbic acid (*trans*, *trans*-2,4-hexadienoic acid) could be decarboxylated by certain species of *Penicillium* isolated from Cheddar cheese preserved with sorbic acid. Fungi capable of this reaction included resistant strains of *Penicillium roqueforti*, *P. notatum*, *P. frequentans* and *P. cyaneofulvum*. The role of *Penicillium* spp., in the production of *trans*-1,3-pentadiene was confirmed by the inoculation of packaged cheese containing sorbic acid with a sorbate resistant strain of *P. roqueforti*. After incubation of 2 weeks at 5°C, *trans*-1,3-pentadiene was detected in the headspace above the cheese. In the same study, samples of kerosene-like tainted cheese were found to be contaminated only with *Penicillium* spp. This finding prompted the authors to claim that only fungi of the genus *Penicillium* were able to produce *trans*-1,3-pentadiene from sorbic acid.

Butter, Yoghurt and Ice cream

Under certain conditions, butter cultures develop a flavour defect described as "green" or "yoghurt-like", which is caused by the production of an excess of acetaldehyde relative to the amount of 2,3-butanedione produced. In desirably flavoured butter, the ratio of these compounds is 1:4; however, when this ratio is less than 1:3, the defect is observed. Subsequent studies with combinations of single-strain starter cultures showed that the green flavour defect occurred when there was an excess number of *Streptococcus lactis* or *S. diacetylactis* in relation to the population of *Leuconostoc citrovorum*.

Yoghurt is another dairy product that can on occasions develop unpleasant flavours as a result of microbial contamination after processing. Incidents have been investigated of vanillin-flavoured yoghurt that developed a smoky or phenol flavours and odour. On each occasion the chemical responsible was shown to be guaiacol. A similar smoky taint has also been recognized in spoiled chocolate ice cream. Here the causes of the problem were guaiacol and 2-ethoxyphenol. The precursors of these compounds were the flavouring additives vanillin and ethylvanillin. It has been shown that species of *Streptomyces* can degrade vanillin to guaiacol in culture media.

FRUITS AND VEGETABLES

Taints caused by the microbiological contamination of foodstuff of plant origin have received far less attention than those occurring in meat, fish or dairy products. A possible explanation is that fruits, vegetables and cereals are generally more resistant to microbial infections than the high-protein animal products.

Fruits

The first reported incidence of a microbial taint in a fruit product was a "buttermilk" off-flavour in single strength orange juice or orange juice concentrate that had resulted from bacterial contamination. The organisms responsible for the taint were strains of acid-tolerant bacteria: *Lactobacillus* spp., including *L. mesenteroides*, and *Leuconostoc* spp., including *Leuconostoc plantarum*. Growth of these microbes in commercial evaporators during the first stage of juice concentration was the major cause of the off-flavour. The compounds responsible for the off-flavours were identified as diacetyl and acetoin, together with 2,3-dihydroxybutane. Carbohydrates present in the orange juice were the most likely precursors of these compounds.

By far the most common cause of citrus fruit decay are *Penicillium* rots due to *P. italicum* and *P. digitatum*, termed blue rot and green rot, respectively. A particularly characteristic odour is associated with both of these rots, and should such fruit be juiced the product has an off-flavour frequently described as "old fruit" or "rotten fruit". The identity of the compound responsible has not been reported. However, 4-vinylguaiacol produces a similar off-flavour in heat abused juice and is formed by the enzymatic or thermal decarboxylation of ferulic acid, a natural component of orange juice. 4-vinyl guaiacol is also formed in beer and worts from the same acid by adventitious wild yeasts and bacteria. Accordingly, it is possible that fungi might contain similar enzymatic systems that could bring about this reaction in infected citrus fruit. The identification of this *Penicillium*-induced taint is long overdue.

Papaya puree prepared by macerating papaya fruit without special treatment invariably develops off-flavours and off-odours as a result of natural enzymatic or microbial activity. Such taints are described as "sulphury", "butyric", "acrid", "pungent" and "sour" and the compounds believed to be responsible were identified as the methyl esters of butanoic, hexanoic and octanoic acids. The same fatty acids were also present in the free form together with pentanoic, heptanoic and nonanoic acids. Bacteria, identified as gram-positive diplococci and gram-positive rods, were present in the puree, and production of the volatile fatty acids was shown to occur with an increase in the numbers of bacteria present in the puree.

Fruits such as apples, pears and cherries infected with the fungus *Penicillium expansum* frequently develop a pungent earthy odour. Investigation of the volatiles produced by this fungus led to the identification of geosmin as the compound responsible.

In 1984, an off-flavour in apple juice was associated with the presence of a heat-tolerant, acid-dependent, spore-forming bacterium, subsequently identified as *Alicyclobacillus acidoterrestris*. Examinations indicated that two compounds were involved: guaiacol was identified in orange juice and 2,6-dibromophenol in apple juice.

In the 1980s, a musty taint in packaged dried vine fruit was a major cause of concern in many producing countries. The taint was sporadic, and only appeared to occur when the dried fruit was transported in general purpose freight containers. The compound responsible was eventually identified as 2,4,6-trichloroanisole. This compound was shown to be formed from the corresponding chlorophenol in the fibre board packaging material by adventitious fungi. The volatile metabolite was then absorbed by the packaged fruit during transportation. Some 17 species of fungi isolated from such materials were shown to produce 2,4,6-trichloroanisole when grown on fibreboard inoculated with 2,4,6-trichlorophenol, species of fungi identified included *Aspergillus*, *Eurotium* and *Penicillium* together with species of *Paecilomyces*, *Merimbla* and *Fusarium*. Species responsible for the highest levels of methylation were *Paecilomyces variotii*, *Fusarium oxysporum* and *Aspergillus flavus*. Recycled newsprint was identified as the major source of the chlorophenols in the fibreboard packaging materials.

Another source of contamination is the timber floors of the freight containers used for the transportation of the packaged fruit. 2,4,6-trichloroanisole was again the compound responsible: this compound and its precursor, 2,4,6-trichlorophenol, were isolated from the container floors, together with 19 species of fungi with the known capacity to biomethylate chlorophenols. The cholorophenols had been introduced into the container floors either as a wood preservative or through accidental spoilage.

2,4,6-trichloroanisole has also been identified as the cause of a phenolic, iodine or musty flavour in some Brazilian coffees, the so-called *"Rio"* flavours. The chloroanisole was produced in the green coffee beans and microbial analysis showed high levels of contamination. Fungi, including *Aspergillus versicolor* and *Wallemia sebi,* were isolated from the surface of the beans, while different strains of *Fusarium* were found in the interior of the beans. *Pseudomonas* were also identified inside the beans.

Another fungal metabolite that has the potential to cause a musty taint in packaged food is 2,4,6-tribromoanisole. This compound can be formed from the fungicide 2,4,6-tribromophenol by microbial methylation. *Paecilomyces variotii* has been identified as one common fungus that can bring about this conversion in fibreboard packaging materials.

Desiccated coconut on storage occasionally develops an off-odour described as ketonic or perfume like. The cause of such off-odours is normally the contamination of the product by xerophilic fungi. Fungi that are known to cause such problems include *Eurotium amstelodami, E. chevaliere, E. herbariorum* and *Penicillium citrinum.* On coconut, these fungi produce odd-numbered methyl ketones (C5-C11). These are derived from even-numbered short-chain fatty acids, with one more carbon atom than the ketones by a modified β-oxidation of the parent fatty acids thus producing 2-pentanone, 2-heptanone, 2-nonanone and 2-undecanone. Other compounds considered to contribute to the off-odour were 2-heptanol and 2-nonanol. The same four ketones responsible for the off-flavour in coconut have recently been shown to cause an "anty" like taint in low salt margarine. The fungus responsible for this problem was identified as *Penicillium solitum.* Taints have also been detected in dessicated coconut which had been previously sterilized either by heat-treatment or gamma irradiation. Storage of such material at high humidity led to the development of a pungent off-odour and the material was heavily contaminated with the bacterium *Bacillus subtilis.* The compounds responsible for the off-odour were subsequently identified as 2,3,5,6-tetramethylpyrazine and 2,3,5-trimethylpyrazine.

Taints in cocoa beans   There is a general agreement among chocolate manufacturers that mouldy taints arising from fungal infections within the cocoa beans are the major causes of mouldy flavours in chocolate products.

Compounds that have been identified as the cause of such problems include 2-methoxy-3-isopropyl-pyrazine (formed by bacteria and actinomycetes) and 2-methylisoborneol (formed by actinomycetes and fungi).

Vegetables

One of the earliest reports of a vegetable taint caused by microorganisms was the occurrence of musty off-flavours in dry white navy beans (*Phaseolus vulgaris*). The compound responsible was geosmin. It was suggested that the likely cause of the problem was actinomycetes, either growing on the beans themselves or in the water supply used during the cultivation of the beans.

Another vegetable off-flavours problem that was assumed to be the result of contamination with actinomycetes was the occurrence of a musty-earthy flavour in canned champignons. The compound responsible was identified as 2-methylisoborneol. Most consumers are aware of the unpleasant odour associated with rotten potatoes, but similar odours have also been observed in processed potato products. Frozen chips with a pig sty like off-flavour were found to contain skatole, indole and *p*-cresol. Examination of potato tubers in storage sheds at the factory producing the chips revealed that the tubers in contact with one of the walls exhibited malodorous soft rots and were heavily contaminated with the above compounds. In addition, tubers not affected by the rot, but stored adjacent to the infected material, were also contaminated to various degrees with the offending compounds. The rotten tubers were found to be infected with the aerobic bacteria *Erwinia carotovora* and *E. chrysanthemi*, together with an anaerobic bacterium believed to be the *Clostridium scatologenes*. This anaerobic bacterium is known to produce skatole, indole and *p*-cresol. Further investigations indicated that the infections had occurred in the field, where the tubers had suffered physiological damage as a result of unusual climatic conditions.

CEREAL AND CEREAL PRODUCTS

Fungal infections of cereals during storage is a major cause of concern in many countries. The principal problem associated with such infections is the production of mycotoxins rather than off-flavours and off-odours. However, many odorous fungal metabolites produced on grain cultures have been identified, but they have been studied as indicator compounds of fungal infections rather than as food taints.

Earthy and musty taints have been observed in some cereal products, and studies have shown that actinomycetes capable of producing geosmin and 2-methylisoborneol can grow in grain bread containing 45–50% moisture, and on wheat grain at 16–18% moisture. However, it was not until 1991 that geosmin was identified as the cause of an earthy taint in a cereal product. A subsequent microbiological examination of the flour led to the isolation of the actinomycete *Streptomyces griseus*. Geosmin has also been identified as the cause of an earthy flavour in corn grain and in corn flour.

A bacterial metabolite that has not been identified in foodstuff as yet is 2,6-dimethyl 3-methoxypyrazine. This compound has an intense odour described as "foul drains" or "sour dishcloths" and is formed by an aerobic, gram-negative bacterium in oil emulsions.

Thus it is interesting to note the change in emphasis that has occurred over the past 40 years with regard to the types of microbes involved in such off-flavour problems. In the 1960's, studies on the role of bacteria in the spoilage of meat, fish and dairy products was paramount. However, in the 1980s and 1990s there has been a significant shift in interest towards problems of fungal origin. Actinomycetes have also been observed to play a more important role in the cause of off-flavours in foodstuff. Only yeasts appear to have been overlooked. Perhaps, in future, these organisms will also be received a greater attention and be shown to produce metabolites deleterious to the flavour quality of food.

1. Define taints.
2. Discuss microbial taints in meat and meat products.
3. What are the microbial taints in seafoods?
4. Discuss microbial taints in milk and milk products.

Unit IV

FOOD POISONING

26

INTRODUCTION

Food poisoning includes ill effects caused by the ingestion of contaminated food by many ways apart from microbial agents. They may be:

1. through the addition of poisons
2. through eating of inherent poisonous substance such as certain mushrooms, fish and molluscs by mistake
3. adulteration of food with poisonous substance such as *Argemone mexicana* in mustard producing epidemic dropsy.

The term "food poisoning" is however restricted only to acute gastroenteritis due to bacterial pollution of food or drink. The term "food-borne" disease is defined as: "A disease, usually either infectious or toxic in nature, caused by agents that enter the body through the ingestion of food". Food-borne diseases may be classified as :

FOOD-BORNE INTOXICATIONS

Food-borne intoxications are caused

1. due to naturally occuring toxins in some foods, including
    i.   lathyrism (beta-oxalyl amino-alanine)
    ii. endemic ascitis (Pyrrolizidine alkaloids)
2. due to toxins produced by certain bacteria, including
    i.   botulism
    ii. staphylococcal toxins
3. due to toxins produced by some fungi, including
    i.   aflatoxin
    ii. ergot
    iii. Fusarium toxins

 4. due to toxins produced by some algae, like
   i. planktonic dinoflagellates
   ii. diatoms
   iii. cyanobacteria
 5. due to food-borne chemical poisoning

FOOD-BORNE INFECTIONS

Food-borne infections include

 1. Bacterial infections such as
   i. salmonellosis
   ii. shigellosis
   iii. *E. coli* diarrhoea
   iv. cholera
   v. streptococcal infection
   vi. brucellosis
   vii. listeriosis
 2. Viral infections such as
   i. viral gastroenteritis
   ii. Hepatitis A
 3. Parasitic infections such as
   i. taeniasis
   ii. trichinellosis

1. Define food-borne disease.
2. Define food poisoning.
3. List the various causes of food-borne intoxications.
4. List the various bacterial food-borne infections.

27

BACTERIAL FOOD POISONING

FOOD-BORNE DISEASE AND FOOD POISONING

Food-borne disease is a disease caused by ingestion of food contaminated by any agent, chemical or biological. Food poisoning is an acute enteritis caused by the ingestion of food, characterized by diarrhoea, vomiting, with or without fever and abdominal pains. Food poisoning is normally associated with the small and large intestine. Certain types of food poisoning are described as intoxications and others as infections.

Intoxications

Intoxication involves food poisoning in which the organism grows in food and releases a toxin from the cells. When the toxin is ingested along with the food, it gives rise to the food poisoning syndrome. The presence of the organism in the food is irrelevant to disease production. It is the toxin that gives rise to the disease. Bacterial toxins that produce intoxications are the exotoxin types of either enterotoxin (affecting the gut) as in staphylococcal intoxication or neurotoxin (affecting the nervous system) as in botulism.

Another category of intoxications are the mycotoxicoses (due to ingestion of mycotoxins) and the diseases caused by algal toxins (shell fish poisoning). Generally, intoxications have short incubation periods.

Infections

These involve food poisoning caused by the ingestion of live organisms. The organisms grow in the gastrointestinal tract to produce the disease. Most microbial food poisonings fall in this category. For example, salmonellosis caused by *Salmonella* species like *Salmonella typhi*. Enteritis associated with food poisoning infections is due to the production of exotoxins or endotoxins that act as enterotoxins.

In certain other types of food poisoning, as in the case of *Clostridium perfringens*, live cells need to be ingested for the disease to occur but the organism does not grow and reproduce in the gut. Vegetative cells sporulate after ingestion and enterotoxin released causing the disease symptoms. Since live cells are needed to be ingested to cause the food poisoning, it can be considered as a food-borne infection.

FOOD POISONING BY *BACILLUS CEREUS*

Bacillus cereus causes two different types of food poisoning: the diarrhoeal type and the emetic type. The diarrhoeal type of food poisoning is caused by an enterotoxin produced during vegetative growth of *B. cereus* in the small intestine while emetic toxin is produced by cells growing in the food. *B. cereus* food poisoning is under reported, as both types of illness are relatively mild and usually lasts less than 24 hours.

Characteristic Features

B. cereus is a gram-positive, spore-forming, motile, aerobic rod but it grows well anaerobically. The genus is divided into six different subgroups and *B. cereus* is classified in the *Bacillus subtilis* group. Four members of this group are closely related: *B. cereus, B. thuringiensis, B. anthracis* and *B. mycoides*. The variation among these four species is mainly due to genes on episomes rather than genes on the chromosome.

Bacillus species sporulate easily after 2–3 days on most media. It grows well on food that has been heat treated (48°C) which usually causes spore germination. The organism is unable to grow below 10°C.

Survivability Characteristics

The organism can survive pasteurization through sporulation. It cannot grow in milk and milk products stored at temperatures between 4°C and 8°C. But the psychrotrophic strains that have developed can grow at temperatures as low as 4–6°C. Since it sporulates, it can survive pH changes (acidity) and it is a problem in home-canned foods.

Spores

The spore of *B. cereus* is an important factor in food-borne illness as it is more hydrophobic than any other *Bacillus* spp. which enables it to adhere to several types of surfaces. Hence, it is difficult to remove from equipments during cleaning. *B. cereus* spores also possess appendages or pili that help in adhesion. These adherence properties not only enable spores to resist normal sanitation procedures, and thus contaminate foods during processing, but also aid in binding to epithelial cells. Spore adhesion to epithelial cells followed by germination and production of enterotoxin may explain the long incubation periods observed in some food-associated utbreaks.

Reservoirs

The organism is widespread being frequently isolated from soil and growing plants. From this natural environment it is easily spread to foods, especially those of plant origin. Through cross-contamination, it may then be spread to other foods such as meat products. The problems in milk and milk products are caused by *B. cereus* which is spread from soil and grass to the udders of the cow and into the raw milk.

Food-borne Outbreaks

The number of outbreaks of *B. cereus* food poisoning is highly underestimated, the main reason being the relatively short duration of both types of diseases. The dominant type of illness caused by *B. cereus* differs from country to country. In Japan, the emetic type is reported about 10 times more frequently than the diarrhoeal type and in Europe and North America, the diarrhoeal type is more frequently reported. Confirmation of *B. cereus* as the cause of food-borne outbreak requires:

1. Isolation of strains of the same serotype from the suspected food and from faeces or vomitus of the patients.

2. Isolation of significant numbers of *B. cereus* serotype known to cause food-borne illness from the suspected food or from the faeces or vomitus of affected individuals.

3. Isolation of significant numbers of *B. cereus* from the suspected food and from the faeces or vomitus of the patients.

Characteristics of the Disease

In some cases, both the illness caused by *B. cereus* can be seen. Although there has been a debate about *B. cereus* food poisoning being an intoxicating type, the long incubation period (>6 hours; average 12 hours) and studies revealing the degradation of the enterotoxin in the gut before reaching the ileum have made this fact unlikely. Infective doses range from 5×10^4 to 10^{11} cells per gram. But foods containing more than 10^4 cells/g may not be safe for consumption.

Pathogenicity

Bacillus cereus toxins   They fall into 4 groups:

1. *The haemolysins*   The first of these is the well defined cerolysin. These toxins are produced in an inactive, oxidized form with sulphydryl groups in

disulphide bonds; reduction frees the sulphydryl groups to produce the active form. The binding site of the thiol activated cytolysins on the eukaryotic cell membrane is cholesterol and the result of binding is a characteristic morphological pitting and micropuncturing of the cell membrane, visible through electron microscope, the result of which is loss of control of free ion exchange and intracellular K^+ ions. But net flow of ions and water into the cell causes swelling and the cell ruptures. After intravenous injection in mice, cereolysin is instantly lethal.

2. *The diarrhoeal enterotoxin*   For many years cereolysin was thought to be the only toxin causing all the diarrhoeagenic episodes. It became clear in the 1970s that it was possible to distinguish and separate an entity, termed enterotoxin because of its ability to produce fluid accumulation in ligated rabbit ileal loops, from the phospho-lipolytic and haemolytic entities. It is regarded as a multi-component protein complex. In addition to eliciting fluid accumulation in ligated loops, it causes severe mucosal damage and these activities reflect in increased vascular permeability with marked necrosis in rabbit or guinea pig skin tests. The degree of production of this necrotic enterotoxin has also been shown to correlate roughly with the severity of infection in non-gastrointestinal *B. cereus* infections. This toxin is responsible for death in severe cases of *B. cereus* infection.

B. cereus* produces at least two different enterotoxins. Although several proteins may be involved in *B. cereus* food-borne illness, only one type of enterotoxin (B component of the haemolysin BL) is likely responsible for the major symptom. The haemolysin is made of three components: L1, L2 and B (the enterotoxin) and all the three must be present for full enterotoxin activity.

Another toxin, enterotoxin T, is composed of a single protein. Much has not been unravelled about the T toxin.

3. *The emetic toxin*   Highly stable compound which is formed at temperatures < 40°C but survives 126°C for 1.5 hours, pH extremes and proteolytic enzymes. It may be associated with sporulation or breakdown products from foods. It is named *cereulide* and it is thought to be an enzymatically synthesized polypeptide.

4. *The phospholipase C group of enzymes*   They are three in numbers, phosphatidylcholine hydrolase, phosphatidylinositol hydrolase and sphingomyelinase. They act on the phospholipases C and sphingomyelin of biological membranes. Under normal circumstances, they probably do not gain access to the phospholipids of cell membranes, but they may possibly act secondarily after exposure of these substrates in the course of other pathological process.

LABORATORY DIAGNOSIS AND IDENTIFICATION

Media Polymyxin pyruvate egg yolk mannitol bromothymol blue agar (PEMBA) is most reliable. Another medium, phenol egg yolk polymyxin agar is also used for selective isolation of *B. cereus* from food samples.

Method Serially diluted food samples can be enumerated (spread plate) using the selective media. After incubation (30°C for 24–48 hours) the specifically coloured colonies (dull peacock blue) confirms the presence of *B. cereus*. In addition, the lecithinase activity can also be detected. They are further confirmed by staining and subculturing onto blood agar for observing β-haemolysis.

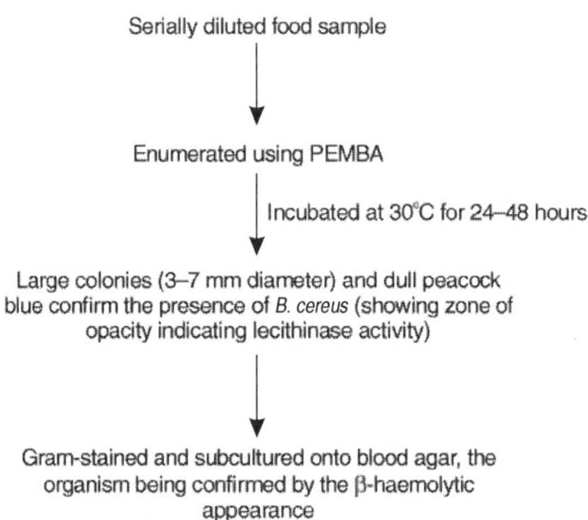

FOOD POISONING BY *CLOSTRIDIUM BOTULINUM*

Clostridium botulinum causes botulism. First recognized as a food-borne disease in the late 1800 and since then it has been a major concern of food processors and consumers. Currently, four categories of human botulism are recognized.

1. Food-borne botulism is caused by eating food contaminated with preformed botulinum neurotoxin (BoNT).

2. Infant botulism is caused by ingestion of viable spores that germinate, colonize, and produce neurotoxin in the intestinal tracts of infants under one year of age.

3. Wound botulism results from infection of a wound with spores of *Clostridium botulinum*, which grow and produce neurotoxin in the wound.

4. Unclassified includes cases of unknown origin and adult cases which resemble infant botulism.

Classification

Gram-positive, anaerobic, rod-shaped, spore-forming bacterium. There are seven types of *C. botulinum*, A, B, C, D, E, F and G, based on the serological specificity of the neurotoxin produced. Food-borne botulism is associated with types A, B, E and very rarely F.

The species is also divided into four groups based on physiological differences as follows:

Group I   All type A strains and proteolytic strains of types B and F produce neurotoxin. Optimal temperature for growth is 37°C with growth occurring between 10 and 48°C. Spores have a high heat resistance (D_{100} = 25 minutes). To inhibit growth, the pH must be below 4.6, salt concentration above 10% and the a_W below 0.94.

Group II   All type E strains and non-proteolytic strains of types B and F. They have a lower optimum growth temperature (30°C) and grows at temperature as low as 3.3°C. Spores have a D_{100} values of less than 0.1 minute. Strains are inhibited by a pH below 5.0, salt concentrations above 5% or a_W below 0.97.

Groups III   All type of C and D strains.

Group IV   Type G strains.

Survivability Characteristics

Temperature, pH, a_W, redox potential, added preservatives and the presence of other microorganisms are the major factors controlling growth of *C. botulinum* in foods.

1. *Low temperature*   Refrigerated storage is used to prevent or inhibit the growth of *C. botulinum*. The established lower limits are 10°C for group I and 3.3 for group II. Production of neurotoxin generally requires weeks at the lower temperature limits for group I and group II organisms. Survival of spores of group II in pasteurized, refrigerated products is of concern because of their ability to grow at refrigeration temperatures.

2. *Thermal inactivation*   Thermal processing is used to inactivate spores of *C.botulinum* and is the most common method of producing shelf stable foods. Spores of types A and B are the most heat-resistant, having D_{121} values of between 0.1 and 0.2 min. These spores are of particular concern in the sterilization of canned low-acid foods. The canning industry has adopted a D value of 0.2 minutes at 121°C as a standard for calculating thermal processes. The Z value (the temperature change necessary to cause a 10-fold change in the D value) for the most resistant strains is approximately 10°C which has also been adopted as a standard. Despite variations in D and Z values, the adoption of a 12 D process as the minimum thermoprocess applied to commercial canned, low-acid foods by the canning industry has ensured the production of safe products.

3. *pH* The minimum pH allowing growth of *C. botulinum* group I is 4.6, for group II is 5.0. Substrate, temperature, nature of the acidulant agent, presence of preservatives, a_w and Eh are the factors that influence the acid tolerance of *C. botulinum*. Acid-tolerant microbes such as yeasts and moulds may grow in acidic products and raise the pH in their immediate vicinity to a level that allows growth of *C. botulinum*.

4. *Salt and a_w* The salt concentration in the aqueous phase, called the brine concentration is critical. The growth limiting brine concentrations are about 10% for group I and 5% for group II. The solute used to control a_w may influence these limits. Generally, NaCl, KCl, glucose and sucrose show similar effects, while glycerol allows growth at lower a_w.

5. *Atmosphere and Eh* Modified atmosphere packaging (MAP) is being increasingly used to extend the shelf life and improve the quality of foods. MAP has been a concern because of creating conditions that might promote growth of *C. botulinum*. CO_2 is used in MAP to inhibit spoilage and pathogenic microbes, but CO_2 may stimulate *C. botulinum*. Only 75% CO_2 inhibited *C. botulinum*. Levels of 15–30% for CO_2 does not inhibit the organism. While it is commonly assumed that *C. botulinum* cannot grow in foods exposed to oxygen, the redox potential (Eh) of most such foods is usually low enough to allow its growth since initial atmospheres containing 20% O_2 does not delay neurotoxin production by *C. botulinum* in pork.

6. *Preservatives* Nitrite has several functions in cured meat products, an important role is the inhibition of *C. botulinum*. Its effectiveness in the inhibition of the organism is dependent on complex interactions among pH, salt, heat treatment, time and temperature of storage and the composition of food. Nitrite is depleted from cured foods and the depletion rate is also dependent on product formulation, pH and time and temperature during processing and storage. A significant contribution of nitrite to the inhibition of *C. botulinum* continues even when nitrite is no longer detectable. Nitrite reacts with many cellular constituents and appears to inhibit *C. botulinum* by more than one mechanism, including reaction with essential iron–sulphur proteins to inhibit energy yielding systems in the cell. The reaction of nitrite, or nitric oxide, with secondary amines in meats to produce nitrosamines, some of which are carcinogenic has led to regulations limiting the amount of nitrite used. Sorbates, parabens, nisin, phenolic antioxidants, fumarates, etc. are also active against *C. botulinum*.

7. *Other microorganisms* The growth of other microorganisms in foods has a very significant effect on the growth of *C. botulinum*. Acid-tolerant yeasts and moulds may make the environment more favourable for growth of *C. botulinum*. Other microbes may inhibit *C. botulinum*, either by changing the environment or by producing specific inhibitory substances or both. For example, lactic acid bacteria including *Lactobacillus, Pediococcus,* and *Streptococcus* can inhibit growth of *C. botulinum* in meat products by reducing the pH and by the production of bacteriocins.

8. Inactivation by irradiation   *C. botulinum* spores are the most radiation-resistant spores. Radappertization is designed to reduce the number of viable spores of the *C. botulinum* by 12 log cycles. Spores are more sensitive in the presence of O_2 or preservatives and at temperatures above 20°C.

Reservoirs

Research shows that spores of *C. botulinum* are commonly present in soils and sediments, but their numbers and types vary depending on the location. Food surveys have largely focused on fish (un-eviscerated salt-cured fish), meats and infant foods, primarily honey. The types most often associated with meats are A and B. These types may also be present in fruits and vegetables, particularly those in close contact with the soil. Different agricultural practices, such as the use of manure as fertilizer may affect the level of contamination. Products in which contamination has often been detected include asparagus, beans, cabbage, carrots, cherries, peaches and tomatoes. A product of particular concern because of the high number of spores found is cultivated mushrooms. The potential presence of spores in honey and other infant foods is problematic because in some infants, the spores can colonize the intestines, produce neurotoxin and cause infant botulism. Only a very low incidence of *C. botulinum* spores has been found in other foods, including dairy products, vacuum-packed products and ready-to-eat foods.

Other important reservoirs could be home-preserved vegetables or meats like ham, fermented sausages and canned products. Temperature abuse of home prepared foods continues to be an important cause of botulism.

Characteristics of the Disease

Food-borne botulism varies from a mild illness, which disregarded could be a serious disease that may be fatal within 24 hours. Symptoms typically appear 12–36 hours after ingestion of neurotoxin. Earlier the symptoms appear, the more serious is the disease. The first symptoms are generally nausea and vomiting (type B and E), followed by neurological signs and symptoms including visual impairment (blurred or double vision, fixed and dilated pupils), difficulty in speaking and swallowing, dry mouth, throat and tongue infections, sore throat as seen frequently in type B strains, general fatigue and lack of muscle coordination, and respiratory impairment which are the main causes of death.

Other gastrointestinal symptoms may include abdominal pain, diarrhoea, or constipation.

Botulism may be confused with other illnesses, including other forms of food-borne poisoning, carbon monoxide poisoning. In botulism, the neurological signs appear first in the cranial nerve area (eyes, mouth and throat) and then descend.

The most common symptom of infant botulism is constipation. The infants usually show a generalized weakness and weak cry. Other symptoms may include feeding difficulty and poor sucking, lethargy, lack of facial expression, irritability, and progressive "floppiness". Respiratory arrests occur frequently but are seldom fatal.

Initially, treatment of food-borne botulism tries to remove or inactivate the neurotoxin by neutralization of circulating neurotoxin with antiserum or to use enema to remove residual neurotoxin from the bowel and gastric lavage or treatment with emetics. Subsequent treatment is mainly to counteract the paralysis of the respiratory muscles by artificial ventilation.

Infective Dose

Little is known concerning the minimum toxic dose of *C. botulinum* and its neurotoxins. There is no tolerance for the presence of neurotoxin or for conditions permitting growth of *C. botulinum*. The mouse LD_{50} for BoNT is approximately 0.1 ng/kg.

Pathogenicity

C. botulinum produces eight antigenically distinct toxins, designated types A, B, C_1, C_2, D, E, F and G. All the toxins except C_2 are neurotoxins. C_2 is an ADP ribosylating enzyme.

Neurotoxins   All seven of the neurotoxins are similar in structure and mode of action. *C. botulinum* neurotoxins are high-molecular eight, two chain proteins which are among the most toxic substances known. Botulinum neurotoxin types A and B possess toxicities of 2×10^8 LD_{50}/mg. The neurotoxins block neurotransmission at peripheral motor nerve terminals by selectively hydrolysing proteins involved in the fusion of synaptic vesicles with the presynaptic plasma membrane, thereby preventing acetylcholine release.

BoNTs are water-soluble proteins produced as a single polypeptide with an approximate molecular weight of 150,000. They are cleaved by a protease (endogenous bacterial proteases or trypsin to produce the active neurotoxin which is composed of one heavy and one light chain linked by a single disulphide bond. The two chains individually are non-toxic.

BoNT's form complexes with non-toxic proteins in naturally contaminated foods and culture supernatants to form progenitor toxin. The non-toxic proteins can be dissociated from BoNT by pH values greater than 7.2 and spontaneously reassociate when the pH is lowered. Three forms of progenitor neurotoxin have been distinguished and are referred to as M (medium sized), L (large) and LL (extra large) toxins. M toxin is produced by all strains producing neurotoxin except those producing type G neurotoxin (neurotoxin + non-toxic non-haemagglutinin (NTNH).

The neurotoxins are arranged as part of a transcriptional unit which includes the genes encoding BoNT as well as genes encoding NTNH components and haemagglutinins. This transcriptional unit is referred to as the BoNT gene complex. The degree of relatedness of the various neurotoxins has been determined on the basis of sequence homologies. The location of the genes coding for BoNTs and the associated non-toxic proteins vary depending on the serotype with the genes coding for BoNTs A, B, E and F are located on the bacterial chromosome. The genes coding for BoNTs C1 and D and the associated non-toxic proteins are encoded by bacteriophages whereas the genes coding for BoNT G and the associated non-toxic proteins are located on a plasmid.

Mode of Action of Neurotoxins   BoNTs block the exocytic release of the excitatory neurotransmitter acetylcholine from synaptic vesicles at peripheral motor nerve terminals. This results in the flaccid paralysis observed in botulism poisoning and has been utilized in the therapy of several neurologic disorders. BoNTs are also being used as tools to study the molecular mechanisms of vesicle docking and membrane fusion mechanisms involved in exocytosis. The H (heavy) chains are responsible for selective binding of the neurotoxin to neurons, internalization of the entire neurotoxin, intraneuronal sorting, and translocation of the L (light) chains into the cytosol. The L chains block exocytosis as soon as they are released into the cytoplasm.

The mechanism of action of BoNTs can be divided into three steps:

1. Binding
2. Internalization
3. Intracellular action

Binding   The potency of the BoNTs results from the specificity of the toxins for neurons. Neurotoxin receptors are located at the motor neuron plasma membrane at the neuromuscular junction. These toxins have a small number of high affinity binding sites and a large number of sites with much lower affinity. While the neurotoxins appear not to share receptors, the receptors may possess at least one *sialic acid* residue. Gangliosides, which are sialic acid containing glycosphingolipids bind to the BoNTs. The neurotoxins first bind to the negatively charged surface of the pre-synaptic membrane, which contains large amount of acidic lipids. After binding to the negatively charged lipids, the neurotoxin may diffuse laterally in the membrane to bind to a protein receptor.

Internalization After the neurotoxin has bound to its receptors at the neuromuscular junctions, the entire neurotoxin is internalized by receptor mediated endocytosis. Once the neurotoxin has been internalized it can no longer be neutralized by antineurotoxin. Studies have shown that the H chain aggregates and forms channels in the endosomal membrane to allow the L chain to pass into the cytoplasm. The L chain exits the endosome by passing through channels created by the H chain.

Intracellular action The L chain is capable of inhibiting neurotransmitter release independently of the H chain. L chains act as zinc-dependent endopeptidases, whose substrates are components of the synaptic vesicle docking and fusion complex. BoNT types B, F, D and G cause selective degradation of vesicle associated membrane protein (VAMP). The zinc-dependent protease activity of the BoNTs appears to be very specific. VAMP, syntaxin and SNAP-25 form the core of a multicomponent complex which mediates fusion of carrier vesicles to target membranes in eukaryotic cells. Proteolysis of these proteins involved in docking and fusion of synaptic vesicles blocks neuro-exocytosis and subsequent neurotransmitter release.

LABORATORY DIAGNOSIS

In view of the severity of the illness, rapid identification of the food source is essential in order to prevent further cases. Suspected foods, together with any containers, should be examined without delay. Serum and faecal specimens should be collected from suspected cases and sent immediately for toxin testing. Suspected foods together with rinses from containers should be examined for the presence of toxin and cultured by direct and enrichment methods using special media. Enumeration of organism is not usually considered necessary.

FOOD POISONING BY *BRUCELLA* SP.

Even though brucellosis is mainly classified under zoonoses (infection transmitted from animal to man), it can be referred as an example for food borne disease since it is also transmitted through milk and milk products. Brucellosis is also named as undulant fever, Malta fever or Mediterranean fever. It is caused by different species of the genus *Brucella* and characterized by intermittent or irregular febrile attacks, with profuse sweating, arthritis and an enlarged spleen.

Taxonomy

The genus *Brucella* can be divided into six species, four of which can infect man by direct or indirect contact with infected animals:

Brucella abortus Primary host cattle, but also transmissible to camels, deer, dogs, horses, sheep, goats, pigs and man.

B. melitensis   Primary host goats and sheep, but can infect many other species, including man. This is most virulent in man, discovered by Bruce (army surgeon in Malta)

B. suis   Primary host is dependent on the biovar, biovars 1,2 and 3 infect pigs, biovar 4 infects reindeers and biovar 5 infects small rodents. All biovars can cause disease in man.

B. canis   Primary host is the dog, man may occasionally be infected.

B. ovis   Primary host is sheep, does not cause disease in man.

B. neotomae   Infection limited to the desert wood rat.

The organism is a gram-negative, round or oval intracellular cocco-bacilli, about 0.4 mm in diameter. Arranged singly, sometimes in pairs, short chains or small clusters. Does not produce capsules, spores or flagella.

Growth characteristics

It requires multiple amino acids and vitamins for growth. The optimum pH range is 6.6–7.4 and temperature range is 20–40°C with an optimum at 37°C.

Survivability

The organism is killed after heating at 60°C for 10 minutes, therefore easily eliminated from milk by pasteurization by holder method and HTST. Brucellae can remain viable in refrigerated milk for 10 days and in cheese for up to 90 days. They may persist in meat for several weeks. The organism is moderately sensitive to acid and hence cannot survive in properly fermented milk products like cheese or yoghurt. It is also sensitive to disinfectants and antibiotics. Brucellosis is most prevalent under conditions of advanced domestication of animals in the absence of hygiene. The organism can travel long distances in milk and dust. It can survive for weeks, or months in favourable conditions of water, urine, faeces, damp soil and manure. Cells of *Brucella* are more resistant to lysozyme but are apparently destroyed by smoking. In raw milk at room temperature, it dies out rapidly with the production of acid. Acid production seems to be the cause of its rapid death in butter and cheese.

Reservoirs

Main reservoirs of human infection are cattle, sheep, goats, swine, buffaloes, horses and dogs. The infected animals excrete the organism in the milk which is one of the important vehicles of food-borne brucellosis. Dairy products prepared from untreated raw milk like butter, cream, soured milk, meat and meat products can also carry the organism. Fresh raw vegetables can also carry infection if grown on soil containing manure from infected farms. Water contaminated with the excreta of infected animals may also serve as a source of infection.

Characteristics of the Disease

It has a highly variable incubation period which usually lasts from 1–3 weeks, but may last for as long as 6 months or more. The disease can vary from an acute febrile disease to a chronic low grade ill defined disease, lasting for several days, months or occasionally years.

The acute phase is characterized by a sudden or insidious onset of illness with

1. Swinging pyrexia especially at nights (40–41°C), rigors and sweating.
2. Arthralgia/arthritis involving larger joints such as hip, knee, shoulder and ankle.
3. Low back pain
4. Headache, insomnia
5. Small firm splenomegaly and hepatomegaly
6. Leucopenia with relative lymphocytosis

The acute phase subsides within 2–3 weeks. The infection being intracellular, may persist giving rise to subacute or relapsing disease.

Pathogenicity

Upon ingestion by the humans, the organisms penetrate gastrointestinal mucosa and in direct contact, organisms enter through breaks in the skin or the conjuctiva. After penetration, the organisms spread via lymphatics through regional nodes and thoracic duct to the bloodstream and get localized in spleen, bone marrow, liver, kidneys, endocardium and elsewhere. In animals, the organism stay confined to the mammary glands, genital organs and in pregnant uterus, foetal fluids and membranes. No exotoxins have been detected.

Host factors Human brucellosis is predominantly a disease of adult males, farmers, shepherds, butchers, laboratory workers and veterinarians are particularly at special risk.

Environmental factors Overcrowding of herds, high rainfall, unhygienic practices in milk and meat production, favour spread of disease. The infection can travel long distances in milk and dust.

Mode of Transmission

Infection transmitted from infected animals to man. Routes of spread are:

1. *Contact infection*   Infection by direct contact with infected tissues, blood, urine, vaginal discharge, aborted foetuses (placenta). Infection occurs through abraded skin, mucosa, conjunctiva. This type of spread is occupational.

2. *Food-borne*   Indirectly by ingestion of raw milk/dairy products (cheese) from infected animals. Fresh raw vegetables can also carry infection if grown on soil containing manure from infected farms. Water may also serve as a source of infection.

3. *Airborne*   The environment in cow shed may be heavily infected. Brucellae may be inhaled in aerosol form in slaughter houses and laboratories. Hence this type is also considered to be occupational.

Incubation Period

It is variable, usually 1–3 weeks to as long as 6 months.

LABORATORY DIAGNOSIS AND ISOLATION

Brucella are hazard group 3 pathogens, hence samples and cultures must be handled accordingly.

Method 1   Direct Culture

Media

A *selective agar*, e.g. Brucella agar base containing dextrose, blood agar or Columbia agar base plus 1% sterile dextrose. These media are suitable for use with the addition of 5% inactivated horse serum (i.e., serum held at 56°C for 30 minutes) and an antibiotic cocktail containing polymyxin, bacitracin, cycloheximide, nalidixic acid, nystatin and vancomycin).

Procedure

1. Transfer the milk sample to sterile test tubes (180 mm × 25 mm) and store overnight at 4°C.
2. Dip a swab into the cream layer and inoculate the surface of a selective agar.
3. Incubate the plates at 37°C in an atmosphere of air containing 10% carbon dioxide.
4. Examine the plates every 2 days for up to 10 days. Colonies are usually visible after 4 to 5 days incubation, and are 1–2 mm in diameter, convex, with round entire edges.

Identification

Brucella sp. can be further identified using antibodies for slide agglutination. Differentiation can also be achieved by the dyes strip method.

1. Impregnate filter paper strips with 1 : 200 basic fuchsin or 1 : 600 thionin and dry.
2. Place a strip of each dye parallel on the surface of serum dextrose agar and cover with a thin layer of the same medium. Allow the medium to solidify.

3. Make streak inoculations of the *Brucella* strains at right angles to the strips.
4. Incubate in 10% carbon dioxide for 2–3 days at 37°C.
5. Examine for growth. Resistant strains grow right across the strip, but sensitive strains show inhibition of growth up to 10 mm from the strip.

	Basic fuchsin (1 : 200)	Thionin (1 : 600)
B. abortus	growth	no growth
B. melitensis	growth	growth
B. suis	no growth	growth

Method 2   Enrichment Culture

Media

Broth bases, e.g. Brucella broth or media suitable for the culture of fastidious organisms such as brain heart infusion broth or tryptone soy broth. Supplement the medium with 5% sterile horse serum and antibiotics as described in method 1.

Procedure

1. Centrifuge 100 ml of the milk for 30 minutes at 1500 rev/min.
2. Transfer the cream layer and deposit from the centrifuged milk to sufficient enrichment broth in a screw capped container to give a ratio of 1 : 10.
3. Incubate the broth, with screwcap loose, in air containing 10% carbon dioxide at 37°C for 5 days.
4. Subculture the broth to selective agar and proceed as described in method 1 (from step 3).

FOOD POISIONING BY *CLOSTRIDIUM PERFRINGENS*

Clostridium perfringens actually causes two quite different human diseases that can be transmitted by food, i.e., *C. perfringens* type A food poisoning and necrotic enteritis.

General Characteristics

C. perfringens is a gram-positive, rod-shaped, encapsulated non-motile bacterium of variable size that is capable of causing a broad spectrum of human and veterinary diseases. The pathogenicity is largely derived from its prolific ability to express protein toxins, including at least two toxins, *C. perfringens* enterotoxin and b-toxin, that are active on the human gastrointestinal tract.

The vegetative cells of *C. perfringens* can double in less than 10 minutes. allowing the organism to multiply very rapidly in food and it forms spores that are highly resistant to environmental conditions such as heat, radiation, and desiccation. *C. perfringens* is an anaerobic but can tolerate some exposure to air and compared with many other anaerobes, requires only relatively modest reductions in oxidation–reduction potential for growth.

Survival Characteristics

Temperature   The heat resistance of *C. perfringens* spores contributes to the organism's ability to cause food poisoning by allowing it to survive in partially cooked foods. The heat resistance properties of *C. perfringens* spores depend on both environmental and genetic factors.

The medium in which a *C. perfringens* spore is heated clearly influences its heat resistance. In a relatively protective medium such as cooked meat medium, many spores will survive boiling for an hour or longer. The involvement of genetic factors in spore heat resistance has been established by observations indicating that spores made by different *C. perfringens* strains vary considerably in their heat resistance properties. Spores made by food poisoning isolates are generally more heat-resistant than spores made by *C. perfringens* isolates from other sources. Partially cooked foods not only fail to rid spores but can actually facilitate the development of *C. perfringens* type A food poisoning, since heating (70–80°C for 20 minutes) is an excellent way to induce the germination of *C. perfringens* spores.

Even the vegetative cells of *C. perfringens* are somewhat heat-tolerant since they have a relatively high optimal growth temperature (43–45°C) and will continue to grow at temperatures up to atleast 50°C. The spores are also resistant to cold temperatures but the vegetative cells are not particularly tolerant of either refrigeration or freezing. Food poisoning may result if viable spores in refrigerated or frozen foods are induced to germinate as this food is warmed for serving.

Water activity   *C. perfringens* is less tolerant of low a_w environments than other common gram-positive food-borne pathogen like *Staphylococcus aureus*. The lowest a_w supporting vegetative growth of *C. perfringens* is reported to be 0.93–0.97, depending on the solute used to control the a_w of the medium.

Redox potential   Relative to other anaerobes, *C. perfringens* does not require an extremely reduced environment for its growth. If the redox potential of the environment is low, *C. perfringens* will then modify the redox potential of its surrounding environment (by producing reducing molecules such as ferredoxin) to produce more optimal growth conditions. The redox potential of many common foods (like raw meats) is often low enough to permit the growth of *C. perfringens*.

pH   Growth of *C. perfringens* is also sensitive of pH extremes, with optimal growth occurring at pH 6–7 (pH that is commonly found in the meat and poultry products that usually serve as food vehicles for *C. perfringens* type A food poisoning and severe inhibition of growth occurring at pH of <5 and > 8.

Chemicals   The effectiveness of curing agents on limiting *C. perfringens* growth in foods has not received much research attention since concentrations of curing salts needed to significantly inhibit survival of *C. perfringens* cells may exceed commercially acceptable levels, i.e., inhibition of *C. perfringens* growth may require at least 6–8% NaCl, 10,000 ppm of $NaNO_3$, or 400 ppm of $NaNO_2$.

1. The simultaneous coapplication of other preservation factors such as heating and non-neutral pH increases the sensitivity of *C. perfringens* to curing salts.

2. The simultaneous use of several curing agents often produces a synergistic inhibition of *C. perfringens* growth.

3. Foods may contain initial burdens of *C. perfringens* cells and spores lower than those used in laboratory studies evaluation the effectiveness of curing agents for inhibition *C. perfringens* growth.

The above-mentioned preservation agents and environmental factors play a major role in inhibiting the outgrowth of germinating *C. perfringens* spores in foods.

Reservoirs for *C. perfringens* type A Food Poisoning

The organism is ubiquitous throughout the natural environment.

(i) soils (10^3–10^4 cfu/g)

(ii) foods (50% of frozen foods)

(iii) dust

(iv) intestinal tracts of humans and domestic animals (human faeces contains about 10^3–10^6 cells per gram)

Surveys have suggested that < 5% of all *C. perfringens* isolates actually carry the *"cpe"* gene, which is considered essential for producing *C. perfringens* type A food poisoning symptoms. Hence, it is suggestive to find out reservoirs for *"cpe"* positive *C. perfringens* which is the actual enterotoxigenic organism.

Food-borne Outbreaks

C. perfringens type A food poisoning annually ranks among the most common food-borne diseases in the developed countries having a 5% mortality rate. This results from at least two factors. First, preparing food in advance and then holding this food for later serving (thus allowing the growth of the organisms in any temperature-abused food). Second, given the relatively mild and non-distinguishing symptoms, its outbreak is not easily identified.

Meat and poultry are the most common food vehicles of *C. perfringens* type A food poisoning and this disease almost always results from temperature abuse during the cooking, cooling, or holding of foods. Partial cooking actually promotes this illness by increasing the germination rates of *C. perfringens* spores present in foods; after outgrowth of these spores into new vegetative cells, the organisms can multiply rapidly in temperature-abused foods that are cooled or stored improperly.

Characteristics of the Disease and Pathogenesis

Symptoms of *C. perfringens* type A food poisoning develop 8–24 hours after ingestion of contaminated food and usually resolve spontaneously within 12–24 hours. Typical symptoms are diarrhoea and severe abdominal cramps. Vomiting and fever are commonly associated with this. While death rates from *C. perfringens* type A food poisoning are low, death is more prevalent in debilitated or elderly individuals afflicted with this illness.

Initially, as a result of temperature abuse, vegetative cells of enterotoxigenic *C. perfringens* multiply rapidly in the food and are consumed when the food vehicle is ingested. Many of the ingested vegetative cells probably die when exposed to stomach acidity but if the food vehicle is sufficiently contaminated, some vegetative cells survive passage through the stomach and enter the small intestine, where they multiply and sporulate. It is during this sporulation in the small intestine that *Clostridium perfringes* enterotoxin (CPE) is expressed. Once released into the intestinal lumen, CPE quickly binds to the intestinal epithelial cells, where it exerts its unique action and produces morphological damage to intestinal epithelial cells. It is this CPE-induced intestinal tissue damage that causes the intestinal fluid loss.

The relatively mild, self limiting nature of this disease, probably stems from two facts:

1. The diarrhoea associated with *C. perfringens* type A food poisoning helps to mitigate the severity of this illness by flushing unbound CPE and many *C. perfringens* cells from the small intestine.
2. CPE preferentially affects the villus tip cells, which are the oldest intestinal cells and can be rapidly replaced in young, healthy individuals by the normal turnover of intestinal cells.

FOOD POISONING BY *ESCHERICHIA COLI* O157 : H7

Escherichia coli strains are a common part of the normal facultative anaerobic microflora of the intestinal tracts of humans and warm-blooded animals. *E. coli* strains that cause diarrhoeal illness are categorized into specific groups based on virulence properties, mechanism of pathogenicity, clinical syndromes and distinct O:H serogroups. These categories include:

1. *Enteropathogenic E. coli strains (EPEC)*   EPEC can cause severe diarrhoea. Humans are an important reservoir.

2. *Enterotoxigenic E. coli strains (ETEC)*   ETEC cause infantile diarrhoea in developing countries. They are also the agents most frequently responsible for traveller's diarrhoea. ETEC colonize the proximal small intestine by fimbrial colonization factors and produce a heat-labile or heat-stable enterotoxin that elicits fluid accumulation and a diarrhoeal response. Humans are the principal reservoir of ETEC.

3. *Enteroinvasive E. coli strains (EIEC)* EIEC cause non bloody diarrhoea and dysentery similar to that caused by *Shigella* spp., by invading and multiplying within colonic epithelial cells. The principal site of bacterial localization is the colon, where EIEC invade and proliferate in epithelial cells, causing cell death. Humans are a major reservoir.

4. *Enteroaggregative E. coli (EAggEC)* Recently, they have been associated with persistent diarrhoea in infants and children in several countries. They are different from the other types of pathogenic *E. coli* because of their ability to produce a characteristic pattern of aggregative adherence on HEp-2 cells. Considerably more epidemiologic information is needed to elucidate its significance as an agent of diarrhoeal disease.

5. *Enterohaemorrhagic E. coli (EHEC)* EHEC were first identified as human pathogens in 1982 when *E. coli* of serotype O157:H7 was associated with two outbreaks of haemorrhagic colitis. All EHEC produce factors cytotoxic to Vero cells. EHEC carries more importance because of its recent origin during 1980s with two food-borne outbreaks of haemorrhagic colitis. It is different from common strains of *E. coli* by their inability to grow well at temperatures of >44.5°C, inability to ferment sorbitol within 24 hours, inability to produce β-glucuronidase, possession of an attaching and effacing gene, carriage of a 60-MDa plasmid and expression of an uncommon 5,000–8,000 MW (outer membrane protein).

Survivability Characteristics

Acid tolerance   *E. coli* O157:H7 is uniquely tolerant to acidic environments thus making this an important food-borne pathogen associated with high-acid foods. It can survive fermentation, drying and storage of fermented sausage for up to 2 months at 4°C.

The organism is not affected by 1.5% of organic acid sprays on food. It survives on mayonnaise (pH 3.6–4.0) for up to 5–7 weeks at 5°C and for 1–3 weeks at 20°C. The mechanism of acid tolerance appears to be associated with a protein that can be induced by pre-exposing the bacteria to acid conditions.

Thermal inactivation   The pathogen has no unusual resistance to heat with D values at 60°C at 45 seconds. The presence of fat increases the thermal tolerance of *E. coli* O157:H7. Pasteurization of milk is an effective treatment. Proper heating of foods of animal origin, i.e., heating foods to an internal temperature of at least 68.3°C is an important critical control point to ensure inactivation of *E. coli* O157:H7.

Reservoirs

Undercooked beef, unpasteurized milk have been implicated in infection. Similarly young calves tend to carry *E. coli* O157:H7 more frequently than adult cattle. Contaminated feedstuff, water, colonized animals in herds, infected wildlife and humans, contaminated facilities and equipment surfaces from contact with faeces also act as important sources of the organism to the reservoirs. Domestic animals like chicks can also act as a reservoir. Person-to-person contact of the infection has also been established since faecal excretion of the pathogen can last for weeks but an asymptomatic long term carrier state has not been identified.

Food-borne Outbreaks

Geographically this infection has been largely concentrated on the North American continent and South Africa. Outbreaks and clusters of *E. coli* O157:H7 peak during the warmest months of the year, because:

1. An increased prevalence of the pathogen in the cattle or other livestock or vehicles of transmission during the summer.
2. Greater human exposure to ground beef or other *E. coli* O157:H7 contaminated foods during the festival season.
3. Greater improper handling or incomplete cooking of products during warm months than other months.

The very young (under 5 years of age) and the elderly (over 65 years) people are at greatest risk of the infection. Although a variety of foods have been implicated in *E. coli* O157:H7 associated illness, most outbreaks have been associated with consumption of raw or undercooked foods of bovine origin.

Characteristics of the Disease

The spectrum of human illness of *E. coli* O157:H7 infection includes non-bloody diarrhoea, haemorrhagic colitis, Haemolytic Uraemic Syndrome (HUS) and thrombotic thrombocytopenic purpura.

Symptoms of haemorrhagic colitis include a crampy abdominal pain followed within 1–2 days by a nonbloody diarrhoea which progresses within 1 or 2 days to bloody diarrhoea that lasts for 4–10 days.

HUS largely affects children and is the leading cause of acute renal failure in children. The syndrome is characterized by a triad of features: acute renal insufficiency, micro-angiopathic haemolytic anaemia and thrombocytopenia. About one-third of patients infected with this organism require hospitalization.

The infective dose is low (1–15 cells/g) and all age groups can be affected by this organism, but infants and young children most frequently experience severe illness. Highest age specific incidence of *E. coli* O157:H7 infection is in the group 2–10 years of age. The high rate of infection in this group is likely due to increased exposure to contaminated foods, contaminated environments and infected animals.

Pathogenicity

The precise mechanism of pathogenicity of *E. coli* O157:H7 has not been fully elucidated. It causes disease by: a) its ability to adhere to the host cell membrane and b) producing Shiga family toxins.

Adherence   Adherence may be through any one of the following methods.

Localized adherence   It involves the initial attachment of the bacterium to epithelial cells and is mediated by bundle-forming pili and other fimbriae.

Signal transduction   It results in an increased level of intracellular calcium, release of inositol phosphates and tyrosine phosphorylation of an epithelial cell protein that leads to the destruction of microvilli.

Intimate adherence   It is mediated by intimin, a 94-kDa OMP which amplifies the accumulation of filamentous actin and other cytoskeletal proteins within the epithelial cell. All the three phenomena together form an attaching and effacing lesion (AE) lesion.

Shiga family toxin   E. coli O157:H7 produces one or two cytotoxins that are cytotoxic to Vero cells and originally called VT1 and VT2. VT1 is immunologically and genetically related to Stx, which is produed by *S. dysenteriae* type 1. Hence these toxins alternately have been named (Shiga like toxins) SLTs. There is no satisfactory animal model for haemorrhagic colitis or HUS.

LABORATORY DIAGNOSIS AND IDENTIFICATION

Isolation of these groups of *E. coli* from food can be done successfully by a combination of direct plating on selective media, such as MacConkey agar or violet red bile-salt lactose agar, and enrichment in minerals modified glutamate or lauryl sulphate tryptose broths. Isolates are confirmed by growth and indole production at 44°C. Demonstration of β-glucuronidase activity is frequently used for confirmation of *E. coli* isolated from food.

Isolation of *E. coli* O157:H7 from foods like red meat and dairy products is achieved by a combination of enrichment in modified trypticase soya broth (MTSB) incubated at 41.5°C for 24 hours, immunomagnetic separation and plating onto Cefixime–tellurite sorbitol MacConkey agar (TC-SMAC). Suspected colonies are confirmed as presumptive positive cells by biochemical (*E. coli* O157:H7 are usually β-glucuronidase negative) and latex agglutination tests. Additional characterization of the organism needs the phage typing, toxin gene typing and pulsed field gel electrophoresis.

FOOD POISONING BY *SHIGELLA* SPP.

Bacillary dysentery or shigellosis is caused by members of the *Shigella* species. These are host-adapted organisms and infect only humans and other primates. With a low infectious dose required to cause disease coupled with oral transmission via faecally contaminated food and water, it is not surprising that dysentery caused by *Shigella* spp. follows in the wake of many natural and man-made disasters.

Classification

There are four species of the genus *Shigella* serologically grouped based on their somatic O antigens: *Shigella dysenteriae* (group A), *S. flexneri* (group B), *S. boydii* (group C) and *S. sonnei* (group D). They are nearly genetically identical to *E. coli* and closely related to *Salmonella* and *Citrobacter.* They are gram-negative, non-motile and oxidase negative rods. Some important biochemical characteristics that distinguish these bacteria from other enteric bacteria are their inability to ferment lactose or utilize citric acid as a sole carbon source, they do not produce H_2S and except for *S. flexneri* they do not produce gas from glucose.

General Characteristics

Shigella species are not particularly fastidious in their growth requirements and can be easily isolated and grown from analytical samples, including water and clinical samples. Identification of shigellae in foods is not as easy as in other sources. Foods have many different physical attributes that may affect the successful recovery of shigellae. These factors include:

1. Composition such as fat content of the food.
2. Physical parameters such as pH and salt.
3. Natural microbial flora of the food in which other microbes in a sample may overgrow shigellae in broth media.

The physiological state of shigellae present in the food is a contributing factor in the successful recovery of these pathogens. The amount of time from the clinical report of a suspected outbreak to the analysis of the food samples can be considerable and thus lessen the chances of identifying the causative agent. Also, shigellae may be present in low populations or in a poor physiological state in the suspected food samples. Under these conditions, special enrichment procedures are required for successful detection of shigellae.

Food-borne outbreaks of shigellosis continue to be a major public health concern and are increasing worldwide. *Shigella* spp. are not associated with any specific foods. Some common foods that were known to be associated with outbreaks caused by shigellae were potato salad, chicken and shellfish.

Whereas epidemiological methods may strongly imply a common food source, *Shigella* spp. are not often recovered from foods and identified by standard bacteriological methods. Also, since shigellae are not commonly associated with any particular food, routine inspections of foods to identify these pathogens are not usually performed. This pathogen is usually introduced into the food supply by an infected person such as a food handler with poor personal hygiene. In some cases, this may occur at the manufacturing site, but more likely it happens at a point between the processing plant and the consumer. Although HACCP (hazard analysis and critical control point) system is a method for regulating food safety and preventing food-borne outbreaks, pathogens such as shigellae that are not indigenous to, but rather are introduced into foods are most likely to be undetected.

From carriers, this pathogen can spread by several routes, including food, fingers, faeces and flies. The highest number of incidences of shigellosis occur during the warmer months of the year. Improper storage of contaminated foods is the second most common factor that accounts for food-borne outbreaks due to shigellae. Other contributing factors are inadequate cooking, contaminated equipment, and food obtained from unsafe sources.

Reservoirs

Humans are the natural reservoir of *Shigella* infections.

Characteristics of the Disease

Disease caused by *Shigella* spp. is distinguished from disease caused by most of the other food-borne pathogens in at least two important aspects: Production of bloody diarrhoea or dysentery and the low infectious dose that can cause clinical symptoms. The clinical picture of shigellosis ranges from a mild watery diarrhoea to severe dysentery, the former usually preceding the latter. The dysentery stage of disease correlates with extensive bacterial colonization of the colonic mucosa. The bacteria invade the epithelial cells of the colon, spread from cell to cell, but penetrate only as far as the lamina propria. The incubation period for shigellosis is 1–7 days. Strains of *S. dysenteriae* type 1 cause the most severe disease, while *S. sonnei* produce the mildest.

An important aspect of *Shigella* pathogenesis is the extremely low infectious dose (as low as 200 organisms). It is a self-limiting disease in normally healthy patients.

Shigellosis can be a very painful and incapacitating disease and is more likely to require hospitalization than other bacterial diarrhoeas. Complications arising from the disease includes severe dehydration, intestinal perforation, toxic mega-colon, septicaemia, seizures, Reiter's syndrome (a form of reactive arthritis, a post-infection sequela to shigellosis. The syndrome consists of three symptoms, urethritis, conjunctivitis and arthritis) and haemolytic uraemic syndrome (HUS- characterized by haemolytic anaemia, thrombocytopenia and acute renal failure) as shown by Shiga toxin produced by *S. dysenteriae* type 1. The Shiga toxin may cause HUS by entering the bloodstream and damaging vascular endothelial cells such as those in the kidney.

While improvements in sanitary and hygienic conditions can help contain secondary spread of shigellosis, the single most effective means of preventing secondary transmission is hand washing.

Pathogenicity

Virulence factors The clinical symptoms of shigellosis can be directly attributed to the hallmarks of *Shigella* virulence: the ability to invade epithelial cells of the intestine, multiply intracellularly, and spread from cell to cell.

Along with the ability to colonize and cause disease, an intrinsic part of a bacterium's pathogenicity is its mechanism for regulating expression of the genes involved in virulence. Virulence in *Shigella* spp. is regulated by growth temperature. The non-invasive phenotype can regain virulence when incubated at 37°C when the bacteria re-expresses its virulence properties. Regulation of gene expression in response to environmental temperature is a useful bacterial strategy. By sensing the ambient temperature of the mammalian host to trigger gene expression, this strategy permits shigellae to economize energy that would be expended on the synthesis of virulence products when the bacteria are outside the host. The system also permits the bacteria to coordinately regulate expression of multiple unlinked genes that are required for the full virulence phenotype.

Shigella virulence is multigenic, involving both chromosomal and plasmid encoded genes. A 180-kb plasmid in *S. sonnei* and a 220-kb plasmid in *S. flexneri* are essential for invasion. Other *Shigella* spp. as well as EIEC contains homologous plasmids that are functionally interchangeable and share significant degrees of DNA homology. In *S. flexneri* 2a, the 37-kb region of the invasion plasmid contains all of the genes necessary to enable shigellae to penetrate into tissue culture cells.

The genes comprising the *ipa BCDA* (invasion plasmid antigens) cluster are required for invasion of mammalian cells. The *ipa* products are associated with the outer membrane of shigellae. *Ipa B* and *ipaC* form a complex on the bacterial cell surface and probably are responsible for transducing the signal that leads to entry of shigellae into the host cells *via* bacterium directed phagocytosis. The product of *ipa B* has also been postulated to be the contact haemolysin that is responsible for lysis of the phagocytic vacuole minutes after entry of the bacterium into the host cell.

A plasmid-encoded virulence gene that is unlinked to the 37-kb region is not required for invasion but is crucial to the intra- and intercellular motility. This gene known as *vir G* or *ics A* (intracellular spread) encodes a protein that catalyses the polymerization of actin in the cytoplasm of the infected cell. The *ics A* protein is unusual in that it is expressed asymmetrically on the bacterial surface, being present only at one pole. The polymerization of actin monomers by *icsA* forms a tail leading from the pole and provides the force that propels the bacterium through the cytoplasm. Hence unipolar expression of *icsA* imparts directionality of movement to the bacterium. This unipolar localization is dependent on the synthesis of a complete lipopolysaccharide (LPS).

In contrast to the genes of the virulence plasmid that are responsible for invasion of mammalian tissues, most of the chromosomal loci associated with *Shigella* virulence are involved in regulation or survival within the host. Aerobactin is a hydroxamate siderophore, which *S. flexneri* uses to scavenge iron, is important for bacterial growth within the mammalian host. Another toxin called the Shiga toxin may account for the generally more severe infections caused by *S. dysenteriae* type 1.

While food-borne infections due to *Shigella* spp. may not be as frequent as those caused by other food-borne pathogens, they have the potential for explosive spread because of the extremely low infectious dose needed to cause overt clinical disease.

LABORATORY DIAGNOSIS OF SHIGELLAE FROM FOOD

Food-borne transmission is usually the result of contamination of ready-to-eat foods by human sewage.

Isolation of the organism from suspect food products is achieved by a combination of direct plating onto a suitable selective medium and by enrichment culture. The enrichment broths currently available are not specific for isolation of *Shigella* spp. but also grow other gram-negative enteric organisms.

Method 1   Enrichment Culture

Media

Selective enrichment medium   Shigella enrichment broth containing peptone 20 g, potassium hydrogen phosphate 2 g, potassium dihydrogen phosphate 2 g, sodium chloride 5 g, glucose 1 g, poly-oxy-ethylene-sorbitan monooleate 1.5 ml and novobiocin 0.55 mg/l.

Selective agar media   XLD, MacConkey agar and Hektoen enteric agar

Non-selective agar   Nutrient agar

Procedure

1. Prepare a 10^{-1} homogenate of food in *Shigella* enrichment broth.

2. Incubate the enrichment broth at 41.5°C under anaerobic conditions (with the container closure loose) for 18 hours.

3. Subculture the enrichment broth to XLD, MacConkey and Hektoen enteric agar. Incubate the plates at 37°C for 20–24 hours.

4. Examine the plates for characteristic colonies, which appear red/cerise on XLD, colourless and lactose negative on MacConkey agar, and green and moist on Hektoen agar. Subculture five suspect colonies from each plate to a non-selective agar, then incubate at 37°C for 18–24 hours.

5. Screen biochemically using TSI slopes, oxidase test and motility. Oxidase negative, non-motile strains that form a yellow butt and red or unchanged slope without production of hydrogen sulphide should be considered as presumptive *Shigella* species. Further biochemical characterization is required to confirm their identity.

6. Perform serological tests on presumptive isolates using *Shigella* agglutinating sera.

FOOD POISONING BY *VIBRIO* SPECIES

There are over 20 species of *Vibrio* which are described including at least 12 capable of causing infection in humans. Of the 12, 8 have been shown to be directly food associated. One of the most consistent aspects of vibrio infections is a recent history of seafood consumption. Vibrios, which are generally the predominant bacterial genus in estuarine waters, are found associated with a great variety of seafoods.

The classical methods of identification are continued as for example, string test, growth on alkaline medium, oxidase test, etc. Molecular techniques to identify the presence of vibrios in foods are becoming more common and are proving to be a very powerful adjunct to more traditional taxonomic methods. The levels of most of the vibrios in both surface waters and shellfish show definite seasonal correlation, generally being greater during the warm weather months. Seasonality is most notable for *V. vulnificus* and *V. parahaemolyticus* infections, whereas those of some vibrios such a *V. fluvialis* occur throughout the year. Unfortunately, because vibrios are part of the normal estuarine microflora and not a result of faecal contamination, vibrio infections will not likely be controlled through shellfish sanitation programs. It is thus essential that raw seafood be adequately refrigerated or iced to prevent significant bacterial growth.

Survival Characteristics

Generally vibrios have been reported to be sensitive to cold but seafoods have also been reported to be protective for vibrios at refrigeration temperatures. Several psychrotrophic strains of vibrios have been isolated like *V. mimicus, V. fluvialis* and *V. parahaemolyticus* from frozen foods and found these to survive well at 10, (4 and –30°C). *V. vulnificus* has been reported to increase in numbers at common refrigerated temperatures of

4–8°C. All the vibrios are sensitive to heat, although a wide range of thermal inactivation rates have been reported. Inactivation times of 15–30 minutes at 60°C and 5 minutes at 100°C seem typical although heat inactivation has been reported to be affected by NaCl levels. Heat sensitivity is also related to the initial load of cells. Doses of 3 kGy of gamma irradiation have been reported to be required for the elimination of vibrios from frozen seafoods. Vibrio species are very sensitive to chemical preservatives like BHA apart from naturally occurring preservatives like spices, essential oils and several organic acids.

VIBRIO CHOLERAE

V. cholerae O1 is the causative agent of cholera, one of the few food-borne diseases with epidemic and pandemic potential. This species is not homogeneous with regared to pathogenic potential. Important distinctions within the species are made on the basis of production of cholera enterotoxin, serogroup and potential for epidemic spread. There are two serotypes namely, Ogawa and Inaba and two biotypes, classical and El Tor type which differ in several characteristics. The El Tor biotype is the most important biotype. *Vibrio cholerae* Non-O1 and Non-O139 have also been implicated in food-borne diarrhoea, found in estuarine environments and infection due to these strains are commonly of environmental origin.

Several tests are suitable for identification of the pathogen and the monoclonal antibody-based coagglutination test is very suitable for the same. DNA probes and PCR techniques have been extremely useful in distinguishing those strains of *V. cholerae* that contain genes encoding cholera toxin (CT) (*ctx*) from those that do not contain these genes. This distinction is particularly important in examining environmental isolates of *V. cholerae* since the great majority of these strains lack *ctx* sequences. This takes as little time as 3 hours from picking the colonies from the agar to final hybridization results. The PCR techniques have been used to detect toxigenic *V. cholerae* O1 in food samples. Restriction fragment length polymorphism (RFLP) analysis can yield valuable epidemiological confirmations.

Reservoirs

Environment is the principal reservoir   *V. cholerae* are part of the normal, free living bacterial flora in estuarine areas. Non-O1/non-O139 strains are much more commonly isolated from the environment than O1 strains, even in epidemic settings in which feacal contamination of the environment might be expected. CT producing *V. cholerae* O1 can persist in the environment in the absence of known human disease. Periodic introduction of such environmental isolates into the human population through ingestion of uncooked or undercooked shellfish appears to be responsible for isolated foci of endemic diseases.

V. cholerae O1 strains are capable of colonizing the surfaces of zooplankton such as copepods, water hyacinths which even promotes its growth. *V. cholerae* produces a chitinase and is able to bind to chitin, which is the principal component of crustacean shells and the organism can grow in media with chitin as the sole carbon source.

Persistence of *V. cholerae* within the environment may be facilitated by its ability to assume survival forms, including a viable but non-culturable state and a rugose(wrinkled morphology) survival form. In this dormant state, the cells are reduced in size and become ovoid. The continued viability of the non-culturable *V. cholerae* can be assessed by a direct viable count procedure in which cells are incubated in the presence of yeast extract and nalidixic acid and examined microscopically for cell elongation. When examined by light microscopy, cells in a rugose culture are small and spherical and are embedded in an amorphous matrix material composed primarily of carbohydrate. The matrix material, or exopolymer, appears to aid aggregation of bacteria in clusters of up to 100 bacteria. In this state, the cells are protected against adverse environmental conditions. These rugose variants survive in the presence of chlorine and other disinfectants and are still capable of causing diarrhoea in volunteers.

Humans and animals form other important reservoirs of the organism. Short term carriage of *V. cholerae* by humans is quite important in transmission of disease. Persons with acute cholera excrete 10^7 to 10^8 cells per g of stool, the asymptomatic carriers being the most important reservoir. *V.cholerae* O1 can be sporadically carried by household animals including cows, dogs and chicken. A dynamic relationship between human and environmental sources of the organism is apparent, with carriage and amplification by human populations playing a critical role in epidemic spread of CT producing *V. cholerae*.

Food-borne Outbreaks

In developing countries, ingestion of contaminated water and food is probably the major vehicle for transmission of cholera, while in developed countries, food-borne transmission is more important. Such distinctions are often difficult to make since contaminated water is frequently used in food preparation. The spectrum of food items implicated in transmission of cholera includes crabs, shrimp, raw fish, mussels, cockles, oysters, clams, rice, raw pork, cooked rice, unhygienic food, frozen coconut milk and raw vegetables and fruit. A common factor in the implicated foods is their neutral or nearly neutral pH. Survival and growth of *V. cholerae* O1 in food are also enhanced by low temperatures, high organic content, high moisture, and absence of competing flora. Survival is increased when foods are cooked before contamination, cooking eliminates competing organisms and has also been suggested to destroy some heat-labile growth inhibitors and produce denatured proteins that the organism uses for growth. Food buffers *V. cholerae* O1 against killing by gastric acid especially crustaceans (presence of chitin).

Disease Symptoms and Infectious Dose

Majority of infections with *V. cholerae* O1 are mild or even asymptomatic. Profuse watery diarrhoea with premonitory symptoms like anorexia, abdominal discomfort, and simple diarrhoea are noted. Initially the stool is brown with feacal matter, but soon the diarrhoea assumes a pale grey colour with an inoffensive, slightly fishy odour. Mucus in the stool imparts the characteristic "rice water" appearance. Vomiting is often present.

In most severe forms, termed "cholera gravis" the rate of diarrhoea may quickly reach 500 to 1000 ml/hour, leading rapidly to tachycardia, hypotension and vascular collapse due to dehydration. Skin turgor is poor, giving the skin a doughy consistency with sunken eyes, hands and feet become wrinkled, as after long immersion ("washerwoman's hands") leading to death. Other symptoms include abdominal cramps and fever with nausea and vomiting.

In healthy volunteers, doses of 10^{11} cfu of *V. cholerae* in buffered saline (pH 7.2) are required to consistently cause diarrhoea.

Pathogenicity

Virulence mechanisms Infection due to *V. cholerae* O1/O139 begins with the ingestion of food or water contaminated with the organism. After passage through the acid barrier of the stomach, vibrios colonize the epithelium of the small intestine by means of one or more adherence factors, invasion into epithelial cells or the lamina propria does not occur. Production of CT disrupts ion transport by intestinal epithelial cells. The subsequent loss of water and electrolytes leads to severe diarrhoea characteristic of cholera.

Mode of action of cholera toxin (CT) The structure of CT is typical of the A-B subunit group of toxins in which each of the subunits has a specific function. The B subunit serves to bind the holotoxin to the eukaryotic cell receptor, and the A subunit possesses a specific enzymatic function that acts intracellularly. CT consists of five identical B subunits and a single A subunit, and neither of the subunits individually has significant secretogenic activity in animal or intact cell systems.

The receptor for CT binding is the ganglioside GM. Binding of CT to epithelial cells is enhanced by a neuraminidase (NANase) produced by *V. cholerae*. This enzyme enhances the effect of CT by catalysing the conversion of higher order gangliosides to GM_1 thereby enhancing the binding of CT and leading to greater fluid secretion.

The intracellular target of CT is adenylate cyclase, which mediates the transformation of ATP to cyclic AMP (cAMP), a crucial intracellular messenger for a variety of cellular pathways. Regulation of adenylate cyclase occurs via G proteins, which serve to link many cell surface receptors to effector proteins at the plasma membrane. The specific G protein involved is the Gsα protein, activation of which leads to increased adenylate cyclase activity. CT catalyses the transfer of the ADP-ribose moiety of NAD to a specific arginine residue in the Gsα protein resulting in the activation of adenylate cyclase and subsequent increase in intracellular levels of cAMP. cAMP activates a cAMP-dependent protein kinase, leading to protein phosphorylation, alteration of ion transport, and ultimately to diarrhoea. The alpha subunit of Gs contains a GTP binding site and an intrinsic GTPase activity. Binding of GTP to the α subunit leads to dissociation of the α, β and γ subunits and subsequent increased affinity of α for adenylate cyclase. The resulting activation of adenylate cyclase continues until the intrinsic GTPase activity hydrolyses GTP to GDP, thereby inactivating the G protein and adenylate cyclase, ADP ribosylation of the a subunit by the A$_1$ peptide of CT inhibits the hydrolysis of GTP to GDP thus leaving adenylate cyclase constitutively activated probably for the life of the cell.

The increased intracellular cAMP concentrations resulting from the activation of adenylate cyclase by CT lead to increased Cl⁻ secretion by intestinal crypt cells and decreased NaCl coupled absorption by villus cells. The net movement of electrolytes into the lumen results in a *trans*-epithelial osmotic gradient which causes water flow into the lumen. The massive volume of water overwhelms the absorptive capacity of the intestine, resulting in diarrhoea.

Prostaglandins and the enteric nervous system are also involved in the response to CT in addition to the cAMP mechanism. The intestine also contains a variety of cells that can produce hormones and neuropeptides such as vasoactive intestinal peptide (VIP) and serotonin that can affect secretion.

Other Toxins

Zonula occludens toxin　Apart from CT, the organism produces various other toxins that are implicated in the disease. The zonula occludens toxin (zot) increases the permeability of the small intestinal mucosa by affecting the structure of the intercellular tight junction, or zonula occludens. Genes coding for this toxin *zot* are located immediately upstream of the *ctx* locus and strains that contain *ctx* sequences almost always contain the *zot* sequences and vice versa. By increasing intestinal permeability, *zot* might cause diarrhoea by leakage of water and electrolytes into the lumen under the force of hydrostatic pressure.

Accessory cholera enterotoxin (Ace) Gene for this is located upstream of *zot* and the product of this gene causes fluid accumulation in rabbits and diarrhoea results from *Ace* monomers aggregating and inserting into the eukaryotic membrane to form an ion channel.

OTHER VIBRIO SPECIES

Vibrio mimicus

It is sucrose negative *V. cholerae* non O1, the important reservoir being water, fishes and prawns. Gastroenteritis due to this organism has been linked only to consumption of seafood. The disease is characterized by diarrhoea, nausea, vomiting and abdominal cramps. This organism does not produce any unique enterotoxins and its level of adherence is also lower when compared to that of *V. cholerae*.

Vibrio parahaemolyticus

This organism has been implicated in food-borne diarrhoea in the 1970s to 1980s. This organism produces a hemolysin termed TDH (thermostable direct haemolysin) or Kanagawa haemolysin.

V. parahaemolyticus occurs in estuarine waters throughout the world with fishes, crabs, oysters, lobsters, shrimps, shellfishes, clams forming major reservoirs. The distribution of this organism is heavily influenced by water temperature, salinity and association with certain planktons. The KP$^+$ strains of *V. parahaemolyticus* are of prime importance in human disease. This organism has a remarkable ability for rapid growth and generation times as short as 8–9 minutes at 37°C have been reported. Eventually the symptoms appear much sooner (4–30 hours) after food consumption. The primary symptoms are diarrhoea and abdominal cramps along with nausea and vomiting and fever. In most severe cases, diarrhoea is watery with mucus and blood. A dose of 10^5–10^7 KP$^+$ cells are sufficient to cause gastroenteritis.

Vibrio vulnificus

This is the most serious in the developed countries like the United States responsible for over 95% of all seafood-related deaths in the country with fatality rates of 60%. The bacterium is unusual in being able to produce wound infections in addition to gastroenteritis and primary septicaemias. Originally termed as the "lactose positive" vibrio, this organism is also sucrose positive and bioluminescent. It is a widespread inhabitant of estuarine environments, from crabs, clams, tarbos, fishes, etc. *V. vulnificus* infections are highly correlated with water and temperature, with most cases occurring during the summer months. The incubation period ranges from 7 hours to several days with a median of 26 hours. The most significant symptoms included fever, chills, nausea and hypotension along with smaller percentage of vomiting, diarrhoea and abdominal pain. An unusual symptom that generally occurs is the development of secondary lesions. These usually occur on extremities, frequently develop into necrotizing fascitis or vasculitis and often necessitate surgical amputation. Gastroinestinal illness with associated diarrhoea is relatively infrequent. The infectious dose of *V. vulnificus* is not known. The polysaccharide capsule produced by some strains of *V. vulnificus* is essential to its ability to initiate infection. The capsule allows these cells to resist phagocytosis. Elevated levels of serum iron appear to be essential for *V. vulnificus* to multiply in the human host. This bacterium can only produce septicaemia in humans with elevated serum iron levels. Another product that may be critical to its virulence is the endotoxic LPS present in the cells. In addition to the role of capsule, iron and endotoxin in the pathogenesis of *V. vulnificus* infections, this bacterium produces a large number of extracellular compounds, including haemolysin, protease, elastase, collagenase, sulphatase, hyaluronidase and fibrinolysin.

Vibrio fluvialis

Referred to as "Group E", there are two biogroups of which biogroup I is anaerogenic and are isolated from aquatic environments and diarrhoeal cases whereas biogroup II was aerogenic and not disease associated. Subsequent studies indicated that the aerogenic strains were a unique species, and these were reclassified as *Vibrio furnisii*.

This organism has been frequently isolated from brackish and marine waters and sediments, fishes, shellfish, oysters, and clams. This organism particularly causes infections in small children. Gastroenteritis with this organism is similar to that of cholera, with diarrhoea and vomiting, moderate to severe dehydration, abdominal pain and fever being common symptoms. A notable difference from cholera is the frequent occurrence of bloody stools in infections due to *V. fluvialis*. Hypotension is absent. Three potential enterotoxins have been described as virulence factors, a Chinese hamster ovarian (CHO) cell cytotoxin, a CHO cell rounding toxin and a CHO cell elongation factor. In addition to the above factors, elastase, mucinase, protease, lipase, lecithinase, chondroitin sulphatase, hyaluronidase, DNase and fibrinolysin have also been reported.

Vibrio furnissii

This organism is similar to *Aeromonas hydrophila* and has been isolated from river and estuarine water, marine molluscs and crustacea. Symptoms included diarrhoea, abdominal cramps, nausea and vomiting. Onset of symptoms occur between 5 to 20 hours with the patients recovering within 24 hours. Haemolysin, CHO elongation factor seem to be the virulence factors for the organism.

Vibrio hollisae

It is unusual among the vibrios in its inability to grow on TCBS agar or MacConkey agar. It grows well on blood agar, and xylose-lysine-desoxycholate agar. There is a strong correlation between *V.hollisae* infections and consumption of raw seafood. Foods most commonly consumed by the patients with gastroenteritis were raw oysters, raw clams, crabs and shrimp. Septicaemia is common in these infections. Other symptoms of gastroenteritis are similar to those caused by non O1 strains of *V. cholerae* and include severe abdominal cramping, vomiting, fever and watery diarrhoea. Virulent factors include a haemolysin, and CHO cell elongation factor.

Vibrio alginolyticus

It was originally classified as a biotype of *V. parahaemolyticus*. It inhabits often at high numbers, seawater and seafood. It is easily isolated from fish, crabs, clams, oysters, mussels and shrimp as well as water. Only rarely has this organism been implicated as a food-borne pathogen. The presence of vibrios, especially *V. cholerae*, *V. vulnificus* and *V. parahaemolyticus* in foods represent a serious and growing public health hazard. Most vibrios demonstrate a seasonality in their ability to be isolated from the environment and foodstuff and in the infections they cause, both being generally greater during the warmer months. Little is known about the susceptibility of the vibrios to food preservation methods. Cold is often considered an effective defense against proliferation of vibrios, although seafoods have been reported to be protective for some of the vibrios at refrigeration temperatures. Thorough heating appears to be the only effective protective measure currently available against proliferation of vibrios.

LABORATORY DIAGNOSIS OF *VIBRIO* SPECIES IMPLICATED IN FOOD POISONING

Traditionally, cholera and other choleraic infections caused by these organisms have not been considered in relation to food microbiology because the mode of transmission is primarily by water either directly or indirectly. The Food Safety Act now embraces water used in the food industry so it is appropriate that mention is made of these organisms with regard to foods that come from an aqueous environment, or are washed in or irrigated with water that may be contaminated.

V. parahaemolyticus, a halophilic gram-negative organism, was first recognized as a food poisoning organism in Japan. It has assumed prominence there and in UK. Enumeration of *V. paraheamolyticus* in foods should be attempted by inoculating and spreading dilutions of food suspensions onto the surface of plates of selective media followed by overnight incubation at 37°C. In addition, a portion (25 g) of the food should be added to a suitable alkaline enrichment broth then incubated at 37°C followed by 41.5°C and subcultured after 6 hours and 18–24 hours onto selective agar plates. Following incubation, suspect colonies are picked from the plates; those which produce oxidase and catalase and are sensitive to vibriostatic agent 0129 (2,4-diamino-6,7-di-iso-propylpteridine) are classed as vibrios.

The main source of pathogenic vibrios is seafood. The presence of any *Vibrio* spp. in cooked food is of significance, as they are easily destroyed by heat.

Method 1 Enrichment culture for pathogenic *Vibrio* spp.

Media

Enrichment broth Alkaline peptone water containing yeast extract 3 g, neutralized peptone 10 g, sodium chloride 20 g/l, pH 8.6.

Selective agar media Thiosulphate citrate bile salt sucrose (TCBS) agar and one other selective agar of choice, e.g. sodium dodecyl sulphate polymixin sucrose (SDS) agar containing proteose peptone 10 g, beef extract 5 g, sucrose 15 g, sodium dodecyl sulphate 1 g, bromothymol blue 0.04 g, cresol red 0.04 g, polymyxin B sulphate 100,000 IU and agar 15g/L with pH 7.6.

Non-selective agar media (containing sodium chloride) e.g. blood agar.

Procedure

1. Prepare a 10^{-1} homogenate of food in alkaline peptone water.

2. For frozen, chilled, salted or otherwise processed food, incubate the homogenate at 37°C for 6 hours. For fresh fish, incubate the homogenate at 41.5°C for 6 hours.

3. After 6 hours, subculture the homogenate to TCBS and SDS agar. Incubate the plates at 37°C for 18–24 hours. Subculture the food homogenate to a new 100 mL volume of alkaline peptone water (secondary enrichment) and incubate at 41.5°C for 18 hours. Reincubate the food homogenate at 41.5°C for 18 hours.

4. At the end of incubation, subculture both the food homogenate and the secondary enrichment broth to TCBS and SDS agar. Incubate the plates at 37°C for 18–24 hours.

5. Examine the TCBS and SDS plates for characteristic colonies.

6. Subculture five suspect colonies from each agar to a non-selective agar containing sodium chloride. Incubate at 37°C for 18–24 hours.

7. Characterize the suspect colonies according to the biochemical tests pertaining to the vibrio group.

Method 2 Enrichment culture for *V. parahaemolyticus*

Media

Enrichment media Salt polymyxin broth (SPB) and either alkaline salt peptone water (ASPW) containing peptone 20 g, sodium chloride 30 g/L, pH8.6 or saline glucose medium with sodium dodecyl sulphate (GST) containing peptone 10 g, meat extract 3 g, sodium chloride 30 g, glucose 5 g, methyl violet 0.002 g, sodium dodecyl sulphate 1.36 g/L with pH8.6.

Selective agar media Thiosulphate citrate bile salt sucrose (TCBS) agar and triphenyl tetrazolium chloride soya tryptone (TSAT) agar.

Procedure

1. Prepare a 10^{-1} homogenate of the food sample with 25 g of food and 225 ml of SPB and ASPW or GST.

2. Incubate the homogenate at 37°C for 18 hours.

3. Subculture to TCBS agar and TSAT agar after 7–8 hours and after 18 hours, incubate the plates for 20–24 hours.

4. Examine TCBS plates for presence of green colonies and TSAT plates for dark red colonies with a diameter of more than 2 mm.

5. Confirm the identity of suspect colonies as described in the above procedure.

Method 3   Direct enumeration of *Vibrio* spp.

Procedure

1. Prepare a 10^{-1} homogenate of the food and further serial dilutions in peptone saline diluent. Use a surface counting method (pour plate/spread plate/drop plate) to enumerate on TCBS agar. If organisms are likely to be stressed, enumeration should also be performed on SDS agar.

2. Incubate the plates at 37°C for 20–24 hours.

3. Count the number of colonies of each type suspected to be vibrios.

4. Confirm suspect colonies as described above.

5. Calculate the count per gram from the proportion of colonies that were confirmed as *Vibrio* spp.

Method 4   Enumeration of *Vibrio* spp. by multiple tube method.

Procedure

1. Prepare a 10^{-1} homogenate and serial decimal dilutions of the food sample as described above.

2. Select a multiple tube method and a liquid medium (peptone saline).

3. Proceed as described above.

4. Calculate the most probable number per gram from the number of tubes that yield growth of *Vibrio* spp.

FOOD POISONING BY *YERSINIA ENTEROCOLITICA*

The genus *Yersinia* comprises 11 species. As members of the family Enterobacteriaceae, yersiniae are oxidase negative, gram-negative, rod-shaped facultative anaerobes which ferment glucose. The genus includes three primary pathogens of humans and several other species which may cause opportunistic infections. The three pathogenic species are *Y. pestis*, the causative agent of bubonic and pneumonic plague, *Y. pseudotuberculosis*, an intestinal pathogen of rodents which occasionally infects humans, and *Y. enterocolitica* a common intestinal pathogen of humans.

Y. enterocolitica first emerged as a human pathogen during the 1930s. It exhibits between 10 and 30% DNA homology with other genera in the Enterobacteriaceae. It is a highly heterogenous, being divisible into a large number of subgroups, chiefly according to biochemical activity and lipopolysaccharide O antigens. Most primary pathogenic strains of humans and domestic animals occur within biovars 1B, 2, 3, 4 and 5. *Y. enterocolitica* strains of biovar 1A are generally obtained from terrestrial and freshwater ecosystems and are often referred to as environmental strains. Not all isolates of *Y. enterocolitica* obtained from soil, water or unprocessed foods can be assigned to a biovar. These strains invariably lack the characteristic virulence determinants of the primary pathogenic yersiniae and may represent a novel non-pathogenic subtypes or even new *Yersinia* species.

Survivability Characteristics

Y. enterocolitica is unusual among pathogenic enterobacteria in being psychrotrophic as evidenced by its ability to replicate at temperatures between 0 and 44°C. The doubling time at the optimum growth temperature is around 34 minutes which increases to 1 hour at 22°C, 5 hours at 7°C and 40 hours at 1°C. Yersiniae withstand freezing and can survive in frozen foods for extended periods even after repeated freezing and thawing. *Y. enterocolitica* is susceptible to heat and is destroyed by pasteurization at 71.8°C for 18s. It is able to grow over a pH range from 4 to 10 with an optimum pH of around 7.6. Tolerance of *Y. enterocolitica* to acid depends on the acidulant agent used, the environmental temperature, the composition of the medium, and the growth phase of the bacteria. Acid tolerance depends on the activity of urease, which catabolizes urea to release ammonia, which elevates the cytoplasmic pH.

Y. enterocolitica is readily inactivated by ionizing and UV radiation and by sodium nitrate and nitrite added to food. It displays relative resistance to these salts in solution and can tolerate NaCl at concentration of up to 5%. It is also resistant to chlorine. This species survives better at room temperature and refrigeration temperatures than at intermediate temperatures. *Y. enterocolitica* persists longer in cooked foods than in raw foods, probably because of increased availability of nutrients in cooked foods and because the presence of other psychrotrophic bacteria, including environmental strains of *Y. enterocolitica*, in unprocessed food may restrict bacterial growth. *Y. enterocolitica* can grow well on cooked beef or pork, in vacuum-packed meat (at refrigeration temperatures), boiled eggs, boiled fish, pasteurized liquid eggs, pasteurized whole milk, and tofu. Proliferation also occurs in refrigerated seafoods such as oysters, raw shrimp and cooked crab-meat. Bacteria may persist for extended periods in refrigerated vegetables and cottage cheese. Although *Y. enterocolitica* tolerates 5% NaCl in culture media, the addition of 5% NaCl to foods slows its growth rate.

Characteristics of Infection

Infection exhibits a wide range of clinical presentation and outcomes. Apart from self limiting diarrhoea, yersiniosis may also give rise to a variety of suppurative and autoimmune complications which is underlined by factors such as age and immune status.

Acute infection *Y. enterocolitica* enters the gastrointestinal tract after ingestion in contaminated food or water. The infective dose for humans is likely to exceed 10^4 cfu. Gastric acid appears to be a significant barrier to infection. Most symptomatic infections with *Y. enterocolitica* occur in children less than 5 years of age. It occurs as diarrhoea, often accompanied by low grade fever and abdominal pain. The character of the diarrhoea varies from watery to mucoid or bloody. The illness lasts for a few days to 3 weeks. Rarely, acute enteritis may progress to intestinal ulceration and perforation to ileocolic intussusception, toxic megacolon, or mesenteric vein thrombosis. Sore throat is a frequent accompaniment and may dominate the clinical picture of older patients.

In children older than 5 years and adolescents, acute yersiniosis often occurs as a pseudo-appendicular syndrome. The usual features of this syndrome are abdominal pain and tenderness localized to the right lower quadrant. These symptoms are usually accompanied by fever, with little or no diarrhoea. This closely resembles appendicitis.

Bacteraemia is a rare complication of infection, except in patients who are immunocompromised or in an iron overloaded state. Infections with *Y. enterocolitica* are noteworthy for the large variety of immunological complications, including reactive arthritis, erythema nodosum, iridocyclitis, glomerulonephritis, carditis and thyroiditis which follow acute infection. Of these, reactive arthritis is the most widely recognized. Arthritis (migratory) typically follows the onset of diarrhoea or the pseudo-appendicular syndrome by 1 to 2 weeks. The joints most commonly involved are the knees, ankles, toes, tarsal joints, fingers, wrists and elbows.

Reservoirs

Y. enterocolitica occupies a broad range of environments and has been isolated from the intestinal tracts of many different mammalian species, from birds, frogs, fish, flies, fleas, crabs, oysters, etc. Foods that may harbour *Y. enterocolitica* include pork, beef, lamb, poultry and dairy products notably milk, cream and ice cream. It is also commonly found in a variety of terrestrial and freshwater ecosystems, including soil, vegetation, lakes, rivers, wells and streams. Several outbreaks of yersiniosis have been linked to the consumption of contaminated cow milk, but cattle do not appear to be an important reservoir of these bacteria. In these outbreaks, milk may have been contaminated with pig or human faeces during or after processing. Contamination of pasteurized milk appears to pose a greater threat of infection than raw milk, probably because *Y. enterocolitica* outgrows other faecal microbes during storage at refrigeration temperatures most readily when competing psychrotrophic microflora have been eliminated.

Food-borne Outbreaks

Most food-borne outbreaks in which a source was identified have been traced to milk. As *Y. enterocolitica* is rapidly destroyed by pasteurization, infection results from the consumption of raw milk or milk that is contaminated after pasteurization. There has been instances when the bacteria were transmitted from raw chitterlings (pig intestine) to affected patients on the hands of food handlers. Other foods which have been responsible for outbreaks of yersiniosis include pork cheese, bean sprouts and tofu. In the latter two foods, contaminated well or spring water was the probable source of the bacteria.

Pathogenicity

Y. enterocolitica is an invasive pathogen which induces an inflammatory response in infected tissues. The distal ileum, in particular the gut-associated lymphoid tissue, bear the brunt of the infection, although adjacent regions of the intestine and the mesenteric lymph nodes are also frequently involved. As investigations in volunteers are precluded by the risk of autoimmune sequelae, most information has been obtained from animal models not taking death as an end point of infection since this is not the usual outcome of infection in humans.

After oral inoculation of mice with a virulent strain of the bacteria, it is seen that most bacteria remain within the intestinal lumen, while a minority adhere to the mucosal epithelium, showing no particular preference for any cell type. Invasion takes place almost exclusively through M cells, which are specialized epithelial cells overlying lymphoid follicles in the intestine. After penetrating the epithelium, yersiniae traverse the basement membrane of the relatively porous dome epithelium of the Peyer's patches. Virulent strains then proliferate in the gut-associated lymphoid tissue and the lamina propria, where they cause localized tissue destruction leading to the formation of microabscesses. They may also spread through the lamina propria to adjacent villi and *via* the lymph to more remote sites. Lesions occur chiefly within intestinal crypts. These lesions are microcolonies of bacteria surrounded by granulocytic and monocuclear inflammatory cells. *Y. enterocolitica* may spread *via* the lymph to the draining mesenteric lymph nodes where the organisms also produce microabscesses.

Virulence Factors

Yersinia is an invasive enteric pathogen. The virulence determinants of *Y. enterocolitica* are classified into those which are chromosomally encoded and those specified by a 70–75-kb virulence plasmid.

Chromosomal determinants of virulence

Invasin All *Y. enterocolitica* strains are able to produce a 91-kDa surface expressed outer membrane protein termed invasin. Analysis of invasin has shown that the amino terminus is inserted in the bacterial outer membrane, and the carboxyl terminus is exposed on the surface. The carboxyl region specifies binding of invasion to specific ligands, known as $\beta1$ integrins on host cells. Integrins are heterodimeric transmembrane proteins that communicate extracellular signals to the cytoskeleton. When invasin binds to its receptors on epithelial cells, a sequence of events which results in cytoskeletal alteration and internalization of the bacteria is initiated. The internalization process mediated by invasion is propagated entirely by the host cells.

Attatchment–invasion locus (ail) Virulent strains of Y. *enterocolitica* produce an outer membrane protein, unrelated to invasin called the *ail* locus (attachment–invasion). This may allow yersiniae to persist in serum by protecting them from non-specific destruction by complement.

Heat-stable enterotoxins The toxin is known as Yst, and it involves binding to and activation of cell associated guanylate cyclase and elevation of intracellular concentration of cyclic GMP. This in turn causes perturbation of fluid and electrolyte transport pathways in intestinal absorptive cells which results in diarrhoea. Yst is encoded by the chromosomal *yst* gene.

Mucoid yersinia fibrillae (Myf) Many enteric pathogens carry distinctive colonization factors on their surface which mediate their adherence to specific sits on the intestinal epithelium. These adhesions are essential virulence determinants of non-invasive enterotoxin secreting bacteria.

LPS Like other gram-negative bacteria, *Y. enterocolitica* may be classified as smooth or rough depending on the amount of 'O' side chain polysaccharide attached to the inner core region of the cell wall LPS. LPS is required for the full expression of the virulence. Smooth LPS may enhance virulence by increasing bacterial hydrophilicity.

Iron acquisition Iron is an essential micronutrient of almost all bacteria. Despite the nutrient rich environment provided to bacteria by mammalian tissues, the availability of iron in some sites may be limited. This is because most iron in tissues is bound to high affinity transport proteins such as transferrin and lactoferrin or is incorporated into organic molecules such as haemoglobin. Several species of pathogenic bacteria produce low-molecular weight, high-affinity iron chelators known as siderophores. These compounds are secreted by the bacteria into the surrounding medium, where they couple with iron. The resultant ferrisiderophore complexes then bind to specific receptors on the bacterial surface and are taken up into the cell. Patients suffering from iron overload show increased susceptibility to severe infections with *Y. enterocolitica* indicating that the availability of iron in tissues may determine the outcome of yersiniosis. More virulent strains of *Y. enterocolitica* produce a novel catechol-containing siderophore termed yersiniabactin. This compound forms a ferrisiderophore complex with iron and then enters the bacteria after binding to a 65-kDa outer membrane protein receptor named FyuA.

Urease All enteric pathogens must negotiate the acid barrier of the stomach to cause disease. In *Y. enterocolitica*, acid tolerance relies on the production of urease which catalyses the release of ammonia from urea and allows the bacteria to resist pH as low as 2.5. Urease is produced chiefly during the stationary phase of growth at temperatures below 37°C.

The virulence plasmid All virulent strains of *Yersinia* spp. carry a 70–75-kb plasmid termed pYV (plasmid for *Yersinia* virulence). Yersiniae which carry this plasmid exhibit a distinctive phenotype known as the low calcium response because it manifests only when pYV bearing bacteria are grown in media containing low concentration of Ca^{2+}. The principal features of the low calcium response are the cessation of bacterial growth after one or two generations and the appearance of at least 11 new proteins on the bacterial surface or in the culture medium. The secreted proteins are referred as (yersinia outer membrane proteins) Yops. They are now given the designations Yop B, D, E, H, M, N, O, P, Q and R.

Yops These *Yops* act directly on host cells to subvert immune defenses. Yop M is a 41-kDa protein which shows homology to a human platelet specific receptor for thrombin. Under normal circumstances, binding of these ligands to platelets cause platelet aggregation and the release of several inflammatory mediators. As Yop M has sufficient thrombin binding activity to inhibit thrombin-induced platelet aggregation, it may interfere with the ability of the host to mount a normal inflammatory response to infection and thus may facilitate bacterial survival in tissues.

Yad A Formerly known as Yop A, it is a 45-kDa outer membrane protein which polymerizes to form fibril like structures on the bacterial surface. YadA acts as an adhesion facilitator by mediating binding to intestinal mucus and to certain extracellular matrix proteins, including collagen and cellular fibronectin. These proteins in turn bind to integrins on epithelial cells and stimulate bacterial internalization by parasite-mediated endocytosis. *Y. enterocolitica* is a versatile food-borne pathogen with a remarkable ability to adapt to a wide range of environments within and outside its host. The bacteria typically access their hosts *via* food or water in which they would have grown to stationary phase at ambient temperature. Under these circumstances, they exhibit factors such as urease and smooth LPS which facilitate their passage through the stomach and the mucous layer of the small intestine. Once *Y. enterocolitica* begins to replicate in the intestine at 37°C, LPS becomes rough exposing the integrins and invasions on the bacterial surface. These factors may promote further invasion while protecting the bacteria from complement-mediated opsonization. When the bacteria make contact with host cells in lymphoid tissue, they are stimulated to synthesize and release Yops, (E and H) which further frustrate the efforts of phagocytes to ingest and remove them. Further bacterial replication may lead to tissue damage and the formation of microabscesses. Some of the more virulent strains produce yersiniabactin so that replication can proceed. Eventually, the cycle is completed when the bacteria rupture microabsecesses in intestinal crypts to re-enter the intestine and regain access to the environment.

LABORATORY DIAGNOSIS AND ISOLATION OF *Y. ENTEROCOLITICA*

The successful isolation of *Y. enterocolitica* from food and environmental samples is achieved by the use of enrichment techniques followed by plating onto selective media. One of the most productive method had been enrichment in simple buffer solutions incubated at low temperatures, 4–9°C, for periods of up to 14 days with periodic subculture of the enrichment broth onto cefsulodin-irgasan-novobiocin (CIN) agar for incubation at 30°C overnight.

Method 1 Enrichment culture

Media

> Enrichment medium   Tris buffered peptone water with peptone 10 g, sodium chloride 5 g, tris (hydroxymethyl) methylamine12.1 g, distilled water 1L, adjusted to pH8.0.
>
> Selective agar   Cefsulodin-irgasan-novobiocin (CIN) agar.

Procedure

1. Homogenize 25 g of food sample in 225 ml of tris buffered peptone water.
2. Incubate the homogenate at 9°C for 2 weeks (pasteurized milk should be incubated at 4°C for 3 weeks with subculture at the end of the incubation period).
3. Subculture after 2 weeks at 9°C as follows: add 1 ml of the incubated homogenate to 9 ml of 0.5% potassium hydroxide or 0.5% of sodium chloride solution and mix. After 15–30 seconds, subculture a loopful of the mixture to CIN agar.
4. Incubate CIN plates at 30°C for 24 hours.
5. Examine CIN plates for the presence of colonies. Typical colonies have a bulls-eye appearance with a red centre surrounded by a transparent border,

and are usually smaller than colonies of other coliforms capable of growth on CIN agar. Colonies may also appear very small and dry, or much large with irregular edges and a large amount of colourless periphery relative to the red centre.

6. Confirm the identity of suspect colonies. *Yersinia* spp. are urease positive that produce an acid, but without production of gas or hydrogen sulphide in TSI agar slopes, and are non-motile at 37°C but motile below 28°C. Some strains produce acid from lactose. *Y.enterocolitica* and related strains can decarboxylate ornithine but not lysine.

7. Colonies may be further characterized using the biochemical reactions specific for *Y. enterocolitica*.

Method 2   Enrichment Procedure

Dual Isolation Procedure

The method uses two enrichment media and different isolation protocols. Enrichment in a buffered peptone sorbitol bile salts broth will recover all strains of *Y. enterocolitica* and related species; the other enrichment procedure is targeted at the pathogenic serotypes.

Media

Enrichment media   Peptone sorbitol bile salts broth (PSB), containing 1% sorbitol and 0.15% bile salts, pH 7.6 and irgasan ticarcillin potassium chlorate broth (ITC).

Selective agar media   Cefsulodin irgasan novobiocin (CIN) agar and salmonella–shigella agar supplemented with 1% sodium desoxycholate and 0.1% calcium chloride (SSDC).

Procedure 1

1. Prepare a 1/10 homogenate of food in PSB broth.
2. Incubate the homogenate at 22–25°C for 3 days with agitation or 5 days without agitation.
3. Subculture directly to CIN agar. Also subculture 1 ml to 9 ml of 0.25% potassium hydroxide or 0.85% sodium chloride solution. After 20 seconds, subculture a loopful to CIN agar.
4. Incubate the plates at 30°C for 24 hours.
5. Examine the plates for suspect colonies and proceed as described in method 1.

Procedure 2

1. Prepare a 1/100 homogenate of food in ITC broth.
2. Incubate the homogenate at 25°C for 48 hours.
3. Subculture to SSDC agar. Incubate at 30°C for 24 hours.
4. Examine the plates for the presence of suspect colonies. These appear small (<1 mm) and grey with an indistinct rim, non-iridescent and very finely granular when examined with oblique transmission.
5. Characterize suspect strains as described in method 1.
6. If pathogenicity tests are required, perform tests for aesculin and or salicin fermentation. Pathogenic strains do not ferment salicin or aesculin.

FOOD POISONING BY *SALMONELLA* SPP.

Of the various food-borne diseases, salmonellosis is the most important one both in terms of the disease intensity, ease of spread and difficulty in prevention and control.

According to WHO, it represents 60–80% of all reported cases of food-borne diseases. The reasons for the increase in occurrence of salmonellosis are:

1. increase in communal feeding
2. increase in international trade in human food
3. higher incidence of salmonellosis in farm animals
4. widespread use of household detergents interfering with sewage treatment
5. wide distribution of prepared foods
6. widespread use of animal feeds containing antimicrobial drugs that favour drug-resistant salmonellae and their potential transmission to human beings.

It is important to learn the taxonomy and morphological features of the organism, growth characteristics, survival characteristics, reservoirs, food-borne outbreaks, disease symptoms, infectious dose, incubation period and mortality rate, pathogenicity and virulence factors, lab diagnosis (sampling, identification with rapid tests).

Salmonella The generic name coined by Lignieres. More than 2300 serovars have been isolated (Kauffmann, White scheme).

Taxonomy

The organism is a facultative anaerobic gram-negative rod belonging to the Enterobacteriaceae. It is motile with peritrichous flagella with two exceptions, *S. pullorum* and *S. gallinarum,* which are non-motile. The organism is chemo-organotrophic (respiratory and fermentative), maximum growth observed at 37°C, can metabolize glucose with acid and gas. It is oxidase-negative, catalase-positive and urease-negative, can utilize citrate, produces H_2S and can decarboxylate ornithine and lysine (to cadavarine). Utilization of lactose and sucrose is plasmid mediated and thus lac^{+}/suc^{-} biotypes have been detected which is of prime concern since they could easily escape detection on Mac plate. The prevalence of *Salmonella* sp. as a biochemically homogenous groups of microbes is rapidly diminishing. Hence there is a need to replace the biochemical tests with molecular techniques.

Certain species encountered as food-borne pathogens are:

S. eastbourne, S. derby, S. newport, S. typhimurium, S. typhimurium PT9, *S. enteritidis, S. enteritidis* PT4, *S. typhimurium* PT204, *S. indiana,*

S. champaign, S. poona, S. saintpaul, S. javiana, S. rubislaw, S. bovismorbificans, S. gallinarum, S. thompson, S. dublin, S. paratyphi A, B *and* C. *S. cholerasuis, S. infantis, S. bongori. S. enterica* is recently added and is found to have six subspecies as *S. enterica* subsp.*enterica, salamae, arizonae, diarizonae, lontenae,* and *indica.*

Growth characteristics

Salmonella can readily adapt to extreme environmental conditions. Thus it can grow or survive at temperatures from 2°–54°C for 24 hours to 2 days. This has proven very difficult to control the survival and growth of the organism in food especially animal based foods. A knowledge about the organisms adaptation towards the extrinsic and intrinsic factors of the food is anticipatory for controlling the survival of the organism.

i. Preconditioning the cells to low temperature increases the growth of *Salmonella* in refrigerated food. Hence the efficacy of chill temperatures in ensuring food safety stands questioned. Incidence of mutant strains capable of growing at temperatures around 48°C and 54°C have been occurring lately.

ii. The physiological adaptability of *Salmonella* sp. is further demonstrated by their ability to multiply at pH values ranging from 4.5–9.5 (optimum pH 6.5–7.5). The growth or enhanced survival of *Salmonella* during fermentative process would result contamination in ready-to-eat product. Presence of acid-tolerant *Salmonella* have lately been solicited and their increased survival in fermented milk further heightened the level of public health hazard since this acquired physiological trait could minimize the antimicrobial action of gastric acidity (pH 2.5) and promote the survival of salmonellae within acidic cytoplasm of the phagocytes.

iii. Foods with $a_w < 0.93$ do not support the growth of salmonellae. Salmonellae are generally inhibited in the presence of 3–4% NaCl, but salt tolerance (food and serovar specific) increases with increasing temperature (in the range of 10–30°C) with slightly decreased growth rate. There is also evidence to point out anaerobiosis and its potentiation of greater salt tolerance in salmonellae (a problem envisaged in modified atmosphere packaged foods) and foods that contain a high salt concentration.

Thus pH, salt concentration and temperature exert profound effects on the growth kinetics of *Salmonella* species.

Survivability of Salmonellae

Survival of salmonellae in unfavorable environment further heightens public health concerns. The composition of freezing liquid, kinetics of freezing process and physiological state of *Salmonella* serovar determine the survival of *Salmonella* in frozen foods. Viability of salmonellae in dry foods stored at > 25°C decreases with increasing storage temperature and with increasing moisture content.

The heat resistance of *Salmonella* species increases as the water activity of heating medium decreases. For example, heating of *S. typhimurium* in a medium adjusted to a_w 0.90 with sucrose and glycerol conferred different levels of heat resistance as calculated by their D values at 57.2°C which was found to vary with 1.8 minutes to 8.3 minutes.

Other important features associated with this adaptive response include:

1. Greater heat resistance of salmonellae grown in nutritionally rich media.

2. Cells derived from stationary phase and previous storage of salmonellae in dry condition.

This adaptive response has serious implications with reference to the safety of thermal processes that expose food products at marginally lethal temperatures (<50°C/15–30 minutes) enhances their heat resistance through rapid synthesis of heat shock proteins. The heat-stressed salmonellae encounter changes in their fatty acid composition of cell membrane to provide a greater proportion of saturated membrane phospholipids which reduce the fluidity of bacterial cell membrane thus increasing membrane resistance to heat damage.

Taking into account the other important factor pH, brief exposure of *S. typhimurium* to mild acid environment, (pH 5.5–6.0-preshock) followed by exposure of the adapted cells to less than or equal to 4.5 (acid shock) triggers a complex acid tolerance response (ATR) that allows the survival of the microorganism under extreme acid environment (pH 3–4). This ATR induces synthesis of 43 acid-shock proteins (ASP) and outer membrane proteins (OMP), reduces the growth rate, pH homeostasis, (i.e., the bacteria maintain internal pH values of 7.0–7.2 upon sequential exposure of cells to external pH of 5).

Important Facts on ATR

1. In the ATR, Mg^{2+}-dependent proton translocating ATPase (encoded by *atp* operon) play an important role in maintaining cellular pH of more than or equal to 5 through an energy dependent transport of intracellular protons to the cell exterior. This is coupled with H^+ coupled ion transport systems (antiport) for the intra- and extracellular transport of K^+, Na^+ and H^+ ions, the electron transport chain dependent efflux of H^+ ions and transport systems committed to symport of H^+ and other solutes.

2. Another factor that makes ATR possible is the Fe^{2+} building regulatory protein encoded by *fur* (ferric uptake regulator) gene (evidence: inability of *fur* mutant strains to survive under highly acidic conditions).

3. The organism's amino acid decarboxylases (acid induced) provide an additional protective mechanisms whereby cadaverine and putrescine (breakdown products of lysine and ornithine respectively) potentiate acid neutralization and enhanced bacterial survival.

4. During the stationary phase of growth, ATR is quite distinct. This phase provides greater acid resistance than log phase ATR. This is induced at pH < 5.5 (maximum ATR seen at pH 4.3). This phase induces synthesis of 15 shock proteins that are not affected by mutations in the *atp* and *fur* genes and this response is independent of external pH, i.e., even though the food is subjected to a more acidic change in pH, the organism can survive in the same pH that has induced the ATR.

The above-mentioned systems operate in the acidic environment that prevails in fermented and acidified foods, and in phagocytic cells of infected host. Acid stress also triggers enhanced bacterial resistance to other environmental conditions. For instance, growth of *S. typhimurium* at pH 5.8 causes

- increased thermal resistance (50°C)
- increased tolerance to high osmotic pressure (2.5 M NaCl)
- induced synthesis of OmpC proteins
- greater surface hydrophobicity
- increased resistance to antibacterial lactoperoxidase system and surface active agents like crystal violet and polymyxin B

Food handlers, food industry managers, quality assurance managers all have to take into account the above-mentioned facts to prevent growth of salmonellae in foods.

Reservoirs

The food poisoning salmonellae (of which *S. typhimurium* is by far the most common) are not natural parasites of man, but are widely distributed in the animal, bird and reptilian kingdoms. Domestic animals (cow, pig, poultry, dog and cat) and rodents may all serve as reservoirs; ducks, rats and mice also are the most frequent hosts of the infection. Spread from the animal reservoirs may occur directly by consumption of naturally contaminated meat which has not been sterilized in the cooking process. Pig meat in various forms has been most incriminated among the processed and made up meats (porkpies, sausages, brawn and cold pork). The incidence of infection among pigs may be greatly increased by cross-infection in the lairage or holding pens before killing or carcasses may be secondarily contaminated in the abattoir.

Another direct source of infection is duck egg, which is contaminated from the oviduct before being shelled and laid. A sample survey indicated that 1.5/1,000 of the 126 million duck eggs sold every year in Britain are contaminated with *S. typhimurium*. Numerous outbreaks and sporadic cases have been traced to duck eggs used in mayonnaise or cakes or as insufficiently cooked eggs. It is therefore recommended that duck eggs to be eaten as such, should be boiled for 10–15 minutes or fried for 4–5 minutes on both sides. If they are used for puddings, cakes, etc., the mixing utensils should be sterilized after use.

A third, and perhaps the most important, direct source of infection is bulked egg or egg products, particularly if imported. Bacteriological examinations by the Public Health Laboratory Service of US have shown that 7–8 % of imported frozen or dried whole egg or egg yolk and 12–19% of imported frozen or dried egg albumen contains human pathogenic salmonellae. The baking processes will usually kill the salmonellae but unfortunately the utensils used for preparing the egg containing mixes may later be used for handling uncooked foods, like cream artificial cream, or mayonnaise, without being sterilized. The salmonellae multiply freely in these latter products, and so outbreaks occur.

Cattle may become infected by *S. dublin.*

S. typhimurium and milk may be contaminated through udder infection, by faecal pollution, or from dirty utensils. It is therefore advisable that milk from tubercle free herds should be pasteurized to prevent the risk of infection with salmonellae, brucellae and the pyogenic cocci.

Another hazard of a more indirect kind is imported animal feeding stuff and fertilizers, bone, fish and meat meals, dried blood, bones, etc. of which up to 30% of samples may harbour salmonellae and these in turn may infect pigs and poultry. It is obvious therefore that greater attention should be directed to imported eggs and animal products and steps taken to minimize the risks attendant on their contamination with salmonellae.

Lastly, there remains the contamination of food in food shops, restuarants, and households by rats and mice, by clinically infected food handlers and symptom-less excretors, and perhaps uncommonly by cats and dogs.

Fruits and vegetables are all-time reservoirs where fertilization of crops with unheated sludge/sewage effluents contaminated with antibiotic-resistant *Salmonella* species.

Food-borne Outbreaks

It is the leading cause of food-borne bacterial disease in humans. Salmonella infections constitute 95% of all bacteriologically proven incidents but only 60% of the general outbreaks. The incidence is higher in warm tropical areas. Salmonellosis can lead to high economic loss in the country by way of medical care, hospitalization, lost time and income, death of animals, decreased production of animals, loss/reduced value of contaminated animal products, control procedure and testing (additional expense).

Even though the mortality rate is less than 1%, regarding children and immunocompromised individuals, this infection becomes a serious problem if not controlled efficiently.

Characteristics of the Disease

The illness presents with four symptoms as

1. Enteric fever (typhoid/paratyphoid)
2. Gastroenteritis
3. Septicaemia
4. Carrier state

Enteric fever has an incubation period of 7–28 days with diarrhoea, fever, abdominal pain, headache and uncomplicated colitis. Treatment is with antibiotics and supportive therapy (fluid replacements).

Systemic infections by non-typhoidal organisms can also degenerate into systemic infections and precipitate chronic conditions with high levels of virulence like *S. dublin* and *S. cholerasuis*.

Salmonella-induced chronic conditions such as aseptic reactive arthritis, Reiters syndrome and ankylosing spondylitis are important since certain strains have the ability to infect mucosal surface with the help of outer membrane LPS and invasiveness.

Several theories have been put forth explaining the onset of chronic conditions: phagocytosis of salmonellae results in chemical alteration and slow phagocyte release of bacterial LPS. These altered LPS moieties are re-inserted into the cell membrane of monocytes and synoviocytes (associated with synovial fluids), thereby rendering these surface-altered host cells susceptible to lysis by antibody and complement, autoimmune response resulting in rheumatoid tissue damage caused by a localized inflammation of host tissue and release of cytokines and enzymes from antigen-specific activated phagocytes.

Infectious Dose

Newborn infants, elders, immunocompromised individuals are susceptible to *Salmonella* infections. Infectious dose is food-dependent.

Food	Microorganism	Quantity (cfu)
Chocolate	*S. napoli*	10^{-1}–10^{2}
	S. typhimurium	$< 10^{1}$
Cheddar cheese	*S. typhimurium*	10^{0}–10^{1}
Potato chips	*S. sinipaul*	$< 4.5 \times 10^{1}$

Generally 1–10 cells can constitute a human infectious dose. Infectious dose also depends on food vehicles like high fat content chocolates (cocoa butter), cheese (milk fat), meat (animal fat) since entrapment of salmonellae within the hydrophobic lipid micelles offers protection against the bactericidal action of gastric acidulants. Following bile-mediated dispersion of the lipid moieties in the duodenum, viable salmonellae would resume their course following attachment and colonization. Infectious dose is inversely related to the incubation period. Generally the incubation period ranges between 5–72 hours/12–36 hours depending on the infectious dose and the immune status of the individual. The duration of the illness is 1–4 days (acute), 7–28 days (severe) and can be over 2 months with intermittent excretion (carrier state).

Pathogenicity For disease production: 1. the organism has to be virulent and 2. the organism has to overcome host immune responses that is divided into specific and non-specific.

Presence of viable salmonellae in human intestinal tract confirms the successful evasion of ingested organisms from non-specific host defenses like antibacterial lactoperoxidase in saliva, gastric acidity, mucosal secretions from intestinal cells, peristalsis, sloughing of intestinal epithelial cells which synergistically oppose bacterial colonization of the intestinal mucosa.

In addition, antibacterial action of neutrophils, macrophages, monocytes coupled with specific T and B lymphocytes, Peyer's patches and complement inactivation mount a formidable defense against the systemic spread of salmonellae.

Attachment and Invasion Whether salmonellosis occurs or not depends on the ability of the organism to attach, enter the enterocytes and specialized M cells overlying Peyer's patches. Salmonellae have to successfully compete with the normal gut flora for attachment sites, evade capture by secretory IgA.

Colonization Interaction of bacterial type 1 (mannose-sensitive) or type 3 (mannose-resistant) fimbriae, surface adhesions, non-fimbriate haemagglutinins or enterocyte-induced polypeptides.

Motility increases the frequency of productive contacts between the pathogen and its target. Proteinaceous appendages (encoded by *inv* C and *inv* G loci) develop on the surface of salmonellae upon contact with epithelial cells.

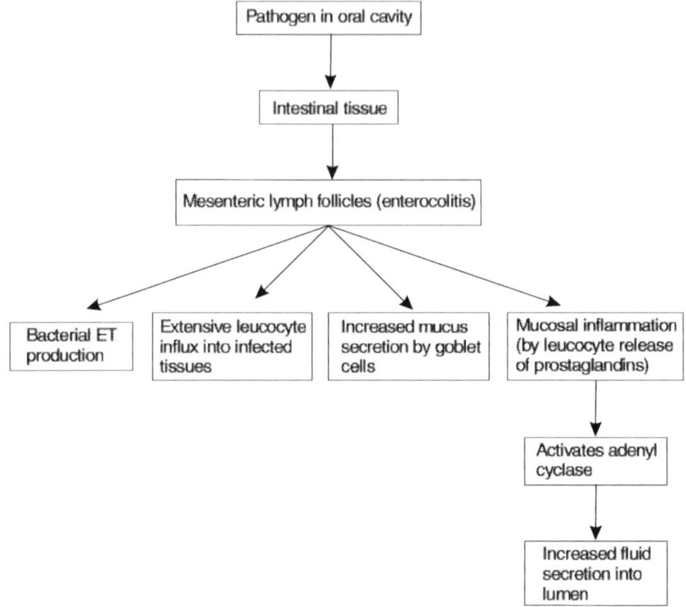

Invasion   The appendages have a role to play during invasion of the bacterium with the host cell. These appendages (0.3–1 nm length and 60 nm in diameter) are short-lived and following attachment, signal transduction between the pathogen and host cell results in culmination of energy dependent invasion of enterocytes and M cells. Important role is played by the *inv* E locus in *Salmonella*. It triggers two profound changes in enterocytes and M cells: a Ca^{2+} influx and cytoskeleton rearrangement in the targeted host cells. The cytoskeleton of eukaryotic cells is a highly organized network of filaments consisting of actin and other protein elements that define the apical basal polity of intestinal epithelial cells and control the movement of intracellular organelles. Gene products from *inv* E of an adherent *Salmonella* cell promote the influx of luminal Ca^{2+} into the mammalian cell. This ionic translocation engenders a polymerization of host cell actin into microfilaments in the vicinity of the invading pathogen. The latter event is seen as an evagination of the apical cytoplasm of epithelial cells around the adherent salmonellae which subsequently mediates the pinocytic uptake of the bacterial pathogen.

Salmonellae can also activate a second host-pathogen system to increase the intracellular Ca^{2+} ion concentration in epithelial cells and enhance the polymerization of actin filaments. Through unknown gene products, adherent salmonellae stimulate phospholipase C activity in the host cells. The enzyme activation results in the breakdown of lipids in the plasma membrane of host cells and in the intracytoplasmic release of inositol triphosphate which in turn, triggers the solubilization of Ca^{2+} bound in the cytoplasmic matrix. Therefore, two distinct systems, the *inv* E dependent translocation of Ca^{2+} from the external medium into the epithelial cells and the inositol phosphate pathway, synergistically contribute to the high cytoplasmic levels of Ca^{2+} required for actin polymerization and subsequent formation of membrane ruffles that are essential for the internalization of salmonellae into epithelial cells.

The appearance of membrane ruffles coincides with the disappearance of the bacterial appendages and with the pathogen directed aggregation of host plasma membrane proteins chiefly the class I MHC proteins which are incorporated into the membrane of infected pinocytotic vacuoles to camouflage the intravacuolar presence of a foreign bacterial antigen thereby pre-empting host-cell-mediated immune responses. This kind of molecular ploy enhances the virulence of the invasive *Salmonella* spp.

Pathogenesis The loci required for invasion and other virulence factors are located within a 40–50 kb segment of chromosomal DNA. The *"inv"* genome is a multigenic locus consisting of as few as 15 genes (*inv*A to O) coding for synthesis of virulence determinants or enzymes responsible for the translocation of virulence determinants to the surface of host cells. The *inv* A, B, C and D are commonly encountered in the genus *Salmonella*. Environmental factors such as high osmolarity and low pO_2 enhance bacterial invasiveness by altering the superhelicity of chromosomal DNA which consequently affects the level of transcription of invasion related genes such as *inv* A. Thus *inv* A triggers signal transduction between the invasive pathogen and the targeted epithelial cell.

Growth and Survival Within Host Cells

Salmonellae are confined to endocytotic vacuoles wherein bacterial replication begins within hours following internalization. The infected vacuoles translocate from the apical to the basal pole of the host cell, where salmonellae are released into the lamina propria. During their multiplication, salmonellae also induce the formation of stable filamentous structures within the epithelial cytosol. These organelles which contain lysosomal membrane glycoproteins and acid phosphatase seem to be connected to the infected vacuoles. These induced filamentous structures are involved in the intravacuolar replication of *Salmonella* spp.

The systemic migration of salmonellae exposes the organism to phagocytosis by macrophages, monocytes and polymorphonuclear leucocytes and to the antibacterial conditions that prevail in the cytoplasm of these host defence cells. For example, the membrane oxidase-dependent formation of toxic oxygen products such as singlet oxygen, super-oxide anion, hydrogen peroxide and hydroxy radicals during the oxidative metabolic burst of phagocytes is countered by the protective activity of several bacterial enzymes, including super-oxide dismutase, peroxidase and catalase. Around 30 bacterial proteins are synthesized in response to the toxic oxygen products of macrophages.

Virulence Factors

The virulence of *Salmonella* spp. as reflected in the ability to cause acute and chronic diseases in humans and in a variety of animal hosts stems from structural and physiological attributes that act synergistically or independently in promoting bacterial adhesion and invasiveness.

The virulence plasmids are large cytoplasmic DNA structures that replicate independently from the bacterial chromosome. These autonomous organelles contain many virulence loci ranging from 30–60 MDa in size, and they occur with a frequency of one to two copies per chromosome. The presence of virulence plasmids within the genus *Salmonella* is limited and has been confirmed in *S. typhimurium, S. dublin, S. gallinarum pullorum, S. enteritidis, S. choleraesuis* and *S. abortus, S. ovis.*

The virulence plasmids enable carrier strains to rapidly multiply within host cells and overwhelm host defense mechanisms. The *Salmonella* plasmid virulence (*"spv"*) region consists of a gene cluster that encodes products for the prolific growth of salmonellae in host reticuloendothelial tissues. Signals that trigger *spv* transcription include the hostile environment within host phagocytes, iron limitation, elevated temperatures, low pH and nutrient deprivation associated with the stationary phase of growth.

Siderophores are yet another facet of the *Salmonella* virulence. These elements retrieve essential iron from host tissues to drive key cellular functions such as the electron transport chain and enzymes associated with iron cofactors.

Toxins are the prominent virulence factor because of its ability to produce overt clinical symptoms in human cases of salmonellosis. *Salmonella* enterotoxin is a thermo-labile protein with a molecular mass of 90 to 110 kDa. It is active at a narrow pH range of 6–8. Thus the functional toxin may be inhibited in the acidic phagolysosomes where the pH is around 4.5–5 or in the epithelial endosomes where the pH is 5–6.5. The *Salmonella* enterotoxin is encoded by a 6.3 kb chromosomal gene which regulates the synthesis of proteins. In addition to expressing enterotoxin, *Salmonella* strains generally produce a thermo-labile cytotoxic protein of 56 to 78 kDa which is localized in the bacterial outer membrane. Other virulence determinants are the capsular polysaccharide Vi antigen which significantly increases the virulence of carrier strains. The length of serotypic LPS that protrudes from the bacterial outer membrane plays an important role in repelling the potentially lytic attack of the host complement system. The primary carbohydrate composition of LPS affects the level of serum complement activation.

Antibiotic Resistance

The prophylactic and growth promoting antibiotic regimens in animal husbandry facilitate the emergence and persistence of drug-resistant *Salmonella* strains and other food-borne bacterial pathogens in farm animals and their meat products. Human salmonellosis attributed to antibiotic resistant strains is linked to the use of antimicrobial agents in farm animals. Such an occurrence is likely to stem from the antibiotic dependent inactivation of resident microflora in the intestinal tract and from the rapid colonization, growth and dissemination into deeper tissues of antibiotic resistant salmonellae to a point beyond therapeutic management. The genetic determinants of antibiotic resistance are encoded in the bacterial chromosome or in cytoplasmic plasmids. Resistance (R) plasmids are autonomous and self replicating fragments of DNA whose transcription products confer resistance to the carrier cell. The propensity for the conjugative intra- and intergeneric transfer of R plasmids carrying single or multiple resistance codons between compatible cells sharing the same ecological niche is a serious public health issue. Such exchanges of genetic materials can favour the emergence and spread of food-borne *Salmonella* spp. that are resistant to one or more medically important therapeutic agents such as chloramphenicol and trimethoprim-sulfamethoxazole. Current practices in animal husbandry further amplify the resistance problem by effectively selecting for resistant strains through the inclusion of sub-therapeutic doses of antibiotics in feed rations.

Several antibiotics such as bacitracin, virginiamycin and apramycin are considered bonafide veterinary drugs because they are reportedly devoid of any therapeutic value in human medicine. However, recent evidence indicates that the agricultural use of apramycin may lead to the emergence of gentamicin resistant *Salmonella* strains in treated livestock. This is because both apramycin and gentamicin-resistance codons lie on a single R plasmid. This leads to a new realization that farm use of any antibiotics may engender major public health problems. The resistance problem is envisaged even in aquatic environment since the aquacultural operations promote the proliferation of *Salmonella* spp. in rearing ponds and basins, and provide favourable conditions for the exchange of R plasmids and cross contamination of reared species.

The use of medically important drugs at subtherapeutic levels in the meat and aquacultural industries should be reassessed in terms of their purported value and of their potentially negative impact on public health and on the safety of the global food supply.

LABORATORY DIAGNOSIS

Foods are sampled principally for the following reasons:

- to check the hygienic production and handling techniques
- to control quality and preserve shelf life performance
- to check the cause of food poisoning
- to verify the quality of imported food

For isolation of *Salmonella,* special care has to be taken while sampling since their very presence in a food is significant. In addition, the levels present may be very low. In these cases it is necessary to use presence or absence procedures rather than relying on enumeration techniques for detection. Presence or absence procedures allow the examination of larger portions of sample, typically 25 g, by use of liquid enrichment procedures in nutrient and selective media formulated to optimize the recovery of the target organism in the presence of other naturally occurring food microflora.

The following foods and feeds have been classified for statutory testing for *Salmonella* in them thus ascertaining the frequency of the presence of the organism in these food items : animal feed, cheese, cream, ice cream, raw milk, pasteurized milk, milk-based drinks, dried milk, eggs, yoghurt, meat and meat products.

In the investigation of outbreaks, efforts should be directed towards demonstrating a common *Salmonella* type in patients and food and in establishing the reason for the presence of salmonellae in food. Incubation time is usually 12–48 hours or longer since multiplication has to occur in the intestine. Isolation of salmonellae from stool specimens in acute cases is usually possible by direct plating on suitable selective agars, although enrichment may be necessary for diagnosis of carriers. No single procedure has been found to be suitable for the recovery of all salmonellae from all types of food. Most of the methods involve primary enrichment in non-selective broth, to allow recovery of sublethally injured organisms followed by secondary enrichment in elective or selective broths. Secondary enrichment broths are subcultured after incubation on to selective agars. Suspect *Salmonella* colonies are checked for purity and their identity confirmed by biochemical and serological tests.

Salmonella typhi, S. paratyphi A, B and C are worthy of special mention since these organisms are categorized as hazard Group 3 pathogens.

ISOLATION OF *SALMONELLA* FROM OTHER FOOD PRODUCTS

Animal Feeds

Mammals and birds reared intensively require large amounts of dehydrated protein feed. This is prepared from meat, bones, blood or feathers or combinations of these. Animal proteins available have high content of salmonellae which depends on the initial contamination of the raw materials and on the hygiene of manufacture. Animals fed with contaminated feed, particularly pigs and poultry often carry these salmonellae in their intestinal tracts with no signs of illness. Meat from such infected animals may become contaminated during slaughter and processing, and the infection passed on to humans during subsequent poor hygiene practices, during preparation or inadequate cooking and storage procedures.

The bacteria in processed food may be damaged as a result of the dehydration process employed during its manufacture, and so a resuscitation step is necessary to ensure the recovery of contaminating organisms.

LABORATORY DIAGNOSIS

Method 1 Pre-enrichment and enrichment culture

Media

Pre-enrichment broth   buffered (1%) peptone water

Enrichment broths   Rappaport Vassiliadis (RV) broth.

Selective agar media   Modified brilliant green agar (BGA), xylose lysine desoxycholate agar (XLD), deoxycholate citrate agar (DCA), mannitol lysine crystal violet brilliant green agar (MLCB).

Procedure

1. Homogenize 25 g of food sample with 225 ml of buffered peptone water. If a large food sample is required maintain a sample to broth ratio of 1:9. Incubate at 37°C for 18 hours.

2. Subculture onto RV broth; add 0.1 ml of RV medium. Incubate at 42°C for 20–24 hours.

3. Subculture a loopful of the broth to two selective agar media.

4. Incubate plates at 37°C for 20–24 hours. Bismuth sulphite agar plates should be incubated for up to 48 hours.

5. Examine plates for typical colonies. Select at least five suspect colonies and subculture to a non-selective agar. Incubate at 37°C for 18–24 hours.

6. Screen biochemically using triple sugar iron (TSI) agar of lysine iron (LI) agar slopes in conjunction with urease and sucrose or lactose media. Incubate at 37°C for 24 hours. Typical strains of salmonellae produce an acid (yellow) butt and an alkaline (red) slope in TSI agar and an alkaline (purple) reaction throughout the LI medium, both with blackening due to hydrogen sulphide production, are urease negative and do not ferment sucrose or lactose.

7. Presumptive isolates of *Salmonella* spp. should be further characterized biochemically and serological tests performed using *Salmonella* agglutinating sera.

Baby Foods

The level of *Salmonella* contamination within a dried powdered formula may be so low that it may be missed by examination of only a 25 g sample. In such instances, where such a product has been implicated in cases of illness in infants, it is recommended that multiple 25 g samples are examined from each individual container.

Method of Isolation

The same as above may be used in addition, selenite cystine (SC) broth can be also included for the enrichment stage. 1 ml of the sample to be added to 10 ml of SC broth and incubated at 37°C for 20–24 hours. Proceed as described in the previous procedure.

Chocolate Products

These have a low water activity and often a high fat content. Chocolate products have now been implicated in a number of *Salmonella* outbreaks. The important factors being low infectious dose and the high fat content of the chocolate which protects salmonellae from the acidity of the stomach.

LABORATORY DIAGNOSIS

Method 1 Pre-enrichment and enrichment culture

Media

Pre-enrichment broth   buffered (1%) peptone water
Enrichment broths   Rappaport Vassiliadis (RV) broth.
Selenite cystine (SC) broth.

Certain additions can be made to the enrichment broths for obtaining better results, for isolation from chocolates, a 10% non-fat dried milk can be added which has shown to reduce the natural bactericidal properties. Proper controls have to be maintained.

Follow the steps (i–vii) given in the previous procedure (animal feed).

Coconut

Salmonellosis has been associated with the consumption of uncooked desiccated coconut. The same procedures that have been highlighted in the previous modules can be followed.

Dairy Products

Milk is at risk of faecal contamination from the cow or other producer species and is subject to potential contamination from equipment, the environment and humans during collection and processing. Milk supports the growth of many pathogens and is a well recognized vehicle for food poisoning.

Cheese   There are different kinds of cheese like the hard-pressed cheese, soft cheese, blue veined mould ripened cheese, etc. One has to follow similar methods that have already been discussed above for the isolation of *Salmonella* spp. with a small modification, i.e.,a surfactant (e.g. tergitol 7, 1% with lactose broth 0.22% with BPW) has to be added to the *Salmonella* enrichment broths so as to aid in dispersion of food since cheese is a high fat food. The same holds good for cream and ice cream.

Yoghurt   Examination of yoghurt for *Salmonella* involves method used for animal feed with a slight modification in the pre-enrichment broth, i.e., addition of di-potassium hydrogen phosphate solution (2%) pH 7.5 in order to neutralize the acidic pH before proceeding further.

Raw milk   Raw milk is a recognized vehicle for food poisoning, where apart from the regular methylene blue dye reduction test, there is a directive specifying the examination for *Salmonella* species. *Salmonella* spp. should be absent in 25 ml as per the Dairy Products Regulations (1995).

One can either go for the method already mentioned in the above schedules or choose ones given below:

LABORATORY DIAGNOSIS

Method 1   Direct enrichment

Media

Enrichment broth   Muller–Kauffmann tetrathionate broth containing no-vobiocin or Rappaport Vassiliadis soya peptone (RVS) broth or Rappaport Vassiliadis (RV) broth.

Selective agars   Any two agars already mentioned in the above schedules.

Procedure

1. Mix well 50 ml of the milk sample in 450 ml of a suitable enrichment medium. Divide the sample into two portions. Incubate one portion at 37°C and the other portion at 415°C for 48 hours.

2. Subculture from the two portions after 24 hours and 48 hours to two selective agar media and proceed as described in steps (3–7) of animal feed.

The same can be followed for pasteurized milk and milk-based drinks.

Meat and Meat Products

Intensive rearing of livestock increases the spread of organisms from animal to animal and in slaughter and processing, from carcass to carcass. Animals may excrete salmonellae without showing symptoms. The prevalence of salmonellae on raw red meats is variable and will differ depending on the type of meat animal and method of rearing. Chicken and other poultry are most often infected with *Salmonella enteritidis* and this organism may infect eggs during their formation in the oviduct. For isolation and enumeration of salmonellae from meat and meat products, one can follow the method given under the animal feed section.

Eggs

Eggs are made use of in different forms like homogenized pasteurized liquid egg, egg powder, egg albumen, crystalline or powdered egg albumen, etc. Before proceeding to the examination, the following data have to be recorded:

- The source of the eggs (shops, supermarket, farm gate, etc.)
- The type and size of the eggs (small, medium, large, etc.)
- The name of the packer or producer
- Data of packing and expiry
- Presence of visible cracks and faecal material adhering to the shell.

LABORATORY DIAGNOSIS

Method 1   **Examination of individual shell eggs without shell disinfection**

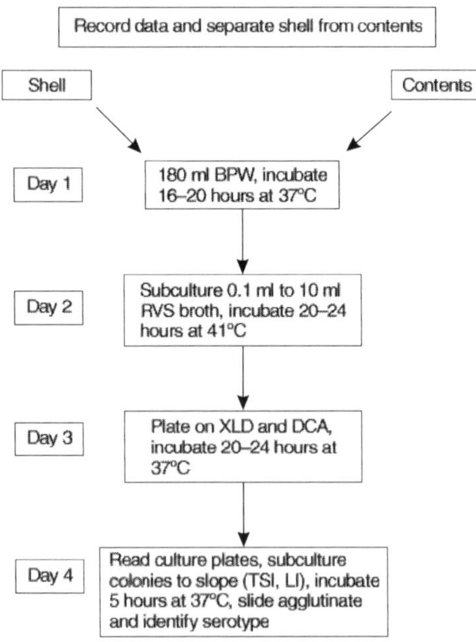

For examination of eggs with shell disinfection, the shell is to be swabbed with BPW soaked swab (sample for shell swab) and disinfected with isopropyl alcohol wipe. The shell has to be separated from the contents and discarded. Follow the steps as given in the previous module.

Raw Liquid Egg

Method 2 Enrichment culture for *Salmonella* spp.

Procedure

1. Add 25 ml of the raw liquid egg to a jar containing 225 ml of BPW plus 5 mg of novobiocin/l and 10 mg of cefsulodin/l.
2. Incubate at 37°C for 18 hours, subculture 0.1 ml to 10 ml of RVS broth.
3. Incubate RVS broth for 20–24 hours at 41.5°C and subculture on XLD and DCA agar. Proceed with isolation and characterization of salmonellae as described earlier.

Pasteurized Liquid Egg

Method 3 Enumeration of *Salmonella* spp.

Procedure

1. Take a sufficient quantity of sample to test 6 × 8 ml by the multiple tube method (addition of 18 ml volumes to each of six 180 ml volumes of the chosen liquid medium (BPW or any selective medium).
2. Add 18 ml to each of six jars containing 180 ml of BPW.
3. Incubate for 18 hours at 37°C.
4. Subculture 0.1 ml from each jar to separate 10 ml volumes of RVS broth.
5. Incubate for 20–24 hours at 41.5°C and subculture on XLD and DCA. Proceed with the isolation and characterization of salmonellae as described earlier.
6. From the number of jars shown to contain salmonellae, calculate the MPN of the organisms/ml of sample from the standard MPN chart.

FOOD POISONING BY *STAPHYLOCOCCUS*

Staphylococcal food poisoning (SFP) ranks as one of the most prevalent causes of gastroenteritis worldwide. It results from ingestion of one or more preformed staphylococcal enterotoxins in *Staphylococcus*-contaminated foods. The aetiological agents of SFP are members of the genus *Staphylococcus*, predominantly *Staphylococcus aureus*. This form of food poisoning is considered as intoxication since it does not require growth of the organisms in the host.

The term "staphylococci" informally describes a group of small, spherical, gram-positive bacteria. They are catalase positive chemo-organotrophs with a DNA composition of 30–40 mol % G+C content.

S. aureus can produce multiple toxins with similar molecular weights as well as similar biological and physico-chemical properties. Thus current classification is based on antigenicity. Initially, differentiation between the multiple antigenic forms of staphylococcal enterotoxin (SE) was based on the observation that many food isolates produce one common antigenic type of toxin, tentatively designated the "F" toxin. Most other enterotoxigenic strains such as those from enteritis patients also produced the "F" toxin in addition to a second antigenic "E" toxin. The discovery of additional isolates that did not conform to this pattern prompted adoption of an improved nomenclature system. The SEs are sequentially assigned a letter of the alphabet in the order of their discovery. The "F" and "E" toxins were designated SEA and SEB respectively. Protein sequencing and recombinant DNA methods have resulted in the current detailed knowledge of the primary sequences of all of the classical SEs. More recent additions to the SE family include SEG and SEH. The SEC serological variant can be further divided into at least three subtypes SEC1, SEC2 and SEC3 based on minor differences in immunological reactivity.

Survival Characteristics

Unique resistance properties of *S. aureus* facilitate its contamination and growth in food. Outside the body, *S. aureus* is one of the most resistant non-spore forming human pathogens and can survive for extended periods in a dry state. Its survival is facilitated by organic material, which is likely to be associated with the organism from an inflammatory lesion.

S. aureus is known for acquiring genetic resistance to heavy metals and antimicrobial agents used in clinical medicine. A remarkable feature of its survivability is its osmotolerance, which permits growth in medium containing the equivalent of 3.5 M NaCl and survival at water activities less than 0.86. This becomes a problem since it does not have any competition from other organisms at this state. Several compounds accumulate in the cells or enhance staphylococcal growth under osmotic stress. Glycine betaine appears to be the most important osmoprotectant for *S. aureus* and its level in the cells increases in response to the NaCl in the environment.

Reservoirs

Humans are the main reservoir for staphylococci involved in human disease, including *S. aureus*. Colonized individuals are carriers and provide the main source for dissemination of the organism to others and to food. In humans, the anterior nares are the predominant site of colonization, although *S. aureus* can be found on other sites such as the skin or perineum. In humans, dissemination can occur by direct contact, indirectly by skin fragments or through respiratory tract droplet nuclei. Most sources of SFP are traced to humans who inoculate food during preparation.

S. aureus may be introduced into food from fomites used in food processing such as meat grinders, knives, storage utensils, cutting blocks and saw blades. Conditions most often associated with food poisoning are:

1) Inadequate refrigeration, 2) Preparation of foods far in advance, 3) Poor personal hygiene, 4) Inadequate cooking or heating of food, 5) Prolonged use of warming plates when serving foods.

Animals are also an important source of *S. aureus* which are often heavily colonized with the organism. One very serious problem for the dairy industry is mastitis, an infectious disease often caused by *S. aureus*. Colonization of animals by this organism is also a public health concern since it may result in contamination of food and milk with *S. aureus* prior to or during processing. Other sources may be air, dust, sewage or water.

Food-borne Outbreaks

SFP occurs as either isolated cases or outbreaks affecting a large number of individuals. Since SFP is self limiting, reporting has not been so great as other food-borne diseases. Isolated cases occurring in the houses are not usually reported. Occurrence of SFP is cyclical in nature. The highest incidence is typically in the summer, when the temperatures are warm and food is more likely to be stored improperly. Approximately one-third of these outbreaks are associated with leftover food.

Characteristics of the Disease

Staphylococcal food poisoning is a food-borne intoxication, and results from the ingestion of food in which *Staphylococcus aureus* has already grown to high numbers and produced exotoxins. The short incubation period and symptoms experienced closely resemble those seen with many chemical intoxicants.

SFP is a self limiting illness presenting with emesis following a short incubation period. Other common symptoms include nausea, abdominal cramps, diarrhoea, headaches, muscular cramping and prostration. Diarrhoea is usually watery but may contain blood as well. There is no fever and other potential symptoms include general weakness, dizziness, chills and perspiration.

Symptoms typically develop within 6 hours after ingestion of contaminated food. The mean incubation rate is 4.4 hours although incubation periods as short as 1 hour have been reported. Death due to SFP is not common and treatment is minimal in most cases although administration of fluids is indicated when diarrhoea and vomiting are severe. Effective doses of SE may be achieved when populations of *S. aureus* are greater than 10^5 organisms per gram of contaminated food.

A basal level of approximately 1ng of SE per gram of contaminated food is sufficient to cause symptoms associated with SFP. Many factors contribute to the likelihood of developing symptoms and their severity. The most important include susceptibility of the individual to the toxin, the total amount of food ingested, the toxin type and the overall health of the affected person. Though SFP outbreaks attributed to ingesting of SEA are much more common, individuals exposed to SEB exhibit more severe symptoms (SEB is generally produced at higher levels than SEA).

Toxins

SEs are single polypeptides of approximately 26 to 28 kDa. Staphylococcal enterotoxins A–J are currently recognized. They are stable molecules in many respects. They are considerably heat-stable. Less extreme elevations of temperature such as those used for pasteurization of milk had little or no effect on SE toxicity. Since temperatures required for inactivating the SEs are much higher than those needed to kill *S. aureus* under the same environmental conditions, toxic food involved in many cases of SFP is devoid of viable organisms at the time of serving.

Another significant property of SEs is their resistance to inactivation by proteases found in the gastrointestinal tract. SEB is susceptible to degradation by pepsin at very low pH levels, but partial neutralization of the gastric acidity by food intake is presumed to temporarily provide a protective environment of the toxin. Presence of disulphide bonds is responsible for an inherent molecular stability of the toxin.

The structural aspects of SEs that enable them to survive degradation by pepsin and other enzymes in the gastrointestinal tract are required for the toxins to induce SFP. Not only this, the SEs must also be able to interact with the appropriate target, leading to emesis, diarrhoea, and other gastrointestinal tract symptoms.

Individually, each SE type and subtype has a sufficient degree of antigenic distinctness to allow its differentiation from other toxins.

Except for rare SFP cases in which massive doses of SEs are consumed, systemic dissemination of the toxins does not contribute significantly to the illness. The site of infection is the abdominal viscera. The emetic response results from a stimulation of local neural receptors in the abdomen which transmit impulses through the vagus and sympathetic nerves, ultimately stimulating the medullary emetic centre.

S. aureus has some unique properties that promote its ability to induce food-borne illness.

LABORATORY DIAGNOSIS

The resemblance of staphylococcal food poisoning with many chemical intoxicants makes it mandatory to look for bacteriological evidence for the confirmation of staphylococcal food poisoning outbreaks. Diagnosis of human cases of staphylococcal food poisoning is straightforward and *S. aureus* can be isolated directly from stool samples by direct plating onto a selective agar. If the food is known or suspected to have been handled or stored in a manner which may have affected the viability of the organisms, then enrichment broth culture followed by plating onto selective agar should be used for isolation. Traditionally, media containing high salt concentrations such as salt meat broth have been used. Presumptive colonies from the selective agar cultures should be confirmed as *S. aureus* by a coagulase test or commercial equivalent and by DNase production.

Isolates from food samples, clinical specimens and from food handlers should be further characterized by special tests like phage typing, toxin typing, etc., carried out at reference laboratories.

Method 1   Enumeration—colony count on Baird–Parker agar

Procedure

1. Prepare a 10^{-1} homogenate and serial decimal dilutions. Select a suitable counting method and enumerate on Baird–Parker agar (this medium allows resuscitation of stressed cells with the formation of characteristic colonies).
2. Select a suitable surface counting method and enumerate on Baird-Parker agar.
3. Incubate plates at 37°C for 48 hours. Examine for the presence of typical colonies, which appear grey black, shiny and convex of diameter 1–1.5 mm surrounded by a zone of clearing.
4. Confirm the identity of the colony types by using the coagulase test and deoxyribonuclease test, subculturing to a non-selective agar medium first if necessary. Include positive and negative controls.

Method 2   Enumeration—colony count using rabbit plasma fibrinogen agar (RPFA)

Medium   Rabbit plasma fibrinogen agar which reads the coagulase reaction directly, eliminating the need for confirmation tests.

Procedure

1. Prepare a 10^{-1} food homogenate and serial decimal dilutions.
2. Use the pour plate method to enumerate using RPFA.
3. Incubate the plates at 37°C for 48 hours. Examine after 18–24 hours for black, grey or white small colonies surrounded by a halo of precipitation, indicating coagulase activity.
4. Count the typical colonies in each plate containing a maximum of 300 colonies and compute per gram.

Method 3 Enumeration—Most probable number technique

Media

Enrichment broth   Giolitti–Cantoni broth containing 0.1% Tween 80, single and double strength, dispensed in 10 ml volumes. Steam and cool just before use, then add 0.1 ml (single strength) or 0.2 ml (double strength) of 1% potassium tellurite solution.

Selective agar   Baird–Parker agar or RPFA.

Procedure

1. Prepare a 10^{-1} homogenate and further decimal dilutions of the food sample.

2. Transfer 10 ml of 10^{-1} homogenate to each of three tubes of double strength enrichment broth, 1 ml of 10^{-1} homogenate to each of three tubes of single strength enrichment broth, and 1 ml of each subsequent dilution to each of three tubes of single strength enrichment broth. Overlay the tubes with liquid paraffin.

3. Incubate the tubes at 37°C for 24 hours. If any tubes show blackening, remove the liquid paraffin and subculture to Baird–Parker agar or RPFA. Reincubate all other tubes for a further 24 hours, then subculture all tubes as before.

4. Incubate the plates at 37°C for up to 48 hours, proceed as described in step (3) of method 2.

5. Use the table to compute the most probable number per gram from the number of tubes that have yielded growth of coagulase positive staphylococci, multiplying the MPN value shown by 10.

LISTERIOSIS

Listeriosis has emerged as one of the major food-borne diseases during the last decade. The emergence of listeriosis is the result of complex interactions between various factors reflecting changes in social patterns. These factors include:

1. Medical progress and consequent demographic changes, such as the increasing proportion of immunocompromised people and the elderly.

2. Change in primary food production, food processing technology, expansion of the agro-food industry, and development of cold storage systems.

3. Changes in food habits (increase consumer demand for convenient food that has a fresh cooked taste, can be purchased ready-to-eat, refrigerated, or frozen, can be prepared rapidly and requires essentially little cooking before consumption).

4. Changes in handling and preparation practices.

Listeriosis is an atypical food-borne disease of major public health concern because of the severity and the non-enteric nature of the disease, a high mortality rate, a long incubation time and a predilection for individuals who have underlying conditions which lead to impairment of T-cell-mediated immunity.

Listeria differs from other food-borne pathogens due to its ubiquity, its resistance to diverse environmental conditions such as low pH and high NaCl concentrations and its microaerobic and psychrotrophic characters.

Listeria is a major concern for the agro-food industry due to the following factors.

- Various ways in which the bacterium can enter into food processing plants.
- Its ability to survive for long periods of time in the environment and on foods and in food processing plants.
- Its ability to grow at very low temperatures (2–4°C).
- Its ability to survive in or on food for prolonged periods under adverse conditions.

Organism

The genus *Listeria* belongs to the *Clostridium* sub-branch together with *Staphylococcus, Streptococcus, Lactobacillus, Brochothrix*. On the basis of DNA–DNA hybridization and 16S rRNA sequencing, the genus *Listeria* presently comprises six species divided into two lines of descent: 1) *L. monocytogenes* and the closely related species, *L. innocua, L. ivanovii, L. welshimeri* and *L. seeligeri*. 2) *L. grayi*.

Only *L. monocytogenes* and *L. ivanovii* are considered virulent. Only *L. monocytogenes* is considered as a public health concern. *L. monocytogenes* is a gram-positive, facultatively anaerobic, catalase-positive, oxidase-negative, non-sporeformer. Cells are coccoid to rod-shaped and motile with peritrichous flagella exhibiting a characteristic tumbling motility. There are 13 serovars which can cause disease, but 955 of human isolates belong to 3 serovars, $\dfrac{1}{2a}, \dfrac{1}{2b}$ and 4*b*.

Growth Characteristics

The organism grows over a wide range of temperature from 0–42°C with an optimum between 30 and 35°C. Growth is extremely slow below 5°C. Growth of all strains is inhibited at pH of below 5.5 and the organism is also salt tolerant being able to grow in 10% sodium chloride. Growth of listeriae is favoured by high humidity and the presence of nutrients.

Survival Characteristics

Heat resistance of *L. monocytogenes* is similar to that of other non-spore-forming gram-positive. The organisms reported from milk are reported to be protected from heat by their intracellular location within milk leucocytes. *L. monocytogenes* can survive for a year in 16% NaCl at pH 6.0 and its survivability in 30% NaCl has also been reported. It can survive at nitrite concentrations that are allowed in foods. According to the WHO, pasteurization is a safe process which reduces the number of *L. monocytogenes* in raw milk to levels that do not pose an appreciable risk to human health.

Freezing and storage at –18°C and even repeated freezing, have little effect and these conditions are more likely to injure than to inactivate *L. monocytogenes.*

Growth of *L. monocytogenes* is not significantly affected by vacuum packaging which augments a new trend in food preservation.

Reservoirs

Listeria monocytogenes is ubiquitous due to its two properties—despite not forming spores, it can survive for long periods of time in many different environments and it is psychrotrophic. There are different types of reservoirs. They are listed below.

Environment Silage (both pasture grass and grass silage), soil, water, sewage, surface waters of canals and lakes, ditches, plants and other crops grown on soil treated with sewage sludge are all contaminated with listeriae.

Food-processing plants In food-processing plants, worker shoes and clothing, transport equipment, raw plant tissue, raw food of animal origin, healthy humans and animal carriers (11–52%, of pigs harbour *L. monocytogenes* in tonsils and 24% of cattle have contaminated internal retropharyngeal nodes), all act as reservoir for *Listeria*.

Also, moist areas such as floor drains, condensed and stagnant water, floors, residues, and processing equipment serve as common sources. It is attached to various kinds of surfaces (stainless steel, glass and rubber), biofilms in meat and dairy processing environments, fingers after hand washing, aerosols and carcasses (contamination by faecal matter during slaughter, the most contaminated area appear to be the cow de-hiding and the pig stunning and hoisting areas apart from the de-feathering machine).

Food *L. monocytogenes* is present in a wide variety of foods, both raw and processed. It is able to survive and grow in many foodstuff during storage. 6% of the 18,000 food samples are contaminated with *L. monocytogenes* and 6% of the positive samples have more than 1000 cfu/g.

- Milk and dairy products are the first and among the most extensively studied foods. Possible sources being feed (10^5 cfu/g), faecal shedding by healthy carrier cows, and the farm environment.
- A wide variety of cheeses especially the soft cheese with a population of *L. monocytogenes* ranging from $10–10^7$ cfu/ml (soft cheeses have a pH that allows the growth of the organism).
- Meat and meat products, including beef, pork, minced meat, smoked and fermented sausages have been associated with *L. monocytogenes.*

Most are surface contamination. *L. monocytogenes* is capable of growing on meat depending on the pH, tissue type (lean or fat), type and amount of indigenous microflora, temperature and curing ingredients.

- Poultry (broiler, ready-to-eat, precooked, chilled or frozen chicken) is also frequently contaminated.
- The organism is often encountered on fresh vegetables (radishes, cucumbers, cabbages and potatoes) but in low numbers. Sources of contamination include soil, water, animal manure, decaying vegetation, and effluents from sewage treatment plants.
- The prevalence of *L. monocytogenes* on raw and ready-to-eat seafood and fish products can be high (10^4 cfu/g in smoked fishes).

Human carriage

Asymptomatic faecal carriage of *L. monocytogenes* is very common including pregnant women, healthy people (2–6%), patients undergoing renal transplantation and patients with symptoms of gastroenteritis.

Food-borne Outbreaks

Food-borne transmission of listeriosis was first demonstrated in 1981. Listeriosis is a rare disease mainly reported from industrialized countries. Various sources that were implicated in the disease were cabbage fertilized with manure from sheep suspected to have had *Listeria* meningitis, pasteurized milk, cheese, pork, etc. Experience during investigations indicate, that several different methods of typing should be used to evaluate relatedness between isolates from patients and foods.

The data obtained during the past 10 years regarding the sources of outbreaks suggest that some foods are more hazardous than others. Highest risk foods are :

1. Ready-to-eat and stored at refrigeration temperatures for a long period of time.
2. Contaminated with a high population of *L. monocytogenes* (>100 cfu/g or ml).

Symptoms of the Disease

Pregnant women are most frequently affected who may be asymptomatic or characterized by a flu-like illness with fever, myalgia or headache. The foetus may be worst affected with foetal death and severe neonatal septicaemia and meningitis.

In non-pregnant adults, meningitis and meningo-encephalitis frequently occur in cases of listeriosis. Foetal infections, including endocarditis, pulmonary infection, septic arthritis, osteomyelitis and peritonitis hepatitis are common.

Infective Dose

The infective dose of *L. monocytogenes* depends on many factors including the immunological status of the host. Microbial characteristics are important risk factors for disease. Published data indicate that the numbers of *L. monocytogenes* in contaminated food responsible for epidemic and sporadic food-borne cases were more than 100 cfu/g. More epidemiological information is needed for an accurate assessment of the infective dose.

Pathogenicity

Virulence factors   During the last decade there have been major advances in studying and identifying the virulence factors of this highly invasive intracellular pathogen.

Once ingested, *L. monocytogenes* crosses the intestinal barrier and then internalized by resident macrophages, in which they can survive and replicate. They are subsequently transported via the blood to regional lymph nodes. When they reach the liver and the spleen, most listeriae are rapidly killed. Depending on the level of T-cell response induced in the first days following initial infection, further dissemination via the blood to the brain may subsequently occur. Hence, infection is not localized at the site of entry but involves entry and multiplication in a wide variety of cell types and tissues. The principal site of infection is the liver. Entry into epithelial cells has been studied by genetic approaches leading to the identification of internalin, a protein required for entry into epithelial cells.

Following invasion of the intestinal barrier, two types of cells, macrophages and hepatocytes are critical to infections. The internalization of the bacterium into the phagocytes is triggered by the bacterium and soon after the entry, bacteria are internalized in membrane bound vacuoles, which are lysed in less than 30 minutes. Intracellular bacteria are released to the cytosol and begin to multiply with a doubling time of about 1 hour. These intracytoplasmic bacteria become progressively covered by a cloud of cell actin filaments which later rearranges into a polarized comet tail up to 40 µm in length. The actin comet tail is made of actin microfilaments that are continuously assembled in the vicinity of the bacterium and are then released and cross-linked. The actin comet tail is stationary in the cytosol and left behind by moving bacteria. The length of the tail is thus proportional to the speed of movement, with faster moving bacteria having longer tails. When bacteria reach the plasma membrane, they put out long protrusions, each with a bacterium at the tip. These protrusions are then internalized by a neighbouring cell, giving rise to a two membrane bound vacuole. After lysis of this vacuole, a new cycle of replication, movement, and spreading of the bacteria begins. The entire cycle is completed in 5 hours.

The strategy of direct cell-to-cell spread allows bacteria to disseminate within host tissues and induce the formation of infectious foci while being sheltered from host defenses such as circulating antibodies.

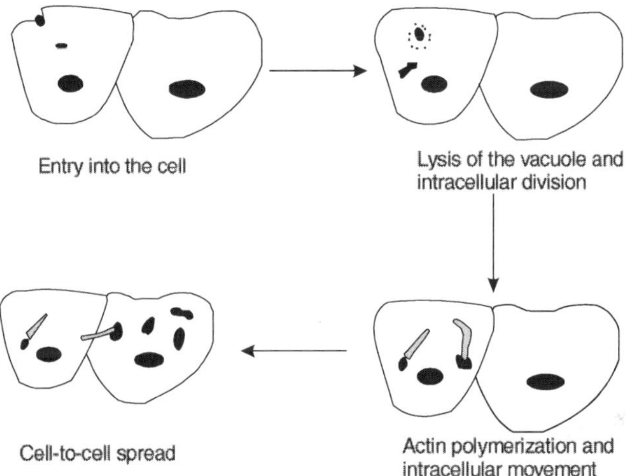

Figure 27.1   Schematic representation of the successive steps in the infectious process

LABORATORY DIAGNOSIS AND IDENTIFICATION

Method 1   Enrichment culture

Media

1. Selective primary enrichment medium   Fraser broth (contains nalidixic acid sodium salt 10 mg/ml and acriflavine hydrochloride 12.5 mg/ml)
2. Selective secondary enrichment medium   Fraser broth (contains nalidixic acid sodium salt 20 mg/ml and acriflavine hydrochloride 12.5 mg/ml)
3. Selective agars   Polymyxin, acriflavin, lithium chloride, ceftazidime, aesculin, mannitol agar and Oxford agar.
4. Non-selective agar   e.g. blood agar, nutrient agar, tryptone soya yeast extract agar.

Procedure

1. Homogenize 25 g of food sample with 225 ml of half Fraser broth.
2. Incubate the half Fraser broth at 30°C for 24 hours.
3. Subculture to Oxford agar and mannitol agar; incubate the plates at 30°C or 37°C for a total of 42–48 hours.

4. Subculture 1 ml of half Fraser broth to 10 ml of Fraser broth; incubate at 37°C for 48 hours.

5. Subculture Fraser broth to Oxford and mannitol agar; incubate the plates at 30°C or 37°C.

6. Examine all agar plates after 24 hours and 42–48 hours of incubation for the presence of typical colonies.

7. Subculture five typical colonies onto a non-selective agar; incubate at 30°C or 37°C for 18–24 hours (strains of *Listeria* species hydrolyse aesculin, producing black zones around the colonies, after 48 hours incubation typical colonies are 2–3 mm diameter with a sunken centre).

8. Confirm the identity of these strains using appropriate biochemical tests:

Reactions of genus *Listeria*

Test	Result
Gram stain	Gram-positive rods
Voges–Proskauer	+
Urease	−
Catalase	+
Oxidase	−
Aesculin hydrolysis	+
D-glucose fermentation	Acid no gas
D-salicin fermentation	Acid no gas
Motility at 22°C	+ tumbling

Method 2 Enrichment culture

Media

1. Selective primary enrichment medium Modified tryptone soya broth containing yeast extract 6 g/L, made selective by the addition of acriflavine hydrochloride (10 mg/L), nalidixic acid sodium salt (40 mg/L) and cycloheximide (50 mg/L).

2. Selective agar Oxford agar/Mannitol agar

3. Non-selective agar e.g. Blood agar, nutrient agar, tryptone soya yeast extract agar.

Procedure

1. Homogenize 25 g of food sample with 225 ml of modified tryptone soya broth.

2. Incubate at 30°C for 48 hours.

3. Subculture from the enrichment broth after 24 hours and 48 hours incubation onto Oxford agar.

4. Incubate plates at 37°C for 48 hours. If species other than *L. monocytogenes* are sought, incubate plates at 30°C. Examine for the presence of typical colonies after 24 hours and 48 hours.

5. Subculture five typical colonies onto a non selective agar, and incubate at 37°C or 30°C for 24 hours.

6. Confirm the identity of these strains by performing various tests.

Method 3   Enumeration by plate method

Procedure

1. Prepare a 10^{-1} homogenate of food sample and serial decimal dilutions or use the homogenate prepared in the previous method, allow to stand for 1hour at 20°C for resuscitation of stressed organisms.

2. Select a surface counting method (plate count, surface drop) and enumerate on Oxford agar. Incubate at 37°C or 30°C for 42–48 hours.

3. Count the number of typical colonies on plates containing up to 150 colonies.

4. Subculture five typical colonies onto non-selective agar and confirm the genus.

Method 4   Enumeration by multiple tube method

If it is likely that the food product contains highly stressed cells of *Listeria,* it may be preferable to use a liquid culture method for enumeration to allow resuscitation.

1. Make serial decimal dilutions of the liquid food sample or 10^{-1} of food homogenate prepared using the minimal recovery diluent (0.85% saline/ 0.1% peptone water).

2. Prepare nine tubes, each containing 10 ml of single strength growth medium appropriate for the organism (any selective enrichment broth).

3. Add 1 ml of the liquid food sample or food homogenate to each of three tubes containing growth medium.

4. Repeat step 3 for each of two subsequent decimal dilutions.

5. Incubate all nine tubes as appropriate for the organisms sought. Refer the MPN chart for the highest dilution showing positive results (similar to coliform tests).

6. After incubation, test those tubes showing the characteristic reactions of the organisms sought by subculture from each tube to a suitable confirmatory agar or liquid medium.

REVIEW QUESTIONS

1. Define food-borne disease.

2. Define food poisoning.

3. List the various causes of food-borne intoxications.

4. List the various bacterial food-borne infections.

5. Discuss the role of *Bacillus cereus* as a food-poisoning organism.

6. Brief the survival characteristics of *Bacillus cereus* in food. Add a note on toxins of *B. cerus.*

7. Describe the isolation procedures for *Bacillus cereus* from food samples.

9. List out the four categories of human botulism.

10. Give the survival characteristics of *Clostridium botulinum.*

11. Discuss the pathogenicity of *Clostridium botulinum.*

12. Discuss the survival characteristics of *Brucella* species.

13. Discuss the laboratory diagnosis of *Brucella* species.

14. Describe the modes of transmission of brucellosis.

15. Discuss the general and survival characteristics of *Clostridium perfringens.*

16. Discuss the pathogenesis and characteristics of the disease caused by *Clostridium perfringens.*

17. List out the various categories of *Escherichia coli* .

18. Discuss the survival characteristics of *E. coli* O157:H7.

19. Discuss the mechanism of pathogenesis of *E. coli* O157:H7.

20. Outline the lab diagnosis of *E. coli* O157:H7.

21. Give the characteristics of shigellosis.

22. Describe the laboratory diagnosis of *Shigella* species from food.

23. Discuss the toxins of *Vibrio cholerae.*

24. Describe the isolation methods of *Vibrio cholerae* and *Yersinia enterocolitica.*

25. Brief out the survival characteristics of *Yersinia enterocolitica* in food.

26. List out the reservoirs of yersiniosis.

27. How does *Listeria* differ from other food-borne pathogens?

28. Discuss the growth and survival characteristics of *Listeria monocytogenes.*

29. Brief out the reservoirs in listeriosis.

30. Explain the isolation methods of *Listeria monocytogenes.*

31. Discuss the pathogenicity of *Salmonella* species.

32. Discuss the pathogenicity of salmonellosis with reference to food-borne attacks.

33. Describe in detail the methods of isolation of *Salmonella* species from various food samples.

34. Discuss the survival characteristics of *Staphylococcus* in food.

35. Describe the laboratory diagnosis of the food-borne pathogen, *Staphylococcus aureus.*

28

FOOD-BORNE VIRUSES

INTRODUCTION

Raw and minimally processed fruits and vegetables are typically sold to the consumer in a ready-to-use or ready-to-eat form. These products do not generally contain preservatives or antimicrobial agents and rarely undergo any heat processing prior to consumption. For many years, raw fruits and vegetables have been implicated as vehicles for transmission of infectious microorganisms. Although fresh products can support the growth and/or survival of many pathogenic bacteria there is little published information on the stability of human pathogenic viruses on these food products. Viruses cannot grow in or on foods but may sometimes be present on fresh products as a result of faecal contamination. This contamination can arise at source in the growth and harvesting area from contact with polluted water and inadequately or untreated sewage sludge used for irrigation and fertilization. Alternatively, fruits or vegetables handled by an infected person might become contaminated with virus and transmit infection. The most frequently reported food-borne viral infections are viral gastroenteritis and hepatitis A, both have been associated with the consumption of fresh fruits or vegetables.

In recent years, it has been recognized that viruses are the important causes of food-borne disease. Unlike bacteria, viruses do not grow or multiply in or on foods, but foods may become contaminated with human viruses and transmit infection. There are many groups of viruses which could contaminate food items, but the major food-borne viral pathogens are those that infect *via* the gastrointestinal tract, such as the gastroenteritis viruses and hepatitis A virus.

Viruses that infect via the gastrointestinal tract are excreted in faeces and may also be present in vomit. Foods become contaminated either directly by infected people or through sewage pollution. Those enteric viruses, which are commonly associated with food-borne outbreaks, either cannot be cultured in the laboratory or can only be cultured with difficulty. Hence information and experimental studies on survival and recovery of viruses from foods often relates to other virus types that are readily cultured. Some of these viruses may be good models for important food-borne viral pathogens, but others may have widely differing characteristics. Enteroviruses, such as coxsackieviruses and vaccine strains of poliovirus, have been used particularly. They infect via the gastrointestinal tract, occur in the environment as a result of sewage contamination and are relatively stable. However, there are very few published reports of food-borne outbreaks of illness caused by enteroviruses.

There is the potential for contamination with other virus groups, such as respiratory viruses, although transmission to humans through foods has not been recognized. Infrequently endogenous contamination of meat and milk can occur with viruses such as tick-borne encephalitis.

EPIDEMIOLOGICAL FEATURES

Viruses are usually transmitted directly from person-to-person, but epidemiological and laboratory investigations indicate that viral diseases can on certain occasions be transmitted via foods, particularly those that are minimally processed. These include seafoods, especially the bivalve molluscan shellfish, and fresh fruits, vegetables and salad items. It might be expected that any virus that infects via the gastrointestinal tract could be food-borne. In practice, however, the most commonly reported food-borne viral infections are viral gastroenteritis and less frequently hepatitis A.

It is believed that the incidence of both is greatly under-reported, but for different reasons. Viral gastroenteritis is a relatively mild disease and most people do not consult a medical practitioner; hence, the majority of cases are not investigated and are not reported. A recent national study funded by the Department of Health of Infectious Intestinal Disease (IID) in the community indicated that for every case of IID detected by National Laboratory Surveillance, there are 136 cases in the community. The study also concluded that a smaller proportion of cases due to common viral pathogens are reported than cases due to common bacterial pathogens. Viral gastroenteritis has a short incubation period of 1–4 days depending on the type of virus. This means that when cases are investigated, the possibility of food-borne transmission is likely to be considered. In contrast, hepatitis A is often a more severe disease and is more likely to be reported. However, the incubation period is 3–6 weeks and hence an association with a food source is unlikely to be made, unless there is a very clearly defined outbreak.

VIRAL GASTROENTERITIS

Several different viruses cause gastroenteritis: the most important include rotavirus, the small round-structured viruses (SRSV) otherwise known as Norwalk-like viruses (NLV), astrovirus and adenovirus types 40 and 41. However, in almost all food-borne outbreaks where a virus is identified, it is an NLV. Rotavirus and astrovirus are only rarely implicated. Adenovirus has not been associated with food- or water-borne transmission.

Viral gastroenteritis is usually regarded as a mild self-limiting disease lasting 24–48 hours. However, people can feel debilitated for 2 to 3 weeks, which has considerable economic implications in terms of working days lost and impaired performance. Symptoms include malaise, abdominal pain, pyrexia, diarrhoea and/or vomiting. A range of symptoms occurring in an outbreak should alert investigators to the possibility of a viral cause. The viruses are usually transmitted by the faecal–oral route, but they are also present in vomitus. Onset of viral gastroenteritis may be sudden and can commence with projectile vomiting. Virus will be disseminated over a wide area in aerosol droplets, which is a particular hazard where food is being prepared and laid out. Although most transmission is directly from person to person, contaminated food and water can give rise to common-source outbreaks. The infective doses are not known, but the evidence from volunteer studies and the typically high attack rates observed in outbreaks suggest that they are very low. For instance, it has been estimated that NLVs have an infective dose of between 10 and 100 virus particles.

Viruses account for 6% of food-borne and 5% of water-borne outbreaks occurring in England and Wales reported to the PHLS and Communicable Disease Surveillance Centre (CDSC). A recent paper prepared by the Public Health Laboratory Service (PHLS) reported on the microbiological status of ready-to-eat fruits and vegetables between 1992 and 1999 in England and Wales. Fruits and vegetables accounted for 4.3% (60/1408) of the total food-borne outbreaks reported during that period. The most commonly identified aetiological agent was the NLVs, which were linked to 20% of the outbreaks. The causative organism of a further 42% of outbreaks was reported as unknown, although the clinical and epidemiological features of these outbreaks suggested that the majority (64%) were also viral.

Norwalk-like Viruses

This group of viruses infects all age groups. There is a variable incubation period of 12–60 hours, which is thought to be dose-dependent. It occurs all year round, although in temperate climates most infections occur over the winter months. These viruses are responsible for both sporadic cases of gastroenteritis in the community and for outbreaks in schools, hospitals, old age homes, hotels and cruise ships. The national IID study indicated that viruses are the most common cause of IID in the community, with NLV the most frequently reported organism. From 1992 to 1997, NLVs accounted for one-third of all gastroenteritis outbreaks reported to the PHLS and Communicable Disease Surveillance Centre (CDSC) and the number of outbreaks of NLV gastroenteritis exceeded the number of outbreaks of salmonellosis. Unlike the salmonellosis outbreaks, however, only 6% of the NLV outbreaks were known to be food- or water-borne. The virus is extremely infectious and secondary cases are a characteristic feature of food-borne outbreaks. Therefore, it is not always possible to determine whether illness is acquired from a food-borne source or by person-to-person transmission. This accounts partly for under-recognition of the extent of food-borne transmission of these viruses.

There have been reports of outbreaks where NLVs have been epidemiologically associated with various fresh products, such as washed salads, imported frozen raspberries, coleslaw, green salads, fresh cut fruits and potato salad.

The virus was discovered in 1972 by electron microscopy. The first virus originated from the town of Norwalk in the United States, and became the prototype of the group. The name small round-structured virus (SRSV), which describes the morphology of the virus particle, was used in the United Kingdom and elsewhere until recently. It has now been agreed informally by virologists working with this group to adopt the name Norwalk-like virus, for the present, so as to bring some conformity to the nomenclature of these viruses. However, formal names have not yet been agreed by the International Committee on Virus Nomenclature. The NLVs form a complex group of viruses. They have formally been classified with the Caliciviridae and are often referred to as human caliciviruses. They are split into two broad genogroups. Most viruses in these two groups have the typical morphology of a 30–35 nm diameter particle with an amorphous surface and ragged outline, as originally described for the SRSV group. Within the Caliciviridae, there is a second genus of human gastroenteritis viruses, provisionally named Sapporo-like viruses (SLVs). The SLVs have the morphology of classical caliciviruses and are genomically distinct from the NLVs. On sequencing, some strains with classical calicivirus morphology fall into the NLV genus. There are suggestions that the epidemiology of SLVs differs from NLVs, in that SLVs mainly cause infections of young children. It is not clear whether SLVs are of significance in food-borne infections and more studies are needed to clarify their role. Genotypic analysis is being used increasingly to investigate the epidemiology of these two groups of viruses. There are several serotypes of human caliciviruses, which correspond broadly with the genotypic groups. Most studies have been carried out with NLVs, and indicate that immunity is complex and short-lived. Volunteer studies have shown that people can be infected repeatedly with the same virus strain.

Rotavirus

Rotaviruses mainly infect young children. It is estimated that they cause one million deaths a year in children under 5 years of age, mostly in developing countries. In developed countries deaths are relatively rare, but rotavirus gastroenteritis is the most frequent reason for admission of young children to hospital. Rotaviruses consistently account for around 80% of all gastroenteritis viruses reported to CDSC, although this figure is probably based on reports of hospitalized children, rather than the occurrence in the community, where NLVs are of greater significance. Food-borne and particularly water-borne spread are probably a significant route of transmission in developing countries, but in developed countries reports are rare.

Astrovirus

The astroviruses form a morphologically distinct group of viruses, and are named from the five- or six-point star seen by electron microscopy on the surface of some particles. Astroviruses have mainly been associated with illness in young children, often under 1 year of age. Reports of astrovirus infection in older children and adults are infrequent, although outbreaks have been reported in the elderly. This may reflect testing policy: detection normally relies on electron microscopy, which is insensitive and often not performed on sporadic samples from adults. The use of more sensitive molecular detection methods is required to assess the incidence and epidemiology of these viruses. Astroviruses have been seen in some adults following the consumption of shellfish or contaminated water, but these incidents appear to be comparatively rare.

HEPATITIS

There are two forms of enterically transmitted hepatitis, hepatitis A and hepatitis E.

Hepatitis A

The most characteristic symptom of hepatitis A is jaundice, but milder symptoms of nausea and general malaise without jaundice are common. Patients may feel unwell for several weeks, but recovery is complete. Deaths are rare. Some infections, particularly in children, may be asymptomatic. Like viral gastroenteritis, transmission is by the faecal–oral route, but the primary site of viral replication is the liver. Virus excretion may commence up to a week before symptoms are apparent, making control difficult.

The epidemiology of food-borne hepatitis A is essentially similar to that of viral gastroenteritis. Food- and water-borne outbreaks have been recognized for over 40 years, but are infrequently reported. Epidemiological evidence to link hepatitis infection to food and water sources is sparse, because of the long incubation period. In the United States, the Centers for Disease Control and Prevention (CDC) placed hepatitis A as the sixth leading cause of food-borne disease from 1988 to 1992. Outbreaks associated with fresh products, particularly soft fruits and salads, have been reported from several countries. Iceberg lettuce, strawberries, diced tomatoes, raspberries and salad items have all been implicated. There is only one serotype of hepatitis A. Following infection immunity is lifelong. An effective vaccine is available. Currently it is used for persons at high risk, such as travellers. The incidence of hepatitis A in developed countries has fallen in recent years and hence a susceptible population has built up. As endemic infection declines, it is possible that an increase in food-borne outbreaks will be seen. The year-round global distribution of fruit and vegetable products poses a risk of infection, particularly when these products are imported from countries with a high incidence of hepatitis A.

Hepatitis E

Hepatitis E has been associated with large water-borne outbreaks in some developing countries, notably in Asia, Africa and Central America. Food-borne transmission has been suggested, but not proved conclusively. Illness appears more severe than hepatitis A, particularly in pregnant women where a death rate of 17–33% has been observed. The primary source of infection appears to be contaminated water rather than person-to-person spread. Secondary person-to-person transmission is estimated at only 0.7–8%. Cases in the United Kingdom are reported infrequently and are mainly imported from endemic areas. With the worldwide distribution of foods, vigilance should be maintained.

PROPERTIES OF FOOD-BORNE VIRUSES

Viruses are very small microorganisms, and basically comprise a nucleic acid core of either DNA or RNA, surrounded by a protein coat. They require living cells in order to replicate and generally have a very restricted host range. Viruses do not multiply in foods or water, or in or on any other environmental sample. However, viruses can survive outside living cells and remain infectious.

Enteric viruses are hardy and survive well in the environment. These viruses survive on inanimate surfaces, on hands and in dried faecal suspensions. Lingering outbreaks have occurred in hospitals, in residential homes and on cruise ships, probably as a result of environmental contamination. NLVs have been detected by PCR in environmental swabs from hospital lockers and hotel carpets supposedly cleaned after incidents of vomiting. The viruses survive just as well on kitchen surfaces and food preparation areas. In one reported outbreak, a kitchen worker vomited into a sink. The following day the sink, which had been cleaned with a chlorine-based disinfectant, was used for washing salad and an outbreak of gastroenteritis associated with NLVs ensued.

Enteric viruses are acid-stable and so are able to survive in the gastrointestinal tract. It is likely that they will survive food processes designed to produce the low pH that inhibits bacterial spoilage organisms (e.g. pickling in vinegar and fermentation processes that produce foods such as yoghurt). Both NLVs and hepatitis A virus retain their activity even after exposure to acidity levels below pH 3.

Most viruses remain infectious after refrigeration and freezing. Frozen foods, that have not received further cooking, have been implicated in a number of incidents of viral gastroenteritis and hepatitis A. Gastroenteritis viruses and hepatitis A virus are inactivated by conventional cooking processes, but retain their infectivity after heating to 60°C for 30 minutes. It is uncertain whether they would be inactivated completely in some pasteurization processes.

DETECTION

NLVs cannot be cultured in the laboratory and until recently detection relied on the use of electron microscopy. This technique is fairly insensitive and requires a minimum of 10^6 virus particles per ml of sample. It has been used widely for detection of virus in faecal samples from patients, but cannot be used for detecting the lower concentration of virus particles present in contaminated food, water and environmental samples. Sequencing of the genome of the Norwalk virus has led to the development of PCR assays, with greatly enhanced sensitivity for virus detection. However, there is great genomic diversity among the NLVs and one set of PCR primers will not detect all strains. PCR assays are being used for the examination of food samples, particularly shellfish, but far more complex nucleic acid extraction techniques are required than when working with clinical specimens from patients. There are also greater problems with naturally occurring inhibitors to the PCR reaction in these types of samples. NLVs have been detected in samples of raspberries associated with an outbreak of gastroenteritis in Quebec. Sequence analysis demonstrated that the strain of NLV identified in the raspberries was identical to that found in the patients. Expression of recombinant virus capsids in yeast and insect cells allows the development of ELISA-based diagnostic assays, but reagents are not widely available and so far the tests only detect a very limited number of NLV strains. Further development and more widespread use of ELISA tests will greatly facilitate the detection of NLVs in clinical samples, although such tests may not be sufficiently sensitive to detect virus in food samples. At present, PCR and

ELISA assays for NLVs are only available in specialist laboratories and are not used for the routine testing of food samples. Commercial test kits are not yet available.

Rotavirus and astrovirus can both be grown in cell cultures in the laboratory. However, it is unreliable and time-consuming for isolation from primary specimens and is not normally used. Rotavirus is frequently detected using commercial ELISA or latex agglutination tests and PCR assays are also available. Rotavirus has been detected in lettuce and shellfish samples, but it is not clear if these were of human origin. Electron microscopy is still the most usual method for the detection of astroviruses, although PCR assays are used in a few laboratories.

Diagnosis of hepatitis A infection in patients is by detection of specific IgM antibody, since virus excretion has largely ceased by the time illness becomes apparent. The virus can be cultured in the laboratory, but this is a long and unreliable procedure. In one outbreak at a summer camp in the United States, virus was isolated from the drinking water supply, but this took 21 weeks. PCR assays have been developed and have been used for detecting virus in water, shellfish and other food and environmental samples. Due to the long incubation period of hepatitis A, food items are not usually available for testing, even if suspected as the source of illness. In particular, minimally processed fruits and vegetables have a short shelf life.

There have been a few experimental studies to investigate seeding and recovery of viruses from fresh products. Transfer of hepatitis A virus to lettuce leaves has been investigated. Rotavirus and poliovirus were recovered from the surface of vegetables. Average recovery rates of 80 and 65%, respectively, were obtained from lettuce; however, recovery of rotavirus from non-leafy vegetables was lower, averaging 44%. Poliovirus and adenovirus were also recovered from vegetable surfaces, obtaining mean efficiencies of approximately 55–58%. There is a need to develop more effective quantitative methods in order to assess the survival of viruses on fresh products and to determine the decontamination efficiencies of current commercial washing systems for fruits and vegetables.

ROUTES OF CONTAMINATION

Fruits and vegetables may become contaminated with viruses in two ways. First, they may be contaminated in their growing area before harvest by coming into contact with inadequately treated sewage or sewage polluted water. Secondly, contamination can arise during processing, storage, distribution or final preparation either directly from infected people or by contact with a contaminated environment. In most outbreaks of food-borne viral disease involving fresh products, it is not known whether contamination took place before, during or after harvest.

Guidelines issued by the World Health Organization state that fruits and vegetables to be eaten raw should not be fertilized with sewage or irrigated with contaminated water. Sewage sludge is sometimes applied to agricultural land, with the benefit that useful plant nutrients and organic matter are recycled to the soil. However, the UK government is proposing more stringent controls for harvesting vegetables from land where conventionally processed sewage sludge is applied. The transmission of viruses is thought to be mainly by surface contamination. There are relatively few reported studies on the possible uptake of viruses within damaged plant tissues during primary growth. Studies with poliovirus report that virus can infiltrate into the roots and body of plants from the soil, but there is no evidence of illness from this source. Viruses from sewage do not bind readily with soil particles and can enter groundwater leading to contamination of water sources.

The viruses causing gastroenteritis and hepatitis A appear to be extremely infectious in very low doses. Large numbers of virus particles can be excreted in the faeces from an infected person. Levels of the order of 10^6–10^{11} infective units per gram have been estimated. Poor personal hygiene is therefore a major route through which viruses can directly reach foods. Virus can be transferred from faecally contaminated fingers to foods or to work surfaces and door handles. There is a significant risk of contamination from field workers who do not have adequate on-site toilet and hand-washing facilities. Even when these facilities are put in place, the workers need to be supervised in such a way as to ensure that the facilities are used. In an outbreak of hepatitis A, associated with frozen raspberries, infection was confirmed in a fruit picker on the farm where the raspberries were cultivated. A number of outbreaks have also been linked to contamination of fresh products from the vomitus of infected food handlers. It has been suggested that between 20 and 30 million virus particles are liberated during vomiting. As well as direct transmission, aerosols produced by vomiting can contaminate exposed food, or surfaces with subsequent transfer to foods.

SURVIVAL

Survival of Viruses in Water

Human enteric viruses will potentially be present in any type of water contaminated by human faecal material and by sewage. Mounting evidence suggests that viruses can survive long enough and in high enough numbers to cause human diseases through direct contact with polluted water or contaminated foods. It was concluded that a thorough and valid assessment of the occurrence and significance of viruses in natural waters is hampered by lack of reliable information. It was also suggested that this is due to inefficient analytical techniques, and because different analytical methods have been used so that comparisons on survival data between different studies cannot be made. However, it does appear that different types of viruses are inactivated at different rates under identical conditions, and therefore no single microorganism can be expected to be an indicator for all viruses.

Temperature is an important factor, with low temperatures favouring viral survival in natural waters. In a study, it was found that there was no significant drop in rotavirus titre after 64 days at 4°C in raw water, treated tap water or filtered water. However, a 99% drop in titre was observed after 10 days at 20°C. Astrovirus survival has been demonstrated in drinking water after 90 days at 4°C. It was concluded in a study that hepatitis A virus and poliovirus could survive in wastewater and groundwater for 90 days or more at 10°C . Hepatitis A virus and poliovirus were shown to survive in excess of one year in mineral water stored at 4°C. Hepatitis A virus can survive in fresh or salt water for up to a year.

Evidences suggest that adsorption of viruses to particulate matter and sediments confers substantial protection against inactivating influences. Salinity and pH do not appear to have a significant direct effect on virus survival under conditions normally found in natural waters, but may have indirect effects by modifying interaction of viruses with particulates. There is some evidence that solar radiation promotes inactivation of viruses, but the effects have not been extensively studied.

Development of PCR-based assays has allowed the detection of NLVs in both river and seawater. In 1997, an outbreak of NLV gastroenteritis occurred among canoeists and was associated with river water at a water sports centre. When hepatitis A virus was detected in lettuce from Costa Rica, it was suggested that the possible source of contamination was the discharge of untreated sewage into river water used to irrigate crops, which is common practice in some less well-developed countries. In another outbreak of hepatitis A, traced to commercially distributed lettuce or tomatoes, it was hypothesized that contamination may have occurred in the fields from dirty water used for growing or irrigation, or possibly from the use of night soil, although this was difficult to prove.

Survival of Viruses in Soil

Enteric viruses may contaminate soil through the land disposal of sewage sludge and dirty irrigation water. A number of research studies have investigated the survival of human pathogenic viruses in soils with conflicting results. Survival appears to depend on a number of different variables, particularly the growing season, soil temperature, rainfall, soil type and composition. For example, viable poliovirus was recovered from spray-irrigated soil after 96 days during the winter season. This was compared to a maximum survival period of only 11 days during the summer, which suggested the higher temperature and solar radiation levels in the warmer seasons accelerated viral inactivation. The possibility also exists that viruses may be mechanically transmitted to fruits and vegetables during harvest. The persistence of inoculated poliovirus in drip irrigation pipes and soil was noted in one study. Moderate environmental conditions and alluvial-type soil, which restricts water infiltration, enhanced viral recoveries in the upper soil layers. It was found in a study that relatively high soil temperature (30°C) and a low moisture content hindered poliovirus survival. Wet soil conditions are frequently associated with low soil temperature. It was reported that a large proportion of outbreaks of water-borne disease in the developed countries resulted from contaminated groundwater. Climate, the nature of the soil and the nature of the resident microflora determine virus survival and retention within soil particles. Both electrostatic and hydrophobic interactions are thought to contribute towards virus adsorption and are controlled by the characteristics of the soil.

Survival of Viruses on Surfaces

Some investigations have focused on the survival of viruses on inanimate environmental surfaces such as stainless steel, glass and plastics. It was observed that a range of enteric viruses, including hepatitis A virus and rotavirus, persisted for extended periods (greater than 30 days) on several types of porous and non-porous surfaces. Greater virus survival was noted at 4°C than at 20°C. The effect of relative humidity on survival of hepatitis A virus on non-porous surfaces has been investigated. At 5°C, relative humidity had little effect on survival time, but at 20°C survival was longest when relative humidity was low. It is apparent from the lingering outbreaks that have occurred on cruise ships that NLVs survive well on environmental surfaces. NLVs have been detected in swabs of lockers in hospital outbreaks and from hotel carpets. A disinfectant formulation is considered to be effective if it is capable of inducing a 1000-fold (99.9%) or greater reduction in the virus titre. Experimental studies have shown that a free chlorine level of 5000 ppm reduced the infectivity titre by more than 99.9% on stainless-steel disks contaminated with the enteric viruses Coxsackievirus B3 and hepatitis A virus, and the respiratory viruses adenovirus type 5, parainfluenza virus type 3 and coronavirus 229E. The virucidal action of sodium hypochlorite (1250 ppm at pH 9.56) was also tested against hepatitis A virus, human rotavirus and a *Bacteroides fragilis* bacteriophage dried on polystyrene. Overall, a less than 1000-fold titre reduction was achieved for all the viruses examined. Chlorine is a common disinfectant used by many fresh product processors to wash fruits and vegetables. However, chlorine levels greater than 200 ppm are thought to cause adverse discolouration (bleaching) and off-flavours in the finished product.

Survival of Viruses on Fruits and Vegetables

Survival times for some enteric viruses have been determined on a range of different fruit and vegetable commodities. It is difficult to draw conclusions from the different studies, since experimental conditions and methods varied. However, most of these studies report viability in excess of the product shelf life. It was hypothesized that under conditions of low or no moisture the aqueous part of the virus inoculum evaporated leaving the virus exposed to air and/or salts. Celery, lettuce and carrots stored at 4°C supported the survival of a range of enteric viruses for up to 8 days.

It was indicated in a study that rotavirus SA-11 (a simian rotavirus that can be cultured) survived on lettuce, radishes and carrots for up to 30 days at 4°C. Viral inactivation at room temperature (25°C) was significantly greater than at 4°C; however, viable rotavirus SA-11 was still detected on lettuce after 25 days. They concluded that the rough or irregular surfaces present on lettuce might offer some additional protection for virus particles. Indeed, protected segments of plants such as the roots, closed leaves and internal fruit parts, may offer favourable conditions that increase survival time up to 60 days. Viable poliovirus was recovered from effluent spray-irrigated lettuce and radishes 23 days after inoculation. Polioviruses was isolated from cucumbers grown in effluent-irrigated soil for a full 8 days after inoculation. The above study investigated the transmission of poliovirus from subsurface drip-irrigated soil to tomato plant leaves and tomato fruits. Virus was not detected in the tomato fruits. However, a number of leaf samples were positive for poliovirus even though the virus was injected 10 cm below the surface of the soil. This result suggests that poliovirus can penetrate into plant tissue through the root system. The lack of viable poliovirus in the tomato fruits may be due to the presence of antiviral substances as witnessed by other authors.

Chilled storage temperatures (2–8°C) typically retard respiration, senescence, product browning, moisture loss, and microbial growth in minimally processed fruits and vegetables, but may contribute to the survival and transmission of viruses to the human host.

The potent antiviral properties of different fruit extracts were described at neutral pH, particularly strawberry and significant differences were found in viral recovery on strawberries, cherries and peaches held in a humid atmosphere at 4°C. Most of the aqueous fruit extracts and infusions demonstrated notable antiviral properties. Although the active compounds were not isolated, these agents could be phenolic in nature. In grapes, these chemicals are thought to be located primarily in the skin. Fruits and vegetables are known to contain a vast array of antimicrobial substances, particularly organic acids, phenolic and sulphur compounds and small polypeptide proteins. Despite this, outbreaks of viral gastroenteritis and hepatitis A have been associated with fruits and fruit juices. In the absence of formal studies it could be inferred that the gastroenteritis and hepatitis A viruses are relatively resistant to these virucidal chemicals.

In view of the increasing use of ready-to-eat vegetables sold in modified atmosphere packaging (MAP), the survival of hepatitis A virus on lettuce in MAP, stored at room temperature and 4°C, has been investigated. Survival at 4°C was the same in MAP as under normal conditions of packaging. At room temperature viral survival was slightly better in MAP containing higher carbon dioxide levels. It was suggested that enhanced virus survival might be due to inhibition of ethylene by carbon dioxide, resulting in reduced physiological spoilage of vegetables such as lettuce and possibly less toxic effects on the virus. Studies have indicated that indigenous microflora in the water environment are deleterious to survival of enteric viruses. These findings highlight the importance of avoiding contamination of food items before packaging in MAP. The survival of herpes simplex virus type 1, suspended in saliva, on the skin of tomatoes and the upper surface of lettuce was studied. Although storage times of only 1 hour were studied, temperature was shown to have a significant effect on virus titre. There was no loss of virus infectivity titre at 2°C compared to a 2-log reduction at room temperature (22–24°C). Although not an enteric virus and not a virus that comes to mind as a possible food-borne infection, herpesviruses could invade through the mucous membranes of the mouth.

DECONTAMINATION OF FRESH PRODUCTS

Most fruit and vegetable washing systems are designed to remove gross contamination such as dirt, insects, and foreign matter. However, they are reported to be less successful at removing microbial contaminants. Vigorous washing of fruits and vegetables with clean potable water typically reduces the number of microorganisms by 10–100-fold and is often as effective as treatment with 100 mg/l chlorine, the current industry standard. Raw materials are typically immersed in cooled sanitized water and then dewatered to remove surface moisture and fruit and vegetable juices from the product. Product agitation is optimized by using water or air jets which enhance surface contact, carriage of product, and suspension of solids and vegetable debris. There has been recent concern over the possible migration of bacterial pathogens into the core tissue of fruits and vegetables during washing. It was found that uptake of bacterial cells was associated with a negative temperature differential between the water and the product. However, the uptake of human viruses by fruits and vegetables under similar conditions has not been studied. Lodging or attachment of microorganisms in tissue crevices may protect cells from direct contact with disinfectants and consequently aid in their survival. Recent studies have found that *Escherichia coli* O157:H7 could survive in the stomata and on cut edges of lettuce following chlorine treatment. Although there are no available data for viruses, cell surface structures are likely to offer some additional protection.

CONTROL

The control of food-borne viral infection was considered in the report of the Advisory Committee on the Microbiological Safety of Food (1998). Sewage pollution is a major factor in the contamination of food and water. This is particularly pertinent to fruits and vegetables that will not be cooked before consumption. Untreated or inadequately treated sewage discharged into natural waters can cause contamination of crops. Sewage sludge is applied to agricultural land, with the benefit that useful plant nutrients are recycled to the soil. In April 2000, the Agricultural Development and Advisory Service (ADAS, UK) published the Safe Sludge Matrix (ADAS, 2000). This document gives clear guidance on the minimum acceptable level of treatment for any sewage sludge applied to agricultural land and provides a framework to ensure microbiological safety. As from 31st December 1999 application of all untreated sludge on agricultural land used to grow food crops has been prohibited in the developed countries. Use on all agricultural land will be prohibited from the end of 2001. Untreated sludge is produced by either the primary settlement or secondary biological stages of sewage treatment. Further processing may be undertaken to produce treated sludge, resulting in improved stability and a reduction in health hazards and odour problems. Currently, treated sludge may be applied to land used for salad crops, but harvesting is not permitted for 30 months. Vegetables cannot be harvested for 12 months. A 10-month harvest interval is required if enhanced treated sludge is used. All applications must comply with the Sludge (Use in Agriculture) Regulations 1989 and DETR Code of Practice 1986. These controls may help to reduce the risks of microbiological contamination of fruits and vegetables in the UK, but still do not address the potential problem of imported products from countries with different standards for organic fertilizers or irrigation water.

The other major source of contamination is from infected people handling food. People with symptoms should be excluded from food handling. However, food-handlers with only very minimal symptoms have been implicated in transmission of viral gastroenteritis. Current recommendations state that food-handlers should be allowed to return to work 48 hours after symptoms have ceased. These recommendations appear to work satisfactorily, but were based on the rapid decrease in virus excretion observed by electron microscopy. Using more sensitive PCR assays, NLVs can be detected for longer periods than electron microscopy and, in some instances, for up to a week after onset of symptoms. It is not clear if people shedding virus detectable by PCR are infectious, but recommendations on how long to exclude people from food-handling need to be kept under review. The main period of excretion for hepatitis A virus is before symptoms become apparent and therefore control is difficult. The wearing of disposable gloves is recommended if foods are to be manipulated by hand, but this does not prevent the transfer of viruses to gloves by touching contaminated surfaces.

If vomiting occurs, virus may spread over a wide area in aerosol droplets. Uncovered foods that will not be cooked should be discarded. The environment should be thoroughly cleaned, including work surfaces, sinks and door handles. Recent studies by the PHLS demonstrated the presence of virus on surfaces and materials that had been cleaned by recommended decontamination methods. This suggests that the current recommendations for the removal of NLVs from contaminated surfaces are inadequate.

Thorough cooking of virus-contaminated foods was advised. However, ready-to-eat fruits and vegetables are unlikely to withstand such harsh treatment and may show deleterious changes in sensory quality. A wash at 80°C for 2 minutes is suggested to ensure the microbiological safety of fresh products. He concluded that this heat treatment does not affect the appearance and taste of most fruits and vegetables provided the food is eaten within 24 hours. However, the majority of minimally processed fruits and vegetables require a shelf life in excess of 1 day and therefore this heating step is not applicable.

Meticulous attention to good food-handling practices and education is essential. There should be provision of adequate toilet and hand-washing facilities, not only in the catering and retail industry, but also for farm workers.

There is an effective vaccine for hepatitis A and it has been suggested that food-handlers should be vaccinated. Epidemiological data currently suggest that food-handlers in the United Kingdom do not pose a significantly greater risk of transmitting hepatitis A infection than other people, and use of vaccine in this group may not be cost-effective. No vaccines have been developed for NLVs, but research studies with NLV recombinant capsid antigens offer the potential for future development of vaccines. There are challenges, however, in designing effective vaccines in that first, there are multiple types of NLVs and secondly, the mechanisms for inducing immunity to these agents is poorly understood.

SUMMARY

Fresh products contribute to the transmission of viral infections. There is a lack of information on the survival of viruses on fresh products related to shelf life and types of packaging. Information is also lacking on the efficiency of current washing and decontamination processes for the removal of viruses. Studies are therefore required to provide this information for NLVs and hepatitis A virus in particular. There is a need for information on rotaviruses and astroviruses. Both these viruses have been implicated in food-borne and water-borne outbreaks in the developed countries, although only very infrequently. The role and extent of food- or water-borne transmission needs to be more clearly defined for both these viruses.

The infectivity of NLVs and hepatitis A virus is difficult to study in the laboratory. NLVs cannot be cultured *in vitro* and culture techniques for hepatitis A virus are time-consuming. Other viruses that can be cultured readily have been used as models for these pathogens in a number of studies. Although only the use of specific human pathogenic viruses will give clear and unequivocal data on survival, inactivation and removal, this is not always safe and practical. Model systems using similar non-pathogenic human or mammalian viruses may be the most satisfactory alternative. For example, feline calicivirus has been used as a model system for NLVs in assessing the efficacy of commercially available disinfectants. Feline calicivirus was also used as a model for NLVs in another study on the heat treatment of shellfish. This organism would appear to be a good candidate for virus studies on fresh products. In order to undertake studies on the survival and removal of viruses, there is a need to develop better methods for inoculation and recovery of virus from a range of fresh products.

There have been recent suggestions that faecal coliforms on fresh products may be an indicator of the probable presence of enteric viruses. However, several authors have found no significant correlation. Despite the shortcomings of indicator organisms such as bacteriophages, there is a need for a safe and convenient model that can be used to help the food industry to assess and optimize new treatment processes. This is not intended as an indicator to monitor food samples directly. Other studies have shown that the behaviour and survival of bacteriophages mimic human viruses more closely than bacteria and would be suitable for such purposes. The optimization of washing and decontamination processes to remove viruses from fruits and vegetables will ultimately contribute to the overall safety of these food products.

REVIEW QUESTIONS

1. Discuss viruses as food-borne pathogens.

29

MYCOTOXIGENIC MOULDS AS AGENTS OF FOOD POISONING

INTRODUCTION

Mycotoxins brought the attention of the scientists in the early 1960s. *Aspergillus* was first described almost 300 years ago and is an important genus in foods. Most *Aspergillus* species occur in foods as spoilage or in biodeterioration. They are extremely common in stored commodities such as grains, nuts and spices.

Aspergillus is a large genus containing more than 100 recognized species, most of which grow well as laboratory culture. There are a number of teleomorphic (ascosporic) genera which have *Aspergillus* conidial states (anamorphs), but the only two of real importance in foods are the xerophilic genus *Eurotium* and *Neosartorya* which produce heat-resistant ascospores and cause spoilage in heat-processed foods. Almost 50 species of *Aspergillus* have been identified as capable of producing toxic metabolites. Chief toxins produced by *Aspergillus* spp. are the aflatoxins, (A. *flavus, A parasiticus, A. nomius*), ochratoxin A (*A. ochraceus*), sterigmatocystin (*A. versicolor*), cyclopiazonic acid (*A. flavus, A. tamarii*), citrinin, patulin and penicillic acids.

ASPERGILLUS FLAVUS AND *A. PARASITICUS*

The most important group of toxigenic aspergilli are the aflatoxigenic moulds. *A. flavus, A. partisiticus* and *A. nomius* are closely related but produce specific toxins. *A. flavus* produces aflatoxins B_1 and B_2 and cyclopiazonic acid. Only certain strains are toxigenic. *A. parasiticus* produces aflatoxins B_1, B_2, G_1 and G_2 but not cyclopiazonic acid and almost all the strains are toxigenic. *A. nomius* also produces B and G aflatoxins but the potential toxigenicity of the organisms are not known.

Occurrence of the Moulds

Aspergillus flavus is widely distributed in nature, but *A. parasiticus* is less widespread. *A. flavus* is the most common species in peanuts and the second most common in corn. Invasion takes place before harvest, not during storage. Peanuts are invaded while still in the ground. In corn, damage in developing kernels is caused by insects that allows entry of aflatoxigenic moulds. Cereals and spices are common substrates for *A. flavus* but aflatoxin production in these products is a result of poor drying, handling or storage.

Factors Affecting Growth and Toxin Production

Aspergillus flavus and *A. parasiticus* have similar growth patterns. Both require growth temperatures of $32–33°C$ and can grow at $10–42°C$. Aflatoxins are produced at $12–40°C$. Optimum water activity required for growth is near 0.99 and aflatoxins are generally produced in greater quantities at higher a_W values with lesser production apparently ceasing at or near a_W 0.85.

Growth occurs at a pH ranging from 2 to 10.5 but aflatoxin production has been reported for *A. parasiticus* between pH 3 and 8 (opt. pH 6). Reduction of available oxygen by modified atmosphere packaging of foods with oxygen scavengers can inhibit aflatoxin formation by both the species.

Aflatoxins

Aflatoxins are difuranocoumarin derivatives. Aflatoxins B_1, B_2, G_1 and G_2 are produced in nature by the moulds. The letters B and G refer to the fluorescent colors (blue and green respectively) observed under UV light and the subscripts refer to their separation patterns on TLC plates.

Aflatoxins are both acutely and chronically toxic in animals and humans, producing acute liver damage, liver cirrhosis, tumour induction and teratogenesis. They also have immunosuppressive effects in combination with other mycotoxins. Immunosuppression can increase susceptibility to infectious diseases. Outbreaks with consumption of rice have been reported with nearly 15% mortality. Another outbreak of hepatitis which affected 400 people, 100 of whom died, almost certainly was caused by aflatoxin that could have been from heavily infected corn that was consumed. The greatest direct impact of aflatoxins on human health is their potential to induce liver cancer where hepatitis B virus also has a role to play. Aflatoxins and hepatitis B virus are co-carcinogens and they increase the probability of human liver cancer. But high aflatoxin intakes are causally related to high incidences of cancer, even in the absence of hepatitis B.

Aflatoxin B1 is metabolized by the microsomal mixed function oxidase system in the liver, leading to the formation of highly reactive intermediates, one of which is 2,3-epoxy aflatoxin B_1. Binding of these reactive intermediates to DNA results in disruption of transcription and abnormal cell proliferation, leading to mutagenesis or carcinogenesis. Aflatoxin also inhibits oxygen uptake in the tissues by acting on the electron transport chain and inhibiting various enzymes resulting in decreased production of ATP.

Cyclopiazonic Acid

This is produced by *A. flavus*, *A. tamari* and *A. versicolor*. The toxin is an indole tetramic acid which can occur in naturally contaminated agricultural commodities and compounded animal feeds. It causes severe gastrointestinal and neurological disorders along with degenerative changes and necrosis in the digestive tract, liver, kidney and heart.

DETECTION OF AFLATOXIGENIC MOULDS

Method 1   Direct Enumeration

Using pour plate method with inclusion of dichloran that helps to restrict the size of mould colonies, facilitating counting. Addition of glycerol reduces the water activity of the medium.

Media

1. Dichloran glycerol chloramphenicol (DG) agar.
2. Dichloran rose Bengal chloramphenicol (DRBC) agar or oxytetracycline glucose yeast extract (OGYE) agar.

Procedure

1. Prepare a 10^{-1} food homogenate and serial decimal dilutions.
2. Use this homogenate and its decimal dilutions with any surface counting method to enumerate on DG agar, DRBC agar or OGYE agar.
3. Incubate the plates at 25°C for 5 days.
4. Count the colonies of yeasts and moulds after 3 and 5 days (to avoid overgrowth) in plates containing up to 150 colonies.
5. Calculate the count per gram.

Method 2   Enumeration of Xerophilic Moulds

This can be done by adjusting the water activity of both the diluent and the isolation medium. The use of a diluent containing 50% glucose is considered satisfactory. An agar medium containing 35–50% glucose will allow recovery of xerotolerant strains. The reduced water activity of the medium may necessitate prolonged incubation for several weeks.

Procedure

1. Prepare a 10^{-1} homogenate and serial decimal dilutions of the food in 50% glucose solution.
2. Use suitable aliquots of this homogenate and dilutions and spread them over the surface of malt extract yeast extract 40% glucose agar.
3. Incubate the plates at 20–25°C for mould isolation up to 3 weeks.

4. Count the colonies and calculate the count per gram.

Identification of *Aspergillus* species requires growth of media developed for this purpose, including Czapek agar, a defined medium based on mineral salts. Growth on Czapek yeast extract +20% sucrose agar can be a useful aid in identifying species of *Aspergillus*. The most effective medium for rapid detection of aflatoxigenic moulds is Aspergillus flavus and parasiticus agar, a medium formulated specifically for this purpose. Under the incubation conditions specified for this medium (30°C for 42–48 hours), *A. flavus*, *A. parasiticus* and *A. nomius* produce a bright orange to yellow colony reverse which is a diagnostic character.

Toxin Detection

Aflatoxins can be detected by chemical or biological methods. In the chemical methods, samples are extracted with organic solvents such as chloroform or methanol in combination with small amounts of water. Extracts are further cleaned up by passage through a silica gel column. The extract is then concentrated, usually by evaporation under nitrogen and separated by TLC or HPLC. TLC is most often used, and aflatoxins are visualized under UV light and quantified by visual comparison with known concentrations of standards or by fluorimetry. The presence of fats, lipids, or pigments in extracts reduces the efficiency of the separation. Immunoassay techniques including enzyme-linked immunosorbent assays and dipstick tests for aflatoxin detection have been developed and commercially made available.

A. OCHRACEUS

A. ochraceus is a widely distributed mould, particularly common on dried foods. There are three toxins, ochratoxin A, B and C. This organism also produces penicillic acid, a mycotoxin of lower toxicity. Other reported metabolites are xanthomegnin and viomellein.

A. ochraceus is widely distributed in dried foods such as nuts, beans, dried fruits and dried fish. It is a xerophile, capable of growth down to a_w 0.79 (opt. 9.9). It grows at temperatures of 8 to 37°C and within a wide pH range of 2.2 to 10.3. Ochratoxin A is produced at quite high temperatures between 25 and 30°C.

Isolation and Detection

This organism is best isolated on reduced a_w media such as dichloran glucose agar. Colonies of *A. ochraceus* are deep ochre-brown to yellow-brown in colour with long stripes bearing radiate *Aspergillus* head. The vesicles are spherical bearing densely packed metulae and phialides with small, smooth pale brown conidia. Ochratoxins can be assayed by routine chromatographic methods like the TLC and HPLC with detection of green fluorescence under UV light at 333 nm. Immunoassay techniques, including ELISA have also been developed and commercialized.

A. VERSICOLOR

This is the most important food spoilage and toxigenic species. It is the major producer of sterigmatocystin, a carcinogenic dihydrofuranoxanthone which is a precursor of the aflatoxins. Sterigmatocystin is also produced by *A. nidulans*.

A. versicolor is a xerophile, with a minimum a_w of 0.74 to 0.78 for growth. It is widely distributed in foods particularly stored cereals, cereal products, nuts, spices and dried meat products. The reported minimum temperature for growth is 9°C at a_w 0.97 and the maximum temperature is 39°C at a_w 0.87. Natural occurrence of sterigmatocystin has been found in rice, wheat and barley. This toxin has low acute oral toxicity because it is relatively insoluble in water and gastric juices. But even low doses can cause tumour and pathological changes.

Isolation and Detection

This organism can be isolated in the low a_w agar such as the dichloran glucose agar. Small grey-green colonies showing pinkish or reddish colour in the mycelium and/or reverse and with mop-like heads are indicative of this organism.

Sterigmatocystin can be detected by TLC with chloroform–methanol as a solvent system. The toxin is visualized as an orange-red spot under UV light. A light yellow fluorescence develops after spraying with acetic acid. An ELISA method for detection of sterigmatocystin has been reported.

A. FUMIGATUS

It is best recognized as a human pathogen causing aspergillosis of the lung. It is a thermophile with a temperature range for growth of between 10 and 55°C and an optimum between 40 and 42°C. It is one of the least xerophilic of the common aspergilli with a minimum a_w of 0.85 for growth. It is best found in decaying vegetation, in which it causes spontaneous heating. It is isolated frequently from foods particularly stored commodities. This organism is capable of producing several toxins which affect the central nervous system, causing tremors. The toxin is called fumitremorgens A, B and C which are toxic cyclic dipeptides.

A. TERREUS

It occurs commonly in soil and foods particularly stored cereals and cereal products, beans, pulses and nuts. They produce a group of tremorgenic toxins known as territrems. These toxins do not contain nitrogen. Territrems are acutely toxic causing whole body tremors within 5 minutes and other neurological symptoms within 20–30 minutes, all of which subside within 1 hour. The toxin appears to act by blocking acetylcholinesterase activity.

The organism produces rapidly growing pale brown colonies, with *Aspergillus* heads bearing densely packed metulae and phialides with minute conidia borne in long columns. The territrems can be detected in chloroform extracts by TLC, exhibiting blue fluorescence under UV light.

A. CLAVATUS

It is found in soil and decomposing plant materials and is easily recognizable by its large blue-green club-shaped heads. It is especially common in malting barley. The organism produces patulin, cytochalasins and the tremorgenic mycotoxins tryptoquivaline, tryptoquivalone and related compounds. *A. clavatus* is a recognized health hazard to workers in the malting industry. Inhalation of large numbers of highly allergenic spores can cause respiratory diseases such as bronchitis, empyema or malt worker's lung, a serious occupational extrinsic allergic alveolitis.

EUROTIUM

Members of this genus are referred to as the *Aspergillus glaucus*. All *Eurotium* species are xerophilic. They are important spoilage moulds in all types of stored commodities like stored grains, spices, nuts and animal feeds. The four most common species are *Eurotium chevalieri, E. repens, E. rubrum* and *E. amstelodami*.

Eurotium chevalieri and *E. amstelodami* have been reported to produce toxic alkaloid metabolites called echinulin and enoechinulins.

PENICILLIUM AS A TOXIGENIC MOULD

The discovery of penicillin in 1929 gave impetus to a search for other *Penicillium* metabolites with antibiotic properties and ultimately to the recognition of citrinin, patulin and griseofulvin as toxic antibiotics or mycotoxins. Retrospectively, over 120 metabolites from common moulds were demonstrably toxic to higher animals of which 42 were produced by one or more *Penicillium* species.

Penicillium is a large genus with 150 species. At least 50 species are of common occurrence. All common species grow and sporulate well on synthetic or semi-synthetic media and are usually readily recognizable to the genus level. Classification of the penicillia is based on microscopic morphology. The genus *Penicillium* is divided into subgenera based on the number and arrangement of phialides and metulae and rami on the main stalk cells. The majority of important toxigenic and food spoilage species are found in subgenus *Penicillium*. Growth of mould does not always mean production of toxin. The conditions under which toxins are produced are often narrower than the conditions for growth. Most toxins can be placed in two broad groups: those that affect liver and kidney function and those that are neurotoxins. The *Penicillium* toxins which affect liver or kidney function are asymptomatic, in contrast, toxicity of the neurotoxins is often characterized by sustained trembling. The mycotoxins, their nature and food source are listed in Table 29.1.

Table 29.1 Major mycotoxins and their food source

Name of toxin	Symptoms	Taxonomy	Nature of toxin	Food source
Ochratoxin A	Causes a serious animal health problem because it plays a major role in the aetiology of nephritis in pigs	*P. verrucosum*. Slow growth on Czapek yeast extract agar (CYA) and malt extract agar (MEA) at 25°C with bright green conidia, clear to pale yellow exudates and rough stipes. Dichloran rose Bengal yeast extract sucrose agar is a selective medium for the enumeration of *P. verrucosum*. It grows most strongly at relatively low temperatures (0°C) with a maximum of 31°C. It is xerophilic (a_W is 0.80).	A fat-soluble metabolite which accumulates in the fat depot of affected animals and from there is ingested by humans eating pork.	Pork, bread made from barley or wheat containing the toxin.
Citrinin	Produces watery diarrhoea, increased water consumption and reduced weight gain due to kidney degeneration	*P. citrinum* *P. verrucosum,* *P. expansum* The penicillus of *P. citrinum* consists of a cluster of three to five divergent metulae, usually apically swollen. The phialides from each metula bear conidia as long columns, producing a distinctive pattern of diagnostic value. They grow well on CYA and MEA at 37°C. A xerophilic mould has a a_W requirement of 0.82.	Renal toxin affecting monogastric domestic animals such as pigs, dogs and domestic birds.	Cereals, especially rice, wheat and corn, milled grains and flour.

(*Contd.*)

Table 29.1   (Continued)

Name of toxin	Symptoms	Taxonomy	Nature of toxin	Food source
Patulin	Immunological, neurological and gastrointestinal effects	*P. expansum* (fruit pathogen) major source being rotting apples and pears. It is a psychrotrophic, growing at -2 to $-3°C$ (opt. $25°C$ and max $35°C$). Minimum a_W is 0.82. It can be grown on Dichloran rose bengal chloramphenicol agar (DRBC) or dichloran 18% glycerol agar (DG18)		Common in apple juice and apple products like cider.
Cyclopiazonic acid	Causes fatty degeneration and hepatic cell necrosis in the liver and kidneys of domestic animals.	*Aspergillus flavus* and *P. cyclopium*, *P. commune* and *P. camemberti* They have the ability to grow at or near $0°C$ and below a_W of 0.85.		Cheese
Citreoviridin	Acute cardiac beriberi. Heart distress, laboured breathing, nausea and vomiting, followed by pain, restlessness and sometimes maniacal behaviour.	*P. citreonigrum* Colonies grow quite slowly on CYA with pale grey-green conidia, and exhibit yellow mycelial, soluble pigment and reverse colours. *Penicillium* consists of small clusters of phialides only. Conidia are spherical smooth walled and tiny. *Eupenicillium ochrosalmoneum* Slow growing and appears bright yellow on both CYA and MEA at $25°C$.		Rice, corn

(Contd.)

Table 29.1   (Continued)

Name of toxin	Symptoms	Taxonomy	Nature of toxin	Food source
Penitrem A	Tremorgenic response-neurotoxins having an emetic effect.	*P. cyclopium,* *P. palitans,* *P. puberulum,* *P. crustosum,* *P. martensii.* *P. crustosum* is a fast-growing species. It produces dull green colonies with a granular texture on both CYA and MEA.	Tremorgens	Cereals, animal feeds, corn, processed meats, nuts, cheese and fruit juices.
PR toxin and Roquefortine	Symptoms are similar to strychnine poisoning.	*P. roqueforti,* *P. crustosum* and *P. chrysogenum.* *P. roqueforti* grows very rapidly on CYA and MEA and produces green reverse colours on one or both media. It is more tolerant to acetic acid, hence can be selectively enumerated on a medium such as malt extract agar (MEA + 0.5% glacial acetic acid). *P. roqueforti* is able to grow in oxygen concentrations below 0.5% and in the presence of 20% carbon dioxide. It is a psychrotrophic, growing vigorously at temperatures as low as 2°C.		Cheese, meats, silage
Secalonic acid D	Toxin due to inhalation	*P. oxalicum.* It grows rapidly on CYA at 25°C and 37°C forming a continuous layer of conidia. It can be enumerated on DRBC and DG18. It grows at temperatures of 8 to 40°C and a_w of 0.86.	Dimeric xanthones	Grain dusts of corn

1. Give a detailed account of toxins of *Aspergillus* species.

2. Describe the detection methods of aflatoxigenic moulds.

3. Discuss *Penicillium* as a food-borne pathogen.

30

ALGAL FOOD POISONING

INTRODUCTION

Seafood toxins can be categorized according to their primary origin: toxins originating from toxic marine algae, principally dinoflagellates, which include ciguatoxins and those toxins associated with the major shellfish poisoning syndromes (diarrhoetic shellfish poisoning (DSP); paralytic shellfish poisoning (PSP); amnesic shellfish poisoning (ASP); and neurotoxic shellfish poisoning (NSP); and contaminants produced by bacteria (tetrodotoxins (TTXs) associated with puffer fish poisoning) or by bacterial decomposition (histamine associated with scombrotoxic fish poisoning). All these syndromes share a lack of organoleptic evidence of contamination in affected fish; most toxic seafoods looks, smell and taste quite normal. Adequate disease prevention requires widespread and efficient monitoring programmes employing accurate and rapid toxin detection methods and also an increased awareness by both the public and clinicians in order to effect swift diagnoses and supportive treatment, where necessary.

TOXIC SYNDROMES ASSOCIATED WITH MARINE ALGAL TOXINS

There are over 4000 species of marine algae (phytoplankton) and about 2% are known to produce toxins (Table 30.1) most of which are *Dinophyceae* spp. (dinoflagellates) with four more genera being responsible for the majority of toxic events, i.e., *Alexandrium* spp., *Gymnodinium* spp., *Dinophysis* spp., and *Prorocentrum* spp. A proportion of these toxic dinoflagellates have a red-brown pigmentation giving rise to the naming of algal blooms as "red-tides". However, not all toxic algae are coloured and incidents of poisoning have occurred in the absence of red blooms. Many toxic species form resting cysts during adverse conditons but produce dense blooms, usually in the summer months, in response to favourable conditions of temperature, pH, salinity, light and the availability of nutrients.

Visible red tides may contain from 20,000 to 30,000 algal cells/ml of sea water; however, concentrations as low as 200 cells/ml may produce toxic shellfish. Algal toxicants are accumulated in the digestive gland of the shellfish, the shellfish themselves remaining unaffected by the toxins.

Both proliferating cells and resting cysts can contain toxins and act as sources of shellfish contamination. The cysts of some dinoflagellates are initially much more toxic than vegetative cells, they sink out of the water column and overwinter at the sediment-water interface. Resting cysts are thought to contribute to shellfish toxicity in the absence of proliferating blooms and to the accumulation of toxins in scavenging crustacea and grazing fin-fish.

Paralytic Shellfish Poisoning

The world wide incidence of paralytic shellfish poisoning (PSP) has been estimated at 1600 cases per year with a possible 300 of these being fatal. Outbreaks have been occurring regularly throughout the world. Blooms of the associated dinoflagellates occur widely throughout Europe.

Time of onset of symptoms varies from 15 minutes to 10 hours, but usually occurs within 2 hours of consumption of the seafood, and is influenced by amount of toxin ingested, age and weight of the patient, nature of the contaminated food and any accompanying alcohol consumption. Dominant clinical features are neurological. Gastrointestinal symptoms are less common. In severe cases, death may occur within 2–25 hours. The toxins associated with PSP are a group of closely related compounds based on saxitoxin (STX), a compound comprising two fused guanidium moieties. PSP toxins are often found in toxic shellfish. The compounds are all water-soluble and heat-stable; a light 5 minutes cook will reduce toxicity by only 30% and increasing this to 20 minutes will only effect a 40% denaturation. All these toxins operate by blocking the sodium channel which traverses excitable cells comprising nerves and muscle fibres. Neurotransmission is facilitated by a Na^+ influx through the channel following depolarization.

Oral intake of 144–1660 µg per person leads to intoxication and may be fatal if the intake is 300–12,400 µg per person.

Diarrhoetic Shellfish Poisoning

Outbreaks of gastrointestinal illness associated with shellfish exposed to dinoflagellates were first reported in 1961 in the Netherlands followed by incidents in Japan in the later 1970s. Areas affected by DSP remain predominantly in Japan and Europe, but now also include North and South America, Australia, Indonesia and New Zealand.

The predominant symptoms of DSP are diarrhoea, nausea, vomiting and abdominal pain lasting up to 3–4 days with a typical onset period of 30 minutes to a few hours after consumption of the toxic seafood. Affected individuals rarely require hospitalization and any treatment prescribed is supportive of the gastrointestinal effects.

Dinophysis spp. and *Prorocentrum* spp. both produce the toxins associated with DSP. In Europe *D. accuta* and *D. accuminata* predominated, *D. fortii* is found occasionally, and *D. norvegica* is common in Norway. The DSP toxin family comprises at least 12 polyether carboxylic acids. Shellfish can create a series of derivatives of each of these toxins during their metabolic processes and these derivatives all vary slightly in their toxicity. The toxins inhibit protein phosphatases which are integral regulatory proteins in metabolism, membrane transport, secretion and cell division. These toxins are potent tumour promoters and are possibly mutagenic and immunotoxic.

Neurotoxic Shellfish Poisoning

Neurotoxic shellfish poisoning has occurred mainly in the USA and Mexico. These toxins are produced by *Gymnodinium* spp. and *Ptychodiscus breve* (also called *G. breve*).

Symptoms usually begin 1–3 hours after toxin ingestion and include paraesthesia (numbness/tingling) in the mouth progressing to the extremities, ataxia, gastrointestinal symptoms and the hot to cold temperature reversal phenomenon which affects the patients' perceptions of touching hot and cold surfaces. Recovery normally occurs in 2–3 days and there is no specific treatment.

The responsible toxins are a family of nine brevetoxins which all act on excitable cells by opening sodium channels causing a sodium ion influx and the release of synaptic neurotransmitters. The toxins are also responsible for killing fishes. The dinoflagellates are easily lysed in surf, releasing toxins directly into the sea water which may cause dermatitis, conjunctivitis and respiratory problems in bathers.

Amnesic Shellfish Poisoning

Amnesic shellfish poisoning was first recognized in 1987 in Canada where 107 cases of human illness were associated with blue mussel consumption. The marine algae *Pseudonitzschia pungens* has been responsible for this toxic syndrome.

The syndrome progresses from typical gastrointestinal symptoms (onset up to 24 hours after consumption of contaminated shellfish) to include a range of neurological effects such as autonomic dysfunction, seizures, focal definits (onset up to 48 hours). Long-term sequelae include short-term memory loss. The causative agent is domoic acid, a potent neurotransmitter with receptor sites in the CNS. It mimics the natural neurotransmitter glutamate and will bind excessively to brain tissue causing necrosis. Recent work has shown that heterotrophic bacteria can enhance the production of domoic acid by *Pseudonitzschia pungens*.

Ciguatera Poisoning

Ciguatera poisoning presents a health risk only in UK due to the importation of exotic fish. There are at least 400 species of tropical, reef feeding fish from oceans between latitudes 35°C on either side of the equator which are potentially ciguatoxic. Those most commonly associated with illness are groupers, was bass, snapper, barracuda, eels, parrot fish and mullet, all of which are regularly consumed in the UK. Globally ciguatera poisoning is the most common fish-borne poisoning syndrome with an estimated 50,000 cases a year.

Onset time varies between a few minutes to 30 hours after consumption of the fish but most patients develop symptoms in less than 6 hours. The most often presented symptoms are tingling and numbness in the mouth, hands and feet and the cold to hot sensory effect (90%), dull aches and sharp shooting pains (85%) and gastrointestinal symptoms (45%). With no known specific antidote, self resolution usually takes about 2–5 days with some neurological effects persisting for weeks, even years in severe cases. Attacks are known to recur when stimulated by stress and hypersensitivity responses induced by foods containing substances which mimic ciguatoxins (e.g. fish, alcohol, nuts). Doses of 0.6 ng/kg body weight have been associated with human illness. Ciguatera poisoning has been shown to be sexually transmissible and toxins may be passed to foetuses *in utero* and to neonates in breast milk.

Dinoflagellates form the genera *Gambierdiscus* (specifically *G. toxicus*), *Ostrepsis, Prorocentrum, Amphidinium* and *Coolia* produce a family of ciguatera associated toxins. Ciguatoxin, a polycyclic quaternary ammonium compound and scaritoxin are lipid-soluble toxins which can be modified as they move up the food chain to produce a range of toxins varying in toxicity and associated clinical features. These toxins facilitate neurotransmission by opening sodium channels in excitatory membranes by the competitive occupation of calcium receptor sites. A range of water-soluble toxins, based on maitotoxin, are found in small herbivorous fish, which play a more minor role in the syndrome and operate via the calcium channel. Maitotoxin has been found to be produced by *G. toxicus*; the fourth toxin group associated with this syndrome is the palytoxin family, like the ciguatoxins these are polycyclic ethers and have been associated with ciguatera poisoning following the consumption of contaminated mackerel. All the toxins are unaffected by typical cooking conditions.

Table 30.1 Toxic syndromes of algal toxin origin

Disease	Clinical symptoms	Toxins	Associated microbes	Shellfish /fish	Occurrence
PSP (paralytic)	Numbness in mouth and extremities, ataxia, dizziness, floating sensation, headache, respiratory distress, paralysis (death)	Saxitoxins, Neo-saxitoxin, Gonyau-toxins	*Alexandrium* spp., *Pyrodinium* spp., *Gymnodinium* spp.	Mussels, cockles, clams, scallops, oysters, crab, lobster	Global
DSP (diarrhoetic)	Diarrhoea, nausea, vomiting, abdominal pain	Okadiac acid, *Dinophysis* toxins, yessotoxin, pectenotoxins	*Dinophysis* spp., *Prorocentrum* spp.	Mussels, cockles, oysters	Europe, Japan, USA South America, Australia, New Zealand, Indonesia
NSP (neurotoxic)	Numbness in mouth and extremities, ataxia, vomiting, diarrhoea, nausea, abdominal pain, hot/cold temperature reversal, paralysis (death)	Brevetoxins, Breve-like toxins	*Ptychodiscus brevis, Psedonitzschia australis*	Oysters, clams, mussels	USA, Caribbean, Spain, New Zealand
ASP (amnesic)	Diarrhoea, vomiting, abdominal cramps, headache, disorientation, short-term memory loss, paralysis (death)	Domoic acid	*Pseudonitzschia* spp.	Oysters, clams	Canada, USA

(Contd.)

Table 30.1   (Continued)

Disease	Clinical symptoms	Toxins	Associated microbes	Shellfish /fish	Occurrence
Ciguatera	Tingling/ numbness in mouth and extremities, hot/cold temperature reversal, cramps malaise, ataxia, temporary blindness (death)	Ciguatoxins Maitotoxins Scaritoxins Palytoxins	*Gambierdiscus toxicus* *Amphidinium* spp. *Ostreopsis* spp. *Prorocentrum* spp.	Groupers, sea bass, snappers, jacks, eels, barracudas, mullet, parrot fish, surgeon fish	Indigenous in tropical regions

REVIEW QUESTIONS

1. What are dinoflagellates?
2. Give the significance of marine algal toxins and their syndromes.
3. Write short notes on
   i. paralytic shellfish poisoning.
   ii. diarrhoetic shell fish poisoning.
   iii. neurotoxic shellfish poisoning.
   iv. amnesic shellfish poisoning.
   v. ciguatera poisoning.

31

FOOD HYGIENE, FOOD REGULATION AND STANDARDS

INTRODUCTION

The basic purpose of food laws and food regulatory agencies is to ensure that all foods reaching the consumer are safe, wholesome and are truthfully labelled. The production and marketing of wholesome foods should be important to everyone affiliated with the food industry.

ADVANTAGES OF THE FOOD LAWS/FOOD CONTROL SERVICES

1. Protection of the consumer against health risks.

2. Reduction of food loss—well trained inspection machinery can provide guidance for the judicious use of pesticides and food additives which can prolong the shelf life of the food.

3. Prevention of gross food adulteration.

4. Control over contaminants and additives—the level of use of pesticides or contaminants (microbial or non-microbial) has to be maintained within the limits of no effect level.

5. Control over dumping of substandard foods. Substandard adulterated or contaminated foods may be dumped in a country which does not have the requisite food laws.

6. Food control can basically contribute to the nutritional improvement of a country.

7. Improvement of foreign exchange—The exporting country can manufacture and label according to the requisite of Food laws of the importing country.

FOOD HYGIENE

Food is a potential source of infection and is liable to contamination by microbes, at any point during its journey from the producer to the consumer. Food hygiene may be defined as the sanitary science which aims to produce food which is safe for the consumer and of good keeping quality. It covers a wide field and includes the rearing, feeding, marketing and slaughter of animals as well as the sanitation procedures designed to prevent bacteria of human origin reaching foodstuff. Food hygiene, in its widest sense, implies hygiene in the production, handling,distribution and serving.

WHO (1984) has defined Food Hygiene or Food Safety as all conditions and measures that are necessary during the production, processing, storage, distribution and preparation of food to ensure that it is safe, sound, wholesome and fit for human consumption. The primary aim of food hygiene is to prevent food poisoning and other food-borne illness. The objective of food control has three aspects— economic, aesthetic and public health.

The laws and regulations are intended to protect the public against injury to health and fraud and deceit. The laws and regulations restrain the sale of foods which are decomposed, adulterated, improperly preserved or misbranded. They provide specifications for safe handling of such perishable and potentially dangerous foods such as milk or milk products and meats. Methods and procedures have been developed to minimize the danger of contamination of foods with poisonous substances or with pathogenic microbes. Prevention requires the team of personnel consisting of an epidemiologist, lab technician and sanitarian.

Different branches of food hygiene include:

1. Milk hygiene
2. Meat hygiene
3. Fish hygiene
4. Egg hygiene
5. Hygiene of vegetables and fruits
6. Food-handlers hygiene
7. Sanitation of eating place

Milk Hygiene

Milk is an efficient vehicle for a great variety of disease agents. Milk gets contaminated by various sources like udder, utensils, personal hygiene of the handlers, storage environment, water, etc. This may lead to various milk-borne diseases that may affect the population. Requisites in the production of clean and safe milk are

1. Healthy and clean animal—Milk from a healthy udder contains only few microorganisms.
2. Sanitary conditions of the dairy farm, i.e., the premises where the animal is housed and milked should be sanitary.

3. Containers and equipment should be sterile and kept covered.

4. Water supply should be bacteriologically safe.

5. Milk handlers must be free from communicable diseases and before milking they must wash their hands and arms. Milking machines should be used as far as possible.

6. Milk must be cooled immediately to 10°C after it is drawn to retard bacterial growth.

7. Milk should be properly pasteurized to increase shelf life.

Meat Hygiene

A number of diseases are transmitted through meat and meat-based foods since animal tissues are important vehicles for transmission of various protozoan diseases like Taeniasis, Trichinellosis and a number of bacterial infections.

Meat inspection is a very important process before being accepted or rejected. Two types of inspection, i.e., Ante-mortem and post-mortem inspection are being carried out.

Ante-mortem rejection   It is based on emaciation, exhaustion, pregnancy, sheep-pox, foot-rot actinomycosis, brucellosis, febrile conditions, diarrhoea and other diseases.

Post-mortem rejection   It is based on *cysticercus bovis*, liver fluke, abscesses, sarcocystis, hydatidosus, septicaemia, parasitic and nodular infection of liver and lungs, tuberculosis and *cysticercus cellulosae*.

Good meat qualities   The meat should be firm and elastic to touch, not be pale pink or deep purple colour.

Slaughter house hygiene   Hygiene of slaughter house is of paramount importance to prevent the contamination of meat during the process of dressing. There is a Model Public Health Act (1955) in India, which standardizes on the location, structure, disposal of wastes, water supply, examination of animal, storage of meat, transportation of meat and miscellaneous other activities connected with meat processing.

Fish Hygiene

Fishes are also important agents for disease transmission since they are perishable foods. It is an intermediate host of tapeworm and consumption of spoilt fishes may lead to fish poisoning. Fresh fish is in a state of stiffness or rigor mortis with bright red gills, and clear and prominent eyes. Fishes also get contaminated through various means during catching, transporting and processing. Prevention of contamination is of utmost importance at each level of processing for increasing the shelf life and quality of fish.

Egg Hygiene

Majority of freshly laid eggs are sterile. Shells of eggs become contaminated by faecal matter from the hens or ducks. The microorganisms can penetrate the shell of the egg, cross the various chemical barriers to reach the yolk leading to spoilage of the egg. Eggs must be stored in a dry condition for preventing spoilage. Eggs can also be pasteurized to increase the shelf life.

Vegetable and Fruit Hygiene

Vegetable and fruits host many pathogens like bacterial, fungal, protozoan which can enter the plant materials during or after their harvest. Standards have been laid down for effective storage of the vegetable and fruits to prevent their spoilage and further disease transmission.

Hygiene for Food-handlers

Food sanitation rests directly upon the state of personal hygiene and habits of the person working in food industries. They may transmit infections like diarrhoea, dysenteries, typhoid, enteroviruses, viral hepatitis, protozoan cysts, eggs of helminthes, staphylococcal/streptococcal infections. Simple rules to be followed in food handling are given below.

1. Medical examination to be carried out at the time of employment. Persons with above diseases or communicable diseases (TB) are not to be employed.
2. Persons with wounds, otitis media, skin infections, etc. should not be permitted to handle food or utensils.
3. The day-to-day health appraisal of food-handlers is important. Those who are ill should be excluded from food handling.
4. Any illness which occurs in a food-handler's family, should be at once notified.
5. Education of food-handlers in matters of personal hygiene, food handling, utensil/dishwashing and insect or rodent control is the best mean of promoting food hygiene.
6. Personal hygiene to be promoted:
   i. Hands—scrubbed and washed with soap and water immediately after visiting lavatory and as often as necessary at other times. Nails to be kept trimmed and free from dirt.
   ii. Hair—to provide covering to the head.
   iii. Overalls—clean white overalls to be worn by all food-handlers.
   iv. Habits—coughing and sneezing in the vicinity of food, licking the fingers before picking up an article of foods, smoking on food premises are to be avoided.

Hygiene in Public Eating Places

The principles of food hygiene in public eating or drinking places are the same as in any other type of food handling activity. The complexity of the problem is mainly due to the ever increasing number of establishments which requires some supervision. At one time, public eating was restricted to special occasions, to the traveller, patient, boarding schools, etc. Due to increased urbanization, distance of working places, mobility and employment, many resort to dining in public eating places. The six minimum essentials for cleanliness and sanitation for public eating or drinking places are as follows:

1. Avoid hand contact with food as far as practicable.
2. Keep perishable food below 40°F or above 140°F.
3. Keep food protected from personal contact and contaminating insects, dust and animals.
4. Discard all food and food products which are not in good quality.
5. Clean and disinfect equipment that comes into direct contact with food.
6. Keep premises presentable and in a sanitary manner at all times.

Sanitation of eating establishments is a challenging problem in India. The Model Public Health Act, Govt. of India (1955) has suggested the following minimum standards for restaurants and eating places in India.

Location   Shall not be near any accumulation of filth or open drain, stable, manure pit and other sources of nuisances.

Floors   To be higher than the adjoining land, made with impervious material and easy to keep clean.

Rooms   Rooms where meals are served shall not be less than 100 sq. feet and shall provide accommodation for a maximum of 10 persons. Walls up to 3 feet should be smooth with rounded corners and should be impervious and easily washable. Lighting and ventilation should be ample with natural and artificial lighting systems along with a good circulation of air.

Kitchen   It should be with ample floor space, window opening, proper flooring and ventilation.

Storage of cooked food   Separate rooms to be provided for storing cooked foods. For long storage, control of temperature is necessary.

Furniture   Furniture should be reasonably strong and easy to keep clean and dry.

Disposal of refuse   Refuse to be collected in covered and impervious bins and disposed off twice a day.

Water supply   It should be an independent source, adequate, continuous and safe.

Washing facilities   Cleaning of utensils or crockery to be done in hot water and followed by disinfection.

FOOD CONTROL ADMINISTRATION

This varies from country to country. The central, state and local bodies are responsible to enact laws and implementations. There should be laws to control raw materials, processing of foods, preservation methods and control of milk or meat hygiene, etc. Laws relating to food control in India are

1. The Prevention of Food Adulteration Act 1954, amended in 1976, 1986.
2. Prevention of Food Adulteration Rules, 1955.
3. Indian Penal Code 1860 as amended.
4. Cantonment Act 1924.
5. The Bombay Municipal Corporation Act, 1888.
6. The Bombay Provincial Municipal Corporations Act, 1949.
7. The Maharashtra Zilla Parishads and Panchayat Samiti's Act, 1961.
8. The Maharashtra Municipalities Act, 1965.
9. Tamil Nadu Public Health Act, 1939, and so on for each State.

Genesis of the PFA Act

Even prior to 1954, many of the states had their own food laws. "First law was followed in Bombay in 1899 to combat the evil of food adulteration. Until 1954, 22 out of the 28 states had their own food laws but they were not of the uniform pattern and hence uniform implementation was not possible. The Central Advisory Board of Health suggested the adoption of Central legislation. In October 1952, the Food Adulteration Bill was introduced in the Lok Sabha by the Health Minister and after approval of both the Houses, it was enacted in 1954 as the Prevention of Food Adulteration Act by the Indian Parliament which came into effect on 1st June, 1955. In 1972, the terms of the PFA Act were extended subsequently to Jammu and Kashmir.

Provisions of the Original PFA Act, 1954

1. Definition of food
2. Definition of Food Adulteration or misbranding
3. Appointment of an Advisory Committee called the Central Committee for Food standards (CCFC)
4. Establishment of the Central Food Laboratory, Calcutta, in 1955 and later in each state to give a final opinion in cases challenged in the court of law.

5. Restriction on imports of adulterated or misbranded foods or other spoilt foods.

6. Power of the State Governments to appoint Public Analysts or Food Inspectors.

7. Procedure for Food Inspectors in drawing and dispatching sample of food to laboratory.

8. Powers to the Central Government for defining the standards of quality, control, over production, distribution, sales, packing, labelling, etc.

9. Penal provisions provided a maximum imprisonment of one year or a minimum fine of Rs. 2000/- in the first instance and imprisonment of 2 years on the second offence and in the third instance, imprisonment up to 4 years.

Food Standards

1. Codex Alimentarius is a collection of international food standards prepared by Codex Alimentarius Commission (Organ of FAO/WHO food standards programme). Indian standards are based on this.

2. PFA standards based on PFA act 1954 "Central Committee of Food Standards" revised periodically to get minimum level of quality of food stuff attainable under Indian conditions.

3. AGMARK (Agricultural Marketing, HO, Faridabad) standards set by the Director of Marketing and Inspection, Government of India. The Agmark gives consumers the assurance of quality in accordance with standards laid down.

4. ISI standards, guarantees good quality in accordance with standard prescribed by Indian Standard Institution for that commodity.

5. ASC specifications for Armed Forces.

1. What are the advantages of food laws?

2. What are the implications of food hygiene?

3. Brief on the different branches of food hygiene.

4. Write a note on hygiene for food handlers.

5. Throw some light on minimum standards for restaurants and eating places in India.

6. What are the laws related to food control administration?

32

INVESTIGATION OF AN OUTBREAK OF FOOD POISONING

INTRODUCTION

Sudden occurrence of an acute food-borne disease in a group within a short period of time among individuals who have consumed one or more foods in common is called an outbreak. It is very rare in household cooking and occurs mostly on festive occasions when large gatherings partake of the meal. Whenever there is an increase in the environmental temperature, there is an increase in outbreaks.

Investigation of an outbreak is carried out by a team of three members: a clinician, a pathologist and an epidemiologist. The investigation must be done faster since delay may result in destruction of valuable evidence due to deterioration of specimens. The aim of any investigation is to implicate the exact item of food and to ascertain circumstances leading to the contamination. The objective of any investigation is to prevent recurrence of the outbreak. There are three stages in investigation: epidemiological, circumstantial and laboratory investigation.

EPIDEMIOLOGICAL INVESTIGATION

The purpose of epidemiological investigation is to

1. Obtain a complete history of the outbreak.
2. Get a complete list of persons affected or unaffected, who consumed the incriminated meal together with a list of various items of food served at the said meal.
3. Interview each person and to examine all cases and to obtain clinical reports.
4. Take a complete food history with a view to identify the food items consumed by them.
5. Write the history of outbreak that includes:

- identity of the persons affected by the illness
- the sequence of events leading to the outbreak
- the time lag between consumption of food and onset of the illness
- signs and symptoms
- duration of illness

6. Do a tentative diagnosis from the above data.

If the number of persons who ate the meal are large or not definitely known and if taking all history is not feasible then a representative sample of the population is taken and investigated.

For example, a list of articles consumed during the meal under suspicion is obtained and a table is constructed as in Table 32.1.

Table 32.1 Basic criteria involved in the epidemiological investigation

Food	Persons who ate the specific item of food			Persons who did not eat the specific item of food			Differences between attack rates
	Number who ate	Number affected	Attack rate %	Number who ate	Number affected	Attack rate %	
Mutton curry	109	80	73	21	14	66	+7
Dal	88	70	79	32	24	75	+4
Pulao (rice)	95	76	80	25	18	72	+8
Raita (curd)	98	84	85	22	2	9	+76
Potato pea curry	78	60	77	42	20	48	+29
Chutney	65	13	20	55	27	48	– 28
Papad	60	6	10	60	24	40	– 30
Sweet	110	77	75	10	6	60	– 15

The general attack rate is 72%. If the attack rates among those who ate individual items of food are examined, it is difficult to identify the food (raita/curd which most probably was responsible for the outbreak), because the attack rates in respect of some of the other items of food were not very different except those for chutney and papad. At the most, only a negative conclusion can be drawn from these attack rates, i.e., chutney or papad was not responsible for the outbreak. But when two groups are contrasted, it becomes easier to identify "raita" as being the most likely food that caused the outbreak. In theory, all those who ate the particular item, which caused the outbreak, should be affected by the illness, but in practice, the association between illness and the food is seldom so perfect and the reasons are

- some persons who ate the food were resistant to the infectious agent and hence not affected.

- one item of food may be contaminated by traces of another during preparation, storage, handling or serving.

- the history of the consumption of individual items of food may be incorrect, due to lapses of memory, misunderstanding of the questions put or any other reason.

LABORATORY INVESTIGATIONS

It is an important part of the investigation. Objectives of the laboratory investigations are:

1. Incriminate the causative agent from stool, vomit or remnants of food by inoculating into appropriate media.

2. Determine the total number of bacteria and the relative numbers of each kind involved. This will give a better indication of the organism involved. Stool samples of the kitchen employees and food-handlers should also be investigated. The samples should be examined aerobically and anaerobically. Phage typing of the organisms should be done to complete the laboratory investigations.

The food samples must be packed in ice and sent by short route with full report of outbreak (brief history, number attacked, time, symptoms and present condition of patients).

Animal experiments   This may be done by feeding rhesus monkeys with the remnants of the food. Protection tests are useful in the case of botulism. Saline infiltrate of the foodstuff is injected subcutaneously into mice protected with antitoxic sera keeping suitable controls.

Blood for antibodies   Blood collected may be useful for retrospective studies.

Environmental study   This includes inspection of the eating places, kitchen(s) and questioning of food-handlers regarding food preparation.

Data analysis   The data should be analysed according to the descriptive methods of time, place and person distribution. Food specific attack rates should be calculated and a case control study may be undertaken to establish the epidemiological association between illness and the intake of a particular food.

CIIRCUMSTANTIAL ENQUIRY

This is done to ascertain the circumstances under which the contamination of the suspected item and the outbreak occurred. The following strategy may be employed:

1. Study the source of infection or contamination.
2. Study the history of the implicated food and source of article suspected to be ascertained.
3. Place of preparation to be inspected for general cleanliness.
4. Ascertain the amount of handling the food before, during and after cooking.
5. Examine the source of supply, method of storage of the food item.
6. Check for the presence of any rodents
7. Check the health of workers—clinically and pathologically.

The above enquiry along with the laboratory results will give complete information pertaining to:

- The aetiological agent
- The source of contamination
- The article of food responsible
- The circumstances leading to contamination
- The incidence, severity and mortality rate and
- Remedial measures recommended

PREVENTION AND CONTROL OF FOOD POISONING

Food Sanitation

Meat inspection   Animals must be free from infection and veterinary staff should examine the animals both before and after slaughter.

Personal hygiene   A high standard of personal hygiene among individuals engaged in handling, preparation and cooking of food is needed.

Food-handlers   Those suffering from infected wounds, boils, diarrhoea, dysentery, throat infection, etc., should be excluded from food handling. The medical inspection of food-handlers is required in many countries but is of limited value in the detection of carriers as it will remove some sources of infection.

Food-handling techniques   The handling of ready-to-eat foods with bare hands should be reduced to a minimum. Time between preparation and consumption of food should be kept short. The importance of rapid cooling and cold storage must be stressed. Food must be thoroughly cooked. The heat must penetrate the centre of the food leaving no cold spots. Most food poisoning organisms are killed at temperatures over 60°C.

Sanitary improvements   Sanitization of all work surfaces, utensils and equipment must be ensured. Food premises should be kept free from rats, mice or flies, and dust.

Health education   Food-handlers should be educated in matters of clean habits and personal hygiene such as frequent and thorough handwashing.

Refrigeration

In the prevention of bacterial food poisoning, emphasis must be placed on proper temperature control. Food should not be left in warm utensils. A few germs can multiply to millions the next morning. Foods not eaten immediately should be kept in cold storage to prevent bacterial multiplication and toxin production. "Cook and Eat the Same Day" golden rule must be followed. When foods are held between 10°C and 49°C, they are in the danger zone—prone to bacterial growth. Cold storage is bacteriostatic at temperatures below 4°C. The refrigeration temperature should not exceed this level.

Surveillance

Food samples must be obtained from food establishments periodically and subjected to laboratory analysis if they are unsatisfactory. Continuing surveillance is necessary to avoid outbreaks of food-borne diseases.

1.   Define an outbreak.
2.   Describe the three stages of a food-borne investigation.
3.   List out the various strategies for the prevention and control of food poisoning.

Unit V

MICROBIAL FOOD FERMENTATION

INTRODUCTION

All food raw materials are contaminated by microorganisms, which take part in the mineralization of organic materials in nature. The microbial reactions mostly resulted in spoilage of the food. However, man learnt some preservation techniques to extend their shelf life. These preservation methods were mainly based on drying or fermentation. Food fermentations are still used to produce so-called fermented foods, but today preservation is not the main objective of the fermentation, but it is rather the specific taste and texture that is the goal of the fermentation.

Microorganisms are the key building blocks of fermented foods and beverages. The final product of fermentation is the result of unique chemical, physical and biological interactions between specific microorganisms and the food. Although fermented foods and beverages have been associated with human culture for several thousand years, their microbiology still holds many mysteries and challenges for development. The functional role of microorganisms in the fermentation process has expanded in recent years, as we have learned more about the linkages between the microbiology and biochemistry of the process, and product quality. Whereas fermentation was once seen as a "generic" process to preserve raw material and transform its sensory characteristics, now, it is intentionally and specifically used to add finer functional assets such as increased nutritional, health, psychological and probiotic benefits, as well as develop regional nuances and individuality with respect to sensory character. Understanding the microbial ecology of the fermentation is the first stage in developing and managing the process to increase product functionality, appeal, safety and value.

Fundamental questions to be answered are: what species and strains are present, which ones are essential, which ones are detrimental, where do they come from, what factors affect their survival and growth, what are their activities and how do these impact on production and product quality? Through the process of ecological discovery, it should be possible to develop starter culture technology, and use this as a means to guide and control the manufacturing operation, and manage product quality. For most fermented products, discovering the ecology presents a scientific challenge because of the diversity of microbial community involved and the chemical and physical complexity of the raw material matrix. The microbiotic is likely to comprise a mixture of bacterial, yeast and fungal species as well as their viruses. Moreover, it is now widely recognized that the influence of any particular organism on the quality of the product extends deeper than species and ecological discovery needs to include strain characterization and differentiation within a species. Molecular analytical technology is now available to determine microbial ecology at the strain level, and to determine which genes are expressed under process conditions.

With this information, it should be possible to optimize the process of production in terms of economic efficiency, product quality, product functionality and product value. The development and use of specific starter culture technology to guide and manage the process have been logical and practical outcomes of ecological research. With this technology, it has been possible to transform processes that once relied on indigenous or "spontaneous" fermentations into well controlled, scaled-up, industrial operations. Notable examples would be beer, wine and cheese production but, even with these modern operations, ecological research has revealed that indigenous flora still impact on process and product.

For most fermented products, discovering their ecology presents a significant scientific challenge because (i) the microbial communities involved are usually quite diverse and (ii) the physical, chemical and biological properties of the food matrix are generally complex. The microbiota is likely to comprise a mixture of bacterial, yeast and fungal species, as well as the viruses that destroy them. Moreover, it is now widely recognized that the influence of any particular organism on product quality extends deeper than species, and ecological discovery must now include strain characterization and differentiation within a species. For many products, the "food matrix" is still biologically active, with endogenous enzymes having a significant input into the total process. This activity adds another layer to the overall complexity.

BASIC PRINCIPLES OF FERMENTATION

The Diversity of Fermented Foods

Numerous fermented foods are consumed around the world. Each nation has its own types of fermented food, representing the staple diet and the raw ingredients available in that particular place. Although the products are well known to the individuals, they may not know the fermentation processes involved. Indeed, it is likely that the methods of producing many of the world's fermented foods are unknown and came about by chance. Some of the more obvious fermented fruit and vegetable products are the alcoholic beverages—beers and wines. However, several more fermented fruit and vegetable products arise from lactic acid fermentation and are extremely important in meeting the nutritional requirements of a large proportion of the world's population. Table 33.1 contains examples of fermented fruit and vegetable products from around the world.

ORGANISMS RESPONSIBLE FOR FOOD FERMENTATIONS

The most common groups of microorganisms involved in food fermentations are bacteria, yeasts and moulds.

Bacteria

Several bacterial families are present in foods, the majority of which are concerned with food spoilage. As a result, the important role of bacteria in the fermentation of foods is often overlooked. The most important bacteria in desirable food fermentations are the Lactobacillaceae which have the ability to produce lactic acid from carbohydrates. Other important bacteria, especially in the fermentation of fruits and vegetables, are the acetic acid producing *Acetobacter* species.

Yeasts

Yeasts and yeast-like fungi are widely distributed in nature. They are present in orchards and vineyards, in the air, the soil and in the intestinal tract of animals. Like bacteria and moulds, yeasts can have beneficial and non-beneficial effects in foods. The most beneficial yeasts in terms of desirable food fermentation are from the *Saccharomyces* family, especially *S. cerevisiae*. Yeasts are unicellular organisms that reproduce asexually by budding. In general, yeasts are larger than most bacteria. Yeasts play an important role in the food industry as they produce enzymes that favour desirable chemical reactions such as the leavening of bread and the production of alcohol and invert sugar.

Table 33.1   Fermented foods from around the world

Name and region	Type of product
Indian sub-continent	
Acar, Achar, Tandal achar, Garam nimboo achar	Pickled fruit and vegetables
Gundruk	Fermented dried vegetable
	Lemon pickle, mango pickle

(Contd.)

Table 33.1 (Continued)

Name and region	Type of product
Asinan, Burong mangga, Dalok, Jeruk, Kiam-chai, Kiam-cheyi, Kong-chai, Naw-mai-dong, Pak-siam-dong, Paw-tsay, Phak-dong, Phonlami-dong, Sajur asin, Sambal tempo-jak, Santol, Si-sek-chai, Sunki, Tang-chai, Tempoyak, Vanilla,	Pickled fruits and vegetables
Bai-ming, Leppet-so, Miang	Fermented tea leaves
Nata de coco, Nata de pina	Fermented fruit juice
East Asia	
Bossam-kimchi, Chonggak-kimchi, Dan moogi, Dongchimi, Kachdoo kigactuki, Kakduggi, Kimchi, Mootsanji, Muchung- kimchi, Oigee, Oiji, Oiso baegi, Tongbaechu- kimchi, Tongkimchi, Totkal kimchi,	Fermented in brine
Cha- ts'ai, Hiroshimana, Jangagee, Nara senkei, Narazuke, Nozawana, Nukamiso-zuke, Omizuke, Pow tsai, Red in snow, Seokbakji, Shiozuke, Szechwan cabbage, Tai-tan tsoi, Takana, Takuan, Tsa Tzai, Tsu, Umeboshi, Wasabi-zuke, Yen tsai	Pickled fruit and vegetables Hot pepper sauce
Africa	
Fruit vinegar	Vinegar
	Hot pepper sauce
Lamoun makbouss, Mauoloh, Msir, Mslalla, Olive	Pickled fruits and vegetables
Oilseeds, Ogili, Ogiri, Hibiscus seed	Fermented fruits and vegetable seeds
Wines	Fermented fruits
America	
Cucumber pickles, Dill pickles, Olives, Sauerkraut	Pickled fruits and vegetables
Lupin seed, Oilseeds,	Pickled oilseed
Vanilla, Wines	Fermented fruits and vegetables
Middle East	
Kushuk	Fermented fruits and vegetables
Lamoun makbouss, Mekhalel, Olives, Torshi, Tursu	Pickled fruit and vegetables
Wines	Fermented fruits
Europe and World	
Mushrooms, Yeast	Moulds
Olives, Sauerkohl, Sauerruben	Pickled fruit and vegetables
Grape vinegar, Wine vinegar	Vinegar
Wines, Citron	Fermented fruits

Moulds

Moulds are also important in the food industry, both as spoilers and preservers of foods. Certain moulds produce undesirable toxins and contribute to the spoilage of foods. The *Aspergillus* species are often responsible for undesirable changes in foods. These moulds are frequently found in foods and can tolerate high concentrations of salt and sugar. However, others impart characteristic flavours to foods and some others produce enzymes, such as amylase for bread making. Moulds from the genus *Penicillium* are associated with the ripening and flavouring of cheeses. Moulds are aerobic and therefore require oxygen for growth. They also have the greatest array of enzymes, and can colonize and grow on most types of food. Moulds do not play a significant role in the desirable fermentation of fruit and vegetable products.

When microorganisms metabolize and grow they release by-products. In food fermentations, the by-products play a beneficial role in preserving and changing the texture and flavour of the food substrate. For example, acetic acid is the by-product of the fermentation of some fruits. This acid not only affects the flavour of the final product, but more importantly has a preservative effect on the food. For food fermentations, microorganisms are often classified according to these by-products. The fermentation of milk to yoghurt involves a specific group of bacteria called the lactic acid bacteria (*Lactobacillus* species). This is a general name attributed to those bacteria which produce lactic acid as they grow. Acidic foods are less susceptible to spoilage than neutral or alkaline foods and hence the acid helps to preserve the product. Fermentations also result in a change in texture. In the case of milk, the acid causes the precipitation of milk protein to a solid curd.

Enzymatic Activity

The changes that occur during fermentation of foods are the result of enzymatic activity. Enzymes are complex proteins produced by living cells to carry out specific biochemical reactions. They are known as catalysts since their role is to initiate and control reactions, rather than being involved in a reaction. Because they are proteinaceous in nature, they are sensitive to fluctuations in temperature, pH, moisture content, ionic strength and concentrations of substrate and inhibitors. Each enzyme will operate most efficiently in certain conditions. Extremes of temperature and pH will denature the protein and destroy enzyme activity. Because they are so sensitive, enzymatic reactions can easily be controlled by slight adjustments to temperature, pH or other reaction conditions. In the food industry, enzymes have several roles—the liquefaction and saccharification of starch, the conversion of sugars and the modification of proteins. Microbial enzymes play a role in the fermentation of fruits and vegetables.

Sequence of Microorganisms in Fermentation

Nearly all food fermentations are the result of more than one microorganism, either working together or in a sequence. For example, vinegar production is a joint effort between yeast and acetic acid forming bacteria. The yeast converts sugars to alcohol, which is the substrate required by the *Acetobacter* to produce acetic acid. Bacteria from different species and the various microorganisms—yeasts and moulds—all have their own preferences for growing conditions, which are set within narrow limits. There are very few pure culture fermentations. An organism that initiates fermentation will grow there until its by-products inhibit further growth and activity. During this initial growth period, other organisms develop that are ready to take over when the conditions become intolerable for the former ones.

In general, growth will be initiated by bacteria, followed by yeasts and then moulds. There are definite reasons for this type of sequence. The smaller microorganisms are the ones that multiply and take-up nutrients from the surrounding area most rapidly. Bacteria are the smallest of microorganisms, followed by yeasts and moulds. The smaller bacteria, such as *Leuconostoc* and *Streptococcus* grow and ferment more rapidly than their close relatives and are therefore often the first species to colonize a substrate. Table 33.2 lists some of the microorganisms found in frutis and vegetables.

Table 33.2 Microorganisms commonly found in fermenting fruits and vegetables

Organism	Type	Optimum conditions	Reactions
Bacteria			
Acetobacter 　　*A. aceti* 　　*A. pasteurianus* 　　*A. peroxydans*	Aerobic rods	$a_W \geq 0.9$	Oxidize organic compounds (alcohol) to organic acids (acetic acid). Important in vinegar production.
Streptococcus 　　*S. faecalis* 　　*S. bovis* 　　*S. thermophilus*	Gram-positive cocci	Acid-tolerant $a_W \geq 0.9$	Homofermentative. Most common in dairy fermentations, but *S. faecalis* is common in vegetable products. Tolerate salt and can grow in high pH media.
Leuconostoc 　　*L. mesenteroides* 　　*L. dextranicum* 　　*L. paramesenteroides* 　　*L. oenos*	Gram-positive cocci		Heterofermentative. Produce lactic acid and acetic acid, ethanol and carbon dioxide from glucose. Small bacteria, therefore have an important role in initiating fermentations. *L. oenos* is often present in wine. It can utilize malic acid and other organic acids.
Pediococcus 　　*P. cerevisiae* 　　*P. acidilactici* 　　*P. pentosaceus*			Saprophytic organisms found in fermenting vegetables, mashes, beer and wort. Produce inactive lactic acid.

(Contd.)

Table 33.2 (Continued)

Organism	Type	Optimum conditions	Reactions
Lactobacillus *L. delbrueckii* *L. leichmannii* *L. plantarum* *L. lactis* *L. acidophilus* *L. brevis* *L. fermentum* *L. buchneri*	Gram-positive rods. Non-motile	Acid-tolerant $a_W \geq 0.9$	Metabolize sugars to lactic acid, acetic acid, ethyl alcohol and carbon dioxide. Saprophytic organisms. Produce greater amounts of acid than the cocci. The genus is split into two types: homo- and heterofermenters. Homofermentative. Produce only lactic acid. *L. plantarum* is important in fruit and vegetable fermentation. Tolerates high salt concentration. Heterofermentative. Produce lactic acid (50%) plus acetic acid (25%) ethyl alcohol and carbon dioxide (25%). *L. brevis* is the most common. Widely distri-buted in plants and animals. Partially reduces fructose to mannitol.
Yeasts			
Saccharomyces *S. cerevisiae* *S. pombe*	Many aerobic, some anaerobes	Tolerate acid, 40% sugar $a_W \geq 0.85$ pH 4–4.5 20–30°C	*S. cerevisiae* can shift its metabolism from a fermentative to an oxidative pathway, depending on oxygen availability. Most yeasts produce alcohol and carbon dioxide from sugars.
Debaryomyces *Zygosaccharomyces rouxii* *Candida species* *Geotrichum candidum*			Tolerant of high salt concentrations and low a_W.

DESIRABLE FERMENTATIONS

It is essential with any fermentation to ensure that only the desired bacteria, yeasts or moulds start to multiply and sgrow on the substrate. This has the effect of suppressing other microorganisms which may be either pathogenic and cause food poisoning or will generally spoil the fermentation process, resulting in an end product which is neither expected nor desired. An everyday example used to illustrate this point is the differences in spoilage between pasteurized and unpasteurized milk. Unpasteurized milk will spoil naturally to produce a sour tasting product which can be used in baking to improve the texture of certain breads. Pasteurized milk, however, spoils (non-desirable fermentation) to produce an unpleasant product which cannot be used. The reason for this difference is that pasteurization (despite being a very important process to destroy pathogenic microorganisms) changes the environment of the microorganism and if pasteurized milk is kept unrefrigerated for some time, undesirable microorganisms start to grow and multiply before the desirable ones. In the case of unpasteurized milk, the non-pathogenic lactic acid bacteria start to grow and multiply at a greater rate than any pathogenic bacteria. Not only do the larger numbers of lactic acid bacteria compete more successfully for the available nutrients, but as they grow they produce lactic acid which increases the acidity of the substrate and further suppresses the bacteria which cannot tolerate an acid environment.

Most food spoilage organisms cannot survive in either alcoholic or acidic environments. Therefore, the production of both of these end products can prevent a food from spoilage and extend the shelf life. Alcoholic and acidic fermentations generally offer cost-effective methods of preserving food in developing countries, where more sophisticated means of preservation are unaffordable.

MANIPULATION OF MICROBIAL GROWTH AND ACTIVITY

The principles of microbial action are identical both in the use of microorganisms in food preservation, such as through desirable fermentations, and also as agents of destruction via food spoilage. The type of organisms present and the environmental conditions will determine the nature of the reaction and the ultimate products. By manipulating the external reaction conditions, microbial reactions can be controlled to produce desirable results. There are several means of altering the reaction environment to encourage the growth of desirable organisms. These are discussed below.

There are six major factors that influence the growth and activity of microorganisms in foods—moisture, oxygen concentration, temperature, nutrients, pH and inhibitors. The food supply available to the microorganisms depends on the composition of the food on which they grow. All microorganisms differ in their ability to metabolize proteins, carbohydrates and fats. Obviously, by manipulating any of these six factors, the activity of microorganisms within foods can be controlled.

Moisture

Water is essential for the growth and metabolism of all cells. If it is reduced or removed, cellular activity is decreased. For example, the removal of water from cells by drying or the change in state of water (from liquid to solid) affected by freezing, reduces the availability of water to cells (including microbial cells) for metabolic activity. The form in which water exists within the food is important as far as microbial activity is concerned. There are two types of water—free and bound. Bound water is present within the tissue and is vital to all the physiological processes within the cell. Free water exists in and around the tissues and can be removed from cells without seriously interfering with the vital processes. Free water is essential for the survival and activity of microorganisms. Therefore, by removing free water, the level of microbial activity can be controlled. The amount of water available for microorganisms is referred to as the water activity (a_w). Pure water has a water activity of 1.0. Bacteria require more water than yeasts, which require more water than moulds to carry out their metabolic activities. Almost all microbial activity is inhibited below an a_w of 0.6. Most fungi are inhibited below a_w of 0.7, most yeasts are inhibited below a_w of 0.8 and most bacteria below a_w 0.9. Naturally, there are exceptions to these guidelines and several species of microorganisms can exist outside the stated range. Table 33.3 provides information on water activity and microbial action. The water activity of foods can be changed by altering the amount of free water available. There are several ways to achieve this—drying to remove water; freezing to change the state of water from liquid to solid; increasing or decreasing the concentration of solutes by adding salt or sugar or other hydrophilic compounds (salt and sugar are the two common additives used for food preservation). Addition of salt or sugar to a food will bind free water and so decrease the a_w. Alternatively, decreasing the concentration will increase the amount of free water and in turn the a_w. Manipulation of the a_w in this manner can be used to encourage the growth of favourable microorganisms and discourage the growth of spoilage ones.

Table 33.3 Water activity for microbial reactions

a_w	Phenomenon	Examples
1.00		Highly perishable foods
0.95	*Pseudomonas, Bacillus, Clostridium perfringens* and some yeasts inhibited	Foods with 40% sucrose or 7% salt
0.90	Lower limit for bacterial growth. *Salmonella, Vibrio parahaemolyticus, Clostridium botulinum, Lactobacillus* and some yeasts and fungi inhibited	Foods with 55% sucrose, 12% salt. Intermediate-moisture foods (a_w = 0.90–0.55)
0.85	Many yeasts inhibited.	Foods with 65% sucrose, 15% salt
0.80	Lower limit for most enzyme activity and growth of most fungi. *Staphylococcus aureus* inhibited	Fruit syrups
0.75	Lower limit for halophilic bacteria	Fruit jams
0.70	Lower limit for growth of most xerophilic fungi	
0.65	Maximum velocity of Maillard reactions	

(Contd.)

Table 33.3 (Continued)

a_W	Phenomenon	Examples
0.60	Lower limit for growth of osmophilic or xerophilic yeasts and fungi	Dried fruits (15–20% water)
0.55	Deoxyribose nucleic acid (DNA) becomes disordered (lower limit for life to continue)	
0.50		Dried foods ($a_W = 0$–0.55)
0.40	Maximum oxidation velocity	
0.25	Maximum heat resistance of bacterial spores	

Oxidation–Reduction Potential

Oxygen is essential to carry out metabolic activities that support all forms of life. Free atmospheric oxygen is utilized by some groups of microorganisms, while others are able to metabolize the oxygen which is bound to other compounds such as carbohydrates. This bound oxygen is in a reduced form.

Microorganisms can be broadly classified into two groups—aerobic and anaerobic. Aerobes grow in the presence of atmospheric oxygen while anaerobes grow in the absence of atmospheric oxygen. In the middle of these two extremes are the facultative anaerobes which can adapt to the prevailing conditions and grow in either the absence or presence of atmospheric oxygen. Microaerophilic organisms grow in the presence of reduced amounts of atmospheric oxygen. Thus, controlling the availability of free oxygen is one way of controlling microbial activity within a food. In aerobic fermentations, the amount of oxygen present is one of the limiting factors. It determines the type and amount of biological product obtained, the amount of substrate consumed and the energy released from the reaction.

Moulds do not grow well in anaerobic conditions, therefore they are not important in terms of food spoilage or beneficial fermentation, in conditions of low oxygen availability.

Temperature

Temperature affects the growth and activity of all living cells. At high temperatures, organisms are destroyed, while at low temperatures, their rate of activity is decreased or suspended. Microorganisms can be classified into three distinct categories according to their temperature preference (Table 33.4).

Table 33.4 Classification of bacteria according to temperature requirements

Type of bacteria	Minimum temperature (°C)	Optimum temperature (°C)	Maximum temperature (°C)	General sources of bacteria
Psychrophilic	0 to 5	15 to 20	30	Water and frozen foods
Mesophilic	10 to 25	30 to 40	35 to 50	Pathogenic and non-pathogenic bacteria
Thermophilic	25 to 45	50 to 55	70 to 90	Spore-forming bacteria from soil and water

Nutritional Requirements

The majority of organisms are dependent on nutrients for energy and growth. Organisms vary in their specificity towards different substrates and usually colonize foods which contain the substrates they require. Sources of energy vary from simple sugars to complex carbohydrates and proteins. The energy requirements of microorganisms are very high. Limiting the amount of substrate available can check their growth.

Hydrogen Ion Concentration (pH)

The pH of a substrate is a measure of the hydrogen ion concentration. A food with a pH of 4.6 or less is termed as high acid or acid food and will not permit the growth of bacterial spores. Foods with a pH above 4.6 are termed low acid and will not inhibit the growth of bacterial spores. By acidifying foods and achieving a final pH of less than 4.6, most foods are resistant to bacterial spoilage.

The optimum pH for most microorganisms is near the neutral value (pH 7.0). Certain bacteria are acid-tolerant and will survive at reduced pH levels. Notable acid-tolerant bacteria include the *Lactobacillus* and *Streptococcus* species, which play a role in the fermentation of dairy and vegetable products. Moulds and yeasts are usually acid-tolerant and are therefore associated with spoilage of acidic foods.

Microorganisms vary in their optimal pH requirements for growth. Most bacteria favour conditions with a near neutral pH (7.0). Yeasts can grow in a pH range of 4 to 4.5 and moulds can grow from pH 2 to 8.5, but favour an acid pH. The varied pH requirements of different groups of microorganisms is used to have a good effect in fermented foods where successions of microorganisms take over from each other as the pH of the environment changes. For instance, some groups of microorganisms ferment sugars so that the pH becomes too low for the survival of those microbes. The acidophilic microorganisms then take over and continue the reaction. The affinity for different pH can also be used to prevent spoilage organisms.

Inhibitors

Many chemical compounds can inhibit the growth and activity of microorganisms. They do so by preventing metabolism, denaturation of the protein or by causing physical damage to the cell. The production of substrates as part of the metabolic reaction also acts to inhibit microbial action.

CONTROLLED FERMENTATION

Controlled fermentations are used to produce a range of fermented foods, including sauerkraut, pickles, olives, vinegar, dairy and other products. Controlled fermentation is a form of food preservation since it generally results in a reduction of acidity of the food, thus preventing the growth of spoilage microorganisms. The two most common acids produced are lactic acid and acetic acid, although small amounts of other acids such as propionic, fumaric and malic acid are also formed during fermentation.

It is highly probable that the first controlled food fermentations came into existence through trial and error and a need to preserve foods for consumption later in the season. It is possible that the initial attempts at preservation involved the addition of salt or seawater. During the removal of the salt prior to consumption, the foods would pass through stages favourable to acid fermentation. Although the process worked, it is likely that the causative agents were unknown. Today, there are numerous examples of controlled fermentation for the preservation and processing of foods. However, only a few of these have been studied in detail—these include sauerkraut, pickles, kimchi, beer, wine and vinegar production. Although the general principles and processes for many of the fermented fruits and vegetable products are the same, relying mainly on lactic acid and acetic acid forming bacteria, yeasts and moulds, the reactions have not been quantified for each product. The reactions are usually very complex and involve a series of microorganisms, either working together or in succession to achieve the final product.

REVIEW QUESTIONS

1. Discuss the basic principles of fermented foods.
2. Broadly classify the organisms important in food fermentations.
3. Give an account of the various fermented foods around the world.
4. Discuss the factors influencing the growth and activity of microorganisms in food.

34

FERMENTATION IN FOOD PROCESSING

INTRODUCTION

Fermented foods are food substrates that are invaded or overgrown by edible microorganisms whose enzymes, particularly amylases, proteases, lipases hydrolyse the polysaccharides, proteins and lipids respectively to non-toxic products with flavours, aromas and textures pleasant and attractive to the human consumer. If the products of enzyme activities have unpleasant odours or undesirable, unattractive flavours or the products are toxic or disease producing, the foods are described as spoiled.

Fermentation plays at least five roles in food processing. They are as follows:

1. Enrichment of the human dietary through development of a wide diversity of flavours, aromas and textures in food.
2. Preservation of substantial amounts of food through lactic acid, alcoholic, acetic acid, alkaline fermentations and high salt fermentations.
3. Enrichment of food substrates biologically with vitamins, protein, essential amino acids and essential fatty acids.
4. Detoxification during food fermentation processing.
5. A decrease in cooking time and fuel requirements.

Evolution of Indigenous Fermented Foods

Philosophy includes theory or investigation of the principles or laws that regulate the universe and underlie all knowledge and reality. Archaeology is the scientific study of the life and culture of ancient peoples. Anthropology is the study of races, physical and mental characteristics, distribution, customs, social relationships, and so on. When we start to study man's foods, we become involved in all the above. In fact, when we study fermented foods, we are studying the most intimate relationships among humans, microbes and foods. There is a never-ending struggle between man and microbes to see which will be first to consume the available food supplies.

Foods invaded by bacteria producing toxins or by fungi producing mycotoxins are dangerous to man. If the products of invasion are ill-smelling, off-flavoured or toxic, human consumers try to avoid them and the foods are described as spoiled. If the microbial products are pleasantly flavoured, have attractive aromas and textures and are non-toxic, the human consumer accepts them and they are designated as fermented foods. Certain flavours such as sweet, sour, alcoholic and meat-like (savoury) appeal to large numbers of humans. Milks sour naturally. Fruit and berry juices rapidly become alcoholic. Over many centuries, people have developed tastes for such products that continue on in modern man. Some anthropologists have suggested that it was stimulation and desire for alcohol that motivated man to settle down and become agriculturists.

CLASSIFICATION OF FOOD FERMENTATIONS

Food fermentations can be classified in a number of ways.

By categories   1. alcoholic beverages fermented by yeasts 2. vinegars fermented with *Acetobacter* 3. milks fermented with lactobacilli 4. pickles fermented with lactobacilli 5. fish or meat fermented with lactobacilli and 6. plant proteins fermented with moulds with or without lactobacilli and yeasts.

By classes   1. beverages, 2. cereal products, 3. dairy products, 4. fish products, 5. fruit and vegetable products, 6. legumes and 7. meat products.

By commodity 1. fermented starchy roots, 2. fermented cereals, 3. alcoholic beverages, (4) fermented vegetable proteins and 5. fermented animal protein.

Sudanese traditionally classify their foods, not on the basis of microorganisms or commodity but on a **functional basis**: 1) Kissar (staples)—porridges and breads such as aceda and kissra, 2) Milhat (sauces and relishes for the staples), 3) marayiss (30 types of opaque beer, clear beer, date wines and meads and other alcoholic drinks) and 4) Akil-munasabat (food for special occasions).

Thus fermentations can be generally classified as:

1. Fermentations producing textured vegetable protein meat substitutes in legume/cereal mixtures. Examples are Indonesian tempe and ontjom.
2. High salt/savoury meat-flavoured/amino acid/peptide sauce and paste fermentations Examples are Chinese soy sauce, Japanese shoyu and Japanese miso, Indonesian kecap, Malaysian kicap, Korean kanjang, Taiwanese inyu, Philippine taosi, Indonesian

tauco, Korean doenjang/kochujang, fish sauces: Vietnamese nuocmam, Philippine patis, Malaysian budu, fish pastes: Philippine bagoong, Malaysian belachan, Vietnamese mam, Cambodian prahoc, Indonesian trassi and Korean jeotkal. These are predominately oriental fermentations but the use of these products is becoming established in many countries.

3. **Lactic acid fermentations** Examples of vegetable lactic acid fermentations are: sauerkraut, cucumber pickles, olives in the Western world; Egyptian pickled vegetables in the Middle East; Indian pickled vegetables and Korean kimchi, Thai pak-sian-dong, Chinese hum-choy, Malaysian pickled vegetables and Malaysian tempoyak. Lactic acid fermented milks include: yoghurts in the Western world, Russian kefir, Middle East yoghurts, liban (Iraq), Indian dahi, Egyptian laban rayab, laban zeer, Malaysian tairu (soybean milk). Lactic acid fermented cheeses in the Western world and Chinese sufu/tofu-ru. Lactic acid fermented yoghurt/wheat mixtures: Egyptian kishk, Greek trahanas, Turkish tarhanas. Lactic acid fermented cereals and tubers (cassava): Mexican pozol, Ghanian kenkey, Nigerian gari; boiled rice/raw shrimp/raw fish mixtures: Philippine balao balao, burong dalag; lactic fermented/leavened breads: sourdough breads in the Western world; Indian idli, dhokla, khaman, Srilankan hoppers; Ethiopian enjera, Sudanese kisra and Philippine puto; Western fermented sausages and Thai nham (fermented fresh pork).

4. **Alcoholic fermentations** Examples are grape wines, Mexican pulque, honey wines, South American Indian chicha and beers in the Western World; wines and Egyptian bouza in the Middle East; palm and jackfruit wines in India, Indian rice beer, Indian madhu, Indian ruhi; in Africa, Ethiopian tej, Kenyan muratina, palm wines, Kenyan urwaga, Kaffir/bantu beers, Nigerian pito, Ethiopian talla, Kenyan busaa, Zambian maize beer; in the Far East, sugar cane wines, palm wines, Japanese sake, Indonesian tape, Malaysian tapuy, Chinese lao-chao, Thai rice wine, Indonesian brem, Philippine tapuy.

5. **Acetic acid/vinegar fermentations** Examples are apple cider and wine vinegars in the West; palm wine vinegars in Africa and the Far East, coconut water vinegar in the Philippines; tea fungus/ Kombucha in Europe, Manchuria, Indonesia, Japan and recently in the United States; Philippine nata de pina and nata de coco.

6. **Alkaline fermentations** Examples are Nigerian dawadawa, Ivory Coast soumbara, African iru, ogiri, Indian kenima, Japanese natto, Thai thua-nao.

7. **Leavened breads** Examples are Western yeast and sourdough breads; Middle East breads.

Lactic Acid Fermentations

A second principle is that fermentations involving production of lactic acid are generally safe. Lactic acid fermentations include those in which the fermentable sugars are converted to lactic acid by lactic-acid organisms such as *Leuconostoc mesenteroides, Lactobacillus brevis, Lactobacillus plantarum, Pediococcus cerevisiae, Streptococcus thermophilus, Streptococcus lactis, Lactobacillus bulgaricus, Lactobacillus acidophilus, Lactobacillus citrovorum, Bifidobacterium bifidus,* and so on.

This single category is responsible for processing and preserving vast quantities of human food and insuring its safety. We are all aware of the excellent safety record of sour milks/yoghurts, cheeses, pickles, and so on. In addition, the lactic acid fermentations provide the consumer with a wide variety of flavours, aromas and textures to enrich the human diet. The most ancient lactic fermentation, likely, is fermented/sour milk. Raw, unpasteurized milk will rapidly sour because the lactic acid bacteria present in the milk ferment milk sugar, lactose to lactic acid. In the presence of acid and a low pH, other microorganisms that might invade the milk and transmit pathogenic organisms are less able to do so.

If the whey is allowed to escape or evaporate, the residual curd becomes a primitive cheese. Fermented cheeses and milks are an extensive topic on their own. Vegetable foods and vegetable/fish/shrimp mixtures are preserved around the world by lactic acid fermentation. The classic lactic acid/vegetable fermentation is sauerkraut. Fresh cabbage is shredded and 2.25% salt is added. There is a sequence of lactic acid bacteria that develop. First, *Leuconostoc mesenteroides* grows producing lactic acid, acetic acid and CO_2 which flushes out any residual oxygen making the fermentation anaerobic. Then *Lactobacillus brevis* grows producing more acid. Finally *Lactobacillus plantarum* grows producing still more lactic acid and lowering the pH to below 4.0. At this pH and under anaerobic conditions, the cabbage or other vegetables will be preserved for long periods of time. Korean kimchi is a fermentation similar to sauerkraut but it includes not only Chinese cabbage but radishes, red pepper, ginger and garlic. It is less acid than sauerkraut and is consumed while still carbonated. It is a staple and makes a major contribution to the Korean diet in which consumption of 100 g or more/capita/day is not uncommon. Kimchi is still a household fermentation in Korea although it can be purchased commercially. Pickled vegetables, cucumbers, radishes, carrots and nearly all vegetables and even some green fruits such as olives, papaya and mango are acid fermented in the presence of salt around the world. It was demonstrated that Indian farmers can safely preserve their surplus vegetables by lactic acid fermentation on the farm. This improves the supply and availability of vegetable foods throughout the year and improves the nutrition of the Indian population.

Pit fermentations Lactic acid fermentations include the "pit" fermentations in the South Pacific Islands. They have been used for centuries by the Polynesians to store and preserve breadfruit, taro, banana and cassava tubers.

The fermented pastes or whole fruits, sometimes punctured, are placed in leaf-lined pits. The pits are covered with leaves and the pits are sealed. It has recently been found that pit fermentations are lactic acid fermentations. The low pH and anaerobic conditions account for the stability of the foods. An abandoned pit estimated to be about 300 years old contained breadfruit still in edible condition. In Ethiopia, pulp of the false banana (*Ensete ventricosum*) is also fermented in pits. It undergoes lactic acid fermentation and is preserved until the pit is opened. Then the mash is used to prepare a flat bread kocho—a staple diet of millions of Ethiopians.

Lactic acid fermented rice/shrimp/fish mixtures Philippine balao balao is a lactic acid fermented rice/shrimp mixture prepared by mixing boiled rice, raw shrimp and solar salt (about 3% w/w), packing in an anaerobic container and allowing the mixture to ferment over several days or weeks. The chitinous shell of the shrimp becomes soft and when the product is cooked, the whole shrimp can be eaten. The process provides a method of preserving raw shrimp or pieces of raw fish (burong dalog). The products are well-preserved by the low pH and anaerobiosis until the containers are opened. Then they must be cooked and consumed.

Yoghurt/cereal mixtures Another household lactic acid fermentation of considerable nutritional importance includes Egyptian kishk, Greek trahanas and Turkish tarhanas. These products are basically parboiled wheat/yoghurt mixtures that combine the high nutritional value of wheat and milks while attaining excellent keeping qualities. The processes are rather simple. Milk is fermented to yoghurt and the yoghurt and wheat are mixed and boiled together until the mixture is highly viscous. The mixture is then allowed to cool, formed into biscuits by hand and sun-dried. Trahanas can be stored on the kitchen shelf for years and used as a base for highly nutritional soups.

In the Egyptian kishk process, tomatoes, onions and other vegetables are sometimes combined with the yoghurt and wheat in the biscuits.

Cereal/legume sour gruels/porridges/beverages Cereal/legume sour gruels/ porridges/beverages include Nigerian ogi, Kenyan uji, South African mahewu/magou, and Malaysian soybean milk yoghurt (tairu). In Nigerian ogi, maize, millet or sorghum grains are washed and steeped for 24 to 72 hours during which time they undergo lactic acid fermentation. They are drained, wet milled and finally wet-sieved to yield a fine, smooth slurry with about 8% solids content. The boiled slurry called "pap", is a porridge. Pap is a very important traditional food for weaning infants and a major breakfast food for adults. Infants (9 months old) are introduced to ogi by feeding once a day as a supplement to breast milk.

Unfortunately the nutritional value of ogi is poor. It is vastly improved by the addition of soybean to produce a soy-ogi. Kenyan uji is a related product except that the grains are ground to a flour before mixing with water and fermenting. The initial slurry is 30% solids. This is fermented for 2 to 5 days yielding 0.3 to 0.5% lactic acid. The slurry is then diluted to 10% solids and boiled. It is then diluted to 4 to 5% solids and 6% sucrose is added for consumption.

South African mageu/mahewu/magou is a traditional sour, non-alcoholic maize beverage popular among the Bantu people of Africa. Corn flour is slurried with water (8 to 10% solids), boiled, cooled and inoculated with 5% w/w wheat flour (household fermentation) which serves as a source of microorganisms and incubated at ambient temperature or the boiled slurry is inoculated with *Lactobacillus delbrueckii* (industrial process) and incubated at 45°C. The slurries are fermented to a pH of 3.5 to 3.9 and then are ready for consumption.

Lactic acid-fermented gruels inhibited the proliferation of gram-negative pathogenic bacteria including toxicogenic *Escherichia coli, Campylobacter jejuni, Shigella flexneri and Salmonella typhimurium.* The mean number of diarrhoea episodes in preschool children over a 9 month period was 2.1 per child using fermented gruels compared with 3.5 per child using non-fermented gruels. It was also reported that using a natural lactic acid culture and flour of germinated seeds (power flour), it was possible to prepare liquid cereal gruels from maize, white sorghum, bulrush millet and finger millet with a 30 to 35% flour concentration.

The energy density of such a lactic acid-fermented gruel was about 1.2 kcal/g, as compared with 0.4 kcal/g in a non-fermented gruel prepared to the same consistency. This is a 3-fold increase in energy density. Also, the *in vitro* protein digestibility of high-tannin cereal varieties was significantly increased from a range of 32 to 40% before fermentation to a range of 41 to 60% after lactic acid-fermentation. Lactic acid fermentation of non-tannin cereals with added flour of germinated sorghum grain or wheat phytase increased iron solubility from about 4% to 9% and 50% respectively. Thus lactic acid fermentation has very profound effects on the nutritive value of cereal gruels for feeding infants and children in the developing world.

Ingestion of foods containing live lactic acid bacteria is likely to improve the resistance of the gastrointestinal tracts of infants and children to invasion by organisms causing diarrhoea.

Cereal/legume steamed breads and pancakes   Indian idli, a sour, steamed bread, and dosa, a pancake are examples of a household fermentation that could be useful around the world. Polished rice and black gram dal in various proportions, that is, 3 : 1 to 1 : 3 are soaked separately during the day. In the evening, the rice and black gram are ground in a mortar and pestle with added water to yield a batter with the desired consistency. The batter is thick enough that requires the use of the hand and forearm to mix it properly. A small quantity of salt is added. The batter ferments overnight, during that time *Leuconostoc mesenteroides* and *Streptococcus faecalis*, naturally present on the grains/legumes/utensils grow rapidly outnumbering the initial contaminants and dominating the fermentation. The organisms produce lactic acid (total acidity as lactic can reach above 1.0%) and carbon dioxide that makes the batter anaerobic and leavens the product. In the morning, the batter is steamed to produce small white muffins or fried as a pancake. Soybean cotyledons, green gram, Bengal gram can be substituted for the black gram. Wheat, maize or kodri can be substituted for the rice to yield Indian dhokla.

Alcoholic Fermentations

A third principle contributing to food safety is that fermentations involving production of ethanol are generally safe foods and beverages. These include wines, beers, Indonesian tape ketan/tape ketella, Chinese lao-chao, South African kaffir/sorghum beer and Mexican pulque. These are generally yeast fermentations but they also involve yeast-like moulds such as *Amylomyces rouxii* and mould-like yeasts such as *Endomycopsis* and sometimes bacteria such as *Zymomonas mobilis*. The substrates include diluted honey, sugarcane juice, palm sap, fruit juices, germinated cereal grains or hydrolysed starch all of which contain fermentable sugars that are rapidly converted to ethanol in natural fermentations by yeasts in the environment. Nearly equal weights of ethanol and carbon dioxide are produced and the CO_2 flushes out residual oxygen and maintains the fermentation anaerobic. The yeasts multiply and ferment rapidly and other microorganisms most of which are aerobic cannot compete. The ethanol is germicidal and, as long as the fermented product remains anaerobic, the product is reasonably stable and preserved. With starchy substrates such as cereal grains, it is necessary to convert some of the starch to fermentable sugar. This is done in a variety of ways, that is, chewing the grains to introduce ptyalin (Andes region of South America) where maize is a staple or germination (malting) of barley or the grains themselves in most of the world where beers are produced.

In Asia, there are at least two additional ways of fermenting starchy rice to alcoholic foods. The first is the use of a mould such as *Amylomyces rouxii* which produces amylases converting starch to sugars and a yeast such as *Endomycopsis fibuligera* which converts the glucose/maltose to ethanol. The sweet/sour alcoholic product of rice fermentation is called tape ketan in Indonesia. It is consumed as a dessert. When cassava is used as substrate, the product is called tape ketella. This process can also be used to produce rice wines, Chinese lao-chao and Malaysian tapuy/tapai. Another method is the Japanese koji process used to ferment rice to rice wine (sake). In this process, boiled rice is overgrown with an amylolytic mould *Aspergillus oryzae* for about 3 days at 30°C. The mould-covered rice called a "koji" is then inoculated with a culture of the yeast *Saccharomyces cerevisiae* and water is added. Saccharification by the mould amylases and alcoholic fermentation by the yeast proceed simultaneously. The result of slow fermentation is high yeast populations and ethanol contents as high as 23% v/v.

Acetic Acid/Vinegar Fermentation

If the products of alcoholic fermentation are not kept anaerobic, bacteria belonging to the genus *Acetobacter* present in the environment oxidize portions of the ethanol to acetic acid/vinegar. Fermentations involving production of acetic acid yield foods or condiments that are generally safe. Acetic acid also is bacteriostatic to bactericidal depending upon the concentration. In most cases, palm wines and kaffir beers contain not only ethanol but also acetic acid. Acetic acid is even more preservative than ethanol. Vinegar is a highly acceptable condiment used in pickling and preserving cucumbers and other vegetables. The alcoholic and acetic acid fermentations can be used to insure safety in other foods.

Over the last few years, there is increasing awareness of a fermented acetic acid beverage called kombucha or tea fungus. Actually kombucha has been known and produced for a long period of time historically in China, Russia and Germany. The fermentation involves sweetened (5 to 15% sugar) tea infusion and a microbial mat containing *Acetobacter* and a variety of yeasts that grows on the surface of the substrate producing primarily acetic and gluconic acids in approximately equal quantities (above 3% total). pH is about 2.5. While various health-promoting qualities have been ascribed to kombucha, the fact appears to be that the acetic acid is the primary microbial inhibitor and, if it is consumed in too large quantities, it can cause indigestion and even death.

Leavened Bread Fermentation

Yeast breads are closely related to the alcoholic fermentation and have an excellent reputation for safety. Ethanol is a minor product in bread because of the short fermentation time; but carbon dioxide produced by the yeasts leavens the bread producing anaerobic conditions and baking produces a dry-surface resistant to invasion by organisms in the environment. Baking also destroys many of the microorganisms in the bread itself.

Yeast-breads are made by fermentation of wheat and rye flour doughs with yeasts, generally *Saccharomyces cerevisiae*. Sour dough breads are fermented with lactic acid bacteria and yeasts.

Alkaline Fermentations

Fermentations involving highly alkaline fermentations are generally safe. Africa has a number of very important foods/condiments that are not only used to flavour soups and stews but also serve as low-cost sources of protein in the diet. Among these are Nigerian dawadawa, Ivory cost soumbara and West African iru made by fermentation of soaked, cooked locust bean Parkia biglobosa seeds with bacteria belonging to genus *Bacillus*, typically *Bacillus subtilis*. Nigerian ogiri is made by fermentation of melon seed (*Citrullus vulgaris*); Nigerian ugba is made by fermentation of the oil bean (*Pentacletha macrophylla*); Sierra Leone ogiri-saro is made by fermentation of sesame seed (*Sesamum indicum*).

This group also includes Japanese natto, Thai thua-nao and Indian kenima all based upon soybean. The essential microorganisms are *Bacillus subtilis* and related bacilli. The organisms are very proteolytic and the proteins are hydrolysed to peptides and amino acids. Ammonia is released and the pH rapidly reaches as high as 8.0 or higher. The combination of high pH and free ammonia along with very rapid growth of the essential microorganisms at relatively high temperatures (above 40°C) make it very difficult for other microorganisms that might spoil the product to grow. Thus, the products are quite stable and well-preserved especially when dried. They are safe foods even though they may be manufactured in an unhygienic environment. These are all household fermentations that depend upon *Bacillus subtilis* spores in the environment. No deliberate inoculum is needed. The seeds are soaked and the seed-coats are removed. The seeds are then cooked and drained. Shallow pots or pans are used for the fermentation. The pans are sources of the required spores, which germinate and overgrow the seeds forming a sticky mucilaginous gum on the surfaces of the beans/seeds. They are very nutritious and a source of low-cost nitrogen for the diet.

Interestingly, the Malaysians ferment locust beans by a similar alkaline process to yield garlic flavoured products. The Japanese ferment soybeans with *Bacillus subtilis* after soaking and cooking to yield a protein-rich food called natto. The Thais also ferment soybeans by a similar process to produce a product called thua-nao consumed in Northern Thailand as a substitute for fish sauce. The Indians also ferment soybeans by similar processes to produce kenima. Thus, these alkaline fermentations involving bacilli fermenting protein-rich beans and seeds are of considerable importance in different parts of the world. They are all household fermentations but Japanese natto has been commercialized.

High Salt Savoury Flavoured Amino/Peptide Sauces and Pastes

Addition of salt in ranges from 13% w/v or higher to protein-rich substrates results in a controlled protein hydrolysis that prevents putrefaction, prevents development of food poisonings such as botulism and yields meaty, savoury, amino acid or peptide sauces and pastes that provide very important condiments particularly for those unable to afford much meat in their diets. Meaty or savoury flavoured amino acid/peptide sauces and pastes include Chinese soy sauce, Japanese shoyu, Japanese miso, Indonesian kecap, Taiwan inyu, Korean kan-jang, Philippine taosi made by fermentation of soybeans or soybean or rice-barley mixtures with *Aspergillus oryzae* and fish sauces and pastes made by fermentation of small fish and shrimp using principally the proteolytic gut enzymes of the fish and shrimp.

The ancient discovery of how to transform bland vegetable protein into meat-flavoured amino acid or peptide sauces and pastes was an outstanding human accomplishment. Like many discoveries, it was probably initially accidental. The earliest substrates were likely meat or fish mixed with a millet koji. In the most primitive process known today, soybeans are soaked and thoroughly cooked. They are mashed, formed into a ball, covered with straw and hung to ferment under the rafters of the house. In 30 days they are completely overgrown with the mould *Aspergillus oryzae*. Packed into a strong salt brine (about 18% w/v) the beans are hydrolysed by the proteases, lipases and amylases in the mould yielding an amino acid or peptide liquid sauce with a meaty flavour and a similar meat-flavoured paste or residue—a primitive miso. One can only guess what a dramatic effect this meat-flavoured sauce and paste had on consumers of the typical bland rice diet in the Orient. The process not only enriches the flavour but it also contains the essential amino acids in the soybean. Because of the relatively high lysine content of the soybean, soy sauce or miso is an excellent adjunct to the nutritional quality of rice. Soy sauce started as an Oriental food; but, in recent years it has been adopted by the West, particularly the Americans.

In the presence of high salt concentration (above 13% w/v and generally in the range of 20 to 23%), fish gut enzymes will hydrolyse fish proteins to yield amino acid or peptide, meaty-flavoured fish sauces and pastes similar to soy sauces or pastes. Halophilic pediococci contribute to the typical desirable fish sauce flavour or aroma. Both soy sauce and fish sauce fermentations are carried out at the household level as well as by large commercial manufacturers.

SAFETY OF FERMENTED FOODS

Fermented foods generally have a very good safety record even in the developing world where the foods are manufactured by people without training in microbiology or chemistry in unhygienic, contaminated environments. They are consumed by hundreds of millions of people every day in both the developed and the developing world. And they have an excellent safety record. What is there about fermented foods that contributes to safety? While fermented foods are themselves generally safe, it should be noted that fermented foods by themselves do not solve the problems of contaminated drinking water, environments heavily contaminated with human waste, improper personal hygiene in food-handlers, flies carrying pathogenic organisms, unfermented foods carrying food poisoning or human pathogens, and even in cooked unfermented foods if handled or stored improperly.

Also, improperly fermented foods can be unsafe. However, application of the principles that lead to the safety of fermented foods could lead to an improvement in the overall quality and the nutritional value of the food supply, reduction of nutritional diseases and greater resistance to intestinal and other diseases in infants.

PRINCIPLES BEHIND SAFETY OF FERMENTED FOOD PROCESSES

The safety of food fermentation processes is related to several principles. The first is that food substrates overgrown with desirable, edible microorganisms become resistant to invasion by spoilage, toxic or food poisoning microorganisms. Other, less desirable (possibly disease producing) organisms find it difficult to compete. An example is Indonesian tempe. The substrate is soaked, dehulled, partially cooked soybeans. During the initial soaking, the soybeans undergo an acid fermentation that lowers the pH to 5.0 or below, a pH inhibitory to many organisms but highly acceptable to the mould. Following cooking, the soy cotyledons are surface dried which inhibits growth of bacteria that might spoil the product. The essential microorganism is *Rhizopus oligosporus,* or related *Rhizopus* species, which knit the cotyledons into a compact cake that can be sliced or cut into cubes and used in recipes as a protein-rich meat-substitute. These moulds have the ability to grow very rapidly at relatively high temperatures, that is, 40 to 42°C, too high for many bacteria and moulds.

The mould utilizes the available oxygen and produces CO_2 which inhibits some potential spoilage organisms. The mould also produces large quantities of spore dust that permeate the environment and contribute to inoculation of the desired microorganisms. The combination of a relatively low pH, no free water and a high temperature in the fermenting bean mass enables *Rhizopus oligosporus* to overgrow the soybeans in 18 hours. Organisms that might otherwise spoil the product are unable to compete with the mould. In addition, the mould also produces some antibiotic activity that inhibits other organisms that might invade and spoil the product. The principle is that, as soon as the substrate is overgrown by the desired organism(s), it is resistant to invasion by other microorganisms. An additional safety factor is that the raw tempe is sliced or cut into cubes and cooked before consumption which destroys vegetative microorganisms that are present.

Soybean tempe (tempe kedelee) has an excellent record of safety in Indonesia and Malaysia where the product is fermented as above and has been used for centuries. In temperate countries including the United States, soybeans do not undergo natural lactic acid fermentation during soaking. The pH remains closer to 6.0 a level that permits growth of contaminating bacteria. It has been shown that unacidified soybeans can be invaded by a variety of food poisoning organisms including *Staphylococcus aureus, Bacillus cereus* and *Clostridium botulinum*.

This can be prevented by artificially acidifying the soybeans or inoculating the soak water with *Lactobacillus plantarum*. In some cases, the unacidified beans or insufficiently acidified soybeans has led to spoilage and could lead to food poisoning. While soybean tempe has an excellent safety record, there is a type of tempe that has a reputation for causing sickness and even death in the consumer. This is tempe bongkrek made from the coconut residue from coconut milk production. Again, if the substrate is properly acidified, it is generally safe and the mould overgrows the substrate knitting it into a compact cake. But, if the substrate is not sufficiently acidified and/or, if the fungal inoculum is insufficient, a bacterium *Burkholdaria (Pseudomonas) cocovenenans* can grow in the coconut residue producing two toxic compounds—bongkrek acid, the most lethal, and toxoflavin. These toxic compounds are inhibitory to the mould and it does not overgrow the substrate properly.

Bongkrek acid is colourless and it can be lethal to the consumer. Toxoflavin is yellow and so, if tempe bongkrek has a yellow colour or if the mould has not overgrown the substrate properly, it should not be consumed. Yet, quite a number of Indonesians in central Java die every year from consumption of improperly fermented tempe bongkrek. This is an example of a fermented food that can become toxic but it is amply documented.

A valid concern is mycotoxins that are present in many cereal grains and legume substrates before fermentation. These are dangerous to the consumer of fermented foods. However, it is not the fermentation that produces mycotoxins. They are produced when the cereal grains or legumes are improperly harvested or stored. It has been found that in the Tempe fermentation, mycotoxin levels are reduced. It was reported that the ontjom mould, *Neurospora* and the tempe mould *Rhizopus oligosporus* could decrease the aflatoxin content of peanut press cake 50% and 70% respectively during fermentation. And during soaking and cooking of raw substrates before fermentation, many potential toxins such as trypsin inhibitor, phytate and haemagglutinin are destroyed. So, in general, fermentation tends to detoxify the substrates.

In summary, food fermentations that improve food safety are as follows:

- Food substrates rapidly overgrown by edible microorganisms leading to desirable flavours, aromas and free of toxins are resistant to development of spoilage, food poisoning or toxin-producing organisms
- Food fermentations involving lactic acid production
- Food fermentations involving ethanol production
- Food fermentations involving acetic acid production
- Food fermentations involve highly alkaline conditions with liberation of free ammonia
- Food fermentations carried out in the presence of high salt concentrations (above 13% w/w) are generally safe.

NUTRITIONAL ASPECTS OF FERMENTED FOODS

"Nutrition" or "nutritional" is the supplying of calories or energy, protein, essential amino acids or peptides, essential fatty acids and vitamins and mineral requirements to satisfy metabolic needs of the consumer.

Two major food problems exist in the world. Starvation (or under-nutrition where there is insufficient food or insufficient economic means to provide the necessary food) and obesity (or over-consumption of food in the wealthy, developed world). There is outright starvation and death in countries such as Ethiopia, Sudan, Somalia and Bangladesh due to poverty, drought, environmental disasters and war combined with lack of economic means to purchase food.

There are a number of nutritional diseases in the developing world today. Kwashiorkor, the result of protein deficiencies, and marasmus, caused by a combination of protein and calorie deficiencies, are found in large numbers of children between the ages of 1 and 3 in the developing world. The immune systems of the children are impaired on restricted protein diets. Such children often develop diarrhoea and the mothers place them on restricted diets such as rice broth. Since the children are already deficient in protein, they quickly develop kwashiorkor with tissue oedema, bloated abdomens, susceptibility to infection and changes in hair pigments. Kwashiorkor can develop in children consuming sufficient calories if their protein is deficient. On the other hand, marasmus is the result of both calorie and protein deficiencies in children under 1 year of age due to early cessation of breast-feeding and the attempted replacement with artificial milks with very low protein contents. The principal features of marasmus are growth retardation, muscle wasting and loss of subcutaneous fat. Both kwashiorkor and marasmus can result in mental retardation if the child survives.

Other nutritional diseases common in the developing world are xerophthalmia, childhood blindness due to vitamin A deficiency; beriberi due to thiamine deficiency including infantile beriberi where sudden death may occur from heart failure in children being nursed by mothers deficient in thiamine; pellagra due to niacin deficiency, riboflavin deficiency; rickets due to vitamin D deficiency; anaemia due to vitamin B_{12} deficiency or insufficient iron in the diet.

The nutritional impact of fermented foods on nutrition-related diseases can be direct or indirect. Food fermentations that raise the protein content or improve the balance of essential amino acids or their availability will have a direct curative effect. Similarly, fermentations that increase the content or availability of vitamins such as thiamine, riboflavin, niacin or folic acid can have profound direct effects on the health of the consumers of such foods. This is particularly true of people subsisting largely on maize where niacin or nicotinic acid is limited and pellagra is incipient and in people subsisting principally on polished rice which contains limited amounts of thiamine and beriberi is incipient. Biological enrichment of foods via fermentation can prevent this.

On the other hand, fermentation does not generally increase calories unless they convert substrates unsatisfactory for humans to human quality foods such as mushrooms produced on straw or waste paper, or converting peanut and coconut presscakes to edible foods such as Indonesian ontjom (oncom). While the Western world can afford to enrich its foods with synthetic vitamins, the developing world must rely upon biological enrichment for its vitamins and essential amino acids. The affluent Western world cans and freezes much of its food but the developing world must rely upon fermentation, salting and solar dehydration to preserve and process its foods at costs within the means of the average consumer. All consumers today have a considerable portion of their nutritional needs met through fermented foods and beverages.

BIOLOGICAL ENRICHMENT BY FERMENTATION

Bioenrichment with Protein

In the Indonesian tape ketan fermentation referred to earlier, rice starch is hydrolysed to maltose and glucose and fermented to ethyl alcohol. The loss of starch solids results in a doubling of the protein content (from about 8% to 16% in rice) on a dry solid basis. Thus, this process provides a means by which the protein content of high starch substrates can be increased for the benefit of consumers needing higher protein intakes. This is particularly important for people consuming principally cassava which has a protein content of about 1% (wet basis). If the tape ketella fermentation is applied to cassava as it is in Indonesia, the protein content can be increased to at least 3% (wet basis)—a very significant improvement in nutrition to the consumer. In addition, the flavour of the cassava becomes sweet or sour and alcoholic, flavours consumers may prefer to the bland starting substrate.

Bioenrichment with Essential Amino Acids

The Indonesian tape ketan/tape ketella fermentation not only enriches the substrate with protein, but the microorganisms also selectively enrich the rice substrate with lysine that the first essential limiting amino acid in rice. This improves the protein quality. Several researchers have reported from 10.6 to 60.0% increase in methionine during the Indian idli fermentation. An increase in methionine, a limiting essential amino acid in legumes greatly improves the protein value.

Bioenrichment with Vitamins

In the wealthy, Western world, nutrients particularly vitamins are added to selected, formulated, manufactured foods as a public health measure. Examples are addition of vitamin D to milk, vitamin A and D to milk and butter and riboflavin to bread. Fruit juices may be fortified with ascorbic acid (vitamin C). While enrichment or fortification are within the means of the Western World, they are far beyond the means of the developing world. Thus, much of the world must depend upon biological enrichment via fermentation to enrich their foods.

Highly-polished white rice is deficient in vitamin B_1 (thiamine). Consumers subsisting principally on polished rice are in danger of developing beriberi, a disease characterized by muscular weakness and polyneuritis leading eventually to paralysis and heart failure. Infants being nursed by mothers suffering from thiamine deficiency may develop infantile beriberi in which sudden death occurs at about 3 months of age due to cardiac failure. The microorganisms involved in the tape ketan fermentation synthesize thiamine and restore the thiamine content to the level found in unpolished rice. This can be a very significant contribution to the nutrition of rice eating people. Tempe, a protein-rich meat substitute in Indonesia is made by overgrowing soaked, dehulled, partially cooked soybeans with *Rhizopus oligosporus* or related moulds. The mould knits the cotyledons into a firm cake that can be sliced and cooked or used in place of meat in the diet. During the fermentation, proteins are partially hydrolysed, the lipids are hydrolysed to their constituent fatty acids, stachyose (a tetrasaccharide indigestible in humans) is decreased decreasing flatulence in the consumer, riboflavin nearly doubles, niacin increases sevenfold and vitamin B_{12} usually lacking in vegetarian foods is synthesized by a bacterium that grows along with the essential mould. The tempe process can be used to introduce texture into many legume or cereal mixtures yielding products with decreased cooking times and improved digestibility. The bacterium responsible for the vitamin

B$_{12}$ production is a non-pathogenic strain of *Klebsiella pneumoniae*. When inoculated into the Indian idli fermentation, it also produces vitamin B$_{12}$.

Mexican pulque is the oldest alcoholic beverage on the American continent. It is produced by fermentation of juices of the cactus plant (Agave). Pulque is rich in thiamine, riboflavin, niacin, pantothenic acid, *p*-amino benzoic acid, pyridoxine and biotin. Pulque is of particular importance in the diets of the low-income group of Mexico.

Kaffir beer is an alcoholic beverage with a pleasantly sour taste and the consistency of a thin gruel. It is a traditional beverage of the Bantu people of South Africa. Alcohol content ranges from 1 to 8% v/v. Kaffir beer is generally made from kaffircorn (*Sorghum caffrorum*) malt and unmalted kaffircorn meal. Maize or millet may be substituted for kaffircorn. During the fermentation, thiamine remains about the same, but riboflavin more than doubles and niacin or nicotinic acid nearly doubles, which is very significant in people consuming principally maize. Consumers of usual amounts of kaffir beer are not in danger of developing pellagra.

Palm sap is a sweet, clear, colourless liquid containing about 10 to 12% fermentable sugar and neutral in reaction. Palm wine is a heavy, milk-white opalescent suspension of live yeasts and bacteria with a sweet taste and vigorous effervescence. Palm wines are consumed throughout the tropics. Palm wine contains as much as 83 mg ascorbic acid per litre.

Thiamine increased from 25 mg to 150 mg/litre, riboflavin increased from 35 to 50 μg/litre and pyridoxine increased from 4 to 18 μg/litre during fermentation. Surprisingly, palm wine contains considerable amounts of vitamin B$_{12}$, 190 to 280 pg/ml. Palm toddies play an important role in nutrition among the economically disadvantaged in the tropics. They are the cheapest sources of B vitamins.

REDUCTION OF TOXINS

During soaking and hydration, that raw substrates undergo in various fermentation processes and the usual cooking, many potential toxins such as trypsin inhibitor, phytate and haemagglutinin and cyanogens in cassava are reduced or destroyed. Even aflatoxin, frequently found in peanut and cereal grains substrates, is reduced in the Indonesian ontjom fermentation. It was found that the ontjom mould, *Neurospora* and the tempe mould *Rhizopus oligosporus* could decrease the aflatoxin content of peanut presscake 50% and 70% respectively during fermentation.

REDUCTION OF COOKING TIMES

Economy of fuel requirements is very important in the developing world where housewives may spend hours every day collecting enough leaves, twigs, wood and dried dung to cook the day's food. Lactic-acid fermented foods generally require little, if any, heat in their fermentation and can be consumed without cooking. Examples are pickled vegetables, sauerkraut, kimchi. Indonesian tempe fermentation converts soybeans that would require as much as 5 to 6 hour cooking, to a product that can be cooked in soup with 5 to 10 minutes of boiling.

SUMMARY

Hence fermented foods provide and preserve vast quantities of nutritious foods in a wide diversity of flavours, aromas, and textures which enrich the human diet. Fermented foods have been with us since humans arrived on earth. They will be with us far into the future, as they are the source of alcoholic foods, beverages, vinegar, pickled vegetables, sausages, cheeses, yoghurts, vegetable protein, amino acid or peptide sauces and pastes with meat-like flavours, and leavened and sourdough breads. All consumers today have a considerable portion of their nutritional needs met through fermented foods and beverages.

1. What is the role of fermentation in food processing?
2. List out the types of food fermentations.
3. Write a note on indigenous fermented foods.
4. Given an account of the safety of fermented foods.
5. Discuss alcoholic fermentation.

35

ROLE OF MICROORGANISMS IN FOOD FERMENTATION

INTRODUCTION

Preservation of foods by fermentation is a widely practiced and ancient technology. Fermentation ensures not only increased shelf life and microbiological safety of a food but also may make some foods more digestible and in the case of cassava, fermentation reduces toxicity of the substrate. Lactic acid bacteria because of their unique metabolic characteristics are involved in many fermentation processes of milk, meats, cereals and vegetables. Although many fermentations are traditionally dependent on inoculation from a previous batch starter cultures for many commercial processes such as cheese manufacture thus ensuring consistency of process and product quality. This chapter outlines the role of lactic acid bacteria in many such fermentations and the mechanism of antibiosis with particular reference to bacteriocins and gives a brief description of some important fermented foods from various countries.

By the middle of the 19th century, two events had occurred which had a very significant impact on the manner in which food fermentations were performed and on our understanding of the process. Firstly, the industrial revolution resulted in the concentration of large masses of populations in towns and cities. This meant that the traditional method of supplying such foods within local communities no longer applied. The need to service these new markets required products to be made in large quantities necessitating the industrialization of the manufacturing process. Secondly, the blossoming of microbiology as a science from the 1850s onwards resulted in the biological basis of fermentation being understood for the first time. Thus, the essential role of bacteria, yeasts and moulds in the generation of fermented foods came to be understood and this ultimately resulted in more controlled and efficient fermentations. Clearly, the traditional approach of backslopping or even natural fermentation of substrate was not an appropriate foundation upon which to base any large-scale industrial process.

The advent of retailing and mass marketing required that products of consistent quality and safety be available. For many fermented foods, but particularly milk-derived products, the characterization of the microorganisms responsible for the fermentation towards the end of the 19th century led to the isolation of starter cultures which could be produced on a large scale to supply factories involved in the manufacture of these products. This significant development had a major impact on the processes used contributed to ensuring consistency of product and reliability of fermentation. These developments parallelled significant technological advances in the handling and processing of milk which has resulted in dairy fermentations being among the most sophisticated and best researched of the food fermentations to this day. For example, research on the lactic acid bacteria, which have a dominant role in the production of many fermented foods, particularly those that are milk-based, has continued to advance at a very impressive rate.

The original and primary purpose of fermenting food substrates was to achieve a preservation effect. However, with the development of the many effective alternative preservation technologies which are now commonly available, particularly in the Western World, this is no longer the most pressing requirement and many of these foods are manufactured because their unique flavour, aroma and texture attributes are much appreciated by the consumer. However, even in these situations, the conditions generated by the fermentation are essential in ensuring the shelf life and microbiological safety of the products. Nevertheless, there are many parts of the world where the preservation role is still the essential one and where the fermentation process is still performed on an artisinal rather than industrial basis.

In many cases, these fermentations have only come under scientific scrutiny in the relatively recent past and thus are still only moderately or poorly understood.

ESSENTIAL ELEMENTS OF FOOD FERMENTATION

Preservation of foods by fermentation depends on the principle of oxidation of carbohydrates and related derivatives to generate end products which are generally acids, alcohol and carbon dioxide. These end products control the growth of food spoilage microorganisms and because the oxidation is only partial, the food retains sufficient energy potential to be of nutritional benefit to the consumer. The term "fermentation" is applied to describe a strictly anaerobic process; however, the general understanding of the term now encompasses both aerobic and anaerobic carbohydrate breakdown processes. Most fermented foods, including the major products that are common in the western world, as well as many of those from other sources that are less well characterized, are dependent on lactic acid bacteria to mediate the fermentation process. The end products of carbohydrate catabolism by these bacteria contribute not only to preservation but also to the flavour, aroma and texture, thereby helping to determine unique product characteristics. Being able to control the specific microorganisms or the succession of microorganisms that dominate the microflora of foods (which is the basis of development of startcr cultures) is therefore very desirable. Fermentation may also increase the nutritional quality of food increasing digestibility as in the fermentation of milk to cheese. In addition, fermented foods would play a major role in future developments. This not only includes traditional activities such as the delivery of probiotic bacteria in products such as fermented milks, but will most probably be extended to the generation of functional components like vitamins, antioxidants and other compounds in a variety of different fermented foods.

METABOLIC ACTIVITY OF LACTIC ACID BACTERIA

Lactic acid bacteria are generally mesophilic but can grow at temperatures as low as 5°C or as high as 45°C. Similarly, while the majority of strains grow at pH 4.0–4.5, some are active at pH 9.6 and others at pH 3.2. Strains are generally weakly proteolytic and lipolytic and require preformed amino acids, purine and pyrimidine bases and B vitamins for growth. All lactic acid bacteria produce lactic acid from hexoses and since they lack functional haem linked electron transport chains and a functional Krebs cycle, they obtain energy via substrate level phosphorylation. The lactic acid produced may be L (+) or, less frequently, D (–) or a mixture of both. It should be noted that D (–) lactic acid is not metabolized by humans and is not recommended for infants and young children, a fact exploited by the marketers of a strain of *L. bavaricus* in Germany for use in the production of speciality L (+) sauerkraut. The pathways by which hexoses are metabolized divides lactic acid bacteria into two groups, homofer) mentative and heterofermentative (Figure 35.1).

Homofermenters such as *Pediococcus, Streptococcus, Lactococcus* and some lactobacilli produce lactic acid as the major or sole end product of glucose fermentation. However, under altered growth conditions and when the initial substrate is a pentose this may change. Homofermenters use the Embden–Meyerhof–Parnas pathway to generate two moles of lactate per mole of glucose and derive approximately twice as much energy per mole of glucose as heterofermenters. Heterofermenters such as *Weisella* and *Leuconostoc* and some lactobacilli produce equimolar amounts of lactate, CO_2 and ethanol from glucose via the hexose monophosphate or pentose pathway. The metabolism of the disaccharide lactose is of primary importance in those lactic acid bacteria used in dairy fermentations. Lactose may enter the cell using either a lactose carrier, lactose permease, followed by cleavage to glucose and galactose or via a phosphoenolpyruvate-dependent phosphotransferase (PTS) followed by cleavage to glucose and galactose 6-phosphate. Glucose is metabolised via the glycolytic pathway, galactose via the Leloir pathway and galactose-6-phosphate via the tagatose 6-phosphate pathway.

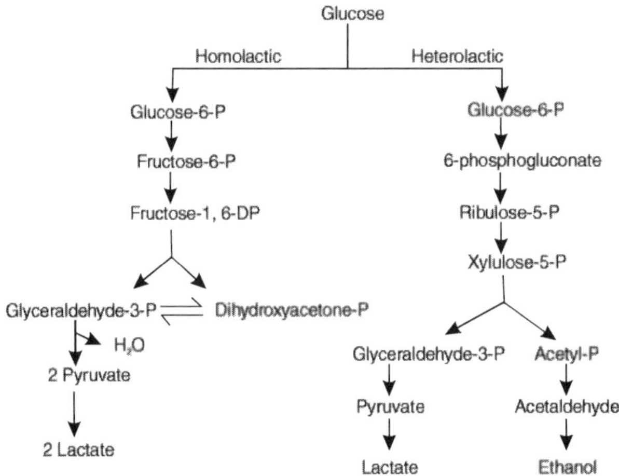

Figure 35.1 Generalized scheme for the fermentation of glucose in lactic acid bacteria

Most *L. lactis* strains used as starters for dairy fermentations use the lactose PTS, the genes for which are plasmid located. Among some thermophilic LAB only the glucose moiety of the sugar is metabolized and galactose is excreted into the medium, although mutants of *S. thermophilus* have been described which metabolize galactose via the Leloir pathway. Citrate metabolism is important among *L. lactis* subsp. *lactis* (biovar *diacetylactis*) and *L. mesenteroides* subsp. *cremoris* strains used in the dairy industry, as it results in excess pyruvate in the cell. The pyruvate may be converted via- acetolactate to diacetyl, an important flavour and aroma component of butter and some other fermented milk products (Figure 35.2). Strategies designed to increase the carbon metabolic flux towards diacetyl production have resulted in mutants which produce large amounts of this compound. The proteolytic system of *Lactococcus* has been investigated in detail due to its pivotal role in allowing growth in milk and the development of flavour and texture in cheese. Casein is degraded by a membrane-anchored serine proteinase (PrtP) with many of the resulting oligopeptides being sufficiently small to allow them to be transported into the cell via an oligopeptide transport system (Opp), where they are further processed by a variety of intracellular peptidases. Amino acid, and di- and tri-peptide transport systems also exist but there is only poor growth in milk when mutants are deficient in PrtP. Commercial culture adjuncts (mesophilic and thermophilic starter cultures) are available to promote proteolysis in cheese and can aid in the development of a consistent cheese flavour. It is notable that over-expression of a lactococcal proteinase did not result in an acceleration of cheese ripening or in an enhancement of flavour.

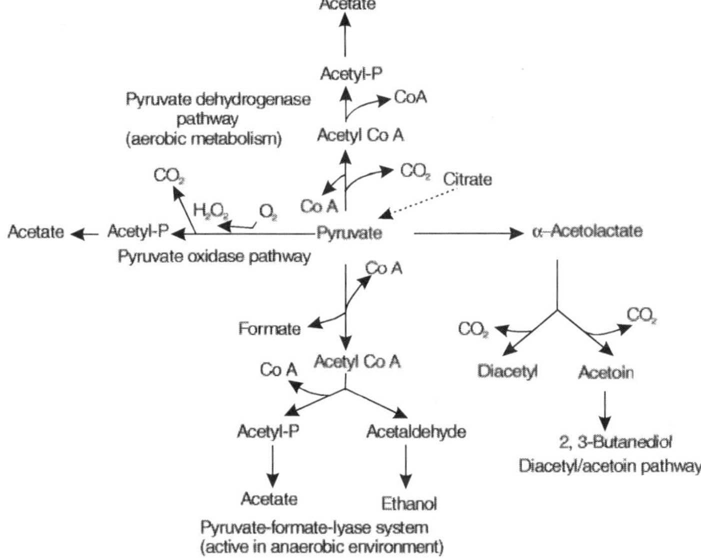

Figure 35.2 Generalized scheme for the formation of important metabolic products from pyruvate in lactic acid bacteria

Lysis of starter cultures during cheese ripening leads to increased proteolytic activity in cheese and thus theoretically results in accelerated ripening and/or a better cheese flavour. A good example of this is found in Swiss cheese where autolytic propionibacteria increase the amount of free proline in the cheese aiding flavour development.

Microbial Interference

General microbial interference is an effective non-specific control mechanism common to all populations and environments including foods. It represents the inhibition of the growth of certain microorganisms by other members of a habitat and was first used to describe the suppression of virulent staphylococci by avirulent strains. In order to operate efficiently the interfering flora, generally the normal background flora of the habitat, needs to outnumber the target host many times. The mechanisms involved include nutrient competition, generation of an unfavourable environment, competition for attachment or adhesion sites and are common to all genera.

Mechanisms of Antibiosis Mediated by Lactic Acid Bacteria

The specific antimicrobial mechanisms of lactic acid bacteria exploited in the biopreservation of foods include the production of organic acids, hydrogen peroxide, carbon dioxide, diacetyl, broad-spectrum antimicrobials such as reuterin and the production of bacteriocins. Many of these factors are thought to play a role in the inhibitory effect of microguard. It is produced by the fermentation of skimmed milk by *Propionibacterium freudenreichii* subsp. *shermanii*, which is subsequently pasteurized and added as a Generally Recognised as Safe (GRAS) food preservative to much of the cottage cheese production in the U.S. to prevent the growth of gram-negative bacteria and moulds.

Organic acids, acetaldehyde and ethanol   The direct antimicrobial effects of organic acids including lactic, acetic and propionic which may be produced by lactic acid bacterial fermentation of foods, are well known. The antagonism is believed to result from the action of the acids on the bacterial cytoplasmic membrane which interferes with the maintenance of membrane potential and inhibits active transport, and may be mediated both by dissociated and undissociated acid. The antimicrobial activity of each of the acids at a given molar concentration is not equal. Acetic acid is more inhibitory than lactic acid and can inhibit yeasts, moulds and bacteria. Propionic acid inhibits fungi and bacteria and is present in microguard as described above and also in another commercial product, bioprofit where the use of a *Propionibacterium freudenreichii* strain along with *Lactobacillus rhamnosus* increases the inhibitory activity against fungi and some gram-positive bacteria. The contribution of acetaldehyde to biopreservation is minor since the flavour threshold is much lower than the levels that are considered necessary to achieve inhibition of microorganisms.

Similarly, although ethanol may be produced by lactic cultures, again the levels produced in food systems are so low that the contribution to antibiosis is minimal.

Hydrogen peroxide   Lactic acid bacteria lack true catalase to break down the hydrogen peroxide generated in the presence of oxygen. It is argued that the H_2O_2 can accumulate and be inhibitory to some microorganisms. Inhibition is mediated through the strong oxidizing effect on membrane lipids and cell proteins. Hydrogen peroxide may also activate the lactoperoxidase system of fresh milk with the formation of hypothiocyanate and other antimicrobials. However, because of the ability of other enzyme systems such as flavoproteins and peroxidases to breakdown H_2O_2 it is not clear what, if any, the *in vivo* contribution of H_2O_2 is to antibacterial activity.

Carbon dioxide   Carbon dioxide, formed from heterolactic fermentation, can directly create an anaerobic environment and is toxic to some aerobic food microorganisms through its action on cell membranes and its ability to reduce internal and external pH. At low concentration, it may be stimulatory to the growth of some bacteria. Production of CO_2 resulting from the use of lactate by propionibacteria in Swiss cheese manufacture is responsible for the characteristic "eyes" of the finished product.

Diacetyl   Diacetyl is a product of citrate metabolism and is responsible for the aroma and flavour of butter and some other fermented milk products. Many lactic acid bacteria including strains of *Leuconostoc, Lactococcus, Pediococcus* and *Lactobacillus* may produce diacetyl, although production is repressed by the fermentation of hexoses. Gram-negative bacteria, yeasts and moulds are more sensitive to diacetyl than gram-positive bacteria and its mode of action is believed to be due to interference with the utilization of arginine. Diacetyl is rarely present in food fermentations at sufficient levels to make a major contribution to antibacterial activity.

Reuterin   Reuterin is produced during stationary phase by the anaerobic growth of *Lactobacillus reuteri* on a mixture of glucose and glycerol or glyceraldehyde. It has a general antimicrobial spectrum affecting viruses, fungi and protozoa as well as bacteria. Its activity is thought to be due to inhibition of ribonucleotide reductase.

Bacteriocins   A recent definition of bacteriocins produced by lactic acid bacteria suggests that they should be regarded as extracellularly released primary or modified products of bacterial ribosomal synthesis, which can have a relatively narrow spectrum of bactericidal activity. They should include at least some strains of the same species as the producer bacterium and against which the producer strain has some mechanism(s) of specific self protection. The possibility of exploiting bacteriocins in food fermentations arises where the inhibitory spectrum includes food spoilage and/or pathogenic microorganisms or gives the producing strain a competitive advantage in the food milieu. The target of bacteriocins is the cytoplasmic membrane and because of the protective barrier provided by the LPS of the outer membrane of gram-negative bacteria, they are generally only active against gram-positive cells. In the context of fermentation, important targets include spoilers such as species of *Clostridium* and heterofermentative lactobacilli and food-borne pathogens including *Listeria monocytogenes, Staphylococcus* spp., *Clostridium, Enterococcus* and *Bacillus* spp. The permeability of gram-negative bacteria can be increased by sublethal injury including that which can occur when using ultrahigh hydrostatic pressure (UHP) and pulsed electric field (PEF) as non-thermal methods of preservation. In addition, disruption of the integrity of the outer membrane through the use of food grade chelating agents such as EDTA and citrate which bind magnesium ions in the LPS layer can increase the effectiveness of bacteriocins against gram-negative bacteria. Many bacteriocins are most active at low pH and there is evidence that bacteriocinogenic strains can be readily isolated from fresh and fermented foods. Strains may naturally produce more than one bacteriocin and heterologous expression of bacteriocins has been demonstrated in constructed strains. Protein engineering has led to the development of nisin derivatives with altered antimicrobial activities or greater solubility at pH 6 than the wild-type nisin. An advantage of bacteriocins over classical antibiotics is that digestive enzymes destroy them. Bacteriocin-producing strains can be used as part of, or adjuncts to starter cultures for fermented foods in order to improve safety and quality. Bacteriocins of lactic acid bacteria, are divided into four major subclasses (Table 35.1).

The majority of those produced by bacteria associated with food belong to classes I and II.

Table 35.1   Classes of bacteriocins produced by lactic acid bacteria

Class	Subclass	Description
I		Lantibiotics—small, heat-stable, containing unusual amino acids
II		Small (30–100 amino acids), heat-stable, non-lantibiotic
	II a	Pediocin-like bacteriocins, with anti-listerial effects
	II b	Two-peptide bacteriocins
	II c	Sec-dependent secretion of bacteriocins
III		Large (> 30 kDa) heat-labile proteins
IV		Complex bacteriocins with glycol- and/or lipid moieties

Class I bacteriocins   Bacteriocins of this class contain post-translationally modified amino acids and are also termed as lantibiotics. It is produced by strains of *L . lactis* subsp. *lactis* and has a broad inhibitory spectrum against gram-positive bacteria, including many pathogens and can prevent outgrowth of *Bacillus* and *Clostridium* spores. It sensitizes spores of *Clostridium* to heat allowing a reduction in thermal processing. Nisin is approved for use, to varying degrees, as a component of the preservation procedure for processed and fresh cheese, canned foods, processed vegetables and baby foods, in more than 50 countries. Typical levels that are used in foods range between 2.5 and 100 ppm. It is most stable in high-acid foods. The addition of a nisin-producing strain of *L. lactis* to the starter culture used in the manufacture of nitrate-free Gouda cheese has been demonstrated to result in the prevention of the outgrowth of *C. tyrobutyricum* spores and it has also been shown to inhibit the growth of *L. monocytogenes* and Camembert cheese. In meats, nisin is not as successful as a preservative, but it may allow a reduction in the levels of nitrite used in cured meat products. Lacticin 3147, produced by a lactococcal isolate from Irish Kefir grains used in the manufacture of buttermilk, is effective against a wide spectrum of gram-positive bacteria. Unlike nisin, lacticin 3147 is effective at neutral pH. The genetic determinants of the lacticin are located on a conjugative plasmid and have recently been transferred to strains used in the manufacture of Cheddar cheese. The resulting cheeses were of normal composition except that they contained no non-starter lactic acid bacteria (NSLAB). This application of lacticin 3147 will prove very useful in studying the role of these latter bacteria in developing flavour and other characteristics in cheeses.

Class II bacteriocins   This class is divided into three subgroups of which the Class IIa is the most common (Table 35.1). This group is composed of the pediocin-like bacteriocins with antilisterial activity. Pediocins are produced by *Pediococcus* spp. and while they are not very effective against spores they are more effective than nisin in some food systems such as meat. Pediococci are the main starter culture used in the manufacture of American-style fermented meats and they are also important in the fermentation of many vegetables. Pediocin PA-1/AcH is the prototype bacteriocin of this class and pediocin-producing cultures are readily isolated from fermented foods. Many studies reported the inhibition of *L. monocytogenes* by pediocins or pediocin-producing cultures in fermented sausages and in Italian salami. Sakacin 674 produced by a *L. sake* isolated from meat and very similar to pediocin PA-1 has been shown to delay or inhibit growth of *L. monocytogenes* in vacuum-packed, sliced Bologna type sausage whether added in purified form or in the form of a bacteriocin producing culture.

Other bacteriocins   In Europe the principal starters used in the manufacture of dry fermented sausage are *L. sake*, *L. curvatus* and *L. plantarum*. Inhibition of *Listeria* by a bacteriocinogenic *L. sake* strain isolated from a naturally fermented sausage has been demonstrated and it was suggested that *L. sake* CTC 494 be employed as a bioprotective culture in fermented meat products.

Aspects to be considered in the use of bacteriocins/bacteriocinogenic cultures in fermented foods   Sensitivity to bacteriocins is strain-dependent and resistance among sensitive cells has been reported with resistance to nisin cited as occurring at a frequency of 10^{-6}. Whether that frequency would be valid in the complex background of a food is unknown. The use of more than one bacteriocin or bacteriocin-producing strain in a specific food system must be carefully controlled so that mutants resistant to one antimicrobial will not be cross-resistant to the others. The implications of resistance arising from general mechanisms such as the alteration of membrane fluidity have to be studied in relation to resistance to other antimicrobial agents. Nisin is the only bacteriocin with GRAS status for use in specific foods and this was awarded as a result of a history of 25 years of safe use in many European countries and was further supported by the accumulated data indicating its non-toxic, non-allergenic nature. Other bacteriocins without GRAS status (which can be based on documented use prior to 1958) will require pre-market approval. Therefore bacteriocinogenic starters, particularly if used in natural fermentations, will most likely afford the best opportunities for the application of bacteriocins in the near future.

FERMENTED FOODS

Fermented foods are now regarded as part of our staple diet. The main substrates used in the commercial production of the most familiar fermented products are milk, meat, cucumber and cabbage. These yield over 400 varieties of cheese of 20 distinct types and a very extensive range of yoghurt and fermented milk drinks, fermented sausages and salamis, pickles and sauerkraut.

Milk Products

The lactic fermentation of milk is required for cheese production. While some cheeses are still made from non-pasteurized milk and may even depend on the natural lactic flora for the fermentation, most are produced on a commercial scale using the appropriate starter culture. These can contain mesophilic *L. lactis* subsp. *lactis and L. lactis* subsp. *cremoris* or thermophilic *S. thermophilus*, *L. helveticus*, and *L. delbrueckii* subsp. *bulgaricus*, depending on the specific application. Thermophilic strains are generally used in cheeses with a high cooking temperature such as Swiss and Italian types.

Secondary microflora are added in some processes to affect texture (e.g. the production of CO_2 by *Propionibacterium* in Swiss cheese) and flavour (e.g. by the production of diacetyl). Moulds, yeasts and bacteria other than lactic acid bacteria are used as secondary microflora in some varieties of cheese (e.g. *Penicillium roqueforti* in blue-veined cheeses). Starter culture improvement with respect to carbohydrate fermentation, proteolysis, production of flavour compounds and protection from the scourge of phage attack have been the subject of much research. The starter used in yoghurt production is a mixed culture of *S. thermophilus* and *Lb. bulgaricus* in a 1:1 ratio. The *Streptococcus*, which is inhibited at pH 4.2–4.4, grows first and the *Lactobacillus*, which adds aroma and flavour (acetaldehyde), can tolerate values as low as pH 3.5–3.8.

Kefir is a fermented milk drink with an alcohol content of up to 1%. The starter consists of characteristic "Kefir grains" which contain the acid- producing *L. lactis* and *L. delbrueckii* subsp. *bulgaricus* and an alcohol-producing *Torula* spp.

Kumiss which is produced in Russia is similar to kefir but uses mares milk. Fermentation can result in an alcohol content of up to 2%. *L. acidophilus* and *Lb. delbrueckii* subsp. *bulgaricus* are used in the production of acidophilus milk and Bulgarian buttermilk where the inoculated sterile milk is held at 37°C until a smooth curd develops. The popularity of fermented milk drinks is increasing, not only because of their attractive taste but also because of the many health benefits associated with them.

Meat Products

Fermented sausages are produced as a result of the lactic acid fermentation of a mixture of comminuted meat mixed with fat, salt, curing agents (nitrate/ nitrite), sugar and spices and these represent traditional foods of central and southern Europe. These sausages are generally classified as dry or semi-dry. Dry sausages have a water activity (a_w) of less than 0.90, are not usually smoked or heat processed and are ate without cooking. The a_w of semi-dry sausages is in the range 0.90–0.95 and they generally receive a heat treatment of 60–68°C during smoking. Fermentation temperatures vary according to the individual product but they are generally less than 22°C for dry and mould-ripened sausages and 22–26°C for semi-dry varieties. Sausages produced without starter have a final pH of 4.6–5.0, while those produced using starter generally have a final pH of 4.0–4.5. The predominant species during lactic acid fermentation of sausages are psychrotrophic *L. sake* and *L. curvatus*. Most fermented sausages formulated with nitrite are produced with added starter culture, generally consisting of lactic acid bacteria (lactobacilli and pediococci) and catalase-positive cocci (*S. carnosus*, *Micrococcus varians*). Yeasts and moulds that are available as starters include *Debaryomyces hansenii*, *Candida famata* and *Penicillium nalgiovense* and *P. chrysogenum*, respectively. The use of starter cultures ensures a good quality, standardized, safe product. Lactic acid bacteria producing bacteriocins have been demonstrated to reduce the count of *L. monocytogenes* by one log early in meat fermentation and this is one particular application where the use of bacteriocinogenic cultures appears to have potential value as an additional inhibitory hurdle.

Vegetable Products

While there are 21 different commercial vegetable fermentations in Europe along with a large number of fermented vegetable juices and blends, the most economically relevant of these are the fermentations of olives, cucumbers (pickles), and cabbage (sauerkraut, Korean kimchi). As raw vegetables have a high microbial load and cannot be pasteurized without compromising product quality, most vegetable fermentations occur as a consequence of providing growth conditions (such as added salt) that favour the lactic acid bacteria. These bacteria are present on fresh vegetables in very low numbers, accounting for only 0.15–1.5% of the total population. The starting material for fermented juice is pasteurized mash or juice and starters of lactobacilli (including *L. plantarum, L. casei, L. acidophilus*), *L. lactis* and *L. mesenteroides*.

Traditional Fermented Foods

Some examples of non-western fermented foods are presented in Table 35.2.

Table 35.2   Examples of traditional fermented foods

Product	Country	Microorganisms	Substrate
Bread	International	*Saccharomyces cerevisiae*, other yeasts, lactic acid bacteria	Wheat, rye, other grains
Bongkrek	Indonesia	*Rhizopus oligosporus*	Coconut press cake
Gari	West Africa	*Corynebacterium menthol*, other yeasts, lactic acid bacteria (*L. plantarum, Streptococcus* spp.)	Cassava root
Idli	Southern India	Lactic bacteria (*L. mesenteroides, E. faecalis, Torulopsis, Candida, Trichosporon pullulans*)	Rice and black gram dhal
Kenkey	Ghana	Unknown	Maize
Kimchi	Korea	Lactic acid bacteria	Cabbage, vegetables, sometimes seafood, nuts
Mahewu	South Africa	Lactic acid bacteria	Maize
Ogi	Nigeria, West Africa	Lactic acid bacteria, *Cephalosporum, Fusarium, Aspergillus, Penicillium* spp., *Saccharomyces cerevisiae, Candida mycoderma, C. valida,* or *C. vint*	Maize
Oncom	Indonesia	*Neurospora intermediate* or *Rhizopus oligosporus*	Peanut press cake
Soy sauce	The Orient (Japan, China, Philippines)	*Aspergillus oryzae* or *A. soyae, Lactobacillus, Zygosaccharomyces rouxii*	Soyabeans and wheat
Tempeh	Indonesia Surinam	*Rhizopus oligosporus*	Soyabeans
Nan	India	*Saccharomyces cerevisiae* Lactic acid bacteria	White wheat flour

(Contd.)

Table 35.2 (Continued)

Product	Country	Microorganisms	Substrate
Cheese	International	Lactic acid bacteria (*L. lactis, S.thermophilus, L. bulgaricus, Propionibacterium shermanii*)	Milk
Yoghurt	International	*S. thermophilus*; *L. bulgaricus*	Milk, milk solids
Fermented sausages	Southern & Central Europe, USA	Lactic acid bacteria (lactobacilli, pediococci), Catalase-positive cocci (*S. carnosus*) *S. xylosus*, *M. varians*) sometimes yeasts and/or moulds	Mammalian meat, generally pork and/or beef, less often poultry
Sauerkraut	International	Lactic acid bacteria *L. mesenteroides, L. brevis, L. plantarum, L. curvatus, L. sake*	Cabbage
Pickles	International	*P. cerevisiae, L. plantarum*	Cucumber
Olives	Mediterranean	*L. mesenteroides, L. plantarum*	Green olives

African Fermented Foods

Pottery linked to wine brewing in Africa has been discovered on excavation sites dating from as early as 690 BC to 560 AD. Many of the foods with the longest recorded pedigree are sour milks and alcoholic beverages. These foods are of particular importance in ensuring adequate intake of proteins and/or calories to consumers living in a climate which favours the rapid deterioration of food. Differing processes with products of different organoleptic properties also extend the attractiveness of what would otherwise be a limited and somewhat monotonous diet. Some fermentations (e.g. cassava) make otherwise inedible foods edible. Breads are also among some of the oldest fermented foods and approximately 60% of the worlds population eats flat breads and gruels made from grains. Sourdough breads are made with starters containing yeasts such as *Saccharomyces* spp. and *Torulopsis* and homofermentative and heterofermentative lactic acid bacteria. Heterofermentative strains such as *L. sanfrancisco*, *L. brevis* and *L. fermentum* are responsible for the characteristic sensory qualities of such breads. "Gari", based on the fermentation of cassava which is one of the most abundant food crops in the tropics, has a long shelf life of about 6 months and is the staple diet of much of the population of West Africa. As a fermented food, it is an example of the rendering safe for human consumption of a food which would otherwise be toxic due to high inherent levels of cyanogenic glucosides. During the fermentation there is a reduction of pH, linamarase activity and total cyanide levels while acid (predominantly lactic acid) levels increase.

Lactic acid bacteria associated with the fermentation include *L. plantarum* and the fermentation also involves other microorganisms including yeasts.

"Ogi" is a fine paste-like sour gruel ate in Nigeria resulting from the submerged fermentation of cereals. It is consumed as a breakfast cereal by adults and is an important traditional weaning food of infants. Similar foods in other African regions are "Koko" or "Kenkey" in Ghana and "Mahewu" in South Africa. The fermentation is dominated by a variety of lactic acid bacteria, particularly *L. plantarum*, while other bacteria such as *Corynebacterium* hydrolyse the corn starch, and yeasts of the *Saccharomyces* and *Candida* species also contribute to flavour development. Mahewu is produced on a large-scale industrial basis and some attempts to develop starter cultures have been made. A starter culture has been used to produce an improved version of ogi called DogiK. The starter strains are lactobacilli isolated from local fermented foods with strong antibacterial activity.

Indian Fermented Foods

"Idli" is fermented steamed cake of rice and dehulled blackgram dhal produced in India. Lactic tetra acid bacteria such as *L. mesenteroides*, *L. delbrueckii*, *P. cerevisiae*, *E. faecalis* and *L. lactis* are responsible for pH reduction and may increase the thiamine and riboflavin content. Yeasts also contribute to the fermentation. "Nan" is leavened flat sourdough bread with a central pocket now prepared worldwide. Organisms involved in the fermentation include *Saccharomyces* yeasts and lactic acid bacteria, particularly *Lactobacillus* species.

Indonesian Fermented Foods

Tempe kedele developed in Indonesia is a soybean fermentation. Soybeans are first soaked in water, generally overnight at ambient temperature, and they are then dehulled, partially cooked and inoculated with moulds of the genus *Rhizopus*. Tempe, which encontains over 40% protein, is a meat substitute and is used in soups or sliced, salted and deep fat fried in coconut oil. Lactic acid bacteria including *L. casei* and *Lactococcus* species dominate the fermentation, which may be initiated by the addition of a commercial culture or a small amount of a previous batch. Following fermentation, beans are bound together with the mycelia of *Rhizopus oligosporus* to form a compact "cake". Tempe is being advocated as a protein-rich source of vitamin B_{12} in the Western vegetarian's diet. The vitamin is thought to be produced by *Klebsiella pneumoniae* and *Citrobacter freundii* and is one of the most frequently cited examples of "bioenrichment". Temhydrolyse peh bongkrek is a coconut press cake, which has led to deaths in its native central Indonesia. Toxicity is due to the production of toxoflavin and bongkrekic acid by the bacterium *Burkholderia cocovenenans* which can grow in the first few days of fermentation if *Rhizopus* growth is not favoured. Oncom (Ontjom) is a fermented peanut press cake. *R. oligosporus* is less frequently used in this product than *Neurospora intermedia*. It is reported that the phytic acid content of peanut press cake is reduced by fermentation.

Oriental Fermented Foods

The characteristic aroma and flavour of soy sauce is due to the enzymatic activities of yeasts, *Tetragenococcus halophilus* and some *Lactobacillus* species. Soy sauce (or shoyu) is a condiment widely used in the cooking and seasoning of Japanese food.

There are five main types of soy sauce in Japan, each with its own distinctive colour, flavour and use. In general, the pH of the sauce is within the range pH 4.6–4.8 and the characteristic high salt concentration is 17–19%. Concentrations of salt less than 16% can result in the development of putrefactive species during fermentation and ageing, and levels greater than 19% interfere with the growth of halophilic bacteria such as *P. halophilus* and osmophilic yeast such as *Zygosaccharomyces rouxii*. Because soybeans contain high levels of protein and oligosaccharides such as stachyose, raffinose, melibiose and sucrose but no significant level of simple sugars, fermentation by lactic acid bacteria and yeast requires the exogenous saccharifying enzymes supplied by the "koji". As a result, the first step in soy sauce fermentation is the production of koji whereby soybeans or a mixture of beans and wheat are inoculated with *Aspergillus oryzae* or *A. soyae* and allowed to stand for 3 days at 25–35°C and 27–37% moisture. This stage is analogous to malting in the brewing process. In the mash stage, called moromi, koji is added to brine to give a salt concentration of 17–19% and fermented at room temperature for 12–14 months (home preparation) or at 35–40°C for 2–4 months (commercial preparation). The bacteria and yeast involved in the fermentation include *P. halophilus*, *L. delbrueckii*, *Z. rouxii* and *Torulopsis* species. The liquid (sauce) is removed from the mash and pasteurized at 70–80°C before bottling.

DEVELOPMENTS IN FOOD FERMENTATIONS

Throughout the world there are many different types of fermented foods in which a range of different substrates are metabolized by a variety of microorganisms to yield products with unique and appealing characteristics. In many of these foods, the biological and microbiological bases of the fermentation processes are poorly understood. What little information is available often deals with the identification and perhaps preliminary characterization of the primary microflora in the finished product. In some instances, there will undoubtedly be a need in the future to produce these foods in circumstances where quality and safety can be guaranteed. This in turn will necessitate a more thorough understanding of the microorganisms involved, in terms of the types and their specific activities, so that the fermentation process can be made more reliable and predictable. It is likely that basic microbiological analyses in conjunction with the appropriate technological developments will, in the first instance at least, be sufficient to achieve these objectives. For those fermentations, where there already exists a considerable body of knowledge regarding the role and activity of the relevant microflora, the challenges enfacing the scientists and technologists are somewhat different. Fermentations involved in the production of many cheeses, yoghurts and some fermented meats in particular, are already quite sophisticated, are generally reliable and predictable and can deliver products of excellent quality. Here, the goals are to further improve reliability and product quality through optimization of starter culture performance and to eliminate those factors that impede the fermentation process. In this regard, it is to be anticipated that the considerable resources that have been devoted to the "biotechnology of lactic acid bacteria" over the past number of years will deliver results with respect to these objectives. This is already evident in the case of the protection of cheese starter cultures of *L. lactis* against bacteriophage infection. This provides an excellent case study as to how a deliberate, programmed research effort, designed to provide a more complete understanding of phage–host interactions, has ultimately yielded strategies for the development of non-recombinant phage-resistant strains for use in industrial fermentations.

The development of the correct flavour characteristics is a critical factor in the production of a range of fermented foods. However, the specific mechanisms by which flavour is generated is not fully understood, although the principal components contributing to flavour, such as protease, peptidase and lipase activities may be known. This is well exemplified in a product such as Cheddar cheese where the fine and subtle flavour attributes of the mature product is an essential element in determining the quality of the cheese. There now exists the ability to manipulate the proteolytic system of certain lactic acid bacteria, albeit in a crude way. Other recent advances related to an understanding of the pathways of amino acid metabolism and the role of cell lysis during product maturation will not only illuminate the biological basis for flavour development but will also allow this characteristic to be controlled and modified according to the needs of the market. There are many other areas where additional functionality can be developed in lactic acid bacteria and in most cases the potential to achieve this is a direct consequence of the fundamental knowledge that has been generated regarding the genetic make-up of these hosts. The prospect of metabolic engineering of strains of lactic acid bacteria to generate derivatives with new attributes is now a real one. It will soon be possible to produce strains that secrete high levels of polysaccharides for use as food-grade texturers, or to elaborate elevated levels of flavour compounds such as diacetyl, acetaldehyde or acetate.

The manipulation of the metabolic flux of these hosts may also yield derivatives that produce health-enhancing compounds such as antioxidants and vitamins. The application of lactic acid bacteria to deliver vaccines is one that is already being intensively investigated and represents a very attractive exploitation of these hosts. The very impressive scientific and technological developments that have been made with lactic acid bacteria over the past number of years are likely to relieve many of the bottlenecks encountered with their full and efficient application in food fermentations in the near future. The availability of information regarding the genetic blueprint of these bacteria (the genome sequence of a strain of *Lactococcus* has already been determined and those of several other lactic acid bacteria will be available in the near future) will also provide new applications that are likely to benefit both the producer and consumer. However, the scientific community and the industrial user need to be aware of consumer concerns regarding recombinant DNA technology, particularly in Europe, and especially when it involves food products. Thus, there is a need to ensure that there will be clear consumer benefits arising from the manipulation of these bacteria and also that the traditional positive attitude associated with fermented foods is not compromised by the exciting biotechnological developments in lactic acid bacteria.

REVIEW QUESTIONS

1. Describe the role of microorganisms in food production and preservation.

36

DYNAMICS OF MICROBIAL POPULATIONS

INTRODUCTION

Industrial control of fermentation processes requires up-to-date knowledge of the physiology, metabolism and genetic properties of such microorganisms. Equally important is the knowledge of their impact on food quality, safety and shelf life. The characters of food fermentation, however, is more complex. Agricultural products of animal or vegetable origin are fermented either by the indigenous microflora or an added starter culture to improve shelf life, nutritional value, health benefit, flavour or texture. Unlike the Western world, Asia has developed many foods based on vegetable proteins using fungi—often in a solid-state fermentation process. In Europe and the US, the main focus of food fermentation has been on food preservation by means of acid fermentation, whereas properties such as taste, nutritional value and health effect are more important in Asian (fungal) fermentations.

THE STARTER BACTERIA

In fermentation, the raw materials are converted by microorganisms to products that have acceptable food qualities. Spontaneous fermentations, i.e., processes initiated without the use of a starter inoculum, have been applied to food preservation for millennia. In a natural fermentation, the conditions are set so that the desirable microorganisms grow preferentially and produce metabolic by-products, which give the unique characteristics of the product. The majority of small-scale fermentation in developing countries and even some industrial processes such as sauerkraut fermentation are still conducted as spontaneous processes. Various types of starter cultures and even back-slopping are widely used in fermentation processes, even in industrialized countries. However, spontaneous food fermentations are neither predictable nor controllable. The natural microflora of the raw material is either inefficient, uncontrollable, and unpredictable, or is destroyed altogether by the heat treatments given to the food.

When the yield is unstable and where the desired microorganisms might not grow, or where pathogenic microbes might also grow, a controlled fermentation is used. In a controlled fermentation, the fermentative microbes are isolated and characterized, then maintained for use. Since the addition of selected bacteria (usually LAB) starts the manufacturing process, they are commonly referred to as starter cultures. Starter cultures are added to the raw materials in large numbers and incubated under optimal conditions. In common, controlled fermented products such as sauerkraut and yoghurt, the primary function of LAB starters is the production of lactic acid from lactose which helps to make the products shelf stable. Other functions of starter cultures may include, flavour, aroma, and alcohol production, proteolytic and lipolytic activities, and inhibition of undesirable organisms. Therefore, microbial starters provide particular characteristics through fermentation is a more controlled and predictable way.

There are two groups of commercial LAB starter cultures:

1. Defined (either single or multiple strains) composed of a known number of strains.
2. Undefined (mixed) in which the number of strains is unknown.

Modern starter cultures are selected either as single or multiple strains, specifically for their adaptation to a substrate or raw material. Suitable cultures for fermentation must be selected at the strain level since not all strains of a species are equally suitable for use as starters, nor are all equally well-adapted to a food substrate. In dairy products, the properties of starter cultures such as phage sensitivity, lactic acid, gas and aroma compounds production are to a large extent, dependent upon the properties of each culture component. Because modern dairy starter cultures are blended empirically for the desired characteristics of the final product, maintenance of the optimal strain balance throughout the cheese fermentation process is important because the complex relationships among microorganisms can be easily altered. Also small variations in microbial composition could have unexpected effects on cheese quality. The largest part of the lactic culture market consists of cultures for the dairy industry. The starter culture industry has been in a phase of consolidation and restructuring; in this context, commercial starter cultures for direct inoculation are increasingly used for their reliability, performance and safety, as well as for their convenience.

THE NON-STARTER BACTERIA

In some fermented dairy products, additional bacteria such as cit(+) *Lactococcus lactis* subsp. *lactis*, *Leuconostoc* spp., *Lactobacillus kefir*, and

Propionibacterium freudenreichii subsp. *germanii*, often referred to as secondary microflora, are often intentionally introduced to produce aroma compounds and carbon dioxide in cultured buttermilk and certain cheeses. Other types of secondary microflora include undefined mixtures of yeasts (*Debaryomyces* spp. and *Geotrichum* spp.), moulds (*Penicillium camemberti* and *Penicillium roqueforti*), and bacteria (*Brevibacterium linens*, *Micrococcus* spp. and mesophilic lactobacilli). The use of micrococci and *B. linens* is usually limited to surface-ripened and mould-ripened cheeses; mesophilic lactobacilli, which can form biofilms and be a source of contamination in a dairy, are believed to contribute considerably to the formation of cheese aroma from amino acids. For example, a useful co-operation between starter *L. lactis* and glutamate-dehydrogenase-positive *L. casei/L. paracasei* strains to stimulate flavour development in Gouda and Cheddar cheeses has been suggested.

THE UNWANTED BACTERIA

Food manufacturing, distribution, and storage rely on well-placed deleterious stresses or hurdles that either inhibit or inactivate contaminating microorganisms in food systems. However, fermented foods are not commercially produced in an environment free of contaminating microbes. For example, dairy products may contain yeasts, moulds, and many other general bacteria whose metabolic activities destroy quality. In cheese, after fermentation is complete, only acid-tolerant bacteria can grow. However if acid development is slow or if the pH does not decrease sufficiently, contaminants which otherwise would have been inhibited may be able to grow. In some fermented foods, the pH can increase during ripening and permits the growth of previously inhibited bacteria. These phenomena, together with the well-known evidences that many food-borne pathogenic bacteria are adapting to sublethal inimical stresses, may allow unwanted organisms (either pathogenic or spoilage) to grow in fermented foods.

MICROORGANISMS IN FOOD ECOSYSTEMS

The importance of ecological concepts in understanding the growth of microbes in foods is well recognized by food microbiologists. These ecological principles are the fundamental to modern quality assurance, predictive modelling, and risk analysis strategies to prevent outbreaks of food spoilage and food-borne disease. They also form the basis for the use of microorganisms in the production of fermented foods and beverages, and for their use as probiotic, starter, and biocontrol agents.

Ecological and Biochemical Factors
Affecting Microbial Growth in Food

In order to effectively manage the growth and activities of microbes in fermented foods, the following points should be raised:

- information on diversity, taxonomic identity, growth cycle, quantitative changes and spatial distribution of microbial species that ferment into the food at every stage of production
- biochemical and physiological data on the food colonization process
- impact of the intrinsic, extrinsic, and processing factors influencing microbial growth, survival and biochemical activity
- relationship between growth and activity of individual microorganisms and product quality and safety

Problems Related to Qualitative and Quantitative Estimation of Microbial Populations in Foods

One of the most important application of ecological concepts to food fermentation ecosystems is the identification, interpretation and manipulation of time-dependent changes in communities of microorganisms. The economic and social consequences of microorganisms in foods depend not only on the species present but, most importantly, on their numbers. The number of microbial cells ultimately determines whether or not the product will cause an outbreak of disease or develop an off-flavour. Unfortunately, the vast majority of ecological studies in food microbiology fall significantly short of providing this quantitative knowledge. Many studies simply describe the isolation and identification of the "most predominant" species at one point in the product's history, while others have provided semi-quantitative data by reporting the frequency of isolation of specific organisms. Population changes are mostly described in reference to microbial groups (e.g. total plate count, coliforms, LAB) rather than data for particular species or strains, which could be more useful to comprehend the final microbiological properties of fermented foods in a better way.

The dynamics of growth, survival and biochemical activity of microorganisms in foods are the result of stress reactions in response to the changing of the physical and chemical conditions into the food microenvironment (e.g. the gradients of pH, oxygen, water activity, salt and temperature) and the ability to colonize the food matrix to grow with spatial heterogeneity (e.g. microcolonies and biofilms). Moreover, because in most food ecosystems microbial populations are generally immobilized and localized in high densities, food production (or degradation) is rarely the result of the activities of an individual but that of a group of microorganisms. Therefore, the growth, survival, and activity of any one species or strain, whether it be an unwanted spoilage or pathogenic organism or a desirable biocontrol of a probiotic agent, will in most cases, be determined by the presence of other microbes and the *in situ* cell-to-cell ecological interactions which often happen in a solid phase. It has been shown that *in situ* solid phase associations similar to those found in food ecosystems could induce unique biochemical and physiological reactions of a microbial population as a whole.

THE SUBLETHALLY INJURED BACTERIA

Stress is any change in the genome, proteome or environment that imposes either reduced growth or survival potential. Such changes lead to attempts by a cell to restore a pattern of metabolism that either fits it for survival or for faster growth. For any stress, the bacterial cell has a defined range within which the rate of increase of colony forming units is positive (growth), zero (survival) or negative (death). In the first two cases, i.e., growth and survival, the cells are sublethally injured, whereas in the case of death (e.g. after a bacteriophage attack) the cells are lethally damaged and rapidly autolyse. The individual values at which the cell moves from one physiological state into the next is conditional on the degree of stress imposed by other environmental conditions.

In microbial populations, viable cells are usually countable on both non-selective and selective media, whereas stressed cells are able to form colonies on non-selective media but are not countable on selective media. In foods, any adverse conditions such as nutrient depleting, low temperature and other stresses can sublethally damage microbes. The viable but non-cultivable (VNC) stage is a kind of stress which induces healthy, cultivable cells to enter a phase in which they are still capable of metabolic activity but do not produce colonies on media (both selective and non selective) that normally support their growth. Sublethally injured cells (including VNC cells) are however, usually able to resume their healthy state.

The VNC stage has been shown in both gram-positive and gram-negative bacteria in the natural environment and it has also been experimentally induced in most food-borne pathogens. Although no molecular technologies have been systematically applied to food ecology, it is conceivable that food-associated microorganisms might enter the VNC state under certain situations, e.g. on the rind of surface-ripened cheese or on the surfaces of fruits and vegetables, where nutrient is limiting. LAB can survive in a VNC state during wine storage and ageing.

1. "Food fermentation is the oldest biotechnology", explain.

2. Discuss about sublethally injured bacteria and its implication in food fermentation.

37

ORGANISMS IMPORTANT IN FOOD FERMENTATION

THE LACTIC ACID BACTERIA

Lactic acid bacteria have the property of producing lactic acid from sugars by a process called fermentation. The genera *Lactobacillus*, *Leuconostoc*, *Pediococcus* and *Streptococcus* are important members of this group. The taxonomy of lactic acid bacteria has been based on the Gram's reaction and the production of lactic acid from various fermentable carbohydrates.

Lactobacilli are gram-positive and vary in morphology from long, slender rods to short coccobacilli, which frequently form chains. Their metabolism is fermentative; some species are aerotolerant and may utilize oxygen through the enzyme flavoprotein oxidase, while others are strictly anaerobic. While spore-bearing lactobacilli are facultative anaerobes, the rest are strictly anaerobic. The growth is optimum at pH 5.5–5.8 and the organisms have complex nutritional requirements for amino acids, peptides, nucleotide bases, vitamins, minerals, fatty acids and carbohydrates.

The genus is divided into three groups based on fermentation patterns:

1. homofermentative which produce more than 85% lactic acid from glucose.
2. heterofermentative which produce only 50% lactic acid and considerable amounts of ethanol, acetic acid and carbon dioxide.
3. less well-known heterofermentative species which produce DL-lactic acid, acetic acid and carbon dioxide.

The species which have been therapeutically used are :

- *L. sporogenes*
- *L. acidophilus*

- *L. plantarum*
- *L. casei*
- *L. brevis*
- *L. delbreuckii*
- *L. lactis*

The metabolic activities of lactobacilli are responsible for their therapeutic benefits.

Lactobacilli cultured in milk medium perform the following activities:

Proteolysis

Proteins are broken down into easily assimilable components.

$$\text{Protein} + H_2O \xrightarrow[\text{from lactobacilli}]{\text{proteinases}} \text{Polypeptide}$$

These activities of lactobacilli in the gastrointestinal tract make protein ingested by the host easily digestible, a property of great value in infant, convalescent and geriatric nutrition.

Lipolysis

Complex fat is broken down into easily assimilable components.

$$\text{Triglycerides (fat)} \xrightarrow[\text{from lactobacilli}]{\text{lipases}} \text{Fatty acids} + \text{glycerol}$$

This property is useful in the preparation of dietetic formulations for infants, geriatrics and convalescents.

Evidences from preclinical and clinical trials have revealed that lactobacilli can breakdown cholesterol in serum lipids. Lactobacilli also assist in the deconjugation of bile salts. Both of these findings have clinical significance.

Lactose metabolism

Lactic acid bacteria have the enzymes β-galactosidase, glycolases and lactic dehydrogenase (LDH) which produce lactic acid from lactose. Lactic acid is reported to have some physiological benefits such as:

(a) Enhancing the digestibility of milk proteins by precipitating them in to fine curd particles.

(b) Improving the utilization of calcium, phosphorus and iron.

(c) Stimulating the secretion of gastric juices.

(d) Accelerating the onward movement of stomach contents.

(e) Serving as a source of energy in the process of respiration.

The levels of optical isomeric forms of lactic acid produced depend upon the nature of the culture. The structural configurations of these isomers are as follows:

$$\underset{\text{D}(-) \text{ Laevorotatory lactic acid}}{\overset{\displaystyle \mathrm{COOH}}{\underset{\displaystyle \mathrm{CH_3}}{\mathrm{H} \blacktriangleright\!\!-\!\!\mathrm{C}\!-\!\blacktriangleleft\mathrm{OH}}}} \qquad \underset{\text{L}(+) \text{ dextrorotatory lactic acid}}{\overset{\displaystyle \mathrm{COOH}}{\underset{\displaystyle \mathrm{CH_3}}{\mathrm{HO} \blacktriangleright\!\!-\!\!\mathrm{C}\!-\!\blacktriangleleft\mathrm{H}}}}$$

In humans, both isomers are absorbed from the intestinal tract. Whereas L(+) lactic acid is completely and rapidly metabolized in glycogen synthesis, D(–) lactic acid is metabolized at a lesser rate, and the unmetabolized acid is excreted in the urine. The presence of unmetabolized lactic acid results in metabolic acidosis in infants. *L. acidophilus* produces the D(–) form and is therefore of disputable clinical benefit, although it has earlier been the probiotic of choice in various therapeutic formulations. *L. sporogenes* on the other hand produces only L(+) lactic acid and hence is preferred.

The ability of lactobacilli to convert lactose to lactic acid is used in the successful treatment of lactose-intolerance. People suffering from this condition cannot metabolize lactose due to lack or dysfunction of the essential enzyme systems. Lactic acid, by lowering the pH of the intestinal environment to 4–5, inhibits the growth of putrefactive organisms and *E. coli*, which require a higher optimum pH of 6 to 7. Some of the volatile acids produced during fermentation also possess some antimicrobial activity under conditions of low oxidation–reduction potential.

Production of Bacteriocins

Bacteriocins are proteins or protein complexes with bactericidal activities directed against species which are closely related to the producer bacterium. The inhibitory activity of lactobacilli towards putrefactive organisms is thought to be partially due to the production of bacteriocins.

Some of the bacteriocins isolated from lactobacilli are listed in Table 37.1.

Table 37.1   Bacteriocins isolated from different *Lactobacillus* species

Substance	Producing species
Acidolin	*L. acidophilus*
Acidophilin	*L. acidophilus*
Lactacin B	*L. acidophilus*
Lactacin F	*L. acidophilus*
Bulgarin	*L. bulgaricus*
Plantaricin SIK-83	*L. plantarum*

(Contd.)

Table 37.1   (Continued)

Substance	Producing species
Plantaricin A	*L. plantarum*
Lactolin	*L. plantarum*
Plantaricin B	*L. plantarum*
Lactolin 27	*L. helveticus*
Helveticin J	*L. helveticus*
Reuterin	*L. reuteri*
Lactobrevin	*L. brevis*
Lactobacillin	*L. brevis*

Production of Other Antagonistic Substances

Lactic acid bacteria also inhibit the growth of harmful putrefactive microorganisms through other metabolic products such as hydrogen peroxide, carbon dioxide and diacetyl.

The metabolites of lactic acid bacteria that exert antagonistic action against putrefactive microorganisms and their mode of action are summarized in Table 37.2.

Table 37.2   Antagonistic activities caused by lactic acid bacteria

Metabolic product	Mode of antagonistic action
Carbon dioxide	Inhibits decarboxylation? Reduces membrane permeability?
Diacetyl	Interacts with arginine-binding proteins.
Hydrogen peroxide/lactoperoxidase	Oxidizes basic proteins.
Lactic acid	Undissociated lactic acid penetrates the membranes, lowering the intracellular pH. It also interferes with metabolic processes such as oxidative phosphorylation.
Bacteriocins	Affect membranes, DNA synthesis and protein synthesis.

Synthesis of B-Vitamins

Experiments on fermented milk products have revealed that lactic cultures require B-vitamins for their metabolic activities. However, some lactic cultures synthesize B-vitamins. Similarly, vitamins are synthesized by the lactic cultures in the gut microflora, in symbiosis with other flora.

It has been observed that the diet of the host influences the nature and levels of beneficial intestinal microflora, such as lactobacilli. The presence of dietary fructo-oligosaccharides was found to enhance the healthful effects of intestinal lactic acid bacteria. These compounds, found naturally in foods such as onion, edible burdock and wheat, are effectively employed as non-nutritive sweeteners (Neosugar, Meiologo). They have the advantage of not being able to be digested by humans and farm animals, rendering them valuable in dietetic products. They are, however, selectively utilized by intestinal lactic acid bacteria, especially bifidobacteria, thereby enhancing the healthful effects of these beneficial intestinal flora.

Lactic Acid Fermentations

The lactic acid bacteria belong to two main groups—the homofermenters and the heterofermenters. They differ in the pathways involved in the lactic acid production. Homofermenters produce mainly lactic acid, *via* the glycolytic (Embden–Meyerhof) pathway. Heterofermenters produce lactic acid plus appreciable amounts of ethanol, acetate and carbon dioxide, *via* the 6-phosphoglucanate/phosphoketolase pathway. The glycolytic pathway is used by all lactic acid bacteria except leuconostocs, group III lactobacilli, oenococci and weissellas. Normal conditions required for this pathway are excess sugar and limited oxygen.

Homolactic fermentation   The fermentation of 1 mole of glucose yields two moles of lactic acid.

$$C_6H_{12}O_6 \rightarrow 2\,CH_3CHOCOOH$$

$$\text{Glucose} \qquad \text{Lactic acid}$$

Heterolactic fermentation   The fermentation of 1 mole of glucose yields 1 mole each of lactic acid, ethanol and carbon dioxide.

$$C_6H_{12}O_6 \rightarrow CH_3CHOCOOH + C_2H_5OH + CO_2$$

$$\text{Glucose} \qquad \text{Lactic acid} \qquad \text{Ethanol} \quad \text{Carbon dioxide}$$

Lactic acid bacteria convert sugars to lactic acid. Flavours are produced as well and the increased acidity of the product helps to prevent the growth of unwanted spoilage bacteria and pathogens. Changes in texture may also occur.

Lactic acid bacteria are among the best studied microorganisms. Important new developments have been made in the research of lactic acid bacteria in the areas of multidrug resistance, bacteriocins, quorum sensing, osmoregulation, autolysins and bacteriophages. Progress has also been made in the construction of food grade genetically modified lactic acid bacteria. These have opened new potential applications for these microorganisms in various industries.

The desirable characteristics of industrial microorganisms are their ability to rapidly and completely ferment cheap raw materials, requiring minimal amount of nitrogenous substances, providing high yields of preferred stereo-specific lactic acid under conditions of low pH and high temperature, production of low amounts of cell mass and negligible amounts of other by-products.

The choice of an organism primarily depends on the carbohydrate to be fermented. *Lactobacillus delbreuckii* subsp. *delbreuckii* are able to ferment sucrose. *Lactobacillus delbreuckii* subsp. *bulgaricus* is able to use lactose. *Lactobacillus helveticus* is able to use both lactose and galactose. *Lactobacillus amylophylus* and *Lactobacillus amylovirus* are able to ferment starch. *Lactobacillus lactis* can ferment glucose, sucrose and galactose. *Lactobacillus pentosus* have been used to ferment sulphite waste liquor.

Lactobacillus have complex nutritional requirements, since groups of microorganisms in this genus have lost their ability to synthesize their own growth factors. They cannot grow solely on carbon source and inorganic nitrogen salts. Organisms such as *Rhizopus oryzae* have less limiting nutritional requirements and can utilize starch feed stocks. They are able to produce pure L (+) lactic acid.

Enzymes for Lactic Acid Fermentation

Lactic acid is produced in the form of L (+) or D (–) lactic acid or as its racemic mixture. Organisms that form the L (+) form or D (–) form have two lactate dehydrogenases (LDH), which differ in their stereospecifity. Some lactobacilli produce L (+) form, which on accumulation induces a racemase, which converts it into D (–) lactic acid until equilibrium is obtained.

Genetically Modified Lactic Acid Bacteria for
Improved L (+) Lactic Acid Bacteria

A few attempts have been made to improve L (+) lactic acid production by metabolic engineering in lactobacilli producing both L (+) and D (–) lactic acids.

In *Lactobacillus helveticus,* inactivation of *ldh*D (D-lactate dehydrogenase gene) led to a two fold increase in the amount of L (+) lactic acid, thereby restoring the total amount of lactic acid to the level in the wild type strain. Two stable *ldh*D negative strains of *Lactobacillus helveticus* were constructed by the gene replacement method. One strain was constructed by an internal deletion of the promoter region thereby preventing the transcription of the *ldh*D gene. The second construct was prepared by replacing the *ldh*D gene with *ldh*L, thus duplicating the gene dosage.

The L-lactate dehydrogenase activity was increased by 53% and 93% respectively in the two modified strains than in the wild type strain. The two D-lactate dehydrogenase negative strains produced only L (+) lactate in an amount equal to the total lactate produced by the wild type strain.

The gene encoding L (+) lactate dehydrogenase was isolated from *Lactobacillus plantarum* and then cloned into *Escherichia coli*. This gene was sequenced and used to construct *Lactobacillus plantarum* strains by either over expressing or not expressing *ldh*L. A multicopy plasmid bearing *ldh*L gene was introduced into *Lactobacillus plantarum* without modification of its expression signals. This increased the L-lactate dehydrogenase activity by 13-fold but it hardly had any effect on the production of L (+) lactate or D (–) lactate. A stable chromosomal deletion in the *ldh*L gene resulted in the absence of L-lactate dehydrogenase activity and in exclusive production of the D-isomer of lactate.

In *Lactococcus lactis,* when the copy number of the *lac* operon in which the *ldh*L gene was increased, it resulted in a slight increase in lactic acid production.

Thus to summarize, the lactic acid bacteria comprise a clade of gram-positive, low-GC, acid-tolerant, non-sporulating, non-respiring rod or cocci that are associated by their common metabolic and physiological characteristics. These bacteria produce lactic acid as the major metabolic end product of carbohydrate fermentation. This trait has historically linked LAB with food fermentations as acidification inhibits the growth of spoilage agents. Proteinaceous bacteriocins are produced by several LAB strains and provide an additional hurdle for spoilage and pathogenic microorganisms. Furthermore, lactic acid and other metabolic products contribute to the organoleptic and textural profile of a food item. The industrial importance of the LAB is further evidenced by their generally regarded as safe (GRAS) status, due to their ubiquitous appearance in food and their contribution to the healthy microflora of human mucosal surfaces. The genera that comprise the LAB are at its core *Lactobacillus, Leuconostoc, Pediococcus, Lactococcus,* and *Streptococcus* as well as the more peripheral *Aerococcus, Carnobacterium, Enterococcus, Oenococcus, Teragenococcus, Vagococcus,* and *Weisella*; these belong to the order Lactobacillales.

ACETIC ACID BACTERIA

These are bacteria that derive their energy from the oxidation of ethanol to acetic acid during respiration. They are gram-negative, aerobic, rod-shaped bacteria. The acetic acid bacteria are found in nature where ethanol is being formed as a result of yeast fermentation of sugars and plant carbohydrates. They can be isolated from the nectar of flowers and from damaged fruit. Other good sources are fresh apple cider and unpasteurized beer which has not been filter sterilized. In these liquids the acetic acid bacteria grow as a surface film due to their aerobic nature and active motility. Vinegar is produced when acetic acid bacteria act on alcoholic beverages such as wine.

Some genera, such as *Acetobacter*, can eventually oxidize acetic acid to carbon dioxide and water using Krebs cycle enzymes. Other genera, such as *Gluconobacter*, doesn't further oxidize acetic acid, as they do not have a full set of Krebs cycle enzymes. Some acetic acid bacteria, notably *Acetobacter xylinum*, are known to synthesize cellulose, something normally only done by plants. As these bacteria produce acid, they are unusually acid tolerant.

The first attempt to classify acetic acid bacteria was made by Hansen in 1894, followed by other investigators. In 1968, a comprehensive account of the taxonomic history of acetic acid bacteria was proposed where the genus *Acetobacter* was divided into two, *Acetobacter* and a new genus *Gluconobacter*. This was based on a study that included isolates from fruits as well as those from the usual source, vinegar. The characteristics of the *Gluconobacter* were the ability to produce large amounts of gluconic acid from glucose, inability to form films in liquid media and poor growth in ethanol-containing substrates such as sake and beer. A similar division based on morphology was proposed in 1954 where those with peritrichous flagella retained the name *Acetobacter*. Those with polar flagella were placed in a new genus *Acetomonas*. Species of the latter also displayed the characteristics ascribed to *Gluconobacter* spp. The *Acetobacter* were classified as being lactophilic in their nutrition. They grow well on lactate, and in many cases with ammonium salts as the sole nitrogen source but relatively poor on gluose.

It was found that strains of *Acetobacter* change their properties right at the time of their isolation. Hence identification of a strain of *Acetobacter* reflects therefore only its properties at the time of isolation and gives no information on its properties at an earlier or later date. Hence *Acetobacter* defy classification. In spite of the propensity of *Acetobacter* for variation, it is possible to perform vinegar fermentations year in year out without interruption using mutation techniques. Some species of *Acetobacter* used in the vinegar industry are *A. curvum, A. aceti, A. schutzenbachii, A. suboxydans* and *A. melanogenum*.

Acetobacter species show sensitivity towards lack of oxygen. The reason for the extreme sensitivity of *Acetobacter* during the fermentation process is the lack of oxygen. *Acetobacter* has high apyrase activity so that ATP which accumulates during the oxidation of ethanol is rapidly hydrolysed and therefore only poorly available for other metabolic activities of the cell. When aeration is interrupted, the ATP pool disappears so quickly that the cell does not have the ability to adjust to the changed conditions. Acetic acid bacteria are damaged if a vinegar fermentation is carried onto the point where all of the ethanol has been oxidized and if the addition of fresh ethanol containing mash is delayed beyond that point. This can cause severe damage to the bacteria. The bacteria are sensitive to changes in temperature. If cooling is stopped during a vinegar fermentation, the temperature rises higher and higher. Damage to the cells of *Acetobacter* increases with the duration of the interruption of cooling, with higher temperatures, and with higher concentrations of acetic acid.

The specific growth rate of *A. aceti* rises sharply with the rate of aeration at an acetic acid concentration of 2%. Increasing concentrations of acetic acid inhibited the uptake of oxygen by the bacterial cells, which was explained by inhibition of the cytochrome system and the organisms are insufficiently adapted to the acetic acid concentrations between 5 and 7%.

Acetic Acid Fermentation

Acetobacter converts alcohol to acetic acid in the presence of excess oxygen.

Oxidation of Alcohol to Acetic Acid and Water

The oxidation of one mole of ethanol yields one mole each of acetic acid and water. The reaction is represented as follows.

$$C_2H_5OH + O_2 \rightarrow CH_3COOH^+ + H_2O$$
$$\text{(Alcohol)} \qquad \text{(Acetic acid)} \quad \text{(Water)}$$

PROPIONIC ACID BACTERIA

The family Propionibacteriaceae, genus *Propionibacterium* and the closely related genus *Corynebacterium,* are classified as members of the Actinomycetaceae groups. *Propionibacterium* species were first described in 1909 by Orla-Jensen who divided them into two principal groups: 1) the classical or dairy propionibacteria and 2) the acnes or cutaneous propionibacteria. The four species of the cutaneous propionibacteria include: *P. lymphophilum, P. granulosum, P. avidum,* and *P. acne.* These organisms are found in soft tissue abscesses, in dental plaque, and on human skin. Five species of dairy propionibacteria are currently recognized: *P. freudenreichii* subsp. *freudenreichii, P. freudenreichii* subsp. *shermanii, P. thoenii, P. acidipropionici, and P. jensenii.*

The classical propionibacteria which are important starter organisms in dairy fermentations, may contribute to natural fermentation of silage and olives and can produce a variety of industrially important products.

Industrial Uses of Propionibacteria

The propionibacteria are probably best known for their role as dairy starter cultures, in which they produce the characteristic eyes and flavour of Swiss type cheeses. The fermentation of lactose to lactic acid by the starter streptococci and lactobacilli provides the substrate for fermentation by the propionibacteria. The characteristic flavour of Swiss cheese is due, to the production of short-chain fatty acids, amino acids, and metabolic intermediates by the propionibacteria. The growth rates and carbon dioxide production by these secondary flora are critical in determining the size and distribution of holes (eyes) in the cheese. In addition to the flavour, the propionic and acetic acids produced by propionibacteria are inhibitory to moulds, yeasts and some bacteria. The presence of these organic acids is known to improve the shelf life of fermented products.

In addition to their use as dairy starter cultures, they play a role in other industrial processes. One such application is the production of vitamin B_{12}. In propionibacteria, the metabolic pathways leading to propionate involve enzymatic reactions that require vitamin B_{12} as a cofactor and the organisms produce enough of this vitamin to be used as a commercial source. Yields of vitamin B_{12} from fermentations with *P. freudenreichii* ATCC 6207 were reported to be 23 mg/ml. Strains of *Pseudomonas* have largely replaced the propionibacteria in commercial vitamin B_{12} fermentations, however, because of their faster growth rates and higher yields of the vitamin.

Propionic and acetic acids, both of which are produced as primary metabolites by the propionibacteria, have many uses as industrial chemicals, including the production of plastics, herbicides, perfumes, etc. In addition, their antimicrobial activity make these acids, particularly propionic acid, as important food and feed preservatives. As a preservative, propionic acid extends the shelf life of food products by inhibiting moulds, yeasts, and some bacteria.

BACTERIA OF ALKALINE FERMENTATIONS

Another group of bacteria are those which bring about alkaline fermentations—the *Bacillus* species. Among them, notable are *Bacillus subtilis, B. licheniformis* and *B. pumilus. Bacillus subtilis* is the dominant species, causing the hydrolysis of protein to amino acids and peptides and releasing ammonia, which increases the alkalinity and makes the substrate unsuitable for the growth of spoilage organisms. Alkaline fermentations are more common with protein-rich foods such as soybeans and other legumes, although there are a few examples utilizing plant seeds. For example watermelon seeds (*Ogiri* in Nigeria) and sesame seeds, (*Ogiri-saro* in Sierra Leone) and others where coconut and leaf proteins are the substrates (Indonesian *semayi* and Sudanese *kawal* respectively).

Although the range of products of alkaline fermentation does not match those brought about by acid fermentations, they are important in that they provide protein rich, low-cost condiments from leaves, seeds and beans, which contribute to the diet of millions of people in Africa and Asia.

CONDITIONS REQUIRED FOR BACTERIAL FERMENTATIONS

Hyrdogen Ion Concentration

Microorganisms vary in their optimal pH requirements for growth. Most bacteria favour conditions with a near neutral pH. The varied pH requirements of different groups of microorganisms is used to have a good effect in fermented foods where successions of microorganisms take over from each other as the pH of the environment changes. Certain bacteria are acid-tolerant and will survive at reduced pH levels. Notable acid-tolerant bacteria include the *Lactobacillus* and *Streptococcus* species, that play a role in the fermentation of dairy and vegetable products.

Temperature

Different bacteria can tolerate different temperatures, which provides enormous scope for a range of fermentations. While most bacteria have an optimum temperature between 20 to 30°C, there are some (the thermophiles) which prefer higher temperatures (50 to 55°C) and those ith colder optimal temperature (15 to 20°C). Most lactic acid bacteria work best at temperatures of 18 to 22°C. The *Leuconostoc* species which initiate fermentation have an optimum of 18 to 22°C. Temperatures above 22°C, favour the *Lactobacillus* species.

Salt Concentration

Lactic acid bacteria can tolerate high salt concentrations. The salt tolerance gives them an advantage over other less tolerant species and allows the lactic acid fermenters to begin metabolism, which produces acid, that further inhibits the growth of non-desirable organisms. *Leuconostoc* is noted for its high salt tolerance and for this reason, it initiates the majority of lactic acid fermentations.

Water Activity

In general, bacteria require a fairly high water activity (0.9 or higher) to survive. There are a few species which can tolerate water activities lower than this, but usually the yeasts and fungi will predominate on foods with a lower water activity.

Oxygen Availability

Oxygen requirements vary from species to species. Some of the fermentative bacteria are anaerobes, while others require oxygen for their metabolic activities. Some, lactobacilli in particular, are microaerophilic. That is they grow in the presence of reduced amounts of atmospheric oxygen. In aerobic fermentations, the amount of oxygen present is one of the limiting factors. It determines the type and amount of biological product obtained, the amount of substrate consumed and the energy released from the reaction. *Acetobacter* requires oxygen for the oxidation of alcohol to acetic acid.

In vinegar production, oxygen has to be made available for the production of acetic acid, whereas with wine it is essential to exclude oxygen to prevent oxidation of the alcohol and spoilage of the wine.

Nutrients

All bacteria require a source of nutrients for metabolism. The fermentative bacteria require carbohydrates—either simple sugars such as glucose and fructose or complex carbohydrates such as starch or cellulose. The energy requirements of microorganisms are very high. Limiting the amount of available substrate can check their growth.

REVIEW QUESTIONS

1. Give a detailed account of lactic acid bacteria and their metabolism.
2. Differentiate between homolactic and heterolactic fermentation.

38

Examples of Lactic Acid Fermentation

LACTIC ACID FERMENTATION

Lactic acid bacteria perform an essential role in the preservation and production of wholesome foods ranging from fermented fresh vegetables such as cabbage (sauerkraut/Korean kimchi) and cucumbers (pickles), fermented cereal yoghurt (Nigerian ogi/Kenyan uji), sour-dough bread and bread-like products without the use of wheat or rye flours (Indian idli/Philippine puto), fermented milks (yoghurts/cheeses), fermented milk/wheat mixtures (Egyptian kishk/Greek trahanas), protein-rich, vegetable protein meat substitutes (Indonesian tempe), amino/peptide meat-flavoured sauces and pastes produced by fermentation of cereals/legumes (Japanese miso/Chinese soy sauce), fermented cereal/fish/shrimp mixtures (Philippine balao/balao; Philippine burong dalag), fermented meats (European salami), etc.

Both the homofermentative and the heterofermentative lactic acid bacteria are generally fastidious on artificial media but they grow readily in most food substrates and lower the pH rapidly to a point where other competing organisms are unable to grow. Leuconostocs and lactic streptococci generally lower the pH to about 4.0–4.5 and some of the lactobacilli and pediococci to about 3.5 before inhibiting their own growth.

In addition to lactic acid, the lactobacilli also have the ability to produce hydrogen peroxide, through oxidation of reduced nicotinamide adenine dinucleotide (NADH) by flavin nucleotides which react rapidly with gaseous oxygen. Flavoproteins such as glucose oxidase also generate hydrogen peroxide and produce an antibiotic effect on other organisms that might cause food spoilage. The lactobacilli themselves are relatively resistant to hydrogen peroxide. For example, it was shown in a study that *Lactobacillus lactis* required a concentration of 125 mg hydrogen peroxide/ml to inhibit an inoculum of 10^4cells/ml while

Staphylococcus aureus was inhibited by a concentration of 4 mg hydrogen perioxide/ml even though the latter contains catalase which is lacking in *L. lactis. Streptococcus lactis* produces the polypeptide antibiotic nisin active against gram-positive organisms including *S. cremoris. S. cremoris,* in turn, produces an antibiotic "diplococcin" active against gram-positive organisms including *S. lactis.*

PRODUCTION OF FERMENTED VEGETABLES

Thus, these two organisms compete in the fermentation of milk products while inhibiting the growth of other gram-positive bacteria. Carbon dioxide produced by heterofermentative lactobacilli also has a preservative effect in foods resulting among others, from its flushing action leading to anaerobiosis if the substrate is properly protected.

Brining and lactic acid fermentation both continue to be highly desirable methods of processing and preserving vegetables because they are cost effective, have low energy requirements and they yield highly acceptable and diversified flavours for humans. Depending upon the salt concentration, salting directs the subsequent course of the fermentation thereby limiting the amount of pectinolytic and proteolytic hydrolysis that occur thus preventing softening and putrefaction. Lactic acid fermentations have some other distinct advantages in that, the foods become resistant to microbial spoilage and to the development of toxins. Acid foods are less likely to transfer pathogenic microorganisms. Acid fermentations also modify the flavour of the original ingredients and often improve the nutritive value.

Since canned or frozen foods are unavailable or too expensive to hundreds of millions of the world's economically deprived and hungry, acid fermentation combined with salting remains one of the most practical methods of preservation often enhancing organoleptic and nutritional quality of fresh vegetables, cereal gruels and milk–cereal mixtures. Lactic acid fermentation is utilized as a major method of processing and preserving vegetables, cereals and legumes throughout the world and particularly in the developing world. Lactic acid fermentations are carried out under three basic types of condition—dry-salted, brined and non-salted. Salting provides a suitable environment for lactic acid bacteria to grow which impart the acid flavour to the vegetable.

Pickling

Pickling is one of the oldest forms of food preservation. It has been traced back to the dawn of civilization, 4500 years ago when people learned to preserve cucumbers by pickling them in a salty brine. However, processors probably have pickled almost every type of vegetable. Today's variety of pickled products include cucumbers, olives, peppers, chutney (an Indian spiced product) and kimchi (a Korean fermented vegetable product).

While not all pickled products undergo fermentation, in many vegetables the natural sugar is converted to lactic acid by specific bacteria during fermentation. This process turns cabbage into sauerkraut and cucumbers into pickles.

Some general principles apply to all types of pickled vegetable products. Factors such as post-harvest handling, quality of ingredients, and proper processing techniques greatly influence the quality of finished pickled products.

The various species of lactic acid bacteria responsible for vegetable fermentation have their own levels of salt tolerance and temperature ranges for growth. Strains of *Leuconostoc mesenteroides* initiate vegetable fermentations, followed by other species of lactic acid bacteria (*Lactobacillus*). The *Leuconostoc* are salt- and sugar-tolerant over a wide range of temperatures and are able to initiate vegetable fermentations more rapidly than any other bacteria.

Carbohydrates in the vegetables furnish energy for the bacteria in the form of sugars and other essential nutrients such as amino acids and peptones, lipids, vitamins and minerals. The bacteria ferment sugar to lactic acid, as well as carbon dioxide, ethyl alcohol, and acetic acid. These products of fermentation rapidly lower the pH, inhibit the growth of undesirable microorganisms, and deactivate vegetable-softening enzymes.

Two general methods of packing are used for ideal fermentation conditions: dry-salting and brine-salting. Usually, shredded or chopped cabbage is dry-salted in the production of sauerkraut. Vegetables such as cucumbers, olives and large vegetable pieces normally use brine.

Basic brines are solutions of salt and water. They may include sugars, flavourings and other ingredients. Salinity is measured with a brine hydrometer, called a "salometer." This instrument has a scale from 0 to 100 degrees. It is calibrated in percentages of saturation with respect to sodium chloride. A saturated solution of sodium chloride, 26.5% salt, reads 100 degrees in a salometer.

In brined vegetables, undesirable aerobic microorganisms are inhibited by the production of carbon dioxide and acid. Carbon dioxide replaces air, creating anaerobic conditions favourable to ascorbic acid stabilization and oxidation inhibition.

Brine or salt draws water from vegetables and decreases the salt concentration in the brine itself. During the pickling process, brine strength is usually raised gradually; otherwise, acid development would be inhibited and the vegetables would become soft. All production facilities recirculate brine to reduce pollution and conserve resources.

Both the concentration of salt in the brine and the salting technique affect the quality of the finished product. Salt reduces the competition from undesirable microorganisms, which encourages lactic acid fermentation. Salt and acid concentration not only control the growth of microorganisms, but influence enzyme activity as well. The temperature also helps determine the rate of acid production and the kinds of bacteria involved in it.

As the salt concentration increases, less acid is produced due to inhibition of lactic acid bacteria. The vegetable's sugar content is directly related to the amount of acid that is produced, so the higher the sugar level, the lesser the salt that is needed. For example, as the amount of fermentable sugars in cabbages is higher than that of cucumbers, less salt is used in sauerkraut production than in pickle processing, and more acid is produced in kraut than in pickles. The salt used in brine production should be free from flow agents and iodine to avoid a cloudy brine and darkened pickles. Pickled and fermented vegetable products may be classified into four general categories based on the process used to create them: 1) unfermented products, such as California-style black and green olives; 2) weak brine fermented products, such as dill pickles; 3) high-salt brine products, such as salt-stock pickles; and 4) dry-salted products, such as sauerkraut.

Sauerkraut is one example of an acid fermentation of vegetables. The name sauerkraut literally translates as acid cabbage. The "sauerkraut process" can be applied to any other suitable type of vegetable product. Because of the importance of this product in the German diet, the process has received substantial research in order to commercialize and standardize production. As a result, the process and the contributing microorganisms are known intimately. Other less well known fermented fruits and vegetables have received less research attention, therefore little is known of the exact process. However, it is safe to assume that the acid fermentation of vegetables is based on this process.

Salt for pickling   For pickling, any variety of common salt is suitable as long as it is pure. Impurities or additives can cause problems. Salt with chemicals to reduce caking should not be used as they make the brine cloudy. Salt with lime impurities can reduce the acidity of the final product and reduce the shelf life of the product. Salt with iron impurities can result in the blackening of the vegetables. Magnesium impurities impart a bitter taste. Carbonates can result in pickles with a soft texture.

DRY-SALTED FERMENTED VEGETABLES

In dry salting, the vegetable is treated with dry salt. The salt extracts the juice from the vegetable and creates the brine. The vegetable is prepared, washed in potable cold water and drained. For every 100 kg of vegetables, 3 kg of salt is needed. The vegetables are placed in a layer of about 2.5 cm depth in the fermenting container (a barrel or keg). Salt is sprinkled over the vegetables. Another layer of vegetables is added and more salt added. This is repeated until the container is three quarters full. A cloth is placed above the vegetables and a weight added to compress the vegetables and assist the formation of a brine which takes about 24 hours. As soon as the brine is formed, fermentation starts and bubbles of carbon dioxide begin to appear. Fermentation takes one to four weeks depending on the ambient temperature. Fermentation is complete when no more bubbles appear, after which the pickle can be packaged in a variety of mixtures.

The "Sauerkraut" Process

Sauerkraut is the clean, sound product of characteristic flavour, obtained by full fermentation, chiefly lactic, of properly prepared and shredded cabbage in the presence of 2–3% salt. It is common in America and Europe. It contains, upon completion of the fermentation, not less than 1.5 % of acid, expressed as lactic acid. Sauerkraut which has been re-brined in the process of canning or repacking contains not less than 1% of acid, expressed as lactic acid. Lactic acid bacteria are the primary group of organisms involved in sauerkraut fermentation. They can be divided into three groups according to their types and end products:

1. *Leuconostoc mesenteroides*   Coccus that produces acid and gas.

2. *Lactobacillus plantarum and L. cucumeris*   Bacilli   that produce acid and a small amount of gas

3. *Lactobacillus pentoaceticus and L. brevis*   Bacilli that produce acid and gas

In addition to the desirable bacteria, there are a range of undesirable microorganisms present on cabbage (and other vegetable material), which can interfere with the sauerkraut process if allowed to multiply unchecked. The quality of the final product depends largely on how well the undesirable organisms are controlled during the fermentation process. Some of the typical spoilage organisms utilize the protein as an energy source, producing unpleasant odours and flavours.

The fermentation process   Sound heads of cabbage are prepared by washing and removing the outer leaves and any defective leaves. The core is removed and the leaves are shredded. Shredding the leaves gives a larger total surface area and allows the extraction of juice. Shredded cabbage or other suitable vegetables are placed in a jar and salt (2.25%)is added. Mechanical pressure is applied to the cabbage and the combination of salt expels the juice (brine), which contains fermentable sugars (3–6%) and other nutrients suitable for microbial activity. The amount of sugar influences the fermentation and the final acidity. When the tank is essentially full, a plastic sheet is used as a cover to keep out dirt and air. The environmental conditions, numbers and kinds of microorganisms, cleanliness of cabbage and tank, salt concentration and distribution, temperature and covering influence the fermentation.

Many types of microbes are associated with raw cabbage. Most of these are not involved in the fermentation. Cabbage contains substances which are inhibitory to gram-negative bacteria. With the inhibitors, salt and anaerobic environment, the lactic acid bacteria tend to dominate. The first microorganisms to act are the gas-producing cocci (*L. mesenteroides*). These microbes produce acids (lactic acid, acetic acid) and alcohol from the sugar, carbon dioxide and other products which contribute to the flavour of sauerkraut. The carbon dioxide helps maintain anaerobic conditions in the fermenting cabbage. When the acidity reaches 0.25 to 0.3% (calculated as lactic acid), these bacteria slow down and begin to die off, although their enzymes continue to function. The activity initiated by the *L. mesenteroides* is continued by the lactobacilli (*L. brevis, Pediococcus cereviseae* and finally *L. plantarum*) until an acidity level of 1.5 to 2% is attained. The high salt concentration and low temperature inhibit these bacteria to some extent. Finally, *L. pentoaceticus* continues the fermentation, bringing the acidity to 2 to 2.5% thus completing the fermentation.

The end products of a normal kraut fermentation are lactic acid along with smaller amounts of acetic and propionic acids, a mixture of gases of which carbon dioxide is the principal gas, small amounts of alcohol and a mixture of aromatic esters. The acids, in combination with alcohol form esters, which contribute to the characteristic flavour of sauerkraut. The acidity helps to control the growth of spoilage and putrefactive organisms and contributes to the extended shelf life of the product. Changes in the sequence of desirable bacteria, or indeed the presence of undesirable bacteria, alter the taste and quality of the product.

Effects of temperature on the sauerkraut process   The optimum temperature for sauerkraut fermentation is around 21°C. A variation of just a few degrees from this temperature alters the activity of the microbial process and affects the quality of the final product. Therefore, temperature control is one of the most important factor in the sauerkraut process. A temperature of 18°C to 22°C is most desirable for initiating fermentation since this is the optimum temperature range for the growth and metabolism of *L. mesenteroides*. Temperatures above 22°C favour the growth of *Lactobacillus* species.

Effects of salt on the sauerkraut process   Salt plays an important role in initiating the sauerkraut process and affects the quality of the final product. The addition of too much salt may inhibit the desirable bacteria, although it may contribute to the firmness of the kraut. The principle function of salt is to withdraw juice from the cabbage (or other vegetable), thus making a more favourable environment for development of the desired bacteria.

Generally, salt is added to a final concentration of 2.0 to 2.5%. At this concentration, lactobacilli are slightly inhibited, but cocci are not affected. Unfortunately, this concentration of salt has a greater inhibitory effect against the desirable organisms than against those responsible for spoilage. The spoilage organisms can tolerate salt concentrations up to 5 to 7%, therefore it is the acidic environment created by the lactobacilli that keep the spoilage bacteria at bay, rather than the addition of salt.

In the manufacture of sauerkraut, dry salt is added at the rate if 1 to 1.5 kg per 50 kg cabbage (2 to 3%). The use of salt brines is not recommended in sauerkraut making, but is common in vegetables that have a low water content. It is essential to use pure salt since salts with added alkali may neutralize the acid.

Use of starter cultures   In order to produce sauerkraut of consistent quality, starter cultures (similar to those used in the dairy industry) have been recommended. Not only do starter cultures ensure consistency between batches, they speed up the fermentation process as there is no time lag while the relevant microflora colonize the sample. Because the starter cultures used are acidic, they also inhibit the undesirable microorganisms. It is possible to add starters traditionally used for milk fermentation, such as *Streptococcus lactis*, without an adverse effect on final quality. Because these organisms only survive for a short time (long enough to initiate the acidification process) in the kraut medium, they do not disturb the natural sequence of microorganisms. On the other hand, if *Leuconostoc mesenteroides* is added in the early stages, it gives a good flavour to the final product, but alters the sequence of subsequent bacterial growth and results in a product that is partially fermented. If gas producing rods (for example *L. pentoaceticus*) are added to the sauerkraut, this disturbs the balance between acetic and lactic acids—more acetic acid and less lactic acid are produced than normal—and the fermentation never reaches completion. If lactic acid, non-gas producing *rods* (*L. cucumeris*) are used as a starters, again the kraut is not completely fermented and the resulting product is bitter and more susceptible to spoilage by yeasts.

It is possible to use the juice from a previous kraut fermentation as a starter culture for subsequent fermentations. The efficacy of using old juice depends largely on the types of organisms present in the juice and its acidity. If the starter juice has an acidity of 0.3% or more, it results in a poor-quality kraut. This is because the cocci which would normally initiate fermentation are suppressed by the high acidity, leaving the bacilli with sole responsibility for fermentation. If the starter juice has an acidity of 0.25% or less, the kraut produced is normal, but there do not appear any beneficial effects of adding this juice. Often, the use of old juice produces a sauerkraut which has a softer texture than normal.

Spoilage and defects in the sauerkraut process The majority of spoilage in sauerkraut is due to aerobic soil microorganisms which break down the protein and produce undesirable flavour and texture changes. The growth of these aerobes can easily be inhibited by a normal fermentation.

Soft kraut may result from many conditions such as large amounts of air, poor salting procedure and varying temperatures. Whenever the normal sequence of bacterial growth is altered or disturbed, it usually results in a soft product. It is the lactobacilli, which seem to have a greater ability than the cocci to break down cabbage tissues, which are responsible for the softening. High temperatures and a reduced salt content favour the growth of lactobacilli, which are sensitive to higher concentrations of salt. The usual concentration of salt used in sauerkraut production slightly inhibits the lactobacilli, but has no effect on the cocci. If the salt content is too low initially, the lactobacilli grow too rapidly at the beginning and upset the normal sequence of fermentation.

Another problem encountered is the production of dark coloured sauerkraut. This is caused by spoilage organisms during the fermentation process. Several conditions favour the growth of spoilage organisms. For example, an uneven distribution of salt tends to inhibit the desirable organisms while at the same time allowing the undesirable salt tolerant organisms to flourish. An insufficient level of juice to cover the kraut during the fermentation allows undesirable aerobic bacteria and yeasts to grow on the surface of the kraut, causing off-flavours and discolouration. A higher fermentation temperature also encourages the growth of undesirable microflora, which results in a darkened colour.

Pink kraut is a spoilage problem. It is caused by a group of yeasts which produce an intense red pigment in the juice and on the surface of the cabbage. It is caused by an uneven distribution of or an excessive concentration of salt, both of which allow the yeasts to multiply. If conditions are optimal for normal fermentation, these spoilage yeasts are suppressed.

Kanji

In Northern India and Pakistan carrots, especially a variety that is deep purple in colour, are fermented to make a traditional ready-to-serve drink known as kanji. Kanji is very popular and considered to have cooling and soothing properties and to be of high nutritional value. After thorough washing, the carrots are finely grated. Each kilogram of grated carrot is mixed with 7 litres of water, 200 g of salt, 40 g of crushed mustard seed and 8 g of hot chilli powder. The mixture is then placed in a glazed earthenware vessel, which is almost entirely sealed, leaving only a tiny hole for gases to escape during fermentation. The mixture is then allowed to ferment for seven to ten days. The type of fermentation that takes place is known as lactic acid fermentation, which must be carried out in the absence of air. Lactic acid bacteria produce lactic acid which reduces the pH (i.e., increases the acidity) to a level that prevents the growth of food poisoning organisms. The final product is slightly acidic in taste and has an attractive purple-red colour. After fermentation, the drink is strained through fine muslin and has to be consumed within 3 or 4 days after which it cannot be used. Each kg of grated carrot yields just over 7 litres of kanji.

Kimchi (Pickled Cabbage)

Kimchi is probably the most important processed food product in Korea. It is an essential dish, eaten at most mealtimes. Production is estimated at over one million tons, mainly household level. Daily consumption as estimated at 150 to 250 grams per person.

Kimchi is a general name for a range of closely related fermented products. It is similar to Sauerkraut in Europe and the United States. There are numerous variations of kimchi depending on the production technique. The main pickled cabbage *kimchis* are *tongbaechu-kimchi tongkimchi* and *bossam*-kimchi.

Appropriate cultivars of Chinese cabbage, with light-green coloured soft leaves and compact structures having no defects, are required for production of kimchi. After removing outer leaves and roots from the cabbage, it is cut into small pieces.

The prepared cabbage is placed in a salt solution (8–15%) for two to seven hours in order to increase the salt content of the cabbage to between 2.0–4.0% (w/w). It is then rinsed several times with fresh water and drained to remove extra water by centrifugation or by allowing it to stand.

Kimchi fermentation is carried out by various microorganisms present in the raw materials and ingredients used in the preparation of kimchi. Among the two hundred bacteria isolated form kimchi, the important microorganisms in kimchi fermentation are known to be *Lactobacillus plantarum, L. brevis, Streptococcus faecalis, Leuconostoc mesenteroides* and *Pediococcus pentosaceus.*

After fermentation, the product can be left to mature for several weeks if refrigeration is available. If stored under warm conditions, the kimchi deteriorates rapidly.

Indian Idli/Dosa

Indian idli is a small, white, acidic, leavened and steam-cooked cake made by lactic fermentation of a thick batter made from polished rice and dehulled black gram dhal, a pulse. The cakes are soft, moist and spongy and have a desirable sour flavour. A closely related product is dosa made from the same ingredients, both finely ground. The batter is generally thinner and dosa is fried like a pancake.

The importance of the idli fermentation is that it is a process by which leavened bread-like products can be made from cereals other than wheat or rye and without yeast. The initial step in the fermentation is to wash both rice and black gram dhal. They are then soaked during the day, generally for 5–10 hours. The ingredients are then drained; and the rice is coarse-ground and the black gram finely ground separately in a mortar or other grinder. To this paste, water and salt is added to make a thick batter. The batter is fermented in a warm place (30–32°C) for overnight during which time acidification and leavening occur. The batter is then placed in small cups and steamed or fried as a pancake.

The proportions of rice to black gram vary from 4:1 to 1:4. Idli or dosa is a product of natural lactic acid fermentation. *L. mesenteroides* and *S. faecalis* develop concomitantly during soaking and then continue to multiply following grinding. Each eventually reaches more than 1×10^9 cells/g is 11–13 hours after formation of the batter. These two species predominate until 23 hours following batter formation. Practically all batters would be steamed by then. If the batter is incubated further, the lactobacilli and streptococci decrease in numbers and *P. cerevisiae* develops. *L. mesenteroides* is the microorganism essential for leavening of the batter and along with *S. faecalis* also responsible for acid production, both being essential for producing a satisfactory idli.

In idli made with a 1:1 proportion of black gram to rice, batter volume increased in 12–15 hours after incubation at 30°C. The pH fell to 4.5 and total acidity rose to 2.8 (as lactic acid). Using a 1:2 ratio of black gram to rice, batter volume increased further and acidity rose to 2.2 in 20 hours. Reducing sugars (as glucose) showed a steady decrease from 3.3 mg/g dry ingredients to 0.8 mg/g in 20 hours reflecting their utilization for acid and gas production. Soluble solids increased while soluble nitrogen decreased. Flatulence-causing oligosaccharides such as stachyose and raffinose are completely hydrolysed.

Philippine Puto

Philippine puto is a leavened, steamed rice cake made from 1-year-old rice grains which are soaked, ground with water and allowed to undergo a natural acid and gas fermentation. Part of the acid is neutralized with sodium hydroxide during the last stage of fermentation. Puto is closely related to Indian idli except that it contains no legume.

Nigerian Ogi (Kenyan Uji)

Nigerian ogi is a smooth-textured, sour porridge with a flavour resembling yoghurt. It is made by lactic acid fermentation of corn, sorghum or millet. Soybeans may be added to improve nutritive value. Ogi has a solid content of about 8%. The cooked porridge known as "pap" is gel-like. The first step in the fermentation is steeping of the cleaned grain for 1–3 days. During this time, the desirable microorganisms develop and are selected. The grain is then ground with water and filtered to remove coarse particles. The pH following steeping should be 4.3. Optimum pH for ogi is 3.6–3.7. The concentration of lactic acid may reach 0.65% and that of acetic acid 0.11% during fermentation. If the pH falls to 3.5, it is less acceptable.

Ogi is a natural fermentation product. A wide variety of moulds, yeasts and bacteria are present initially. *L. plantarum* appears to be the essential microorganism in the fermentation. Following depletion of the fermentable sugars it is able to utilize dextrins from the corn. *S. cerevisiae* and *Candida mycoderma* contribute to the desirable flavour.

Nigerian Gari

Nigerian gari is a granular starchy food made from cassava *(Manihot utilisima* or *M. esculenta)* by lactic acid fermentation of the grated pulp, followed by a dry-heat treatment to gelatinize and semi-dextrinize the starch and drying. Cassava tubers are washed, peeled and grated. An inoculum of 3-day-old cassava juice or fermented mash liquor is added. The pulp is then placed in a cloth bag, excess water is squeezed out and the pulp undergoes an anaerobic acid fermentation for 12–96 hours. Optimum temperature is 35°C. When the pH of the mash reaches 4.0 with about 0.85% total acid (as lactic acid), the gari has the desired sour flavour and a characteristic aroma. Further moisture may be removed and then the pulp is toasted (semi-dextrinized) in shallow iron pots and dried to less than 20% moisture in village processes. Village-processed gari has a carbohydrate content of about 82% with 0.9% protein.

For consumption, the gari is added to boiling water in which it increases in volume by 300% to yield a semi-solid plastic dough. The stiff porridge is rolled into a ball (10–30 g wet wt.) and dipped into stew for consumption. *Leuconostoc* reaches populations of 10^8/g in 24 hours while yeasts reach 10^5–10^6/g. Lactic, acetic, propionic, succinic, and pyruvic acids are identified in gari with aldehydes and esters providing the characteristic aroma.

Philippine Balao Balao

Balao balao is a lactic acid fermented rice shrimp mixture, generally prepared by blending cooked rice, whole raw shrimp and solar salt, and then allowing the mixture to ferment for several days or weeks depending upon the salt content. The chitinous shell becomes soft and when the fermented product is cooked, the whole shrimp can be eaten.

With a salt concentration of 3% added to the rice/shrimp mixture, the pH falls to an organoleptically desired value of 4.08 with titratable acidity reaching 1.32% acid (as lactic acid) in 4 days. Total counts of lactobacilli reach a peak of 1.1×10^9 cells/g and yeast counts reach 78×10^6 cells/g in 6 days.

Balao balao made with 3% salt is best in colour, odour, flavour, texture, general acceptability and is least salty. Balao balao offers a basic method of preservation for cereal/shrimp/fish mixtures. When properly packed to exclude air, sufficient acid is produced to preserve the products without resorting to high-temperature cooking.

Mexican Pulque

Mexican pulque is a white, acidic, alcoholic beverage made by fermentation of juice of the *Agave,* mainly *A. atrovirens* or *A. americana,* the century plants. It has been a national Mexican drink since the time of the Aztecs. Pulque plays an important role in the nutrition of the low-income groups in the semi-arid regions of Mexico. The essential microorganisms in the pulque fermentationare *L. plantarum,* a heterofermentative *Leuconostoc* sp., *S. cerevisiae,* and *Zymomonas mobilis.* The heterofermentative *Leuconostoc* plays an essential role in producing dextrans, that contribute a characteristic viscosity to pulque, and also increases the acidity of the Agave juice very rapidly inhibiting the growth of other less desirable bacteria. *L. plantarum* contributes to the final acidity of pulque. *S. cerevisiae* appears to be a major producer of ethanol. Under anaerobic conditions, *Zymomonas* uses the Entner–Douderoff pathway transforming 45% of the glucose to ethanol and CO_2. It also produces some acetic acid, acetylmethylcarbinol and some slimy gums which may contribute to the viscous nature of traditional pulque. The pH falls from 7.4 to 3.5 4.0. Total acid increases from 0.029% to 0.4–0.7% (as lactic acid). Sucrose decreases from 18.6% to less than 1%. Ethanol increases from 0% to 4–6 % (v/v). The B vitamins are present in nutritionally important quantities.

BRINE-SALTED FERMENTED VEGETABLES

Brine is used for vegetables which inherently contain less moisture. A brine solution is prepared by dissolving salt in water (a 15 to 20% salt solution). Fermentation takes place well in a brine of about 20° salometer. As a general rule, a fresh egg floats in a 10% brine solution. Properly brined vegetables will keep well in vinegar for a long time. The duration of brining is important for the overall keeping qualities. The vegetable is immersed in the brine and allowed to ferment. The strong brine solution draws sugar and water out of the vegetable, which decreases the salt concentration. It is crucial that the salt concentration does not fall below 12%, otherwise conditions do not allow fermentation. To achieve this, extra salt is added periodically to the brine mixture.

Once the vegetables have been brined and the container sealed, there is a rapid development of microorganisms in the brine. The natural controls which affect the microbial populations of the fermenting vegetables include the concentration of salt and temperature of the brine, the availability of fermentable materials and the numbers and types of microorganisms present at the starting time of fermentation. The rapidity of the fermentation is correlated with the concentration of salt in the brine and its temperature.

Most vegetables can be fermented at 12.5° to 20° salometer salt. If so, the microbial sequence of lactic acid bacteria generally follows the classical sauerkraut fermentation. At higher salt levels of up to about 40° salometer, the sequence is skewed towards the development of a homofermentation, dominated by *Lactobacillus plantarum*. At the highest concentrations of salt (about 60° salometer) the lactic acid fermentation ceases to function and if any acid is detected during brine storage it is acetic acid, presumably produced by acid-forming yeasts which are still active at this concentration of salt.

Dill Pickles

Dill pickles are fermented cucumbers. "Dill" pickles get their name because spices and herbs like dill are added to the fermentation vats to give flavour. There are many pickle products produced from cucumbers. Because of their general acceptance, pickles were the first fermented vegetable product made on a large commercial scale. The fermentation process is very similar to the sauerkraut process, only difference is brine is used instead of dry salt. The washed cucumbers are placed in large tanks and salt brine (15 to 20%) is added. The cucumbers are submerged in the brine, ensuring that none float on the surface and this is essential to prevent spoilage. The strong brine draws the sugar and water out of the cucumbers, which simultaneously reduces the salinity of the solution. In order to maintain a salt solution so that fermentation can take place, more salt has to be added to the brine solution. If the concentration of salt falls below 12%, it will result in spoilage of the pickles through putrefaction and softening.

A few days after the cucumbers have been placed in the brine, the fermentation process begins. The process generates heat which causes the brine to boil rapidly. Acids are also produced as a result of the fermentation.

In natural fermentation of cucumber, the natural microflora plays a major role. The microflora includes bacteria, yeasts and moulds. The salt and a lowered redox potential favour the growth of facultatively anaerobic organisms. The vats are covered with plastic, and UV rays from sunlight or UV lamps are used to prevent surface growth of film-forming yeasts. There is a miscellaneous group of yeasts at the beginning of the fermentation. Yeasts can affect the fermentation by utilizing sugars that would otherwise be metabolized to lactic acid by the LABs. The yeast can also utilize the produced lactic acid, raise the pH and allow other potential spoilage types of microbes to grow. The yeasts produce large amounts of gas which can be associated with spoiled cucumbers. These problems can be tackled with controlled fermentations. Here the cucumbers are washed and sanitized with a chlorine solution which removes most of the undesirable microbes. After brining (20°–25° salometer), the cover brine is acidified with acetic acid and buffered with sodium acetate or sodium hydroxide. The brine is purged with nitrogen to remove dissolved CO_2. A culture of *L.plantarum* is added for the fermentation. With controlled fermentation, the need to add more salt during storage is reduced. During fermentation, visible changes take place which are important in judging the progress of the process. The colour of the cucumber surface changes from bright green to a dark olive green as acids interact with the chlorophyll. The interior of the cucumber changes from white to a waxy translucent shade as air is forced out of the cells. The specific gravity of the cucumbers also increases as a result of the gradual absorption of salt and they begin to sink in the brine rather than floating on the surface.

Microbes Involved in the Fermentation Process

As with the sauerkraut process, the gram-positive coccus *Leuconostoc mesenteroides* predominates in the first stages of pickle fermentation. This species is more resistant to temperature changes and tolerates higher salt concentration than the subsequent species. As fermentation proceeds and the acidity increases, lactobacilli start to take over from the cocci. The active stage of fermentation continues for between 10 to 30 days, depending upon the temperature of the fermentation. The optimum temperature for *L. cucumeris* is 29 to 32°C. During the fermentative period, the acidity increases to about 2% and the strong acid producing types of bacteria reach their maximum growth. If sugar or acetic acid is added to the fermenting mixture during this time it increases the production of acid.

Problems in Pickles

The production of excessive amounts of acid during the fermentation, results in shrivelling of the pickles, possibly due to over-activity of the *L. mesenteroides* species. If the brine is stirred, it may introduce air, which makes conditions more favourable for the growth of spoilage bacteria. In general, if the pickles are well covered with brine, the salt concentration is maintained and the temperature is at an optimum, it should be quite simple to produce good quality pickles.

There are two main defects of fermented pickles—bloaters and softening.

Bloaters Bloaters are those naturally fermented and cured pickles, that float on the brine or are hollow or have large air spaces in the interior. Bloater formation is due to the accumulation of gas inside the cucumber during fermentation. Several factors are responsible for this defect. In the early stages of the fermentation, the coliforms and certain halophilic bacteria can produce hydrogen and cause bloaters. The fermentation of sugar by yeasts results in formation of much gas. Some bloaters are formed in fermentations without yeasts being present. The respiration of cucumber tissue plus the fermentation by the homofermentative *P. cerevisiae* or *L. plantarum* produces enough CO_2 to cause bloater formation. Piercing of the fruits prior to brining controls the defect. Purging of CO_2 from cucumber brines with nitrogen can reduce the amount of bloater damage.

Softening Softening of the pickles is attributed to pectinolytic enzymes that degrade the cucumber tissue. The main source of these enzymes is moulds that enter the vat with the cucumbers, especially with portions of flowers that may remain attached to the cucumbers. Pectin degrading enzymes are naturally present in the cucumber. Softening can be essentially controlled by removing any flowers attached to the cucumbers. Pasteurization of the fermented pickles at about 82°C for 25–30 minutes will inactivate the pectinolytic enzymes.

Olives

Olives are stone fruits with a very tough skin. They are inedible until fermented. Olives are brined and fermented in a manner similar to cucumbers.

There are two main types of olives—the Spanish-style green and the Greek black. To make them suitable for human consumption, olive fruits need to undergo three processing steps. Oleuropeine is a bitter glucoside that makes the fresh fruits unedible, and therefore, it is necessary to deplete the olive flesh of this compound. Treatments with solutions of sodium hydroxide (lye) are known to hydrolyse the oleuropeine. If this treatment is performed three or more times with continuous aeration, "black ripe olives" are obtained. However, if after the lye treatment the olives are placed in a solution of sodium chloride (brine) there a lactic acid fermentation takes place and "Spanish green olives" are obtained. Texture is one of the organoleptic characteristics most affected by these treatments. Sodium chloride either improves texture by reducing the electrostatic repulsion of acidic groups (as happens at low values of pH), or it has the opposite effect on texture by competing with calcium. Endogenous enzymatic activity produces losses in firmness, as occurs during ripening and in the case of fermentations.

The green olives are picked unripe and treated with a sodium hydroxide solution (lye) to break down a compound called oleuropein which has a very bitter taste and also inhibits lactic acid bacteria. It also softens the olive skins.This takes several hours, after which the caustic solution is washed off thoroughly. The olives are placed in brine and the natural fermentation sequence begins. Sometimes sugar is added to the brine that starts the microbial growth. Due to the low pH and salt, lactic acid bacteria and some yeast replace the natural flora of the fruits. Air is kept out of the vat so that spoilage moulds and yeasts cannot grow. The process takes some weeks. The product contain 1% lactic acid and has a pH about 4.

The black olives are ripe and are not treated with lye. The fermentation is slower because it takes time for nutrients to diffuse through the olive skins into the brine for the microbes to grow on. A variety of yeasts are responsible for the fermentation process, with significant numbers of lactic acid bacteria present only if the salt content is less than 6–7%. The final product is much less acid than green olives (pH 4.5–4.8) and so more salt is usually added to prevent the growth of spoilage microbes on storage.

During fermentation, the numbers of Enterobacteriaceae and *Pseudomonas* spp. in the brine decreases while lactic acid bacteria and yeast populations increase. Scanning electron microscopy showed that a yeast-rich biofilm developed on the epicuticular wax of the olive skin during fermentation. Yeasts also predominated in the stomatal openings, but bacteria were more numerous in intercellular spaces in the sub-stomatal flesh. Citric, malic and tartaric acids were the major organic acids accumulating in the brine during fermentation.

A defect known as sloughing spoilage include severe softening, skin rupture and flesh sloughing. The defect is caused by gram-negative pectinolytic bacteria (belonging to the strains of *Xanthomonas* and various coliforms) and occurs during the washing to remove the lye prior to brining.

NON-SALTED, LACTIC-ACID-FERMENTED VEGETABLES

Some vegetables are fermented by lactic acid bacteria, without the prior addition of salt or brine. Examples of non-salted products include gundruk (consumed in Nepal), sinki and other wilted fermented leaves. The detoxification of cassava through fermentation includes an acid fermentation, during which time the cyanogenic glycosides are hydrolysed to liberate the toxic cyanide gas.

The fermentation process relies on the rapid colonization of the food by lactic acid producing bacteria, which lower the pH and make the environment unsuitable for the growth of spoilage organisms. Oxygen is also excluded as the lactobacilli favour an anaerobic atmosphere. Restriction of oxygen ensures that yeasts do not grow.

Gundruk (Pickled Leafy Vegetable)

Gundruk is particularly popular in Nepal. The annual production of gundruk in Nepal is estimated at 2,000 tons and most of the production is carried out as the household level.

Gundruk is obtained from the fermentation of leafy vegetables in Nepal. It is served as a side dish with the main meal and is also used as an appetizer. Gundruk is an important source of minerals particularly during the off-season when the diet consists of mostly starchy tubers and maize which tend to be low in minerals.

In the months of October and November, during the harvest of the first broad mustard, radish and cauliflower leaves, large quantities of leaves accumulate—much more than can be consumed fresh. These leaves are allowed to wilt for one or two days and then shredded with a knife or sickle.

Shredded leaves are tightly packed in an earthenware pot and warm water (at about 30°C) is added to cover all the leaves. The pot is then kept in a warm place. After five to seven days, a mild acidic taste indicates the end of fermentation and the gundruk is removed and sun-dried. This process is similar to sauerkraut production except that no salt is added to the shredded leaves prior to gundruk fermentation. The ambient temperature at the time of fermentation is about 18°C.

Pediococcus and *Lactobacillus* species are the predominant microorganisms during gundruk fermentation. The fermentation is initiated by *L. cellobiosus* and *L. plantarum*, and other homolactics make a vigorous growth from the third day onwards. *Pediococcus pentosaceus* increases in number on the fifth day and thereafter declines. During fermentation, the pH drops slowly to a final value of 4.0 and the amount of acid (as lactic) increases to about 1% on the sixth day.

It has been found that a disadvantage with the traditional process of gundruk fermentation is the loss of 90% of the carotenoids, probably during sun-drying. Improved methods of drying might reduce the vitamin loss.

Sinki (Pickled Radish)

Sinki is a sour pickle prepared from radish taproots. It is consumed traditionally in India, Nepal and parts of Bhutan, where it is used as a base for soup or eaten as a pickle. It is one of the most popular pickles in Nepal. Fresh radish roots are harvested, washed and wilted by sun-drying for one to two days. They are then shredded, re-washed and packed tightly into an earthenware or glass jar, which is sealed and left to ferment. The optimum fermentation time is twelve days at 30°C. Sinki fermentation is initiated by *L. fermentum* and *L. brevis*, followed by *L. plantarum*. During fermentation the pH drops from 6.7 to 3.3. After fermentation, the radish substrate is sun-dried to a moisture level of about 21%. There is a second processing method involving fermentation in a clay lined pit for two to three months. For consumption, sinki is rinsed in water for two minutes, squeezed to remove the excess water and fried with salt, tomato, onion and green chilli. The fried mixture is then boiled in rice water and served hot as soup along with the main meal.

Sunki

Sunki is a non-salted and fermented vegetable product prepared from the leaves of "Otaki-turnip" in Kiso district, Nagano prefecture, Japan. Sunki is eaten with rice and in miso soup. The Otaki-turnip is boiled, inoculated with "Zumi" (a wild small apple), dried Sunki from the previous year and allowed to ferment for one to two months. Sunki is produced under low temperature (in winter season). Microorganisms involved include *Lactobacillus plantarum*, *L. brevis*, *Bacillus coagulans* and *Pediococcus pentosaceus*.

PIT FERMENTATIONS

South Pacific pit fermentations are an ancient method of preserving starchy vegetables without the addition of salt. The raw materials undergo an acid fermentation within the pit, to produce a paste with good keeping qualities. Pit fermentations are also used in other parts of the world, for example in Ethiopia, where the false banana (*Ensete ventricosum*) is fermented in a pit to produce a pulp known as kocho. Foods preserved in pits can last for years without deterioration, therefore pits provide a good, reliable cheap means of storage.

Kocho (Pickled False Banana)

It is common in Ethiopia. False banana (*Ensete ventricosum*) is fermented in a pit to produce a pulp known as *kocho*.

The type of soil and its drainage are important in the selection of a pit site. Pits are often lined with stones to prevent the soil from the side walls falling into the bottom. A family pit may be 0.6 to 1.5 metres deep and 1.2 to 2 metres wide with a capacity of about fifty breadfruits. A community pit is usually much larger, with the capacity to hold up to 1000 breadfruits. A family pit requires at least 1000 green banana leaves and four sacks of dried banana leaves for lining the walls and top. It is essential that proper attention is given to hygiene of the pit and the fruits to be stored in it.

The central stems are removed from fresh banana leaves and they are wilted in the sun until they become soft and pliable. The pit is lined with dry leaves, then green leaves are folded and arranged, overlapping each other, around the sides of the pit and extending over the top. At least two or three layers of banana leaves are used to seal the pit and prevent contamination by the soil. Washed, peeled food is placed in the pit, green banana leaves are folded over the top of the food and heavy stones are placed on top to weigh down the leaves.

Root crops and bananas are peeled before placing in the pit, breadfruit are scraped and pierced. Food is left to ferment for three to six weeks, after which time it becomes soft, has a strong odour and a paste-like consistency. The fermented paste can be left in the pit and removed as required. Usually, it is removed and replaced with a second batch of fresh food to ferment. The fermented food is washed and fibrous material removed. It is then dried in the sun for several hours to remove the volatile odours. It is then pounded into a paste with grated coconut or coconut cream and sugar, the mixture is wrapped in banana leaves and either baked or boiled.

During fermentation, carbon dioxide builds up in the pit, creating an anaerobic atmosphere. As a result of bacterial activity, the temperature rises much higher than the ambient temperature. The pH of the fruits within the pit decreases from 6.7 to 3.7 within about four weeks. Inoculation of the fruit in the pit with lactic acid bacteria greatly speeds up the process.

In the South Pacific, pit fermentations are an ancient method of preserving starchy vegetables such as banana, plantain, breadfruit, cassava, taro, sweet potato, arrowroot and yams. The products undergo an acid fermentation, to produce a paste with good keeping qualities. It is usually pounded with a little sugar, coconut cream or fresh coconut and boiled or baked to make a type of pudding.

Pak-Gard-Dong (Pickled Leafy Vegetable)

Pak-Gard-Dong is a fermented mustard leaf (*Brassica juncea*) product made in Thailand. The mustard leaves are washed, wilted in the sun, mixed with salt, packed into containers for 12 hours. The water is then drained and a 3% sugar solution added. They are again allowed to ferment for three to five days at room temperature. Microorganisms associated with the fermentation include *L. brevis, L. cerevisiae* and *L. plantarum.*

A similar product (Hum choy) is made in the South of China. This is produced by fermenting a local leafy vegetable. The leaves are washed and drained. They are then covered in salt and hung on racks and sun-dried. The wilted leaves are placed in earthenware pots and covered with rice water, obtained after washing rice grains. The pots are sealed and the leaves allowed to ferment for four days. The product can be stored for up to two months if the seal is not broken.

ALKALINE BACTERIAL PRODUCTS OF FERMENTATION

Kawal

It is commonly eaten in Sudan. Kawal is a strong smelling Sudanese, protein-rich food prepared by fermenting the leaves of a wild African legume, *Cassia obtusifolia* and is usually cooked in stews and soups. It is used as a meat replacer or a meat extender. Its protein is of high quality, rich in sulphur amino acids which are usually obtained from either fish or meat.

The sickle pod plant (*Cassia obtusifolia*) is a wild legume that grows in Sudan. The leaves are collected late in the rainy season when the plant is fully grown. All the stems, pods and flowers are removed. The leaves are not washed, since it is thought that natural microorganisms on the leaves are important for proper fermentation.

The leaves of the leguminous plant are pounded into paste without releasing the juice. The paste is placed in an earthenware jar and covered with sorghum leaves. The whole jar is sealed with mud and buried in the ground up to the neck in a cool place. Every three days the contents are mixed by hand.

Flow Diagram

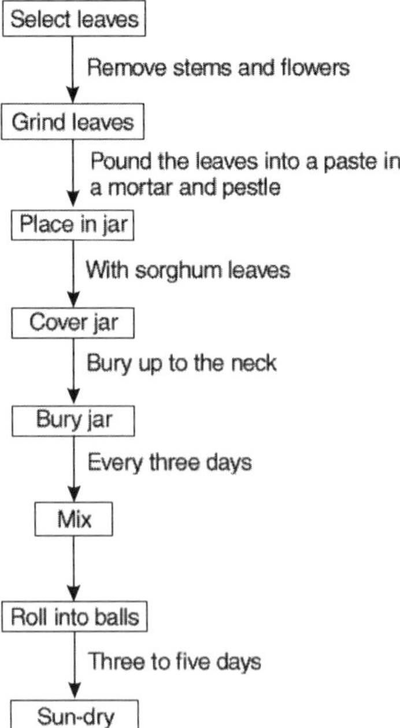

The fermentation takes about fourteen days. The fermentation is extremely complex. The main microorganisms are *Bacillus subtilis* and *Propionibacterium* species. Lactic acid bacteria including *Lactobacillus plantarum;* yeasts including *Candida krusei* and *Saccharomyces* spp. and moulds including *Rhizopus* spp. are also involved.

After about fourteen days, the strong-smelling black fermented paste is made into small balls and sun-dried for five days. The entire process is represented in the above flow diagram.

Ombolo Wa Koba

In Zaire, cassava leaves are fermented to produce ombolo wa koba which is traditionally eaten with boiled cassava and plantain bananas. Cassava leaves are allowed to wilt and turn black. This takes about three to four days. The cassava leaves are then chopped up and placed in a pot of boiling water for about one hour. During this processing stage, a water-soluble extract of ash is produced by placing the ash of burnt dried banana skins and palm tree flowers in a strainer and pouring water through it. The extract is then added to the boiled cassava leaves. The extract is alkaline and neutralizes the cyanohydric acid liberated when the leaves are chopped up. Salt and dried fish or meat is also added. After allowing the cassava leaf mixture to cool a little, acid palm oil is then added. This reacts with the excess alkali and neutralizes it. The product is now ready to be eaten.

1. Discuss the role of beneficial and spoilage microbes in the manufacture of pickles.

2. Give the importance of salt in pickling.

3. Define sauerkraut.

4. Discuss the spoilage defects in the sauerkraut process.

39

FERMENTED MEAT AND MEAT PRODUCTS

MEAT STARTER CULTURES

The use of starter cultures in fermented meat and meat products is a relatively recent practice. It is usually a combination of lactobacilli, pediococci, micrococci and staphylococci. Proper inoculation and incubation procedures are the most critical steps for the uniform production of safe, flavoured and wholesome fermented meat products. Inoculation is accomplished by one of the three methods: **natural fermentation**, which relies on indigenous microflora in the meat to serve as the inoculum; **back inoculation**, which involves transfer of a portion of raw meat from a previous batch of sausage to the present batch, i.e., transfer from a mother batch of raw sausage; **the use of starter cultures**, i.e., inoculation of unfermented meat with a pure strain or strains of lactic acid bacteria.

In fresh meat, lactic acid bacteria are a minor component of the microflora, but when meat is packaged and stored under vacuum, the resulting microenvironment facilitates the growth of lactic acid bacteria. An essential requirement for starter cultures is that they should be produced and preserved in a viable and metabolically active form suitable for commercial distribution. Currently, the predominant genera of bacteria used either singly or as mixed starter cultures are *Pediococcus, Lactobacillus, Micrococcus* and *Staphylococcus*. These cultures are available either fresh, frozen, freeze-dried or in a low temperature stabilized liquid form. Cultures with optimum low temperature are preferred for flavour development and red colour enhancement.

Bacterial starter cultures for meat products have been selected on the basis of being homofermentative, capable of rapidly converting glucose or sucrose to DL-lactic acid anaerobically with sustained growth to pH 4.5, tolerant of salt brines up to 6% and capable of growth in the presence of sodium or potassium nitrate or nitrite. These cultures are aciduric, capable of growth in the range of 21 to 43°C, non-proteolytic, non-lipolytic, inactivated at 60.5–63.2°C, resistant to phage infection and mutation, and able to outgrow and suppress pathogens.

Specific advantages favouring the use of commercial starter cultures over natural fermentations (back inoculations) are consistency of the inoculum, reduced risk of bacterial cross-contamination, uniformity of lactic acid development, production of desirable flavour components, predictability of pH end point and reduced risk of proteolytic and pathogenic bacteria outgrowth. Other benefits include the acceleration of fermentation time, to increase commercial plant throughput and the reduction of product defects such as off-flavours, lack of tangy flavour, excessive softness, crumbly texture, gas pockets and pinholes all of which can be attributed to heterofermentative bacteria.

An inoculum population of 10^7 cfu/g of raw product is sufficient for rapid lactic acid development within 6–18 hours under controlled temperature, airflow and relative humidity conditions, but incubation time is also dependent upon carbohydrate type and concentration, spices composition, exclusion of oxygen, and product diameter or thickness. Single species of *Lactobacillus plantarum, L. acidilactici, L. curvatus, L. sake, L. pentosus, P. pentosaceus,* or a combination of these bacteria can start the fermentation process.

Use of starter cultures assures dominance of desirable microbes, production of acids consisting of >90% lactic acid, and modulation of fermentation based on combinations of inoculum level, glucose concentration and incubation time and temperature intervals. Thus the combined effects of low pH, increased acidity, concomitant loss of moisture during drying, reduction of a_w, concentration of curing salts such as sodium chloride and sodium nitrite, bacterial inhibition of spoilage or pathogenic microbes and heat-processed fermented meat and meat products against spoilage by inactivating indigenous tissue and bacterial enzymes.

Mould and yeast starter cultures are not widely used. Their effect is primarily cosmetic, but it has been reported that the mycelial coat can reduce moisture loss and facilitate uniform drying. Catalase produced by moulds may serve as an antioxidant by reacting with surface oxygen to prevent it from entering the product, while nitrate reductase promotes the development of red surface colour. Moulds are capable of decomposing lactic acid, which increases pH and results in a milder flavour. *Penicillium nalgiovense* is most commonly used for this purpose, the sue of non-toxigenic moulds may reduce the risk of mycotoxin production by other moulds. Yeasts such as *Candida famata* and *Debaryomyces hansenii* are used alone or in combination with bacterial cultures to produce a powdery surface or a fruity or alcoholic aroma which may be consumed as spoiled if uncontrolled.

MEAT MICROBIOLOGY

Although unaware of the process, early sausage makers knew that once the animal was killed, it was a race between external preservation techniques and the decomposition of the raw meats to decide the ultimate fate of the tissue. We now realize that once the inherent defense mechanisms of the live animal are destroyed, the meat tissue is subjected to rapid decay. The slaughtering process affords extensive contamination of the sterile tissue with primarily gram-negative microorganisms. This includes enteric bacteria from animal intestines (Enterobacteriaceae, including *E. coli* and *Salmonella*), as well as contaminants such as *Pseudomonas* and gram-positive lactic acid bacteria and staphylococci associated with humans, animals and the environment. As a result, fresh meat spoilage is usually associated with gram-negative, proteolytic bacteria which decompose the protein with the production of offensive odours (putrid, rotten egg) and flavours. Salting, curing and drying of the fresh meat have proven to be effective means to control the fresh meat microflora and thus preserve the tissue for later consumption. This is possible since the microbiology of salted meats is entirely different from that of fresh meats. The curing salts (sodium chloride, sodium nitrite, sodium nitrate) and subsequent proper handling methods favour the growth of gram-positive bacteria while inhibiting the proliferation of gram-negative bacteria. This "microbial inversion," occurring during the salting and curing process, provided the "accidental" origin of fermentation cultures for all salted meats and fermented sausages. Once the raw meat was salted and spiced, the mixture was stuffed tightly into sausage casings or skins, which excluded air and held the sausage during ageing. The casing also provided a convenient medium to hang the mixture for smoking and drying. Excluding air also prevented mould growth, except on the casing surface. Consequently, a microenvironment was provided for those microorganisms that were not only salt tolerant, but which also could grow in the absence of air. These gram-positive, fermentative types of microorganisms included the lactic acid bacteria, staphylococci, micrococci, and yeasts.

As the product dried, a fermentation ensued whereby residual meat sugar, or added carbohydrates, was fermented to a variety of end products, including various organic acids, carbon dioxide, alcohols, etc., that contributed to a variety of flavours and textural properties. The primary fermentation product, lactic acid, served to lower the pH and contributed to the stability of these sausages against food-borne pathogens and other undesirable microorganisms. The lower pH also aids in drying since the meat proteins are less able to bind water under acidic conditions. Although actually a spoilage condition, the "souring" of the product could be tolerated to a much greater extent than the fresh meat type of proteolytic spoilage, and this served to preserve the meat. If properly controlled, the resulting flavour could make the meat more acceptable to the palate, through a "tangy" sensation in combination with the salt, spices, smoke, and other imparted formulation and processing characteristics. Lactic acid resulting from fermentation has a tendency to enhance the "salty" and "smoky" flavours. Lactic acid is non-toxic to man and moderately pleasant to the palate, but it is inhibitory to most undesirable microorganisms, including most pathogenic microorganisms, which tend to die off in the final product during storage. Similar lactic acid fermented foods include cheeses, sauerkraut, yoghurt and pickles.

Lactic Acid Bacteria (LAB)

The most common lactic acid microorganisms found in fermented meats are various strains of lactobacilli, leuconostocs, pediococci, and streptococci. *Lactobacillus plantarum* is mostly found in sausages fermented at higher temperatures (30°C) and *L. sake* and *L. curvatus* dominate the flora at mild temperatures (20–24°C). Pediococci involved include *P. acidilactici* at higher temperatures (<29°C) and *P. pentosaceus* at lower temperatures (>29°C). These harmless, beneficial types of microbes are ubiquitous in the environment and are highly competitive. They are fastidious microorganisms, requiring many nutrients for growth (which meats can provide). They primarily convert various sugars to lactic acid and other end products. They can grow with or without air, but are very rapid acid producers in the absence of air, since the fermentation process produces low energy and large quantities of acid must be produced for growth. This anaerobic (no air) environment is the characteristic condition within the internal portion of the sausage. The lactic acid microorganisms are also very salt tolerant and thrive within the typical sausage formulation. Typically, the initial counts of LAB in raw meat mixes, depending upon the raw materials and the environment, are around 1,000–10,000 cells/g. During fermentation, those numbers increase to 10,000,000–100,000,000 cells/g. Lactic acid type microorganisms are commonly used commercially today as starter cultures to produce a variety of fermented food products, including cheese, yoghurt, sour cream, pickles, and olives, as well as dry salami, pepperoni, summer sausage, and fermented meat snacks. They are also used for human and animal nutrition, establishing a beneficial intestinal microflora (i.e., probiotic cultures).

Micrococcaceae and Staphylococcaceae

Other types of common microbes in salted meats are members of the Micrococcaceae and Staphylococcaceae, including *Kocuria* and *Staphylococcus*, respectively. The predominant types are coagulase-negative staphylococci that are very salt tolerant and can also grow with or without oxygen. So far, the most common strains belong to the species *Staphylococcus carnosus*, *S. xylosus*, and *K. varians*. Previously, many of these strains were classified as *Micrococcus* species. These bacteria generally are harmless and do not represent a microbial hazard. However, *Staphylococcus aureus* is a known food pathogen. The *Kocuria* and staphylococci are more salt tolerant than the lactic acid microorganisms, and thus survive and grow in much lower water activity environments, as is the case with highly salted products and as the sausage loses water. Fortunately, these microorganisms are not very acid-tolerant and tend to die off at lower pH during and after the fermentation process. They produce catalase, which reduces hydrogen peroxide, and some strains contribute to flavours as a result of lipolytic and proteolytic activity. A characteristic of the micrococci and staphylococci is that they can reduce nitrate (NO_3) to nitrite (NO_2), which provides the active component (nitric oxide, NO) that initiates the typical meat curing reactions. This characteristic was very important in the early days of meat processing, before the availability of commercial cures, since the nitrate was found as a natural contaminant in the salt and/or saltpeter (KNO_3) that was added. Without this nitrate reduction to nitrite, the meat will not show the characteristics of cured meat. Since the staphylococci are sensitive to acid, this conversion would occur early in the process (prior to fermentation) before the lactic acid microorganisms became the dominant microflora of the meat mix. With "microbial succession," the staphylococci would gradually die off as the lactic acid microorganisms produces lactic acid. Today, most processors, just add the nitrite directly to the meat mix, thus eliminating the need for nitrate reduction by these microbial types. Initially, these strains are present at 100–1000 cells per gram, increasing to 1–10 million cells per gram in some fermentations.

Undesirable Microorganisms

Not all the lactic acid bacteria that occur in meat are desirable. For example, *L. viridescens* can cause greening of the meat due to the production of hydrogen peroxide. *L. brevis* and *L. mesenteroides* can cause gas production and unacceptable souring. *Brochothrix* can cause souring, off-flavours and odours. Yeasts and moulds can also be present in salted meats and can survive and grow at lower pH. Both prefer to grow under aerobic conditions at the outer edge and external surface of the meat product (although yeast can grow anaerobically). Yeast can result in gas production and fruity flavours. Some moulds can attack proteins and produce ammonia, which raises the pH on the sausage surface. Others may be lipolytic, attacking fats, or cellulytic, attacking casings. There is also the potential for production of mycotoxins, although this has not yet been shown to be a problem in sausage production. Although these microorganisms are often undesirable, they can be controlled via fermentation and proper drying.

Many pathogenic microorganisms can also be present in the raw sausage mix, including *S. aureus, E. coli* O157: H7*, Listeria monocytogenes,*

Salmonella, and clostridia. Fortunately these pathogens also are not very competitive with the lactic acid microorganisms and are inhibited by low pH, nitrite and lower water activity. The parasite *Trichinella spiralis* may also be present in pork.

DRIED MEAT CURING AND MICROBIAL FERMENTATION

Traditionally, dried meat microbiology involves a natural development or "wild" fermentation in which a "microbial succession" occurs. Without a starter culture, initially all types of microorganisms increase. The total microbial count, which is primarily LAB (when starter cultures are used), rises to a high level (between 10^6 and 10^7 cfu/g) and then remains fairly constant. Enterobacteriaceae and psychrotrophs decrease once LAB reach high levels. Micrococcaceae increase more slowly and eventually decrease.

As mentioned previously, Micrococcaceae (*Kocuria*) and Staphylococcaceae (*Staphylococcus*) produce nitrate reductases that convert nitrate to nitrite. Lactic acid microorganisms ferment the product and lower the pH. The fermentation of the sugars results in lactic acid formation and the acidity of the meat increases. In chopped meat, the salting process also favours the same types of microorganisms, but the greater amount of surface area and the exposure to oxygen resulted in less fermentation activity and lactic acid production. In these types of dried meats (e.g. prosciutto, basturma), uniform salting over the entire surface is most critical to inhibit pathogens and spoilage microorganisms. Since the internal meat tissue is sterile, the high salt levels applied to the surface retard most microbes as the salt gradually penetrates the tissue, which should be kept cool during penetration or equilibration. These products have higher pH and greater risk of causing illness if the brine content is not sufficient.

Historically, some dry and semi-dry sausage processors would add the salt (or saltpeter), sugars and spices to the meat and hold the resulting meat mix refrigerated for 1–2 weeks in shallow pans. This process was called "pan curing" or "panning." The salt inhibited gram-negative bacteria and allowed the lactic acid bacteria, and some micrococci and staphylococci to develop and reduce the naturally occurring nitrate to nitrite. Although rare today, some processors still use a straight "nitrate cure" containing only nitrate—a process that is generally not effective without the presence of the nitrate-reducing microorganisms. A "mixed cure" of added nitrate and nitrite is very common for dried meats that use a combined starter culture (nitrate-reducing micrococci and LAB) and that are not fully cooked.

Nitrate reduction to nitrite provides additional cure colour development and the nitrate provides a reservoir for the production of nitrite during the shelf life of the product.

Starter Cultures

Use of starter cultures resulted from an understanding of the natural development or "wild" fermentation. Gram-negative bacteria yield to gram-positive bacteria (micrococci, staphylococci and LAB), resulting in microbial inversion. Micrococci and staphylococci convert nitrates to nitrites and yield to LAB (microbial succession). Fermentation results in lower pH. However, relying on indigenous microorganisms to properly ferment products was a risky business. Thus the "backslop," "back inoculation" or "mother batch" method evolved, which depended on using the batter from a previous batch to inoculate the next one—a potentially uncontrolled starter culture where undesirable, as well as desirable, microorganisms are recycled. However, this method has been successfully used by some processors who have implemented appropriate controls to prevent contamination or loss of the effective fermentation culture. Ultimately this process evolved into controlled development with prepared starter cultures. These cultures resulted in controlled "spoilage" through addition of staphylococci and LAB, if nitrate is added to the batch, or LAB only if nitrite is added. Today, most fermented meat processors either add lactic acid starter cultures and/or harmless staphylococci to the raw meat mix. The microbial basis for fermentation was first determined about 1900, but the first starter cultures were only patented in the 1940's. Meat starter cultures were first commercialized in 1958, but by 1973 only about 33% of all fermented meat processors in North America used commercial starter cultures. Today, over 95% of all processors of these products use meat starter cultures.

Commercial starter cultures Commercial meat starter cultures are selected microorganisms that have been isolated from the meat, purified, grown to large numbers under controlled conditions, concentrated, and then preserved by freezing or drying prior to use. Eventually, the level of bacteria in many meat products is the same, whether a culture has been added or not. However, when a bacterial culture is added to meat, the microbial growth is controlled by bacteria with well-known characteristics. The addition of high levels (1–10 million bacterial cells per gram) of appropriate microorganisms such as in an LAB starter culture to the initial fermented sausage mix assures microbial dominance over the potential pathogenic and other undesirable microorganisms that might be present. The lactic acid microbes reduce product pH *via* fermentation while the staphylococci assure more efficient curing through nitrate or nitrite reduction and oxygen scavenging ability. There are two primary reasons to add such an overpowering population of the desired starter culture. First, they instantly become the dominant microflora, inhibiting all other microorganisms; and, second, the large number of microbial cells provides a tremendous amount of surface area, which results in high metabolic performance (e.g. desired end products such as lactic acid and/or nitrate reduction to nitrite).

Commercial meat starter cultures are not all the same. They differ in form, strain, purity, activity, and consistency, depending on the manufacturer. The functions of commercial meat starter cultures are to allow fermentation with acid development and reduced pH, which has preservative effects for safety and shelf life, and to produce more drying efficiency. Commercial meat starter cultures are also used for flavour development (proteolytic, lipolytic), and for colour development and stability (nitrate reduction, oxygen scavenger). There are cultures of moulds and yeasts that are used on the surface. Starter cultures can also show antioxidant characteristics.

The commercial cultures used in the U.S. differ from those used internationally. Those that are most common (and the fermentation temperatures used) are *Pediococcus acidilactici* (32–46°C), *P. pentosaceus* (21–38°C), *Lactobacillus plantarum* (21–38°C), and *Staphylococcus carnosus* (formerly classified as *Micrococcus varians or M. halobius* (21–38°C). Internationally, the most common are *Staphylococcus carnosus*, *Lactobacillus curvatus, L. sake, S. xylosus, L. plantarum, Penicillium* spp., (mould), *P. pentosaceus, P. acidilactici*, and *Debaryomyces* spp. (yeast).

Specific starter cultures The specific starter cultures used in meat processing and whether a manufacturer uses a single culture or a combination of cultures will depend on the product, process conditions, desired fermentation temperature, final product characteristics desired, and regulatory requirements. Some of the specific starter cultures are

- Traditional fermented sausages—European-style, lower temperature, slower fermentations with enhanced colour and flavour development
- Fast fermented products—U.S. style, higher temperature, rapid fermentations with emphasis on acid development
- Flavour and colour enhancing cultures—strictly for flavour and colour development with less emphasis on acid development
- Surface treatment cultures—mould and yeast cultures for surface appearance and flavour
- Whole muscle cultures—whole muscle products with emphasis on colour and bioprotection
- Bioprotection cultures—food safety and shelf life function
- Probiotic cultures—for nutritional purposes

Frozen cultures There are different forms of cultures as well. They can consist of frozen liquid or pellets, dry (freeze-dried) cultures, or frozen "syrup." Frozen pellet cultures are the same as frozen liquid, but are easier to handle. They are measurable cultures, but they must be kept at (–40°C) or below before use (culture freezer). They are easily customized for specific culture blends. Dry meat cultures are used more worldwide. They are easy to distribute and are used primarily by small-scale processors. Their primary use is for slower fermentations at lower temperatures (21–27°C). These cultures should be kept frozen prior to use and added directly to sausage mix or diluted in water and then added. Cans of frozen liquid culture have been most common, but this is changing to frozen pellets and dry cultures. Regardless of the specific starter culture, detailed product specifications should accompany the cultures. These specifications should include:

- specific microbial strain

- function of the starter culture

- purity of the product, including absence of pathogenic microorganisms and maximum number of "non-type" harmless strains

- minimum number of specific microbial strain per gram in culture

- minimum activity of the culture in terms of pH reduction and/or colour development

Meat cultures are alive and need to be handled appropriately to ensure they do not lose viability and activity. Proper receipt, storage, and stock rotation according to the manufacturer's instructions are essential to maintain optimal performance.

Fermentation Process

Fermentation results in an increase in acid along with a concomitant decrease in pH due to the fermentation of sugar (usually dextrose). Acid development is monitored by the drop in pH. This phase of the process has historically been referred to as the "holding," "greening," or "dripping of the product." In all cases, this "fermentation phase" is where the conditions are established to effect the most efficient fermentation of added sugars to lactic acid. In this phase of the process, it is critical to measure product pH. Product pH is the negative logarithm of the hydrogen ion concentration and is indicative of the acid concentration. The product pH can be affected by many factors, including the "buffering capacity" (the resistance to change in pH) of the meat mix. Being a logarithmic measurement, pH is not linear and may not be directly proportional to the acid concentration. Total acidity is a linear measurement (titration) and is directly indicative of the acid concentration (taste or "tang"), but is more difficult to measure in the processing environment. Generally, measuring pH is sufficient as an indicator of the progress of the fermentation phase. When measuring pH with the direct probe method (i.e., insertion directly in the meat mix), it is very important to routinely clean the probe of protein and fat residue, both of which can result in a false pH reading. Additionally, the pH reading from a direct probe should be routinely correlated with the standard water dilution method (i.e., meat slurry) to assure accuracy. The water used for the dilution should be distilled water of neutral pH and do not contain any buffering agents. Appropriate fermentation rate and final pH in the meat reflect that starter cultures are viable and have been handled correctly. The fermentation rate depends on meat type and condition: pork, beef, poultry, temperature, percentage of fat and moisture, meat age, dominant microflora, and pH. A salt level >3.0% slows down growth of the starter culture, but can be overcome with higher temperatures and the use of an appropriate culture. The usual added salt for dried sausages is 3.3%. Fermentation depends on the sugar types and levels. In general, dextrose is universally the most fermented carbohydrate, followed by corn syrup, sucrose, lactose, maltodextrins, starches and other more complicated carbohydrates.

In general, increasing sugar levels up to 1% decreases pH proportionately. In specific fermented meat products (e.g. pepperoni), limiting the added sugar to 0.5–0.75% achieves adequate fermentation with no residual carbohydrate present after fermentation. This prevents the "charring" of the product due to the reaction of reducing sugars with protein (i.e., Maillard reaction) during heating. A lower pH is obtained with increasing temperature at the same sugar level. Spice types can increase or decrease culture activity. This usually is dependent upon the manganese content of the respective spice. For example, black and white pepper increase fermentation rate while the antimicrobial properties of mustard and garlic inhibit fermentation. The following additives can adversely affect the fermentation rate due to microbial inhibition: curing ingredients, antioxidants, phosphates, smoke, liquid smoke, non-fat dry milk, starch, and soy products. Obviously, the starter culture type, activity, handling, and age will affect the culture's performance. It is critical that the optimum starter culture be used for the desired meat product and process. The casing diameter will also affect the fermentation rate and final pH by affecting heat penetration and moisture migration in, and then out. Generally, large diameter products ferment slower due to slower heat penetration, but they result in a lower final pH for the same reason and/or slower drying. As expected, the specific process affects the fermentation rate and final pH. In general, the higher the fermentation process temperature and humidity, the faster the fermentation; however, the fermentation temperature should be at the optimum growth temperature of the added starter culture. Added smoke will sometimes inhibit fermentation at the product surface, but the significance will depend upon the product diameter. The final pH will be affected by the added carbohydrate, the heating temperature after fermentation, and the drying conditions. Fermentation temperature affects the time to reach pH 5.3, which is critical for control of *S. aureus* in fermented products. At 18°C, it takes 36 hours; at 24°C, it takes 19 hours; at 28°C, it takes 13 hours; and at 38°C, it takes only 7 hours to reach pH 5.3.

A simplified flow chart illustration of the fermentation method of sausage is seen in Figure 39.1.

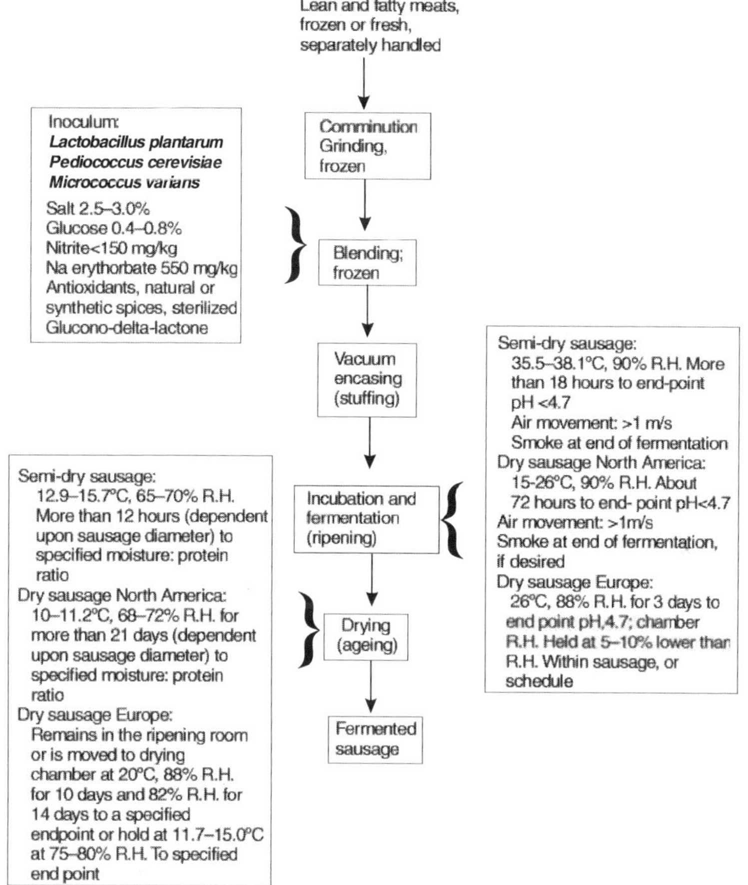

Figure 39.1   Steps in preparation of fermented sausage

Problems in Fermentation

If the fermentation to a desired pH does not occur within the normal time period, it can be due to a variety of reasons. Generally, if a fermentation problem does occur, it is the result of a total lack of fermentation (the pH does not change from its initial value of 5.6–6.0), a partial fermentation (the pH drops slightly to 5.4–5.6), and/or inconsistent fermentation (variation in fermentation activity from piece to piece or from location to location). The following lists some typical causes for inadequate and/or inconsistent meat fermentations.

No Fermentation

- No starter culture added
- No fermentable sugars added
- Excessive salt added
- Starter culture mishandled or thawed and refrozen or culture premixed with cure, salt, chemicals
- Antimicrobial agents added to formulation
- Antibiotic residues in raw meat

Inconsistent or partial fermentation

- Inadequate distribution of starter culture
- Insufficient fermentable sugars added
- Inconsistent internal product temperature and/or processing temperature and/or humidity
- Reduced fermentative activity of the starter culture or out of code product/improper stock rotation or mishandled culture
- Antimicrobial agents added to formulation
- Antibiotic residues in raw meat

FERMENTED SAUSAGE PRODUCTION

The broad group of fermented or raw sausages include a large number of products whose characteristic properties are partially or completely dependent on fermentative action of certain types of bacteria. The comminuted meat mass may be submitted to curing either prior to or after stuffing. The stuffed sausages are processed by smoking, drying and ageing which make a product entirely suitable for eating without further cooking.

The main subgroups (Table 39.1) of fermented sausages are semi-dry or quickly fermented and dry or slowly fermented (semi-fermented) sausages. There are both hard and soft types in both subgroups.

Semi-dry Sausages

Semi-dry sausages differ greatly from dry sausages by their pronounced "tangy" flavour of forced fermentation resulting in lactic acid accumulation and a bulk of other products of fermentative breakdown. The addition of starter cultures for a number of semi-dry sausages is particularly successful.

Semi-dry sausages are usually stuffed in medium- and large-diameter natural or artificial casings. The duration of production (smoking and fermentation) of these sausages depends upon their type but rarely exceeds several days.

The pH of semi-dry sausages is explicitly acid (i.e., 4.8 to 5.4); although they are often finely chopped and spreadable, many of them can be cut to thin slices; their water content reaches 35 per cent or more.

Semi-dry sausages are regularly smoked and only exceptionally slightly cooked by the heat applied in the smokehouse at various temperatures, mostly not exceeding 45°C and very occasionally rising to nearly 60°C for a strictly limited time; after smoking, the sausages are usually air-dried for a relatively short time.

Semi-dry sausages usually contain an important proportion of beef. Their shelf life is surprisingly good due to low water activity, accumulation of acids and smoke compounds, counteracting the effect of lactic acid bacteria on spoilage microorganisms, etc. A high level of hygiene and the ability to perform dexterously all operations in the manufacturing process are basic prerequisites for the good keeping quality of semi-dry sausages. Semi-dry sausages have improved stability if stored in the chiller, protected from humidity rather than at room temperature.

This category of sausages is popular in many European countries and North America. As these sausages need only little refrigeration, they can be successfully produced in many subtropical countries.

The main sausages of this group are: summer sausages (with a series of varieties in many countries), different types of cervelats and metwursts, lebanon bologna (in USA), etc.

Dry Sausages

The organoleptic and other properties of dry sausages depend not only upon the products of sugar bacterial fermentation but are also strongly influenced by biochemical and physical changes occurring during the long drying or ageing process. The use of starter cultures for this category of raw sausages is less successful than for the semi-dry varieties. The length of production, either with or without smoking, and drying periods depends upon a multiplicity of factors, such as diameter and physical properties of casings, sausage formulation, choice and methods of preparing meat, conditions of drying, etc. but overall processing time may require up to 90 days. The final pH of dry sausages is usually somewhat higher (5.0–5.5) than in semi-dry sausages, and it increases during the second part of this long ageing process.

Dry sausages are made from selected, mainly coarsely chopped, meat (some Italian salamis, some types of sucuk); often they are moderately chopped (majority of small-diameter dry sausages) and very occasionally finely chopped. They are cut in thin slices, their water content is under 35 per cent, but normally less than 30 per cent. Most varieties of dry sausages are subjected to cold smoking (12 to 18°C) but sometimes not; in some countries they are often heavily spiced with red pepper or garlic or sometimes heavily smoked and strongly salted. In principle, they are processed by long, continous air-drying, sometimes after a comparatively short period of smoking.

The formulation, degree of grinding, level of fermentation, smoking intensity, temperature of ageing and type and size of casing as well as other factors determine the properties of the final product. Dry sausages are stuffed in both natural and artificial casings of varying diameters. In the preparation of dry sausages, natural casings are preferred because they adhere closely to the sausages as sausages shrink. Sausages stuffed in casings with a diameter exceeding 4.5 cm are often called "salami"; salamis are chiefly made from coarsely ground meat, predominantly of pork and are not smoked. The shelf life of dry sausages is excellent, which may be especially attributed to the high salt-to-moisture ratio. These sausages are normally kept without refrigeration.

Raw sausages, which are not submitted to the smoking process, are known as air-dried sausages. This variety of dry sausage is characterized by a highly attractive appearance and by its yeasty-cheesy flavour. Air-dried sausages are marked with or without mould overlay.

Table 39.1 Classification of fermented sausages

Category	Ripening times	Final water activity	Application of smoke	Examples
Dry, mould-ripened	> 4 weeks	< 0.90	No	Genuine Italian Salami French saucisson sec
Dry, mould ripened	> 4 weeks	< 0.90	Yes (during fermentation)	Genuine Hungarian salami
Dry, no mould-growth	> 4 weeks	< 0.90	Yes or no	German "Dauerwurst"
Semi-dry, mould-ripened	< 4 weeks	0.90–0.95	No	Various French and Spanish raw sausages
Semi-dry, no mould growth	< 4 weeks (usually 10–20 days)	0.90–0.95	Yes (with exceptions)	Most fermented sausages in Germany, the Netherlands, Scandinavia, USA, etc
Undried, spreadable	2 days to 2 weeks	0.94–0.97	Yes or No	German "Streichmettwurst" Spanish "sobrasada"

Traditional and Modern Methods of Sausage Fermentation

In raw sausage manufacture, amongst the many factors involved, development of microflora and its effect on product quality play a major part. The fermentation process in particular, results in the desired flavour characteristics and tang.

In the traditional production process, fermentation is accomplished by natural flora. In order to achieve safe fermentation of the raw sausage, it is of importance to give the microflora the proper growth conditions as well as the appropriate type of the meat. One of the number of methods offered for choice are procedures requiring extended incubation times. For instance, the raw sausage mixture, containing meats, curing salts and sugar, can be placed in 15–18 cm deep pans and kept for 2–4 days at 3–4°C. After remixing, the mix is stuffed into casings and the drying process continued at 12–15°C with or without simultaneous smoking. A number of alternative procedures are found in practice.

The inherent bacterial sausage microorganisms use various sugar substances as energy sources, whereby they produce acids and contribute to the flavour of the raw sausages. While dextrose is degraded by almost all kinds of bacteria, lactose or starch products of higher molecularity are converted only by some of them. In the second case, the speed of acidification is delayed. Acidification, i.e., with a sufficiently low pH (below 5.2–5.3) is indispensable for adequate binding and colour development in the sausage. A pH value of 4.8 or below influences the taste, but does not contribute to better binding properties of the final product. However, this low pH gives fermented sausages excellent keeping qualities.

In general, addition of sugar varies from 0.3 to 2.0 per cent. If dextrose or other easily degradable sugars are used, acidification is fast and the amount of sugar added should be somewhat lower. In opposition, corn syrup solid must be added at somewhat higher levels in order to compensate for its lower acidification properties.

A sufficient level of acidification can also be obtained by the addition of glutamine or some other acids and particularly by the addition of glucono-delta-lactone (GDL). When GDL is used, acidification (release of gluconic acid) occurs at 21–23°C. At GDL levels under 0.5 per cent, the curing agent should be nitrate, but at GDL levels higher than 0.5 per cent nitrite is preferred since too high rate of acidification does not permit the process of denitrification performed by microorganisms.

Bacterial starter cultures used in the production of dry sausages are lactic acid producing and belong mainly to the genera *Lactobacillus*, *Pediococcus* and *Streptococcus*. Besides a few French (ferments lactiques) and American preparations of lactobacilli (Lactacell MC, etc.), and Spanish mixed-culture preparations, those mainly used on the European market are the micrococci and lactobacilli. As a matter of fact, the market today is dominated by three types of products: starter cultures containing micrococci, starter cultures containing a mixture of micrococci and lactobacilli and starter cultures containing lactic acid-producing cocci. The reason for applying these bacteria in the production of raw sausages is their ability to produce a consistent and controlled acidification able to inhibit growth of undesirable microorganisms and as an aid in obtaining the desired structure and colour of the final product.

Using starter cultures in current production practices, desired acidity can be achieved within 24 hours at a high (35–41°C) incubation temperature. This is in contrast to traditional manufacturing processes which utilized the natural flora of the meat as the source of lactic acid bacteria and required extended incubation times.

Microorganisms Involved in Sausage Fermentation

The microorganisms that are primarily involved in sausage fermentation include species of LAB, gram-positive catalase-positive cocci, moulds, and yeasts (Table 39.2). In spontaneously fermented European sausages, facultative homofermentative lactobacilli constitute the predominant flora throughout ripening. *L. sakei* and/or *L. curvatus* generally dominate the fermentation process. *L. sakei* appears to be the most competitive of both strains, frequently representing half to two-thirds of all LAB isolates from spontaneously fermented sausage, whereas *L. curvatus* is frequently found in amounts up to one-fourth of all LAB isolates. Other lactobacilli that may be found, albeit generally at minor levels, include *L. plantarum*, *Lactobacillus bavaricus* (now reclassified as *L. sakei* or *L. curvatus*), *Lactobacillus brevis*, *Lactobacillus buchneri*, and *Lactobacillus paracasei*. Recently, the new species *Lactobacillus versmoldensis* has been isolated from German, quick-ripened, salami-style sausages where it was present in numbers of up to 10^8 cfu/g.

Pediococci are less frequently isolated from European-fermented sausages but occasionally occur in small percentages. They are more common in fermented sausages from the United States where they are deliberately added as starter cultures to accelerate acidification of the meat batter.

Enterococci are sometimes associated with fermented meat products, in particular artisan products from Southern Europe, where they increase during early fermentation stages and can be detected in the end product at levels of 10^2–10^5 cfu/g. They are ubiquitous in food processing establishments and their presence in the gastrointestinal tract of animals leads to a high potential for contamination of meat at the time of slaughter. Opinions about their significance vary, as they may enhance food flavour but also compromise safety if opportunistic pathogenic strains proliferate or antibiotic resistance is spread.

Coagulase-negative staphylococci and kocuriae are GCC that participate in desirable reactions during ripening of dry fermented sausages, such as colour stabilization, decomposition of peroxides, proteolysis, and lipolysis .

They are poorly competitive in the presence of actively growing aciduric bacteria, often not growing more than one log cfu/g during ripening. The non-pathogenic, coagulase-negative staphylococci are dominated by *Staphylococcus xylosus*, *Staphylococcus carnosus*, and *Staphylococcus saprophyticus*, but other species occur too. In addition to staphylococci, *Kocuria varians*, formerly known as *Micrococcus varians*, or other kocuriae are sometimes isolated in small quantities from naturally fermented sausage.

Moulds, usually *Penicillium nalgiovense* and *Penicillium chrysogenum*, are used in mould-ripened sausages, particularly in Southern Europe. A yeast population, dominated by *Debaryomyces hansenii*, may also be found on the sausage surface and originates from the house flora or is sometimes added as starter culture.

Probiotic Meat Products

The use of probiotic lines is an attractive approach for designing functional meat products. Probiotics is defined as "live microorganisms which, when administered in adequate amounts confer a health benefit on the host". Representative probiotic bacteria are intestinal strains of *Lactobacillus* and *Bifidobacterium*. According to the scientific basis for isolation and defining productive bacteria, desirable properties of such bacteria are:

- human origin
- resistance to acid and bile toxicity
- adherence to human intestinal cells
- colonization of the human gut
- antagonism against pathogenic bacteria
- production of antimicrobial substances
- immune modulation properties
- history of safe use in humans

Development of Probiotic Meat Products

In recent years, the possibility of development of probiotic meat products has been discussed in the field of meat science and industry. By using probiotic bacteria, potential health benefits can be introduced to meat products. Target products with probiotic bacteria are mainly dry sausages, which are processed by fermentation without heat treatment. Technically, it has already become possible to produce probiotic meat products. German and japanese producers have developed meat products containing human intestinal lactic acid bacteria. In 1998, a German producer launched a salami product containing three intestinal lactic acid bacteria strains (*Lactobacillus acidophilus*, *Lactobacillus casei*, *Bifidobacterium* spp.). This product is claimed to have health benefits and is thought to be the first probiotic like salami products to be marketed.

Studies have shown that the consumption of such meat products with *L. paracasei* brought some probiotic effects. After 4 week ingestion of the product, the values of CD4 (T-helper cells) were elevated and phagocytosis index increased. Also, the expression of CD54 (glycoprotein responding to inflammatory regulators) decreased and the titre of antibodies against oxidized LDL increased.

Prebiotics and Synbiotics

Prebiotics are defined as "non-digestible food ingredients that beneficially affect the host by selectively stimulating the growth and/or activity of one or a limited number of bacteria in the colon and thus improve the health of the host". Several prebiotic substances are known to enhance the activity of probiotic bacteria. Oligosaccharides and dietary fibres are representative prebiotic substances used for processed foods. Such prebiotics have been utilized for several meat products in Japan. Some desireable attributes in functionally enhanced prebiotics listed are:

- targeting at specific probiotics (*Lactobacillus* and/or *Bifidobacterium*)
- active at low dosage and lack of side effects
- persistence through the colon
- protection against colon cancer
- enhance the barrier effect against pathogens
- inhibit adhesion of pathogens

A mixture of probiotics and prebiotics, known as synbiotics is utilized for many foods such as fermented dairy products in European countries and Japan. Synbiotics are foods containing both probiotic bacteria and prebitoic ingredients to provide a diet in which the intestinal growth of the probiotic bacteria is enhanced by the presence of the prebiotic, thus promoting the chance of the probiotic bacteria becoming established in the gut, and conferring a health benefit.

FUNCTIONAL STARTER CULTURES FOR A MORE TASTY PRODUCT

Bacteria Versus Meat Enzymes

The flavour of fermented sausage is influenced by several factors, primarily the source, quantity and type of ingredients (e.g. meat, salt, and spices), but also the temperature, processing time, smoking, and choice of starter culture. Basic flavour results from the interaction of taste (mainly determined by lactic acid production and the pattern of peptides and free amino acids resulting from tissue-generated proteolysis) and aroma (mainly determined by volatile components derived from bacterial metabolism and lipid autoxidation.) In Northern Europe, smoking is important for flavour and lactic acid is believed to be the main taste component. However, acetic acid is also present and is actually needed in small amounts for full dry sausage flavour.

Too high concentrations of acetic acid, however, produce a prickly, astringent flavour. In Southern European sausage, predominant acidity is not sought for and may be rejected. In the latter type of sausage, where acidity is mild and smoking is rare, flavour is primarily generated by proteolytic and lipolytic activities from tissue enzymes, but the starter culture influences flavour too because of its aroma-generating metabolic activities. For instance, flavour variability between chorizo manufacturers has been related to ester formation, attributed to different starter cultures employed, followed by the effects of smoking (phenols) and spices (terpenes) if applied. Usually, mediterranean sausages are also inoculated on the surface with moulds or yeasts that contribute to the sensory properties as well.

Proteolytic enzymes, principally those endogenous to the meat, are of major importance for flavour. Meat proteases, particularly cathepsin D-like enzymes, seem to be responsible for proteolysis and peptide formation during fermentation, while microbial enzymes rather act on the released oligopeptides during the later stages of ripening. Microbial proteolytic activity against meat proteins is low under conditions found in fermented sausages, but a certain activity, albeit to a minor extent and in a strain dependent manner, may partly contribute to initial protein breakdown. More importantly, the peptides generated by muscle proteolysis can be taken up by bacteria that further split them intracellularly into amino acids and may convert them to aroma components.

Lipolysis is believed to play a central role in aroma formation. It leads to the release of free fatty acids and is mainly due to tissue lipases, although bacterial lipolytic activity has been described too, in particular by staphylococci. Short-chain fatty acids (C<6) lead to strong cheesy odours, whereas medium - and long-chain fatty acids can act as precursors. Lipolysis is only the first step in the process and is followed by further oxidative degradation of the liberated fatty acids into alkanes, alkenes, alcohols, aldehydes, ketones, and furanic cycles. The bacterial flora plays a role in the oxidation of free fatty acids although non-enzymatic reactions also occur.

With the exception of post-mortem glycolysis, carbohydrate catabolism is mainly of bacterial origin (both LAB and GCC) and compounds such as lactic acid, acetic acid, ethanol, acetoin, and diacetyl, all may play a role in the complexity of sausage flavour. The use of selected strains that produce interesting aroma components as functional starter cultures could lead to more tasty sausages as well as to a reduction of the ripening time.

It should be noted that the raw material (e.g. type of meat and spices) and the sausage technology (e.g. fermentation temperature, salting, and ripening time) will interact with the metabolism of the starter culture. Developing of functional starters for improved flavour needs thus to take into account knowledge about raw materials, technology, and sensory quality, and should be targeted to specific applications. Moreover, starters should not possess undesirable characteristics, such as the formation of toxic compounds or too high quantities of acetic acid or acetoin. Acetoin production, stimulated by low pH and low sugar availability at the end of the ripening period, may lead to fermented sausages with a dairy product odour.

Lactic Acid Bacteria (LAB)

The primary contribution of LAB to flavour generation is ascribed to the production of large amounts of lactic acid and some acetic acid, although they also produce volatiles through fermentation of carbohydrates. They usually do not possess strong proteolytic or lipolytic properties, although a degree of peptidase and lipase activity has been observed for some meat strains. Exopeptidases from meat lactobacilli contribute, in conjunction with muscle aminopeptidases, to the generation of free amino acids, contributing to flavour. LAB isolated from Greek sausages exhibited high *in vitro* leucine and valine aminopeptidase activities. However, lactobacilli and pediococci display low catabolism of branched-chain amino acids, and hence do not play a major role in the formation of typical sausage aroma compounds such as 3-methyl butanal, as it is the case for staphylococci.

Little information is available about the lipolytic activity of lactobacilli during sausage fermentation, but some *in vitro* activity has been documented for *L. sakei*, *L. curvatus*, and *L. plantarum*. However, lipases from lactobacilli often display little or no activity under conditions found in fermented sausages, although for some the production of lipase appears to be significant under conditions relevant for sausage ripening. In addition to lactobacilli, other LAB may be added to the sausage batter to influence flavour. Enterococci display several metabolic activities such as proteolytic, lipolytic, and esterolytic activities. They play an important role in the ripening and the development of flavour in several traditional Mediterranean cheeses, and it is speculated that they may also be of importance in traditional fermented sausages. Also, the addition of *Lactococcus lactis* subsp. *cremoris* could result in a higher amount of free amino acids. Finally, a strain of

Carnobacterium piscicola has been suggested as a new starter culture because of its high formation of aroma compounds derived from leucine metabolism, similar to the ones formed by staphylococci. Besides the necessary amino acid converting enzymes, strains need to possess specific peptide uptake mechanisms to transfer the required amino acids intracellularly, since peptides, rather than free amino acids, are the preferred substrate for lactic acid bacteria. Indeed, no relation could be found between volatile products of amino acid degradation and free amino acid concentration.

The major contribution of LAB to flavour seems, however, to be limited to their carbohydrate catabolism, mainly the production of organic acids, whereas gram-positive catalase-positive cocci (GCC) appear to be more appropriate for the generation of specific aroma compounds.

Gram-positive, Catalase-positive Cocci (GCC)

To ensure sensory quality of fermented sausages, the contribution of GCC is needed. Staphylococci, in particular *S. xylosus* and *S. carnosus*, modulate the aroma through the conversion of amino acids (particularly the branched-chain amino acids leucine, isoleucine, and valine) and free fatty acids.

Aroma generation depends however on sausage technology and variety. For fast-ripened sausages, increasing inoculum levels of staphylococci may increase methyl-branched aldehyde production, whereas in slow-ripened sausages the situation is more complex. In the latter case, aroma production is particularly pronounced and high inoculation levels favour the formation of methyl-branched acids and sulphites, whereas low levels favour diacetyl and ethyl ester production. It is thus possible to modify sausage aroma profiles by changing the inoculation level of the *Staphylococcus* starter culture. In addition, additives such as nitrate, nitrite, or ascorbate, precultivation parameters, and environmental factors clearly influence the generation of aroma compounds. The use of well-selected strains that generate high amounts of aroma components could permit to achieve improved sensory qualities and/or to accelerate the meat fermentation process. Selection of appropriate staphylococci in view of the application will be crucial. Strains of *S. xylosus*, for instance, predominate in Southern European salamis, which are characterized by a rounded aroma and a less acidic taste, and have been recommended when production of very aromatic sausage is intended. The species has been shown to produce, amongst others, 3-methyl-1-butanol, diacetyl, 2-butanone, acetoin, benzaldehyde, acetophenone, and methyl-branched ketones. In comparison, *S. carnosus* is believed to direct more of the branched-chain amino acids into methylbranched aldehydes (e.g. 3-methyl-butanal, 2-methyl-butanal, and 2-methyl-propanal) and their acids. The compounds 3-methylbutanal and 3-methyl-butanoic acid, are derived from leucine (Figure 39.2).

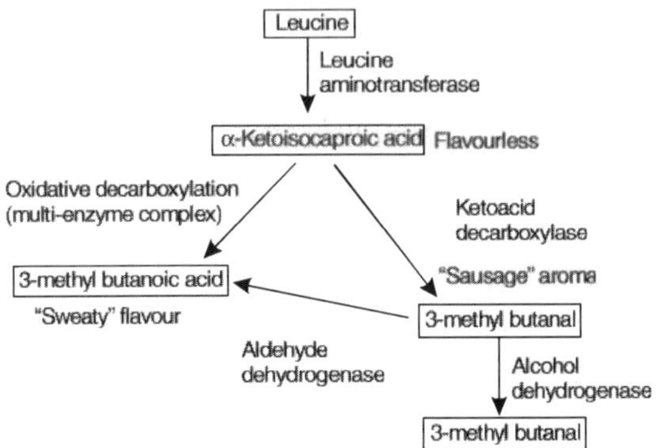

Figure 39.2   Hypothetical pathway for the conversion of leucine into aromatic compounds have been linked with sausage aroma

Moreover, they can further be converted into fruity esters (e.g. ethyl 2- or 3-methylbutanoate), which are believed to occur mainly in products with high contents of acids and alcohols, and where microbial esterase activity is low. The presence of an incomplete β-oxidation pathway in staphylococci explains the formation of methyl ketones (2-pentanone and b-heptanone) as being derived from intermediates of this pathway. β-Ketoacyl-CoA esters are deacylated into β-ketoacids by a thioesterase and then decarboxylated to the methyl ketone. Besides contributing to flavour, GCC also prevent the formation of off-flavours and can be used to control the oxidation of unsaturated fatty acids, due to their nitrate reductase and antioxidant activities. In conclusion, selected *S. carnosus* or *S. xylosus* strains, with specific peptide uptake systems and branched-chain amino acid converting and fatty acid oxidizing activities, could be used as functional starter cultures to obtain a tastier end product. Maturity correlates significantly with higher amounts of branched-chain aldehydes and alcohols arising from the breakdown of branched-chain amino acids, and branched- and straight-chain methyl ketones in turn derived from microbial β-oxidation of fatty acids.

Moulds and Yeasts

Moulded sausages are very common in the Mediterranean area. It has been shown that the superficial inoculation of the sausage with atoxigenic moulds, e.g. *Penicillium* or *Mucor* species, contributes to sensory quality. This contribution is mediated by lactate oxidation, proteolysis, degradation of amino acids, lipolysis, lipoxidation, the delay of rancidity, and reduced water loss due to slower evaporation. Moreover, moulds contribute to the overall attractiveness of the end product due to their characteristic white or greyish appearance, to the stabilization of colour through catalase activity, oxygen consumption and protection against light, and to easy skin peeling. A characteristic popcorn odour in mould-fermented sausages has been ascribed to 2-acetyl-1-pyrroline, which may be caused by conversion of proline, often found in sausage collagen casings, by the moulds.

However, as with bacterial starter cultures, the selection of mould starter strains should be done carefully since the proteolytic and lipolytic capabilities, and hence the effect on the end product, can significantly differ between strains and depend on the applied technology. The role of yeasts in sausage flavour formation is not well characterized. In model sausage minces, it has been observed that *Candida utilis* is able to produce several volatile compounds, in particular esters and alcohols, of which many are probably derived from branched-chain amino acids, whereas *D. hansenii* has very little effect on the production of volatile compounds. In other studies, *Debaryomyces* spp. affects proteolysis and volatile generation . Appropriate amounts of *Debaryomyces* influence volatile production by inhibiting lipid oxidation due to its antioxidative effect and by promoting the generation of ethyl esters. However, too large amounts of *Debaryomyces* may generate high amounts of acids (e.g. 2-methyl-propanoic and 2- and 3-methyl-butanoic acid) that mask the positive effect. Also, due to the fungistatic effect of garlic or the presence of other starter cultures, yeast cultures may die out before the ripening process ends, and fail to improve sensory quality.

FUNCTIONAL STARTER CULTURES FOR A SAFER PRODUCT

Bacteriocin Production

The main antimicrobial effect responsible for safety is evidently the rate of acidification of the raw meat. Nevertheless, certain antimicrobials such as bacteriocins may also play a role, in particular in slightly acidified products or to eliminate undesirable microorganisms that display acid tolerance (e.g. *L. monocytogenes*).

Bacteriocins produced by LAB are antibacterial peptides or proteins that kill or inhibit the growth of other gram-positive bacteria. They often have narrow inhibitory spectra and are most active towards closely related bacteria likely to occur in the same ecological niche. LAB produce a diversity of bacteriocins that are generally active towards other LAB, contributing to the competitiveness of the producer, but also towards food-borne pathogens such as *L. monocytogenes*. The so-called class IIa bacteriocins are particularly active towards *Listeria*. The application of bacteriocin-producing LAB in the meat industry offers therefore a way of natural food preservation.

Lactobacilli sausage isolates frequently produce bacteriocins or bacteriocin-like compounds, as has been shown for *L. sake*, *L. curvatus*, *L. plantarum*, *L. brevis* and *L. casei*.

The use of bacteriocin-producing *L. sake* as starter cultures permits to decrease *Listeria* levels in fermented sausage. Antilisterial effects have also been demonstrated with bacteriocinogenic *L. curvatus* and *L. plantarum* sausage starter cultures. In certain products, e.g. in American-style sausages fermented at elevated temperatures (27–38°C versus 20–25°C), bacteriocin-producing pediococci rather than lactobacilli are used. Many enterococci from food origin are known to produce bacteriocins with high antilisterial activity, including enterococci isolated from fermented sausage. The use of bacteriocin-producing enterococci is well documented in cheese manufacture, but information on their potential as bioprotective cultures in the meat industry is scarce. Bacteriocin-producing *Enterococcus faecium* strains have been included into the starter culture in Spanish-style dry fermented sausage to effectively inhibit Listeria. The enterococci were partially competitive in the early stages of the fermentation, reducing listerial counts, but disappeared during further ripening. In addition, kinetic studies have indicated that bacteriocin production by *E. faecium* RZS C5 occurs in the very early growth phase. These observations suggest that bacteriocin producing enterococcal strains may be used as co-cultures that are effective during the first stage of the fermentation process.

Also, the bacteriocin-producing *Enterococcus casseliflavus* IM 416K1 has been applied in the production of cacciatore sausages, the habitat from which the strain was isolated, leading to the elimination of *L. monocytogenes*. As an alternative to using enterococci, application of enterococcal bacteriocins as additives to the sausage mixture may reduce *Listeria* counts. The addition of enterocins A and B significantly reduced *L. innocua* counts, while the application of the producer strain *E. faecium* CTC 492 did not, probably due to an insufficient enterocin production. Bacteriocin-producing lactococci and a bacteriocin-producing *Leuconostoc mesenteroides* have been isolated from fermented sausage too. Bacteriocin-producing *L. lactis* strains have been used as new, functional starters for fermented sausage manufacture, despite the fact that they are not particularly adapted to sausage technology, e.g. displaying sensitivity to nitrite.

Based on the above, the use of bacteriocin producers as new starter cultures may offer considerable food safety advantages, without risk for human health due to toxicological side effects. It represents a safe way of natural food preservation that, most likely, has always occurred in fermented foods for centuries. Moreover, *in situ* bacteriocin production generally does not lead to organoleptic or flavour imperfections, as has been shown through taste panels with different strains. Still, some disadvantages have to be considered related to the fact that bacteriocin activity *in situ* is lower than may be expected from *in vitro* experiments. Activity may be less effective in the sausage due to low production, genetic instability, the inability to uniformly distribute bacteriocin throughout the product, low solubility of the bacteriocin, inactivation by meat proteases, resistance of the target strain, and interference by meat components, in particular adsorption to fat and meat particles. It is recommended to use strains that are well adapted to the sausage environment, preferably sausage isolates, for optimal performance and bacteriocin production. For instance, the results obtained during sausage fermentation with *L. plantarum* are less convincing than the results that are generally obtained with *L. sake* or *L. curvatus*. In this context, it has been shown kinetically that the sausage isolates *L. sake* CTC 494, *L. curvatus* LTH 1174, and *L. curvatus* L442 optimally produce bacteriocin under conditions of pH and temperature that prevail during European sausage fermentation. Also, appropriate cultures are to be selected according to the specific formulation and the technology used. For instance, pediococci are more suitable than lactobacilli when a fast fermentation at a high temperature is desired. Also, the performance of bacteriocinogenic LAB as starter or co-cultures depends on the sausage ingredients and the recipe used. For instance, performance is different if a nitrate or a nitrate/nitrite recipe is followed and depends on the use of additives such as sodium chloride and pepper.

The development of heterologous expression systems for bacteriocins may offer advantages over native systems, such as facilitating the control of bacteriocin gene expression, achieving higher production levels, producing multiple bacteriocins, or acquisition of antimicrobial properties by industrial strains. Also, in some cases the engineering of bacteriocin molecules could lead to more active compounds or increase their stability. However, such techniques are not easily acceptable by the consumer.

Evidently, bacteriocins are not meant to be used as the sole means of food preservation, but should be appropriately integrated in a multihurdle preservation system, at all times respecting good manufacturing practice.

Other Antimicrobials

As an alternative to bacteriocin-producing starter cultures, strains that produce other specific antimicrobial compounds may be proposed. The introduction of the lysostaphin gene of *Staphylococcus simulans* biovar. *staphylolyticus* into meat starter lactobacilli or *P. nalgiovense* can be used to prevent the growth of *S. aureus*. Lysostaphin is an endopeptidase that specifically cleaves the glycine–glycine bonds unique to the interpeptide cross-bridge of the *S. aureus* cell wall.

Another point of interest to improve food safety could be the use of *Lactobacillus reuteri*-containing starter cultures that produce reuterin or reutericyclin. Reuterin is a mixture of monomeric, hydrated monomeric and cyclic dimeric forms of b-hydroxy-propionaldehyde that has a broad spectrum of activity, including fungi, protozoa and a wide range of gram-positive and gram-negative bacteria. Reutericyclin is a tetrameric acid antibiotic that is active towards gram-positive bacteria. Application of purified reuterin on the surface of Turkish-style beef sausage was able to inhibit growth of *L. monocytogenes* but not *Salmonella*. New antimicrobials with application possibilities are still being discovered. For instance, strains of *L. plantarum* produce a number of interesting compounds, including a mixture of low-molecular-mass molecules that act synergistically with lactic acid , 3-hydroxy fatty acids, antifungal cyclic peptides, phenyllactic acid, and 4-hydroxyphenyllactic acid. Most of these compounds are active towards moulds and yeasts, but some of them also towards gram-positive and gram-negative bacteria, including *Listeria* and *Salmonella*, respectively.

Microbial inhibitory action is not always ascribed to specific metabolites and remains frequently unspecified. A commercial starter culture to which strains of *Lactobacillus acidophilus*, *L. paracasei*, or *Bifidobacterium lactis* from intestinal or dairy origin were added, increased the safety of Hungarian salami, because these cultures displayed strong inhibition of *L. monocytogenes* and *E. coli* O111 during sausage fermentation. The latter reduction was one to two log units higher than when the commercial starter was used alone, but the mechanism of action has not been studied. Also, staphylococci and lactobacilli isolated from spontaneously fermented Greek sausage exhibited unprecised activity towards *L. monocytogenes* and *S. aureus*. Application of a *S. xylosus* sausage isolate that produces an antagonistic substance resulted in *Listeria*-free Italian sausage. The bioprotective cultures *Lactobacillus rhamnosus* E-97800, *L. rhamnosus* LC-705, and *L. plantarum* ALC-01 displayed antilisterial activity in Northern-European type dry sausages.

FUNCTIONAL STARTER CULTURES FOR A MORE RELIABLE PRODUCTION PROCESS

The use of bacteriocin-producing starter cultures may not only contribute to food safety, but also to the prevention of food spoilage. In particular, the growth of LAB that produce hydrogen peroxide or cause sliminess, off-odours or off-flavours may be inhibited. Because bacteriocin producers are more competitive than non-producing variants, the application of such strains as new starter cultures may improve the competitiveness of the starter and lead to a more controlled and standardized fermentation process. *L. curvatus* LTH 1174, for instance, achieved a dominance of 97% after inoculation at 10^5 cfu g^{-1} in German-style fermented sausage, whereas rep-PCR has confirmed the competitiveness of *L. sake* CTC 494 and *L. curvatus* LTH 1174 in Belgian-type fermented sausage. The use of strains with antioxidant properties due to catalase or superoxide dismutase, for instance *S. carnosus*, may help to inhibit lipid oxidation and prevent deterioration of colour and texture as well as the formation of toxic compounds.

FUNCTIONAL STARTER CULTURES WITH A TECHNOLOGICAL ADVANTAGE

The use of functional starter cultures may be useful to reduce levels of nitrite and nitrate, which are under discussion because of their contribution to the formation of health-affecting nitrosamines. Nitrite and nitrate are required in sausage fermentation technology as curing agents for microbial stability and colour formation. Bacteriocin-producing strains may be used to replace part of the preservative action, whereas the use of strains that are able to generate nitrosylated derivatives of myoglobin, i.e., to convert brown metmyoglobin into red myoglobin derivatives, could be useful to partially take over the colouration function of the curing agents. The latter possibility has been demonstrated with strains of *Lactobacillus fermentum* in smoked sausages.

Also, the use of strains that lead to accelerated ripening of the sausage may yield technological advantages, resulting in a reduction of storage time and an increased profit margin and competitiveness of the end product.

Bacterial strains, such as lactococci, that display strong proteolytic activity or autolysis during maturation could be applied as new starter cultures.

FUNCTIONAL STARTER CULTURES FOR A HEALTHIER PRODUCT

Probiotic Starter Cultures

In view of the recent interest for healthier meat products, probiotic starter cultures seem promising. Probiotics consist of live microorganisms as food or feed supplement, which, when ingested in sufficient amounts and when applied to animal or man, beneficially affect the host by improving the properties of the indigenous microbiota in the intestinal tract. Some strains of *Lactobacillus* are good candidates as probiotic cultures because they are normal components of the gut flora and display health-promoting properties *in vivo*. Enterococci are also marketed as probiotic preparations in health care. The use of probiotic strains is already very common in the dairy industry. In the future, probiotic LAB strains may be used in functional, probiotic meat products too. In some countries, however, meats do not have a reputation of being "health foods", which may compromise their marketing potential.

A first strategy consists of checking existing commercial starter cultures or sausage isolates for probiotic properties. For instance, the commercial meat starter strains *L. sake* Lb3 and *P. acidilactici* PA-2 can be regarded as potentially probiotic starter cultures because of their survival capacities under simulated gastrointestinal conditions.

Alternatively, it may be investigated if strains with probiotic properties, for instance intestinal isolates, perform well in a meat environment during sausage fermentation. Such strains should be strong competitors against the natural meat microbiota and grow to numbers that might have health-promoting effects. Several lactobacilli of intestinal origin have been shown to survive the sausage manufacturing process and can be detected in high numbers in the end product. It is certainly an asset if these new meat starter cultures also contribute to food safety. In this matter, the intestinal lactobacilli *L. rhamnosus* FERM P-15120 and *L. paracasei* subsp. *paracasei* FERM P-15121 have been shown to inhibit the growth and enterotoxin production of *S. aureus* to the same extent as a commercial *L. sake* starter culture in sausages fermented at either 20°C or 35°C. On the other hand, the intestinal strain *L. acidophilus* FERM P-15119 could not satisfactorily decrease *S. aureus* numbers, indicating the importance of careful strain selection.

In all cases, it should be checked that the sensory properties are not negatively affected when strains from non-meat origin are used. The (potential) probiotic strains *L. rhamnosus* GG, *L. rhamnosus* LC-705, *L. rhamnosus* E-97800 and L. plantarum E-98098 have been tested as functional starter culture strains in Northern European sausage fermentation without negatively affecting the technological or sensory properties, with a (minor) exception for *L. rhamnosus* LC-705. Similarly, the intestinal isolates *L. paracasei* L26 and *B. lactis* B94 had no negative impact on the sensory properties of the product when applied in conjunction with a traditional meat starter culture.

Ultimately, human studies should confirm the functionality of such probiotic fermented sausages. The daily consumption of 50 g of probiotic sausage by healthy volunteers, containing *L. paracasei* LTH 2579, during several weeks has been shown to modulate various aspects of host immunity but there was no significant influence on serum concentration of different cholesterol fractions and triacylglycerides. In faecal samples, there was a statistically significant increase in the numbers of *L. paracasei* LTH 2579, but not in the faeces of all volunteers.

Nutraceutical and Micronutrient Producers

The use of strains that produce micronutrients and nutraceuticals, e.g. vitamins and conjugated linoleic acid (CLA), as starter cultures may lead to health-promoting meat products. A nutraceutical is defined as any substance that may be considered as a food or part of a food that provides medical or health benefits, including the prevention and treatment of disease.

Most LAB have limited biosynthetic capacities for vitamin production. However, careful selection could reveal strains that produce vitamins in considerable amounts. Moreover, through metabolic engineering it could be possible to develop starter cultures for the *in situ* production of vitamins, such as folic acid and riboflavin. Several health-promoting properties have been ascribed to CLA, including anti-atherogenic action, inhibition of carcinogenesis, anti-diabetic effects, enhancement of immunological function, and reduction of body fat. Generally, CLA is found in meat from ruminants because of the bacterial hydrogenation of dietary linoleic acid in the rumen. However, meat from ruminants is not so often used in fermented sausages as in pork, in particular in Europe and North America. In addition to rumen-derived bacteria, it has been shown that other bacterial species, including lactobacilli, bifidobacteria and propionibacteria, can produce CLA. This opens perspectives to increase the nutritional value of fermented sausages when such strains are used as new starter cultures.

Reduction of Undesirable Compounds

It is important during strain selection that no undesirable compounds such as toxins, biogenic amines, or D(–) lactic acid, that could aversely affect health, are formed. If surface mould growth is desirable, it must be checked whether the mould starter culture produces mycotoxins or antibiotics. The use of moulds that are free of mycotoxin production as starter cultures could be useful in outcompeting mycotoxin-producing strains from the house flora. Moulds may also produce green or dark spots that are not acceptable to most consumers or have a negative impact on flavour and taste. During the ripening of fermented sausages, biogenic amines such as tyramine, histamine, tryptamine, cadaverine, putrescine, and spermidine, may be formed by the action of microbial decarboxylases on amino acids that originate from meat proteolysis. Microbial decarboxylation reactions may be ascribed to both the microorganisms that were introduced via the starter culture and the ones that constitute part of the natural population of the meat. In general, starter bacteria have limited tyrosine-decarboxylating activity, but contaminant non-starter LAB, in particular enterococci, are believed to be responsible for tyramine production. The use of decarboxylase-negative starter cultures that are highly competitive and fast acidifiers prevents the growth of biogenic amine producers and leads to end products nearly free of biogenic amines, as long as the raw material is of sufficient quality. Also, the introduction of starter strains that possess amine oxidase activity might be a way of further decreasing the amount of biogenic amines produced *in situ*. Although it is known that the superficial inoculation with *P. camemberti* in cheeses increases the concentration of certain amines, the production of amines by moulds in fermented sausages does not appear significant but has not been fully studied yet.

The nature of the lactic isomer produced by the LAB strains is of concern, since high levels of the D(–) lactic acid isomer are not hydrolysed by lactate dehydrogenase in humans and are thus capable of causing acidosis. Therefore, strains producing L(+) lactic acid should be preferably selected.

CONCLUSION AND FUTURE PERSPECTIVES

Fermented sausages not only contribute to regional pride and culture identity, they are also highly appreciated specialities with gastronomic value. Examples in Europe include varieties of Italian salami, Spanish chorizo, French saucisson, German Mettwurst, and many others. Products from different origins clearly posses different sensory qualities. Moreover, traditional fermented sausages are a rich source of interesting wild-type microorganisms. The deliberate use of such microorganisms in large-scale fermentations could help to introduce advantages of artisan specialities in industrial products. Also, a better understanding of the microbiota and population dynamics of artisan fermented sausages could improve the fermentation process and the safety and standardization of such products and hence reduce economic losses, in this way reinforcing their position in a market characterized by ever increasing globalization and industrialization. Useful information on the characterization of the microbiota and the study of population dynamics and strain dominance in meat products is to be expected from molecular techniques, e.g. denaturing gradient gel electrophoresis (DGGE).

Previous studies have aimed at applying mathematical modelling, biokinetics, and predictive microbiology, which are classically used to study the negative aspects related to food pathogens, toxin production, and food spoilage, to simulate and predict the behaviour of beneficial microorganisms, in the case of functional starter cultures. For instance, the kinetics of *L. sake* CTC 494 and *L. curvatus* LTH 1174 have been studied in detail in function of the sausage environment. In the future, rather than studying the food ecology of a single population, research will have to focus on a synecological approach, not exclusively taking into account interactions with the food environment but also with other microbial populations.

Recent advances in the field of metabolic engineering, genomics, and bioinformatics are expected to contribute to the future development of new, functional starter cultures. Exploration studies of the natural diversity of wild strains occurring in traditional, artisan foods, including comparative genomics, microarray analysis, transcriptomics, proteomics, and metabolomics, may generate a framework leading to the generation of new, industrial starters with increased diversity, stability, and industrial performance. It will permit rapid, high-throughput screening of promising wild strains with interesting functional properties and lacking negative characteristics, as well as the construction of genetically modified starter cultures with a tailored functionality.

Bioinformatic tools can be used to search in genomes for essential components, for instance with regard to flavour development, such as peptidases, aminotransferases, enzymes for biosynthesis of amino acids, and transport systems for peptides and amino acids. Genetic engineering of starters can be performed for a variety of purposes where a suitable starter cannot easily be found in nature. However, it is currently difficult to predict in what direction the future regulatory requirements will influence innovation through biotechnology, in particular genetic modification. One of the most promising applications is self-cloning, i.e., if the resulting strain does not contain DNA originating or derived from any other species than the host organism, which is currently exempted from the regulation on genetically modified organisms by the European Union (regulations 90/219 and 90/220). However, the use of genetically modified microorganisms in food and food processing remains controversial due to a lack of acceptance by consumers, especially in Europe.

Besides the focus on functional characteristics of potential new starter cultures, negative aspects such as antibiotic resistance, virulence genes, and undesirable metabolite formation should not be overlooked. The control of antibiotic resistance, for instance, is a topic of importance since the high load of endogenous bacteria in meat raw material and the inoculation with starters may represent a problem concerning the spreading of antibiotic resistance.

In addition, it should be emphasized that the performance of a selected starter culture should be seen in a context of the application, since functionality will depend on the type of sausage, the technology applied, the ripening time, and the ingredients and raw materials used. For instance, interactions between animal variability in tissue enzymes and bacterial peptidase or oxidative activity should be investigated. Moreover, fermentation temperature and ripening time will affect microbial ecology and hence performance. Nevertheless, our knowledge in the fields of raw materials, technologies and quality characteristics of fermented meat products is rapidly expanding and industrial application, possibilities of new, functional meat starter cultures are becoming more concrete.

Table 39.2 Some important properties of lactic acid bacteria involved in sausage fermentations

Property	*L. curvatus*	*L. sake*	*L. plantarum, L. pentosus*	*P. pentosaceus*	*P. acidilactici*
Cell shape	Rods	Rods	Rods	Cocci in tetrads	Cocci in tetrads
Growth at a_w 0.93	–(+)	+(–)	+(–)	V	+
Growth at 4°C	+	+	–	–	–
Growth at 50°C	–	–	–	–	+

(Contd.)

Table 39.2   (Continued)

Property	*L. curvatus*	*L. sake*	*L. plantarum,* *L. pentosus*	*P. pentosaceus*	*P. acidilactici*
Peptidoglycan type	Lys-D-Asp	Lys- D -Asp	Meso-DAP	Lys- D -Asp	Lys- D -Asp
Lactic acid enantiomer	DL(L)	DL	DL	DL	DL
Nitrate reductase	–	–	V	–	–
Nitrite reductase					
(+) haem	–	–	V	V	–
(–) haem	–	V	V	–	–
Catalase					
(+) haem	–	+	V	–	+
(–) haem	–	V	V	+	–
Ammonia from arginine	–(+)	+	–	+	+
Fermentation of maltose	V	V	+(–)	+	–
Sucrose	+(–)	+	+(–)	V	V
Lactose	V	V	+(–)	V	V
Melibiose	–(+)	+	+(–)	V	V
Mannitol	–	–	+	–	–
Gluconate	–(+)	+	+(–)	V	–
Minimal pH in MRS broth	4.0	4.0	3.7	3.8	3.8

L—*Lactobacillus*; P—*Pediococcus*.
+(–) postive with exceptions; – (+) negative with exceptions; V—variable (strain dependent).
L. pentosus is similar to *L. plantarum* but capable of fermenting D-xylose.
Strains forming L-lactate only have been formely classified as *L. bavaricus*.
All strains ferment glucose and ribose.

REVIEW QUESTIONS

1. Discuss the role of various classes of bacteria useful in meat and meat product preservation.

2. Write about dried meat curing and microbial fermentation.

3. Discuss the fermentation process in detail.

4. What are the reasons for fermentation failure in meat?

5. Outline the development of probiotic meat products.

40

YEAST FERMENTATION AND ITS PRODUCTS

YEASTS

A yeast is a unicellular fungus which reproduces asexually by budding or division, especially the genus *Saccharomyces* which is important in food fermentations. Yeasts and yeast-like fungi are widely distributed in nature. They are present in orchards and vineyards, in the air, the soil and the intestinal tract of animals. Like bacteria and moulds, they can have beneficial and non-beneficial effects in foods. Most yeasts are larger than most bacteria. The most well known examples of yeast fermentation are in the production of beverages and the leavening of bread. Some yeasts are chromogenic and produce a variety of pigments, including green, yellow and black. Others are capable of synthesizing essential B group vitamins.

Although there is a large diversity of yeasts and yeast-like fungi, (about 500 species), only a few are commonly associated with the production of fermented foods. They are all either ascomycetous yeasts or members of the genus *Candida*. Varieties of the *Saccharomyces cerevisiae* species are the most common yeasts in fermented foods and beverages based on fruit and vegetables. All strains of this genus ferment glucose and many ferment other plant derived carbohydrates such as sucrose, maltose and raffinose. In the tropics, *Saccharomyces pombe* is the dominant yeast in the production of traditional fermented beverages, especially those derived from maize and millet.

CONDITIONS NECESSARY FOR FERMENTATION

Most yeasts require an abundance of oxygen for growth, therefore by controlling the supply of oxygen, their growth can be checked. In addition to oxygen, they require a basic substrate such as sugar. Some yeasts can ferment sugar to alcohol and carbon dioxide in the absence of air but require oxygen for growth.

They produce ethyl alcohol and carbon dioxide from simple sugars such as glucose and fructose.

$$C_6H_{12}O_6 \xrightarrow{\text{Yeast}} 2C_2H_5OH + 2CO_2$$

Glucose　　　　　　　　Ethyl alcohol　　Carbon dioxide

In conditions of excess oxygen (and in the presence of *Acetobacter*) the alcohol can be oxidized to form acetic acid. This is undesirable if the end product is a fruit alcohol, but is a technique employed for the production of fruit vinegars.

Yeasts are active in temperature range from 0 to 50°C, with an optimum temperature range of 20° to 30°C.

The optimum pH for most microorganisms is near the neutral point (pH 7.0). Moulds and yeasts are usually acid tolerant and are therefore associated with the spoilage of acidic foods. Yeasts can grow in a pH range of 4 to 4.5 and moulds can grow from pH 2 to 8.5, but favour an acid pH.

In terms of water requirements, yeasts are intermediate between bacteria and moulds. Bacteria have the highest demands for water, while moulds have the least need. Normal yeasts require a minimum water activity of 0.85 or a relative humidity of 88%.

Yeasts are fairly tolerant of high concentrations of sugar and grow well in solutions containing 40% sugar. At concentrations higher than this, only a certain group of yeasts—the osmophilic type—can survive. There are only a few yeasts that can tolerate sugar concentrations of 65–70% and these grow very slowly in these conditions. Some yeasts—for example the *Debaryomyces*—can tolerate high salt concentrations. Another group which can tolerate high salt concentrations and low water activity is *Zygosaccharomyces rouxii*, which is associated with fermentations in which salting is an integral part of the process.

PRODUCTS OF YEAST FERMENTATION

The major products of yeast fermentation are alcoholic drinks and bread. With respect to fruits and vegetables, the most important products are fermented fruit juices and fermented plant saps. Virtually any fruit or sugary plant sap can be processed into an alcoholic beverage. This process is an alcoholic fermentation of sugars to yield alcohol and carbon dioxide.

It should be noted that alcohol production requires special licenses or is prohibited in many countries.

Alcohol and acids are two primary products of fermentation, used in the preservation of foods. Several alcohol-fermented foods are preceded by an acid fermentation and in the presence of oxygen and *Acetobacter*, alcohol can be fermented to produce acetic acid. Most food spoilage organisms cannot survive in either alcoholic or acidic environments. Therefore, the production of both these end products can prevent a food from undergoing spoilage and extend its shelf life.

Although primitive wines and beers have been produced, with the aid of yeasts, for thousands of years, the microorganisms associated with the fermentation were observed and identified nearly four hundred years ago. It was not until the 1850s that Louis Pasteur demonstrated unequivocally the involvement of yeasts in the production of wines and beers. Since then, the knowledge of yeasts and the conditions necessary for fermentation of wine and beer has increased to the point where pure culture fermentations are now used to ensure consistent product quality. Originally, alcoholic fermentations would have been spontaneous event that resulted from the activity of microorganisms naturally present. These non-scientific methods are still used today for the home preparation of many of the traditional beers and wines.

Alcoholic drinks fall into two broad categories: wines and beers. Wines are made from the juice of fruits and beers from cereal grains. The principal carbohydrates in fruit juices are soluble sugars; the principal carbohydrate in grains is starch, an insoluble polysaccharide. The yeasts that bring about alcoholic fermentation can attack soluble sugars but do not produce starch-splitting enzymes. Wines can therefore be made by the direct fermentation of the raw material, while the production of beer requires the hydrolysis of starch to yield sugars fermentable by yeast, as a preliminary step.

Raw fruit juice is usually a strongly acidic solution, containing 10 to 25 per cent soluble sugars. Its acidity and high sugar concentration make it an unfavourable medium for the growth of bacteria but highly suitable for yeasts and moulds. Raw fruit juice naturally contains many yeasts, moulds, and bacteria, derived from the surface of the fruit. Normally the yeast used in alcoholic fermentation is a strain of the species *Saccharomyces cerevisiae*.

The fermentation may be allowed to proceed spontaneously, or can be "started" by inoculation with a must that has been previously successfully fermented by *S. cerevisiae* var. *ellipsoideus*. Many modern wineries eliminate the original microbial population of the must by pasteurization or by treatment with sulphur dioxide. The "must" is then inoculated with a starter culture derived from a pure culture of a suitable strain of wine yeast. This procedure eliminates many of the uncertainities and difficulties of older methods. At the start of the fermentation, the "must" is aerated slightly to build up a large and vigorous yeast population; once fermentation sets in, the rapid production of carbon dioxide maintains anaerobic conditions, which prevents the growth of undesirable aerobic organisms, such as bacteria and moulds. The temperature of fermentation is usually from 25 to 30°C, and the duration of the fermentation process may extend from a few days to two weeks. As soon as the desired degree of sugar disappearance and alcohol production has been attained, the microbiological phase of wine making is over. Thereafter, the quality and stability of the wine depend very largely on preventing further microbial activity, both during the "ageing" in wooden casks and after bottling.

At all stages during its manufacture, fruit juice alcohol is subject to spoilage by undesirable microorganisms. Pasteur, whose descriptions of the organisms responsible and recommendations for overcoming them are still valid today, first scientifically explored the problem of the "diseases" of wines. The most serious aerobic spoilage processes are brought about by film-forming yeasts and acetic acid bacteria, both of which grow at the expense of the alcohol, converting it to acetic acid or to carbon dioxide and water. The chief danger from these organisms arises when access of air is not carefully regulated during ageing. Much more serious are the diseases caused by fermentative bacteria, particularly rod-shaped lactic acid bacteria, which utilize any residual sugar and impart a mousy taste to the wine. Such wines are known as turned wines. Since oxygen is unnecessary for the growth of lactic acid bacteria, wine spoilage of this kind can occur even after bottling. These risks of spoilage can be minimized by pasteurization after bottling.

Although yeasts are the principal organisms involved, filamentous fungi, lactic acid bacteria, acetic acid bacteria and other bacterial groups all play a role in the production of alcoholic fruit products.

GRAPE WINE

Grape wine is perhaps the most common fruit juice alcohol. Because of the commercialization of the product for industry, the process has received most research attention and is documented in detail.

The production of grape wine involves the following basic steps: crushing the grapes to extract the juice; alcoholic fermentation; malolactic fermentation if desired; bulk storage and maturation of the wine in a cellar; clarification and packaging. Although the process is fairly simple, quality control demands that the fermentation is carried out under controlled conditions to ensure a high-quality product.

The distinctive flavour of grape wine originates from the grapes as raw material and subsequent processing operations. The grapes contribute trace elements of many volatile substances (mainly terpenes) which give the final product the distinctive fruity character. In addition, they contribute non-volatile compounds (tartaric and malic acids) which impact on flavour and tannins that give bitterness and astringency. The latter are more prominent in red wines as the tannin components are located in the grape skins.

Normal grapes harbour a diverse microflora, of which the principal yeasts (*Saccharomyces cerevisiae*) involved in desirable fermentation are less in number. Lactic acid bacteria and acetic acid bacteria are also present. The proportions of each and total numbers present are dependent upon a number of external environmental factors including the temperature, humidity, stage of maturity, damage at harvest and application of fungicides. It is essential to ensure proliferation of the desired species at the expense of the non-desired ones. This is achieved through ensuring fermentation conditions that encourage *Saccharomyces* species.

The fermentation may be initiated using a starter culture of *Saccharomyces cerevisiae* in which case the juice is inoculated with populations of yeast of 10^6 to 10^7 cfu/ml juice. This approach produces a wine of generally expected taste and quality. If the fermentation is allowed to proceed naturally, utilizing the yeasts present on the surface of the fruits, the end result is less controllable, but produces wines with a range of flavour characteristics. It is likely that natural fermentations are practiced widely around the world, especially for home production of wine.

During alcoholic fermentation, yeasts are the prominent species. The composition of fruit juice— its acid and sugar level and low pH—favour the growth of yeasts and production of ethanol that restricts the growth of bacteria and fungi.

In natural fermentations, there is a progressive pattern of yeast growth. Several species of yeast, including *Kloeckera, Hanseniaspora, Candida* and *Metschnikowia*, are active for the first two to three days of fermentation. The build up of end-products (ethanol) is toxic to these yeasts and they die off, leaving *Saccharomyces cerevisiae* to continue the fermentation to the end. *S. cerevisiae* can tolerate much higher levels of ethanol (up to 15% v/v or more) than the other species who only tolerate up to 5 or 8% alcohol. Because of its tolerance of alcohol, *S. cerevisiae* dominates wine fermentation and is the species that has been commercialized for starter cultures.

Traditionally, fermentation was carried out in large wooden barrels or concrete tanks. Modern wineries now use stainless steel tanks as these are more hygienic and provide better temperature control. White wines are fermented at 10 to 18°C for about seven to fourteen days. The low temperature and slow fermentation favours the retention of volatile compounds. Red wines are fermented at 20 to 30°C for about seven days. This higher temperature is necessary to extract the pigment from the grape skins.

RED GRAPE WINE

Location of Production

Red grape wines are made in many African, Asian and Latin American countries including Algeria, Morocco and South Africa.

Product Description

Red grape wine is an alcoholic fruit drink of 10 to 14% alcoholic strength. The colour ranges from a light red to a dark red. It is made from the fruit of the grape plant (*Vitis vinifera*). There are many varieties of grape used including Cabernet, Sauvignon, Grenache, Nebbiolo, Pinot Noir, and Torrontes. The skins of the grape are allowed to be fermented in red wine production, to allow for the extraction of colour and tannins, which contribute to the flavour. The grapes contribute trace elements of many volatile substances, which give the final product the distinctive fruity character.

Raw Material Preparation

Ripe and undamaged grapes should be used. Crushed grape juice with skin is known as must.

Processing

The crushed grapes are transferred to fermentation vessels. The ethanol formed during this fermentation assists with the extraction of pigments from the skins. This takes 24 hours to three weeks depending on the colour of the final product required (Figure 40.1).

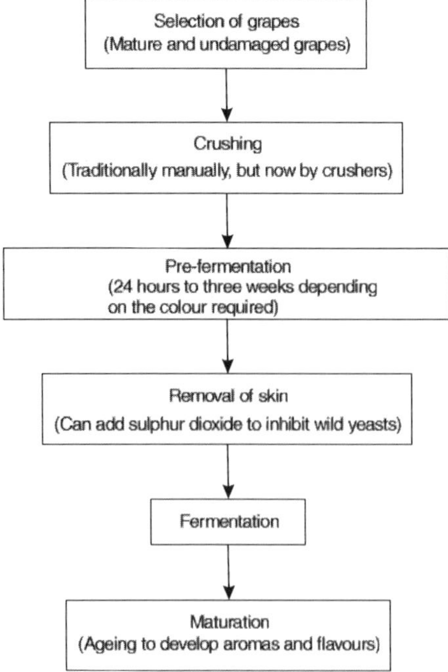

Figure 40.1   Flow chart of red grape wine production

The skins are then removed and the partially fermented wine is transferred to a separate tank to complete the fermentation. The fermentation can be carried out by naturally occurring yeasts on the skin of the grape or by using a starter culture of *Saccharomyces cerevisiae* in which case the juice is inoculated with populations of yeast. This approach produces a wine of generally expected taste and quality. If the fermentation is allowed to proceed naturally, utilizing the yeasts present on the surface of the fruits, the end result is less controllable, but produces wines with a range of flavour characteristics. Traditionally, fermentation was carried out in large wooden barrels or concrete tanks. Modern wineries now use stainless steel tanks as these are more hygienic and provide better temperature control.

Fermentation ends naturally when all the fermentable sugars have been converted to alcohol or when the alcoholic strength reaches the limit of tolerance of the yeast strain involved. Fermentation can be stopped artificially by adding alcohol and by sterile filtration or centrifugation. Some wines can be drunk immediately. However, most wines develop distinctive flavours and aromas by ageing in wooden casks.

Packaging and Storage

Traditionally wine was delivered to the point of sale in casks. The product is traditionally packaged in glass bottles with corks, made from the bark of the cork oak (*Quercus suber*). The bottles should be kept out of direct sunlight. During storage, wines are prone to non-desirable microbial changes. Yeasts, lactic acid bacteria, acetic acid bacteria and fungi can all spoil or taint wines after the fermentation process is completed.

WHITE GRAPE WINE

Location of Production

White grape wines are made in many African, Asian and Latin American countries including Algeria, Morocco and South Africa.

Product Description

White grape wine is an alcoholic fruit drink of 10 to 14% alcoholic strength. It is prepared from the fruit of the grape plant (*Vitis vinifera*), and is pale yellow in colour. There are many varieties used including Airen, Chardonnay, Palomino, Sauvignon Blanc and Ugni Blanc. The main difference between red and white wines is the early removal of grape skins in white wine production. The distinctive flavour of grape wine originates from the grapes used and subsequent processing operations. The grapes contribute trace elements of many volatile substances (mainly terpenes) which give the final product the distinctive fruity character.

Preparation of Raw Materials

Ripe and undamaged grapes should be used. The grapes are crushed to yield the juice, and the skins are removed and separated out. Sometimes the juice is clarified by allowing it to stand for 24 to 48 hours at 5 to 10°C, by filtering or centrifugation. Pectinolytic enzymes may be added to accelerate the breakdown of cell wall tissue and to improve the clarity of juice. Excessive clarification removes many of the natural yeasts and flora. This is beneficial if a tightly controlled induced fermentation is desired, but less so in natural fermentation. Long periods of settling out, however, encourage the growth of natural flora, which can contribute to the fermentation.

Processing

The clarified juice is transferred to a fermentation tank where fermentation either begins spontaneously or is induced by the addition of a starter culture. Traditionally, fermentation was carried out in large wooden barrels or concrete tanks. Modern wineries now use stainless steel tanks as these are more hygienic and provide better temperature control. White wines are fermented at 10 to 18°C for about seven to fourteen days. The low temperature and slow fermentation favour the retention of volatile compounds.

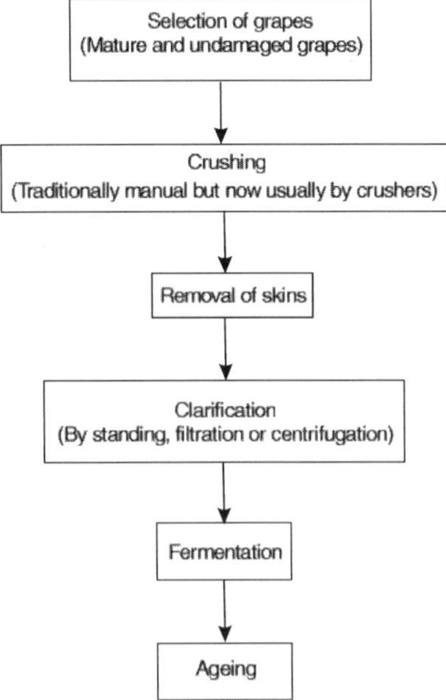

Figure 40.2 Flow chart of white grape wine preparation

The fermentation results from naturally occurring yeasts on the skin of the grape or using a starter culture of *Saccharomyces cerevisiae.* This approach produces a wine of generally expected taste and quality. If the fermentation is allowed to proceed naturally, utilizing the yeasts present on the surface of the fruits, the end result is less controllable, but produces wines having a range of flavour characteristics. It is likely that natural fermentations are practised widely around the world, especially for home production of wine. (Figure 40.2).

Packaging and Storage

Traditionally wine was delivered to the point of sale in casks. The product is traditionally packaged in glass bottles with corks, made from the bark of the cork oak (*Quercus suber*). The bottles should be kept out of direct sunlight. During storage, wines are proned to non-desirable microbial changes. Yeasts, lactic acid bacteria, acetic acid bacteria and fungi can all spoil or taint wines after the fermentation process is completed.

BANANA BEER

Location of Production

Throughout Africa.

Product Description

Banana beer is made from bananas, mixed with a cereal flour (often sorghum flour) and fermented to an orange, alcoholic beverage. It is sweet and slightly hazy with a shelf life of several days under proper storage conditions. There are many variations in the production process made. For instance *Urwaga* banana beer in Kenya is made from bananas and sorghum or millet, and *Lubisi* is made from bananas and sorghum.

Preparation of Raw Materials

Ripe bananas (*Musa* spp.) are selected. The bananas should be peeled. If the peels cannot be removed by hand, then the bananas are not sufficiently ripe.

Processing

The first step is the extraction of banana juice. Extraction of a high yield of banana juice without excessive browning or contamination by spoilage microorganisms and proper filtration to produce a clear product is of great importance. Grass is used as an aid in obtaining clarified juice.

One volume of water is added to every three volumes of banana juice. This makes the total soluble solids low enough for the yeast to act. Cereals are ground and roasted, and added to improve the colour and flavour of the final product. The mixture is placed in a container, which is covered in polythene to ferment for 18 to 24 hours (Figure 40.3). The raw materials are not sterilized by boiling and therefore provide an excellent substrate for microbial growth. It is essential that proper hygienic procedures are followed and that all equipment is thoroughly sterilized to prevent contaminating bacteria from competing with the yeast and producing acid instead of alcohol. This can be done by cleaning with boiling water or with chlorine solution. Care is necessary to wash the equipment free of residual chlorine as this would interfere with the action of the yeast. Strict personal hygiene is also essential.

For many traditional fermented products, the microorganisms responsible for the fermentation are unknown to scientists. However, there has been research to identify the microorganisms involved in banana beer production. The main microorganism involved is *Saccharomyces cerevisiae* which is the same organism involved in the production of grape wine. However, many other microorganisms associated with the fermentation have been identified. These varied according to the region of production.

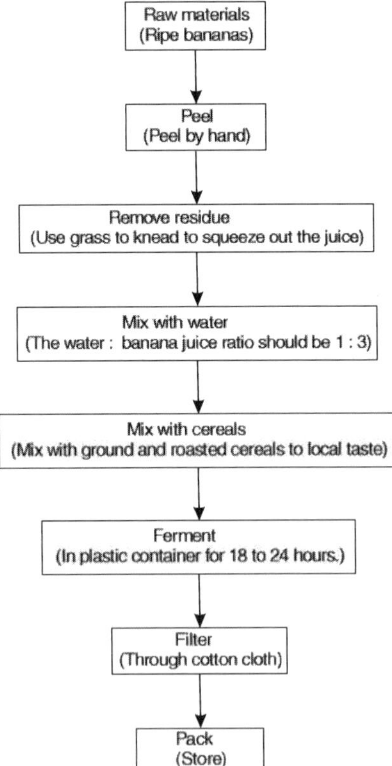

Figure 40.3   Flow diagram of banana beer production

After fermentation, the product is filtered through cotton cloth.

Packaging and Storage

Clean glass or plastic bottles are used. The product is kept in a cool place away from direct sunlight.

CASHEW WINE

Location of Production

Cashew wine is made in many countries in Asia and Latin America.

Product Description

Cashew wine is a light yellow-coloured alcoholic drink prepared from the fruit of the cashew tree (*Ancardium occidentale*). It contains an alcohol content of 6–12%.

Preparation of Raw Materials

Cashew apples are sorted and only mature and undamaged cashew apples should be selected. These should be washed in clean water.

Processing

The cashew apples are cut into slices to ensure a rapid rate of juice extraction when crushed in a juice press. The fruit juice is sterilized in stainless steel pans at a temperature of 85°C in order to eliminate the wild yeast. The juice is filtered and treated with either sodium or potassium metabisulphite to destroy or inhibit the growth of any undesirable types of microorganisms—acetic acid bacteria, wild yeasts and moulds.

Wine yeast (*Saccharomyees cerevisiae* var *ellipsoideus*) are added. Once the yeast is added, the contents are stirred well and allowed to ferment for about two weeks. The wine is separated from the sediment. It is clarified by using fining agents such as gelatin, pectin or casein which are mixed with the wine. Filtration is carried out with filter-aids such as fullers earth. The filtered wine is transferred to wooden vats.

The wine is then pasteurized at 50°–60°C. Temperature should be controlled, so as not to heat it to above 70°C, since its alcohol content would vaporize at a temperature of 75°–78°C. It is then stored in wooden vats and then subjected to ageing. At least six months should be allowed for ageing.

If necessary, wine is again clarified prior to bottling. During ageing, and subsequent maturing in bottles many reactions, including oxidation, occur with the formation of traces of esters and aldehydes, which together with the tannin and acids already present enhance the taste, aroma and preservative properties of the wine.

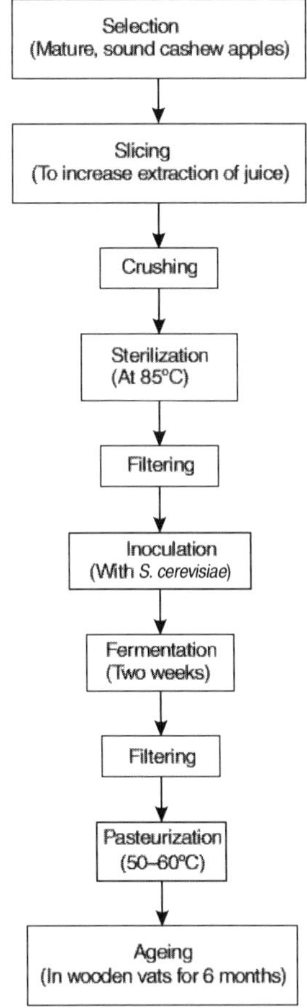

Figure 40.4   Flow chart of cashew wine production

Packaging and Storage

The product is packaged in glass bottles with corks. The bottles should be kept out of direct sunlight.

TEPACHE

Tepache is a light, refreshing beverage prepared and consumed throughout Mexico. In the past, tepache was prepared from maize, but nowadays various fruits such as pineapple, apple and orange are used. The pulp and juice of the fruit are allowed to ferment for one or two days in water with some added brown sugar. The mixture is contained in a lidless wooden barrel called a *"tepachera"*, which is covered with cheese cloth. After a day or two, the *tepache* is a sweet and refreshing beverage. If fermentation is allowed to proceed longer, it turns into an alcoholic beverage and later into vinegar. The microorganisms associated with the product include *Bacillus subtilis*, *B. graveolus* and the yeasts—*Torulopsis insconspicna, Saccharomyces cerevisiae* and *Candida queretana*.

COLONCHE

Colonche is a sweet, fizzy beverage produced in Mexico by fermenting the juice of the fruits of the prickly pear cacti—mainly *Opuntia* species. The procedure for preparing colonche is essentially the same as has been followed for centuries. The cactus fruits are peeled and crushed to obtain the juice, which is boiled for 2–3 hours. After cooling, the juice is allowed to ferment for a few days. Sometimes old colonche or tibicos may be added as a starter. Tibicos are gelatinous masses of yeasts and bacteria, grown in water with brown sugar. They are also used in the preparation of tepache.

FORTIFIED GRAPE WINES

Fortified wines are made in the Republic South Africa and North Africa. Fortified wines are made by adding spirits to wines, either during or after fermentation, resulting in an increased alcohol content, to about 20 per cent, i.e., approximately double that of table wines.

DATE WINE

Date wines are popular in Sudan and North Africa. They are made using a variety of techniques. Dakhai is produced by placing dates in a clean earthenware pot. To every one volume of dates, two and four volumes of boiling water is added, This is allowed to cool and is then sealed for three days. More warm water is then added and the container is sealed again for seven to ten days. Many variations of date wine exist: *El madfuna* is produced by burying the earthenware pots underground. *Benti merse* is produced from a mixture of sorghum and dates. *Nebit* is produced from date syrup.

SPARKLING GRAPE WINE

Sparkling grape wines are made in the South Africa. Sparkling wines can be made in one of three ways. The cheapest method is to carbonate wines under pressure. Unfortunately, the sparkle of these wines quickly disappears, and the product is considered inferior to the sparkling wines produced by the traditional method of secondary fermentation. This involves adding a special strains of wine- yeast (*S. cerevisiae* var. *ellipsoideus*)—a champagne yeast—to wine that has been artificially sweetened. Carbon dioxide produced by fermentation of the added sugar gives the wine its sparkle. In the original champagne method, which is still widely used today, this secondary fermentation is carried out in strong bottles, capable of withstanding pressure but early in the nineteenth century a method of fermenting the wine in closed tanks was devised, this being considerably cheaper than using bottles.

JACK-FRUIT WINE

Jack-fruit wine is an alcoholic beverage made by ethnic groups in the eastern hilly areas of India. As its name suggests, it is produced from the pulp of jackfruit (*Artocarpus heterophyllus*). Ripe fruit is peeled and the skin is discarded. The seeds are removed and the pulp is soaked in water. Using bamboo baskets, the pulp is ground to extract the juice, which is collected in earthenware pots. A little water is added to the pots along with fermented wine inoculum from a previous fermentation. The pots are covered with banana leaves and allowed to ferment at 18 to 30°C for about one week. The liquid is then decanted and drunk. During fermentation, the pH of the wine reaches a value of 3.5 to 3.8, suggesting that an acidic fermentation takes place at the same time as the alcoholic fermentation. Final alcohol content is about 7 to 8% within a fortnight.

FERMENTED PLANT SAPS

Virtually any sugary plant sap can be processed into an alcoholic beverage. The process is essentially an alcoholic fermentation of sugars to yield alcohol and carbon dioxide. Many alcoholic drinks are made from the juices of plants including coconut palm, oil palm, wild date palm, nipa palm, raphia palm and kithul palm.

PALM WINE

Location of Production

Palm "wine" is an important alcoholic beverage in West Africa, where it is consumed by more than 10 million people.

Product Description

Palm wine can be consumed in a variety of flavours varying from sweet unfermented to sour fermented and vinegary alcoholic drinks. There are many variations and names including *emu* and *ogogoro* in Nigeria and *nsafufuo* in Ghana. It is produced from sugary palm saps. The most frequently tapped palms are raphia palms (*Raphia hookeri* or *R. vinifera*) and the oil palm (*Elaeis guineense*). Palm wine has been found to be nutritious. The fermentation process increases the levels of thiamine, riboflavin, pyridoxin and vitamin B_{12}. Like many African alcoholic beverages, palm wine has a very short shelf life. The product cannot be preserved for more than one day. After this time accumulation of an excessive amount of acetic acid makes it unconsumable. The bark of a tree (*Saccoglottis gabonensis*) may be added as a preservative. The alkaloid and phenolic compounds, which are extracted from the wine have antimicrobial effect.

Preparation of Raw Materials

Sap is collected by tapping the palm. Tapping is achieved by making an incision between the kernels and a gourd is tied around to collect the sap which is collected in a day or two. The fresh palm juice is a sweet, clear, colourless juice containing 10–12 per cent sugar and is neutral. The quality of the final wine is determined mostly by the conditions used in the collection of the sap. Often the collecting gourd is not washed between collections and the residual yeasts in the gourd quickly begin the fermentation.

Processing

The sap is not heated and the wine serves an excellent substrate for microbial growth. It is therefore essential that proper hygienic collection procedures are to be followed, to prevent contaminating bacteria from competing with the yeast and producing acid instead of alcohol.

Fermentation starts soon after the sap is collected and within an hour or two, the sap becomes reasonably high in alcohol (up to 4%). If allowed to continue to ferment for more than a day, the sap begins turning into vinegar. Organisms responsible include *S. cerevisiae*, and *Schizosaccharomyces pombe*, and the bacteria *Lactobacillus plantarum* and *L. mesenteroides*. There are reports that the yeasts and bacteria originate from the gourd, palm tree, and tapping implements. However, the high sugar content of the juice would seem to selectively favour the growth of yeasts which might originate from the air. This is supported by the fact that fermentation also takes place in plastic containers. Within 24 hours the initial pH is reduced from 7.4–6.8 to 5.5 and the alcohol content ranges from 1.5 to 2.1 per cent. Within 72 hours, the alcohol levels increase from 4.5 to 5.2 per cent and the pH is 4.0. Organic acids present are lactic acid, acetic acid and tartaric acid. The main control points are extraction of a high yield of palm sap without excessive contamination by spoilage microorganisms, and proper storage to allow natural fermentation to take place.

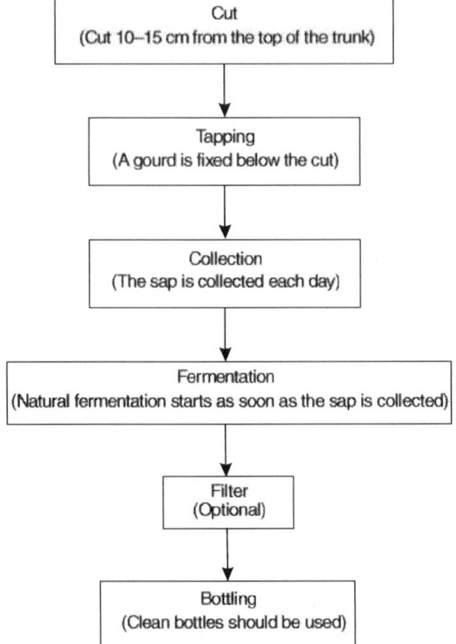

Figure 40.5   Flow chart of palm-wine production

Packaging and Storage

Clean glass or plastic bottles should be used. The product should be kept in a cool place away from direct sunlight.

TODDY

Location of Production

Throughout Asia, particularly India and Sri Lanka.

Product Description

Toddy is an alcoholic drink made by the fermentation of sap from a coconut palm. It is white and sweet with a characteristic flavour. The alcoholic content is 4–6% and has a shelf life of about 24 hours.

Preparation of Raw Materials

The sap is collected by slicing off the tip of an unopened flower. The sap oozes out and can be collected in a small pot tied underneath the flower.

Processing

The fermentation starts as soon as the sap is collected in the pots, particularly if a small amount of toddy is left in the pots. The toddy is fully fermented in six to eight hours. The product is usually sold immediately due to its short shelf life.

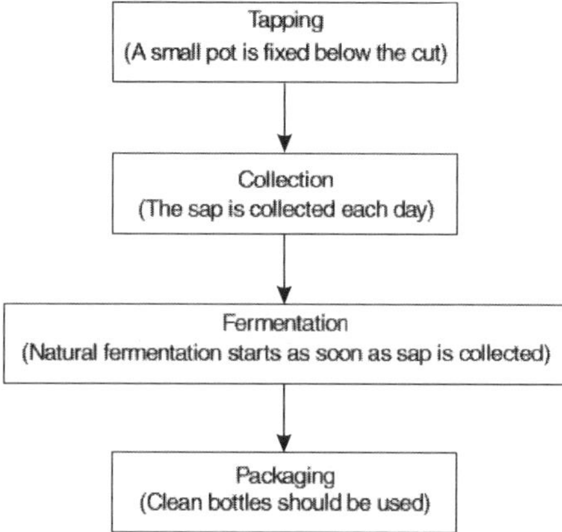

Figure 40.6   Production process of toddy

Packaging and Storage

Clean glass or plastic bottles are used to store. The product should be kept in a cool place away from direct sunlight.

PULQUE

Location of Production

Pulque is the national drink in Mexico, where it is claimed, and it originated with the early Aztecs. Pulque is a traditional beverage that now forms the basis of a national industry, together with the spirits mezcal and tequila that are obtained from it. Pulque plays an important role in the nutrition of low income people in Mexico with B vitamins being present in nutritionally important levels.

Product Description

Pulque is a milky, slightly foamy, acidic and somewhat viscous beverage. It is obtained by fermentation of aguamiel, which is the name given to the juices of various cacti, notably *Agave atrovirens* and *A. americana* which are often called the "Century plant" in English. The alcohol content of pulque varies between six to seven per cent. The beverage obtained upon distilling pulque is called "Mezcal", and if manufactured in the Tequila region from a numbered distillery, it is referred to as "Tequila". The drink is often considered an aphrodisiac. The name Ticyaol is given to a good strain that makes one particularly virile. Pulque is frequently the potion of choice used by women during menstruation.

Preparation of Raw Materials

The juices are extracted from the plants of eight-to-ten years old and fermentation takes place spontaneously, although occasionally the juices are inoculated with a starter from previous fermentations.

Processing

The juice is allowed to ferment naturally through a mixed fermentation, although yeast (*Saccharomyces carbajali*) is the main actor. *Lactobacillus plantarum* produces lactic acid and the viscosity of pulque is caused by the activity of two species of *Leuconostoc* which produce dextrans (Wood and Hodge). During fermentation of the juices of the plant, the soluble solids are reduced from between 25–30% to 6%; the pH falls from 7.4 to 3.5; the sucrose content falls from 15% to 1% and vitamin levels are increased. For instance the vitamin content (mg of vitamins per 100 g of product), increases from 5 to 29 for thiamine, 54 to 515 for niacin and 18 to 33 for riboflavin.

Packaging and Storage

Clean glass or plastic bottles should be used. The product should be kept in a cool place away from direct sunlight.

ULANZI (BAMBOO WINE)

Location of Production

East and Southern Africa.

Product Description

Ulanzi is a fermented bamboo sap obtained by tapping young bamboo shoots during the rainy season. It is a clear, whitish drink with a sweet and alcoholic flavour.

Preparation of Raw Materials

The bamboo shoots should be young in order to obtain a high yield of sap. The growing tip is removed and a container is fixed in place to collect the sap. The container should be clean in order to prevent contamination of the fresh sap.

Processing

The raw material is an excellent substrate for microbial growth, and fermentation begins immediately after collection. Fermentation takes five to twelve hours depending on the strength of the final product desired.

Packaging and Storage

Packaging is usually only required to keep the product for its relatively short shelf life.

BASI (SUGAR CANE WINE)

Basi is a sugar cane wine made in the Philippines by fermenting boiled, freshly extracted, sugar cane juice. A dried powdered starter is used to initiate the fermentation. The mixture is allowed to ferment for up to three months, and to age for up to one year. The final product is light brown in colour and has a sweet and a sour flavour. A similar product called shoto sake is made in Japan.

MURATINA

Muratina is an alcoholic drink made from sugar cane and muratina fruit in Kenya. The fruit is cut in to half, sun dried and boiled in water. The water is removed and the fruit is again sun dried. The fruit is added to a small amount of sugar cane juice and incubated in a warm place for 24 hours, after which it is removed and sun dried. The dried fruit is then added to a barrel of sugar cane juice which is allowed to ferment for one to four days. The final product has a sour alcoholic taste.

RICE WINES

Rice wines are produced from the hydrolytic breakdown products of cereal starches and other polysaccharides. They range from simple Thai rice wine to highly sophisticated Japanese sake. The moulds involved in alcohol production of Asian rice wines include *A. rouxii*, and the yeasts *S. cerevisiae*, *Saccharomycopsis* burtonii, *S. fibuligera*, *C. lactosa* and related yeasts. Rice and/or cereal wines are produced on both a cottage and a commercial scale in most Asian countries, especially Japan, China, Korea, Thailand, the Philippines and Vietnam. Wines may be distilled to obtain a liquor or spirit, for instance the famous Indonesian brem bali, an alcoholic liquor produced in Bali from the liquid portions of tape ketan. These liquors can also be used to fortify rice wine. Sake is a pale yellow rice wine of Japanese origin with an alcohol content of 15–16% (weight in volume, w/v) or higher. Steamed rice is mixed with tane-koji (*A. oryzae*) and allowed to ferment for 5–6 days, after which it is mixed with yeast moto or ragi starter and water to form the main mash, or moromi. Moto dominates the moromi fermentation. Wild yeasts tend to die off at the early stages of moromi fermentation due to nitrite produced by nitrate-reducing bacteria. Malaysian rice wine or tapai is lighter in colour, ranging from red to pink. It is made from cooked gelatinized rice and red pulverized ragi (yeast cake, or jui-piang). Yakju and takju are Korean alcoholic beverages originally made from rice, but which are now made from wheat, barley, corn or millet. In the traditional yakju process, steamed, cooled rice is mixed with nuruk amylolytic starter, and yeast inoculum is added. Takju is made by diluting fresh yakju liquor prior to filtration. In comparing the traditional and industrial rice wine brewing processes in Korea, it was noted that nuruk is used in the traditional method, normally carried out at low temperatures (5–10°C). The industrial brewing follows the Japanese sake brewing method and undergoes relatively high temperature fermentation (25°C). The alcohol content in the industrial rice wine tends to be higher but the wine is lower in esters than the traditional brew. In the Philippines, tapuy is an acidic but sweet alcoholic rice wine and is known by other names such as binubudan (Ifugao), binuburan (Ilocano) or purad (Tagalog). The Thai rice wines such as sato and krachae are cloudy yellow liquids made from glutinous rice. In Vietnam, ruou nep and ruou nep than are made by steaming white or purple glutinous rice, respectively, and inoculation with men, an amylolytic starter. Fermentation is carried out in a two-stage process of which the first stage is aerobic mould fermentation in a solid-state condition. The main alcoholic fermentation occurs during the second stage after water has been added and lasts for approximately 5 days. At the final stage, the wine is a turbid suspension of pink-red colour, containing 8–14% (w/v) alcohol and some residual sugars. The wine may be clarified and/or strengthened by blending with distilled alcohol, depending on local demand.

FACTORS AFFECTING WINE FERMENTATION

There are several variables which can affect the fermentation process and final quality of wine. Factors which are most important to control are:

Clarification and pretreatment of juice   Excessive clarification removes many of the natural yeasts and flora. This is beneficial if a tightly controlled induced fermentation is desired, but less so if the fermentation is a natural one. Long periods of settling out, however, encourage the growth of natural flora, which can contribute to the fermentation.

Chemical composition of juice   The main consituents of grape juice are glucose (75 to 150 g/l), fructose (75 to 150 g/l), tartaric acid (2 to 10 g/l), malic acid (1 to 8 g/l) and free amino acids (0.2 to 2.5 g/l). The main reaction is the fermentation of glucose and fructose to ethanol and carbon dioxide. However, the presence of nitrogenous and sulphurous products also contributes to the fermentation. The addition of sulphur dioxide to the juice delays the growth of yeast, but does not necessarily inhibit the growth of non-*Saccharomyces* strains. Fruits generally contain sufficient substrates—soluble sugars—for the yeast to ferment and convert into an acceptable concentration of alcohol. Sugar can be added to fruit juices with a low sugar content, to increase the amount of fermentable substrate.

Temperature   Temperature has an impact on the growth and activity of different strains of yeast. At temperatures of 10 to 15°C, the non-*Saccharomyces* species have an increased tolerance to alcohol and therefore have the potential to contribute to the fermentation.

Influence of other microorganisms   Other microorganisms have the potential to influence wine production at all stages of the process. Prior to harvest, yeasts grow on the surface of grapes. Fungicides are used to control their growth, but these disturb the natural balance of flora, thus making it difficult to carry out a "natural" fermentation. Overuse of fungicides can lead to the development of resistant strains of yeast which have the potential to produce toxins which destroy the desirable yeast species. These yeasts are known as "killer" strains. Other microbes have further chances to influence the fermentation during the clarification process, after fermentation and during maturation and bottling when *Acetobacter* species can oxidize the alcohol and produce acetic acid.

About two to three weeks after the alcoholic fermentation is finished wines often undergo a malolactic fermentation. This occurs naturally and lasts for about four weeks. It is a lactic acid fermentation, initiated by lactic acid bacteria resident in the wine. Inoculating the fermented wine with cultures of *Leuconostoc oenos* can start the process if it is desired. The main reaction of these bacteria is the decarboxylation of L-malic acid to L-lactic acid, which decreases the acidity of the wine and increases its pH by about 0.3 to 0.5 units. Wines produced from grapes grown in colder climates tend to have a higher concentration of malic acid and a lower pH (3.0 to 3.5) and the taste benefits from this slight decrease in acidity. The benefits of this process are that it imparts a more mellow flavour to the wine. The growth of malolactic bacteria also contributes to the taste of the wine. Wines that have undergone a malolactic fermentation appear to be less susceptible to any further damage from other bacteria. This could be because *L. oenos* has used up all available substrate, or it may have secreted bacteriocins which prevent the growth of other species. Although the malolactic fermentation seems to be a useful process, not all wines benefit from it. Wines produced from grapes in warmer climates tend to be less acidic (pH > 3.5) and a further reduction in acidity may have adverse effects on the quality of the wine. Decreasing the acidity also increases the pH to values which can allow spoilage organisms to multiply. It is difficult to prevent the malolactic fermentation from taking place naturally, especially later on after the wine has been bottled. In low-acid wines, the acidity may be adjusted after this fermentation has taken place. The malolactic fermentation can be prevented by controlling several factors: the wine pH (< 3.2); ethanol content (> 14%) and levels of sulphur dioxide (>50 mg/l). The bacteriocin nisin can also be used to control the growth of malolactic bacteria. However the subtle blend of aromas and flavours that contribute to the final taste may be lost by such stringent control.

The conversion of malic acid to lactic acid is one of the main reactions carried out by wine lactic acid bacteria. *L. oenos* needs to be present in significant numbers (greater than 10^6 cfu/ml) for the reaction to take place at a suitable pace. The bacteria use residual pentose and hexose sugars in the wine as a substrate for growth. The main reaction is the deacidification (or decarboxylation) of malic acid. In addition to this, the by-products of the reaction impart flavours and aromas to the wine.

During storage, wines are prone to non-desirable microbial changes. Yeasts, lactic acid bacteria, acetic acid bacteria and fungi can all spoil or taint wines after the fermentation process is completed. The changes that occur are increased acidification through the formation of acetic and other acids from alcohol; increased carbonation through a secondary fermentation of residual sugars and flavour changes through the metabolism of numerous compounds.

OTHER MISCELLANEOUS PRODUCTS

Kombucha

Kombucha from Central and East Asia is a beverage obtained by fermentation of sweetened boiled tea with a mixed culture of yeasts and acetic acid bacteria. Other names for kombucha, or 'tea fungus', include 'fungus japonicus', 'tee kwass', 'tea kvass', 'champignon de longue vie', 'Indo-Japanese tea fungus' and 'Manchurian mushroom'. Kombucha is a symbiosis of *Acetobacter* spp.— mainly *Acetobacter xylinum*—and various yeasts. The mixed yeast–bacterial culture growing on sugary tea extract accumulates lactic (0.1%), acetic (traces) and gluconic (0.01–0.3%) acids, and some ethanol (0.3%). The pH decreases steadily to about 2.5. The resulting beverage also contains vitamins and minerals and is considered to be a healthy product.

The yeast flora of commercial tea fungus includes the genera *Brettanomyces* (56%), *Zygosaccharomyces* (29%) and *Saccharomyces* (26%). *Saccharomycodes ludwigii* and *Candida kefyr* (anamorph of *Kluyveromyces marxianus*) in isolated cases, as well as a pellicle-forming yeast, *C. krusei* and apiculate yeasts, *Kloeckera* spp. and *Hanseniaspora* spp. have also been reported. However, *Zygosaccharomyces kombuchaensis* is the dominant yeast species now known to be commonly associated with kombucha tea. In another study of four commercial kombucha products, the yeasts found included *Brettanomyces bruxellensis* (anamorph of *Dekkera bruxellensis*), *Candida stellata*, *S. pombe*, *T. delbrueckii* and *Zygosaccharomyces bailii*. Comparing these findings, it appears that the fermentation is initiated by osmotolerant yeasts and is then succeeded and ultimately dominated by acid-tolerant species.

CONDIMENTS

Wadis

Wadis, traditionally consumed in Punjab and Bengal of India, are now popular in many places of India, Pakistan and Bangladesh. These dried, hollow, brittle cones or balls (3–8 cm diameter, 15–40 g in weight) are used as a spicy condiment or an adjunct for cooking vegetables, grain legumes or rice. To prepare wadi, dals generally of blackgram, are soaked, drained, ground into a smooth soft dough, left to ferment for 1–3 days, moulded into cones or balls, deposited on bamboo or palm mats smeared with oil, and sun-dried for 4–8 hours. The surface of the cones or balls becomes covered with a mucilaginous coating which helps to retain the gas formed during their fermentation. The wadis look hollow, with many air pockets and yeast spherules in the interior and a characteristic surface crust. Initially the microflora is diverse and contains lactic acid bacteria, bacilli, flavobacteria and yeasts. Gradually, *L. mesenteroides*, *L. fermentum*, *S. cerevisiae* and *T. Cutaneum* become dominant. *Candida vartiovaarae* and *K. marxianus* are also often found. The development and prevalence of microflora are affected by the seasons, summer being more favourable for bacteria and winter for yeasts. The production of acid and gas results in a fall of pH from 5.6 to 3.2 and two-fold rise in the volume of the dough. The lactic acid bacteria are mainly responsible for the acidification of dough, favourable conditions for the yeasts to grow and become active for leavening. The fermentation brings about a significant increase in soluble solids, non-protein nitrogen, soluble nitrogen, free amino acids, proteolytic activity and B vitamins including thiamine, riboflavin and cyanocobalamin. On the other hand, the levels of reducing sugars and soluble proteins decrease. Amylase activity increases initially, but declines thereafter. Wadis prepared by inoculating sterilized ingredients with a mixed culture of *C. krusei* (anamorph of *Issatchenkia orientalis*) and *L. mesenteroides* resemble the marketed ones. In contrast, the uninoculated controls were hard and compact and, when broken, had a glistening surface.

Papads

Papad (papadam) is another important condiment or savoury food in India, Pakistan and Bangladesh. This thin, usually circular, wafer-like product is used to prepare curry or is eaten by itself as a crackly snack or appetizer with meals after roasting or deep-frying in oil. Papad-making under controlled conditions has already developed into a cottage or small-scale industry. Black gram flour or a blend of blackgram with Bengal gram, lentil (*Lens culinaris*), red gram or green gram (*Vigna radiata*) flour is hand-kneaded with a small quantity of peanut oil, common salt (about 8%, w/w), 'papad khar' (saltworts produced by burning a variety of plant species, or from very alkaline deposits in the soil) and water, and then pounded into a stiff paste. The dough (sometimes with a backslop and spices added) is left to ferment for 1–6 hours. The fermented dough is shaped into small balls which are rolled into thin, circular flat sheets (10–24 cm diameter, 0.2–1.2mm thick) and generally dried in the shade to 12–17% (w/w) moisture content. *Candida krusei* and *S. cerevisiae* are involved in the preparation of papad.

Soysauces

Soy sauces are light to dark brown liquids with a meat-like salty flavour used in cooking and as a table condiment. Traditionally made in China, Japan, Korea, Thailand, the Philippines, Indonesia and Malaysia, soy sauce is now also produced in Europe and the Americas. There are two specific fermentation stages involved in soy sauce production: aerobic koji fermentation, which involves the use of *A. oryzae* or *Aspergillus sojae*, and an anaerobic moromi or salt mash, which undergoes lactic acid bacteria and yeast (*Zygosaccharomyces rouxii*) fermentations. The two main groups of enzymes produced by *A. oryzae* during koji fermentation are carbohydrases (α-amylases, amyloglucosidase, maltase, sucrase, pectinase, β-galactosidase, cellulase, hemicellulase and pentosan-degrading enzymes) and proteinases. Lipase activity has also been reported. These major enzymes hydrolyse carbohydrates and proteins to sugars and amino acids and low molecular weight peptides, respectively. These soluble products are essential for yeast and bacterial activities during moromi fermentation.

In this fermentation *Tetragenococcus halophila* initially proliferates and produces lactic acid, which lowers the pH to 5.5 or less. Acid-tolerant dominant yeasts, notably *Z. rouxii*, grow and produce about 3% (w/v) alcohol and several compounds which add characteristic aroma to soy sauce. Although *Z. rouxii* is the dominant moromi yeast which produces alcohol and several compounds that add characteristic aromas to soy sauce, other yeasts such as *Candida versatilis* and *Candida etchellsii* also produce phenolic compounds, i.e., 4-ethylguaiacol and 4-ethylphenol, which contribute to soy sauce aroma. Nearly 300 types of flavour compounds have been identified in Japanese soy sauce. *Zygosaccharomyces rouxii* produces flavour compounds including alcohols, glycerol, esters, 4-hydroxy-5-methyl-3(3 H)-furanone (HMMF), 4-hydroxy-2 (or 5)-ethyl-5(or 2) -methyl-3 (2 H)-furanone (HEMF) and 4-hydroxy-2,5-dimethyl-3(2 H)-furanone (HDMF). Of the furanones, HEMF produced by *Z. rouxii* and *Candida* spp. gives Japanese-type soy sauce its characteristic flavour. This compound is also reported to have antitumour and antioxidative properties. Higher alcohols such as isobutyl alcohol, isoamyl alcohol and 2-phenyl ethanol, produced by *C. versatilis*, are also important flavour constituents of soy sauce. Certain strains of yeasts have deleterious effects on soy sauce. Film-forming yeasts, mainly belonging to the genera *Zygosaccharomyces*, *Hansenula and Pichia*, cause spoilage in moromi fermentation.

Miso

Miso is a salty paste with a meat-like flavour made by fermenting soybean, with or without the addition of rice or barley, using A. oryzae and a yeast, Z. rouxii. Sometimes, *Tetragenococcus halophila* and *Enterococcus faecalis* are also involved in the fermentation. Miso is a seasoning agent and is also used in the preparation of miso soup. Heat-treated rice and/or soybeans are used to prepare 'shinshu' or ricesoybean miso. After the initial solid-substrate fermentation dominated by *A. oryzae*, salt (38% of the original weight of dry soybeans) is added to the koji and mixed thoroughly. The mixture is backslopped or inoculated with *Z. rouxii* and allowed to ferment for up to 15 days for sweet miso and 2–12 months for salty miso. Although other halophilic yeasts such as *Torulopsis versatilis* may be present, only *Z. rouxii* produces the desired metabolites for an acceptable product. Flavour components in miso are similar to those of soy sauce. Furanones, HEMF and HDMF, produced by *Z. rouxii*, have been identified as important flavour components in miso. Miso also contains B vitamins (riboflavin and cyanocobalamine) as a result of yeast fermentation.

Future Perspectives

Upgrading of traditional home-scale processes is needed so that they can continue to maintain and strengthen the cultural heritage and can compete successfully with imported products. Whereas small-scale manufacture has the advantages of short distribution lines, income generation for families, etc., urbanization and the resulting growing demand for ready-to-consume high-quality foods requires larger-scale controlled industrial production. Examples of industrialized traditional fermented foods are: i) alcoholic snacks such as tapai, which are now produced at a small cottage scale in Malaysia using commercially available pure culture starters of the starch-degrading mould *A. rouxii* and the yeast *S. fibuligera*; ii) rice wines such as Japanese sake, using A. oryzae for rice saccharification and sake yeasts (*S. cerevisiae* strains selected for reduced foam production, or killer properties if required); iii) condiments such as soy sauce inoculated with *Z. rouxii* and *Candida* spp. and miso in which similar halo-tolerant yeasts are used for flavour development. Yeast products such as enzymes, B vitamins, trace elements (selenium, chromium), glycans, flavour components and carotenoid pigments occur in traditional foods, but could be exploited more effectively as purified substances and food ingredients. Yeasts have a relatively high content of protein, lipids and micronutrients. In view of the widespread micronutrient deficiencies in regions that depend predominantly on plant-based diets, the addition of yeast derived food products could contribute to improved nutritional status. In conclusion, a wide variety of yeasts are involved in traditional fermented foods. Although the occurrence of various yeasts has been reported, knowledge and understanding of their ecology including aspects such as microbial successions and competitiveness, and of their genetic and physiological properties remain to be acquired. In particular, yeasts that contribute to desirable product properties require more precise characterization, using genomics, proteomics and physiological approaches for more efficient identification and exploitation, while developing consumer friendly strategies to control fermentations and safeguard hygiene.

REVIEW QUESTIONS

1. Describe the conditions necessary for yeast fermentations.
2. Describe the production of fruit alcohols.
3. List out the various products of yeast fermentation.

4-1

ENZYMES OF LACTIC ACID BACTERIA IN VINIFICATION

INTRODUCTION

Two key groups of organisms are involved in the production of red, white, and sparkling wine. The yeasts, typically strains of *Saccharomyces cerevisiae*, carry out the primary or alcoholic fermentation, in which sugars are converted to ethanol and CO_2. Lactic acid bacteria (LAB), especially *Oenococcus oeni* (formerly *Leuconostoc oenos*, conduct the secondary or malolactic fermentation (MLF) of wine by decarboxylating L-malic acid to L-lactic acid and CO_2. Apart from these two crucial reactions in grape vinification, a myriad of other changes occur to complete the transformation of grape juice to wine. Compounds that stimulate our visual, olfactory, gustatory, and tactile senses are either released from the various ingredients or are synthesized, degraded, or modified during vinification. Many of these processes involve the action of enzymes. Such enzymes can be free or cell associated and originate from sources that include enzyme addition, the grapes themselves, the grape microflora (fungi, yeast, or bacteria), the inoculated microbes, or microbes associated with winery equipment and storage vessels to which the wine is exposed during production.

Current viticultural practices and vinification processes are essentially protocols for favouring the activities of certain enzymes while discouraging the activities of others. Thus, winemakers can broadly achieve desirable outcomes during fermentation by using a selected wine yeast strain characterized by desirable physiological and hence enzymatic properties. Conversely, adverse reactions, such as the browning associated with polyphenoloxidases, can be minimized by excluding oxygen from the grape juice or through addition of sulphur dioxide (SO_2) to inhibit enzyme activity.

A more recent strategy in the history of winemaking is the addition of a microbial culture or enzyme preparation to juice or wine that confers a specific or selective group of enzymatic activities. These activities can either amplify the effect of indigenous enzymes or be novel. Initially, such additives addressed issues of juice-processing efficiency and wine recovery. Thus, for many decades the gelling seen in many fruit juices as a result of pectins has been reduced or eliminated with pectinase enzymes, most often derived from *Aspergillus* fungi, which increase juice extraction or minimize filter blockage. Enzyme-based solutions that provide a broader range of benefits, such as flavour enhancement or manipulation of colour, have now become available.

In the development of new enzyme treatments, efforts have often been centered on desirable activities identified in the microorganisms used or encountered during vinification, especially the yeast. In part, this approach has been taken because of legal restrictions on the nature of additives that can be added to wine. It is unlikely that the use in winemaking of wine yeast with a novel enzymatic capability would require regulatory approval, whereas the addition of an enzyme extract or purified enzyme preparation may require such approval. Despite the appeal of this approach, extensive efforts have yielded only a small number of technologically important enzymes, and even fewer of these enzymes perform satisfactorily under winemaking conditions, which include a high sugar (glucose and fructose) content, a low pH (pH 3.0 to 4.0), low temperatures (–15°C), and the presence of ethanol (up to 15% [v/v] or more) or SO_2.

Interestingly, the LAB that grow and thrive in grape juice or wine under conditions that interfere with the production and activity of desirable enzymes in yeast or fungi have been poorly studied as a source of enzymes with potential usefulness in vinification. Young wine can be a nutritionally deficient environment that could be expected to lead to the elaboration by LAB of numerous enzymatic activities for nutrient scavenging. The activities of greatest interest are those conferred by a single enzyme, ideally one with an extracellular localization. Such enzymes are most amenable to separation from the cell biomass or preparation as an enzyme-enriched extract, which may be desirable when the originating organism is difficult to grow or is not wanted in grape juice or wine.

THE MALOLACTIC ENZYME

The LAB most commonly associated with wine belong to *O. oeni* and selective *Lactobacillus* and *Pediococcus* spp. The major function of LAB is the conversion of L-malic acid to L-lactic acid during the MLF. This conversion may be achieved by one of three pathways. Most wine-borne LAB decarboxylate L-malic acid to L-lactic acid and carbon dioxide, in a reaction catalysed by the malolactic enzyme, without the release of intermediates. One exception to this is observed in *Lactobacillus casei* and

Lactobacillus faecalis, which use a malic enzyme (malate dehydrogenase) to metabolize L-malic acid to pyruvate. L-lactate dehydrogenase then acts on pyruvate to produce L-lactic acid. A second exception is evident in *Lactobacillus fermentum*, in which metabolism of L-malic acid yields D-lactic acid, L-lactic acid, acetate, succinate, and carbon dioxide. Despite the importance of the MLF, its occurrence is both highly unpredictable and difficult to control or manipulate. Consequently, techniques that facilitate the efficient and complete conversion of L-malic acid to L-lactic acid in grape juice and wine have been sought. Such techniques aim to separate this central enzyme-driven conversion from the problematic growth of the source LAB in the wine. Examples from the beverage and food industries include bioreactor systems comprising LAB cells immobilized alone, LAB cells co-immobilized with yeast, or free *O. oeni* cells or enzymes and cofactors.

The ability of the malolactic enzyme, as a single enzyme, to conduct the conversion of L-malic acid to L-lactic acid has made it the activity of choice for such bioreactor systems, as well as heterologous expression studies. A bioreactor containing NAD, manganese ions, and the malolactic enzyme from *L. oenos* strain 84.06 achieved a 62 to 75% conversion rate for L-malic acid to L-lactic acid in wine. Incomplete conversion was attributed to enzyme inactivation and instability of the cofactor NAD at the wine pH. The expression of the malolactic enzyme encoded by the *mleS* gene from *Lactococcus lactis* in an *S. cerevisiae* wine yeast enabled it to effect the MLF and alcoholic fermentation simultaneously. Whether achieved via such recombinant methods or via bioconversions with cells or enzyme preparations, the potential benefits of enhanced application of malolactic enzyme warrant further research. Identification of a malolactic enzyme that is more resilient under wine conditions and improved delivery systems is of foremost interest.

PROTEOLYTIC AND PEPTIDOLYTIC ENZYMES

Nitrogen compounds in grape juice include compounds that are variously essential or detrimental to successful fermentation and wine quality. The bulk of the nitrogenous fraction is comprised of the alpha amino acids and ammonium, which along with peptides containing up to five amino acid residues represent the assimilable nitrogen that is vital for yeast growth and fermentative activity and suppression of hydrogen sulphide. Conversely, the proteins of grapes are considered a nuisance as they become unstable in the finished wine and can precipitate to produce a haze. Bentonite fining remains the most common and effective method for removal of haze-forming proteins from wine despite the unwanted effects of removing some assimilable nitrogen, modifying the flavour, and changing the kinetics of fermentation. Proteases have been sought from a variety of sources, and they have been evaluated as an alternative to bentonite treatment to remove unwanted proteins while possibly also liberating assimilable nitrogen for exploitation by yeast. Commercial proteolytic preparations, such as trypsin and pepsin, do not function optimally at the low temperatures and pH used during winemaking. Proteases from *Aspergillus niger* have similarly been unsuccessful under winemaking conditions. Researchers have investigated wine and beer yeasts as alternate sources of such enzymes, reasoning that these organisms would be more suited to the conditions of the corresponding fermentations, but generally the results have been disappointing. Alternate enzymes or alternate sources are clearly called for, and thus the proteolytic and peptidolytic activities of LAB are receiving greater attention. LAB are fastidious in their amino acid requirements, and there is clear evidence that some LAB produce the activities needed to procure peptides and amino acids to meet these requirements. The potential importance of these activities for winemaking is in part linked to the nature of the enzyme, its cellular location, and how it is applied to the wine.

Activities that are lost is the culture supernatant or associated with whole cells have been reported for most species. As a result, such activities could be evident in intact cells of a LAB when it is grown in grape juice or wine. Organisms whose growth mirrors that of *O. oeni* and is most apparent after or toward the end of the primary fermentation are unlikely to have an impact on yeast growth. At this time yeast needs minimal assimilable nitrogen; therefore, any proteolytic or peptidolytic activity of LAB is beneficial mainly for haze reduction. Conversely, the sensitivity of *Lactobacillus* and *Pediococcus* to ethanol relegates their growth in mixed cultures with yeast to the early stages of the primary fermentation. Degradation of proteins and peptides at this early stage might not only affect protein haze formation in the finished wine but also release assimilable nitrogen to benefit yeast growth.

The application of a cell-free enzyme extract is one way to dissociate a desired enzymatic activity from the need to grow a particular LAB in grape juice or wine. This approach also introduces the possibility of exploiting the considerable cohort of intracellular enzymes identified to date, but it might be necessary to consider the stability of these enzymes under wine conditions. In considering the importance of individual enzyme types, proline-specific peptidases might be less important in providing assimilable nitrogen since the liberation of proline has little nutritional value to yeast cells because of their inability to exploit this amino acid under oenological conditions.

In the absence of extensive studies of wine LAB, the nature and frequency of proteolytic and peptidolytic activities identified by dairy researchers strongly suggest that similar activities also exist in wine LAB. It is recognized that the levels of individual peptides and amino acids can increase or decrease during LAB growth in wine, and the only general point of agreement is that the arginine concentration decreases while the ornithine concentration increases during MLF. When applied to sterile grape juice, a concentrated, purified exoprotease is thought to degrade proteins at a high rate. As encouraging as these findings are, there are some questions that remain to be answered. For example, it is not known whether the observed degradation of grape proteins releases peptides and amino acids in amounts that provide a nutritional benefit to the yeast or bacteria involved in the winemaking process and whether these activities are able to reduce the potential for haze formation in wine in which protein is unstable.

GLYCOSIDASES

The sensory properties of wine are the result of a multitude of individual compounds. Four groups of these compounds, the monoterpenes, C13-norisoprenoids, benzene derivatives, and aliphatic compounds, all can occur linked to sugars to form glycosides. Monoterpenes and some benzene derivatives and C13-norisoprenoids play an important role in determining wine aroma, particularly for varieties such as Muscat, Gewurztraminer, and Riesling. Aliphatic compounds, which include the aliphatic alcohols, carboxylic acids, lactones, and ethyl esters, are more related to the flavour of a wine. The remaining benzene derivatives include the anthocyanins, which contribute to wine colour. Importantly, the characteristics of the glycosides differ from those of the corresponding aglycones.

Generally, the glycosides are water-soluble and less reactive and volatile than the aglycones, possibly explaining why plants store a great number of compounds in the glycosidic form. In wine, volatile, aromatic compounds that are otherwise detectable by human senses are non-volatile and undetectable when they are in the glycosidic form. Accordingly, because as much as 95% of such aromatic compounds is present in the glycosidic form, most of the aromatic potential of these compounds is not realized. Conversely, monoglucoside anthocyanins represent the principal form in which the anthocyanins that contribute to colour in red wines are found. When these colour compounds are deglycosylated, the corresponding anthocyanidin is less stable and is readily converted to a brown or colourless compound. While this outcome may be undesirable in a red wine, these enzymes have been proposed as a means to reduce the color intensity in white or rose wines produced from red grapes.

The glycosidase enzymes that cleave the sugar moiety from glycosides can therefore have a major impact on the sensory profile of a wine. The occurrence of many types of such enzymes is a reflection of the complexity of their glycoside substrates, which can contain either mono- or disaccharides. The terminal sugar can be either D-glucopyranoside, L-rhamnopyranoside,

L-arabinofuranoside, D-apiofuranoside, or D-xylopyranoside, and the additional central sugar in disaccharides is always D-glucopyranoside. Removal of these sugars requires a glycosidase specific for the terminal sugar, followed by, in the case of a disaccharide, a D-glucopyranosidase. The latter enzyme is essential for liberation of aglycones from all diglycosides and D-glucopyranosides; With the aim of increasing the aromaticity of wines, glycosidases have been widely studied in several organisms, including both wine-related and non-wine-related organisms. Grapevines produce glycosidases, although these enzymes have little activity against wine glycosides. Given its importance in winemaking, much attention has been paid to *S. cerevisiae*, but this yeast shows very limited production of glycosidases, much of which is intracellular. Studies of other wine yeasts, including the apiculates and the spoilage yeasts, have yielded wide-ranging levels of activities, primarily D-glucosidase (D-glucopyranosidase) activities.

Aspergillus is a common source of commercial enzyme preparations that have glycosidic activities; however, these preparations are often impure, requiring resolution before characterization in the laboratory, and they have undesirable effects on the wine. More importantly, the enzymes of fungi are frequently ineffective in wine. The same is true for many of the glycosidic activities from the various source organisms examined to date, which can be limited by sensitivity to one or more of the following key wine parameters: low pH (pH 3.0 to 4.0), ethanol content (9 to 16%, v/v), or residual sugar content (10 g/l). Interestingly, the LAB, which can thrive under these conditions, have received little attention as a potential source of glycosidic enzymes.

While the glycosidases of some LAB have been studied, wine isolates have only recently been included. Limited data have been reported for *O. oeni*, and no data are available for wine *Lactobacillus* and *Pediococcus* species. No enzymatic activity by *O. oeni was observed* against arbutin, an artificial glycosidic substrate. In another study, changes in the glycoside content of Tannat wines during MLF indirectly supported the existence of such activities in the commercial *O. oeni* strains used. More specific data have come from examinations of commercial wine *O. oeni* isolates, which were shown to have the potential for high glycosidase activity against nitrophenyl glycosides. D-Glucosidase was the predominant activity, and some D-xylopyranosidase and L-arabinopyranosidase activities were also detected.

Notably, these activities were only partially inhibited under wine-like conditions. At pH 3.5 and in the presence of glucose (20 g/l) and ethanol (12%, v/v), one isolate retained 50% of the activity seen under optimized conditions. More recent work demonstrated that some *O. oeni* strains are able to act on glycosides extracted from the highly aromatic Muscat variety or the non-aromatic Chardonnay variety. In agreement with results obtained with synthetic substrates, the pattern of hydrolysis of selected glycosides from the Chardonnay variety showed that strain EQ 54 had little activity other than a D-glucosidase activity, and greater hydrolysis of the mixture occurred only after addition of commercial L-rhamnopyranosidase and L-arabinofuranosidase preparations. While the use of enzymes and/or selected cultures to liberate aroma compounds from natural grape aroma glycosides is still in the early stages of development, the findings to date for LAB and synthetic or natural glycosides are very encouraging and justify further investigation. LAB appear to possess the full array of glycosidases needed to hydrolyse many of the glycosides found in grapes and wine, although some enzymes have limited activity. Determining the precise sensory significance of glycosidic activities, as well as the longevity of any positive effects, is an important objective of future studies.

POLYSACCHARIDE-DEGRADING ENZYMES

The polysaccharides of higher plant cell walls and middle lamellae that affect wine production include cellulose (primarily glucans), hemicellulose (primarily xylans), and pectic substances. Such compounds are present in grape juice as a result of berry disruption or release through the action of degradative enzymes from the grapes. In fruits infected with the mould *Botrytis cinerea*, glucans are excreted by this pathogen directly into the berry, and fungal enzymes release grape polysaccharides, particularly type II arabinogalactan and rhamnogalacturonan II. While fungal enzymes appeared not to enhance the release of polysaccharides (mannoproteins) from yeast, these compounds can be released during yeast cell growth, through exposure to shear (e.g. during pumping and centrifugation), and particularly upon autolysis.

Collectively, polysaccharides reduce juice extraction and are primarily responsible for fouling of filters during clarification steps. Wine quality also can be affected through changes in clarity, while an effect on viscosity may influence mouth feel and the perception of tastes and aromas. Enzymes capable of degrading polysaccharides therefore have the potential to improve juice yields and wine processability through the removal of problem colloids, to increase wine quality via breakdown of grape cell walls to yield better extraction of colour and aroma precursors, and to alter the perception of wine components. These complex macromolecules are hydrolysed by a number of distinct enzymes, including pectinases (protopectinase, pectin methylesterase, polygalacturonase, and pectin and pectate lyase activities), cellulases (endoglucanase, exoglucanase, and cellobiase activities), and hemicellulases (-D-galactanase, -D-mannase, and -D-xylanase activities). There have been few reports of attempts to specifically identify polysaccharide-degrading enzymes in LAB, despite the importance of these enzymes to winemaking.

The pectinolytic activities of LAB have largely been addressed in studies of fermentation processes other than winemaking, in which their significance remains unclear. For example, early work on silage microflora suggested that cellulases and hemicellulases are produced, whereas more recent work indicated that combinations of *Lactobacillus plantarum* and *Pediococcus cerevisiae* (*Pediococcus damnosus*) had negligible ability to degrade plant cell walls. Pectin methylesterase and polygalacturonate lyase activities have been detected in the spontaneous fermentation of cassava roots, but the study neither confirmed nor discounted the involvement of the LAB present in the fermentation. At the very least, *L. plantarum* is able to liberate reducing sugars from polymeric carbohydrates during corn straw ensiling.

An extracellular glucanase that is produced early in the stationary phase of cell growth has been demonstrated in *O. oeni*. This enzyme was determined to be b-1,3 in nature and to be capable of hydrolysing yeast cell wall macromolecules; thus, it was proposed that the enzyme plays a role in yeast cell autolysis following alcoholic fermentation. Further work is required to confirm the significance of this activity along with its efficacy at temperatures below 10°C, at which currently available glucanases are insufficiently active. Similarly, the absence of additional polysaccharide-degrading enzymes cannot be assumed until a comprehensive and specific search for such activities, such as the search conducted for wine yeasts and fungi, has been completed for LAB.

ESTERASES

Esters are a large group of volatile compounds that are usually present in wine at concentrations above the sensory threshold. Most wine esters are produced by yeast as secondary products of sugar metabolism during alcoholic fermentation. Esters can also be derived from the grape and from the chemical esterification of alcohols and acids during wine ageing. The importance of esters in winemaking lies in their prominent role in determining the aroma and, by extension, the quality of wine. Esters are responsible for the desirable, fruity aroma of young wines, although they can also have a detrimental effect on wine aroma when they are present at excessive concentrations. Quantitatively, the most important wine esters are mainly yeast derived and include (i) ethyl esters of organic acids, (ii) ethyl esters of fatty acids, and (iii) acetate esters. Ethyl acetate is usually the predominant ester in wine, and with a low sensory threshold, it is often an important contributor to wine aroma. At low concentrations, ethyl acetate aroma is desirable and described as fruity, but at higher concentrations it imparts an undesirable nail polish remover character to wine. Other important wine esters and their aromas include isoamyl acetate (banana), ethyl hexanoate (fruity, violets), ethyl octanoate (pineapple, pear), and ethyl decanoate (floral). Esterolytic activity during wine production could result in either an increase or a decrease in wine quality, depending on the ester involved. In addition, the compounds liberated by the esterases (for example fatty acids and higher alcohols) could contribute to wine aroma.

While the esterases of yeast have been extensively researched, there has been little work focusing on the esterases of wine LAB. Current knowledge of LAB esterases is based primarily on work carried out in the dairy industry, in which such enzymes contribute to the characteristic flavours and defects of cheeses. Most of this work has focused on the metabolism of esters by LAB, and it is now suspected that these enzymes have the ability to both synthesize and hydrolyse esters. Thus, dairy LAB synthesize esters, including ethyl butanoate and ethyl hexanoate, while ester hydrolysis is also supported by abundant experimental evidence. The appearance of esterase activity in association with whole cells or in culture supernatants of some LAB implies that growth in grape juice or wine of these species may modify the ester profile of the beverage. Where intracellular esterase activities are reported, cell disruption would presumably be required in order for these activities to have an impact on wine.

Some wine flavour studies have reported changes in the concentration of individual esters during MLF. For example, increases in ethyl acetate, isoamyl acetate, and ethyl lactate levels, while some studies have reported a decrease in the levels of some esters following MLF. These results suggest that like the esterases of dairy isolates, esterases of wine LAB are involved in both the synthesis and hydrolysis of esters. No further investigation has been reported. Therefore, further research into the esterase systems of wine LAB should help determine the precise nature of these enzymes and their effects on the sensory properties of wine.

UREASES

Ethyl carbamate is a known carcinogen and is formed in wine *via* the spontaneous acid ethanolysis of certain carbamyl precursors, including urea and citrulline. Due to the health risks associated with elevated levels of ethyl carbamate in wine, the sources of the precursor compounds have been studied extensively. The pathways by which LAB can contribute to the ethyl carbamate precursor pool in wine have also been investigated.

Ureases produced by microorganisms are substrate-specific enzymes that catalyze the hydrolysis of urea. Urea-degrading enzymes are a potential tool for reducing the concentration of urea in wine in order to avoid dangerous and illegal concentrations of ethyl carbamate. Most commercial urease products are derived from nonwine sources (e.g. beans) and are unsuitable for use in wine, which has a pH that typically is well below the neutral pH optima. Ureases derived from LAB have been investigated as alternate enzymes for the removal of excess urea in wine. The acid ureases produced by *L. fermentum* and *Lactobacillus reuteri* were very effective over the pH range from 2.0 to 4.0, which included typical wine pH values. Wine components, such as phenolic compounds (e.g. grape seed tannins), sulphur dioxide, ethanol, and organic acids (e.g. malic acid, lactic acid, and pyruvic acid), did, however, inhibit the activity of the urease from *L. fermentum*. Whether such inhibition limits the usefulness of ureases, at least in wine of a certain type or composition, remains to be determined through further study.

PHENOLOXIDASES

Laccases (*p*-benzendiol:oxygen oxidoreductase) and tyrosinases (monophenol monooxygenase) are two groups of phenoloxidases which are widely distributed in nature. They have been found in bacteria, filamentous fungi, insects, and higher plants. Both groups of enzymes catalyse the transformation of a large number of (poly)phenolic and nonphenolic aromatic compounds and thus have potential uses in bioremediation processes in the paper and pulp, tanning, and food industries (olive mill and brewery wastewater). One of the main applications of laccases in the food industry is product stabilization in fruit juice, beer, and wine processing.

A myriad of phenolic compounds are found in musts and wine; these compounds range from simple hydroxybenzoic acid and cinnamic acid derivatives to more complex molecules, such as catechins, anthocyanins, flavonols, flavanones, and tannins. Such compounds are responsible for the desirable attributes of colour, astringency, flavour, and aroma of wine, as well as unwanted attributes, including browning, flavour and aroma alterations, and some forms of haze, which are the consequence of enzymatic and chemical oxidoreduction in white musts and wines. In vinification two main approaches have been taken to combat oxidative decolorization and flavour alteration (madeirization).

One of these approaches is inhibition of enzymes in the must using sulphur dioxide as a reductant and inhibitor; alternatively, the polyphenol substrate content of white wine is reduced by limiting maceration of the must. Where introduction has not been avoided or reduced, polyphenols are removed from the must or wine with fining agents, such as polyvinylpolypyrrolidone, gelatin, casein, and egg albumin, in addition to bentonite. More recently, chitosan, a polymeric adjuvant, was used as a fining agent and was found to be comparable to potassium caseinate in terms of wine stabilization.

Treatment of the must with enzymes such as laccases, tyrosinases, tannases, and peroxidases has been considered an alternative to treatment with physical-chemical adsorbents. Laccases are thought to be the most promising enzymes, because they have broader specificity for phenolic compounds, are more stable at the pH of must and wine, and are less affected by sulphur dioxide. With these enzymes, wine stabilization is achieved through pre-fermentative treatment to bring about oxidation of polyphenol substrates normally involved in the madeirization process. The oxidized products polymerize and precipitate, and they are subsequently removed by conventional clarifiers and filtration. To date, laccases from lignin-degrading fungi, such as *Trametes versicolor*, *Coriolus versicolor*, and *Agaricus bisporus*, and tyrosinases from mushrooms have been used in wine processing with promising results.

When the enzymes were applied either singly or in combination and immobilized on supports ranging from metal chelate affinity matrices to molecular sieves, good results were observed in terms of polyphenol removal, retention of enzyme activity, and reuse of the carrier support.

The occurrence of (poly)phenoloxidases in LAB, particularly those commonly associated with winemaking, is also of interest. These enzymes are present in fermentative LAB. Specifically, *L. lactis* is able to reduce humic acids, a constituent of soil humus containing substituted phenols and polyphenols, which implies that a polyphenoloxidase is involved. Studies of wine LAB, particularly *O. oeni*, have revealed that the growth and rate of MLF of these organisms are influenced by phenolic compounds. Gallic acid and anthocyanins, metabolized by growing cells, have a positive effect on growth and malolactic activity, whereas tannins are inhibitory. Other workers have observed that growth of *O. oeni* is inhibited by phenolic acids, including *p*-coumaric, caffeic, ferulic, *p*-hydroxybenzoic, protocatechuic, gallic, vanillic, and syringic acids, while growth of *Lactobacillus hilgardii* was stimulated. These findings are indicative of the evolution of polyphenoloxidases (and possibly tannases) from some of these organisms. If such activities do exist, their contribution to browning or decolorization is yet to be determined, but it might be expected to be most relevant to species that grow in grape juice early in fermentation. This could include LAB which are inoculated early, such as *Lactobacillus*, or contaminating LAB. Spectrophotometric assays analogous to those described for plant polyphenoloxidases, in which substrates such as catechol, catechin, and *p*-courmaric acid are used, should facilitate extended surveys of wine LAB for such activities.

LIPASES

Wine lipids can be derived from the grape berry or can be released from yeast during autolysis. Grape lipids can originate from a number of sources within the berry, including the skin, seeds, and berry pulp. The lipid profiles of each of these sources have been shown to be different, due to variation in both the concentration and the fatty acid composition of neutral lipids, glycolipids, and phospholipids. The grape lipid profile also varies with grape maturation, climate, and variety; red varieties tend to have greater total lipid concentrations than white varieties. Yeast autolysis following fermentation releases many different types of lipids, including tri-, di-, and monoacylglycerols and sterols, in amounts and proportions which vary with the yeast strain. Such lipids are known to influence not only the sensory profile of sparkling wine but also the foam characteristics.

The action of lipases on wine lipids could yield a range of volatile compounds, including fatty acids. The low aroma thresholds of fatty acids allows them to contribute to wine aroma, but since their odours are described as vinegar, cheesy, and sweaty, their impact might not be desirable. A more positive contribution to the aroma profile of wine can develop when volatile compounds such as esters, ketones, and aldehydes are derived from these fatty acids. The lipolytic activity of wine LAB has not been thoroughly investigated, but preliminary work with nonwine substrates suggests that some LAB may produce lipases. In a study of LAB isolated from wines, lipase activity was observed in several strains of *L. oenos* (*O. oeni*) and one species of *Lactobacillus*. By contrast, a more recent study failed to find lipolytic activity in wine isolates comprising 32 *Lactobacillus* strains, two *Leuconostoc* strains, and three *Lactococcus* strains. The lipolytic activity of LAB has been more extensively researched in other areas of food production. In dairy foods, lipases can contribute to flavour and processability. On the basis of this work, LAB are now generally acknowledged to be weakly lipolytic, and their lipases display substrate specificity which is both strain- and species-dependent. By utilizing numerous substrates and various degrees of cell fractionation, several such studies have provided information about activities in genera that are of interest in winemaking, namely, *Lactobacillus* and *Pediococcus*. Since lipases are located extracellularly or are associated with the whole cells, LAB have the potential to influence the wine lipid content when they are grown in grape juice or wine. The ability of any of these enzymes to attack membranes of yeast and grape cells and to influence wine aroma remains to be determined.

CONCLUSION

A considerable amount of research has been conducted to determine the enzymatic properties of LAB specifically isolated from wine or, more commonly, the enzymatic properties of analogous species isolated from other foods or beverages. From this work, it is clear that these organisms possess an extensive collection of enzymatic activities, many of which have the potential to influence wine composition and therefore the processing, organoleptic properties, and quality of wine. In many cases the precise nature and extent of this influence has yet to be delineated for the LAB that are associated with winemaking and grow under the conditions encountered during grape juice fermentation and wine maturation. Ideally, the potential for these organisms that has been highlighted in this review will stimulate fuller characterization of wine LAB so that at the very least these strains will be able to be applied by the winemaker in a more informed manner. In other cases, these organisms may serve as a source for the preparation of enzyme extracts that are better able to function under the harsh and changing environmental conditions of wine fermentation.

1. Explain in detail the various enzymes of lactic acid bacteria.

PRODUCTS OF MIXED FERMENTATIONS

INTRODUCTION

Most traditional fermented food products are made by a complex interaction of different microorganisms. This chapter deals with the products made when there is more than one set of microbial succession involved.

VINEGARS

Vinegar is the product of a mixed fermentation of yeast followed by acetic acid bacteria. Vinegar, literally translated as sour wine, is one of the oldest fermented products used by man. It is the acetic acid produced by the fermentation of alcohol (ethanol) which gives the characteristic flavour and aroma to vinegar.

It can be made from almost any fermentable carbohydrate source, for example, fruits, vegetables, syrups and wine. The basic requirement for vinegar production is a raw material that will undergo an alcoholic fermentation. Apples, pears, grapes, honey, syrups, cereals, hydrolysed starches, beer and wine are all ideal substrates for the production of vinegar. To produce a high-quality product it is essential that the raw material is mature, clean and in good condition.

Indigenous vinegars can be made by the spontaneous fermentation of a fruit or alcohol. All that is necessary is an alcoholic substrate, strains of acetic acid forming bacteria (*Acetobacter*) and oxygen to enable the oxidation of alcohol. However, this process is very slow and vinegars produced by this method tend to be of inferior quality. Controlled fermentation conditions produce a more acceptable product. A wide range of vinegar are made using different raw materials.

COCONUT WATER VINEGAR

Location of Production

Throughout Asia particularly Philippines and Sri Lanka.

Product Description

A clear liquid with a distinctive acetic acid taste with a hint of coconut flavour.

Raw Material Preparation

Coconut water is a waste product, which is abundant in Philippines, Sri Lanka, Thailand and other countries. Its conversion into vinegar therefore presents an attractive option for decreasing wastage and producing a valuable product.

Processing

Coconut water is a good base for vinegar, but its sugar content is too low (only about 1%). Sugar is added to bring the level of sugar up to 15%. After the addition of sugar, the coconut juice is allowed to ferment for about seven days, during that time the sugar is converted to alcohol. An alternative method is to pasteurize the coconut water and sugar mixture, and add yeast.

After this initial fermentation, strong vinegar (10% v/v) is added to stimulate the growth of acetic acid bacteria and discourage further yeast fermentation. The acetic acid fermentation takes approximately one month, yielding a vinegar with approximately 6% acetic acid. The fermentation will take less time than this if a generator is used.

After fermentation, the vinegar must be stored in anaerobic conditions to prevent spoilage by the oxidation of acetic acid. Clarification can be achieved by stirring with a well-beaten egg white, heating until the egg white coagulates and filtering.

PINEAPPLE PEEL VINEGAR

Location of Production

Latin America and Asia.

Product Description

This product enables the utilization of pineapple peels, which are usually discarded during the processing or consumption of the fruit. The product has a distinct, very light pineapple flavour and has the same uses as any commercial vinegar.

Raw Material Preparation

The peels should be from very well-washed ripe pineapples (damaged, rotten or infected fruits should not be used as a source of peels). Use only the peels, not the leaves or stems. The water used should be potable water, boiled if necessary. All the equipment should be well cleaned, as well as the bottles, which should also be steam-sterilized before use.

Processing

The peels should be cut into thin strips and put into clay or pewter pots. Aluminium or iron pots should not be used. Sugar and clean water are added. Each pot is then inoculated and covered with a clean cotton cloth, held around the pot with an adhesive tape, to prevent contamination by insects or dust. The inoculated pineapple is fermented at room temperature (about 20–22°C) for about eight days. The acidity should be checked daily. The water level should be maintained during this period. The product should be increasingly acid and by the eighth day it should have the required concentration of 4 per cent acetic acid in vinegar. If higher acidity is desired the product is left to ferment for another one or two days. The development of acidity should be checked by tasting the product during fermentation. The residual bacteria removed may be reused as a residue inoculum two or three times more. The processing steps are illustrated in Figure 42.1.

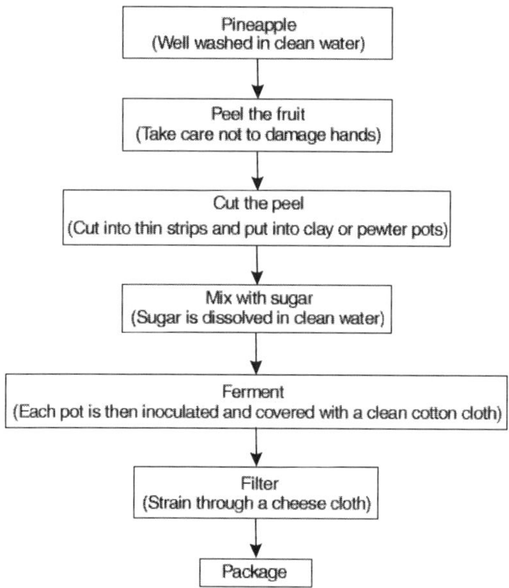

Figure 42.1   Production process of pineapple peel vinegar

The traditional process may be improved by a two-stage fermentation in which alcohol is first formed by yeast (*Saccharomyces cerevisiae*) and the "must" is then inoculated with acetic acid bacteria (*Acetobacter pasteurianus*). The process involves liquidizing the peels and diluting with water (water : pulp is 4:1), adjusting the pH to 4.0 using sodium bicarbonate and adding yeast nutrient (ammonium phosphate) at 0.14 g per litre. A starter culture is added at 2.7 g per litre and the fermentation allowed to take place at 25°C for two days. The "must" is then filtered and inoculated with acetic acid bacteria and allowed to ferment for eleven days with aeration of the "must". Other parts of the process are similar. Additional equipment includes a pH meter, refractometer, liquidizer, fermentation locks and equipment for preparing the starter cultures.

Packaging and Storage

The vinegar is bottled in clean glass bottles and stored in a cool dark place.

PALM WINE VINEGAR

Palm wine vinegar is produced across West Africa. It is a vinegar containing about 4% acetic acid, produced from the oxidation of palm wine. It is mainly consumed by people in urban areas as a salad dressing and meat tenderizer, although it also has medicinal uses and is valued in certain rituals. Palm wine is fermented using the same process as for grape wine vinegar—the oxidation of alcohol to acetic acid. The spontaneous process takes about four days. The optimum fermentation temperature is 30°C.

COCONUT TODDY VINEGAR

Coconut toddy vinegar is produced throughout South Asia particularly in Sri Lanka. It is a clear liquid with a strong acetic acid flavour and a hint of coconut flavour. The fresh toddy is strained, prior to allowing yeast fermentation to occur naturally for 48 to 72 hours. The yeast cells and debris are then removed by progressive sedimentation. After two to four weeks of settling, the fermented toddy is placed in barrels. The alcohol is then converted into acetic acid by acetic acid bacteria which are naturally present. The process can be hastened by adding vinegar as a starter. The fermented toddy is converted into vinegar in about three months. Ageing for six months, results in a pleasantly flavoured final product.

NIPA PALM VINEGAR

In East Asia, particularly Papua New Guinea, a vinegar is made from the sap of the *Nipa* palm (*Nypa fruticans*).

QUICK PROCESS PICKLES

Quick process pickles are easy to make but do not really constitute a fermented food product. For this technique, vegetables are soaked in a low salt solution for a few hours. They are then drained and placed in a container. The container is filled with a hot vinegar and spice mixture or a hot oil and spice mixture. There are hundreds of different recipes utilizing locally available fruit and vegetables such as mango sour pickle, sliced mango pickle, sweet olive pickle, hot olive pickle, sweet tamarind pickle, chalta pickle and green chilli.

COCOA POWDER

Location of Production

Africa, Asia and Latin America particularly Cote d'Ivoire, Ghana, Indonesia and Brazil.

Product Description

A fine brown powder with the characteristic taste of cocoa. It is a major ingredient in the confectionery and bakery industries. The product has a short shelf life. "Drinking chocolate" is a mixture of cocoa powder and sugar.

Raw Material Preparation

Cocoa beans are the seeds of the cocoa plant (*Theobroma cacao*). Cocoa pods are cut from the cocoa tree. The pods are cut and the beans removed. Only fully ripe and undamaged beans should be selected. It is important that the beans are processed quickly.

Processing

It was formerly believed that cocoa beans were fermented to remove the adhering pulp. However, a good flavour in the final cocoa or chocolate is dependent on good fermentation. Fermentation is carried out in a variety of ways but all depend on heaping a quantity of fresh beans with their pulp and allowing microorganisms to produce heat. The majority of beans are fermented in heaps although better results are obtained using boxes, which result in a more even fermentation.

Fermentation lasts from five to six days. During the first day, the adhering pulp is liquified and drained away with the rising temperature steadily. The initial alcoholic fermentation gives way to acetification. This along with other chemical changes cause the temperature to rise above 50°C and the beans die. It was thought in the past that death was mainly due to increasing temperature. It is now known that acetic acid at a concentration of 1 per cent in the bean is the cause of death and that it is only enhanced by heat, lactic acid and ethanol. The pH value of the cotyledon drops from 6.45 to 4.5 over 120 hours and during the same period the acetic acid content increased from 0 to 1.36 per cent, while the lactic acid content increased from 0.005 to 0.12 per cent. When the bean dies maceration of the tissue takes place, allowing enzymes and substrate to mix freely. The possible substrates for enzymes are carbohydrates, lipids, phenolics and amino acids. In addition it is known that the bacteria can metabolize alcohols and organic acids of various kinds. Possible major substrates for microorganisms are carbohydrates, lipids, phenolics and amino acids. Unlike some flavours and aromas, that of chocolate is not attributable to a single compound.

During fermentation the external appearance of the beans changes. At first they are pinkish with a covering of white mucilage. Gradually the colour darkens and the mucilage disappears. The beans on the surface are always darker than those deeper in the heap or box, indicating that the colour change is oxidative. As the beans are mixed, their colour becomes a more uniform orange-brown and they are only slightly sticky. At this stage they are ready for drying.

The beans need to be dried to a moisture content of less than 7.5%. The beans are dried by either being spread out in the sun in layers a few centimetres thick or in artificial dryers. There are numerous types of dryers but it is important to control any smoky products of combustion come in contact with the beans otherwise taints will appear in the final product. The beans are cleaned to remove the extraneous matter.

Cocoa beans consist of an outer skin that needs to be removed and inner "nib". The shell is sometimes removed before roasting and sometimes after roasting.

For cocoa powder, roasting temperatures of 120 to 150°C are used. There are many designs of roasters: both batch and continuous systems. The operation is controlled so that the cocoa is heated to the required temperature without burning the shell or the cotyledon. The heat is applied evenly over a long period of up to 90 minutes to produce uniform roasting. The bean must not be contaminated with any combustion products from the fuel used and provision must be made for the escape of any volatile acids, water vapour and decomposition products of the bean. After roasting, the beans are cooled quickly to prevent scorching. The roasted nibs are ground into a powder in a plate mill. The resulting powder is sieved through fine silk, nylon or wire mesh.

To produce cocoa powder, some of the cocoa butter needs to be removed. With low fat cocoa powder, more than 90% of the cocoa butter is removed. With medium fat cocoa powder, more than 78% of the cocoa butter has been removed. Finally, high fat cocoa powder has less than 78% of the cocoa butter removed. Extrusion, expeller, or screw presses are used in the cocoa industry to remove the cocoa.

The cake from the mill is ground in a hammer mill to produce the cocoa powder. The production of cocoa powder is summarized in Figure 42.2.

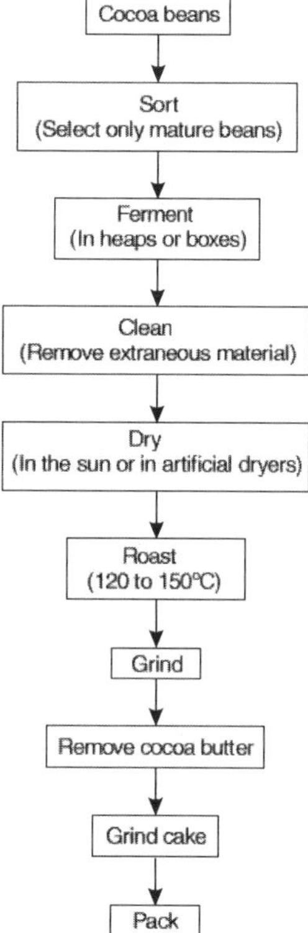

Figure 42.2   Production process of cocoa powder

Packaging and Storage

Cocoa powder is hygroscopic (picks up moisture from the air) and should be protected, especially in humid climates. Lidded tins or sealed polythene bags should be used.

COFFEE

Location of Production

Throughout Africa, Asia and Latin America particularly Brazil, Colombia, Indonesia, Mexico and Cote d'Ivoire.

Product Description

A fine dark brown powder made from roasted coffee beans which is brewed with boiling water and consumed as a drink.

Raw Material Preparation

Coffee beans are harvested from two plants *Coffea arabica and Coffea canephora* variety *robusta*. Only ripe berries should be used in coffee production. Berries can be placed in water so that immature berries which float can be identified and discarded.

Processing

Dry processing is the simpler of the two processing methods and is popular in Brazil for the processing of *robusta* coffee and in Sri Lanka for processing *arabica* coffee. The coffee cherries are dried immediately after harvest by sun drying on a clean dry floor or on mats. The dried berry is then hulled to remove the pericarp. This can be done by hand using a pestle and mortar or in a mechanical huller. The mechanical hullers usually consist of a steel screw, the pitch of which increases as it approaches the outlet thereby removing the pericarp. The hulled coffee is cleaned by winnowing.

Wet processing (Figure 42.3) involves squeezing the berry in a pulping machine or pounding in a pestle and mortar to remove the outer fleshy material (mesocarp and exocarp) and leave the bean covered in mucilage. This mucilage is removed by fermentation. Fermentation involves placing the beans in plastic buckets or tanks and allowing them to sit, until the mucilage is broken down. Natural enzymes in the mucilage and yeasts and bacteria in the environment work together to break down the mucilage. The coffee should be stirred occasionally and every so often a handful of beans should be tested by washing in water. If the mucilage can be washed off and the beans feel gritty rather than slippery, the beans are ready.

There is much debate about the fermentation of coffee beans. Some researchers feel that the mucilage breakdown is caused by enzymatic breakdown. If these "pulped" beans are piled up or put in a container and protected from any bacterial or other contamination the fermentation will progress. After a number of hours the enzymes of the pulp will have acted on the torn tissues, gorged with starches, sugars and pectins, in such a manner that, without any microbial intervention, the remaining pulp will be easily detached from the beans and washed off in water. However most investigators acknowledge the necessity for the presence of microorganisms for the depectinization of the beans.

The following microorganisms have been isolated: *Leuconostoc mesenteroides, Lactobacillus plantarum, Lactobacillus brevis, Streptococcus faecalis, Aerobacter (Enterobacter)* and *Escherichia,* pectinolytic species of *Bacillus, Saccharomyces marscianus, S. bayanus* and *Flavobacterium* sp., *Erwinia dissolvens, Fusarium spp., Aspergillus spp.,* and *Penicillium.*

The beans should then be washed immediately as off-flavours develop quickly. To prevent cracking, the coffee beans should be dried slowly to 10% moisture content (wet basis). Drying should take place immediately after to prevent off-flavours developing. The same drying methods can be used for this as for the dry processed coffee. After drying, the coffee should be rested for 8 hours in a well ventilated place. The thin parchment around the coffee is removed either by hand, in a pestle and mortar or in a small huller. The hulled coffee is cleaned by winnowing.

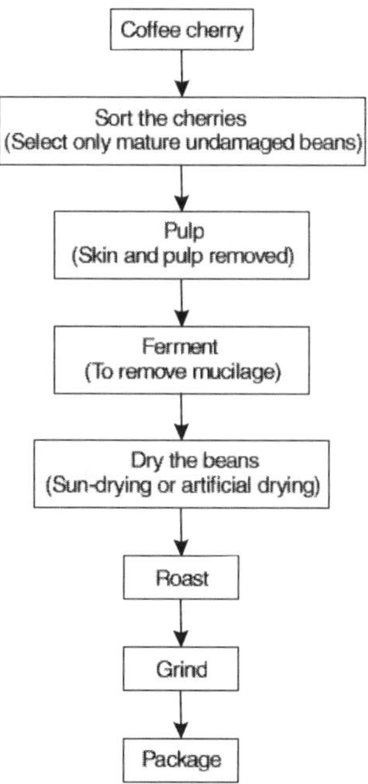

Figure 42.3   Production process of coffee

The final flavour of the coffee is heavily dependent on how the beans are roasted. Roasting is a time- and temperature-dependent process. The roasting temperature needs to be about 200°C. The degree of roast is usually assessed visually. One method is to watch the thin white line between the two sides of the bean. When this starts to go brown, the coffee is ready. Coffee beans can be roasted in a saucepan as long as they are continually stirred. A small improvement is made by roasting the coffee in sand, as this provides a more uniform heating. A roaster will produce a high-quality product.

Grinding is a means of adding value to a product. However, it is fraught with difficulties. It is easy to make an assessment of an intact bean, while a ground product presents some difficulty. The fear of adulteration and the use of low-quality product is justified. Because of this there is a great deal of market resistance to ground coffee. This market resistance can only be overcome by consistently producing a good product. There are basically two types of grinders—manual grinders and motorized grinders.

Packaging and Storage

Roasted beans can be stored in sacks. Milled beans need to be packaged quickly to prevent the loss of volatile flavour components. The packaging material should be airtight. Polythene is not suitable as it is a low barrier to loss of aroma.

1. List out the products of mixed fermentations and highlight their significance.
2. Explain about quick-process pickles.
3. Write about fermented cocoa and coffee products.

43

MICROBIOLOGY OF VINEGAR PRODUCTION

INTRODUCTION

Vinegar is the product obtained by acetic acid fermentation of alcohol-containing solutions. It is a clear aqueous liquid which is either colourless or has the colour of the raw material. It is also known as white vinegar, spirit vinegar, alcohol vinegar or grain vinegar. The vinegar is designated according to the particular raw material used. For example, wine vinegar is produced by vinegar fermentation of grape wine. Cider vinegar is produced from fermented apple juice. Malt vinegar is the product made by the alcoholic and subsequent acetous fermentation of an infusion of barley malt or cereals whose starch has been converted by the malt. Whey vinegar is produced by the alcoholic and subsequent acetous fermentation of concentrated whey. Fruit vinegar is made from fruits which are available in surplus like dates, citrus fruits and bananas. Sugar vinegar is made by the alcoholic and subsequent acetous fermentation of sugar syrup and molasses. Glucose vinegar is made by the glucose solutions. Similarly rice vinegar is made by the saccharification of rice starch. Vinegar is the product of a two-stage fermentation. In the first stage, yeasts convert sugars into ethanol anaerobically, while in the second, ethanol is oxidized to acetic acid aerobically by bacteria of the genera *Acetobacter* and *Gluconobacter*. This second process is a common mechanism of spoilage in alcoholic beverages.

TYPES OF VINEGAR

The predominant type of vinegar is white or distilled vinegar. Vinegar is usually described in terms of grain strength, the grain being ten times the acid percentage. For example, 10% acid is referred to as 100 grain. Some of the most popular vinegars and their characteristics are shown below:

Balsamic vinegar   It is brown in colour with a sweet-sour flavour. It is made from the white Trebbiano grape and aged in barrels of various woods. Some gourmet Balsamic vinegars are over 100-years old.

Cane vinegar   It is made from fermented sugar cane and has a very mild, rich-sweet flavour. It is most commonly used in Phillippine cooking.

Champagne vinegar   It has no bubbles. It is made from a still, dry white wine of Chardonnay or Pinot Noir grapes (both of which are used to make champagne).

Cider vinegar   It is made from apples and is the most popular vinegar used for cooking.

Coconut vinegar   It is low in acidity, with a musty flavour and a unique aftertaste. It is used in many Thai dishes.

Distilled vinegar   It is harsh vinegar made from grains and is usually colourless. It is best used only for pickling.

Malt vinegar   It is very popular in England. It is made from fermented barley and grain mash, and flavoured with woods such as beech or birch. It has a hearty flavour and is often served with fish and chips.

Rice wine vinegar   It has been made by the Chinese for over 5,000 years. There are three kinds of rice wine vinegar: red (used as a dip for foods and as a condiment in soups), white (used mostly in sweet and sour dishes), and black (common in stir-fries and dressings).

Sherry vinegar   It is aged under the full heat of the sun in wooden barrels and has a nutty-sweet taste.

Wine vinegar   It can be made from white, red, or rose wine. These vinegars make the best salad dressings.

VINEGAR BACTERIA

Acetic acid bacteria (Figure 43.1) are gram-negative, ellipsoidal to rod-shaped cells that have a required aerobic metabolism with oxygen as the terminal electron acceptor. The identification of the acetic acid bacterial species has been traditionally performed by studying physiological and chemotaxonomic properties. Taxonomic studies based on partial sequence comparisons of 16S rRNA have shown that *Gluconoacetobacter* can be considered as a new genus which is present along with other species during wine fermentations. Bacterial 16S rRNA sequences are attractive targets for developing identification methods because they represent conserved regions in all bacteria.

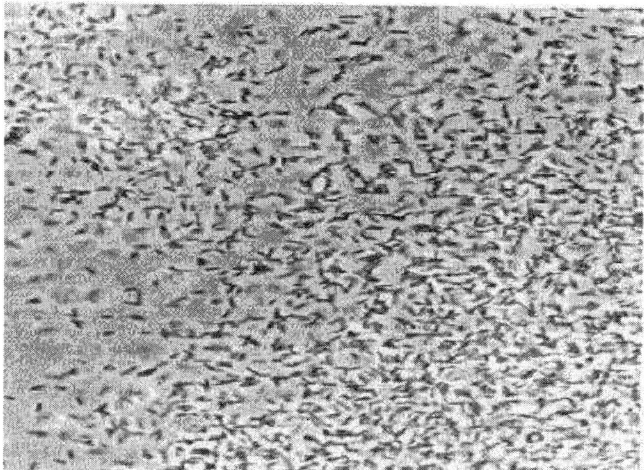

Figure 43.1   Acetic acid bacteria

The restriction fragment length polymorphisms (RFLPs) of the gene coding for rRNAs show interspecies and intraspecies differences in bacteria. The PCR-RFLP method is used for the rapid identification of acetic acid bacteria at the genus level and the identification of *Acetobacter*, *Gluconobacter* and *Gluconoacetobacter* species. PCR has been shown to be a suitably accurate technique for identifying bacterial strains and for determining the taxonomic relationships between bacterial species.

FERMENTATION PROCESS

The vinegar fermentation is an oxidative fermentation in which diluted solutions of ethanol (mash) are oxidized by *Acetobacter* in the presence of oxygen to acetic acid and water. All mashes must contain ethanol, water and nutrients of the acetic acid bacteria.

Apart from nutrients like potassium, magnesium, calcium, and ammonium salts, water is another important raw material. Water used must be clear, colourless, odourless and without any sediment or suspended particles. For a good fermentation, it is essential to have an alcohol concentration of 10 to 13%. If the alcohol content is much higher, the alcohol is incompletely oxidized to acetic acid. If it is lower than 13%, there is a loss of vinegar because the esters and acetic acid are oxidized. In addition to acetic acid, other organic acids are formed during the fermentation which become esterified and contribute to the characteristic odour, flavour and colour of the vinegar.

Biochemistry of Vinegar Fermentation

Initially, alcohol is dehydrogenated to form acetaldehyde and releases two hydrogen ions and two electrons. In the second step, two hydrogen ions bind with oxygen to form water that hydrates acetaldehyde to form aldehyde. During step three, aldehyde dehydrogenase converts acetaldehyde to acetic acid and releases 2 hydrogen ions and 2 electrons.

Ethanol is oxidized by the alcohol cytochrome-553 reductase (with the help of a haem cofactor) to acetaldehyde which is further oxidized by the coenzyme independent aldehyde dehydrogenase.

1. Formation of Acetaldehyde

$$CH_3-\underset{H}{\overset{OH}{C}}-H \xrightarrow{\text{Alcohol dehydrogenase}} CH_3-\underset{H}{\overset{O}{C}} + 2H^+ + 2e^-$$

Acetaldehyde

2. Hydration of Acetaldehyde

$$CH_3-\underset{H}{\overset{O}{C}} + H_2O \longrightarrow CH_3-\underset{H}{\overset{OH}{C}}-OH$$

Intermediate product

3. Formation of Acetic acid

$$CH_3-\underset{H}{\overset{OH}{C}}-OH \xrightarrow{\text{Aldehyde dehydrogenase}} CH_3-\underset{OH}{\overset{O}{C}} - 2H^+ + 2e^-$$

Acetic acid

4. Electron transfer

$$4H^+ + 4e^- + O_2 \xrightarrow{\text{Cytochrome system}} 2H_2O$$

The yeasts and bacteria exist together in a form known as commensalism. The acetobacter are dependent upon the yeasts to produce an easily oxidizable substance (ethyl alcohol). It is not possible to produce vinegar by the action of one type of microorganism alone. Most acetifications are run on a semicontinuous basis, i.e., when acetification is nearly complete and acetic acid levels are around 10–14%, a proportion of the fermenter's contents is removed and replaced with an equal volume of fresh alcoholic vinegar stock. As a result, this helps conserving the culture by maintaining a relatively high level of acidity thus preventing contamination. Besides, it also prevents against overoxidation as the bacterium *Acetobacer europaeus*, a species commonly used for vinegar production will not overoxidize when the acetic acid concentration is more than 6%.

Microorganisms Involved in the Fermentation of Vinegar

The organisms involved in vinegar production usually grow at the top of the substrate, forming a jelly like mass. This mass is known as "mother of vinegar". The mother is composed of both *Acetobacter* and yeasts, that work together. The principal bacteria are *Acetobacter aceti, A. xylinum* and *A. ascendens*. The main yeasts are *Saccharomyces ellipsoideus* and S *cerevisiae*. It is important to maintain an acidic environment to suppress the growth of undesirable organisms and to encourage the presence of desirable acetic acid producing bacteria. It is a common practice to add 10 to 25% by volume of strong vinegar to the alcoholic substrate, in order to attain a desirable fermentation.

The alcoholic fermentation of sugars should be completed before the solution is acidified because any remaining sugar will not be converted to alcohol after the acetic acid is added. Incomplete fermentation of the juice results in a "weak" product. The acetic acid strength of good vinegar should be approximately 6%.

FERMENTATION METHODS

Small-scale Production

Vinegar can be made at home by introducing oxygen into barrels of wine or cider and allowing fermentation to occur spontaneously. This process is not very rigorously controlled and often results in poor quality.

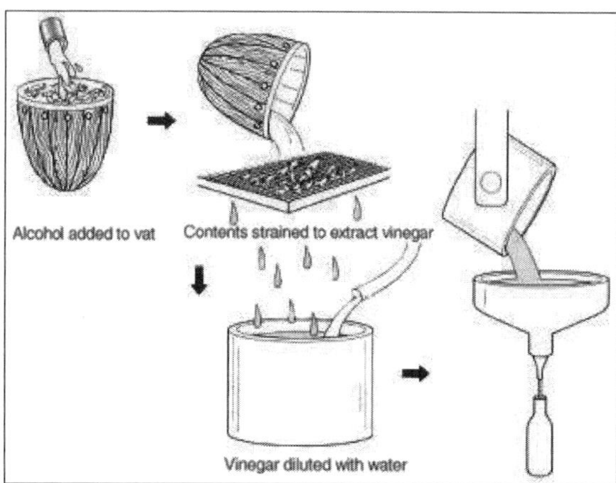

Figure 43.2   Orleans process

The Orleans Process

The Orleans process is one of the oldest and well known methods for the production of vinegar (Figure 43.2). It is a slow, continuous process, which originated in France. High grade vinegar is used as a starter culture, to which wine is added at weekly intervals. It is a surface culture technique where the bacteria form a surface film at the interface between the acetifying medium and air. First, holes are drilled at the ends of the barrel a few inches above the liquid surface. The holes are left open and covered with a fine screen. Thus the Orleans process consists of wood barrels filled with alcoholic liquid fermented for about 1 to 3 months at 21°C to 29°C. After fermentation, part of the vinegar is drawn off for bottling purposes and an equivalent amount of alcoholic liquid or mash is added. The vinegar is fermented in large (200 litre) capacity barrels. Approximately 65 to 70 litres of high grade vinegar is added to the barrel along with 15 litres of wine. After one week, a further 10 to 15 litres of wine are added and this is repeated at weekly intervals. After about four weeks, vinegar can be withdrawn from the barrel (10 to 15 litres per week) as more wine is added to replace the vinegar. The room temperature is kept at approximately 29°C. Samples are taken periodically by inserting a spigot into the side holes and drawing liquid off. When the alcohol has converted to vinegar, it is drawn off through the spigot. About 15% of the liquid is left in the barrel to blend with the next batch.

One of the problems encountered with this method is the addition of more liquid to the barrel without disturbing the floating bacterial mat. This can be overcome by using a glass tube which reaches to the bottom of the barrel. Additional liquid is poured in through the tube and therefore does not disturb the bacteria. Wood shavings are sometimes added to the fermenting barrel to help and support the bacterial mat.

Generator Fermentation

Early in the nineteenth century, a vinegar-making system called the trickle method [now called generator fermentation or quick process (Schnellessig)] was developed by German chemist Schutzenbach in 1832. In this process, the bacteria were grown and formed a thick slime coating around a non-compacting material like beech wood shavings, charcoal or coke. The non-compacting material was packed into large upright wooden tanks (Figure 43.3) of 2000 cubic feet capacity above a perforated wood grating floor. The wood shavings are generally made of air-dried beech wood sliced to form a coil of about 2 inches long and 1¼ inches in diameter. The generators must be closely monitored to prevent overoxidation or unacceptable temperatures. Oxygen is allowed into the vats in two ways. One is through bungholes that have been punched into the sides of the vats. The second is through the perforated bottoms of the vats. An air compressor blows air through the holes. The process takes about 3 to 7 days. Two-thirds of the final vinegar product is withdrawn from the tank and fresh mash added. Replacement mash is slowly poured into the tank until the working level for acetification of the solution and a beginning temperature of 21.1°C are reached. The optimum temperature for generator operation is 32.2°C. Each gallon of 190 proof alcohol oxidized to acetic acid releases 3.2×10^7 to 3.7×10^7 Joules of energy. The optimum temperature for *Acetobacter* is about 30°C. A temperature control system is necessary to prevent overheating and consequent inactivation of the bacteria. The vinegar produced in this method has a very high acetic acid content, often as high as 14%, and must be diluted with water to bring its acetic acid content to a range of 5–6%.

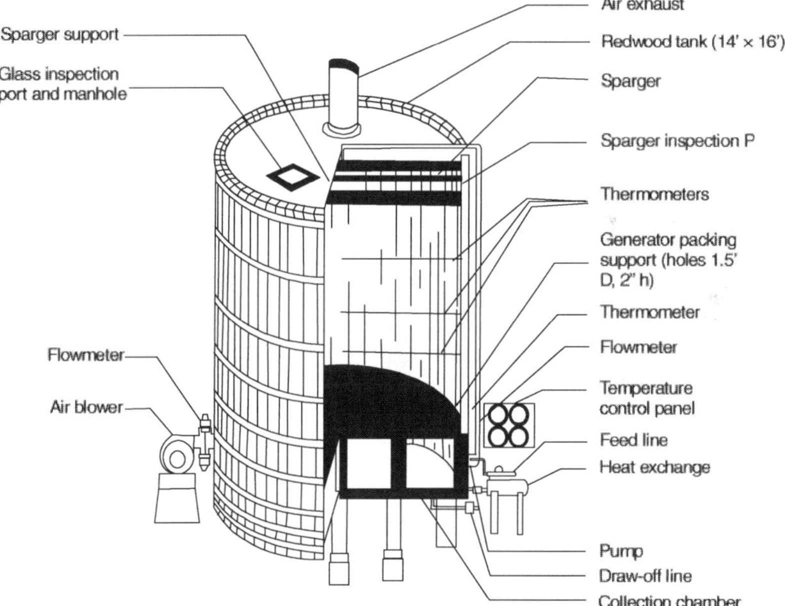

Figure 43.3 Vinegar generator

Submerged Vinegar Fermentation

Today, the most common production method is submerged culture which improves the general fermentation conditions like aeration, stirring, heating, etc. As generator culture systems are slow and expensive, submerged culture fermenters have become widely used at industrial process. The fastest rates of acetification are achieved using submerged acetification in which acetic acid bacteria grow suspended in a medium which is oxygenated by sparged air. The most commercially successful technique is the Frings acetator (Figure 43.4).

Figure 43.4   The Frings acetator

Submerged culture is very efficient and rapid, a semicontinuous run normally takes 24–48 hours. Hence it is also called "quick fermentation method". It does however require far more careful control methods than simpler processes. In this process, the mash is stirred and aerated frequently. The fermenters are usually fitted with a heat exchanger for the maintenance of the optimum temperature during the fermentation process. The typical operation mode in industrial submerged cultures is semicontinuous (Figure 43.5).

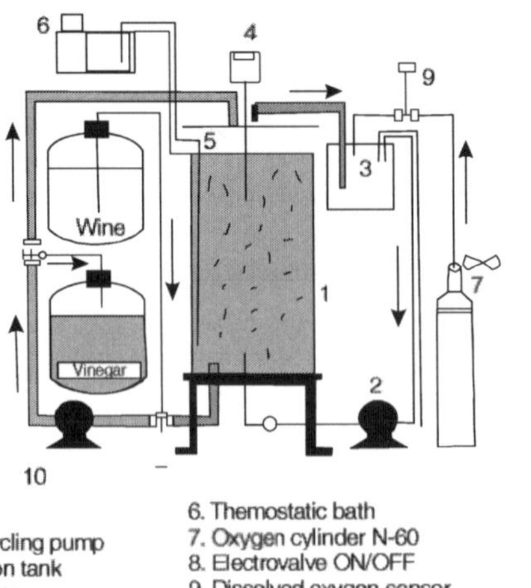

1. Reactor
2. Gas recycling pump
3. Expansion tank
4. Air diffuser
5. Heat exchanger
6. Themostatic bath
7. Oxygen cylinder N-60
8. Electrovalve ON/OFF
9. Dissolved oxygen sensor
10. Feed inlet and effluent outlet pump

Figure 43.5   The semicontinuous fermentation system

This operation consists of the development of successive discontinuous cycles of acetification. At the end of every cycle, a given volume of acetic acid is discharged and refilled with mash. The best temperature for industrial production of 11 to 12% vinegar is 30°C. Damage to the bacteria may occur above 30°C. In addition, the bacterial condition also affects the concentration of acetic acid produced.

The Bourgeois Process

This process is based on the use of two fermenters with alternating fermentation cycles. In one of the fermenters, the fermentation proceeds until the ethanol concentration has dropped to zero. A thermostat stops the aeration at this time. In the meantime, fermentation proceeds in the other fermenter whose contents are used for the inoculation of the first fermenter. A simple turbo mixer with air compressor and an automatic device for temperature control complete the equipment. The lack of ethanol in the finished product lessens its aroma. Therefore, the addition of ethanol after completion of the fermentation is recommended. Total production is not favourable.

Processing Steps After Fermentation

A resting period of several months is recommended for vinegar produced from natural raw materials since storage improves the quality of vinegar. Vinegar should have a clean aroma which is related to the raw material that has been used. Raw vinegar contains acetic acid bacteria which make it opaque since a portion of the acetic acid bacteria settles during storage. To clarify the vinegar an aqueous suspension of bentonite is added to the vinegar, and mixed with it intimately. The suspended particles are permitted to settle for several hours, so that the supernate is usually clear and easy to filter.

Filtration is carried out with suspensions of diatomaceous earth. Filtration is continued until a concentrate is obtained which constitutes only about 0.5% of the filtered volume.

Vinegar is pasteurized by short heating just before bottling. Sulphiting of vinegar up to 50 mg SO_2 per litre is also in use, instead of pasteurization.

1. What is vinegar?
2. List out the various types of vinegars.
3. Write a note on vinegar bacteria.
4. Explain the biochemistry of vinegar fermentation.
5. Write a short note on the microbes involved in the production of vinegar.
6. Give a detailed account of the production of vinegar.
7. Write the steps involved in processing of vinegar.

44

BEER PRODUCTION

OVERVIEW

The exact origin of beer is unclear. It is apparent, however, that beer has been existing for centuries. In the 18th century BC owners of brewpubs in Babylonia were fined for overcharging their customers. Humans developed their own versions of beer made from available grains. Sake was developed long ago in Asia and Pre-Columbian Native Americans made beer from corn. Beer may have been important because it provided an alternative to drinking contaminated water, or a palatable use for grain that had gotten wet and would otherwise have spoiled. Beers are made from corn, wheat, rice, and barley but they are classified by the type of yeast used in their production. Beers brewed in Colonial America were all ales, made with top-fermenting yeast, like those found in England. In the mid 1800s, German emigrants brought bottom-fermenting yeast to the United States and produced German lagers. The mellow lager quickly replaced ale as the beer of choice. The majority of breweries in the world still produce lagers. During the 19th century, several American beers called common beers that do no fit the ale or lager styles developed. One such example, California steam beer which originated during the 1849 California Gold Rush, uses lager fermentation and ale yeast. "Steam" refers to the high concentration of CO_2 produced during secondary fermentation.

A variety of flavours have been used in beer to balance the sweetness of the grains. An ancient Peruvian recipe that is still used, uses strawberries for flavour. Around 300 years ago, hops (*Humulus lupulus*) became the flavour of choice because they also contributed antibacterial action and clarifying action. The a-acids (humulone, cohumulone, and adhumulone) in hops provide the bitter flavour and large proteins settle out with tannins to clarify the beer. Other molecules in hops including *trans*-isohumulone are responsible for the antibacterial properties.

BARLEY MALT AS SOURCE OF BEER

Barley malt is to beer as grapes are to wine. It is ideally suited to brewing for many reasons. Malted barley has a high compelement of enzymes for converting its starch supply into simple sugars and contains protein, which is needed for yeast nutrition. Of course, one important element is its flavour. There are two types of barley: six row and two row.

Six-row barley malt   Generally, six-row barley has a higher enzyme content, more proteins, less starch, and a thicker husk than two row barley. The higher level of diastatic enzymes makes six-row barley desirable for conversion of adjunct starches (those that lack enzymes) during mashing. On the down side, the higher protein content can result in greater break material (hot and cold), as well as possibly increase problems with haze in the finished beer. The husk is high in polyphenols (tannins) that results not only in haze, but also imparts an astringent taste.

Two-row barley malt   Generally, two-row barley has a lower enzyme content, less protein, more starch and a thinner husk than six-row barley. Of the first two of these characteristics, the protein content of two-row barley depends greatly on the barley strain, and enzyme content depends very much on the strain and degree of kilning. In comparison to six-row barley, two-row has a higher starch content—the principal contributor to extract. The thinner husk associated with two-row barley makes for mellower (less astringent) beers due to lower levels of polyphenols.

Barley Malt Identification

The number of rows of kernels helps in easy identification of two- and six-row varieties. In six-row varieties, two-thirds of the kernels are twisted in appearance because of insufficient space for symmetrical development. Since they must overlap, they twist as they grow. In two-row barley there are no lateral kernels; all kernels are straight and symmetrical. The kernels of two-row barley are broader than the central kernels of six-row barley and do not taper as sharply.

Malting

Malting serves the purpose of converting insoluble starch to soluble starch, reducing complex proteins, generating nutrients for yeast development, and the development of enzymes. The three main steps of the malting process are steeping, germination and kilning.

Barley steeping   Steeping begins by mixing the barley kernels wih water to raise the moisture level and activate the metabolic processes of the dormant kernel. The water is drained, and the moist grains are turned several times during steeping to increase oxygen uptake by the respiring barley. Generally, the barley spends about 40 hours in tanks of fresh, clean water, with three intervals during which the water is allowed to drain. Draining is done to remove dissolved carbon dioxide and to reintroduce oxygen rich water. Steeping is complete when the white tips of the rootlets emerge, which is known as chitting. At this point the grains will be swollen one-third times their original size.

Barley germination In the next step, the wet barley is germinated by maintaining it at a suitable temperature and humidity level until adequate modification has been achieved. Germination is done on floor, drum, or in boxes. Floor malting is an old process in which the chatted malt is spread on the floor to a height of 10–20 cm. Germination in drums is still done, but is not very economical.

Malt kilning The final step is to dry the green malt in the kiln. Malts are kilned at different temperatures. The temperature regime in the kiln determines the colour of the malt and the amount of enzymes which survive for use in the mashing process. Low temperature kilning is more appropriate for malts when it is essential to preserve enzymatic (diastatic) power. These malts are high in extract but low in colouring and flavouring compounds. Pilsner and pale ale malts are examples of malts kilned at low temperatures. Malts kilned at intermediate temperatures, such as Munich and Vienna malts, are lower in enzymes but higher in colouring and flavouring compounds. Malts kilned at high temperatures, such as crystal and chocolate malts, have little amount of enzymes, thus are lower in extract.

Malt Modification

In general, modification refers to the extent to which the endosperm breaks down. During malting, enzymes break down the cell structure of the endosperm, releasing the nutrients necessary for yeast growth and making the starch available for enzyme degradation during mashing. Modification of the endosperm correlates with growth of the acrospire. As the acrospire grows, chemical changes are triggered that result in the production of numerous enzymes, which are organic catalysts. Their function is to break down the complex starches and proteins of the grain.

Malt Constituents

Malt is largely made up of carbohydrates, which are composed of insoluble cellulose, soluble hemicellulose, dextrins, starch, and sugars. Cellulose constituents do not contribute to fermentable extract or desirable flavours in the malt. Hemicellulose is a constitutent of the endosperm cell walls, which consist largely of β-glucan. Dextrins are residual, unfermentable fractions of amylopectin. Starch, accounting for about 60–65% of the malt's weight, is composed of amylose, which is reduced to maltose and maltotriose, and amylopectins that decompose into glucose. Glucose, a monosaccharide, accounts for about 1–2% of the total starch found in a barley kernel. Maltose, a disaccharide or double-sugar, is most closely associated with brewing and is formed by two molecules of glucose. Another disaccharide present in malt is sucrose. The only significant trisaccharide or triple-molecule sugar in brewing is maltotriose which is slowly fermentable by most strains of brewing yeast.

Malt Analysis

Malt analysis provides guidance on the effectiveness of the malting process and the suitability of the malt for brewing. The brewer judges malt quality by referring to the malt analysis provided by the maltster. A malt analysis provides very useful information, listing a number of parameters.

Wheat Malt

Wheat malt, for obvious reasons, is essential in making wheat beers. Wheat is also used in malt-based beers (3–5%) because its protein gives the beer a fuller mouthfeel and enhanced beer head stability. On the down side, wheat malt contains considerably more protein than barley malt, often 13 to 18% and consists primarily of glutens that can result in beer haze. Compared to barley malt it has a slightly higher extract, especially if the malt is milled somewhat finer than barley malt.

Malt Extracts

Malt extracts can be used as a sole source of fermentable sugar, or they can be combined with barley malt. The malt extract comes in the form of syrup to dried powder. If the final product is a dried powder, the malt extract has undergone a complete evaporation process by means of "spray-drying", thus removing almost all water content. Syrups are more popular than dried malt extract, possibly because they are easier to store. A common problem noticed in malt extract beers is the thin, dry palate, which correlates with a low terminal gravity. Another common problem is the lack of a true "dark malt" flavour in dark beers.

HOPS

Hops, a minor ingredient in beer, are used for their bittering, flavouring, and aroma-enhancing powers (Figure 44.1). Hops also have pronounced bacteriostatic activity that inhibits the growth of gram-positive bacteria in the finished beer and, when in high enough concentrations, aids in precipitation of proteins.

Figure 44.1 Hop cones

The hop cones resemble those of pine but are smaller and have a softer texture. The hops are added to the wort and boiled for 50–90 minutes, which has the incidental advantages of sterilizing and concentrating the wort and purging the wort of harsh grainy flavours.

Hop Constituents

Hops contain hundreds of components, but of particular interest are resins, oils and polyphenols.

Hop resins Hop resins are subdividied into hard and soft, based on their solubility. Hard resins are of little significance as they contribute nothing to the brewing value, while soft resins contribute to the flavouring and preservative properties of beer. Alpha and beta acids are two compounds present in the soft resins and are responsible for bitterness. Alpha acids are responsible for about 90% of the bitterness in beer. Magnesium, carbonate and chloride ions also can accentuate hop bitterness.

Alpha acids Alpha acids are the precursors of beer bitterness since they are convereted into iso-alpha acids in the brew kettle. The three major components of alpha acids are humulone, cohumulone and adhumulone.

Beta acids Hops also contain a second group of acids known as the beta acids. The beta acids (lupulone, colupulone and adlupulone) are only marginally bitter.

Hop oils Although hops that have high alpha acid content are preferred for their bittering and flavouring properties, hops are also selected for the character of their oils. Oils are largely responsible for the characteristic aroma of hops and either directly or indirectly, for the overall perception of hop flavours. Hops selected with characteristic oil content are often referred to as aroma or "noble" type hops. Oils also tend to increase a beer's bitterness and also enhance the body or mouth-feel of the beer.

Hop polyphenols   Polyphenols found in hops include the anthocyanogens, tannins, and catechins. Some polyphenols act as antioxidants, protecting beer against oxidation, while others contribute to beer colour and haze formation. Polyphenols may also cause an unpleasant astringency. Significant proportions are removed during boiling by precipitation with proteins.

Hop Varieties

Although there is only one hop species (*Humulus lupulus*) that is useful for brewing, there are a number of varieties in that species, each with its own spectrum of characteristics. Varieties of hops are chosen for the properties of bitterness, flavour or bouquet that they will lend to the beer. Hop varieties can be roughly divided into two classes, bittering hops and aroma hops. Bittering hop varieties are those that impart bitter flavour to beer and have high alpha acid levels. Aroma hops, with low to medium alpha levels, mainly impart characteristic hop aromas to beer. Generally, bittering hops are added at the beginning of the boil, a process referred to as "kettle hopping" or "bitter hopping", while aroma hops are added in the final stages of the boil for their aromatic and flavouring properties, a process referred to as "late hopping".

Whole Hops

When hops are used in their raw or unprocessed form, directly from the bale, they are designated as whole, raw or leaf hops. There is a belief that whole hops provide the best aroma. The biggest disadvantages of whole hops are the amount of space required, susceptibility to deterioration, poor utilization and brewhouse equipment requirements.

Dry Hopping

Dry hopping is the process of adding hops to the primary fermenter, the maturation tank, or the casked beer to increase the aroma and hop character of the finished beer. Some brewers believe dry hopping should not be done during primary fermentation because of the risk of contaminating the beer with microrganisms. Dry hopping adds no bitterness to the beer, and any lingering bitterness will dissipate in a few weeks. This is because alpha acids are only slightly soluble in cold beer.

YEASTS

Yeasts are single-celled microbes that reproduce by budding. They are biologically classified as fungi and are responsible for converting fermentable sugars into alcohol and other by-products. There are literally hundreds of varieties and strains of yeast. In the past, there were two types of beer yeast: ale yeast (the "top-fermenting" type, *Saccharomyces cerevisiae*) and lager yeast (the "bottom-fermenting type, *Saccharomyces uvarum*, formerly known as *S. carlsbergensis*). Today, as a result of recent reclassification of *Saccharomyces* species, both ale and lager yeast strains are considered to be members of *S. cerevisiae*.

Ale yeast   Ale yeast strains are best used at temepratues ranging from 10–25°C though some strains will not actively ferment below 12°C. Ale yeasts are generally regarded as top-fermenting yeasts since they rise to the surface during fermentation, creating a very thick, rich yeast head. That is why the term "top-fermenting" is associated with ale yeasts. Fermentation by ale yeasts at these relatively warmer temperatures produces a beer high in esters, which many regard as a distinctive character of ale beers.

Lager yeast   Lager yeast strains are best used at temperatures ranging from 7–15°C. At these temperatures, lager yeasts grow less rapidly than ale yeasts, and with less surface foam they tend to settle out to the bottom of the fermenter as fermentation nears completion. Therefore they are often referred to as "bottom" yeasts. The final flavour of the beer will depend on the strain of lager yeast and the temperatures at which it was fermented.

Yeast Nutritional Requirements

To grow successfully, yeast requires an adequate supply of nutrients— fermentable carbohydrates, nitrogen sources, vitamins and minerals—for healthy fermentation. These nutrients are naturally present in malted barley or developed by enzymes during the malting and mashing process.

Carbohydrates   Only low molecular weight sugars such as the mono- di- and oligosaccharides are available for yeast growth. Polysaccharides are not used by the yeast. The sugars are in order of concentration, maltose, maltotriose, glucose, sucrose and fructose, which together constitute 75–85% of the total extract. The other 15–20% consists of non-fermentable products such as dextrins, β-glucans, pentosans, and oligosaccharides. Regardless of concentration, fermentable carbohydrates are usually assimilated by yeast in the following order: sucrose, glucose and fructose are consumed most rapidly (24–49 hours); followed by maltose (70–72 hours); then maltotriose (after 72 hours). Some overlap in assimilation does occur. A majority of the strains leave maltotetrose and dextrins unfermented.

Nitrogen   Nitrogen is available for yeast growth in wort as amino acids, peptides and ammonium salts. Yeast prefers to use ammonium salts, but these are present in wort only in very small amounts. Amino acids and peptides are therefore the most important wort constitutents. Amino acids collectively referred to as "free amino nitrogen (FAN)" are the principal nitrogen source in wort and are an essential component of yeast nutrition. It is the amino acids that the yeast cells use to synthesize more amino acids and, in turn to synthesize proteins.

Vitamins Vitamins such as biotin, pantothenic acid, thiamine, and inositol are essential for enzyme function and yeast growth. Biotin is obtained from malt during mashing and is involved in carboxylation of pyruvic acid, nucleic acid synthesis, protein synthesis, and synthesis of fatty acids. Biotin deficiencies will result in yeast with high death rates. Pantothenic acid is required by many strains of fermentating yeast and is an essential factor in carbohydrate and lipid metabolism and in cell memebrane function. Pantothenic acid deficiencies can lead to the accumulation of hydrogen sulphide. Thiamine is essential in oxo-acid decarboxylation. Inositol is required for cell division; deficiencies will decrease the rate of carbohydrate metabolism.

Minerals Yeasts are unable to grow unless provided with a source of a number of minerals. These include phosphate, potassium, calcium, magnesium, sulphur, and trace elements. Phosphate is involved in energy conservation, is necessary for rapid yeast growth, and is part of many organic compounds in the yeast cell. Potassium ions are necessary for the uptake of phosphate. Calcium improves the flocculation properties of yeast and should be present in a concentration greater than 50 mg/l. Magnesium is required for yeast growth and acts as an enzyme activator. Yeast requires sulphur for the synthesis of methionine and for cysteine, which is incorporated into protein, glutathione, coenzyme A, and thiamine. The elements zinc, copper, and manganese are required in trace amounts.

Yeast By-products

The flavour and aroma of beer is very complete, being derived from a vast array of components that arise from a number of sources. Not only do malt, hops and water have an impact on flavour, so does the synthesis of yeast, which forms by-products during fermentation and maturation. The most notable of these by-products are ethanol and carbon dioxide; but in addition, a large number of other flavour compounds are produced such as:

Aldehydes There are many flavour-active aldehydes present in beer. These are formed at various stages in the brewing process and are produced by oxidation of alcohols and various fatty substances. Aldehyde levels reach a maximum during primary fermentation or immediately after kraeusening, and then decrease. Aldehyde is reduced to ethanol by the end of the primary fermentation. If oxygen is introduced back into the process, the ethanol is oxidized back into acetaldehyde.

Esters Esters are considered the most important aroma compounds in beer. They make up the largest family of beer aroma compounds and in general impart a "fruity" character to beer. Esters are more desirable in ales than in lagers. Ester production is increased by 1) high fermentation temperatures, 2) restricting wort aeration, 3) increasing the attenuation limit, and 4) increasing the wort concentration to above 13%. In addition, the type of yeast affects ester levels. Most of the esters are formed during primary fermentation, and some ester formation occurs during maturation. However, the level of esters could double with a long secondary fermentation.

Diacetyl Diacetyl and 2,3-pentanedione, which are classified as ketones, are important in contributing flavour and aroma to beer. Often these two ketones are grouped and reported as the vicinal diketone(VDK) content of beer, which is the primary flavour in differentiating aged beer from green beer. Of the two, diacetyl is more significant because it is produced in larger amounts and has a higher flavour impact than 2,3-pentanedione. A buttery or butterscotch flavour usually indicates the presence of diacetyl, while 2,3-pentanedione has a honey flavour.

Dimethyl sulphide Another major compound responsible for sulphury flavours in beer is dimethyl sulphide (DMS) which is a desirable flavour component in lager beer but not in ales. In lagers it will lead to a malty/sulphury note. The taste threshold for DMS is considered to be from 50 to 60 µg/L. DMS also enhances the malt character of beer.

Fatty acids Fatty acids are minor constituents of wort and increase in concentration during fermentation and maturation. They give rise to goaty, soapy, or fatty flavours and are recognized as common flavour characteristics in both lagers and ales; but they are more prevalent in lagers because of the tendency of some lager yeast strains to produce greater quantities of fatty acids than do strains of ale yeast.

Fusel alcohols Fusel alcohols are a group of by-products that are sometimes called "higher alcohols." They contribute directly to beer flavour but are also important because of their involvement in ester formation. Fusel alcohols have strong flavours, producing an "alcoholic" or "solvent-like" aroma. They are known to have warming effect on the palate. About 80% of fusel alcohols are formed during primary fermentation. The yeast strain is very important, with some being able to produce up to three times as much fusel alcohols as others.

Nitrogen compounds Yeast also excretes some nitrogen compounds during fermentation and maturation as amino acids and lower peptides which contribute to the rounding of the taste and an increase in palate fullness. Harvesting of the yeast too soon can therefore produce empty, dry beers even when they are subsequently lagered for a long time. The beginning of autolysis can be detected by an excessive increase in the amino acid content.

Organic acids Some of these organic acids are derived from malt and are present at low levels in wort, with their concentrations increasing during fermentation. Other acids are produced solely as a result of yeast metabolism. Organic acids can directly affect the flavour of beer by lowering its pH.

Sulphur compounds   Volatile sulphur compounds such as hydrogen sulphide, dimethyl sulphide, sulphur dioxide, and thiols make significant contributions to beer flavour. When present in small concentrations, sulphur compounds may be acceptable or even desirable, but in excess they give rise to unpleasant off-flavor, (e.g. rotten egg flavours). Three main sources of sulphur compounds in beer are raw materials (malt and hops), yeast metabolism and spoilage organisms—in particular *Zymomonas anaerobia, Enterobacter aerogenes* and *Hafnia protea*.

Yeast Selection

Selection of a yeast with the required brewing characteristics is vital for both product quality and economic standpoint. The criteria for yeast selection will vary according to the requirements of the brewing equipment and the beer style, but they are likely to include the following:

Rapid fermentation   A rapid fermentation without excessive yeast growth is important, as the objective is to produce a beer with the maximum attainable ethanol content consistent with the overall flavour balance of the product.

Yeast stress tolerance   The yeast strain should be tolerant to alcohol, osmotic shock, and temperature. Another stress point for yeast can be the collection, separation (centrifuging/pressing), and transfer (pumping) throughout the plant.

Flocculation   The flocculation characteristics of yeasts are of great importance. The term "flocculation" refers to the tendency to form clumps of yeast called flocs. The flocs of yeast cells descend to the bottom in the case of bottom-fermenting yeasts or rise with carbon dioxide bubbles to the surface in the case of top-fermenting yeasts. The flocculation characteristics need to be matched to the type of fermentation vessel used. A strongly cropping strain will be ideal for skimming from an open fermenter but unsuitable for a cylindroconical fermenter.

Attenuation   Attenuation referes to the percentage of sugars converted to alcohol and carbon dioxide, as measured by specific gravity. Most yeasts ferment the sugars—glucose, sucrose, maltose and fructose. To achieve efficient conversion of sugars to ethanol (good attenuation) requires the yeast to be capable of completely utilizing the maltose and maltotriose. Brewing yeasts vary significantly in the rate and extent to which they use these sugars. Lager strains are often better at utilizing maltotriose than their ale counterparts. The degree of attenuation obtainable exerts a great influence on the organoleptic properties of the resultant beer and consequently is one of the determinant factors in the process of yeast selection.

Flavour component   The selection of the yeast strain itself is perhaps one of the most important contributors to beer flavour. Different strains will vary markedly in the by-products they produce: esters, higher alcohols, fatty acids, hydrogen sulphide and dimethyl sulphide. The yeast strain must also be capable of reproducible flavour production.

Storage characteristics The storage characteristics of a yeast are very important for maintaining viability during storage between fermentations and rapid attenuation when repitched.

Yeast mutations Yeast mutations are a common occurrence in breweries, but their presence may never be detected. Usually the mutant has no adverse effect since it cannot compete with normal yeast and generally disappears rapidly. In some cases, though mutant yeast will overcome the normal brewing yeast and may express itself in many different ways. For example, a mutation could affect the fermentation of maltotriose, or there could be a continuous variation in the fermentation rate. Reportedly, lager yeast mutates more rapidly than ale yeast. Most commonly, mutations are due to poor handling of brewing yeast.

Yeast degeneration Yeast degeneration refers to the gradual deterioration in performace of the brewing yeast. Yeast degeneration has a harmful effect on the course of brewing fermentations. It is characterized by some of the following symptoms: sluggish fermentations, premature cessation of fermentation (resulting in high residual fermentable levels in beer), gradual lengthening of fermentation times, and poor foam or yeast head formation. Some brewers have noticed that the flavour of beer becomes increasingly "dry" as a result of yeast degeneration.

Yeast Pure Cultures

The process of culturing yeast strains involves isolation of a single yeast cell, maintenance of yeast cultures, and the propagation of the yeast until an amount sufficient for pitching is obtained.

The isolation of pure cultures Pure yeast cultures are obtained from a number of sources. Most often, the yeast is already in use in the brewery, but it can also be obtained from other breweries, commercial distributors, or culture collections. Various procedures are used to collect pure cultures, including culturing from a single colony, a single cell, or a mixture of isolated cells and colonies.

The maintenance of pure cultures Once yeast has been selected, accepted and fully proven for use in brewing, it is essential that a pure culture is maintained in the laboratory yeast bank for prolonged periods. Some of the most common methods used for maintaining the purity and characteristics of the yeast culture are subculturing, desiccation and lyophilization.

The propagation of pure cultures The objective of propagation is to produce large quantities of yeast with known characteristics for the primary role of fermentation, in a short time as possible. Most brewers use a simple batch system of propagation, starting with a few milliliters of culture and scaling up until there is enough yeast to pitch a commercial brew. Scale-up introduces actively growing cells to a fresh supply of nutrients in order to produce a crop of yeast in the optimum physiological state.

Stock Maintenance of Harvested Yeast

Yeast concentration　The yeast slurry or "barm" removed directly from the fermenter usually has the consistency of thin cream, of which a major proportion is beer. There are a number of disadvantages in direct reuse of barm yeast for repitching. One problem is the possibility of carrying infection forward and another is that the yeast concentration is variable within the slurry. Consequently, the barm is washed and concentrated prior to its use as pitching yeast. Bacterial infective levels can be controlled by yeast washing. Rotary vacuum filters and simple mesh strainers can be used to separate yeast; however, the most common system of yeast separation is the plate and frame filter, yeast press.

Culture contamination　It frequently happens that brewing yeasts carry a persistent low level of contaminants such as *Obesumbacterium proteus*, acetic acid bacteria, and slow growing *Torula* type yeasts. These organisms are generally regarded as harmless because their numbers never reach a point where they are likely to have adverse effects on the beer. On the other hand, *L. pastorianus*, *Z. anaerobia*, and *S. carlsbergensis* are strains considered harmful at low levels. Lager fermentations seem to be more susceptible to bacterial contamination than do ale fermentations primarily because the pH drops more slowly for lagers than for ales and bacteria are suppressed by falling in pH.

Yeast washing　Pitching yeasts collected from brewery fermentations are never absolutely free of microbiological infection. In spite of whatever care and sanitary precautions are taken, some bacteria and wild yeasts will contaminate the pitching yeast. To minimize microbiological infection, brewery yeast can be washed using different procedures.

Yeast storage　In most breweries, yeast is stored during the period between cropping and repitching. Pitching yeast may be stored within the brewery as a slurry or as a "barm" in a yeast collection vessel, or it may be stored under a layer of water or beer, under a layer of wort or as pressed cake.

Yeast Viability and Replacement

Yeast viability　Viability is a measure of yeast's ability to ferment. Yeast viability is determined by the standard culture methods, by selective straining, or by more advanced methods such as the slide viability method, flocculation tests and fermentation tests.

Yeast replacement Most brewers discard yeast after a successive number of fermentations because it may be intermixed with other yeasts, contaminated with wild yeast and bacteria, or mutated to less desirable strains. Depending on their yeast handling facilities and procedures, some brewers use their yeast in production for less than three generations, while others discard their yeast only after 5–10 successive fermentations. However, there are exceptions to the rule, with some brewers routinely discarding yeast after 30 brewery fermentations. Those brewers who employ high pitching rates will probably have to replace their yeast more often because continuous use and high pitching rates tend to increase the average age of the yeast, thus reducing their vigour.

BEER ADJUNCTS

Adjuncts are nothing more than unmalted grains such as corn, rice, rye, oats, barley and wheat. Adjuncts are used mainly because they provide extract at a lower cost. Use of adjuncts results in beers with enhanced physical stability, superior chill proof qualities, and greater brilliancy. The greater physical stability has to do with the fact that adjuncts contribute very little proteinaceous material to wort and beer, which is advantageous in terms of colloidal stability. Rice and corn adjuncts contribute little or no soluble protein to the wort, while other adjunct materials such as wheat and barley, have higher levels of soluble protein. Except for barley, adjuncts also contribute little or no polyphenolic substances.

Adjuncts can be used to adjust fermentability of a wort. Many brewers add sugar and/or syrup directly to the kettle as an effective way of adjusting fermentability, rather than trying to alter mash rest times and temperatures. Adjuncts are often used for their flavour contribution, for example, rice has a very neutral aroma and taste, while corn tends to impart a fuller flavour to beer. Wheat tends to impart a dryness to beer. Semi-refined sugars add flavour to ales that has been described as imparting a luscious character. Adjuncts will also alter the carbohydrate and nitrogen ratio of the wort, thereby affecting the formation of by-products such as esters and higher alcohols.

Adjuncts are used for colour adjustment, as in the case of dark sugars. On the other hand, aduncts such as rice and pure starches and sugars are used to dilute malt colours to produce light-coloured beers.

Some adjuncts are used for their chemical properties; raw barley and wheat, which contribute glycoproteins to enhance foam stability. Other adjuncts, low in protein, are used to improve colloidal stability since they will dilute the amount of potential haze forming proteins.

Finally, the use of adjuncts can result in increased brewing capacity, reduced labour costs, improved hot and cold breaks, and shorter brewing cycles.

If the level of adjuncts used is too high, there is a risk of producing wort with insufficient insoluble nitrogen for yeast growth.

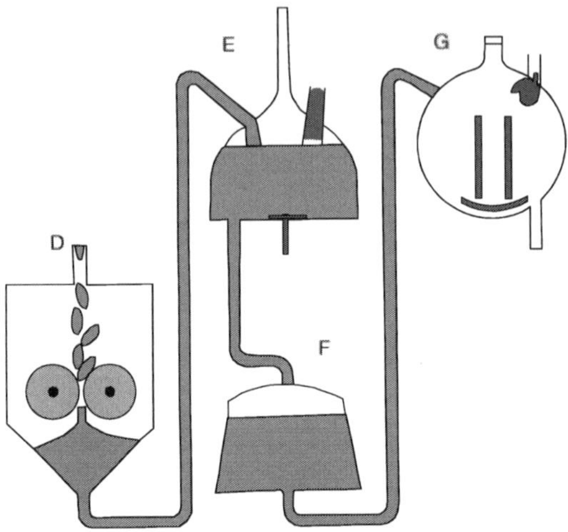

Figure 44.2   Simple illustration of brewing malt

The brewhouse consists of brewery buildings housing machinery and equipment for the production of wort. Processes taking place here include milling of the kiln dried malt (D), mashing (E), filtration (F) and wort boiling (G) (Figure 44.2).

Malt Milling

The objective of milling is to split the husk, preferably lengthwise, in order to expose the starchy endosperm for milling and allow for efficient extraction and subsequent filtration of the wort. Malt milling is usually done by either dry or wet milling.

The most commonly used mills in breweries are dry mills. A refinement to dry milling is conditioning of the malt with steam or warm water. This practice minimizes the risk of fracturing the malt husks, thus leaving the husks intact for lautering. The effect is to produce a more open bed of spent grains on the mash. In a wet milling operation, the whole uncrushed malt is pre-steeped in hot water to the point where the husks reach a water content of approximately 20% and the endosperm remains nearly dry which results in a semi-plastic, pasty consistency. The steeped malt is then passed through a roll mill in which the endosperm is squeezed out of the surrounding husk, leaving it fully available to the subsequent actions of mashing. The husk remains tough and intact for its eventual role as the filter medium for wort separation.

Mashing

Mashing is the process of converting starch form the milled malt and solid adjuncts into fermentable and non-fermentable sugars to produce wort of the desired composition. The composition of the wort will vary according to the style of beer. Mashing involves mixing milled malt and solid adjuncts with water at a set temperature and volume to continue the biochemical changes initiated during the malting process. Mashing involves four stages. (Figure 44.3).

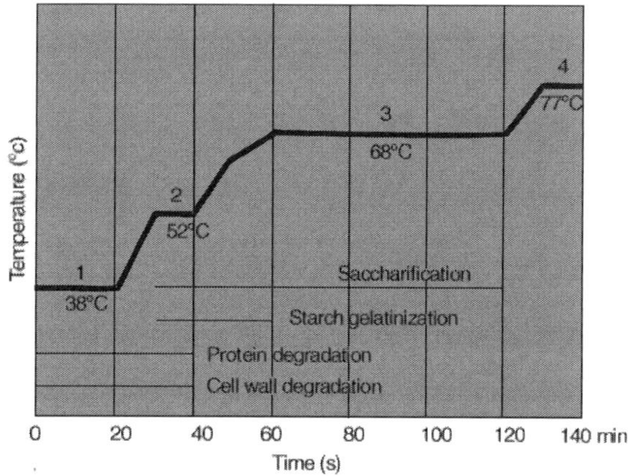

Figure 44.3   Representation of the four stages of mashing

1. *Mashing-in*   Mixing of malt and water
2. *Protein pause*   Release of peptides and amino acids
3. *Sugar pause*   Release of maltose and dextrins
4. *Mashing-off*   Degradation of residual starch, and inactivation of enzymes

Mashings are normally performed at pH 5.5 where most malt-derived enzymes exhibit high activity. Conditions include a controlled stepwise increase in temperature that preferentially favours one enzyme over the other, eventually degrading cell wall, proteins and starch (Figure 44.3). Since the enzymes which degrade cell walls and proteins are rather heat-labile, it is important for their function that mashing begins at a low temperature. Subsequent mashing at 65°C, or higher, is particularly geared to control conversion of gelatinized starch into fermentable sugars using malt-derived starch-degrading enzymes.

Principal mashing enzymes include (1-3,1-4)-β-glucanase and xylanase for cell wall degradation, endopeptidase and carboxypeptidase for protein degradation; and amylases, limit dextrinase and α-glucosidase for starch degradation

Wort Separation

After mashing, when all the starch has been broken down, it is necessary to separate the liquid extract (the wort) from the solids (spent grain particles and adjuncts). Wort separation (Figure 44.4) is important because the solids contain large amounts of protein, poorly modified starch, fatty material, silicates, and polyphenols (tannins). The objectives of wort separation(lautering) include the following to produce clear wort, to obtain good extract recovery and to operate within the acceptable cycle time.

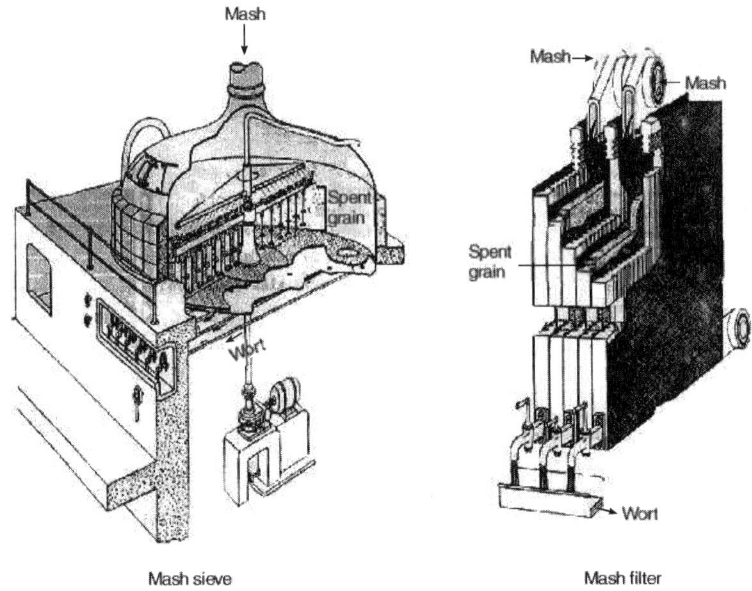

Figure 44.4   Wort separation

The quality of the grist form the mill can greatly affect wort clarity, extraction recovery and overall lauter times.

Wort separation may be carried out by a number of different methods like the mash tun, the lauter tun, the mash filter and the strain master. The wort is transferred to the kettle or to a vessel, i.e., a wort collection vessel. Bottom entry into the kettle allows for a gentler handling of the wort with less chance of oxidation. Holding the wort at about 77°C is recommended in the interest of preventing enzyme activity. Worts must not be allowed to cool below 55°C because of the possibility of contamination by bacteria (e.g. *Lactobacillus delbrueckii*).

Wort Boiling

Following extraction of the carbohydrates, proteins, and yeast nutrients from the mash, the clear wort must be conditioned by boiling the wort in the kettle. The purpose of wort boiling is to stabilize the wort and extract the desirable components from the hops. The principal biochemical changes that occur during wort boiling are as follows:

- sterilization
- destruction of enzymes
- protein precipitation
- colour development
- isomerization
- dissipation of volatile constituents
- concentration
- oxidation

Wort Cooling

After boiling and clarification, the wort is cooled in preparation for the addition of yeast and subsequent fermentation. The principal changes that occur during wort cooling are as follows:

- cooling the wort to yeast pitching temperature
- the formation and separation of cold break and
- oxygenation of the wort to support yeast growth

Wort is usually cooled through plate heat exchangers. Wort enters the heat exchanger at approximately 96–99°C and exits cooled to pitching temperature. The first stage utilizes water to remove the bulk of the heat, cooling the incoming water to within 3°C of fermentation temperature. In the second stage, the wort is cooled to the fermentation temperature by a secondary refrigerant (e.g. glycol).

Aeration of the chilled wort is needed in order to provide the yeast with sufficient oxygen for growth during fermentation. The amount of oxygen required depends on yeast strains, wort temperature and a number of other factors.

BEER FERMENTATION

Fermentation is the process by which fermentable carbohydrates are converted by yeast into alcohol, carbon dioxide and numerous by-products. The by-products have a considerable effect on the taste, aroma and other characteristic properties of the beer. Fermentation is dependent upon the composition of the wort, the yeast and fermentation conditions.

Wort composition influences fermentation by the presence and concentration of various nutrients, pH and degree of aeration and temperature. These factors can affect the rate of fermentation, the extent of fermentation, the amount of yeast produced, and the quality of beer produced.

The factors that affect fermentation conditions are time, wort temperature, volume, fermenter design, pressure, agitation and currents in the wort.

The brewer's ability to pitch the correct number of yeast cells to initiate fermentation is crucial to consistently producing a product of superior and constant quality. Pitching rates are governed by a number of factors, including wort gravity, wort constituents, fermentation temperature, degree of wort aeration, fermentation capacity of the yeast, yeast viability, flocculation characteristics, and previous history of the yeast. For example, highly flocculent yeast strains may settle prematurely, requiring the brewer to either overpitch or to mix and aerate by "rousing".

Low pitching rates can result in long lag and reproductive phases. This in turn, can increase the production of by-products such as higher alcohols, esters, and diacetyl. Furthermore, since many sugars are not metabolized with low pitching rates, the resulting beer tends to have a residual sweetness (lower attenuation). Insufficient yeast growth can eventually lead to slower and even stop the fermentations. Poor yeast growth can also lead to high sulphur dioxide levels.

Uninoculated wort is liable to contamination by many types of mould; bacteria such as *Pediococcus* spp. and *Lactobacillus* spp; and wild yeasts such as *Hansenula, Dekkera, Brettanomyces, Candida* and *Pichia.* Other *Saccharomyces* species may be present. However, once the yeast inoculum is added, the selective effect against microbes other than brewing yeast begins. This is due to falling pH, the antimicrobial effect of hop compounds, the developing anaerobic conditions in the wort, and the production of carbon dioxide and ethanol.

Fermentation Process

The fermentation stages in traditional ale fermentation systems can be described as follows. After the wort is run into the fermenter and the yeast added, the yeast begins building cell walls and reserves, which is referred to as the "lag phase". At the end of the lag phase, the yeast begins to divide. The first visible sign of fermentation is the appearance of fine bubbles on the wort surface at the sides of the fermenter. These bubbles gradually spread, until the surface of the wort is completely covered. After about 18 hours, the bubbles thicken and their colour changes to light brown and the cold break rises to the surface of the wort. At the same time the specific gravity starts to fall as does the pH, with a subsequent rise in temperature and yeast count.

Eventually, maximum fermentation is reached (after 36–48 hours) and the yeast surface is covered with a white foam of constantly moving pinnacles and crevasses as carbon dioxide rises to the surface. Gradually, the yeast activity slows and the colour of the head changes from white to pale cream as yeast starts to rise to the surface to replace the foam. The yeast rises for some time, eventually forming a thick covering.

Traditional lager brewing involves pitching the yeast beween 5 to 6°C and allowing the temperature to rise between 8–9°C. This generally results in a better quality beer because low fermentation temperature retards the development of by-products—esters, fusel alcohols and diacetyl—all of which are inappropriate in lagers. The yeast is pitched between 7 and 8°C and after a couple of days the temperature is increased to 10–11°C. Some brewers are known to pitch between 12 and 14°C and then increase the temperature to as high as 18°C.

Use of high temperatures at the end of primary fermentation serve to reduce vicinal diketones (VDKs). This procedure is known as a diacetyl rest. VDKs, which are assimilated by yeast towards the end of fermentation, are responsible for off-flavours in beer. The diacetyl rest reinvigorates the yeast culture so that it metabolizes those by-products such as diacetyl and 2,3-pentanedione that are excreted early in the fermentation, thereby removing them from solution. Depending upon the yeast type, the medium and the physical environment, this process is variable in time and temperature.

Following fermentation and maturation, the beer is diluted with cool carbonated water to a prescribed original gravity or to a prescribed alcohol concentration. There are a number of advantages associated with high-gravity brewing. It results in beers that are more consistent and more physically stable since the compounds responsible for haze are more easily precipitated at higher concentrations. The disadvantages are longer fermentation times, different flavour characteristics and poorer hop utilization than normal gravity fermentations.

Removal of excess yeast from beer at the end of primary fermentation is necessary to minimize the load on the filter medium, to prevent yeast metabolism and autolysis, and to provide a source of yeast for pitching subsequent brews.

The time it takes for autolysis to set in depends on the temperature and the strain of yeast. Yeast will stay in good condition at cold temperatures. Thus, separating the beer from the layer of sediment yeast at the end of primary fermentation is particularly important, especially when fermentation is carried out at elevated temperatures. Many of the off-flavours in beer that are described as "yeasty" are due to the excretion of fatty acids by the yeast, a process which is accelerated at high temperatures.

BEER CONDITIONING

Following primary fermentation, many undesirable flavours and aromas are present in the "green" or immature beer. Conditioning reduces the levels of these undesirable compounds to produce a more finished product. The component processes of conditioning are maturation, clarification and chillproofing.

Beer maturation   Maturation of "green beer" involves four general schemes: traditional lagering, bottle conditioning, casking and accelerated lagering.

Lagering beer involves secondary fermentation of remaining fermentable extract at a reduced rate controlled by low temperatures and low yeast count. The low temperatures also aid in setting the remaining yeast and precipitating haze-forming material (polyphenol complexes). The evolution of carbon dioxide during secondary fermentation not only carbonates the beer, but also it reduces by-products (including sulphur compounds and other volatiles). Bottle conditioning involves secondary fermentation and clarification in the bottle induced by adding yeast and sugar to the beer. Cask conditioned beers involve secondary fermentation and clarification in the cask induced by adding yeast, sugars, hops, and finings. In accelerated lagering, the beer is fully attenuated, virtually free of yeast, and stored at higher temperatures. Unlike beers that have undergone traditional or accelerated lagering, beers conditioned in a bottle or cask are neither filtered nor pasteurized.

Beer clarification   Following fermentation, beer is quite turbid due to the presence of yeast, protein or polyphenol complexes, and other insoluble material, all of which are responsible for haze formation in beer. Extended lagering periods at low temperatures, the addition of finings to the beer, and centrifugation are some of the techniques that brewers use to reduce these substances.

Lagering   Lagering involves holding the beer in shallow vessels at temperatures between -1 and $2°C$. The cold temperatures reduce fermentation by-products by reducing their solubility and encouraging precipitation of yeasts and other haze loading material. Natural sedimentation yields beer that is slightly turbid.

Fining   Although good clarity can be obtained from simple sedimentation by prolonged storage at low temperatures, better results can be obtained in less time by adding fining agents either during transfer from the fermenter or during storage. Isinglass and wood chips can be used to improve clarity by precipitating negatively charged yeast cells.

Centrifugation   Centrifuging is a popular method of reducing the solids content of beers, and takes the place of lagering or fining.

Beer chillproofing   In certain circumstances, fining alone cannot provide satisfactory clarity because of the presence of proteins and polyphenols (tannins). Positively charged proteins combine with polyphenols during conditioning to form a colloidal haze which is somewhat soluble at warmer temperatures but insoluble at cooler temperatures. To remove proteins or polyphenols and to improve its physical stability, the beer undergoes a procedure often referred to as "chillproofing".

Proteolytic enzymes, tannic acid, hydrolysable tannins, silica gels and bentonite are used to improve the beer's physical stability or to "chillproof" it.

Beer Filtration

Extended lagering periods and the addition of flocculation aids in reducing yeast and haze loadings. Centrifuges are mainly used in the preliminary reduction of suspended particles, primarily in yeast before sending to the conditioning tanks. Although these methods are very effective in prefiltering the beer, a final filtration is needed to remove residual yeast, other turbidity-causing materials and microorganisms in order to achieve colloidal and microbiological stability.

Beer Carbonation

The next major process which takes palce after filtration and prior to packaging is carbonation. Carbon dioxide not only contributes to perceived "fullness" or "body" and enhances foaming potential, it also acts as a flavour enhancer and plays an important role in extending the shelf life of the product. Typically, carbon dioxide levels range from 1.2 to 1.7 volumes of carbon dioxide per volume of beer (v/v). The time required to reach a desired carbon dioxide concentration depends on a number of physical factors. Temperature and pressure play an important role in determining the equilibrium concentration of carbon dioxide in solution. Fixing the temperature and pressure at appropriate settings will bring about the desired carbon dioxide concentration.

Pasteurization

Pasteurization is an alternative to sterile filtration for reducing the number of harmful microbes in beer. The basis for pasteurization is the heating of the beer for a predetermined period of time at specific temperatures, thereby assuring the microbiological stability of the beer. Flash pasteurization is used for continuous treatment of bulk beer prior to filling the bottles, cans, or keg. The beer is heated to atleast 71.5°C to 74°C and held at this temperature between 15 and 30 seconds.

Bottling Beer

Once the final quality of the beer has been achieved, it is ready for packaging. The packaging of beer is one of the most complex aspects of brewery operations and the most labour intensive of the entire production process.

REVIEW QUESTIONS

1. List out the various steps involved in beer production.
2. What are hops?
3. What is dry hopping?
4. Differentiate between ale yeast and lager yeast.
5. Discuss the nutritional requirements of yeast.
6. Explain in detail the yeast by-products.
7. Define flocculation.
8. Write short notes on beer adjuncts.
9. What is wort?
10. Describe the beer fermentation in detail.
11. What is lagering?
12. What is chillproofing?

45

FERMENTED MILK PRODUCTS

INTRODUCTION

The origin of fermented milk is observed in antiquity. The Rigveda of the Hindus—known to be more than 3000 years old—mentions about *dahi* the popular fermented milk in India which is produced and consumed in every Indian home even today.

The fermented milks may be described as products prepared from milks (whole, partially or fully skimmed milk, concentrated milk or milk reconstituted from partially or fully skimmed milk), homogenized or not, pasteurized or sterilized and fermented by means of specific organisms. Metchnikoff advocated consumption of sour milk to prolong life. Table 45.1 lists some fermented products of milk.

Table 45.1 Fermented milk products

Fermented product	Microorganisms responsible for fermentation	Description
Sour cream	*Streptococcus* sp. *Leuconostoc* sp.	Cream is inoculated and incubated until the desired acidity develops.
Cultured buttermilk	*Streptococcus* sp. *Leuconostoc* sp.	Made with skimmed or partly skimmed pasteurized milk.
Bulgarian buttermilk	*Lactobacillus bulgaricus*	Product differs from commercial buttermilk in having higher acidity and lacking aroma.
Acidophilus milk	*Lactobacillus acidophilus*	Milk for propagation of *L. acidophilus* and the milk to be fermented are sterilized and then inoculated with *L. acidophilus*. This milk product is used for its medicinal therapeutic value.

(Contd.)

Table 45.1 (Continued)

Fermented product	Microorganisms responsible for fermentation	Description
Yoghurt	*Streptococcus thermophilus, Lactobacillus bulgaricus*	Made from milk in which solids are concentrated by evaporation of some water and addition of skim milk solids. Product has consistency resembling custard.
Kefir	*Streptococcus lactis, Lactobacillus bulgaricus*, yeasts	A mixed lactic acid and alcoholic fermentation; bacteria produce acid, and yeasts produce alcohol.

NUTRITIVE VALUE

Fermentation of milk increases the nutritive value to some extent. The microbial cells consumed along with the fermented milk are likely to be hydrolysed partially in the digestive tract, releasing many essential amino acids to be utilized by the body. Fermented milk nutrients are easily assimilable than liquid milk. It has good effect in controlling various pathogens in the intestine. The importance of fermented milk as therapeutic agents in the treatment of gastrointestinal disorders has been recognized in the Ayurvedic systems of medicine in India since early times.

YOGHURT

Yoghurt is a fermented milk produced and is characterized by its viscous consistency, a strong acidulous taste due to high acidity (pH 4.6) and a distinct aroma caused mainly by acetaldehyde. The steps in the preparation of yoghurt is described in Figure 45.1.

Pretreatment of raw milk The milk used for yoghurt must be of highest bacteriological quality. It must have low count of organisms and substances which suppress the development of starter microbes. Hence, the milk must not contain penicillin, bacteriophages, residues of detergents/sanitizers. The milk must be carefully analysed at the dairy. The visible extraneous matter present in milk are to be removed by passing it through cloth filters. Clarification can also be done in a specially designed centrifugal clarifier.

Standardization of fat The methods employed for standardization of fat in milk are removal of part of the fat content of milk, mixing full cream milk with skim milk or by using standardizing centrifuge.

Increasing the milk solids to 14–16% is done by

1. addition of skim milk powder

2. vacuum evaporation of milk
3. use of concentrates prepared by ultra filtration or reverse osmosis

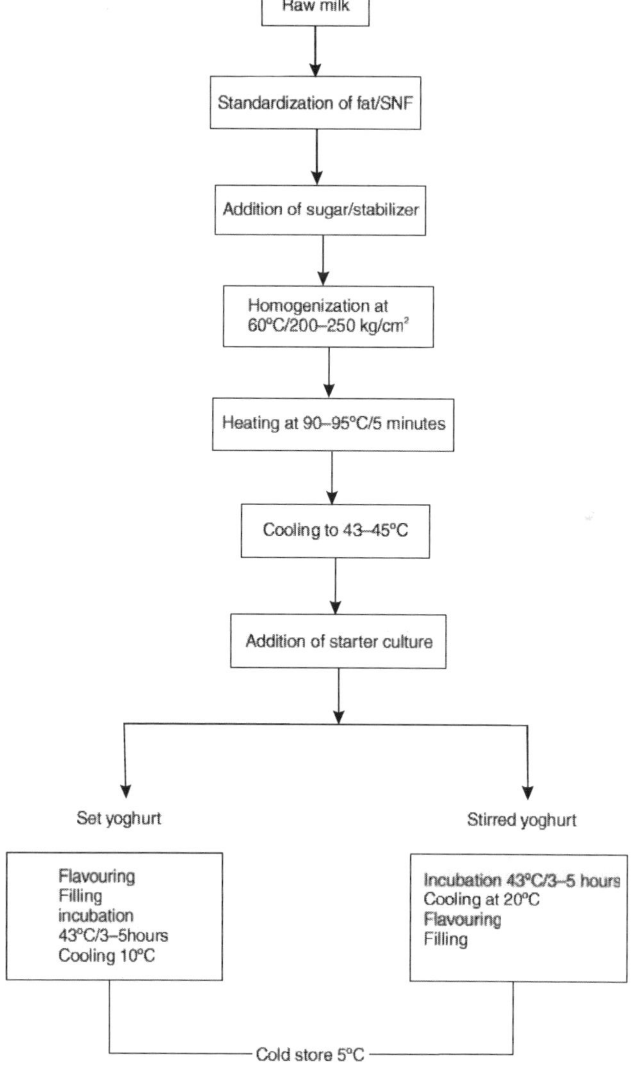

Figure 45.1   Steps in the preparation of yoghurt

Homogenization   Homogenization of milk is done by first pre-treating it
to 60–65°C and passing it through an homogenizer at 100–200 kg/cm² to

stabilize the product and provide smooth texture. Stabilizers or emulsifiers like agar agar, alginate, gums, etc. are added if necessary at low concentrations to increase the viscosity of the product and avoid syneresis from the coagulum. Nearly 5% of sugar is added to the basic mix. Sugar inhibits the de7velopment of lactic acid. This explains the reason for the addition of a part of the sugar to the base mix and the desired amount can come from sweetened fruit preparations added later after fermentation.

Heat treatment   Heat treatment is the next important stage. This involves heating the milk at 90–95°C for 5 minutes to improve the properties of the milk suited as a substrate for the microbes of the starter to ensure firm body and texture in the yoghurt and reduce whey separation. The removal of air in heating produces a more suitable environment for the growth of microaerophilic lactic cultures. The high temperature renders the milk suitable for subsequent growth of starter lactobacilli which have fastidious nutritional requirements.

Cooling   After the heat treatment, the substrate is cooled to incubation temperature (42–45°C). The cultures contain two strains of microorganisms, *Streptococcus salivarius* subsp. *thermophilus* and *Lactobacillus delbrueckii* subsp. *bulgaricus* which coexist in symbiosis and together produce the desired characteristics of the yoghurt. Best yoghurt is obtained when the ratio of cocci to bacilli is between 1:1 and 2:1. The bacilli should not grow faster because the flavour will become sharply acid. One of the factors which affect the relative bacterial counts, is the incubation temperature. At 40°C the ratio is 4:1 and at 45°C it is about 1:2. The optimum temperature for inoculation (and incubation) in yoghurt production is between 42–43°C.

As soon as inoculation has taken place and the starter organisms mixed into the substrate, the organisms multiply rapidly and ferment lactose to lactic acid. It is important that the temperature is regulated throughout and that no contaminants are allowed to come in contact with the product. The incubation temperature is 42–43°C. A ratio of 1:1 or 2:1 is obtained at a dosage of 2.5–3% and the incubation time is 2.5–3 hours. During this time, the acidity increases from 0.88 to 1.10% lactic acid. At the end of this period the inoculated milk should contain the largest possible number of active microorganisms. Cooling during incubation keeps the bacterial activity at the optimum level and the bacterial count will be the highest at the end of the incubation period. The cooling period is restricted to 3–3.5 hours. The fermentation is done in large multipurpose tanks and the cooling of the coagulum to 20°C is done within about 30 minutes to prevent yoghurt becoming unpalatable through excess acidity. The level of lactic acid required in the end product is around 1.2–1.4%. In the set process coagulation of milk takes place in retail containers. In this process the yoghurt milk is cooled to inoculation temperatures, starter added and directly filled in retail container, placed in the incubation cabinet to get desired consistency, cooled to 7°C and transferred to cold store for final reduction in temperature to 4.5°C. The heating and cooling can also be done in water bath or air tunnels with conveyor systems.

Conventional yoghurt producers use a multi-strain starter comprising two thermophilic organisms in approximately equal proportions. The two types of organisms possess complementary relationship known as symbiosis when used together. They can grow separately but are usually grown as a mixed population under controlled conditions to ensure a correct balance between the two microorganisms. Initially streptococci having an optimum growth at 45°C dominate the bacterial culture producing lactic acid and carbonyl compounds which are responsible for creamy/buttery aroma and flavour of the finished yoghurt. The streptococci remove oxygen from the milk (if not, may lead to the production of toxic hydrogen peroxide). Growth may continue until the pH is 5.5 when the medium becomes suitable for lactobacilli to proliferate.

Lactobacillus delbrueckii subvar. *bulgaricus* has an optimum temperature of 45°C. It produces considerable amounts of lactic acid and acetaldehyde which gives the characteristic yoghurt flavour. Certain amino acids liberated by lactobacilli have a stimulating effect on the growth of streptococci. The starter organisms hydrolyse lactose through enzyme b-D-galactosidase, use glucose for producing lactic acid with accumulation of galactose; de-stabilize calcium-caeseinate–phosphate complex to form coagulum; produce flavour compounds like acetaldehyde, acetone, acetoin and diacetyl; produce peptides and amino acids which act as precursors for enzyme and chemical action leading to flavour compounds; the peptidase enzymes cause bitter peptides and free amino acids, and lipases act on fat and produce short chain triglycerides giving flavour besides increasing vitamins like folic acid and niacin concentrations.

At 5°C, the yoghurt has a shelf life of about 2 weeks, after which bacterial growth although restricted will increase in acidity and impair flavour and make it unpalatable. The bacteria may be destroyed and the yoghurt may separate into curd and whey. Since yoghurt is susceptible to attack by yeasts and moulds, great care is necessary that the starter is free from these organisms and do not gain access during packing.

ACIDOPHILUS MILK

The therapeutic value of acidophilus milk was stressed many years ago. The organisms concerned is *Lactobacillus acidophilus* and this grows slowly in milk. This poor development perhaps resulted in many batches being severely contaminated. Now, selective strains have been isolated which provide desired acidity in milk in around 12 hours. The original system for getting high degree of microbiological cleanliness, required heat treatment of milk at 95°C for one hour. The milk is then cooled to 37°C and held at this temperature for 3–4 hours to allow germination of spores and then reheated to 95°C to destroy the resultant cells. The milk is then homogenized at 150 kg/cm² at 60°C and cooled to 37°C. The milk is inoculated with *L. acidophilus*. Because of the difficulty in growing this organism in the laboratory, direct to vat inoculation (DVI) cultures are usually employed. The milk is held for 3–4 hours until coagulation takes place. For medicinal use, the acidity is restricted to 0.6–0.7% lactic acid prior to chilling and bottling. The organisms survive best at lower level and since viable cells are essential for any therapeutic activity the incubation is restricted to get the desired acidity. The coagulum is broken to improve mouthfeel. The organisms retain their viability for around 7 days of storage at 5°C.

L. acidophilus are able to get themselves implanted in the large intestine of human beings through regular consumption of the product and thereby control gastrointestinal disorders such as diarrhoea, dyspepsia, constipation, flatulence, colitis in adults and children, unlike the organisms in dahi which cannot be implanted in the intestine and control intestinal disorders. Some times the *L. acidophilus* is used in conjunction with 25% of yoghurt culture. This reduces fermentation time to around 4 hours. Acidophilus yoghurt can be made by simply adding a concentrated culture of *L. acidophilus* to normal yoghurt. In terms of cell activity the accelerated growth of selected strains could prove advantageous.

Acidophilus paste   It is made by concentrating milk to 23% total solids and fermenting with 5% *L. acidophilus* (Table 45.2).

Table 45.2 Various formulations of acidophilus milk

Product	Microorganism
Acidophilus milk	*L. acidophilus*
Acidophilus paste	*L. acidophilus*
Acidophilus buttermilk	*L. acidophilus* + mesophilic starters
Acidophilus natural buttermilk	*L. acidophilus*+mesophilic starters
Acidophilus yoghurt	*L. acidophilus*+*S. salivarius* subsp. *thermophilus*+*L. delbrueckii* subsp. *bulgaricus*
Acid yeast milk	*L. acidophilus* +lactose fermenting yeasts
Bioghurt	*L. acidophilus*+ *S. salivarius thermophilus*
Acidophilus	*L. acidophilus*+*Lactococcus lactis* subsp. *lactis* +Kefir culture

BIFIDUS MILK

Fermented milks containing bifidobacteria are made either by pure strains of these organisms alone or in combination with other lactic acid bacteria. Bifidium organisms were discovered in faeces of breast-fed infants. The products *of L. acidophilus* dominate in the small intestine whereas *Bifidobacterium* dominates in the large intestine. The preparation involves lactic acid fermentation with a culture of *Bifidobacterium bifidum* and in specific cases *B. longum* in addition. The product has a slightly acetic flavour, pH of 4.3–4.7 with around 10^8/ml bifidobacteria. It can be stored for a week at 5°C. The heat treatment of milk is similar to other products. Inoculum is added at 2% and incubated at 37°C/24 hours. For bifighurt, 6% starter is used and fermented at 42°C/4 hours.

Table 45.3   Various formations of bifidus mik

Product	Microorganisms
Bifidus milk	*B. bifidum / B. longum*
Bifighurt	*B. bifidum + S. salivarius* subsp. *thermophilus*
Biogarde	*B. bifidum + S. salivarius + L. acidophilus*
Biokys	*B.bifidum + L. acidophilus + P. acidilactici*
Cultura	*B. bifidum + L. acidophilus*
Mil mil	*B. bifidum + B. breve + L. acidophilus*

CULTURED BUTTERMILK

It is a fermented milk product made by using mesophilic starters. It is highly nutritious and ideally suited as a supplement to local foods. First grade milk free from antibiotics and detergents has to be used. Milk is homogenized at 200 kg/cm^2 at 15°C and it is heat treated at 90°C/ 13minutes to get satisfactory end product. It is then cooled to 23°C and inoculated with a mixed culture having *Lactococcus lactis* subsp. *lactis* and *L. lactis* subsp. *cremoris* for acid production and *Lactococcus lactis* subsp *diacetylactis, Leuconostoc citrovorum* as primary producers of aroma and flavour. The inoculum may range from 1–2%. When acidity reaches 0.9% the product is gently mixed, cooled, bottled and stored at 5°C. The fermentation time may be around 16–20 hours. The final product is a viscous drink with pleasing aroma and flavour. Fermentation predigests several milk constituents, synthesizes water-soluble vitamins of the B complex and makes a nutritionally upgraded milk. Cultured cream is also made in a similar manner.

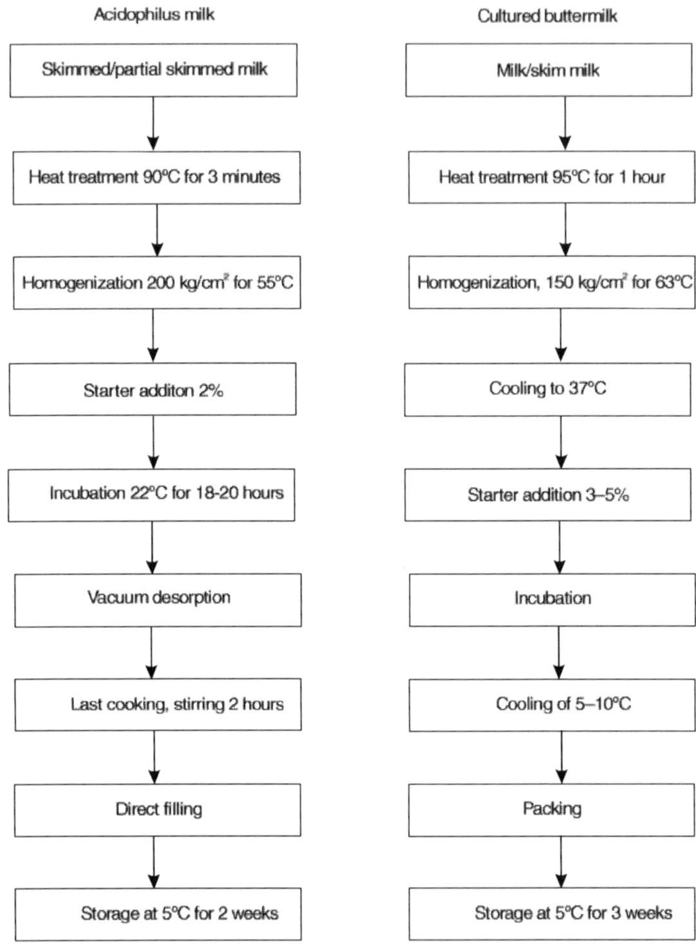

Figure 45.2 Preparation of fermented milk products

MANUFACTURE OF CHEESE

Cheese making is an ancient craft that evolved centuries ago as a method of preserving milk. Currently, there are enormous number of cheese varieties available. (Table 45.4). Many are still produced on a small scale using traditional methods but the most popular cheese, e.g. Cheddar, are manufactured on a large scale using highly technical semi-automated processing.

Table 45.4 Classification of some cheese

Cheese	Microorganisms
Soft, unripened	
Cottage	*Lactococcus lactis*
	Leuconostoc citrovorum
Cream	*Streptococcus cremoris*
Neufchatel	*Streptococcus diacetilactis*
Soft, ripened, 1–5 months	
Brie	*Lactococcus lactis*
	Penicillium candidium
	Streptococcus cremoris
	Penicillium camemberti
	Brevibacterium linens
Camembert	*Lactococcus lactis*
	Streptococcus cremoris
	Penicillium candidium
	Penicillium camemberti
Limburger	*Lactococcus lactis*
	Brevibacterium linens
	Streptococcus cremoris
Semisoft, ripened, 1–12 months	
Blue	*Lactococcus lactis*
	Penicillium roqueforti
	Streptococcus cremoris
	Penicillium glaucum
Brick	*Lactococcus lactis*
	Brevibacterium linens
	Streptococcus cremoris
Gorgonzola	*Lactococcus lactis*
	Penicillium roqueforti
	Streptococcus cremoris
	Penicillium glaucum
Monterey	*Lactococcus lactis*
	Streptococcus cremoris
Muenster	*Lactococcus lactis*
	Brevibacterium linens
	Streptococcus cremoris
Roquefort	*Lactococcus lactis*
	Penicillium roqueforti

(Contd.)

Table 45.4   (Continued)

Cheese	Microorganisms
	Streptococcus cremoris
	Penicillium glaucum
Hard, ripened, 3–12 months	
Cheddar	*Lactococcus lactis*
	Lactobacillus casei
	Streptococcus cremoris
	Streptococcus durans
Colby	*Lactococcus lactis*
	Lactobacillus casei
	Streptococcus cremoris
	Streptococcus durans
Edam	*Lactococcus lactis,*
	Streptococcus cremoris
Gouda	*Lactococcus lactis*
	Streptococcus cremoris
Gruyere	*Lactococcus lactis*
	Lactobacillus helveticus
	Streptococcus thermophilus
	Propionibacterium shermanii or
	Lactobacillus bulgaricus and
	Propionibacterium freudenreichii
Swiss	*Lactococcus lactis*
	Lactobacillus helveticus
	Propionibacterium shermanii or
	Lactobacillus bulgaricus and
	Propionibacterium freudenreichii'
	Streptococcus thermophilus
Very hard, ripened, 12–16 months	
Parmesan	*Lactococcus lactis*
	Lactobacillus bulgaricus
	Streptococcus cremoris
	Streptococcus thermophilus
Romano	*Lactobacillus bulgaricus*
	Streptococcus thermophilus

Pre-treatment of milk   A good quality milk is generally more important for cheese making because it is not possible to pasteurize cheese milk since the growth conditions for the harmful microbes are far better in cheese than any other milk based product. The bacterial content of milk to be used for cheese making must be low, because microbes growing in milk may develop unwanted flavour and enzymes. Besides, some of the organisms may survive pasteurization and cause faults in cheese. The content of psychrotrophs should be low. This is important when milk has to be stored at 5°C for one or more days. The raw milk on arrival at cheese factories may have counts ranging from 10^5 to 10^7/ml depending on the levels of hygiene. Not only the total counts, but composition of the bacterial flora is also very important. The organisms may be mostly pseudomonads, *Aeromonas, Alcaligenes*, some lactic acid bacteria, spore-forming gram-positive rods, coryneforms, micrococci and coliforms. Of these, psychrotrophs may multiply during transport and storage: if the temperature of the insulated tankers increase, may lead to the production of extracellular thermoduric lipases and proteinases. These organisms are usually killed by low temperature pasteurization but the enzymes may cause off-flavour during ripening of cheese. The number of thermoduric organisms should be low. The butyric acid bacterial spores are not killed by ordinary pasteurization. This may cause extensive gas production and develop unwanted flavours. Above all hygienic milk handling becomes a necessity.

Basic stages involved in cheese making are as follows:

1. *Standardization* It is done to adjust fat or to have balanced ratio of fat to casein or increasing fat through cream for making cream cheese.

2. *Clarification* The clarifier is an effective alternative to filtration for the removal of extraneous matter, leucocytes and some bacteria. It is carried out at 32°C–35°C centrifugally.

3. *Bacterofugation* If centrifuged milk is passed through the unit a second time, about 90% of the remaining 10% bacteria is removed. Thus a total of 99% bacteria can be removed but it cannot be relied upon to remove all pathogenic organisms. This process is very much used in preparation of cheese where most of bacteria, yeast and mould can be removed.

4. *Homogenization* In making Feta cheese, milk is homogenized at low pressure. The purpose is to reduce the whey exudation from the coagulum, to make cheese whiter and promote fat hydrolysis. This treatment is useful for soft cheese and blue-veined cheese also.

5. *Thermization* When raw milk must be stored for a few days before using it for cheese making, it is subjected to a heat treatment (63°C/10–15 seconds) and cooled to 5°C prior to storing in insulated silos. It is not a substitute for pasteurization.

6. *Ultrafiltration* By membrane filtration under low pressure, milk can be separated into retentate with fat, casein, whey proteins and bacteria, and a permeate containing lactose, water and mineral salts. The retentate can be used in making UF-Feta, Camembert and other cheeses.

7. *Ripening of milk*   This is achieved by heating the milk to 30–33°C and adding starter bacteria which has the effect of speeding up the natural process of milk souring. The organisms grow using lactose, as an energy source and this is converted to lactic acid by a complex series of reactions involving many different enzymes. Selected strains of lactic acid organisms with predictable acid development and production of flavour products are used as starters to get steady rate of acids throughout the curd making process. Starter, a) ensures consistent acidity development b) aids rennet action and subsequent coagulation by the developed acidity, c) helps expulsion of whey from the curd and d) contributes to flavour, body and texture of cheese during ripening, eye formation in Swiss cheese, besides suppressing the growth of undesirable organisms.

The microorganisms used for cheese making are given in Table 45.5.

Table 45.5   Microorganisms involved in cheese making

Type	Function
Lactococcus lactis subsp. *lactis*	Acid formation
L. *lactis* subsp. *cremoris*	Acid formation
L. lactis subsp. bivar *diacetylactis*	Acid, gas and flavour products
S. *salivarius* subsp. *thermophilus*	Acid in high scald cheese
Lactobacillus delbrueckii subsp. *bulgaricus*	Acid in high scald cheese
Lactobacillus helveticus	Acid in high scald cheese
Streptococcus faecalis	Acid and flavour in high scald cheese
Propionibacterium shermanii	Gas and flavour production

Many non-starter organisms are also used in making soft and semi-soft cheese. They are given in Table 45.6.

Table 45.6   Non-starter organisms used in cheese making

Microorganisms	Function
Penicillium roqueforti	Blue-veined cheese flavour
P. candidium	White mould surface growth
P. camemberti and *P. album*	White/blue mould flavour
Brevibacterium linens	Colour/flavour production

(Contd.)

Table 45.6   (Continued)

Microorganisms	Function
Micrococcus varians	Flavour production
Debaryomyces hansenii	Yeast/flavour production
Geotrichum candidum	Mould/flavour production
Bifidobacterium	Product stabilization

All additives required are added and mixed separately before rennet addition. For example, addition of 15 g of salt petre/100 kg of cheese milk prevents blowing (development of too much gas in the cheese) caused by coliform bacteria or butyric acid bacteria and possibly propionic acid bacteria.

A simplified flowchart for the production of cheddar cheese is given in Figure 45.3.

8. *Rennetting* After an increase in acidity of milk created by starter, rennet extract is added to milk and uniformly distributed to effect coagulation of milk.

The coagulating enzymes available are given in Table 45.7.

Table 45.7   Coagulating enzymes

Type	Source of enzymes
Animal	Calf (chymosin)/(pepsin) Pig (pepsine)
Bacteria	*Bacillus subtilis* *B. mesenteroides* *B. polymyxa*
Fungi	*Mucor meihei* *M. pusilus* *Endothia parasitica*

9. *Cutting the coagulum* Cutting has the greatest effect on moisture expulsion. It releases a large volume of whey, establishes the size of the curd particles and increases the area of curd from which moisture can escape. Stirring maintains the curd particulate form in the whey and makes it possible to cook uniformly and control the expulsion of moisture. Low temperature cooking leads to softer type of cheese. High cooking temperature produces drier curd for slow maturing with long keeping quality. Special knives are used for getting different sizes of cubes.

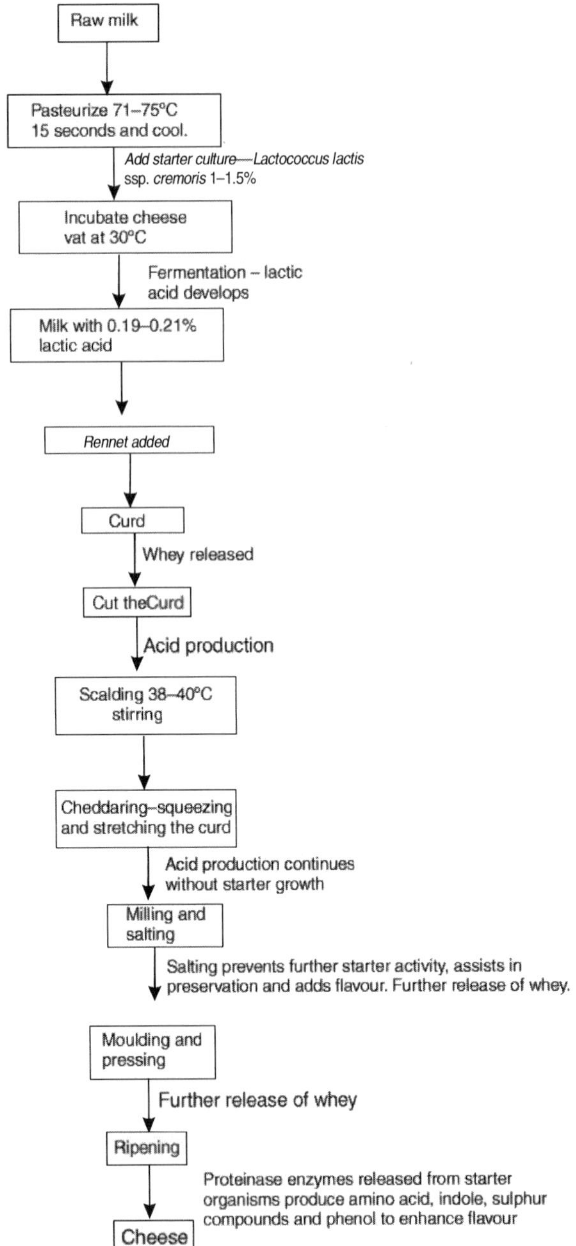

Figure 45.3 Manufacture of cheddar cheese

10. *Scalding* In high scald, the temperature may be 52–58°C, in medium scald 30–42°C and in low scald around 30–35°C. During combined action of stirring and heat, lactic acid within the curd particle is formed by the starter organisms, embedded in cheese particles and the curd cubes shrinks in size. When the desired development in the curd has reached, whey is drained for texturizing the curd. The texture may be close as in Cheddar, plastic as in Emmenthal and elastic as in Edam or Gouda.

11. *Draining the whey* The curd is allowed to settle, acidity is measured and when it has reached the desired level, the whey is run off until a compact mass of curd forms in the vat. This process varies from cheese to cheese.

12. *Milling* Milling of the texturized matted curd is done at 0.6–0.7% acidity (lactic) in Cheddar. Milling helps uniform distribution of salt (1–2%). Salt acts as a preservative and flavour enhancer.

13. *Pressing* The curd is filled in moulds and pressed. The degree of pressing and length of time varies with the type of cheese.

14. *Dressing and package* The green cheese is waxed or packed in heat shrink barrier film (impermeable to oxygen and moisture and permeable to CO_2).

15. *Churning* It is done under controlled humidity and temperature conditions. Occasional turning of the blocks is necessary to avoid deformation of shape, rotting due to condensation between cheese block and shelf and aid in even distribution of moisture throughout the block of cheese.

1. List out the various fermented milk products.
2. Describe the nutritive value of fermentative milk products.
3. Describe the production of yoghurt.
4. Describe the production of acidophilus milk.
5. Give a detailed account of bifidus products.
6. Describe the microbiology of cheese.
7. Describe the basic stages involved in cheese making.

MICROBIOLOGY OF STARTER CULTURES

INTRODUCTION

Starters are carefully selected microorganisms which are deliberately added to milk to initiate and carry out the desired fermentation in the production of fermented milk products. The starter organisms bring about specific changes in the appearance, body, flavour and texture characteristics of the finished products. Selected cultures were available from special laboratories from 19th century onwards.

Starters can be single strain starters, multiple strain starters, mixed strain starters or multiple mixed strain starters depending on the number of strains and on the proportion in which they are present on subculturing the starters. The most common starters are composed of lactic acid bacteria, but other bacteria, yeasts and moulds are in some cases included with them in the production of different fermented products.

ROLE OF STARTERS IN DAIRY FERMENTATIONS

The cultures of bacteria used in the manufacture of cheese and other fermented milk products are known as starters. Starters play a vital role in the manufacture of these products; they produce the lactic acid that influences important quality characteristics such as texture, moisture content, freeing the pathogenic microorganisms and their toxins, and taste. The rate of acid production is critical in the manufacture of certain products, e.g. Cheddar cheese. Depending on the product, especially in mechanized cheese production units, starters may also be required to produce acid at a consistently fast rate through out the manufacturing period. The negative redox potential created by starter growth in cheese also aids in preservation and the development of flavour in Cheddar and similar cheeses. Additionally antibiotic substances, now referred to as bacteriocins produced by starters, e.g. nisin may also have a role in preservation.

ECOLOGY OF STARTER BACTERIA

Most starters in use today have originated from lactic acid bacteria originally present as part of the contaminating microflora of milk. These bacteria have probably originated from vegetation in the case of lactococci or the intestinal tract in the case of *Bifidobacterium* spp. and *Lactobacillus acidophilus*.

Modern starter cultures are developed from the practice of retaining small quantities of whey or cream from the successful fermented product of previous day and using this as the inoculum or starter for the next day's production. This practice has been called by various names but the term "back-slopping" is used widely.

CLASSIFICATION OF STARTER BACTERIA

The bacteria used in the manufacture of fermented dairy products are generally lactic acid bacteria (LAB); however, *Propionibacterium shermanii* and *Bifidobacterium* spp. and, which are not lactic acid bacteria, although *Bifidobacterium* species do produce lactic acid, are also used. In addition, other bacteria including *Brevibacterium linens*, responsible for the flavour of Limburger cheese; and moulds (*Penicillium* species) are used in the manufacture of Camembert, Roquefort and Stilton cheeses.

As molecular biology has developed, genetic methods have increasingly been employed to study the genetic similarities between different bacterial groupings. These techniques have resulted in significant changes in the classification of the LAB. There are now 11 genera, and starter bacteria are found in five groups namely, *Enterococcus, Lactococcus, Lactobacillus, Leuconostoc,* and *Streptococcus*. While there is, some doubt whether bacteria classified within the *Enterococcus* group should be used as starters, they are present in some artisanal cultures and therefore, in practice, starters. There are concerns about enterococci because some are pathogens. However, it is their ability to exchange antibiotic resistance genes, particularly for glycopeptide antibiotics (vancomycin and teicoplanin), that perhaps raises most concern. Vancomycin is one of the antibiotics that may be effective against multiple antibiotic-resistant bacteria.

Enterococcus

These organisms are gram-positive, catalase-negative cocci that tend to form chains of varying length. They are normal inhabitants of the intestinal tract of man and other animals, and are often used in microbiology as indicators of faecal contamination; some species of the genus are pathogens. Apart from their ability to grow at 45°C, at pH 9.6 and in the presence of high concentrations of salt they are similar to lactococci.

Prior to recent taxonomic research, the *Enterococcus* species used as starters were classified as faecal streptococci and Group D streptococci.

Leuconostoc

Leuconostoc species are important flavour producers in some fermented dairy products. There is general agreement that two species, *Leuconostoc mesenteroides* subsp. *cremoris* and *L. lactis* are important in starter cultures. Unlike lactococci, leuconostocs grow on Rogosa agar and are heterofermentative producing carbon dioxide from glucose and fructose. While the carbon dioxide production is undesirable in Cheddar cheese, gas production is desirable in some varieties, e.g. Emmenthal.

On microscopic examination, leuconostocs generally appear as gram-positive cocci similar in size and shape (occur in pairs and in usually short chains) to lactococci. However, small rods can often be found and since leuconostocs grow on Rogosa agar, it may be assumed that these cultures are contaminated, with lactobacilli for example. Unlike lactococci, leuconostocs do not produce ammonia from arginine and produce the D-isomer of lactic acid. With some exceptions leuconostocs grow weakly in milk, and are not capable of reducing litmus before coagulation in litmus milk medium.

Isolation and identification of leuconostocs in starters is time-consuming and laborious. Carbohydrate fermentation and identification of the lactic acid isomer are useful elements in an identification protocol.

Streptococcus

S. thermophilus is the only species of this genus found in starter cultures. This streptococcus is classified as a thermophile, growing at 45°C, and higher, and is widely used in the manufacture of yoghurt, Mozzarella, and in some other cheeses. Now it has been used widely in the manufacture of Cheddar cheese. It is a component, along with lactococci, in some DVI/DVS cultures to produce acid rapidly during scalding and may confer an additional measure of bacteriophage (phage) protection. Its incorporation in Cheddar-cultures also increases the profitability of DVI/DVS cultures to culture-suppliers.

Like lactococci and many leuconostocs, strains of *S. thermophilus* are catalase-negative; cocci-shaped and occur in pairs and chains. Generally, most strains produce long chains only. L-lactic acid only is produced and carbon dioxide is not produced from glucose. Some strains produce urease and have the potential to produce CO_2 from urea. Since *S. thermophilus* can grow in the regeneration section of pasteurizers, high levels can occasionally occur in cheese. Urease-producing strains have the potential to cause openness in cheese.

Strains differ in their ability to utilize galactose. Use of non-galactose fermenting strains will result in high levels of this reducing sugar in products. Since galactose and other reducing sugars react with amino acids in the Maillard reaction it is necessary to select only galactose-utilizing strains to reduce the probability of undesirable colour changes occurring in heated products.

S. salivarius, commonly found in saliva, has been shown by DNA–DNA hybridization studies to be similar to *S. thermophilus*.

S. thermophilus is sensitive to low levels of salts and to high osmotic strength media. M17 medium widely used in studies with lactococci is not an ideal medium for the growth of some strains, unless modified to reduce its osmotic strength by the reduction of its glycerophosphate content.

Lactobacillus

This genus consists of a large group of gram-positive, catalase-negative, rod-shaped bacteria. Some species are homofermentative while others are heterofermentative. While some species produce mainly L-lactate from glucose, others produce D-lactate. Since some strains exhibit significant racemase (an isomerase enzyme) activity, D/L lactic acid is also produced. Strains may also exhibit coccoid morphology and this can lead, as discussed previously, to confusion with leuconostocs and perhaps even lactococci.

Lactobacilli are used as starters in the manufacture of yoghurt, and Mozzarella cheese. They are also used as starter adjuncts to promote faster ripening of Cheddar and similar cheeses, to reduce the incidence of bitterness, and as probiotics in yoghurt type products. *L. delbrueckii* subsp. *bulgaricus* is widely used along with *S. thermophilus* as a starter in yoghurt manufacture. This subspecies is homofermentative, produces almost 2% w/v lactic acid in milk, has an optimum temperature of 42°C and grows at temperatures of 45°C and higher. It will not grow in low concentrations of salt and is sensitive to bile salts.

L. acidophilus, which is normally present in the intestine, is generally not used as a starter; it is widely used as a probiotic. This bacterium, is homofermentative, produces high concentrations of D-lactic acid in milk, has an optimum temperature of 37°C, and is relatively tolerant of oxygen, compared to *Bifidobacterium* species that are frequently used in conjunction with this organism. Little growth occurs at temperatures less than 20°C and most strains show no growth at 15°C. Because *L. acidophilus* produces D-lactate there have been some concerns about its use in infant nutrition.

L. casei is also a normal inhabitant of the small intestine and is resistant to bile. It is used as a probiotic although it is found in some starter cultures and is commonly one of the members of non-starter lactic acid bacteria (NSLAB) found in Cheddar cheese. L-lactate is the main isomer of lactose produced, although some strains produce small concentrations of D-lactate due to weak racemase activity. Rogosa agar is widely used as a general isolation medium for lactobacilli. *L.helveticus* is frequently used along with other thermophilic lactic acid bacteria in the manufacture of a range of fermented milk products including Emmenthal cheese, Mozzarella and yoghurt. One advantage of including this species along with *L. delbrueckii* subsp. *bulgaricus* is that *L.helveticus* utilizes galactose and this can be useful if products free of reducing sugars are required. Since many strains have been shown to possess proline-iminopeptidase-like activity, *L. helveticus* has been used to produce modified "Cheddar-type cheese" with some of the 'sweetness' characteristics of Swiss cheeses like Emmenthal. More recently, designated strains have been used as starter adjuncts to reduce bitterness in a range of cheeses, to improve flavour and/or to accelerate ripening. Bitterness is reduced due to peptidase action on starter-derived hydrophobic peptides. The species is homofermentative and produces high concentrations of D/L lactic acid in milk. Many strains grow, at 45°C although lower temperatures 42–43°C generally give higher recoveries when enumerated using selective media such as Rogosa or modified MRS agars. Most strains show no or little growth at 15°C (some atypical strains may take several weeks to grow at 15°C or below).

Lactococcus

Originally, the bacteria in this group were classified as members of the genus *Streptococcus*. They were differentiated from other streptococci, some of which are pathogens, by their specific reaction with Group N antiserum and by their tolerance to temperature, salt and dyes. It is now known that serotyping LAB has limited value in species differentiation; strains of the same species may react with different sera and some strains may exhibit no group antigen.

Lactic Streptococci

The lactic group of the genus *Streptococcus* originally included the species *S. lactis* and *S. cremoris* and a subspecies of *S. lactis*, *S. lactis* subsp. *diacetylactis*. However, even in the 1970s it was suggested that *S. lactis* strains might be variants of *S. diacetylactis* that were unable to ferment citric acid, since citrate permease-negative strains of *S. diacetylactis* had been described. Bacteria in this group were designated as the lactic streptococci. The designation "lactic" was used by Sherman (1937) for mainly historical reasons, including the use of the term by Lister (1878) to describe a bacterium that we now know as *L. lactis* subsp. *lactis*.

Evidences suggest that *S. cremoris*, *S. lactis* and *S. diacetylactis* formed a phenotypically and genotypically continuous spectrum of variation. The three so-called species were known to be similar in DNA base-composition indicating that they are closely related. Genetic relationships between the genus *Streptococcus* and the genera *Lactobacillus* and *Leuconostoc* were becoming apparent.

Rationale for assigning lactic streptococci to Lactococcus genus   The lactic streptococci have been assigned to the *Lactococcus* genus because of molecular and chemotaxonomic studies. Essentially the extent of DNA–DNA and DNA–rRNA hybridization, similarity between the profiles produced by restriction mapping of chromosomal DNA and the nucleotide sequence of the 16S and 32S RNAs have formed the basis for the creation of the genus. Antisera against purified superoxide dismutase have been used to demonstrate similarity between lactococci but not streptococci or enterococci.

Because citrate utilization is plasmid DNA coded and not coded on the cell chromosome , citrate utilization and the associated diacetyl production are variable traits and the former *S. diacetylactis* has not been given the subspecies status.

This genus, contains one important starter species *Lactococcus lactis*. This species consists of gram-positive, catalase-negative, homofermentative cocci that normally form short to long chains. Only L-lactic acid is formed from lactose. *Lactococcus lactis* is divided into two subspecies *L. lactis* subsp. *lactis* and *L. lactis* subsp. *cremoris* and a biovariant of *L. lactis* subsp. *lactis*, designated variously as either *L. lactis* subsp. *lactis* biovar. *diacetylactis* or cit(+). *L. lactis* subsp. *lactis*. The latter is identical to *L. lactis* subsp. *lactis* apart from its ability to utilize citrate in the presence of carbohydrate.

TYPES OF STARTER CULTURE

Single-strain Starters

Pure cultures consist of only one species of lactic acid bacteria, whereas mixed cultures have several species. Pure cultures may have one or more strains of the same species. A pure culture can be used in conjunction with a mixed culture. Some organisms like *Lactococcus lactis* only ferment the lactose to lactic acid, some others, besides, produce aroma compounds.

Multi-strain Starters

Defined mixtures of three or more single strains of *S. cremoris* and/or *S. lactis*. *Leuconostoc* and *S. diacetylactis* strains may also be used. Multiple starters are composed of strains which are mutually competitive with each other and should have strains that can be maintained in the proper ratio of lactic acid to flavour production. Lactic streptococci are homofermentative in nature and they produce lactic acid from lactose. On the other hand leuconostoc species are heterofermentative and produce acetic acid, lactic acid, ethanol and CO_2 besides aroma compounds.

Mesophilic Starter Cultures

Mesophilic lactic cultures growing at an optimum of 30°C contains group N streptococci and/or leuconostocs. They are composed of mesophilic organisms which possess acid production, gas formation and production of enzyme activity for cheese ripening, i.e., proteases and peptidases. They play an important role in ripened cream butter where diacetyl production is an important requirement. A variety of fermented milks are produced worldwide by using mesophilic starters made up of streptococci and leuconostocs by themselves or in mixtures with other microbes and yeasts. Types of starter cultures are either lactic acid producing microorganisms and/or flavour producers. The types of starters are referred to as K, L-(B), or DL(BD) cultures. O type cultures contain no aroma and are mainly used for cheese making. Traditional mesophilic starters of unknown composition are still used in many countries, the reason being that they are less sensitive to bacteriophages. The stability of these multiple strain starters due to symbiotic relationships has also been considered as an advantage.

Thermophilic Starters

Thermophilic starters were developed many years ago in the Middle East area with its subtropical climate, where, in summer the temperature can be 40°C or more. Under adverse conditions of milk production, the quality of fermented product varied very much.

The thermophilic cultures used in the production of yoghurt and similar fermented milks are composed of obligate homofermentative lactobacilli and streptococci. The organisms used are *Lactobacillus delbrueckii* subsp. *bulgaricus* and *Streptococcus salivarius* subsp. *thermophilus*. These two LAB grow in association with milk and form typical yoghurt starter culture. This growth is symbiotic because the rate of acid development is greater when the two microbes are grown together as compared to the single strain.

L. delbrueckii subsp. *bulgaricus* stimulate *S. salivarius* subsp. *thermophilus* by releasing several amino acids (leucine, lysine, cystine, valine and histidine) while *S. thermophilus* produces formic acid like compounds which promote the development of *L. delbrueckii* subsp. *bulgaricus*. Other stimulatory compounds for yoghurt starter cultures are peptides, purine, oxaloacetic acid, and fumaric acid, orthophosphates, denatured whey proteins, CO_2 and partially hydrolysed casein.

Kefir Starters

Certain dairy products are produced using natural microflora which contains yeasts as an essential component. These products differ considerably in physical, biological and microbiological properties from those produced with pure cultures of LAB. Yeasts exert a favourable effect on the activity of LAB by providing them the growth stimulants as well as by metabolizing some of the lactic acid. Yeasts isolated from traditional cultured dairy products manifest a high antibiotic activity against tuberculosis. Kefir has high nutritional, biological and dietetic value and are widely recommended for people, particularly patients with gastrointestinal and metabolic diseases.

Kefir Grains

The basic microflora of the kefir grains is composed of LAB, acetic acid bacteria and yeasts. Non-lactose fermenting yeasts are found in the deep layers of the kefir grains while lactose-fermenting yeasts are present mainly in the peripheral layers. The surface microflora of kefir grains comprises mesophilic lactic acid streptococci, mesophilic and thermophilic lactobacilli and acetic acid bacteria. Coliforms are inhibited by natural kefir microflora and pathogenic *Shigella* and *Salmonella* do not grow when they are introduced in milk together with kefir starters.

The microflora of kefir starter, which is prepared by growing kefir grains in milk, consists of different groups of microorganisms. Mesophilic homofermentative lactic acid streptococci (*Lactococcus lactis*, *Lactococcus lactis* subsp. *cremoris*) form the longest and highly active part of kefir starter microflora. This produces rapid acidity development during the first few hours of fermentation. It also contains *S. durans*.

Lactobacilli are constantly present in kefir grains. The number of mesophilic lactobacilli is less than 10^2–10^3 cells/ml. Species most frequently found are obligate heterofermentative *L. brevis*, the facultative heterofermentative *L. casei* subsp. *rhamnosus,* and obligate homofermentaive *L. delbrueckii* subsp. *bulgaricus* and *L. helveticus*. Mesophilic heterofermentative lactic acid streptococci like the *Leuconostoc mesenteroides* and *Leuconostoc mesenteroides dextranicum* provide the specific taste and aroma of kefir. Yeasts like *Kluyveromyces marxianus* subsp. *marxianus* are found in kefir grain. Yeasts play an important role by promoting symbiosis among microbes, in CO_2 formation and the development of characteristic taste and aroma. Acetic acid bacteria like *Acetobacter aceti* and *Acetobacter rasens* play an important role in maintaining symbiosis among the kefir grain microflora.

1. What are starter cultures?
2. What is the role of starter cultures?
3. Brief out on the classification of starter bacteria.
4. What are the types of starter cultures?
5. What are kefir grains?

GLOSSARY

Attenuation Refers to the percentage of sugars converted to alcohol and carbon dioxide, as measured by specific gravity.

Avidin A glycoprotein present in egg albumen.

Bacteriocin Large and diverse group of ribosomally synthesized extracellular antimicrobial proteins or peptides which have a bactericidal or bacteriostatic effect on other closely related bacteria.

Baroduric Those capable of enduring high pressures but grow well at atmospheric pressures.

Biopreservation Extending the storage life and enhancing the safety of food products by using natural or controlled microflora, mainly LAB and/or their antibacterial products.

Botulinum cook A heat process which reduces the population of the most resistant *C. botulinum* spores by an arbitrarily established factor of 12 decimal values, i.e., killing of cells by 12 log cycles.

Chilling Storage at temperatures above freezing, i.e., 16°C down to –2°C.

Cryoprotective agents Substances that protect the microbial cell during freezing and thawing, e.g. glycerol and dimethyl sulphoxide.

D value Time taken to kill 90% of the microorganisms at a certain temperature.

Electronic asteurization Treatment of food using either X-rays or electron beams.

F value The time necessary to destroy a given number of microorganisms at a standard temperature (usually 121°C for spores, 60°C for vegetative cells) having a specific Z value.

Fermented foods Food substrates that are invaded or overgrown by edible microorganisms whose enzymes, particularly amylases, proteases and lipases hydrolyse the polysaccharides, proteins and lipids respectively to non-toxic products with flavours, aromas and textures pleasant and attractive to the human consumer.

Flat A can with both ends concave; it remains in this condition even when the can is brought down sharply on its end on a solid, flat surface.

Flipper A can that normally appears flat; when brought down sharply on its end on a flat surface, one end flips out. When pressure is applied to this end, it flips in again and the can appears flat.

Food antimicrobials Chemical compounds added to or present in foods that retard microbial growth or kill microorganisms thereby resisting deterioration in safety or quality.

Food-borne disease A disease, usually either infectious or toxic in nature, caused by agents that enter the body through the ingestion of food.

Food hygiene The sanitary science that aims to produce food which is safe for the consumer and of good keeping quality.

Food irradiation Microbial destruction using ultraviolet, ionizing radiation and microwave heating without the generation of high temperature and is also referred to as cold sterilization.

Food poisoning Acute gastroenteritis due to bacterial contamination of food or drink.

Food safety The conditions and measures that are necessary during the production, processing, storage, distribution and preparation of food to ensure that it is safe, sound, wholesome and fit for human consumption.

Gamma rays Radioactive fission products of cobalt-60 and cesium-137.

Good manufacturing practice A preventive system with specifications, standards and procedures according to which the manufacturing process is to be carried out.

Hard swell A can bulged at both ends, and so tightly that no indentation can be made with thumb pressure. A hard swell will generally "buckle" before the can bursts. Bursting usually occurs at the double seam over the side seam lap, or in the middle of the side seam.

Heat penetration Ability to transfer heat to every part of the food such that it becomes sterile inside the container.

Heat resistance The capacity to withstand heat. It is generally expressed as thermal death time.

Heterofermentative LAB Bacteria which ferment glucose to produce only 50% lactic acid and considerable amounts of ethanol, acetic acid and carbon dioxide.

High hydrostatic pressure The technology by which a product is treated at or above 100 MPa.

Homofermentative LAB Bacteria which ferment glucose to produce more than 85% lactic acid.

Implicit factors Mutual influences, synergistic or antagonistic, among the primary selection of organisms resulting from the influence of intrinsic parameters of the food.

Intrinsic factors Factors that are inherent to the food itself.

Lagering A process that involves secondary fermentation of remaining fermentable extract at a reduced rate controlled by low temperatures and low yeast count.

Malting A process that converts insoluble starch to soluble starch, reducing complex proteins, generating nutrients for the development of yeast and its enzymes.

Manothermosonication A new preservation technique that combines heat and ultrasound to decrease the intensity of heat treatments.

Mashing The process of converting starch from the milled malt and solid adjuncts into fermentable and nonfermentable sugars to produce wort of the desired composition.

Metabiosis The tendency of an organism to rely on another to produce a favourable environment.

Metabiotic spoilage association Two or more microbial species contributing to spoilage through exchange of metabolites or nutrients.

Metacryotic fluid The unfrozen concentrated solutes of sugars and salts oozing out from packages as a viscous material.

Microwave heat treatment Heating due to electromagnetic radiowaves that generate internal heat.

Modified atmospheric packaging The enclosure of food products in gas-barrier materials, in which the gaseous environment has been changed.

Ohmic heating Direct electric heating of food particles to destroy microbes.

Pasteurization A term given to heat processes in the range of 60–80°C for few minutes followed by immediate cooling.

Pulsed electric field The passage of high voltage in short bursts into foods placed between two electrodes.

Quick freezing Frigid air (cool air) blown across the materials being frozen.

Quorum sensing The ability to regulate gene expression as a function of cell density.

Radicidation Low-level irradiation treatment to destroy viable non-spore-forming pathogenic organisms to reduce the problem of food-borne illness.

Radurization The use of low doses of a radiation to destroy a sufficient number of microbes and enhance the shelf life of the product.

Sauerkraut The clean, sound product with characteristic flavour, obtained by full fermentation, chiefly lactic, of properly prepared and shredded cabbage in the presence of 2–3% salt.

Soft swell A can bulged at both ends, but not so tightly that the ends cannot be pushed in somewhat with thumb pressure.

Specific spoilage organism Spoilage microflora composed of microorganisms that have contributed to the spoilage i.e., single species being responsible for spoilage.

Spoilage potential The ability of a pure culture to produce the metabolites that are associated with the spoilage of a particular product.

Springer A can with one end permanently bulged. When sufficient pressure is applied to this end, it will flip in, but the other end will flip out.

Starter cultures Carefully selected microorganisms which are deliberately added to food substrate to initiate and carry out the desired fermentation in the production of fermented food products.

Still air freezing Freezing in air with only natural air circulation or electric fans.

Thermoradiation Combination of heat and radiation.

Vinegar The product of a mixed fermentation of yeast followed by acetic acid bacteria.

X-rays The rays generated when electrons from an electron beam bombard a heavy metal target such as tungsten.

Yoghurt A fermented milk product having viscous consistency, a strong acidulous taste due to high acidity (pH 4.6) and a distinct aroma caused mainly by acetaldehyde.

Z value The temperature required to reduce the D value by tenfold.

REFERENCES

A´ lvarez, I., Man˜ as, P., Condo´N, S. and Raso, J. (2003a). "Resistance variation of *Salmonella enterica* serovars to pulsed electric field treatments." *J. Food Sci.* 68: 2316–2320.

A´ lvarez, I., Man˜ as, P., Sala, F.J. and Condo´N, S. (2003b). "Inactivation of *Salmonella enteritidis* by ultrasonic waves under pressure at different water activities."*Appl. Environ. Microbiol.* 69: 668–672.

A´ lvarez, I., Raso, J., Sala, F.J. and Condo´n, S. (2003c). "Inactivation of *Yersinia enterocolitica* by pulsed electric fields." *Food Microbiol.* 20: 691–700.

Ababouch, L.H., Grimit, L. Eddafry, R.and Busta, F.F. (1995). "Thermal inactivation kinetics of *Bacillus subtilis* spores suspended in buffer and in oils." *J. Appl. Bacteriol.* 78: 669–676.

Abee, T. and Wouters, J.A. (1999). "Microbial stress response in minimal processing." *Int. J. Food Microbiol.* 50: 65–91.

Abee, T., Krockel, L. and Hill, C. (1995). "Bacteriocins: modes of action and potentials in food preservation and control of food poisoning." *International Journal of Food Microbiology.* 28: 169–185.

Acott, K., Sloan, A.E. and Labuza, T.P. (1976). "Evaluation of antimicrobial agents in a microbial challenge study for an intermediate moisture dog food." *J. Food Sci.* 41: 541–546.

ACSH. 1988. *Irradiated food.* American Council on Science and Health, New York.

Aersten, A., Vanoirbeek, K., De Spiegeleer, P., Sermon, J., Hauben, K., Farewell, A., Nystro¨m, T. and Michiels, C.W. (2004). "Heat shock protein-mediated resistance to high hydrostatic pressure in *Escherichia coli.*" *Appl. Environ. Microbiol.* 70: 2660–2666.

Ahmed, F.I.K. and Russell, C. (1975). "Synergism between ultrasonic waves and hydrogen peroxide in the killing of micro-organisms." *Journal of Applied Bacteriology* 39: 31–40.

Allison, D.G., D'Emanuele, A., Egington, P. and Williams, A.R. (1996). "The effect of ultrasound on *Escherichia coli* viability." *J. Basic Microbiol.* 36: 3–11.

Alpas, H., Kalchayanand, N., Bozoglu, F. and Ray, B. (2000). "Interactions of high hydrostatic pressure, pressurization temperature and pH on death and injury of pressure-resistant and pressure-sensitive strains of foodborne pathogens." *Int. J. Food Microbiol.* 60: 33–42.

Alpas, H., Kalchayanand, N., Bozoglu, F., Sikes, A., Dunne, C.P. and Ray, B. (1999). "Variation in resistance to hydrostatic pressure among strains of food-borne pathogens." *Applied and Environmental Microbiology.* 65: 4248–4251.

Alpas, H., Kalchayanand, N., Bozuglu, F. and Ray, B. (1998). "Interaction of pressure, time and temperature of pressurization on viability loss of *Listeria innocua.*" *World J. Microbiol. Biotech.* 14: 251–253.

Ames, J.M. (1998). "Applications of the Maillard reaction in the food industry." *Food Chemistry.* 62(4): 431–439.

Ananth, V., Dickson, J.S., Olson, D.G. and Murano, E.A. (1998). "Shelf life extension, safety, and quality of fresh pork loin treated with high hydrostatic pressure." *J. Food Prot.* 61: 1649–1656.

Ananth, V., Murano, E.A. and Dickson, J.S. (1995). "Shelf life extension and safety of fresh pork treated with high hydrostatic pressure." *J. Food Prot.* 58 (suppl. book of abstracts). p. 8.

Andrews, L.S., Ahmedna, M., Grodner, R.M. et al. (1998). "Food preservation using ionizing radiation." *Rev. Environ. Contam. Toxicol.* 154: 1–53.

Arce-Garcia, M.R., Jimenez-Munguia, M.T., Palou, E. and Lopez-Malo, A. (2002). Ultrasound treatments and antimicrobial agents effects on *Zygosaccharomyces rouxii*. IFT Annual Meeting Book of Abstracts, 2002, Session 91E-18.

Aronsson, K. and Ro¨nner, U. (2001). "Influence of pH, water activity and temperature on the inactivation of *Escherichia coli* and *Saccharomyces cerevisiae* by pulsed electric fields." *Innovat. Food Sci. Emerg. Technol.* 2: 105–112.

Asaka, M. and Hayashi, R. (1991). "Activation of polyphenoloxidase in pear fruits by high pressure treatment." *Agr. Biol. Chem.* 55: 2439–2440.

Ashie, I.N.A. and Simpson B.K. (1996). "Application of high hydrostatic pressure to control enzyme related fresh seafood texture deterioration." *Food Res. Int.* 29: 569–575.

Balaban, M.O., Kincal, D., Hill, S., Marshall, M.R. and Wildasin, R. (2001). The synergistic use of carbon dioxide and pressure in nonthermal processing of juices. IFT Annual Meeting Book of Abstracts, 2001, Session 6-3.

Balassa, G., Milhaud, P., Raulet, E., Silva, M.T. and Sousa, J.C. (1979). "A *Bacillus subtilis* mutant requiring dipicolinic acid for the development of heat-resistant spores." *J. Gen. Microbiol.* 110: 365–379.

Balasubramaniam, V.M., Balasubramaniam, S. and Reddy, N.R. (2001). "Effect of pH and temperature on spore inactivation during high pressure processing." IFT Annual Meeting Book of Abstracts, 2001, Session 59H-5.

Balny, C. and Masson, P. (1993). "Effects of high pressure on proteins." *Food Reviews International.* 9: 611–628.

Banga, J.R., Alonso, A.A., Gallardo, J.M. and Perez-Martin, R.I. (1993). "Mathematical modelling and simulation of the thermal processing of anisotropic and non-homogeneous conduction-heating canned foods: application to canned tuna." *J. Food Eng.* 18: 369–387.

Barbosa-Canovas, G.V., Gongora-Nieto, M.M., Pothakamury, U.R. and Swanson, B.G. (1999). *Preservation of Foods with Pulsed Electric Fields.* Academic Press, San Diego.

Barbosa-Canovas, G.V., Pothakamury, U.R., Palou, E. and Swanson, B.G. (1998). *Nonthermal Preservation of Foods.* Marcel Dekker, New York.

Bargiota, E., Rico-Muñoz, E. and Davidson, P.M. (1987). "Lethal effect of methyl and propyl parabens as related to *Staphylococcus aureus* lipid composition." *Int. J. Food Microbiol.* 4: 257–266.

Beaman, T.C. and Gerhardt, P. (1986). "Heat resistance of bacterial spores correlated with protoplast dehydration, mineralization, and thermal adaptation." *Appl. Environ. Microbiol.* 52: 1242–1246.

Beaman, T.C., Pankratz, H. S. and Gerhardt, P. (1988). "Heat shock affects permeability and resistance of *Bacillus stearothermophilus* spores." *Appl. Environ. Microbiol.* 54: 2515–2520.

Benito, A., Ventoura, G., Casadei, M., Robinson, T. and Mackey, B. (1999). "Variation in resistance of natural isolates of *Escherichia coli* O157 to high hydrostatic pressure, mild heat and other stresses." *Appl. Environ. Microbiol.* 65: 1564–1569.

Beuchat, L.R. (1981). "Microbial stability as affected by water activity." *Cereal Foods World.* 26: 345–349.

Blumeathal, D. (1990). *Food irradiation: Toxic to bacteria, safe for humans.* FDA Consumer, v. 24, Department of Health and Human Services.

Bower, C.K. and Daeschel, M.A. (1999). "Resistance responses of microorganisms in food environments." *Int. J. Food Microbiol.* 50: 33–44.

Brown, K.L., Ayres, C.A., Gaze, J.E. and Newman, M.E. (1984). "Thermal destruction of bacterial spores immobilized in food/alginate particles." *Food Microbiol.* 1: 187–198.

Brudzinski, L. and Harrison, M.A. (1998). "Influence of incubation conditions on survival and acid tolerance response of *Escherichia coli* O157:H7 and non-O157:H7 isolates exposed to acetic acid." *J. Food Protect.* 61: 542–546.

Brul, S. and Coote, P. (1999). "Preservative agents in foods. Mode of action and microbial resistance mechanisms." *Int. J. Food Microbiol.* 50: 1–17.

Brul, S., Coote, P., Oomes, S., Mensonides, F., Hellingwerf, K. and Klis, F. (2002). "Physiological actions of preservative agents: prospective of use of modern microbiological techniques in assessing microbial behaviour in food preservation. *Int. J. Food Microbiol.* 79: 55–64.

Brul, S., Klis, F.M.S., Oomes, J.C.M., Montijn, R.C., Schuren, F.H.J., Coote, P. and Hellingwerf, K.J. (2002). "Detailed process design based on genomics of survivors of food preservation processes." *Trends Food Sci. Technol.* 15: 325–333.

Buchanan, R.L., Edelson, S.G. and Boyd, G. (1999). "Effects of pH and acid resistance on the radiation resistance of enterohaemorrhagic *Escherichia coli.*" *J. Food Prot.* 62: 219–228.

Buchanan, R.L., Edelson-Mammel, S.G., Boyd, G. and Marmer, B.S. (2004). "Influence of acidulant identity on the effects of pH and acid resistance on the radiation resistance of *Escherichia coli* O157:H7." *Food Microbiol.* 21: 51–57.

Burgos, J., Ordonez, J.A. and Sala, F. (1972). "Effect of ultrasonic waves on the heat resistance of *Bacillus cereus* and *Bacillus licheniformis* spores." *Applied Microbiology.* 24: 497–498.

Butz P., Funtenberger, S., Haberditzl, T. and Tauscher, B. (1996). "High pressure inactivation of *Byssochlamys nivea* ascospores and other heat resistant moulds." *Lebensm. Wiss. u Technol.* 29: 404–410.

Butz, P. and Tauscher, B. (2002). "Emerging technologies: chemical aspects." *Food Research International.* 35: 279–284.

Butz, P., Edenharder, R., Fister, H., Tauscher, B. 1997. "The influence of high pressure processing on antimutagenic activities of fruit and vegetable juices." *Food Res. Int.* 30: 287–291.

Calderon-Miranda, M.L., Barbosa-Canovas, G.V. and Swanson, B.G. (1999a). "Inactivation of *Listeria innocua* in skim milk by pulsed electric fields and nisin." *International Journal of Food Microbiology.* 51: 19–30.

Calderon-Miranda, M.L., Barbosa-Canovas, G.V. and Swanson, B.G. (1999b). "Inactivation of *Listeria innocua* in liquid whole egg by pulsed electric fields and nisin." IFT Annual Meeting Book of Abstracts, Session 83A-5.

Cano, M.P., Hernandez, A. and Ancos, B.De. (1997). "High pressure and temperature effects on enzyme inactivation in strawberry and orange products." *J. Food. Sci.* 62: 85–88.

Capellas, M., Mor, M.M., Sendra, E., Pla, R. and Guamis B. (1996). "Populations of aerobic mesophils and inoculated *E. coli* during storage of fresh goat's milk cheese treated with high pressure." *J. Food Prot.* 59: 582–587.

Carlez, A., Cheftel, J.C., Rosec, J.P., Richard, N., Saldana, J.L. and Balny, C. (1992). "Effects of high pressure and bacteriostatic agents on the destruction of *Citrobacter freundii* in minced beef muscle." In: *High Pressure and Biotechnology.* Balny, C., Hayashi, R., Hermans, K.,Masson P. (eds.). John Libbey and Co. Ltd., London. pp 365–368.

Carlez, A., Rosec, J.P., Richard, N. and Cheftel, J.C. (1993). "High pressure inactivation of *Citrobacter freundii, Pseudomonas fluorescens* and *Listeria innocua* in inoculated minced beef muscle." *Lebensm. Wiss. u Technol.* 26: 357–363.

Carlez, A., Rosec, J.P., Richard, N. and Cheftel J.C. (1994). "Bacterial growth during chilled storage of pressure-treated minced meat." *Lebensm. Wiss. u Techno.* 27: 48–54.

Carlez, A., Veciana, N.T. and Cheftel, J.C. (1995). "Changes in color and myoglobin of minced beef due to high pressure processing." *Lebensm. Wiss. u Technol.* 28: 528–538.

Carpi, G., Gola, S., Maggi, A., Rovere, P. and Buzzoni M. (1995). "Microbial and chemical shelf-life of high-pressure treated salmon cream at refrigeration temperatures." *Ind. Conserve.* 70: 386–397.

Casadei, M.A., Man˜ as, P., Niven, G.W., Needs, E. and Mackey, B.M. (2002). "Role of membrane fluidity in pressure resistance of *Escherichia coli* NCTC Z8164." *Appl. Environ. Microbiol.* 68: 5965–5972.

Castellari, M., Matricardi, L., Arfelli, G., Rovere, P. and Amati, A. (1997). "Effects of high pressure processing on phenoloxidase activity of grape musts." *Food Chem.* 60: 647–649.

Castro, A.J., Barbosa-Canovas, G.V. and Swanson, B.G. (1993). "Microbial inactivation of foods by pulsed electric fields." *Journal of Food Processing and Preservation.* 17: 47–73.

Cazemier, A.E., Wagenaars, S.F. and ter Steeg, P.F. (2001). "Effect of sporulation and recovery medium on the heat resistance and amount of injury of spores from spoilage bacilli." *J. Appl. Microbiol.* 90: 761–770.

Cerf, O. (1977). "Tailing of survival curves of bacterial spores." *J. Appl. Bacteriol.* 42: 1–19.

Cheah, P.B. and Ledward, D.A. (1996). "High pressure effects on lipid oxidation in minced pork." *Meat Sci.* 43: 123–134.

Cheah, P.B. and Ledward, D.A. (1997). "Inhibition of metmyoglobin formation in fresh beef by pressure treatment." *Meat Sci.* 45: 411–418.

Cheftel, J.C. (1995). "Review: High pressure, microbial inactivation and food preservation." *Food Sci. Technol. Int.* 1: 75–90.

Cheftel, J.C. (1996). "Changes in color and myoglobin of minced beef meat due to high pressure processing." In: Process optimisation and minimal processing of foods CIPA-CT94-0195, Dec. Warsaw Agr. Univ., Poland. pp. 149–150.

Cheftel, J.C. and Culioli, J. (1997). "Effects of high pressure on meat: a review." *Meat Sci.* 46: 211–236.

Cherry, J.P. (1999). "Improving the safety of fresh produce with antimicrobials." *Food Technol.* 53(11): 54–59.

Chipley, J.R. (1993). "Sodium benzoate and benzoic acid." In: *Antimicrobials in Foods,* 2nd edn. Davidson, P.M. and Branen, A.L. (eds.). Marcel Dekker, Inc., New York. pp. 11–48

Chung, W. and Hancock, R.E.W. (2000). "Action of lysozyme and nisin mixtures against lactic acid bacteria." *International Journal of Food Microbiology.* 60: 25–32.

Cleveland, J., Montville, T.J., Nes, I.F. and Chikindas, M.L. (2001). "Bacteriocins: safe, natural antimicrobials for food preservation." *International Journal of Food Microbiology.* 71: 1–20.

Clouston, J.G. and Wills, P.A. (1969). "Initiation of germination and inactivation of *Bacillus pumilus* spores by hydrostatic pressure." *J. Bacteriol.* 97: 684–690.

Cords, B.R. and Dychdala, G.R. (1993). In: *Antimicrobials in Foods,* 2nd edn. Davidson, P.M. and Branen, A.L. (eds.). Marcel Dekker, Inc., NewYork. pp. 469–537.

Crawford, Y.J., Murano, E.A., Olson, D.G. and Shenoy, K. (1996). "Use of high hydrostatic pressure and irradiation to eliminate *Clostridium sporogenes* spores in chicken breast." *J. Food Prot.* 59: 711–715.

Crawford, Y.J., Murano, E.A., Olson, D.G. and Shenoy, K. (1996). "Use of high hydrostatic pressure and irradiation to eliminate *Clostridium sporogenes* spores in chicken breast." *Journal of Food Protection.* 59: 711–715.

Cross, A.S. (1990). "The biological significance of bacterial encapsulation." In: *Current Topics in Microbiology and Immunology: Bacterial Capsules.* Vol.150. Jann, K. and Kann, B., (eds.). Springer-Verlag, Berlin. pp. 86–95.

Cruess, W.V. (1966). *Commercial Fruit and Vegetable Products,* 4th edn. McGraw-Hill, New York.

Drnenburg, H., Fenselau, K., Wiesner, P. and Knorr, D. (1996). "Effect of high pressure treatment on the activity of pectinase, polyphenol oxidase and peroxidase of a plant cell culture from *Lycopersicon esculsentum* (abstr.)." In: High Pressure Processing of Foods Symp., 8-9, Nov., 3. Cologne, Germany.

Dantzer, W.D., Hermawan, N. and Zhang, Q.H. (1999). "Standardized procedure for inactivation of spores by pulsed electric fields." IFT Annual Meeting Book of Abstracts, 1999, Session 83A-9.

Datta, N. and Deeth, H.C. (1999). "High pressure processing of milk and dairy products." *Australian Journal of Dairy Technology.* 54: 41–48.

Davidson, P.M. (2001). "Chemical preservatives and natural antimicrobial compounds." In: *Food Microbiology: Fundamentals and Frontiers,* 2nd edn. Doyle, M.P., Beuchat, L.R. and Montville, T.J., (eds.). ASM Press, Washington, D.C.

Davidson, P.M. and Branen, A.L. (1993). *Antimicrobials in Foods,* 2nd edn. Marcel Dekker, Inc., N.Y.

DeRuiter, F.E. and Dwyer, J. (2002). "Consumer acceptance of irradiated foods: dawn of a new era?" *Food Science Technology.* 2: 47– 58.

Desrosier, N.W. (1970). *The Technology of Food Preservation.* The AVI Publishing Company, Inc., Connecticut. pp. 123–159.

Deuchi T. and Hayashi, R. (1992). "High pressure treatments at sub-zero temperature: application to preservation, rapid freezing and rapid thawing of foods." In: *High Pressure and Biotechnology.* Balny, c., Hayashi R., Hermans, K., Masson, P. (eds.). John Libbey and Co. Ltd., London. pp. 353–355.

Deuchi, T. and Hayashi, R. (1990). "A new approach for food preservation: use of non-freezing conditions at subzero temperature generated under moderate high pressure." In: *Pressure Processed Food Research and Development.* Hayashi, R. (ed.). San-Ei Shuppan Co., Kyoto, Japan. 37–51.

Deuchi, T. and Hayashi, R. (1991). "Pressure-application to thawing of frozen foods and to preservation under subzero temperature." In: *High Pressure Science for Food.* Hayashi, R. (ed.). San-Ei Shuppan Co., Kyoto, Japan. pp. 101–110.

Dickson, J.S. (2001). "Radiation inactivation of microorganisms." In: *Food irradiation: principles and applications.* Molins, R.A., (ed.). John Wiley, New York. pp. 23–36.

Dickson, J.S. (2001). "Radiation inactivation of microorganisms." In: *Food Irradiation: Principles and Applications.* Molins, R., (ed.). Wiley, New York. pp. 23– 35.

Diehl, J.F. (1983). "Radiolytic effects on foods." In: *Preservation of Foods by Ionising Radiation.* Vol.1. *Josephson,* E.S. and Peterson, M.S., (eds.). CRC Press Inc. Boca Raton, Florida. pp.279–357

Diehl, J.F. (1992). "Food irradiation: Is it an alternative to chemical preservatives?" *Food Additives and Contaminants.* 9: 409–416.

Diehl, J.F. *Safety of irradiated foods,* 2nd edn. Marcel Dekker, New York.

Dong, U.L., Jiyong, P., Jungil, K. and Ick, H.Y. (1996). "Effect of high hydrostatic pressure on the shelf life and sensory characteristics of *Angelica keiskei* juice." *Korean J. Food Sci. Technol.* 28: 105–108.

Donsi, G., Ferrari, G. and Matteo, M.Di. (1996). "High pressure stabilization of orange juice: evaluation of the effects of process conditions." *Ital. J. Food Sci.* 8: 99–106.

Doyle, M.P., Beuchat, L.R. and Montville, T.J. (1997). *Food Microbiology: Fundamentals and Frontiers.* ASM Press, Herndon, Va.

Drake, M.A., Harrison, S.L., Asplund, M., Barbosa canovas, G. and Swanson, B.G. (1997). "High pressure treatment of milk and effects on microbiological and sensory quality of cheddar cheese." *J. Food Sci.* 62: 843–845.

Dumay, E.M., Kalichevsky, M.T. and Cheftel, J.C. (1994). "High-pressure unfolding and aggregation of beta-lactoglobulin and the baroprotective effects of sucrose." *J. Agr. Food Chem.* 42: 1861–1868.

Earnshaw, R. (1996). "High pressure food processing." *Nutrition and Food Science.* 2: 8–11.

Earnshaw, R.G. (1992). "High pressure as a cell sensitiser: new opportunities to increase the efficacy of preservation process." In: *High Pressure and Biotechnology.* Balny, C., Hayashi, R., Hermans, K., Masson, P. (eds.). John Libbey and Co. Ltd., London. pp. 261–266.

Earnshaw, R.G. (1998). "Ultrasound: a new opportunity for food preservation." In: *Ultrasound in Food Processing.* Povey, M.J.W. and Mason, T.J. (eds.). Blackie Academic and Professional, London. pp. 183– 192.

Earnshaw, R.G., Appleyard, J. and Hurst, R.M. (1995). "Understanding physical inactivation processes: combined preservation opportunities using heat, ultrasound and pressure." *International Journal of Food Microbiology.* 28: 197–219.

Eklund, T. (1989). "Organic acids and esters." In: *Mechanisms of Action of Food Preservation Procedures.* Gould, G.W. (ed.). Elsevier Applied Science, London. pp. 161–200.

El Moueffak, A., Cruz, C., Antoine, M., Montury, M., Demazeau, G., Largeteau, A., Roy, B. and Zuber, F. (1996). "High pressure and pasteurization effect on duck foie gras." *Int. J. Food Sci. Technol.* 30: 737–743.

Erickson, L.E. (1982). "Recent developments in intermediate moisture foods." *J. Food Prot.* 45: 484–491.

Erkman, O. (2001). "Antimicrobial effect of pressurized carbon dioxide on *Yersinia enterocolitica* in broth and foods." *Food Science and Technology International.* 7: 245– 250.

Erkmen, O. and Karatas, S. (1997). "Effects of high hydrostatic pressure on *Staphylococcus aureus* in milk." *J. Food Eng.* 33: 257–262.

Eshtiaghi, M.N. and Knorr, D. (1993). "Potato cubes response to water blanching and high hydrostatic pressure." *J. Food Sci.* 58: 1371–1374.

Eskin, N.A.M., Henderson, H.M. and Townsend, R.J. (1971). *Biochemistry of Foods.* Academic Press, New York.

Evrendilek, G.A. and Zhang, Q.H. (2001). "Effect of acid and sublethal stresses on the inactivation of E. coli O157:H7 treated by PEF." IFT Annual Meeting Book of Abstracts, 2001, Session 28-8.

Ezaki, S. and Hayashi, R. (1992). "High Pressure effects on starch: structural change and retrogradation." In: *High Pressure and Biotechnology.* Balny, C., Hayashi, R., Hermans, K., Masson P., (eds.). John Libbey and Co. Ltd., London. pp. 261–266.

Farber, J.M. and Brown, B.E. (1990). "Effect of prior heat shock on heat resistance of *Listeria monocytogenes* in meat." *Appl. Environ. Microbiol.* 56:1584–1587.

Farkas, J. and Andrassy, E. (1993). "Interaction of ionising radiation and acidulants on the growth of the microflora of a vacuum-packed chilled meat product." *International Journal of Food Microbiology*. 19: 145–152.

Farr D. (1990). "High pressure technology in the food industry." *Trends Food Sci. Technol.* 1: 14–16.

FDA. (1998). Direct food substances affirmed as generally recognized as safe; egg white lysozyme. Food and Drug Administration, Washington, D.C. Fed. Reg. 63: 12421–12426.

FDA. 2000. Sodium diacetate, sodium acetate, sodium lactate and potassium lactate; Use as food additives. Food and Drug Administration, Washington, D.C. Fed. Reg. 65: 17128-17129.

Felipe, X., Capellas, M. and Law, A.J.R. (1997). "Comparison of the effects of high pressure treatments and heat pasteurization on the whey proteins in goat's milk." *J. Agr. Food Chem.* 45: 627–631.

Fellows, P. (1988). Food Processing Technology, Principles and Practice. VCH, 306–310.

Fenemma, O.R. (1975). *Principles of Food Science*. Marcel Dekker, Inc., New York.

Fernandez-Molina, J.J., Barbosa-Canovas, G.V. and Swanson, B.G. (2001a). "Inactivation of *Pseudomonas fluorescens* in skim milk by combination of pulsed electric fields and organic acids." IFT Annual Meeting Book of Abstracts, 2001, Session 59H-30.

Fernandez-Molina, J.J., Barbosa-Canovas, G.V. and Swanson, B.G. (2001b). "Inactivation of *Listeria innocua* by combining pulsed electric fields and acetic acid in skim milk." IFT Annual Meeting Book of Abstracts, 2001, Session 59H-31.

Finol, M.L., Marth, E.H. and Lindsay, R.C. (1982). "Depletion of sorbate from different media during growth of *Penicillium* species." *J. Food Protect*. 45: 398–404.

Food Standards Agency Consumer Committee (2004). *Food Irradiation Cons. Comm D030/04.*

Fornari, C., Maggi, A., Gola, S., Cassara, A. and Manachini, P.L. (1995). "Inactivation of *Bacillus* endospores by high-pressure treatment." *Ind. Conserve*. 70: 259–265.

Fox, J.B. Jr., Thayer, D.W., Jenkins, R.K., Phillips, J.G., Ackerman, S.A., Beecher, G.R., Holden, J.M., Morrow, F.D. and Quirbach, D.M. (1989). "Effect of gamma irradiation on the B vitamins of pork chops and chicken breasts." *International Journal of Radiation Biology*. 55: 689–703.

Fox, J.B., Thayer, D.W., Jenkins, R.K. *et al.* (1989). "Effect of gamma irradiation on the B vitamins of pork chops and chicken breasts." *Internat. J. Radiat. Biol*. 55: 689–703.

Fox, P.F. (1991). *Food Enzymology*. Vol. 1. Elsevier Applied Sci., London.

Frank, J.F. and Koffi, R.A. (1990). "Surface-adherent growth of *Listeria monocytogenes* is associated with increased resistance to surfactant sanitizers and heat." *J. Food Protect*. 53: 550–554.

Franz, C.M.A.P. and von Holy, A. (1996). "Thermotolerance of meat spoilage lactic acid bacteria and their inactivation in vacuum-packaged vienna sausages." *Int. J. Food Microbiol*. 29: 59–73.

Fuchigami, M., Ka to, N. and Teramoto, A. (1998). "High-pressure-freezing effects on textural quality of Chinese cabbage." *J. Food Sci*. 63: 122–125.

Fuji, T., Satomi M., Nakatsuka G., Yamaguchi T. and Okuzumi, M. (1994). "Changes in freshness indexes and bacterial flora during storage of pressurized mackerel." *J. Food Hyg. Soc. Jpn.* 35: 195–200.

Galazka, V.B., Ledward, D.A., Dickinson, E. and Langley, K.R. (1995). "High pressure effects on emulsifying behavior of whey protein concentrate." *J. Food Sci.* 60: 1341–1343.

Ganzle, M.G., Weber, S. and Hammes, W.P. (1999). "Effect of ecological factors on the inhibitory spectrum and activity of bacteriocins." *International Journal of Food Microbiology.* 46: 207–217.

Garcia, M.L., Burgos, J., Sanz, B. and Ordonez, J.A. (1989). "Effect of heat and ultrasonic waves on the survival of two strains of *Bacillus subtilis.*" *Journal of Applied Bacteriology.* 67: 619–628.

Garcia-Graells, C., Masschalck, B. and Michiels, C.W. (1999). "Inactivation of *Escherichia coli* in milk by high-hydrostatic-pressure treatment in combination with antimicrobial peptides." *Journal of Food Protection.* 62: 1248–1254.

Garza, S., Ibarz, A., Pagan, J. and Giner, J. (1999). "Non-enzymatic browning in peach puree during heating." *Food Research International.* 32: 335–343.

Gaucheron F., Famelart, M.H., Mariette, F., Raulot, K., Michel, F. and Graet, Y. (1997). "Combined effects of temperature and high pressure treatments on physicochemical characteristics of skim milk." *Food Chem.* 59: 439–447.

Gee, M., Farkas, D. and Rahman, A.R. (1977). "Some concepts for the development of intermediate moisture foods." *Food Technol.* 32(4): 58–63.

Gervilla, R., Capellas, M., Ferragut, V. and Guamis, B. (1997a). "Effect of high hydrostatic pressure on *Listeria innocua* 910 CECT inoculated into ewe's milk." *J. Food Prot.* 60: 33–37.

Gervilla, R., Felipe, X., Ferragut, V. and Guamis, B. (1997b). "Effect of high hydrostatic pressure on *Escherichia coli* and *Pseudomonas fluorescens* strains in ovine milk." *J. Dairy Sci.* 80: 2297–2303.

Gomes, M.R.A. and Ledward, D.A. (1996). "Effect of high pressure treatment on the activity of some polyphenol oxidases." *Food Chem.* 56: 1–5.

Gongora-Nieto, M.M., Seignour, L., Riquet, P., Davidson, P.M., Barbosa-Canovas, G.V. and Swanson, B.G. (1999). "Hurdle approach for the inactivation of *Pseudomonas fluorescens* in liquid whole egg." IFT Annual Meeting Book of Abstracts, 1999, Session 83A-2.

Goodner, J.K., Braddock, R.J. and Parish, M.E. (1998). "Inactivation of pectinesterase in orange and grapefruit juices by high pressure." *J. Agr. Food Chem.* 46: 1997–2000.

Gould, G.W. (1995). "Biodeterioration of Foods and an Overview of Preservation in the Food and Dairy Industries." *International Biodeterioration and Biodegradation.* pp. 267–277.

Gould, G.W. and Jones, M.V. (1989). "Combination and synergistic effects." In: *Mechanisms of Action of Food Preservation Procedures.* Gould, G.W. (ed.). Elsevier Applied Science, London. pp. 401–421.

Gould, G.W. and Sale, A.J.H. (1970). "Initiation of germination of bacterial spores by hydrostatic pressure." *J. Gen. Microbiol.* 60: 335–346.

Gow, C.Y., Hsin, T.L. (1996). "Comparison of high pressure treatment and thermal pasteurization effects on the quality and shelf life of guava puree." *Int. J. Food Sci. Technol.* 31: 205–213.

Grahl, T. and Markl, H. (1996). "Killing of microorganisms by pulsed electric fields." *Applied Microbiology and Biotechnology.* 45: 148–157.

Haas, G.J. and Herman, E.B. (1978). "Bacterial growth in intermediate moisture food systems." *Lebensm.Wiss.Technol.* 11: 74–78.

Haas, G.J., Bennett, D., Herman, E.B. and Collette, D. (1975). "Microbial stability of intermediate moisture foods." *Food Prod. Dev.* 9(3): 86.

Haas, G.J., Prescott, H.E, Dudley, E., Dik, R., Hintlian, C. and Keane, L. (1989). "Inactivation of microorganisms by carbon dioxide under pressure." *Journal of Food Safety.* 9: 253–265.

Haas, G.J., Prescott, H.E.JR. and D'initio, J. (1972). "Pressure freeze-air drying: a new technique to reduce deterioration in drying time." *J. Food Sci.* 37: 430–433.

Hansen, N.H. and Riemann, H. (1963). "Factors affecting the heat resistance of nonsporing organisms." *J. Appl. Bacteriol.* 26: 314–333.

Hashizume, C., Kimura, K. and Hayashi, R. (1995). "Kinetic analysis of yeast inactivation by high pressure treatment at low temperatures." *Biosci. Biotech. Biochem.* 59: 1455–1458.

Hauben, K.J.A., Bartlett, D.H., Soontjens, C.C.F., Cornelis, K., Wuytack, E.Y. and Michiels, C.W. (1997). "*Escherichia coli* mutants resistant to inactivation by high hydrostatic pressure." *Appl. Environ. Microbiol.* 63: 945–950.

Hauben, K.J.A., Wuytack, E.J., Soontjens, C.F. and Michiels, C.W. (1996). "High-pressure transient sensitization of *Escherichia coli* to lysozyme and nisin by disruption of outer-membrane permeability." *Journal of Food Protection.* 59: 350–355.

Hausam, J.G., Crandall, P.G., O'Bryan, C., Hettiarachchy, N.S. and Ahn, D.U. (2002). "Ascorbic acid and sodium chloride effects on microbial stability and quality characteristics of irradiated poultry breast meat." IFT Annual Meeting Book of Abstracts, 2002, Session 96-3.

Hayakawa, I., Kanno, T., Yoshiyama, K. and Fujio, Y. (1994). "Oscillatory compared with high pressure sterilization on *Bacillus stearothermophilus* spores." *J. Food Sci.* 59: 164–167.

Hayashi, K., Takahashi, S., Asano, H. and Hayashi, R. (1990). "Effect of pressure on the acid hydrolysis of proteins and polysaccharides." In: *Pressure Processed Food Research and Development.* Hayashi, R. (ed.). San-Ei Shuppan Co., Kyoto, Japan. pp. 288–293.

Hayashi, R. (1989). "Application of high pressure processing and preservation: philosophy and development." In: *Engineering and Food.* Vol.2. Spiess W.E.L., Schibert, H., (eds.). Elsevier Applied Science, London, England. pp. 815–826.

Hayashi, R. and Hayashida, A. (1989). "Increased amylase digestibility of pressure-treated starch." *Agr. Biol. Chem.* 53: 2543–2544.

Hayashi, R., Kawamura, Y., Nakasa, T. and Okinaka, O. (1989). "Application of high pressure to food processing: pressurization of egg white and yolk, and properties of gels formed." *Agr. Biol. Chem.* 53: 2935–2939.

Heinz, V. and Knorr, D.W. (1995). "Inactivation of *Bacillus subtilis* endospores by ultra-high-pressure in combination with other treatments." IFT Annual Meeting Book of Abstracts. p. 268.

Hendrickx, M., Silva, C., Oliveira, F. and Tobback, P. (1993). "Generalized (semi)-empirical formulae for optimal sterilization temperatures of conduction-heated foods with infinite surface heat transfer coefficients." *J. Food Eng.* 19: 141–158.

Heremans, K. (1992). "From living systems to biomolecules." In: *High Pressure and Biotechnology.* Balny, C., Hayashi, R., Hermans, K., Masson, P. (eds.). John Libbey and Co. Ltd., London. pp. 37–44.

Heremans, K. (1995). "High pressure effects on biomolecules." In: *High Pressure Processing of Foods.* Ledward, D.A., Johnston, D.E., Earnshaw, R.G., Hasting, A.P.M. (eds.). Nottingham University Press, Nottingham, England. pp. 81–97.

Hite, B.H. (1899). The effect of pressure in preservation of milk.West Virginia University Agricultural Experiment Station Bulletin 85, 15. (Cited in Hoover, D.G., Metrick, C., Papineau, A.M., Farkas D.F. and Knorr D. (1989). "Biological effects of high hydrostatic pressure on food microorganisms." *Food Technology.* 43: 99–107).

Hite, B.N., Giddings, N.J. and Weakly, C.E. (1914). "The effect of certain microorganisms encountered in the preservation of fruits and vegetables." *Bull. West Virginia Agr. Exp. Sta.* 146: 3–67.

Honma, K. and Haga, N. (1991). "Effects of high pressure treatment on sterilization and physical effect of high pressure treatment on sterilization and physical properties of egg white." In: *High Pressure Science for Food.* Hayashi, R., (ed.). San-Ei Pub. Co., Kyoto, Japan. pp. 317–324.

Hoover, D.G., Metrick, C., Papineau, A.M., Farkas, D.F. and Knorr, D. (1989). "Biological effects of high hydrostatic pressure on food microorganisms." *Food Technol.* 43: 99–107.

Huisman, G.W., Siegele, D.A., Zambrano, M.M. and Kolter, R. (1996). "Morphological and physiological changes during stationary phase." In: *Escherichia coli* and *Salmonella Cellular and Molecular Biology.* Neidhardt, F.C. (ed.). ASM Press, Washington. pp. 1672–1682.

Humpheson, L., Adams, M.R., Anderson, W.A. and Cole, M.B. (1998). "Biphasic thermal inactivation kinetics in *Salmonella enteritidis* PT4."*Appl. Environ. Microbiol.* 64: 459–464.

Ibarz, A., Sangronis, E., Barbosa canovas, G.V. and Swanson, B.G. (1996). "Inhibition of polyphenoloxidase in apple slices during high hydrostatic pressure treatments." In: Institute of Food Technologists Ann. Meet., Book of abstracts, p. 100.

Ibrahim, H.R., Yamada, M., Kobayashi, K. and Kato, A. (1992). "Bactericidal action of lysozyme against Gram-negative bacteria due to insertion of a hydrophobic pentapeptide into its C-terminus." *Bioscience, Biotechnology and Biochemistry.* 56: 1361–1363.

ICGFI. (1991). *Facts about food irradiation.* International Atomic Energy Agency, Vienna.

Iu, J., Mittal, G.S. and Griffiths, M.W. (2001). "Reduction in levels of *Escherichia coli* O157:H7 in apple cider by pulsed electric fields." *Journal of Food Protection.* 64: 964–969.

Jacobs, S.E. and Thornley, M.J. (1954). "The lethal action of ultrasonic waves on bacteria suspended in milk and other liquids." *Journal of Applied Bacteriology.* 17: 38–56.

Jaenick, R. (1991). "Protein stability and molecular adaptation to extreme conditions." *Euro. J. Biochem.* 202: 715–728.

Jagus, R., Terebiznik, M.R., Cerrutti, P., Pilosof, A.M.R. and Huergo, M.S. (1999). "Combined effects of pulsed electric field technology and nisin on *E. coli* inactivation." IFT Annual Meeting Book of Abstracts, 1999, Session 83A-6.

Jayaram, S., Castle, G.S.P. and Margaritis, A. (1992). "Kinetics of sterilization of *Lactobacillus brevis* cells by the application of high voltage pulses." *Biotechnology and Bioengineering*. 40: 1412–1420.

Jeyamkondan, S., Jayas, D.S. and Holley, R.A. (1999). "Pulsed electric field processing of foods: a review." *Journal of Food Protection*. 62: 1088–1096.

Jia, M., Zhang, Q.H. and Min, D.B. (1999). "Pulsed electric field processing effects on flavour compounds and microorganisms of orange juice." *Food Chemistry*. 65: 445–451.

Jin, Z.T., Su, Y., Tuhela, L., Singh, B. and Zhang, Q.H. (1998). "Inactivation of *Bacillus subtilis* using high voltage pulsed electric fields and ultrasonication. IFT Annual Meeting Book of Abstracts, 1998, Session 59C-15.

Johnston, D.E., Austin, B.A. and Murphy, R.J. (1992). "The effects of high pressure treatment on skim milk. In: *High Pressure and Biotechnology.*" Balny, C., Hayashi, R., Hermans, K., Masson P., (eds.). John Libbey and Co. Ltd., London. pp. 243–247.

Joint FAO/IAEA/WHO Study Group on High Dose Irradiation (1998). Weekly Epidemiological Record.

Jolibert, F., Tonello, C., Sagegh, P. and Raymond, J. (1995). "Effects of high pressure on polyphenol oxidases in fruit." *Imbottigliamento*. 18: 122–137.

Jones, J.M. (1992). *Food Safety*. Eagan Press, St. Paul, MN.

Josephson, E.S. (1983). "An historical review of food irradiation." *J. Food Safety*. 5: 161–189.

Josephson, E.S., Thomas, M.H. and Calhoun, W.K. (1978). "Nutritional aspects of food irradiation: An overview." *Journal of Food Processing and Preservation*. 2: 299–313.

Kalchayanand, N., Hanlin, M.B. and Ray, B. (1992). "Sublethal injury makes Gram-negative and resistant Gram-positive bacteria sensitive to the bacteriocins pediocin AcH and nisin." *Letters in Applied Microbiology*. 15: 239–243.

Kalchayanand, N., Sikes, A., Dunne, C.P. and Ray, B. (1994). "Hydrostatic pressure and electroporation have increased bactericidal efficiency in combination with bacteriocins." *Applied and Environmental Microbiology*. 60: 4174–4177.

Kalchayanand, N., Sikes, A., Dunne, C.P. and Ray, B. (1998). "Interaction of hydrostatic pressure, time and temperature of pressurization and pediocin AcH on inactivation of foodborne bacteria." *Journal of Food Protection*. 61: 425– 431.

Kalichevsky, M.T., Knorr, D. and Lillford, P.J. (1995). "Potential food applications of high-pressure effects on ice-water transitions." *Trends Food Sci. Technol.* 6: 253–258.

Kamihara, M., Taniguchi, M. and Kobayashi, T. (1987). "Sterilization of microorganisms with supercritical carbon dioxide." *Agricultural and Biological Chemistry*. 51: 407–412.

Kanda, Y. and Aoki, M. (1992). "Development of pressure-shift freezing method: Part I, observation of ice crystals of frozen 'Tofu'." In: *High Pressure Bioscience and Food Science*. Hayashi, R., (ed.). San-Ei Shuppan Co., Kyoto, Japan. pp. 24–26.

Kaplow, M. (1970). "Commercial development of intermediate moisture foods." *Food Technol.* 24: 889–893.

Karatzas, A.K., Kets, E.P.W., Smid, E.J. and Bennik, M.H.J. (2001). "The combined action of carvacrol and high hydrostatic pressure on *Listeria monocytogenes* Scott A." *J. Appl. Microbiol.* 90:463–469.

Karel, M. (1973). "Recent research and development in the field of low-moisture and intermediate-moisture foods." *Crit. Rev. Food Technol.* 3: 329–373.

Karmas, E. and Chen, C.C. (1975). "Relationship between water activity and water binding in high and intermediate moisture foods." *J. Food Sci.* 40: 800–801.

Kempkes, M.A. (2001). Personal Communication. Diversified Technologies, Bedford, MA.

Khadre, M.A. and Yousef, A.E. (2002). "Susceptibility of human rotavirus to ozone, high pressure, and pulsed electric field." *J. Food Prot.* 65: 1441–1446.

Kilcast, D. (1994). "Effect of irradiation on vitamins." *Food Chemistry.* 49: 157–164.

Kim, A.Y. and Thayer, D.W. (1996). "Mechanism by which gamma irradiation increases the sensitivity of *Salmonella typhimurium* ATCC 14028 to heat." *Appl. Environ. Microbiol.* 62: 1759–1763.

Kimura, K., Ida, M., Yosida, Y., Ohki, K., Fukumoto, T. and Sakuin. (1994). "Comparison of keeping quality between pressure-processed jam and heat-processed jam: changes in flavor components, hue, and nutrients during storage." *Biosci. Biotech. Biochem.* 58: 1386–1391.

Kingsley, D.H., Hoover, D.G., Papafragkou, E. and Richards, G.P. (2002). "Inactivation of hepatitis A virus and a calicivirus by high hydrostatic pressure." *J. Food Prot.* 65: 1605–1609.

Kinsloe, H., Ackerman, E. and Reid, J.J. (1954). "Exposure of microorganisms to measured sound fields." *Journal of Bacteriology.* 68: 373–380.

Kloczko, I. and Radomski, M. (1996). "Preservation of fruits, vegetables and juices by high hydrostatic pressure." *Przemysl Spozywczy.* 50: 25–26.

Knorr, D. (1993). "Effects of high-hydrostatic-pressure processes on food safety and quality." *Food Technology.* 47: 156–161.

Knorr, D. (1994). "Hydrostatic pressure treatment of food: microbiology." In: *New Methods of Food Preservation.* Gould, G. W. (ed.). Blackie Academic and Professional, London. pp. 159–175.

Knorr, D. (1995). Advantages and limitations of non-thermal food preservation methods. In: *VTT Symp.* 148 New Shelf-Life Technologies and Safety Assessments, 7–12. Technical Research Center of Finland, Finland.

Knorr, D. (2001). "Combination treatments to improve food safety and quality." IFT Annual Meeting Book of Abstracts, 2001, Session 6-4.

Knorr, D., Geulen, M., Grahl, T. and Sitzmann, W. (1994). "Food application of high voltage field pulses." *Trends in Food Science and Technology.* 5: 71–75.

Knorr, D., Schlueter, O. and Heinz, V. (1998). "Impact of high hydrostatic pressure on phase transitions of food." *Food Technol.* 52: 42–45.

Kooiman, W.J. (1973). "The screw cap tube technique: a new and accurate technique for the determination of the wet heat resistance of bacterial spores." In: *Spore Research*. Barker, A. N., Gould, G. W. and Wolf, J. (eds.). Academic Press, London, United Kingdom. pp. 87–92.

Kotrola, J.S., Conner, D.E. and Mikel, W.B. (1997). "Thermal inactivation of *Escherichia coli* O157:H7 in cooked turkey products." *J. Food Sci.* 62: 875–905.

Krokida, M.K. and Maroulis, Z.B. (1999). "Effect of microwave drying on some quality properties of dehydrated products." *Drying Technology*. Marcel Dekker, Inc. pp. 449–465.

Krokida, M.K. and Maroulis, Z.B. (2001). "Structural properties of dehydrated products during rehydration." *International Journal of Food Science and Technology*. 36: 529–538.

Krokida, M.K. and Maroulis, Z.B. and Saravacos, G.D. (2001). "The effect of the method of drying on the colour of dehydrated products." *International Journal of Food Science and Technology*. 36: 53–59.

Krokida, M.K., Kiranoudis, C.T., Maroulis, Z.B. and Marinos-Kouris, D. (2000). "Effect of pretreatment on colour of dehydrated products." *Drying Technology*. 18(6):1239–1250.

Krokida, M.K., Kiranoudis, C.T., Maroulis, Z.B. and Marinos-Kouris, D. (2000). "Drying related properties of apple." *Drying Technology*. 18(6): 1251–1267.

Labuza, T.P. (1980). "The effect of water activity on reaction kinetics of food deterioration." *Food Technol.* 34(4): 36–41.

Lacroix, M. and Ouattara, B. (2000). "Combined industrial processes with irradiation to assure innocuity and preservation of food products—a review." *Food Research International*. 33: 719–724.

Lado, B.H. and Yousef, A.E. (2003). "Selection and identification of a *Listeria monocytogenes* target strain for pulsed electric field process optimization." *Appl. Environ. Microbiol.* 69: 2223–2229.

Lambert, A.D., Smith, J.P. and Dodds, K.L. (1991). "Shelf life extension and microbiological safety of fresh meat—a review." *Food Microbiology*. 8: 267–297.

Lambert, A.D., Smith, J.P., Dodds, K.L. and Charbonneau, R. (1992). "Microbiological changes and shelf life of MAP, irradiated fresh pork." *Food Microbiology*. 9: 231–244.

Leadley, C.E. and Williams, A. (1997). "High pressure processing of food and drink-an overview of recent developments and future potential." In: *New Technologies*. Bull. No. 14, Mar., CCFRA, Chipping Campden, Glos, UK.

Lee, F.A. (1983). *Basic Food Chemistry*, 2nd edn. The AVI Publishing Co., Westport, Connecticut. pp. 283-302.

Leistner, L. (1992). "Food preservation by combined methods." *Food Research International*. 25: 151–158.

Leistner, L. (1994). "Further developments in the utilization of hurdle technology for food preservation." *Journal of Food Engineering*. 22: 421–432.

Leistner, L. (2000). "Basic aspects of food preservation by hurdle technology." *International Journal of Food Microbiology*. 55: 181–186.

Leistner, L. and Gould, G.W. (2002). *"Hurdle Technologies: Combination Treatments for Food Stability, Safety and Quality."* Kluwer Academic/Plenum Publishers, New York.

Lemaître, J.P., Echchannaoui, H., Michant, G., Divies, and Rousset. (1998). "Plasmid-mediated resistance to antimicrobials among Listeriae." *J. Food Protect.* 61: 1459–1464.

Lewicki P.P. (1998). "Some remarks on rehydration of dried foods." *Journal of Food Engineering.* 36: 81–87.

Leyer, G.J. and Johnson, E.A. (1992). "Acid adaptation promotes survival of *Salmonella* spp. in cheese." *Appl. Environ.Microbiol.* 58: 2075–2080.

Liang, Z., Mittal, G.S. and Griffiths, M.W. (2002). "Inactivation of *Salmonella typhimurium* in orange juice containing antimicrobial agents by pulsed electric field." *Journal of Food Protection.* 65: 1081–1087.

Licciardello, J.J., Ravesi, E.M., Tuhkunen, B.E. and Racicot, L.D. (1984). "Effect of some potentially synergistic treatments in combination with 100 Krad irradiation on the iced shelf life of cod fillets." *Journal of Food Science.* 49: 1341–1375.

Liewen, M.B. and Marth, E.H. (1985). "Growth and inhibition of microorganisms in the presence of sorbic acid: A review."*J. Food Protect.* 48: 364–375.

Lillard, H.S. (1994). "Decontamination of poultry skin by sonication." *Food Technology.* 48: 72–73.

Linton, M., McClements, J.M.J. and Patterson, M.F. (1999). "Survival of *Escherichia coli* O157:H7 during storage of pressure-treated orange juice." *J. Food Protect.* 62: 1038–1040.

Liu, X., Yousef, A.E. and Chism, G.W. (1997). "Inactivation of *Escherichia coli* O157:H7 by the combination of organic acids and pulsed electric field." *Journal of Food Safety.* 16: 287–299.

Loaharanu, P. (1994). "Status and prospects of food irradiation." *Food Technology.* 48(5): 124–131.

Loaharanu, P. (2003). *Irradiated foods,* 5th edn. Rev. American Council on Science and Health, New York.

Lopez, F.R., Carrascosa, A.V. and Olano A. (1996). "The effects of high pressure on whey protein denaturation and cheese-making properties of raw milk." *J. Dairy Sci.* 79: 929–1126.

Ludikhuyze, L.R., Broeck I.van.den., Weemaes, C.A., Herremans, C.H., Impe J.F.van, Hendrickx M.E. and Tobback, P.P. (1997). "Kinetics for isobaric-isothermal inactivation of *Bacillus subtilis* alpha-amylase under dynamic conditions. *Biotechnol. Prog.* 13: 617–623.

MacDonald, A.G. (1993). "Effects of high hydrostatic pressure on natural and artificial membranes." In: *High Pressure and Biotechnology.* Balny, C., Hayashi, R., Heremans, K. and Masson, P., (eds.). John Libbey and Company Ltd., London. pp. 67–75.

Macfarlene, J.J. (1985). "High pressure technology and meat quality." In: *Developments in Meat Science,* Vol.3. Lawrie, R. (ed.). Elsevier Applied Science Publishers, London. pp. 155–184.

Macfarlene, J.J., Mckenzie, I.J., Turner, R.H. and Jones, P.N. (1984). "Binding of comminuted meat: effect of high pressure." *Meat Sci.* 10: 307–320.

Mackey, B., Forestie're, K., Isaacs, N.S., Stenning, R. and Brooker, B. (1994). "The effect of high hydrostatic pressure on *Salmonella thompson* and *Listeria monocytogenes* examined by electron microscopy." *Lett. Appl. Microbiol.* 19: 429–432.

Mackey, B.M. (2000). "Injured bacteria." In: *The Microbiological Safety and Quality of Foods*. Vol. I. Lund, B.M., Baird-Parker, T.C. and Gould, G.W. (eds.). Aspen Publishers, Gaithersburg, Inc. pp. 315–341.

Mackey, B.M. and Derrick, C.M. (1987). "Changes in the heat resistance of *Salmonella typhimurium* during heating at rising temperatures." *Lett. Appl. Microbiol.* 4: 13–16.

Maggi, A., Gola, S., Spotti, E., Rovere, P. and Mutti, P. (1994). "High-pressure treatments of ascospores of heat-resistant moulds and patulin in apricot nectar and water." *Ind. Conserve.* 69: 26–29.

Maggi, A., Cassara, A., Rovere, P. and Gola, S. (1995). "Use of high pressure for inactivation of butyric clostridia in tomato serum." *Ind. Conserve.* 70: 289–293.

Maggi, A., Gola, S., Spotti, E., Rovere P., Miglioli L., Dall'aglio G. and Lonneborg, N.G. (1996). "Effects of combined high pressure-temperature treatments on *Clostridium sporogenes* spores in liquid media." *Ind. Conserve.* 71: 8–14.

Mallidis C.G. and Drizou, D. (1991). "Effect of simultaneous application of heat and pressure on the survival of bacterial spores." *J. Appl. Bacteriol.* 71: 285–288.

Mallidis, C.G. and Scholefield, J. S. (1985). "The release of dipicolinic acid during heating and its relation to the heat destruction of *Bacillus stearothermophilus* spores." *J. Appl. Bacteriol.* 59: 479–486.

Man͂ as, P. and Mackey, B.M. (2004). "Morphological and physiological changes induced by high hydrostatic pressure in exponential- and stationary-phase cells of *Escherichia coli*: relationship with cell death." *Appl. Environ. Microbiol.* 70: 1545–1554.

Man͂ as, P., Paga´n, R., Raso, J., Sala, F.J. and Condo´n, S. (2000). "Inactivation of *S. typhimurium*, *S. enteritidis* and *S. senftenberg* by ultrasonic waves under pressure." *J. Food Prot.* 63: 451–456.

Manvell, C. (1997). "Minimal processing of food. Food Science and Technology Today." 11: 107–111.

Manzocco, L., Calligaris, S., Mastrocola, D., Nicoli, M.C. and Lerici, C.R. (2001). "Review of non-enzymatic browning and antioxidant capacity in processed foods." *Trends in Food Science and Technology.* 11: 340–346.

Marquez, V.O., Mittal, G.S. and Griffiths, M.W. (1997). "Destruction and inhibition of bacterial spores by high voltage pulsed electric field." *Journal of Food Science.* 62: 399– 409.

Marth, E.H., Capp, C.M., Hasenzahl, L., Jackson, H.W. and Montville, T.J., Winkowski, K. and Chikindas, M.L. 2001. "Biologically based preservation systems." In: *Food Microbiology: Fundamentals and Frontiers,* 2nd edn. Doyle, M.P., Beuchat, L.R. and Montville, T.J., (eds.). American Society for Microbiology, Washington, DC. pp. 629–647

Martins, S.I.F.S., Jongen, W.M.F and van Boekel, M.A.J.S. (2001). "A review of Maillard reaction in food and implications to kinetic modelling." *Trends in Food Science and Technology.* 11: 364–373.

Mason, T.J. (1998). "Power ultrasound in food processing—the way forward." In: *Ultrasound in Food Processing.* Povey, M.J.W., Mason, T.J. (eds.). Blackie Academic and Professional, London. pp. 105–126.

Masschalck, B., Van Houdt, R. and Michiels, C.W. (2001a). "High pressure increases bactericidal activity and spectrum of lactoferrin, lactoferricin and nisin." *International Journal of Food Microbiology.* 64: 325–332.

Masschalck, B., Van Houdt, R., Van Haver, E.G.R. and Michiels, C.W. (2001b). "Inactivation of Gram-negative bacteria by lysozyme, denatured lysozyme, and lysozyme-derived peptides under high hydrostatic pressure." *Applied and Environmental Microbiology.* 67: 339–344.

Master, A.M., Krebbers, B., Van den Berg, R.W. and Bartels, P.V. (2004). "Advantages of high pressure sterilisation on quality of food products." *Trends Food Sci. Technol.* 15: 79–85.

Masuda, M., Saito, Y., Iwanami, T. and Hirai, Y. (1992). "Effects of hydrostatic pressure on packaging materials for food." In: *High Pressure and Biotechnology.* Balny, C., Hayashi, R., Hermans, K., Masson, P. (eds.). John Libbey and Co. Ltd., London. pp. 545–547.

Mermelstein, N.H. (1997). "High-pressure processing reaches the U.S. market." *Food Technology.* 51: 95–96.

Mertens, B. (1992). "Recent developments in high pressure processing." In: New Technologies for the Food and Drink Industries. Symp. Proc., Part 1. CCFRA, Chipping Campden, Glos, UK.

Mertens, B. and Deplace, G. (1993). "Engineering aspects of high-pressure technology in the food industry." *Food Technology.* 47: 164–169.

Metrick, C., Hoover, D.G. and Farkas, D.F (1989). "Effects of high hydrostatic pressure on heat." *J. Food Protect.* 61: 432–436.

Miyao, S., Shindoh, T., Miyamori K. and Arita, T. (1993). "Effects of high pressure processing on the growth of bacteria derived from surimi (fish paste)." *J. Japanese Soc. Food Sci. Technol.* 40: 478–484.

Moio, L., Pietra, L.L.A., Cacace, D., Palmieri, L., Martino, E.DE., Carpi, G., Dall'aglio, G. and Masi, P. (1995). "Application of high pressure treatment for microbiological stabilization of must." *Vignevini.* 2: 3–6.

Molina-Hoppner, A., Doster, W., Vogel, R.F. and Ganzle, M.G. (2004). "Protective effect of sucrose and sodium chloride for *Lactococcus lactis* during sublethal and lethal high-pressure treatments." *Appl. Environ. Microbiol.* 70: 2013–2020.

Molins, R.A., Motarjemi, Y, Käsferstein, F.K. (2001). "Irradiation: a critical control point in ensuring the microbiological safety of raw foods." *Food Contr.* 12: 347–356.

Monk, J.D., Beauchat, L.R. and Doyle, M.P. (1995). "Irradiation inactivation of food-borne microorganisms." *Journal of Food Protection.* 58: 197–208.

Moorman, J.E., Toledo, R.T. and Schmidt, K. (1996). "High-pressure throttling (HPT) reduces microbial population, improves yogurt consistency and modifies rheological properties of ultrafiltered milk." In: Institute of Food Technologists Ann. Meet. Book of abstracts, p. 49.

Morgan, S.M., Ross, R.P., Beresford, T. and Hill, C. (2000). "Combination of hydrostatic pressure and lacticin 3147 causes increased killing of *Staphylococcus* and *Listeria*." *Journal of Applied Microbiology.* 88: 414–420.

Moseley, B.E.B. (1989). "Ionizing irradiation: action and repair." In: *Mechanisms of Action of Food Preservation Procedures.* Gould, G.W. (ed.). Elsevier Applied Science, London. pp. 43–70.

Mozhaev, V.V., Heremans, K., Frank, J., Masson, P. and Balny, C. (1994). "Exploiting the effects of high hydrostatic pressure in biotechnological applications." *Trends in Biotechnology*. 12: 493–501 (Cited in Datta, N., Deeth, H.C., 1999. "High pressure processing of milk and dairy products." *Australian Journal of Dairy Technology*. 54: 41–48).

Muhr, A.H. and Blanshard, J.M.V. (1982). "Effect of hydrostatic pressure on starch gelatinization." *Carbohydr. Polym.* 2: 61–64.

Muhr, A.H., Wetton, R.E., Blanshard, J.M.V. (1982). "Effect of hydrostatic pressure on starch gelatinization, as determined by DTA." *Carbohydr. Polym.* 2: 91–102.

Mulet-Powell, N., Lacoste-Armynot, A.M., Vinas, M., and Simeon de Buochberg, M. (1998). "Interactions between pairs of bacteriocins from lactic bacteria." *J. Food Protect.* 61: 1210–1212.

Nachmanson, J. (1995). "Packaging solutions for high quality foods processed by high hydrostatic pressure." In: *Proceedings of Europak*, vol. 7, 390–401.

Niemira, B.A. (2001). "Citrus juice composition does not influence radiation sensitivity of *Salmonella enteritidis*." *Journal of Food Protection*. 64: 869–872.

Nishi, K., Kato, R. and Tomita, R. (1994). "Activation of *Bacillus* spp. spores by hydrostatic pressure." *Nippon Shokuhin Kogyo Gakkaishi.* 41: 542–549.

Nishiwaki, T., Ikeuchi, Y. and Suzuki, A. (1996). "Effects of high pressure treatment on Mg-enhanced ATPase activity of rabbit myofibrils." *Meat Sci.* 43: 1445–1155.

Niven, G.W., Miles, C.A. and Mackey, B.M. (1999). "The effects of hydrostatic pressure on ribosome conformation in *Escherichia coli*: an *in vivo* study using differential scanning calorimetry." *Microbiology SGM.* 145: 419–425.

O'Brien, J.K. and Marshall, R.T. (1996). "Microbiological quality of raw ground chicken processed at high isostatic pressure." *J. Food Prot.* 59: 146–150.

Ogawa, H. (1992). "Effect of hydrostatic pressure on sterilization and preservation of citrus juice." In: *High Pressure and Biotechnology*. Balny, C., Hayashi, R., Hermans, K., Masson, P. (eds.). John Libbey and Co. Ltd., London. pp. 269–278.

Ohlsson, T. (1994). "Minimal processing- preservation methods of the future: an overview." *Trends in Food Science and Technology*. 5: 341–344.

Ohmori, T., Shigehisa, T., Taji, S. and Hayashi, R. (1991). "Effect of high pressure on the protease activities in meat." *Agr. Biol. Chem.* 55: 357–361.

Ohshima, T., Ushio, H. and Koizumi, C. (1993). "High pressure processing of fish and fish products." *Trends Food Sci. Technol.* 4: 370–375.

Okamoto, M. and Hayashi, R. (1990). "Application of high pressure to preferential enzymatic degradation of lactoglobulin in bovine milk whey." In: *Pressure Processed Food Research and Development*. Hayashi, R., (ed.). San-Ei Pub. Co., Kyoto, Japan. pp. 67–71.

Okamoto, M., Kawamura, Y.and Hayashi R. (1990). "Application of high pressure to food processing: textural comparison of pressure- and heat-induced gels of food proteins." *Agr. Biol. Chem.* 54: 183–189.

Okos, M.R., Narsimhan, G., Singh, R.K., Weitnauer, A.C., Heldman, D.R. and Lund, D.B. (1992). "Food Dehydration." In: *Handbook of Food Engineering*. Heldman, D.R. and Lund, D. B. (eds.). Marcel Dekker, Inc. pp. 475–480.

Ouattara, B., Giroux, M., Smoragiewicz, W., Saucier, L. and Lacroix, M. (2002). "Combined effect of gamma irradiation, ascorbic acid, and edible coating on the improvement of microbial and biochemical characteristics of ground beef." *Journal of Food Protection.* 65: 981–987.

Pagan, R., Esplugas, S., Gongora-Nieto, M.M., Barbosa-Canovas, G.V. and Swanson, B.G. (1998). "Inactivation of *Bacillus subtilis* spores using high intensity pulsed electric fields in combination with other food conservation technologies." *Food Science and Technology International.* 4: 33–44.

Pagan, R., Manas, P., Alvarez, I. and Condon, S. (1999). "Resistance of *Listeria monocytogenes* to ultrasonic waves under pressure at sublethal (manosonication) and lethal (manothermosonication) temperatures." *Food Microbiology.* 16: 139–148.

Palou, E., Lopez-malo, A., Barbosa-canovas, G.V., Swanson, B.G. and Welti, J. (1996). "Combined effect of high hydrostatic pressure and water activity on *Zygosaccharomyces bailii* inhibition." In: Institute of Food Technologists Ann. Meet., Book of Abstracts, p. 57.

Palou, E., Lopez-malo, A., Barbosa-canovas, G.V., Welti-chanes, J., Davidson, P.M. and Swanson, B.G. (1998). "High hydrostatic pressure come-up time and yeast viability." *J. Food Prot.* 61: 1657–1660.

Papineau, A.M., Hoover, D.G., Knorr,D. and Farkas, D.F. (1991). "Antimicrobial effect of water-soluble chitosans with high hydrostatic pressure." *Food Biotechnol.* 5: 45–47.

Patterson, M. (1999). "High-pressure treatment of foods." In: *The Encyclopedia of Food Microbiology.* Robertson, R.K., Batt, A. and Patel, P.D. (eds.). pp. 1059–1065. Academic Press, London.

Patterson, M.F. (2005). "Microbiology of pressure-treated foods." *Journal of Applied Microbiology* 98(6): 1400.

Patterson, M.F. and Kilpatrick, D.J. (1998). "The combined effect of high hydrostatic pressure and mild heat on inactivation of pathogens in milk and poultry." *J. Food Sci.* 54: 1547–1564.

Patterson, M.F. and Kilpatrick, D.J. (1998). "The combined effect of high hydrostatic pressure and mild heat on inactivation of pathogens in milk and poultry." *J. Food Prot.* 61: 432–436.

Patterson, M.F., Quinn, M., Simpson, R. and Gilmour, A. (1995). "Sensitivity of vegetative pathogens to high hydrostatic pressure treatment in phosphate-buffer saline and foods." *J. Food Prot.* 58: 524–529.

Patterson, M.F., Quinn, M., Simpson, R. and Gilmour, A. (1995). "Effects of high pressure on vegetative pathogens." In: *High Pressure Processing of Foods.* Ledward, D.A., Johnston, D.E., Earnshaw, R.G., Hasting, A.P.M. (eds.).Nottingham University Press, Nottingham. pp. 47–64.

Paul, P., Chawla, S.P., Thomas, P., Kesavan, P.C., Fotedar, R. and Arya, R.N. (1997). "Effect of high hydrostatic pressure, gamma-irradiation and combination treatments on the microbiological quality of lamb meat during chilled storage." *J. Food Safety.* 16: 263–271.

Pehrsson, P.E. (1996). "Application of high-pressure pasteurization to citrus processing." In: Institute of Food Technologists Ann. Meet., Book of abstracts, p. 108.

Pellegrino, P.M., Fell, N. F. and Gillespie, J. B. (2002). "Enhanced spore detection using dipicolinate extraction techniques." *Anal. Chim. Acta.* 455: 167–177.

Pittia, P., Wilde, P.J., Husband, F.A. and Clark, D.C. (1996). "Functional and structural properties of beta-lactoglobulin as affected by high pressure treatment." *J. Food Sci.* 61: 1123–1128.

Pol, I.E., Mastwijk, H.C., Bartels, P.V. and Smid, E.J. (2000). "Pulsed electric field treatment enhances the bactericidal action of nisin against *Bacillus cereus*." *Applied and Environmental Microbiology.* 66: 428–430.

Ponce, E., Pla, R., Mor, M.M., Gervilla, R. and Guamis, B. (1998). "Inactivation of *Listeria innocua* inoculated in liquid whole egg by high hydrostatic pressure." *J. Food Prot.* 61: 119–122.

Popper, L. and Knorr, D. (1990). "Applications of high-pressure homogenization for food preservation." *Food Technol.* 44: 84–89.

Pothakamury, U.R., Barbosa-Canovas, G.V. and Swanson, B.G. (1993). "Magnetic field inactivation of microorganisms and generation of biological changes." *Food Technology.* 12: 68–72.

Potter, N.N. (1973). *Food Science.* The AVI Publishing Company, Inc., Westport, Connecticut. pp. 238–254.

Qin, B., Pothakamury, U.R., Vega-Mercado, H., Martin, O., Barbosa-Canovas, G.V. and Swanson, B.G. (1995). "Food pasteurization using high intensity pulsed electric fields." *Food Technology.* 49: 55–60.

Qiu, X., Sharma, S., Tuhela, L., Jia, M. and Zhang, Q.H. (1998). "An integrated PEF pilot plant for continuous non-thermal pasteurization of fresh orange juice." *Transactions of ASAE.* 41: 1069–1074.

Rademacher, B., Pfeiffer, B. and Kessler, H.G. (1998). "Inactivation of microorganisms and enzymes in pressure-treated raw milk." In: *High Pressure Food Science, Bioscience and Chemistry.* Isaacs, N.S. (ed.). The Royal Society of Chemistry, Cambridge. pp. 145–151.

Raso, J. and Barbosa-Ca´novas, G. (2003). "Nonthermal preservation of foods using combined processing techniques." *Crit. Rev. Food Sci. Nutr.* 43: 265–285.

Raso, J., A´lvarez, I., Condo´n, S. and Sala, F.J. (2000). "Predicting inactivation of *Salmonella senftenberg* by pulsed electric fields." *Innov. Food Sci. Emerg. Technol.* 1: 21–29.

Raso, J., Gongora, M., Calderon, M.L., Barbosa-Canovas, G.V. and Swanson, B.G. (1998c). "Resistant microorganisms to high intensity pulsed electric field pasteurization of raw skim milk." IFT Annual Meeting Book of Abstracts, 1998, Session 59D-20.

Raso, J., Paga´n, R., Condo´n, S. and Sala, F.J. (1998a). "Influence of temperature and pressure on the lethality of ultrasound."*Applied and Environ. Microbiol.* 64: 465–471.

Raso, J., Palop, A., Paga´n, R. and Condon, S. (1998b). "Inactivation of *Bacillus subtilis* spores by combining ultrasonic waves under pressure and mild heat treatment." *J. Appl. Microbiol.* 85: 849–854.

Ray, B. (1993). "Sublethal injury, bacteriocins, and food microbiology." *ASM News.* 59: 285–291.

Ray, B. (2001). "Bacteriocins, mild heat and high pressure for preserving low acid meat products." IFT Annual Meeting Book of Abstracts, 2001, Session 6–7.

Reed, J.M., Bohrer, C.W. and Cameron, E.J. (1951). "Spore destruction rate studies on organisms of significance in processing of canned foods." *Food Res.* 165: 383–408.

Resurreccion, A.V.A., Galvez, F.C.F., Fletcher, S.M. and Misra, S.K. (1995). "Consumer attitudes toward irradiated food: results of a new study." *Journal of Food Protection.* 58: 193–196.

Roberts, C.M. and Hoover, D.G. (1996). "Sensitivity of *Bacillus coagulans* spores to combinations of high hydrostatic pressure, heat, acidity and nisin." *Journal of Applied Bacteriology.* 81: 363–368.

Roberts, C.M. and Hoover, D.G. (1996). "Sensitivity of spores of *Clostridium sporogenes* PA3679 and *B. subtilis* 168 to combination of high hydrostatic pressure, heat, acidity, and nisin." In: Institute of Food Technologists Ann. Meet., Book of abstracts, pp. 174–175.

Roberts, T.A., Baird-Parker, A.C. and Tompkin, R.B. (1996). *Micro-organisms in foods. 5. Microbiological specification of food pathogens.* Chapman and Hall, London, United Kingdom.

Robinson, R.A. and Stokes, R.H. (1970). *Electrolyte Solutions,* 2nd edn. Butterworths, London.

Rockland, L.B., and Nishi, S.K. (1980). "Influence of water activity on food product quality and stability." *Food Technol.* 34(4): 42–52.

Rockville, M.D., Clavero, M.R.S., Monk, D. J., Beuchat, L.R., Doyle, M.P. and Brackett, R.E. (1994). "Inactivation of *Escherichia coli* 0157:H7, salmonellae, and *Campylobacter jejunum* raw ground beef by gamma irradiation." *Applied and Environmental Microbiology.* 60(6): 2069–2075.

Ross, K.D. (1975). "Estimation of water activity in intermediate moisture foods." *Food Technol.* 29(3): 26–34.

Rovere, P., Carpi, G., Maggi, A., Gola, S. and Dall'aglio G. (1994). "Stabilisation of apricot puree by means of high pressure treatments." *Prehram Technol. Biotechnol. Rev.* 32: 145–150.

Rovere, P., Sandei, L., Colombi, A., Munari, M., Ghiretti, G., Carpi, G. and Dall'aglio G. (1997). "Effects of high pressure treatments on chopped tomatoes." *Ind. Conserve.* 72: 3–12.

Rovere, P., Tosoratti, D. and Maggi, A. (1996). "Sterilizing trials to 15 000 bar to obtain microbiological and enzymatic stability." *Ind. Alimentari.* 35: 1062–1065.

Russell, A.D. (1991). "Mechanisms of bacterial resistance to non-antibiotics: Food additives and food and pharmaceutical preservatives." *J. Appl. Bacteriol.* 71: 191–201.

Russell, A.D. (1997). "Plasmids and bacterial resistance to biocides." *J. Appl. Microbiol.* 83: 155–165.

Russell, A.D., Furr, J.R. and Maillard J.Y. (1997). "Microbial susceptibility and resistance to biocides." *ASM News.* 63: 481–487.

Sablon, E., Contreras, B. and Vandamme, E. (2000). "Antimicrobial peptides of lactic acid bacteria: mode of action, genetics and biosynthesis." *Applied Microbiology and Biotechnology.* 68: 21–60.

Sala, F.J., Burgos, J., Condon, S., Lopez, P. and Raso, J. (1995). "Effect of heat and ultrasound on microorganisms and enzymes." In: *New Methods of Food Preservation.* Gould, G.W. (ed.). Blackie Academic and Professional, London. pp. 176–204.

Sale A.J.H., Gould G.W. and Hamilton, W.A. (1970). "Inactivation of bacterial spores by hydrostatic pressure." *J. Gen. Microbiol.* 60: 323–334.

Sale, A.J.H. and Hamilton, W.A. (1967). "Effects of high electric fields on microorganisms I. Killing of bacteria and yeast." *Biochim Biophys Acta.* 148: 781–788.

Sale, A.J.H. and Hamilton, W.A. (1968). "Effects of high electric fields on micro-organisms: III. Lysis of erythrocytes and protoplasts." *Biochemica et Biophysica Acta.* 163: 37–43.

Sale, A.J.H., Gould, G.W. and Hamilton, W.A. (1970). "Inactivation of bacterial spores by hydrostatic pressure." *Journal of General Microbiology.* 60: 323–334.

Salunkhe, D.K. (1974). *Storage, Processing and Nutritional Quality of Fruits and Vegetables.* CRC Press, Ohio. pp. 29–30.

Satin, M. (1993). *Food Irradiation A Guidebook.* Technomic Publ. Co., Inc., Lancaster, PA.

Satin, M. (2002). "Use of irradiation for microbial decontamination of meat: situation and perspectives." *Meat Science.* 62: 277–283.

Scott, W.J. (1957). "Water relations of food spoilage microorganisms." *Adv. Food Res.* 7: 83–127.

Sepulveda, D.R., Gongora-Nieto, M.M., San-Martin, M.F., Swanson, B.G. and Barbosa-Canovas, G.V. (2002). "Pulsed electric fields and its induced heating inactivates Listeria spp." IFT Annual Meeting Book of Abstracts, 2002, Session 96–10.

Severini, C., Romani, S., Dall'aglio, G., Rovere, P., Conte, L. and Lerici, C.R. (1997). "High pressure effects on lipid oxidation of extra virgin olive oils and seed oils." *Ital. J. Food Sci.* 9: 183–191.

Seyderhelm, I. and Knorr, D. (1992). "Reduction of *Bacillus stearothermophilus* spores by combined high pressure and temperature treatments." *ZFL (J. Food Ind.).* 43: 17–20.

Seyderhelm, I., Boguslawski, S., Michaelis, G. and Knorr, D. (1996). "Pressure induced inactivation of selected food enzymes." *J. Food Sci.* 61: 308–310.

Shapero, M., Nelson, D.A. and Labuza, T.P. (1978). "Ethanol inhibition of *Staphylococcus aureus* at limited water activity." *J. Food Sci.* 43: 1467–1469.

Shearer, A.E.H., Dunne, C.P., Sikes, A. and Hoover, D.G. (2000). "Bacterial spore inhibition and inactivation in foods by pressure, chemical preservatives, and mild heat." *Journal of Food Protection.* 63: 1503–1510.

Sherry, A.E., Patterson, M.F. and Madden, R.H. (2004). "Comparison of 40 *Salmonella enterica* serovars injured by thermal, high-pressure and irradiation stress." *J. Appl. Microbiol.* 96: 887–893.

Shigehisa, T., Ohmori, T., Saito, A., Taji, S. and Hayashi, R. (1991). "Effects of high hydrostatic pressure on characteristics of pork slurries and inactivation of microorganisms associated with meat products." *Int. J. Food Microbiol.* 12: 207–216.

Shigehisa, T., Ohmori, T., Saito, A., Taji, S. and Hayashi, R. (1991). "Effects of high hydrostatic pressure on characteristics of pork slurries and inactivation of microorganisms associated with meat and meat products." *Int. J. Food Microbiol.* 12: 207–216.

Shimada, K. (1992). "Effect of combination treatment with high pressure and alternating current on the lethal damage of *Escherichia coli* cells and *Bacillus subtilis* spores." In: *High Pressure and Biotechnology*. Balny, C., Hayashi, R., Hermans, K., Masson, P. (eds.). John Libbey and Co. Ltd., London. pp. 49–51.

Shimada, K. and Shimahara, K. (1981). "Factors affecting the survival fractions of resting *Escherichia coli* B and K-12 cells exposed to alternating current." *Agricultural and Biological Chemistry*. 45: 1589–1595.

Shimada, K. and Shimahara, K. (1985). "Leakage of cellular contents and morphological changes in resting *Escherichia coli* B cells exposed to an alternating current." *Agr. Biol. Chem.* 49: 3605–3607.

Shimada, K. and Shimahara, K. (1987). "Effect of alternating current exposure on the resistivity of resting *Escherichia coli* B cells to crystal violet and other basic dyes." *J. Appl. Bacteriol.* 62: 261–268.

Shimada, K. and Shimahara, K. (1991). "Decrease in high pressure tolerance of resting cells of *Escherichia coli* K-12 by pretreatment with alternating current." *Agricultural and Biological Chemistry*. 55: 1247–1251.

Shoji, T., Saeki, H., Wakameda, A. and Nakamura, M. (1990). "Gelation of fish meat paste 'Surimi' by HPP." In: *Pressure Processed Food Research and Development*. Hayashi, R. (ed.). San-Ei Pub. Co., Kyoto, Japan. pp. 99–102.

Singh, R.K., Lund, D.B. and Buelow, F.H. (1983). "Storage stability of intermediate moisture apples: Kinetics of quality change." *J. Food Sci.* 48: 939–944.

Singh, R.K., Lund, D.B. and Buelow, F.H. (1984). "Computer simulation of storage stability in intermediate moisture apples." *J. Food Sci.* 49: 759–764.

Sinskey, A.J. (1976). "New developments in intermediate moisture foods: Humectants." In: *Intermediate Moisture Foods*. Davies, R., Birch, G.G. and Parker, K.J. (eds.). Applied Science Publ., London. pp. 260–277.

Sitzmann, W. (1995). "High-voltage pulse techniques for food preservation." In: *New Methods of Food Preservation*. Gould, G.W. (ed.). Blackie Academic and Professional, London. pp. 236–251.

Sizer, C.E. and Balasubramaniam, V.M. (1999). "New intervention processes for minimally processed juices." *Food Technology*. 53: 64–67.

Sloan, A.E. and Labuza, T.P. (1976). "Prediction or water activity lowering ability of food humectants at high a_W." *J. Food Sci.* 41: 532–535.

Sloan, A.E., and Labuza, T.P. (1975). "Investigating alternative humectants for use in foods." *Food Prod. Dev.* 9(7): 75–88.

Sloan, A.E., Waletzko, P.T. and Labuza, T.P. (1976). "Effect of order-of-mixing on a_W lowering ability of food humectants." *J. Food Sci.* 41: 536–540.

Smelt, J.P., Hellemons, J.C., Wouters, P.C. and van Gerwen, S.J. (2002). "Physiological and mathematical aspects in setting criteria for decontamination of foods by physical means." *Int. J. Food Microbiol.* 78: 57–77.

Smelt, J.P.P.M. (1998). "Recent advances in the microbiology of high pressure processing." *Trends Food Sci. Technol.* 9: 152–158.

Smelt, J.P.P.M., Hellemons, J.C., Wouters, P.C. and van Gerwen, S.J.C. (2002). Physiological and mathematical aspects in setting criteria for decontamination of foods by physical means." *Int. J. Food Microbiol.* 78: 57–77.

Smith, K., Mittal, G.S. and Griffiths, M.W. (2002). "Pasteurization of milk using pulsed electric field and antimicrobials." *Journal of Food Science.* 67: 2304–2308.

Sofos, J.N. and Busta, F.F. (1993). "Sorbic acid and sorbates." In: *Antimicrobials in Foods,* 2nd edn. Davidson and Branen, A.L., (eds.). Marcel Dekker, New York. pp. 49–94.

Sofos, J.N., Beuchat, L.R., Davidson, P.M., and Johnson, E.A. (1998). Naturally occurring antimicrobials in food. Task Force Report No. 132, 103 pp., Council for Agricultural Science and Technology, Ames, Iowa.

Sonoike, K., Setoyama, T., Kuma, Y.and Kobayashi, S. (1992). "Effect of pressure and temperature on the death rates of *Lactobacillus casei* and *Escherichia coli.*" In: *High Pressure and Biotechnology.* Balny, C., Hayashi, R., Hermans, K., Masson, P. (eds.). John Libbey and Co. Ltd., London. pp. 297–301.

Stevens, K.A., Sheldon, B.W., Klapes, N.A. and Klaenhammer, T.R. (1992). "Effect of treatment conditions on nisin inactivation of Gram-negative bacteria." *Journal of Food Protection.* 55: 763–766.

Su, Y., Zhang, Q.H. and Yin, Y. (1996). "Inactivation of *Bacillus subtilis* spores using high voltage pulsed electric fields." IFT Annual Meeting Book of Abstracts, 1996, Session 26A-14.

Suzuki, A., Kim, K., Honma, N., Ikeuchi, Y. and Saito, M. (1992). "Acceleration of meat conditioning by high pressure treatment." In: *High Pressure and Biotechnology.* Balny, C., Hayashi, R., Hermans, K., Masson, P. (eds.). John Libbey and Co. Ltd., London. pp. 219–227.

Takahashi, T. and Haga, S. (1997). "Application of high pressure during the processing of uncooked and fermented hams with pickle curing." *Ann. Sci. Technol.* 68: 414–419.

Taki, Y., Awaao, T., Toba, S. and Mitsuura, N. (1991). "Sterilization of *Bacillus* spp. spores by hydrostatic pressure." In: *High Pressure Science for Food.* Hayashi, R. (ed.). San-Ei Pub. Co., Kyoto, Japan. pp. 217–224.

Tamagawa, K., Endo, Y., Osada, K. and Komiyama, Y. (1996). "Effects of hydrostatic pressure treatment on sterilization, viscosity and browning of grated yam named 'tororo'." *J. Japanese Soc. Food Sci. Technol.* 43: 194–202.

Teixeira, A.A., Dixon, J.R., Zahradnik, J.W. and Zinsmeister, G.E. (1969). "Computer determination of spore survival distributions in thermally-processed conduction-heated foods." *Food Technol.* 23: 78–80.

Terebiznik, M.R., Jagus, R.J., Cerrutti, P., De Huergo, M. and Pilosof, A.M.R. (2000). "Combined effect of nisin and pulsed electric fields on the inactivation of *Escherichia coli.*" *Journal of Food Protection.* 63: 741–746.

Tewari, G., Gill, C.O., Jayas, D.S., Jeremiah, L.E. and Holley, R.A. (1999). "Oxygen absorption kinetics of oxygen scavengers." *Intl. J. Food Sci. Technol.* Submitted for publication.

Thakur, B.R. and Nelson, P.E. (1998). "High pressure processing and preservation of foods." *Food Reviews Int.* 14: 427–447.

Thayer, D.W. (1990). "Food irradiation: Benefits and concerns." *Journal of Food Quality.* 13: 147–169.

Thayer, D.W. (1992). "Irradiation for control of foodborne pathogens on meats and poultry." In: *Safeguarding The Food Supply Through Irradiation Processing Techniques.*

Thayer, D.W., Christopher, J.P, Campbell, L.A. et al. (1987). "Toxicology studies of irradiation-sterilized chicken." *J. Food Prot* .50: 278–288.

Thevelein, J.M., Assche, J.A.V., Heremans, K. and Gerlsma, S.Y. (1981). "Gelatinization temperature of starch, as influenced by high pressure." *Carbohydr. Res.* 93: 304–307.

Tilbury, R.H. (1976). "The microbial stability of intermediate moisture foods with respect to yeasts." In: *Intermediate Moisture Foods.* Davies, R., Birch, G.G. and Parker, K.J. (eds.). Applied Science Publ., London. pp. 138–165.

Timson, W.J. and Short A.J. (1965). "Resistance of microorganisms to hydrostatic pressure." *Biotechnol. Bioeng.* 7: 139–159.

Timson, W.J. and Short, A.J. (1965). "Resistance of microorganisms to hydrostatic pressure." *Biotechnology and Bioengineering.* 7: 130–159.

Torres, J.A. and Karel, M. (1985). "Microbial stabilization of intermediate moisture food surfaces. III. Effects of surface preservative concentration and surface pH control on microbial stability of an intermediate moisture cheese analog." *J. Food Process. Preserv.* 9: 107–119.

Torres, J.A., Bouzas, J.O. and Karel, M. (1985b). "Microbial stabilization of intermediate moisture food surfaces." II. Control of surface pH." *J. Food Process Preserv.* 9: 93–106.

Torres, J.A., Motoki, M. and Karel, M. (1985a). "Microbial stabilization of intermediate moisture food surfaces. I. Control of surface preservative concentration." *J. Food Process Preserv.* 9: 75–92.

Troller, J.A. (1980). "Influence of water activity on microorganisms in foods." *Food Technol.* 34(5): 76–80.

Troller, J.A. (1985). "Effects of a_w and pH on growth and survival of *Staphylococcus aureus*." In: *Properties of Water in Foods.* Simatos, D. and Multon, J.L. (eds.). NATO Adv. Sci. Inst. Ser., Ser. E, Appl. Sci., No. 90. Martinus Nijhoff Publ., Dordrecht, Netherlands. pp. 247–257.

Troller, J.A. and Christian, J.H.B. (1978). *Water Activity and Food.* Academic Press, New York.

U.S. Food and Drug Administration, 2000. Kinetics of microbial inactivation for alternative food processing technologies: pulsed electric fields. [Accessed 21 June 2000],

Ulmer, H.M., Heinz, V., Ganzle, M.G., Knorr, D. and Vogel, R.F. (2002). "Effects of pulsed electric fields on inactivation and metabolic activity of *Lactobacillus plantarum* in model beer." *J. Appl. Microbiol.* 93: 326–335.

USDA. (1992). FSIS Backgrounder: *Poultry irradiation and preventing food-borne illness.* Food Safety and Inspection Service, United States Department of Agriculture, Washington, D.C.

van Camp, J. and Huyghebaert, A. (1995). "High pressure induced gel formation of a whey protein and haemoglobin protein concentrate." *Lebensm. Wiss. u Technol.* 28: 111–117.

van Den Berg, C. and Bruin, S. (1981). "Water activity and its estimation in food systems:Theoretical aspects." In: *Water Activity: Influences on Food Quality.* Rockland, L. and Stewart, G. F. (eds.). Academic Press, New York. pp. 1–64.

VanArdsel, W.B., Copley, M.J. and Morgan, A.I. (1973). *Food Dehydration,* 2nd edn. Vol.2. The AVI Publishing Co., Westport, Connecticut.

Vega-Mercado, H., Pothakamury, U.R., Chang, F.J., Barbosa-Canovas, G.V. and Swanson, B.G. (1996). "Inactivation of *Escherichia coli* by combining pH, ionic strength and pulsed electric field hurdles." *Food Research International.* 29: 117–121.

Walkenstrom, P. and Hermansson, A.M. (1997). "High pressure treated mixed gels of gelatin and whey proteins." *Food Hydrocol.* 11: 195–208.

Walker, H.W. and Matches, J.R. (1965). "Release of cellular constituents during heat inactivation of endospores of aerobic bacilli." *J. Food Sci.* 30: 1029–1036.

Warmbier, H.C., Schnickles, R.A. and Labuza, T.P. (1976). "Effect of glycerol on non-enzymatic browning in a solid intermediate moisture model food sytem." *J. Food Sci.* 41: 528–531.

Warth, A.D. (1985). "Resistance of yeast species to benzoic and sorbic acids and to sulfur dioxide." *J. Food Protect.* 48: 564–569.

Warth, A.D. (1988). "Effect of benzoic acid on growth yield of yeasts differing in their resistance to preservatives." *Appl. Environ. Microbiol.* 54: 2091–2095.

Wedzicha, B.L. (1984). *Chemistry of Sulphur Dioxide in Foods.* Elsevier Applied Sci., London.

Weemaes, C., Ludikhuyze, L., van Den Broeck, I. and Hendrickx, M. (1998). "High pressure inactivation of polyphenoloxidases." *J. Food Sci.* 63: 873–877.

Welch, T.J., Farewell, A., Neidhardt, F.C. and Barlett, D.H. (1993). "Stress response of *Escherichia coli* to elevated hydrostatic pressure." *J. Bacteriol.* 175: 7170–7177.

Wilkins, K.M. and Board, R.G. (1989). "Natural antimicrobial systems." In: *Mechanisms of Action of Food Preservation Procedures.* Gould, G.W. (ed.). Elsevier Applied Science, London. pp. 285–362.

Williams, A. (1994). "New technologies in food preservation and processing, part II." *Nutrition and Food Science.* 1: 20–23.

Williams, A. (1994). "New technologies in food processing: Part II." *Nutrition and Food Science.* 1: 20–23.

Williams, J.C. (1976). "Chemical and nonenzymic changes in intermediate moisture foods." In: *Intermediate Moisture Foods.* Davies, R., Birch, G.G. and Parker, K.J. (eds.). Applied Science Publ., London. pp. 100–119.

Woese, C. and Morowitz, H.J. (1958). "Kinetics of the release of dipicolinic acid from spores of *Bacillus subtilis.*" *J. Bacteriol.* 76: 81–83.

Wood, O.B. and Bruhn, C.M. (2000). "Position of the American dietetic association: food irradiation." *Journal of the American Dietetic Association.* 100: 246–253.

Woodroof, J.G., and Luh, B.S. (1986). *Commercial Fruit Processing,* 2nd edn. The AVI Publishing Co., Westport, Connecticut.

World Health Organisation (1994). *Safety and nutritional adequacy of irradiated food.* WHO, Geneva.

Wouters, P.C, A´ lvarez, I. and Raso, J. (2001a). "Critical factors determining inactivation kinetics by pulsed electric field food processing." *Trends Food Sci. Technol.* 12: 112–121.

Wouters, P.C., Bos, A.D. and Ueckert, J. (2001b). "Membrane permeabilization in relation to inactivation kinetics of *Lactobacillus* species due to pulsed electric fields." *Appl. Environ. Microbiol.* 67: 3092–3101.

Wouters, P.C., Glaasker, E. and Smelt, J.P.P.M. (1998). "Effects of high pressure on inactivation kinetics and events related to proton efflux in *Lactobacillus plantarum*." *Appl. Environ. Microbiol.* 64: 509–514.

Ye H.Y., Wang, H., Wang, J.F., Xu, Q., Li, Y.H., (1996). "Pressure tolerance of microbes and techniques for high pressure disinfection of acid food (jams etc.) at room temperature." *Food Sci.* 17: 30–34.

Yousef, A.E. (2001). "Efficacy and limitations of non-thermal preservation technologies." IFT Annual Meeting Book of Abstracts, 2001, Session 9-1.

Zhang, Q., Monsalvez-Gonzalez, A., Barbosa-Canovas, G.V. and Swanson, B.G. (1994). "Inactivation of *E. coli* and *S. cerevisiae* by pulsed electric fields under controlled temperature condition." *Transactions of ASAE.* 37: 581–587.

INDEX

Made in the USA
Monee, IL
07 July 2026